PHYSICS OF UNSTABLE NUCLEI

PROCEEDINGS OF THE INTERNATIONAL
SYMPOSIUM ON PHYSICS OF UNSTABLE NUCLEI

ISPUN07

PHYSICS OF UNSTABLE NUCLEI

PROCEEDINGS OF THE INTERNATIONAL SYMPOSIUM ON PHYSICS OF UNSTABLE NUCLEI

ISPUN07

HOI AN, VIETNAM 3 – 7 JULY 2007

edited by

Dao Tien Khoa
INST Hanoi, Vietnam

Peter Egelhof
GSI Darmstadt, Germany

Sydney Gales
GANIL, France

Nguyen Van Giai
IPN Orsay, France

Tohru Motobayashi
RIKEN, Japan

World Scientific

NEW JERSEY · LONDON · SINGAPORE · BEIJING · SHANGHAI · HONG KONG · TAIPEI · CHENNAI

Published by

World Scientific Publishing Co. Pte. Ltd.

5 Toh Tuck Link, Singapore 596224

USA office: 27 Warren Street, Suite 401-402, Hackensack, NJ 07601

UK office: 57 Shelton Street, Covent Garden, London WC2H 9HE

British Library Cataloguing-in-Publication Data
A catalogue record for this book is available from the British Library.

PHYSICS OF UNSTABLE NUCLEI
Proceedings of the International Symposium on PHYSICS OF UNSTABLE NUCLEI ISPUN07

ISBN-13 978-981-277-614-3
ISBN-10 981-277-614-1

Printed in Singapore.

PREFACE

The *International Symposium on Physics of Unstable Nuclei 2007* (ISPUN07) was held from July 3 to 7, 2007, at a sea side resort on the central coast of Vietnam, near the ancient town of Hoi An which is protected by UNESCO as a World Heritage. The present volume contains the texts of most of the invited talks delivered at the conference, as well as the oral and poster presentations.

The first meeting of this series, ISPUN02, took place in 2002 at Halong Bay, another famous World Heritage in Vietnam, with the aim to encourage nuclear physics research in that part of the continent and also to give an opportunity to the new generation of Vietnamese physicists to meet their colleagues from abroad. An early attempt with a conference in Hanoi in 1994 already showed a good response from the international nuclear physics community. The success of ISPUN02 led to the idea of having a regular cycle of such meetings in Vietnam, focusing on current issues of nuclear structure and reaction studies with unstable nuclei.

ISPUN07 was organized by Institute for Nuclear Science and Technique (INST) and Institute of Physics and Electronics (IPE) in Hanoi, and it has strongly benefited from the encouragement and support from various organizations both at home and abroad: the Natural Science Council of Vietnam, Vietnam Atomic Energy Commission, GANIL, CEA Saclay, GSI Darmstadt, RIKEN, and the EU Asia-Link Network on Nuclear Physics and Astrophysics. In particular, the involvement of the latter network has allowed the attendance of a number of Asian students from China and Vietnam. The symposium organizers have been very much impressed by the fact that the Chinese students did not hesitate to travel all the way by train from Beijing or Lanzhou (a 3-day trip each way) to Hoi An to attend this meeting. The attendance at ISPUN07 was quite satisfactory with a total of 90 participants, about 80% coming from abroad and the rest from Vietnamese universities and institutes in Hanoi, Ho Chi Minh City, and Dalat.

The scientific program of the symposium contains three main parts:

- Construction of the new facilities and instruments. The present status and future plans of the major experimental facilities in the world were the objects of many presentations: the RIBF at RIKEN, SPIRAL2 project at GANIL, FAIR facility at GSI, Radioactive Ion Beams facility in Brazil (RIBRAS), and the REX-ISOLDE physics program with an outlook of the HIE-ISOLDE upgrade.
- Present status and new problems in the experimental studies of unstable nuclei. The main emphasis is of course on the structure studies of nuclei far from stability:

the low lying states, pygmy dipole resonances at low energies, halo structure, changes in subshell closures around $N = 28$ and $Z = 50$, $N = 82$, superheavy nuclei produced in ^{48}Ca induced reactions etc. Other important issues discussed at ISPUN07 include the symmetry term of the nuclear Equation of State, stellar nucleosynthesis, high precision nuclear reaction studies with unstable beams etc.
- Status and progress of theoretical approaches. The presentations covered the work done in the AMD and GCM approaches, the shell model description of medium mass exotic nuclei, self-consistent mean field approaches (including the continuum effects) in both the non-relativistic and relativistic frameworks for ground states and excited states, the microscopic study of the Equation of State of asymmetric nuclear matter and recent progress in theoretical nuclear reaction studies.

The ISPUN07 symposium will leave a long lasting memory to all its participants. Especially, for the young Chinese and Vietnamese students, it has been definitely a unique opportunity to learn about the latest developments in the studies of unstable nuclei, to meet and discuss physics with many world experts of the field. For others, it was not only just a scientific conference but also a unique occasion to discover the beauty of a far away country as well as the hospitality of its friendly people. An interesting scientific program and active contributions by all participants are the key to the success of ISPUN07 and we would like, therefore, to express our best thanks to the members of the international advisory board for their help in suggesting the scientific program and excellent speakers for this symposium, as well as all participants of ISPUN07 for their valuable contributions.

Dao Tien Khoa, *INST Hanoi*
Nguyen Van Giai, *IPN Orsay*
Sydney Galès, *GANIL*
Peter Egelhof, *GSI Darmstadt*
Tohru Motobayashi, *RIKEN*

INTERNATIONAL ADVISORY COMMITTEE

for the International Symposium on Physics of Unstable Nuclei ISPUN07

ORGANIZING COMMITTEE

of the International Symposium on Physics of Unstable Nuclei ISPUN07

EDITORIAL BOARD

for the Proceedings of ISPUN07

CONTENTS

PART A

Invited Talks and Oral Presentations

STUDIES at the RIKEN RI BEAM FACTORY

T. MOTOBAYASHI*

RIKEN Nishina Center,
2-1 Hirosawa, Wako, Saitama 351-0198, Japan
**E-mail: motobaya@riken.jp*
http://www.riken.jp/engn/r-world/research/lab/wako/heavyion/index.html

The RIKEN RI Beam Factory (RIBF) project has reached a milestone in Dec. 2006 by extracting the first beam. Its basic part, accelerators and a superconducting RI beam separator called BigRIPS are now in operation. The first experimental result, production of a new neutron-rich isotope ^{125}Pd, was obtained in May 2007 with 345 MeV/nucleon ^{238}U beams. The RIBF project was launched about ten years ago, encouraged by success of nuclear physics and nuclear astrophysics studies using fast RI beams produced by the accelerator facility which has been running since 1986. At RIBF, radioactive isotope (RI) beams are produced via the projectile fragmentation of heavy ions or in-flight fission of uranium ions. Most of the expected path of the r-process nucleosynthesis can be reached with the intensity higher than one particle per day. In view of its outstanding performance on the variety and intensity of RI beams, the RIBF is one of the new-generation facilities competitive to those planned in the United States (FRIB) and Europe (FAIR at GSI), and is their first realization.

1. Studies with fast RI Beams

The word "RI beams" means "Radioactive Isotopes beams", that is beams with unstable nuclei. RI beams at RIKEN have been produced by the projectile fragmentation scheme since 1990. Heavy ion beams with intermediate energy (a few tens MeV/nucleon) bombard a production target, and produced projectile fragments are separated by a magnetic analyzer system to be used to bombard a secondary target. Due to the high performance of the fragment separator RIPS,[1] the RI beam intensities are highest in the world for many light neutron-rich nuclei. Thanks to this high-intensity capability, the very neutron-rich fluorine isotope ^{31}F was found to be particle-stable.[2] Besides these intermediate-energy RI beams, beams of light unstable nuclei at lower energies, typically 5 MeV/nucleon, are also available in the CRIB facility constructed by CNS.[3] The CRIB uses in-flight direct reactions, such as (p,n) and (^{3}He,n), to produce RI beams, which are mainly used for studying low-energy reactions of astrophysical interest.

The RI beams from the RIPS have been used for various experiments. Studies of interaction cross section have been performed, and neutron halo and neutron skin structures have been established in some light neutron-rich nuclei.[4] The neutron halo or the neutron skin indicates that some neutrons can be decoupled from protons despite of the strong p-n interaction. Disappearance of magic numbers in some neutron-rich regions in the nuclear chart is another phenomena of interest. The first Coulomb excitation experiment with fast RI beams has been performed at RIKEN for the neutron-rich ^{32}Mg nucleus with the N=20 magic number.[5] The extracted large B(E2) value indicates that the N=20 shell closure disappears in ^{32}Mg. A new method for studying properties of unstable nuclei, measurement of γ-rays from fast-moving excited nuclei in coincidence with reaction products with particle identification, was applied to this experiment, and many experiments using this technique have been performed so far for studying nuclear structures of nuclei around the shell closure at N=8 and N=20. Recently, the decoupling of protons and neutrons has been revealed also in excitation of the ^{16}C nucleus to its 2^+ state by measuring the 2^+ lifetime,[6,7] inelastic scattering with ^{1}H,[8] and the one with ^{208}Pb.[9] Neutrons strongly contribute to the 2^+ state excitation, while protons have little contribution. This decoupling picture might be related to the low electric quadrupole moments measured for ^{15}B and ^{17}B using the β-NMR technique.[10] Studies of particle-unbound states in unstable nuclei are another highlight. The Coulomb dissociation to simulate astrophysical (p,γ) reactions[11] and the neutron halo structure[12] have been extensively studied. The nuclear reactions used in these studies involve no mass transfer and match the fast RI beam, since the their good kinematical matching persist even in the relevant energy range.

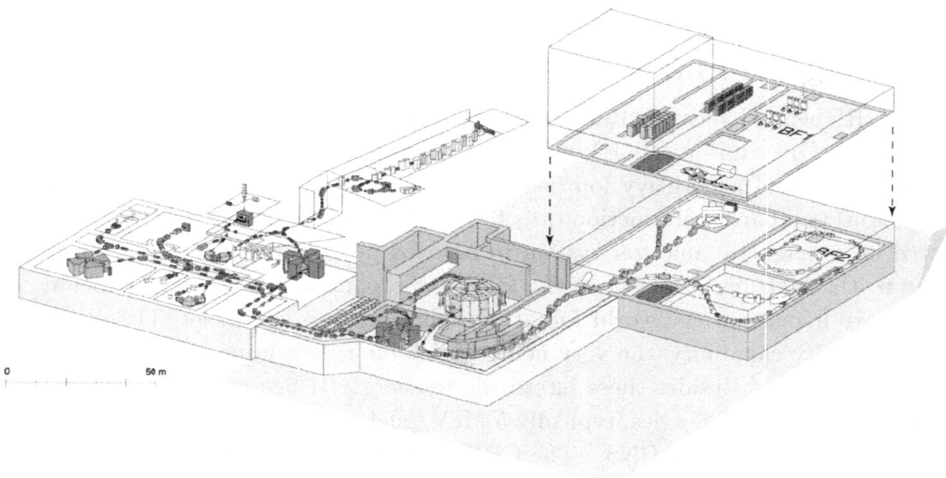

Fig. 1. Schematic view of the RIKEN accelerator complex.

2. RI Beam Factory

2.1. *Overview*

Encouraged by the success of research using the fast RI beams, the RI Beam Factory (RIBF) project was planned. The RIBF new facility consist of three cyclotrons, the fixed-frequency Ring Cyclotron (fRC) with K=570 MeV, Intermediate stage Ring Cyclotron (IRC) with K=980 MeV, and Superconducting Ring Cyclotron (SRC) with K=2600 MeV, which successively boosts the beam energy up to 345 MeV/nucleon.[13] The 16 MV variable-frequency linear accelerator (RILAC) and four-sector ring cyclotron (RIKEN Ring Cyclotron, RRC) with K=540 MeV, which have been in operation since 1987, are used as their injectors. A schematic view of the facility is shown in Fig. 1.

To provide intense beams the new RIBF accelerators, several improvements have been made around the injector RILAC. Thus, intense ions at several MeV/nucleon energies became available. In 2004, the first event indicating the production of the isotope $^{278}113$ by the $^{209}Bi(^{70}Zn,n)$ fusion reaction has been observed.[14] The beams have been used also for various applications to nuclear chemistry, bio and medical science, and materials science.

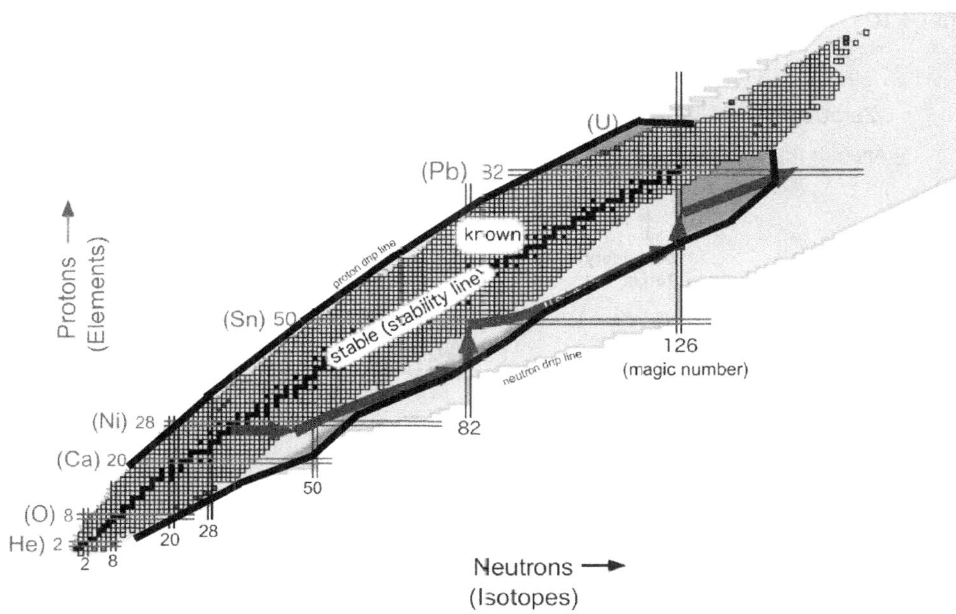

Fig. 2. Nuclear chart covered by the RIBF. The thick solid curves indicate the limit of RI productions of 1 particle per day with 1 pμA primary beam intensity.

RI beams are produced via the projectile fragmentation of heavy ions or in-flight fission of uranium ions by a superconducting fragment separator BigRIPS[15] shown in Fig. 3. It consists of fourteen magnets of superconducting quadrupole triplets and six room-temperature dipoles magnets. Separation of reaction products is made by a two-stage separation scheme. The first stage serves to produce and separate RI beams with a wedge-shaped degrader inserted at the momentum-dispersive focus F1. The second-stage identifies RI beam species event-by-event and tags the secondary beam that still contains various different ions. The horizontal and vertical angular acceptances are designed respectively to 80 mrad 100 mrad vertically, while the momentum acceptance is 6%. These angular and momentum acceptances enable one to collect about half of the fission fragments produced by a 350 MeV/nucleon uranium beam. With the goal intensity of the primary beams, 1 pμA for ions up to uranium, the RI beam yield estimated by the code EPAX2[16] is shown in Fig. 2. For example, the intensities of the doubly magic nuclei ^{78}Ni, ^{132}Sn and ^{100}Sn are expected to be 10, 100 and 1 particles/s, respectively, encouraging detailed studies of nuclei far from the stability valley. Most of the expected path of the r-process nucleosynthesis can be reached with the intensity higher than one particle per day.

The Zero Degree spectrometer (ZDS), also shown in Fig. 3, analyzes secondary reaction products emitted in the beam direction. Its typical use is for γ-ray measurements in coincidence with fast-moving excited nuclei. The ZDS is used to identify the reaction product. For other applications, such as β-decay measurements with stopped RIs, the ZDS is also useful.

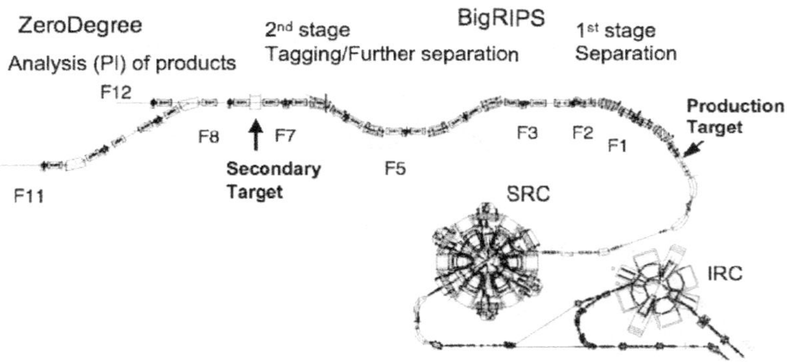

Fig. 3. Plan view of the Big RIPS and Zero Degree spectrometer together with the two cyclotrons IRC and SRC of the RIBF accelerate complex.

2.2. *Status of RIBF*

The parts completed so far are the cyclotron complex, the Big RIPS, and the Zero Degree spectrometer. After the first extraction of the primary beam of 345 MeV/nucleon ^{27}Al^{10+}, on December 28 in 2006, the accelerators have been tuned, and ^{238}U ions were successfully accelerated in Mar. 2007.

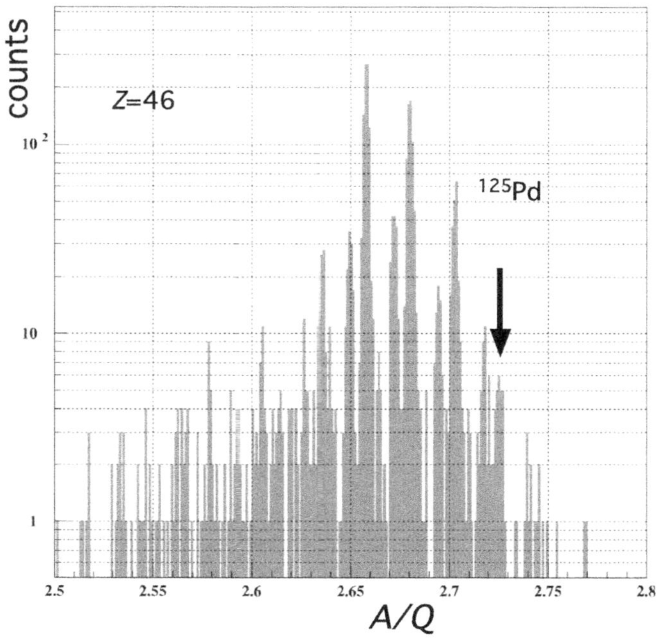

Fig. 4. Yield distribution for the atomic number Z=46 obtained by the in-flight fission on uranium beams with the energy of 345 MeV/nucleon.. The arrow indicates the peak position for the new isotope ^{125}Pd.

Commissioning of RI beam production started in March and the first attempt of new-isotope production[17] was performed with a 345 MeV/nucleon $^{238}U^{86+}$ beam delivered to a beryllium production target of 7mm thick. The parameters of the BigRIPS were matched to neutron-rich fission products with the atomic number around 50. The atomic number Z and the mass-to-charge ratio A/Q were obtained event-by-event by measuring the velocity (or time-of-flight), energy loss, total energy, and magnetic rigidity using beam-line counters. The yield for Z=46 (palladium) is plotted as a function of A/Q in Fig. 4. Observed double peak structures are due to the mixture of ions not fully stripped. A peak corresponding to the new isotope ^{125}Pd, indicated by the arrow, is clearly seen with a good statistics, and some indication for ^{126}Pd is also seen. The data were taken with the primary beam current of 4×10^7 particle/sec on an average, which is about 10^{-5} of the goal intensity, 1 particle μA, and the data acquisition time is about one-day. This indicates high potential of the Big RIPS separator coupled with the RIBF accelerator complex, and future possibility in accessing a large amount of unknown unstable nuclei is foreseen. Since the RI beam intensities achieved so far is not enough for secondary reaction studies, various efforts are on-going including the improvement of the beam transmission in every stage of the accelerator complex. The use of the ^{48}Ca and ^{86}Kr beams is currently assumed for the first series of experiments.

2.3. *Planned Experimental Installations*

Besides experimental studies in the first periods using the Big RIPS and the Zero Degree Spectrometer, the nuclear-physics and nuclear-astrophysics programs in the RIBF includes construction of various kinds of new apparatuses.

The SHARAQ spectrometer[18] is of a QQ-D-Q-D configuration. By employing dispersion-matching optics with a specially designed beam line from the BigRIPS, the momentum resolution of 15,000 will be achieved. The SHARAQ spectrometer is under construction and will be commissioned in 2008.

The SLOWRI[19] aims at conducting various experiments using slow RI beams. Fast secondary beams from the Big RIPS will be efficiently stopped and extracted by a gas-catcher system with the RF ion-guide technique.

The SAMURAI is a spectrometer with a large solid angle and a large momentum acceptance dedicated to particle-correlation studies. It is of a QQQ-D configuration, where the dipole (D) is a superconducting magnet with 6.7 Tm rigidity. The large gap of 80 cm is useful for measurements of projectile-rapidity neutrons.

The system for electron-RI scattering experiments employs a Self-Confining Radioactive Ion Target (SCRIT[20]). RI ions are transversely confined by the attractive force caused by the electron beam itself. A mirror potential is applied externally to achieve longitudinal confinement. Test experiments to examine the confinement mechanism are successfully made at an existing electron ring.

The rare RI mass ring[21] is to measure the mass of rarely produced (1 particle per day, for example) exotic nuclides in 10^{-6} accuracy. Each ion is injected individually to the ring by a trigger signal provided from a counter in the Big RIPS. The ring is tuned to achieve the isochronous condition, and a time-of-flight of the ion in the ring is measured.

To allow for running an RI-beam based experiment and a super heavy element search simultaneously, construction of a new injector[22] is being considered. Another possibility to extend the research opportunity is to build a beam line that brings back the IRC beam to the RIPS separator. Various nuclear moment studies and condensed matter research are planned.

3. Summary

The RIKEN RI Beam Factory (RIBF), one of the new-generation facilities for beams of unstable nuclei, has started its operation. RI-beams are produced by the RIBF cyclotron complex coupled with the superconducting RI beam separator Big RIPS. Their potential was demonstrated by production of the new isotope ^{125}Pd. When the RIBF reaches to its full performance, almost all nuclides along the r-process nuclear synthesis can be created and experimentally studied. Continuous attempts to improve the primary beam intensity together with construction of various experimental equipments will open a new domain of nuclear physics, and will create a new view on atomic nucleus as well as on the element genesis in the universe, incorporated with the world efforts for realizing RI-beam facilities being constructed or planned.

References

1. T. Kubo *et al.*, *Nucl. Instr. Meth. B* **70**, 309 (1992).
2. H. Sakurai *et al.*, *Phys. Lett. B* **448**, 180 (1999).
3. Y. Yanagisawa *et al.*, *Nucl. Instr. Meth. A* **539**, 74 (2005).
4. T. Suzuki *et al.*, *Phys. Rev. Lett.* **75**, 3241 (1995).
5. T. Motobayashi *et al.*, *Phys. Lett. B* **346**, 9 (1995).
6. N. Imai *et al.*, *Phys. Rev. Lett.* **92**, 062501 (2004).
7. H.J. Ong *et al.*, to be published.
8. H.J. Ong *et al.*, *Phys. Rev. C* **73**, 024610 (2006).
9. Z. Elekes *et al.*, *Phys. Lett. B* **686**, 34 (2004).
10. H. Ueno *et al.*, *Nucl. Phys. A* **738**, 211 (2004).
11. for example, T. Motobayashi, *Nucl. Phys. A* **693**, 258 (2001).
12. for example, T. Nakamura, *et al. Phys. Rev. Lett.*, **96**, 252502 (2006).
13. Y. Yano, *Nucl. Instr. Meth. B* **261**, 1009 (2007).
14. K. Morita *et al.*, *J. Phys. Soc. Jpn.* **73**, 2593 (2004).
15. T. Kubo *et al.*, *IEEE TRANS. APPLIED SUPERCONDUCTIVITY* **17**, 1069 (2007).
16. K. Suemmerer and B. Blank, *Phys. Rev. C* **61**, 034607 (2000).
17. Y. Yano, presented at the Int. Symp. Nucl. Phys. (INCP07), Tokyo, July 2007.
18. T. Uesaka *et al.*, *CNS Annual Report 2004*, 42 (2005); H. Sakai, this symposium.
19. M. Wada *et al.*, *Nucl. Instr. and Meth. A* **532**, 40 (2004).
20. M. Wakasugi *et al.*, *Nucl. Instr. and Meth. A* **532**, 216 (2004).
21. Y. Yamaguchi *et al.*, *CNS Annual Report 2004*, 83 (2005).
22. O. Kamigaito *et al.*, *RIKEN Accel. Prog. Rep.* **39**, 261 (2005).

DILUTE NUCLEAR STATES

MARTIN FREER*

School of Physics and Astronomy, University of Birmingham
Birmingham, B15 2TT, UK
**E-mail: M.Freer@bham.ac.uk*

An examination of dilute, cluster, states in the nuclei ^{12}C and ^{10}Be is made. Decay correlations are used to characterise the spins of excited states in both nuclei. In addition, measurements of the breakup into its cluster constituents, $2\alpha+2p$, of the super-Borromean nucleus ^{10}C nucleus is reported.

Keywords: Nuclear clustering; ^{12}C, alpha-decay; ^{10}Be and ^{10}C clustering.

1. Introduction

The cluster structure of light nuclei has been well studied and well documented. There are a few cases in which the clustering is remarkably well enhanced. Here, examples of these in the nuclei ^{12}C and ^{10}Be are examined. We also investigate a new type of nuclear structure which can in principle exist in ^{10}C.

1.1. *Background to dilute states in ^{12}C*

The structure of the nucleus ^{12}C remains one of the most important questions in nuclear physics. The ground state possesses the triangular 3α symmetry, but resides so far below the α-decay threshold (7.27 MeV) that the cluster structure is strongly suppressed. The ground-state does, however, possess an oblate deformation and the first excited state at 4.44 MeV (2^+) and the 14.08 MeV (4^+) would be the corresponding rotational excitations. The state which is of greatest interest is the 7.65 MeV, 0^+, state. This is the portal for the synthesis of ^{12}C in stellar nucleosynthesis, and is named after Hoyle who proposed its existence.[1] The importance of this state is arguably unparalleled in nuclear science. As a consequence one would anticipate that the level of knowledge its structure would be also be well documented. This is far from the case and remains an open question.

For many years it was believed, in accordance with the Ikeda picture and the ideas of Morinaga[2] and Brink,[3] that the 7.65 MeV should correspond to a linear chain structure. This interpretation was always problematic as a 2^+ member of the rotational band built on the 3 α-chain, is predicted by the Alpha Cluster Model to lie at $\sim$8.4 MeV.[4] The closest 2^+ candidate is some 3 MeV higher in energy, and thus cannot be rotationally linked with the 7.65 MeV state if the configuration possesses a significant deformation. This feature alone casts considerable doubt on the connection of this state with the 3 α-chain. The energy 8.4 MeV lies between the 7.65 MeV state and a 3^- state at 9.64 MeV. In this region a further state can unambiguously excluded. However, the same is not true above the 3^- state. Thus, an extended configuration with a large moment of inertia, perhaps a bent chain,[5–7] cannot be excluded.

There is, however, a rather intriguing possible alternative. Stimulated by developments in the field of cold-atom physics it has been speculated that the Hoyle-state corresponds to a dilute gas of α-particles. The α-particle possesses a bosonic nature and the proposed large radial extent of the excited state would in principle permit the internal fermionic degrees of freedom to be neglected. This invites the possibility that the three α-particles condense into the lowest (s-state) of their interaction potential - a Bose-Einstein condensate (BEC). In order to describe such a possibility, a condensate wave-function has been developed to reflect the possible character of the state.[8–10] One of the main successes of this model is that it manages to reproduce the form factor for the electron inelastic excitation to the Hoyle-state without any arbitrary normalisation[12] (see Fig. 1). These calculations reveal that

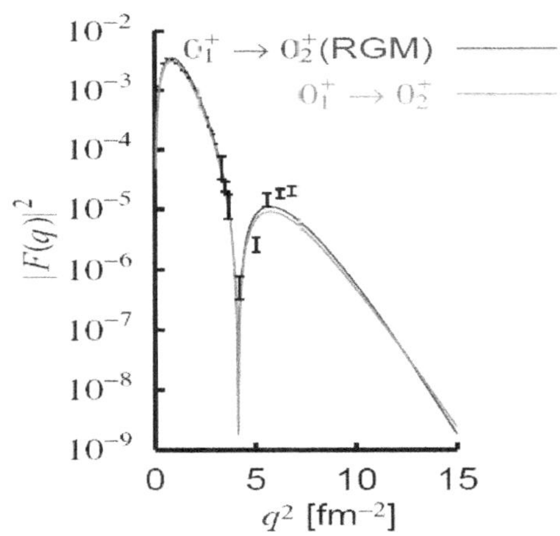

Fig. 1. The calculated inelastic form factor for electron inelastic scattering from the 0_1^+ ground state to the 0_2^+ excited state,[12] compared with the experimental data from.[13]

the spatial extent of the 0_2^+, 7.65 MeV, excited state is such that the volume of the Hoyle-state is ~3-4 times that of the ground-state. It is this large volume which permits the realisation of the clustering and the liberation of the α-particle gas.

The question then arises as to the nature of the excitations of the 0_2^+ state. If it is a condensate then the next level in the potential containing the 3α-particles would be a d-state, and thus the promotion of an α-particle from the s to d-state would result in a 2^+ excitation. The model predicts this to lie ~2.5 MeV above the 0_2^+ state, i.e. close to 10 MeV. From an experimental perspective this is a very challenging region to untangle. Aside from the narrow 9.64 MeV 3^- and 10.84 MeV 1^- states, there is a well documented broad 0_3^+ resonance close to 10.3 MeV. The width is such ($\Gamma \sim 3$ MeV) that it extends across the whole region of interest. The experimental situation remains unclear at present. Measurements at RCNP using the ^{12}C(α,α') reaction indicate the possibility of a 2^+ component close to the 9.64 MeV 3^- state.[14] On the other hand, $\beta^{\pm}$-decay studies suggest that the 2^+ strength in this region is small.[15]

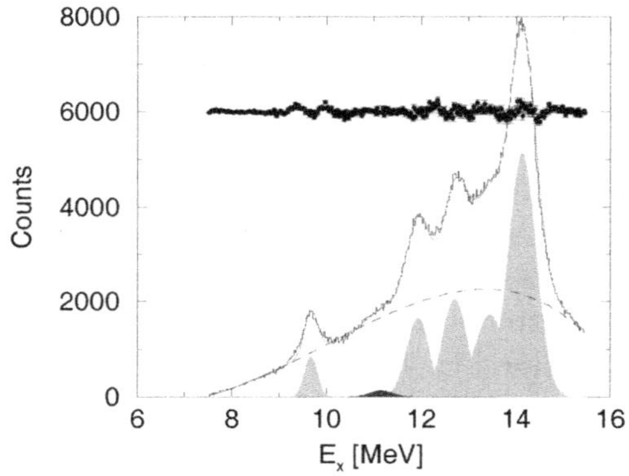

Fig. 2. ^{12}C excitation energy spectrum for decays to α+^{8}Be*. The histogram corresponds to the experimental data, the solid line the fit, and the dashed line the estimated background. The peaks included in the fit are shown shaded, and the magnitude of the contribution from a state at 11.16 MeV is shown in a darker shade. The residuals of the fit to the data are shown across the top of the spectrum (filled circles).

If a 2^+ excitation does exist then is there a collective 4^+ excitation also? If a vibrational model is employed then the 4^+ should lie close to 7.65+2×2 = 11.65 MeV (taking the energy of the 2^+ state to lie under/close to the 9.64 MeV state). There is no known state of this character close to this energy. If however, a rotational model is used then the energy of the 4^+ would reside at 7.65+3.33×2= 14.31 MeV. This could place the 4^+ state beneath the 4^+ collective excitation of the ground state. This would make detection difficult. Thus, it is possible that the collective excitations of the Hoyle-state is of of the best hidden secret in nature.

2. Present measurements of ^{12}C

It is clear that the region of ^{12}C excitation energy 7-15 MeV is of prime importance in terms of the determination of the structure of this all important nucleus. In order to add to our understanding in this region, a study of the ^{12}C(^{12}C, 3α)^{12}C reaction has been performed in which the α-decay to the ^{8}Be proceeds via the 2^+ excited state as opposed to the ground state. This selection of the decay path could in principle highlight states of higher spin in the 10 MeV region. It also has the effect of selecting unnatural parity states.

2.1. *Correlation measurements*

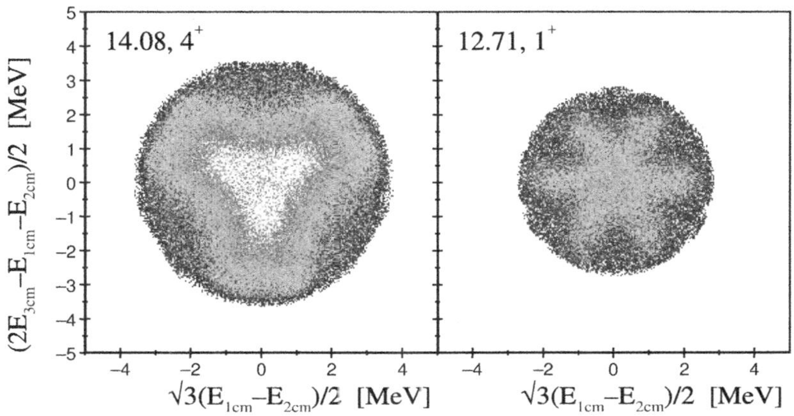

Fig. 3. Dalitz plots for the decays of the 14.08 and 12.71 MeV states via the 2^+ excited state in ^{8}Be.

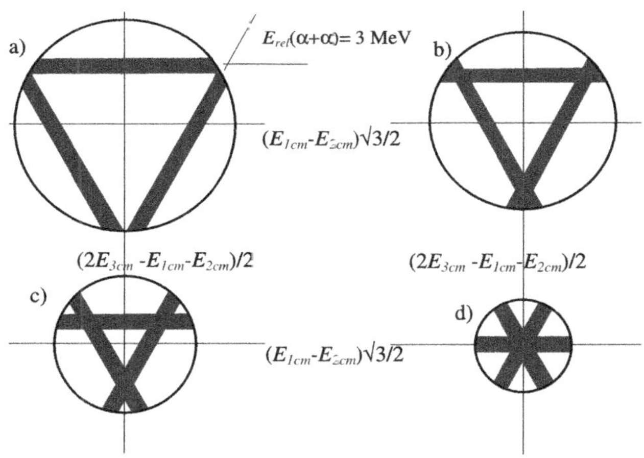

Fig. 4. Illustration of the evolution of the Dalitz structure with ^{12}C excitation energy. The bands show the location of the ^{8}Be(2^+) resonance. a) corresponds to an excitation energy close to 20 MeV and d) to 12.7 MeV. b) and c) to energies of ~16 and ~14 MeV, respectively. The 12.71 MeV state corresponds to case d) and the 14.08 MeV state approximately to c).

The spectrum of states populated in this instance is shown in Fig. 2. The data have been fitted with all of the states known to decay to this final state, taking into account the experimental resolution. The important states in the decay to ^{8}Be excited states are; 9.64 MeV (3^-), 10.84 MeV (1^-), 11.83 MeV (2^-), 12.71 MeV (1^+), 13.35 MeV ([2^-]) and 14.08 MeV (4^+). There appears to be a small but necessary contribution from a state at 11.16 MeV reported in ref.[17] As judged from the residuals, the fit to the spectrum includes all of the important components from narrow resonances in the present excitation energy region.

In order to evaluate the spins of the states observed in Fig. 2, the decay correlations have been analysed. These correlations correspond to the emission pattern of the α-particles in the ^{12}C centre-of-mass frame. These are shown for the well known states at 12.71 and 14.08 MeV, which have spins 1^- and 4^+, respectively, in Fig. 3. The decay patterns observed for these two states are characteristically different. In principle, it might be anticipated that this difference might be attributed to the two spins of the states. Rather, the changing pattern is really a reflection of the changing decay energy with respect to the ^{8}Be excited state. As illustrated in Fig. 4, the locus of the final state correlation of the two α-particles arising from the decay of the 3 MeV excited state moves towards the centre of the plot as the excitation energy decreases. Monte Carlo simulations of the reaction and decay process coupled with the acceptance and resolution of the detection system reproduce the patterns observed for the two states in Fig. 3. A similar analysis for the state at 13.35 MeV, which is tentatively assigned a spin of 2^- in Ref.[17] is shown in Fig. 6. The correlations are not definitive but appear to show that rather than a 2^- state,

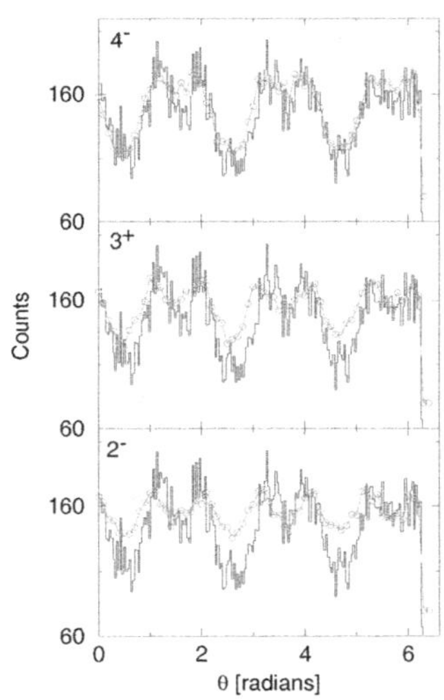

Fig. 5. The angular dependence of the yield in the Dalitz analysis for the 13.35 MeV state, compared with the simulations (open-circles) for 2^-, 3^+ and 4^- decays. Note, it has been assumed that there is a 50 % contribution to the experimental spectrum from an isotropic background.

the 13.35 MeV state may in fact be associated with a higher spin – possibly 4^-. A similar analysis for the lower energy 11.83 MeV state which is assigned a spin and parity 2^-,[17] may in fact also have a different assignment (4^-), see Ref.[18] for details.

These measurements demand that other studies are performed to examine the structure in this region. Having stated that, no new evidence for excitations of the Hoyle state have emerged and this challenge remains.

3. Clustering in ^{10}Be

An alternative way to study the nature of the Hoyle state is to examine the structure of neighbouring nuclei, i.e. those in which correspond to the addition or removal of nucleons. In the case of ^{10}Be two protons are removed. The Hoyle state in the shell model limit would correspond to 2 protons and 2 neutrons in the *sd*-shell. The corresponding state in ^{10}Be would possess two neutrons in the same orbits. In the framework of the Antisymmetrised Molecular Dynamics (AMD)[19] this corresponding state lies at 6.1793 MeV [0^+], and possesses a rather remarkable cluster structure. The calculations suggest that the state is associated with an α+2n+α cluster structure. Moreover, the two neutrons are found to intercede between the α-particles in a molecular orbit which possesses σ-type characteristics. The location of the neutrons servers to promote the deformation. There is a well known 2^+ state at 7.542 MeV which has been found to have a very large α-decay reduced width.[20] A recent study of resonant scattering of ^{6}He+^{4}He has found evidence for a state at 10.15 MeV with spin and parity 4^+, again with a very large reduced α-width.[21] If these three states are indeed linked then the moment of inertia that is associated with the underlying structure is extremely large, making it the most deformed state yet observed in nature.

It is thus possible that the dilute nature of the ^{12}C Hoyle state is manifest in the rather unusual structure of the 6.1793 MeV [0^+] state in ^{10}Be. It would be interesting to find similar states in the intermediate systems ^{11}C and ^{11}B.

4. The super-Borromean nucleus ^{10}C

The nucleus ^{10}C occupies a rather unique position in the chart of nuclides, as illustrated in Fig. 6. It resides at the top of a loop of particle stable nuclei, caused by the presence of two nuclei which are particle unstable in the ground state; ^{9}B and ^{6}Be. The removal of a proton from ^{10}C results in the formation of ^{9}B, which proton decays into ^{8}Be, which in turn decays into two α-particles. In other words, the removal of the proton causes the dissembly of the whole nucleus. Similarly, the removal of an α-particle produces ^{6}Be, which then decays into 2p+α via an intermediate unbound ^{5}Li ground state. The removal of p-p, α-α or p-α pairs also results in the system "falling apart". Nuclei which are composed of three constituents which are all required to bind the system have become known as Borromean, after the symbolic three interlocking Borromean rings. Thus, the ^{10}C nucleus should be termed a super-Borromean nucleus since it has 4 constituents all of which are required to produce a particle-stable ground state. In fact the Borromean rings are a member

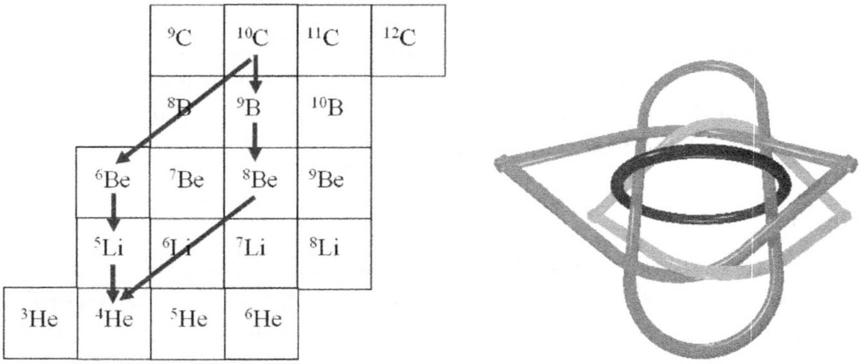

Fig. 6. Illustration of the decay properties of excited, unbound, states of ^{10}C. The white squares indicate nuclei which are particle-unstable in the ground state, others correspond to either stable nuclei, or those which undergo weak decays. The right hand part shows the 4 interlocking rings of the super-Borromean, Brunnian system.

of a set of non-trivial mathematical knots which become trivial if any link is broken. These are known as Brunnian knots after the mathematician Brunn.[22] Thus, ^{10}C nucleus corresponds to a class of Brunnian nuclei. The symbolic 4 interlocking rings representing the structure of ^{10}C are shown in Fig. 6.

The structure of this nucleus has recently been probed in break-up measurements at GANIL.[23] These have measured the cross sections for the breaking each one of the links, in order to determine the influence of the Brunnian structure in the ground state. Moreover, they also probe the excited states in ^{10}C, and in principle are sensitive to the states which are analogs of the Hoyle state and the states in ^{10}Be with the α+2n+α cluster structure.[23]

5. Summary

The study of the structure of light nuclei offers exciting challenges as they, in principle, possess a wide range of exotic structures. The appearance of clustering in some cases gives rise to states with striking properties. The 7.65 MeV 0^+ state in ^{12}C is a prime example, as it is believed to possess an exceptionally large charge radius. The associated dilute nature would allow the nucleons to precipitate into an α-particle gas, possibly an α-condensate. The understanding of the structure of this state remains one of the greatest challenges for contemporary nuclear physics. The nucleus ^{10}Be also has an excited 0^+ state with a rather unusual α+2n+α cluster structure. In this instance, the neutrons reside in a *molecular* orbital between the two α-particles. Finally, ^{10}C holds the promise of containing an even more exotic Brunnian structure. Here, all four of the $2\alpha+2p$ cluster constituents are required to bind the ground state. This, of course, possesses a strong link with Borromean systems such has ^{6}He and ^{11}Li. It is clear, that much more effort is required to fully characterise such systems, but what is known at present is intriguing.

Acknowledgments

I would like to thank my colleagues at the Universities of Birmingham and Surrey, in the UK, and LPC, Caen, France for their essential contribution to the work described in these proceedings.

References

1. F. Hoyle, The Astrophysical Journal, Supplement Series, Vol. 1, 12 (1954).
2. H. Morinaga, *Phys. Rev.* **101**, 254 (1956).
3. D.M. Brink *Proceedings of the International School of Physics "Enrico Fermi"*, Varenna, 1965, Course 36, edited by C. Bloch (Academic Press, New York 1966), p. 247 (1966).
4. A.C. Merchant and W.D.M. Rae, *Nucl. Phys. A* **549**, 431 (1992).
5. N. de Takacsy and S. das Gupta, *Phys. Lett. B* **33**, 556 (1970).
6. H. Friedrich, L. Satpathy and A. Weiguny, *Phys. Lett. B* **36**, 189 (1971).
7. N. de Takacsy, *Nucl. Phys. A* **178**, 469 (1972).
8. A. Tohsaki, H. Horiuchi, P. Schuck and G. Röpke, *Phys. Rev. Lett.* **87**, 192501 (2001).
9. Y. Funaki, A. Tohsaki, H. Horiuchi, P. Schuck and G. Röpke, *Phys. Rev. C* **67**, 051306 (2003).
10. T. Yamada and P. Schuck, *Phys Phys. C* **69**, 024309 (2004).
11. G. Röpke and P. Schuck, *Mod. Phys. Letts. A* **21**, 2513 (2006).
12. Y. Funaki, A. Tohsaki, H. Horiuchi, P. Schuck and G. Röpke, *Eur. Phys. J. A* **28**, 259 (2006).
13. I. Sick and J.S. McCarthy, *Nucl. Phys. A* **150**, 631 (1970); A. Nakada, Y. Torizuka and Y. Horikawa, *Phys. Rev. Lett.* **27**, 745 (1971); and 1102 (erratum); P. Strehl and Th. H. Schucan *Phys. Lett. B* **27**, 641 (1968).
14. M. Itoh *et al*, *Proc. of the 8th Int. Conf. on Clustering Aspects of Nuclear Structure and Dynamics*, Nara, Japan, 2003, ed. K. Ikeda, I. Tanihata and H. Horiuchi (*Nucl. Phys. A* **738**, 268 (2004)).
15. H.O.U. Fynbo *et al*, *Nature* **433**, 136 (2005).
16. Y. Funaki, A. Tohsaki, H. Horiuchi, G. Röpke and P. Schuck, *Mod. Phys. Letts. A* **21**, 2331 (2006).
17. F. Azjenberg-Selove, *Nucl. Phys. A* **506**, 1 (1990).
18. M. Freer, *et al.*, accepted for publication in Phys. Rev. C.
19. Y. Kanada-En'yo, H. Horiuchi and A. Doté, *Phys. Rev. C* **60**, 064304 (1999).
20. J. A. Liendo *et al*, *Phys. Rev. C* **65**, 034317 (2002).
21. M. Freer *et al*, *Phys. Rev. Lett.* **96**, 042501 (2006).
22. H. Brunn, Über Verkettung, *S.-B.Math.-Phys.Kl. Bayer Akad. Wiss.* 22, 77 (1892).
23. N. Curtis, *et al.*, submitted for publication.

STUDIES OF EXOTIC SYSTEMS USING TRANSFER REACTIONS AT GANIL

D. BEAUMEL* , S. FORTIER, E. RICH, E. TRYGGESTAD, E. BECHEVA, Y. BLUMENFELD, F. DELAUNAY, N. FRASCARIA, S. GALES[1], L. GAUDEFROY, J. GUILLOT, F. HAMMACHE, E. KHAN, V. LIMA, J.A. SCARPACI, O. SORLIN

Institut de Physique Nucléaire, Université Paris-Sud-11-CNRS/IN2P3, France
** E-mail: beaumel@ipno.in2p3.fr*

A. FOMICHEV, S.V. STEPANTSOV, G.W. TER-AKOPIAN, R. WOLSKI[2]

FLNR/JINR, Dubna, Russia

A. GILLIBERT, N. KEELEY, V. LAPOUX, L. NALPAS, E.C. POLLACCO

CEA-Saclay DSM/DAPNIA/SPhN, F-91191 Gif-sur-Yvette,France

K.W. KEMPER

Department of Physics,Florida State University, Tallahassee,Florida 32306-4350,USA

P. ROUSSEL-CHOMAZ

Ganil, CEA/DSM-CNRS/IN2P3, Caen (France)

U. DATTA PRAMANIK

Saha Institute of Nuclear Physics, Kolkata ,India

D. SANTONOCITO

LNS Catania, INFN, Italy

Note: [1] Present address: GANIL, CEA/DSM-CNRS/IN2P3, Caen, France

Note: [2] Also: Institute of Nuclear Physics, Cracow, Poland

Investigation of unbound nuclear systems ^{9}He, ^{7}H and ^{4}n was performed at GANIL-SPIRAL via transfer reactions using the ^{8}He beam at 15.4 A MeV. The missing mass spectra were deduced from kinetic energies and emission angles of light ejectiles detected by the Silicon array MUST. The ^{9}He spectrum was obtained from the d(^{8}He,p)^{9}He reaction. In addition to previously known low-lying narrow resonant states, the neutron transfer reaction populates a peak just above neutron threshold, which may be identified with the $s_{1/2}$ ground state of ^{9}He. The d(^{8}He,^{3}He) and d(^{8}He,^{6}Li) reactions were used to search for resonant states in the t+4n (^{7}H) and 4n systems, respectively. The missing mass spectrum of the 4n system does not give evidence for the existence of a bound "tetraneutron". However the comparison with the results of 5-body phase-space calculations emphasizes the existence of correlations in the 4n system. The broad structure observed at $\simeq$1 MeV above the t+4n emission threshold is proposed to be the ground state of ^{7}H.

Keywords: transfer reactions with exotic beams, ^{9}He, ^{7}H, tetraneutron

1. Introduction

Reactions induced by light neutron-rich ion beams offer the opportunity to investigate unbound exotic few-nucleon systems with particularly high N/Z ratio, including eventual multineutron clusters (Z=0). Search for resonant states and subsequent determination of their energies and widths constitute an important challenge for the existing models of nuclear forces, as these properties may bring strong constraints on the poorly-known neutron-neutron and 3-and 4-body interaction parameters. The present paper reports on experimental results about the very neutron-rich unbound nuclear systems ^{9}He and ^{7}H and the tetraneutron 4n, which could be reached by means of transfer reactions induced by the 15.4 MeV/u ^{8}He beam delivered by the SPIRAL facility at GANIL. The experimental methodology consists in using a thin target of CD2 and detect the recoiling charged particle using the large acceptance Silicon array MUST[1] to deduce the energy of the populated states as well as the scattering angle. An identification of the quasi-projectile is also performed in coincidence in order to better select the reaction channel.

2. Shell inversion in ^{9}He

A spectacular effect of the evolution of shell structure as function of neutron-proton asymmetry is observed in N=7 isotones. The shell energy gap E(s1/2)-E(p1/2) decreases from ^{15}O to ^{11}B, and becomes negative in ^{11}Be, with ground state J_π being equal to 1/2+ instead of the normally expected 1/2- value. The existence of neutron halo in ^{11}Be and in ^{11}Li is in fact closely related to this parity inversion phenomenon, considering the small binding energy of the last neutron. In a first GANIL experiment using a ^{11}Be beam produced by fragmentation, we have studied the ^{11}Be(p,d)^{10}Be neutron-pickup reaction. Results have confirmed that the wave function of ^{11}Be ground state was predominantly s1/2.[2] In a second step, the ^{11}Be(d,^{3}He)^{10}Li proton-pickup reaction, which selectively excites s1/2 neutron states in ^{10}Li, was studied using the MUST array.[3] This experiment has demonstrated that the s1/2 single-particle state was located just above the ^{9}Li+n emission threshold, thus being the ^{10}Li ground state.

The main motivation of the experiment using the ^{8}He beam from SPIRAL, was to search for a possible p1/2-s1/2 shell inversion in the next N=7 nucleus ^{9}He, as previously observed in ^{11}Be and ^{10}Li. This very neutron-rich unstable nucleus was previously studied using (π^-,π^+),[4] (^{13}C,^{13}O)[5] and (^{14}C,^{14}O)[5] double-charge exchange reactions on a ^{9}Be target. A narrow resonant state at about 1.3 MeV above neutron threshold was identified to the ^{9}He ground state. On the other hand, results from a ^{11}Be fragmentation experiment[6] were found to be consistent with the existence of a s1/2 ground state at 0.1 MeV above threshold. In the present experiment, we used the (d,p) reaction, well-established as a classical and powerful spectroscopic tool for studying single-particle neutron states. The d(^{8}He,p)^{9}He reaction was studied at E=15.3 MeV/u using a thin (0.9mg/cm2) CD2 target. Energies of resonant states in ^{9}He were obtained by measuring kinetic energies and emission angles of protons detected at backward angles using MUST. Background could be suppressed by gating with ^{8}He (from ^{9}He decay) stopped in a plastic telescope placed downstream the target. The missing mass spectrum of ^{9}He is presented in fig. 1 (left). Proton angular distributions were compared with DWBA predictions for neutron transfer to unbound single-particle states (fig.1 (right)) in order to get information on transferred angular momentum and spectroscopic factor. Somewhat unexpected were the small natural widths (well below single-particle expectations) of ^{9}He states previously observed in charge-exchange reactions,[45] in contrast with neighbouring unbound nuclei which display very broad resonances. The present data confirms the existence of the two narrow low-energy resonant states (E=1.34 and 2.38 MeV) of refs,[45] The measured L=1 spectroscopic factor of the 1.34 MeV resonance is much lower than 1, indicating a strong fragmentation of the p1/2 single-particle state. As for the s1/2 neutron shell, the present (d,p) experiment also indicates that the ^{9}He ground state is located just above neutron threshold (E=0.1 MeV) and has $J^{\pi} = 1/2+$,in agreement with the results of ref.[6] The ^{9}He ground state is thus, like $^{10}Li_{gs}$, a s1/2 virtual state. Therefore the s1/2-p1/2 shell inversion observed in ^{11}Be and ^{10}Li also exists in the unbound nucleus ^{9}He. A theoretical explanation can be found in ref.[7] Such an inversion can be understood considering the properties of the spin-isospin dependent part of the nuclear interaction: the neutron $p_{1/2}$ orbit is pushed at higher energies as the nucleus loses protons in its spin-flip patner $p_{3/2}$ and dynamical correlations finally invert +1/2 and -1/2 states.

3. Search for neutral clusters ("tetraneutron")

The search for neutron clusters at SPIRAL has been triggered by the very intriguing result obtained in an experiment performed at GANIL[8] , studying the breakup of ^{14}Be. In this experiment, breakup neutrons were detected using the DEMON array, in coincidence with charged fragments measured using a Si-CsI telescope placed around zero degrees. The analysis revealed the presence of anomalous events in neutron detectors, that is events for which the deposited energy exceeds by far the energy deduced from the time-of-flight (assuming that a single neutron is detected). Moreover, these events were found to be in coincidence with ^{10}Be fragments. None were found in the other channels. A detailed study of free neutrons pileup lead to the conclusion that such process cannot be invoked to explain the data. It was then suggested that these events could be due to bound tetraneutrons. Later on, the same approach was used at SPIRAL to investigate the breakup of ^{8}He[9] , a good candidate for tetraneutron search. It was argued that strong clustering effect would imply a ground-state wave function of ^{8}He containing a large fraction of the alpha+4n component, which would favour the formation of a bound tetraneutron, and its observation after a breakup process. The presence of 13 anomalous events in neutron detectors is reported, as illustrated on Fig.2 (left). All these events are in coincidence with ^{4}He fragments (that is, in the A(beam)-4 channel) and none in coincidence with ^{6}He, providing another evidence in favour of tetraneutron observation. A very detailed analysis using 4 different methods shows that these events cannot be due to the pileup of free breakup neutrons. The analysis of the neutron-fragment relative angle spectrum built with these 13 events shows that these events cannot correspond to a bound tetraneutron. The situation is less clear concerning an eventual 4-neutron resonant state. The data are compatible with a resonance located below 1 MeV above the 4n threshold, but the poor statistics do not allow a definite conclusion.

Our group has proposed another approach based on a missing mass measurement to investigate this issue. Starting from the same 15 A.MeV ^{8}He beam, the alpha-transfer reaction ^{8}He(d,^{6}Li) was studied using the MUST array to which thin (65 μm thickness) Silicon layers were added in order to provide an unambiguous identification of ^{6}Li particles. Neutrons were also detected using thick plastic detectors so

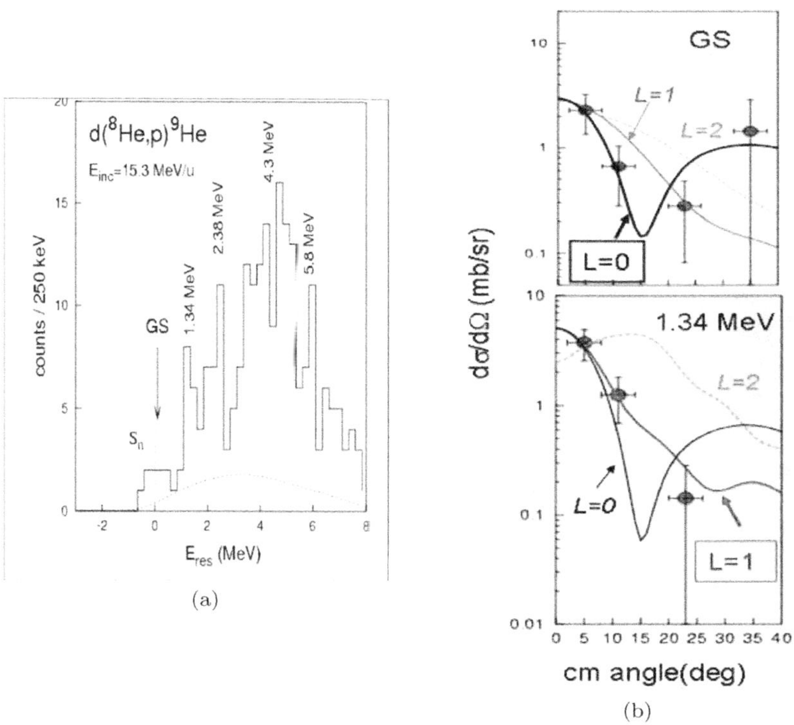

Fig. 1. (a) Scatter plot of the fragment identification parameter PID vs Ep/En (ratio of the deposited energy in neutron detector to the energy deduced by time of flight). (b) Excitation energy spectrum in the 4n system deduced from the ^{6}Li energy and gated by the plastic detectors.

as to improve the reaction channel selection. The excitation energy spectrum in the 4n system deduced from the ^{6}Li energy and angle is represented on Fig.2 (right). This spectrum is gated by the detection of at least one neutron in plastic detectors placed in the forward direction to detect recoil neutrons. Counts observed below the alpha+4n threshold can be accounted for the background originated from ^{12}C in the CD2 target. Namely, no bound tetraneutrons are observed in the limits of the statistics. Above the alpha+4n threshold, a smooth increase of the cross-section is observed. The data cannot be satisfactorily reproduced by a pure 5-body phase-space calculation, which represents a clear evidence of correlations between the 4 neutrons in the exit channel. These correlations has been investigated using simple simulations of 2n-2n interaction in the final-state. More sophisticated analysis is to be done in the near future.

On the theory side, state-of-the-art quantum Monte-Carlo calculations lead to the conclusion that a bound tetraneutron would not be compatible with our present knowledge on nuclear forces.[10] These calculations however cannot yet make reliable predictions concerning the unbound case. Recently, Faddeev method calculations could be achieved by Lazauskas and Carbonell[11] in which no resonant state were found. In conclusion, both experiment and theory seem to agree at the present time that bound four-neutron clusters do not exist. The situation is less clear concerning resonances, such state being more difficult to observe due to their width.

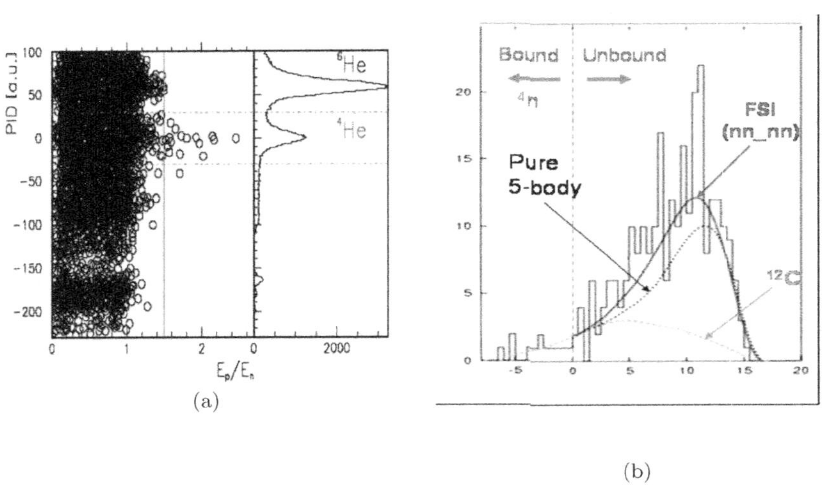

Fig. 2. (a) Scatter plot of the fragment identification parameter PID vs Ep/En (ratio of the deposited energy in neutron detector to the energy deduced by time of flight). (b) Excitation energy spectrum in the 4n system deduced from the ^{6}Li energy and gated by the plastic detectors.

4. The heavy hydrogen ^{7}H

Search for resonant states in heavy unbound Hydrogen nuclei and subsequent determination of their energies and widths constitute an important challenge for the existing models of nuclear forces, as their properties may bring strong constraints on the poorly-known neutron-neutron and 3- and 4-body interaction parameters. The heavy hydrogen ^{7}H represents an extremely neutron rich system, with N/Z=6, comparable to neutron star case.

The first evidence for a low-lying resonance in the p+6n system was obtained at RIKEN using the p(^{8}He,2p)^{7}H reaction at 61 A.MeV.[12] A structure located just above the t+4n threshold superimposed on a large background was observed. However the poor center of mass energy resolution (1.9 MeV) did not allow extracting accurate information on its energy and width.

Experimental information on this very exotic ^{7}H nucleus can be provided by missing mass measurements using a secondary beam of ^{8}He. The setup of the tetraneutron experiment described above allowed a simultaneous investigation of the ^{7}H production channel through the proton pick-up reaction (d,^{3}He), using the ^{8}He SPIRAL beam at 15.3 A.MeV. Figure 3 displays the missing mass spectrum of ^{7}H, obtained after subtracting the contribution of Carbon atoms of the CD2 target. The dotted line shows the 6-body phase space convoluted with efficiency. The existence of a low-lying resonance in the ^{7}H system above t+4n threshold is clearly confirmed by the present data. The resonance energy 1.56±0.27 MeV and width $\Gamma = 1.74 \pm 0.72$ MeV are deduced by fitting a Breit-Wigner curve to the experimental points.

A second SPIRAL experiment has investigated ^{7}H using a similar missing-mass method but with different experimental setup.[14] The one-proton transfer reaction ^{12}C(^{8}He,^{13}N)^{7}H induced by the 15.4 A.MeV ^{8}He beam of SPIRAL on a ^{12}C gas target was measured. The main part of the experimental setup was the Time-Charge Projection Chamber MAYA,[13] used as a detector for the reaction products as well as a gas target cell. By detecting the recoiling ^{13}N gated by tritons, the ^{7}H spectrum could be obtained. The statistics is low but the spectrum is clean. A resonant state of ^{7}H is observed , with an estimated resonance position at 0.57 +0.42-0.21 MeV above the ^{3}H+4n threshold, and a width $\Gamma = 0.09 + 0.94 - 0.06$ MeV.

The resonance characteristics found in the two SPIRAL experiments are somewhat different but, together with the early RIKEN experiment of Korshenninikov, the three set of data are pointing to the very important result that ^{7}H exists as a resonance near the t+4n threshold.

5. The MUST2 array

A major upgrade of the MUST array, called MUST2, has been achieved recently, with the aims of significantly increasing the angular coverage, improving the detection capabilities of multiparticle events, and allowing easy coupling with gamma-ray detectors such as EXOGAM. This new device relies on the same structure as MUST, namely a modular set of three-stages telescopes composed of Double Sided Silicon

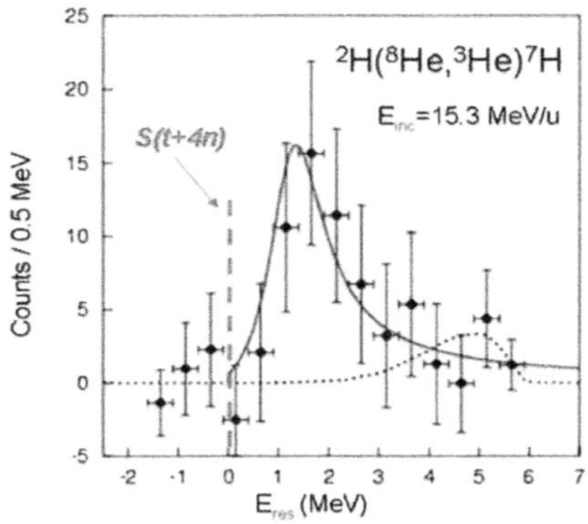

Fig. 3. Excitation energy spectrum obtained from the recoiling ^{3}He. Zero energy corresponds to the ^{3}H+4n threshold. The resonance energy and width are deduced by fitting a Breit-Wigner curve (solid line) to the data (see text). The dotted line corresponds to a 6-body phase-space convoluted with the detection efficiency.

Strips detectors (DSSSD), Si(Li) and CsI crystals with photodiode readout. The first layer consists of new DSSSD, 300mm thick, offering a large active surface of 10x10cm2, with customized PCB frames and 128 strips on each side. These detectors are manufactured by MICRON SC company from 6" silicon wafers. For the second layer, 5mm thick Lithium drifted Silicon detectors of two different types have been used. The first type, built by FZ Julich laboratory, has a surface 5x9 cm2 and is segmented in 8 pads, while the second, made at IPN Orsay, has reduced size of 5.5x5.5 cm2 and 4 pads. Two detectors of the first type are needed per telescope, and four of the second type. Dead areas are equivalent for both cases. Finally, the last stage is equipped with 16 tallium doped CsI crystals provided by SCIONIX company, with photodiode readout. These modules are in geometrical correspondence with the Si(Li) detector pads. The thickness of each module is 4cm, which allows the energy meaurement of protons up to 120 MeV (taking into account the Si and Si(Li) layers). To handle the increase of number of channels as compared to MUST, a completely new electronics was designed and built. Its basic component is an new innovative ASIC (Application Specific Integrated Circuit) comprising 16 bipolar channels, with energy and time measurement. The 16 channels are multiplexed and require only one analogic line. The ASICs are mounted on motherboards, 9 units per board, placed in a liquid-cooled cupper block at the back of each telescope, operating under vaccum. Time and energy informations are sent to a specifically developped VXI board integrating 14 bits ADC, performing the coding and a pretreatment of the events. The readout, control, and storage of the data is taken in charge by the GANIL data acquisition system. At the present time, the MUST2 array comprises 6 telescopes and is operational since the end of 2006.

6. Conclusion

The light neutron-rich ^{8}He beam delivered by SPIRAL at GANIL has allowed access to the neigbouring very exotic systems ^{9}He, ^{7}H and 4n, by means of transfer reactions performed in inverse kinematics. Data on the ^{8}He(d,p) reaction show evidence of the occurence of 2s1/2-1p1/2 shell inversion in the ground state of ^{9}He, as observed in other N=7 isotones such as ^{10}Li and ^{11}Be. The heavy hydrogen ^{7}H could be searched through the ^{8}He(d,^{3}He) reaction. The observed broad resonance above the t+4n threshold is attributed to its ground state. Finally the alpha-transfer reaction (d,^{6}Li) has allowed to obtain a missing mass spectrum of the 4n system. No evidence in favour of a bound state is deduced. On the instrumental side, a new generation array, MUST2, is now operational to perform further direct reaction studies with improved capabilities.

References

1. Y. Blumenfeld et al., *Nucl. Instr. Meth. Phys. Res. A* **421**, 471 (1999).
2. S. Fortier et al., *Phys. Lett. B* **461**, 22 (1999); J.S. Winfield et al., *Nucl. Phys. A* **683**, 48 (2001).
3. S. Pita, Ph.D. Thesis, University of Paris VI (2000)
4. K.K. Seth et al., *Phys. Rev. Lett.* **58**, 19 (1987).
5. H.G. Bohlen et al., *Z. Phys. A* **330**, 227 (1988); H.G. Bohlen et al., *Prog. Part. Nucl. Phys* **42**, 17 (1999).
6. L. Chen et al., *Phys. Lett. B* **505**, 21 (2001).
7. T. Otsuka et al., *Phys. Rev. Lett.* **87**, 082502 (2001).
8. F.M. Marquès et al., *Phys. Rev. C* **65**, 044006 (2002).
9. V. Bouchat, Ph.D. thesis, Université libre de Bruxelles (2005)
10. S. Pieper, *Phys. Rev. Lett.* **90**, 252501 (2003).
11. R. Lazauskas and J. Carbonell, *Phys. Rev. C* **72**, 034003 (2005).
12. A.A. Korsheninnikov et al., *Phys. Rev. Lett.* **90**, 08251 (2003).
13. W. Mittig et al., *Nucl. Phys. A* **722**, 10c (2003).
14. M. Caamano et al., *Phys. Rev. Lett.* **99**, 062502 (2007).

FIRST RESULTS FROM THE MAGNEX LARGE-ACCEPTANCE SPECTROMETER

A. CUNSOLO[A,B], F. CAPPUZZELLO[A,B], M. CAVALLARO[A,B], A. FOTI[B,C], A. KHOUAJA[A], S.E.A. ORRIGO[A,B], T. BORELLO-LEWIN[D], M.R.D. RODRIGUES[D], M.D.L. BARBOSA[D], H. PETRASCU[D], J.S. WINFIELD[E],

a) INFN Laboratori Nazionali del Sud, Catania, Italy

b) Dipartimento di Fisica e Astronomia, Università di Catania, Italy

c) INFN Sezione di Catania, Catania, Italy

d) IFUSP, São Paulo, Brazil

e) GSI, Darmstadt, Germany

f) NIPNE, Bucharest, Romania

Abstract: *The MAGNEX large-acceptance spectrometer was commissioned with beams from the LNS Tandem. The optics was tested with elastically scattered ^{7}Li, ^{16}O and ^{48}Ti beams with apertures mounted after the target. A demonstration of the particle identification capabilities of the PSD start detector and the focal plane detector was given by a measurement of the $^{19}F(^{7}Li,^{7}Be)^{19}O$ charge exchange reaction at $25° \pm 5°$. The charge state distribution of scattered ^{48}Ti ions was also measured and found to be in good agreement with predictions. Preliminary results of the software reconstruction of excitation energy and angles, which used matrices based on a 3D-interpolation of the measured field maps, are encouraging.*

1 Introduction

The MAGNEX large-acceptance spectrometer [1,2] was designed particularly for RNBs from the EXCYT facility at the LNS, although there are several advantages that carry over to experiments induced by stable beams from either the Tandem or SC cyclotron. The spectrometer relies partly on hardware (especially the shape of the entrance and exit dipole pole tips) and partly on software (ray reconstruction) to overcome the strong aberrations inherent in the 50-msr acceptance.

Apart from the final scattering chamber, the spectrometer was complete and ready for the first beam to target and scattered particles through to the focal plane in April 2005. The Focal Plane Detector (FPD) [3] continued to be tested with an α-source during the summer and was further tested with scattered beams from the LNS Tandem in December.

The spectrometer was ready for initial commissioning in February 2006. "Initial" because the small scattering chamber ("*Cameretta*") with a fixed exit angle port at 25° was used. Furthermore, the fields for the magnetic elements – Quadrupole, Dipole and α surface coils – were set according to estimated values rather than from those derived from the measured field maps. The β surface coil, for which the effect on the focal surface is rather complicated, was not energized in general, apart from a brief test to show that it did have some effect.

2 Experimental details

The optics in general, was tested with elastically scattered ^{16}O, ^{7}Li and ^{48}Ti beams on gold and nickel targets. Various apertures were mounted about 10 cm after the target, including one corresponding to the nominal 50-msr full acceptance and an aperture with a set of 1-mm holes. The microchannel-based start detector (PSD) described in ref. [3] was installed in a small chamber after the "*Cameretta*".

The FPD was fixed at a position roughly corresponding to the intermediate kinematic factor $K = -0.25$. Normally, the FPD should be moved closer to or further from the dipole exit, in order to compensate the kinematic shift of the focal plane for different reactions. However, since the remote control system for this movement was not finished, we decided to compensate changes in K by use of the αsurface coil.

2.1 FPD Parameters

Suitable isobutane gas pressures and proportional wire voltages were studied for the different heavy ions detected. The choice of gas pressure is essentially a compromise between producing enough primary electrons through ionization and minimizing the straggling of the ions passing through the detector. The anode wire voltages need to be sufficiently high to induce charges to produce signals above noise on at least three consecutive strips (pads), while keeping in the proportional regime. However, in practice the voltage is often limited by the saturation of the preamplifier for the energy-loss signal, which is taken from the wire. Some typical pressures and voltages used during the tests are given in Table 1.

Table 1: Typical FPD isobutane gas pressures and proportional and the silicon detector that triggered the event. The final wire voltages for various detected ions. "Energy" is after the scattering at 25° and after energy-loss in the PSD foil.

Ion	Energy (MeV)	Pressure (mbar)	Voltage (V)
^{7}Li	50	30	750
^{16}O	75	20	650
^{48}Ti	100	10	490

2.2 FPD counting rate and data acquisition dead time

With such a large gas-filled detector (dimensions 160 cm long, 25.5 cm tall, 50 cm deep), the acceptable counting rate to maintain good particle identification is mostly limited by the drift time of the electrons and the build up of positive ions ("space charge"). Since we have a rather long acquisition readout time (224 multiplexed signals must be processed by the VME CRAMS modules), the data acquisition dead time may also be significant. These two count-rate dependent parameters were measured by setting the spectrometer to detect the elastic scattering of ^{48}Ti on a gold target and, run by run, increasing the beam current. Table 2 gives the measured counting rate in the detector (the number of silicon detector triggers) and the resulting dead time of the acquisition system (the ratio of the number of accepted events to triggers). Only at the highest beam current studied, corresponding to a trigger rate of 1.75 kHz, was there a noticeable deterioration of the ΔE and horizontal position signals.

Table 2: Event rate and dead-time for various beam currents (electrical) of ^{48}Ti measured in our Faraday cup.

Ibeam (nA)	**Rate (Hz)**	**Dead time (%)**
1.8	170	9
4.4	390	18
8.5	750	29
15.0	1320	42
20.0	1750	49

2.3 Particle identification

A demonstration of the particle identification capabilities of the PSD and FPD was given by the selection of ^{7}Be after the ^{7}Li beam impinged on a target of AlF_3 (this was a test of a future $^{19}F \rightarrow {}^{19}O$ charge exchange reaction, discussed also in Section 3). First a cut (condition) was applied to the $Z = 4$ line in a spectrum of ΔE *vs*. Er. Here ΔE is the energy-loss signal in the gas under the 68-mm deep anode section of the ionization chamber, corrected by $\cos(\theta_{foc})$ for the angular variation in track length, and Er is the "OR" of any of the (calibrated) silicon energies.

The second step is to select the mass by applying a condition to a matrix of TOF *vs*. X_{foc} (gated by $Z = 4$), where the TOF is the time difference between the PSD and the silicon detector that triggered the evenet. The final result, passed through the software reconstruction procedure to correct for flight-path differences, is shown in Fig.1.

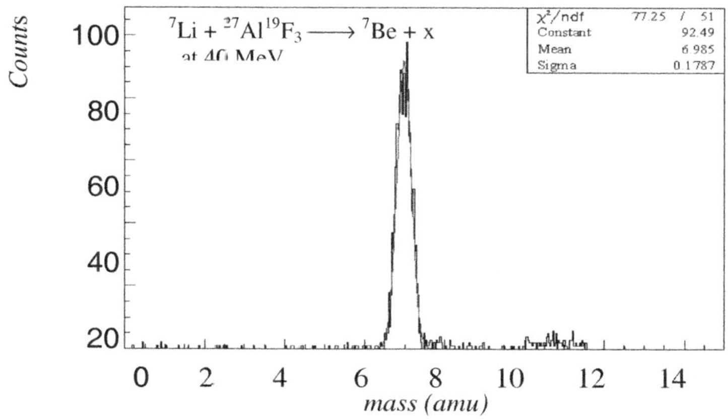

Fig. 1. Reconstructed mass spectrum of ^7Be based on TOF gated by a $\Delta E - Er$ cut.

2.4 Charge state measurement

The charge state distribution of 120 MeV ^{48}Ti ions elastically scattered from a gold target was measured by taking separate runs with the magnets set to the appropriate rigidity in turn. The integrated counts of the peaks, scaled by the Faraday cup integrated charge and corrected for the acquisition dead time are plotted in Fig. 2. They are in good agreement with predictions of algorithms taken from the code INTENSITY [5].

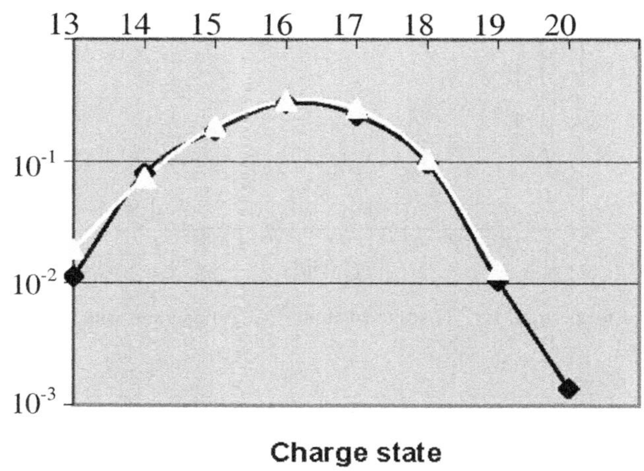

Fig. 2. Charge state distribution of scattered 120 MeV ^{48}Ti ions. Diamonds: measurements by MAGNEX. Triangles: predictions by algorithms from ref. [5].

3 Angle and Position Spectra

The horizontal position X_{foc} in the FPD is taken from the drift chamber (DC) closest to the gas window. The vertical position Y_{foc} is obtained from the drift time of the electrons. The horizontal and vertical angles, θ_{foc} and φ_{foc}, are calculated from the first and subsequent DCs. Normally, the initial vertical angle φ_i is taken directly from the PSD. For the initial commissioning this was not implemented, so φ_i had to be reconstructed from φ_{foc}. Because of the small φ_{foc}/φ_i, the reconstruction of φi from is compromised by straggling and detector resolution [2].

The software reconstruction [6] to obtain the initial (scattering) angle and energy uses the COSY INFINITY program [7] to calculate the transport matrices and their inverses. Although it was not available at the time of the experiment, the complicated 3D-interpolation of the maps of the measured fields is now complete and we are able to use realistic fringe fields in the reconstruction. This analysis is still in progress. Nevertheless, the preliminary results are encouraging: see the reconstructed θ_i spectrum shown in Fig. 3. For this measurement an aperture with 1 mm diameter holes was mounted after the target; the horizontal separation between holes subtending 73 mrad. The width of the first peak in the figure is consistent with the size of the holes.

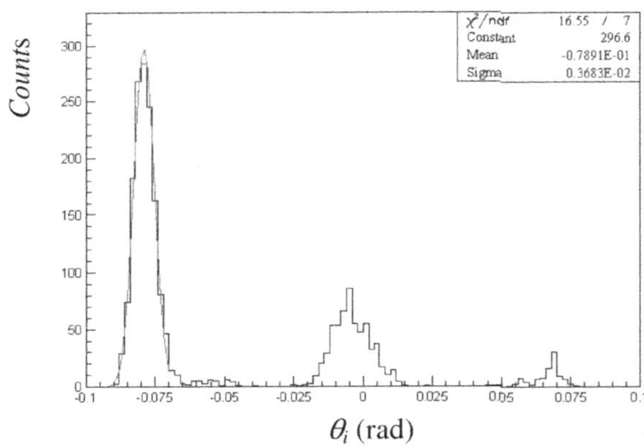

Fig 3: Reconstructed θ_i spectrum for ^{16}O scattered from ^{58}Ni. An aperture with 1 mm holes was located after the target.

To check the momentum dispersion of the spectrometer, two measurements of X_{foc} were made with the same magnet settings but with the energy of the ^{48}Ti beam from the Tandem changed by a few percent. After adjusting the energies for the loss in the target and PSD foil, we obtained a dispersion of 3.68 cm/% for a K of -0.104, which agrees well with the ion optical design.

Figure 4 shows a comparison between the focal plane position spectrum X_{foc} and the reconstructed excitation energy spectra *Exc* for the $^{19}F(^{7}Li,^{7}Be)^{19}O$ charge exchange reaction mentioned in Section 3. The doublet near 1.5 MeV in *Exc* corresponds to the ground state of ^{19}O with the ejectile ^{7}Be in its ground and 0.429 MeV excited state (there also may be the unresolved 96 keV state of ^{19}O contributing to both ^{7}Be peaks). In the X_{foc} spectrum the doublet is not resolved. Because of the large scattering angle, as well as the PSD (3.5 μm) and FPD gas-window (6 μm) foil thicknesses – both of which were much thicker than would be used in the actual experiment – the resolution obtained does not reflect the true capabilities of the spectrometer (the energy-loss in the PSD foil as well as that in the target explains why the ground state of ^{19}O is not at 0 MeV, since those effects were not included in the reconstruction). The figure is intended only as a preliminary indication of the potential of the ray reconstruction procedure.

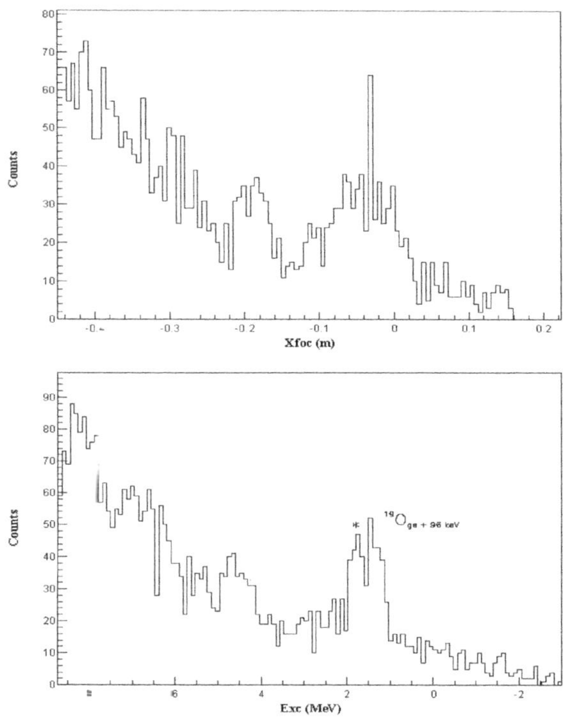

Fig 4. Focal plane position spectrum (top) and software-reconstructed excitation energy spectrum (bottom) for the (^{7}Li, ^{7}Be) charge-exchange reaction on an AlF_3 target. The peak marked with an asterisk is associated to the ejectile in its 0.429 MeV excited state.

References

[1] A.Cunsolo et al., *Nucl. Instr. Meth. A* **481**, 48 (2002).

[2] A.Cunsolo et al., *Nucl. Instr. Meth. A* **484**, 56 (2002).

[3] A.Cunsolo et al., LNS Activity Report 2004, p. 113.

[4] A.Cunsolo et al., *Nucl. Instr. Meth. A* **495**, 216 (2002).

[5] J.A.Winger et al., *Nucl. Instr. Meth. B* **70**, 380 (1992).

[6] A.Lazzaro et al., Proc. of the 7th Int. Computational Accelerator Physics Conf., East Lansing, Michigan, Oct. 2002 (M. Berz and K. Makino, eds.) IOP conf. series 175 (2005) p.171.

[7] M.Berz and K.Makino, COSY INFINITY version 8.1.

THE ICHOR PROJECT AND SPIN-ISOSPIN PHYSICS WITH UNSTABLE BEAMS

H. SAKAI for ICHOR collaboration

Department of Physics, University of Tokyo,
7-3-1 Hongo, Bunkyo, Tokyo 113-0033, Japan
**E-mail: sakai@phys.s.u-tokyo.ac.jp*
www.nucl.phys.s.u-tokyo.ac.jp

The ICHOR project is introduced. It aims to explore the unexplored spin-isospin responses of nuclei such as IVSM or DGTGR by using the exothermic heavy-Ion charge-exchange reactions employing radioactive beams.

Keywords: isovector spin giant resonances, exothermic heavy-ion charge-exchange reaction

1. Introduction

The spin-isospin interaction produces various interesting responses such as the Gamow-Teller (GT) β-decay, GT giant resonance (GTGR), isovector-spin monopole giant resonance (IVSM), spin dipole resonance (SDR), double GT giant resonance (DGTGR) etc. depending on momentum and/or energy of nuclear system.[1] The knowledge of those responses often plays an essential role in understanding the astrophysical phenomena. So far the spin-isospin responses have been extensively studied mainly by using the charge-exchange (CE) reactions with stable beams leading to endothermic reactions.

Generally the spin-isospin responses under consideration ($\Delta L = 0, \Delta S = 1$) are most strongly excited at small momentum transfer (preferably $q = 0$) by the reaction. However, the reaction kinematics due to the endothermic reaction is such that the excitation of nucleus is always associated with the finite momentum transfer ($\omega < q$ (space-like region)). Therefore it becomes extremely difficult to search for spin-isospin responses in the highly excited region due to the increasing momentum transfer.

The exothermic Heavy-Ion Charge-Exchange (HICE) reaction employing radioactive projectiles having a large positive reaction Q-value allows us to access the so far unexplored $\omega > q$ region (time-like region), even realizing $q = 0$ condition. The energy range of 100-300 MeV/A is most suited for the study of spin-isospin responses.

We initiated the ICHOR project[a] (Isospin-spin responses in CHarge-exchange exOthermic Reactions) in 2005 which aims to explore these spin-isospin responses by using the HICE reactions.

The ICHOR project consists of two parts.

(1) Study of the intermediate states of double-beta decay nuclei by (p, n) and (n, p) reactions. We identify the GT strength distributions going to/from intermediate states which can be used to analyze the two neutrino double-beta decay process which is a part of the double Gamow-Teller transitions.
(2) Search for the new spin-ispspin responses with the exothermic HICE reactions by construction of the magnetic spectrometer SHARAQ. It is designed exclusively for the radioactive beam experiments having a capability of full dispersion matching with a radioactive beam production beam line. It is under construction at the RI Beam Facility (RIBF) at RIKEN.

In the following sections, we explain the exothermic HICE reaction and show two examples of research by the exothermic HICE reactions Then we explain the status of the magnetic spectrometer SHARAQ which is under construction.

[a]Grant-in-Aid for Specially Promoted Research of MEXT, Japan. Spokesperson : H. Sakai.

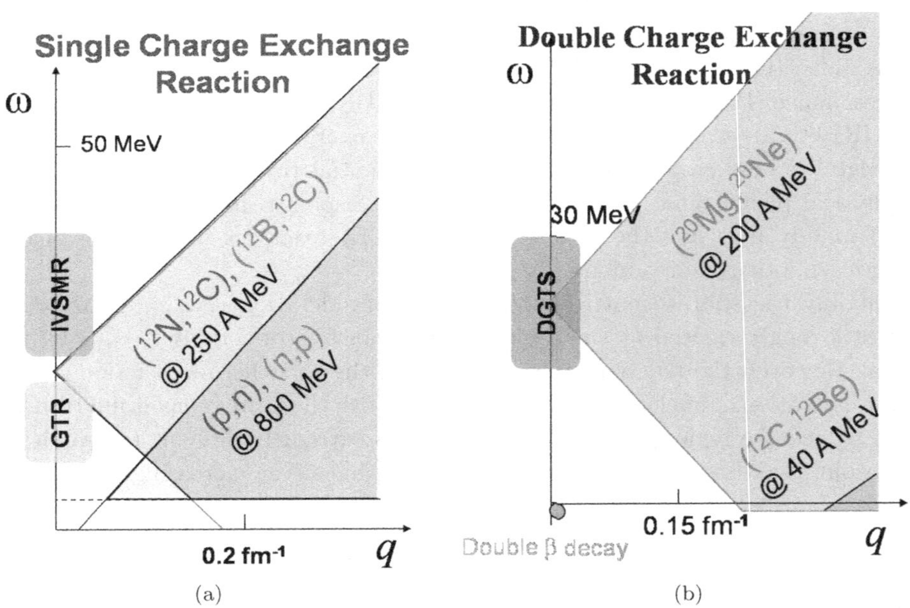

Fig. 1. Kinematical condition for (a) the study of the IVSMR and (b) the study of the DGTGR using unstable and stable nuclear beams indicated in the figure.

2. Exothermic HICE reaction

For the nuclear reaction of $A(a,b)B$, momentum transfer q, excitation energy ω of the particle B and the mass difference $\Delta m(\Delta M)$ between the particle $a(A)$ and $b(B)$ are expressed by

$$\begin{aligned}\vec{q} = \vec{p}_B - \vec{p}_A &= \vec{p}_a - \vec{p}_b, \\ \omega &= M'_B - M_B \\ \Delta m &= M_a - M_b \quad , \\ \Delta M &= M_A - M_B,\end{aligned}$$

where $\vec{p}_i$ and M_i denote the momentum and the rest mass of particle i, respectively. It is often defined $Q_{gg} = \Delta m + \Delta M$ in a heavy ion reaction. For charge exchange reaction, $|\Delta m| \ll M_a$, with small energy and momentum transfer compared to the rest mass of the relevant particle at small scattering angle ($q \ll M_A$, $q \ll \sqrt{M_a^2 + p_a^2}$, and $\theta \ll 1$), the relationship between ω and q is expressed as:

$$\begin{aligned}\omega &\simeq \frac{\Delta m}{\gamma} + \Delta M \pm \beta\sqrt{q^2 - (p_a\theta)^2}, \\ (\beta q)^2 &\simeq \left(\omega - \frac{\Delta m}{\gamma} - \Delta M\right)^2 + (\beta p_a \theta)^2 \geq \left(\omega - \frac{\Delta m}{\gamma} - \Delta M\right)^2 \quad ,\end{aligned} \tag{1}$$

where β and γ denote the velocity and the γ factor of the incident particle a. Because of $\Delta m < 0$ and $\Delta M < 0$ for the charge exchange reaction with a stable beam and stable target combination, the momentum transfer q is always greater than ω/β, i.e. restriction to the space-like region (see Eq. (1)). However, if we use a charge exchange reaction from an unstable nuclear projectile with a large mass excess to a stabler ejectile, namely $\Delta m \gg 0$, we can access to the time-like region ($q < \omega$) when $\Delta m/\gamma + \Delta M > 0$ in the right hand side of Eq. (1) even though $\Delta M < 0$.

3. Excitations of IVSMR and DGTR by exothermic HICE reactions

To show the merit of the exothermic HICE reactions, two examples of research are briefly introduced.

3.1. *IVSMR*

The first example is the study of the isovector spin monopole resonances (IVSMR) in both β^- and β^+ directions. There are some experimental hints on IVSMR(β^-)[2,3] but none exists on IVSMR(β^+). Excitation energies and widths provide us very important information on the nuclear matter compressibility with spin degrees of freedom.

In addition the precise neutron skin thickness $\delta_{np} = \sqrt{< r^2 >_n} - \sqrt{< r^2 >_p}$ can be extracted reliably from the IVSMR study. The δ_{np} value gives a strong constraint on the neutron matter equation-of-state (EOS) and consequently on neutron stars.

The δ_{np} value was experimentally extracted by Yako[4] for the first time employing the model independent spin-dipole sum-rule,

$$S_- - S_+ = \frac{9}{4\pi}[N < r^2 >_n - Z < r^2 >_p], \quad (2)$$

where $S_-(S_+)$ is the total spin-dipole strength for the $\beta^-(\beta^+)$ direction which can be obtained experimentally. The result was $\delta_{np} = 0.07 \pm 0.04$ fm for ^{90}Zr. Unfortunately, the uncertainty of ± 0.04 fm is not enough to constrain the effective interactions which are used for the neutron matter EOS calculations.[4,5]

However there is a chance to extract more precise δ_{np} value by using the model independent spin sum-rule for the spin monopole transitions which is

$$S_- - S_+ = 3[N < r^4 >_n - Z < r^4 >_p], \quad (3)$$

where $< r^4 >_n$ and $< r^4 >_p$ are the mean quadruple radii (mqr) of the neutron and proton distributions, respectively. Because of the r^4 dependence, it is highly sensitive to the neutron/proton distribution. Please be aware it is the r^2 dependence for SDR (see Eq. 2). Thus one can probably deduce more reliably the $\delta_{np} = \sqrt[4]{< r^4 >_n} - \sqrt[4]{< r^4 >_p}$ value from the measured S_- and S_+ values. For this purpose one needs to derive the total iso-vector spin-mpnopole strengths for both β^- and β^+ directions from the experiments.

For the study of IVSMR, the exothermic (^{12}N,^{12}C) and (^{12}B,^{12}C) reactions should be a good choice for β^- and β^+ directions, since they have large positive Δm values, namely +16.83 MeV and +13.88 MeV, respectively together with $\Delta S = 1$. A kinematic relation between ω and q is schematically plotted in Fig. 1(a). The IVSMR is expected to be located at around $\omega \sim 35(\sim 15)$ MeV for $\beta^-(\beta^+)$ direction. It is clear from Fig. 1(a) that one can achieve small momentum transfer for exciting those IVSMRs in the high excitation energies.

3.2. *DGTR*

The double Gamow-Teller giant resonance (DGTR), which is a two-phonon state, the GT giant resonance built on the GT giant resonance and which is of particular importance in relevance to double beta decay, has never been observed, while other multi-phonon giant resonances, such as double-isobaric analogue and double-giant dipole resonances, have been observed via the (π^+,π^-) reaction. There are several attempts to search for DGTR using the endothermic (^{18}O,^{18}Ne)[6] or (^{11}B,^{11}Li) reactions, which were not successful. However the exothermic (^{20}Mg,^{20}Ne) reaction, for example, with a large mass excess of Δm =25.6 MeV can achieve easily an excitation energy of ~20 MeV with $q = 0$, which could be an essential condition to excite DGTR as can be seen in Fig. 1(b). This is in striking contrast to the (^{18}O,^{18}Ne) and the (^{11}B,^{11}Li) reactions where larger momentum transfers than 0.75 fm^{-1} are unavoidable.

4. Magnetic Spectrometer SHARAQ

The SHARAQ spectrometer is a high resolution magnetic spectrometer designed for radioactive isotope (RI) beam experiments at the RI Beam Factory (RIBF). The spectrometer will be used in nuclear physics experiments in combination with a variety of radioactive isotope (RI) beams from BigRIPS.[7]

The SHARAQ spectrometer can analyze particles with magnetic rigidity of 1.8–6.8 Tm, corresponding to an energy of 40–440 MeV/A for $A/Z = 2$ particles. It is designed for momentum and angular resolutions of $\delta p/p \sim 1/15000$ and $\delta\theta \sim$ 1 mrad, respectively. The SHARAQ can be rotated around its target position in the range from $-2°$ to $15°$ to allow 0 degree and forward angle measurements. The designed parameters of the SHARAQ magnetic spectrometer are indicated in Table 1. Fig. 2 shows (a) an overview of the SHARAQ spectrometer and (b) the particle trajectories.

RI beams, produced by the primary heavy-ion beams in BigRIPS, are transported to the SHARAQ target position through the BigRIPS beam-line and a SHARAQ beam-line which is in preparation.[8] The beam-line and the spectrometer are designed to fulfill dispersion matching conditions both in both lateral and angular dispersions simultaneously which are crucial to achieve the high spacial and angular resolving powers of the spectrometer because of the large intrinsic momentum spread of RI beams. The beam-line is also designed to operate in an achromatic mode to provide a small beam spot at the target position for diagnostic and experimental purposes.

The SHARAQ spectrometer consists of three quadrupole (Q) and two dipole (D) magnets in the order of QQDQD.[9] The first quadrupole doublet (SDQ) consists of two superconducting quadrupoles (Q1 and Q2). The following normal-conducting

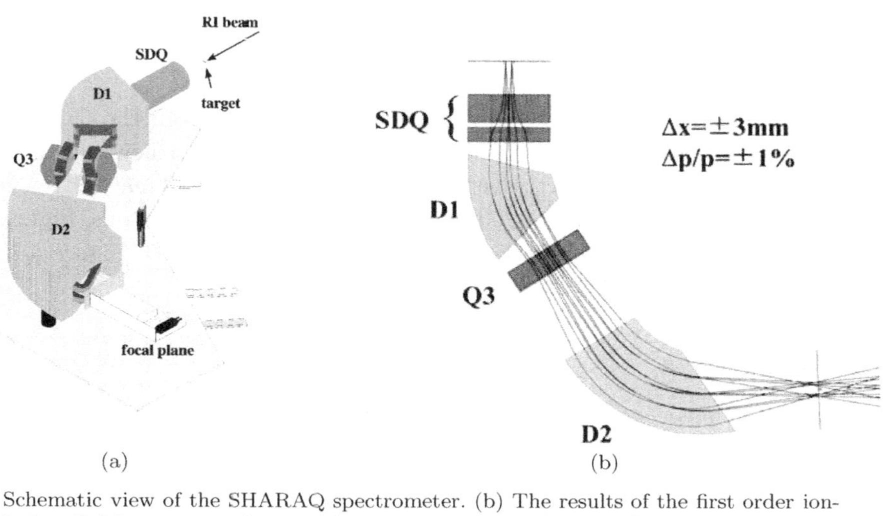

(a) (b)

Fig. 2. (a) Schematic view of the SHARAQ spectrometer. (b) The results of the first order ion-optical calculation. Solid curves are the particle trajectories in the horizontal plane.

Fig. 3. Photograph of the D2 magnet.

Table 1. Design parameters of the SHARAQ spectrometer.

Dispersion (D) [m]	5.86
Horizontal magnification (M_x)	0.40
D/M_x [m]	14.7
Resolving power (for image size of 1 mm)	14700
Vertical magnification	0.0
Angular resolution	<1 mrad
Vertical acceptance [mrad]	±50
Horizontal acceptance [mrad]	
in dispersion matching mode	±17
in achromatic beam line mode	±30
Solid angle [msr]	
in dispersion matching mode	2.7
in achromatic beam line mode	4.8

dipole (D1) and quadrupole (Q3) magnets are recycled from the decommissioned spectrograph SMART. The second dipole magnet, D2, is a new magnet specially designed to achieve the spectrometer specifications.[10] The D2 magnet is a 60° bending magnet with a pole gap of 200 mm and it weighs about 300 tons. Figure 3 shows the assembled D2 magnet at the factory in the spring of 2007. Figure 4 shows the location of the SHARAQ spectrometer in the RIBF.

5. Summary

The ICHOR project which has started in 2005 is briefly introduced. It is an abbreviation of the Isospin-spin responses in CHarge-exchange exOthermic Reactions and it consists mainly of two parts: study of the intermediate states of double beta decay nucleus by the (p, n) and $(n.p)$ reactions and investigation of IVSMR and/or DGTR by the exothermic HICE reactions. The merit of using the exothermic HICE reac-

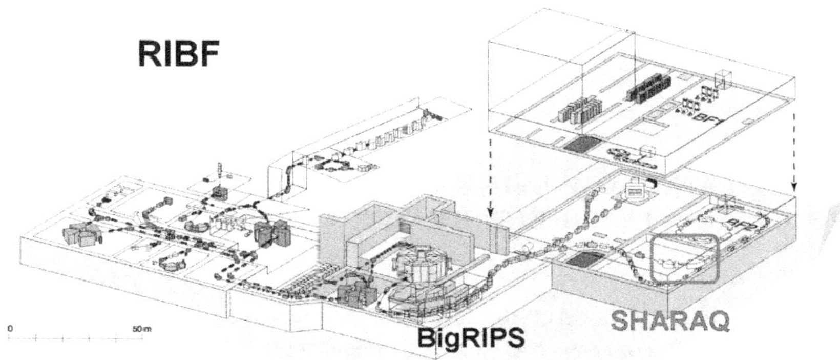

Fig. 4. Schematic bird view of RIBF and the SHARAQ spectrometer.

tions is stressed. For the latter purpose, the high-resolution SHARAQ spectrometer dedicated to the secondary RI beam is now under construction at RI Beam Factory (RIBF) of RIKEN. It is designed to achieve a high momentum resolution of $\delta p/p$=1/14700 and a high angular resolution of $\delta\theta \sim 1$ mrad. The spectrometer is scheduled to be operational in 2008.

The author thanks Profs. Shimoura and Uesaka for allowing to use some materials from Refs. [11,12]. This work is supported by the Grant-in-Aid of Specially Promoted Research (Grant No. 17002003) of the Ministry of Education, Culture, Sports, Science, and Technology of Japan.

References

1. M. Ichimura, H. Sakai and T. Wakasa, *Prog. Part. Nucl. Phys.* **56**, 446 (2006).
2. D.L. Prout *et al.*, *Phys. Rev. C* **63**, 014603 (2000).
3. R.G.T. Zegers *et al.*, *Phys. Rev. Lett.* **90**, 202501 (2003).
4. K. Yako, H. Sagawa and H. Sakai, *Phys. Rev. C* **74**, 051303(R) (2006).
5. H. Sagawa *et al.*, *Phys. Rev. C* **76**, 024301 (2007).
6. J. Blomgren *et al.*, *Phys. Lett. B* **362**, 34 (1995).
7. T. Kubo, *Nucl. Instrum. Methods Phys. Res. B* **204**, 97 (2003).
8. T. Kawabata *et al.*, in XIth International Conference on Electromagnetic Isotope Separators and Techniques Related to their Applications (EMIS2007), Deauville, 2007.
9. T. Uesaka *et al.*, CNS Annual Report 2005.
10. G. P. Berg, "Design Study of a New Dipole Magnet D2 for the SHARAQ Spectrometer", *unpublished.*
11. S. Shimoura, *ibid.* in EMIS2007.
12. T. Uesaka, *ibid.* in EMIS2007.

STRUCTURE AND LOW-LYING STATES OF THE ^{8}HE EXOTIC NUCLEUS VIA DIRECT REACTIONS ON PROTON

V. LAPOUX*, N. KEELEY, F. SKAZA, N. ALAMANOS, F. AUGER, A. DROUART, A.GILLIBERT, L. NALPAS, E. C. POLLACCO and J-L. SIDA

DSM/DAPNIA/SPhN, CEA-Saclay,
Gif-sur-Yvette, F-91191, France
**E-mail: vlapoux@cea.fr*

D. BEAUMEL, E. BECHEVA, Y. BLUMENFELD, F. DELAUNAY and J-A. SCARPACI

IPN-Orsay, CNRS-IN$_2$P$_3$, Orsay, F-91406 France

L. GIOT and P. ROUSSEL-CHOMAZ

GANIL, Bld Henri Becquerel, BP 5027, Caen, F-14021, France

K. W. KEMPER

Department of Physics, Florida State University,
Tallahassee, Florida 32306-4350, USA

R.S. MACKINTOSH

Department of Physics and Astronomy, The Open University,
Milton Keynes, MK7 6AA, UK

A. PAKOU

Dpt of Physics, The University of Ioannina, Ioannina, 45110, Greece

R. RAABE

IKS, University of Leuven, Leuven, B-3001, Belgium

K. RUSEK

DNR, The Andrzej Soltan Institute for Nuclear Studies,
Warsaw, PL-00681, Poland

S. STEPANTSOV and R. WOLSKI

Flerov Laboratory of Nuclear Reactions, JINR,
Dubna, RU-141980, Moscow region, Russia

The structure of the ^{8}He exotic nucleus was investigated using direct reactions of the ^{8}He SPIRAL beam on a proton-rich target. The (p,p') scattering to the 2_1^+ state, the (p,d)^{7}He and (p,t)^{6}He transfer reactions, were measured at the energy E_{lab} = 15.7 A.MeV. The excitation spectrum of ^{8}He was extracted. Above the known 2_1^+ excited state at 3.6 MeV, a second resonance was found around 5.4 MeV. Within the coupled-reaction channels framework, the cross sections were analyzed and the spectroscopic factors for the (p,d)^{7}He$_{gs}$, the (p,t)^{6}He(0^+), and the (p,t)^{6}He(2^+) were deduced. It is inferred that the ^{8}He ground state has a different neutron-skin structure from the one suggested by previous $\alpha + 4n$ models assuming a pure $(1p_{3/2})^4$ configuration.

Keywords: (p,p') scattering; weakly-bound nuclei; resonances; coupled-reaction channel analysis.

1. Goals

In order to test the validity of the nuclear models and to improve their predictive power, the properties of the exotic nuclei such as very diffuse nuclear surfaces, low-lying resonances or new magic shells are experimentally investigated using radioactive ion beams. With the facilities developed worldwide, we can begin the exploration of the effect of large variation of isospin along isotopic chains. Using the particle spectroscopy technique, direct reactions such as (p,p') scattering and one-nucleon transfer reactions are measured to obtain the signatures for the new properties.

In this scope we have studied the wave function of the ^{8}He exotic nucleus. This nucleus has the largest N/Z ratio amongst the known nuclei and the nuclear correlations producing this weakly-bound structure at the drip line have to be clarified. Our experimental goals were to explore the structure and to obtain the low-lying resonant states using (p,p') scattering.

In the Helium chain, large spatial extensions of the neutron distributions have been found for ^{6}He and ^{8}He, producing a halo in the case of ^{6}He[1] and a possible neutron skin for ^{8}He.[2] From reaction cross sections[3] and high energy proton elastic scattering,[4] it is shown that ^{8}He has a matter rms radius of 2.5±0.1 fm. It means that it has almost the same size as ^{6}He, although having two neutrons more. The 4-neutron skin structure expected in the ^{8}He nucleus is generally described as the 5-body COSMA model[2] does, with parameters adjusted to reproduce the 4-neutron separation energy: an inert Alpha core surrounded by four valence neutrons. In this model, the neutron $1p_{3/2}$ subshell is fully occupied, which results in the neutron-skin structure and in a rather compact nucleus compared to ^{6}He.

During our complete experiment, all the reaction channels (elastic, inelastic and transfer) were simultaneously measured in the same conditions. From the analysis of the (p,p') angular distributions, it is possible to test the validity of the various microscopic calculations proposed for this nucleus. All previous data were not conclusive about the resonant states above the first excited 2^+ state measured at RIKEN.[5] This state was obtained from the inelastic ^{8}He(p,p') scattering measured at 73 A.MeV,[5] but there was not enough statistics to provide information on the ^{8}He structure and on other resonant states. We carried out the measurement in inverse kinematics using the ^{8}He beam delivered by the SPIRAL facility at GANIL and an optimized particle detection set-up.

2. Experimental set-up

The direct reactions of ^{8}He on a proton-rich target of polypropylene foil were measured at the incident energy of 15.7 A.MeV. We used the dedicated tools to study direct reactions induced by radioactive beams: the beam tracking detectors CATS,[6] to define the impact of the beam on the target, and the particle detector array MUST.[7] The MUST array was a wall of 8 modules located at 15 cm from the target. Each MUST module was a telescope (active area of 6x6 cm^2), with a 300 μm thick position-sensitive Si-strip followed by a 3 mm-thick Si(Li) stage. The emitted light charged particles [proton from (p,p') reaction, deuteron from (p,d) or triton from (p,t)] were detected and identified by the MUST telescopes. The heavy ejectiles were measured in coincidence in a wall of plastic scintillator located at forward angles. Mean intensity during the experiment was 5000/s (maximum value was 10^4/s). With the CATS and MUST devices, the scattering angle in the laboratory (lab) frame and the energy of the recoiling nuclei were reconstructed event by event; the angular resolution was of 0.4° (lab). In the excitation spectrum obtained from proton scattering, using the thin (1.48 mg/cm^2) to measure the angles below $40^\circ_{c.m}$, and the thick one (8.25 mg/cm^2), the energy resolution was ranging from 600 keV to 1.2 MeV.

3. Results

Elastic, inelastic and transfer reaction cross-sections were extracted from the ^{8}He+p data and compared to theoretical calculations yielding accurate results on the structure of the ^{8}He.[8–10] We summarize here the main findings. The angular distribution for elastic scattering is shown in Fig. 1 (left). The entrance channel potential was calculated with the microscopic complex JLM nucleon-nucleus potential,[11] using the no-core shell model (NCSM) ^{8}He ground state densities.[12] The usual parameter set, for the normalization of the real and imaginary parts of the potential for the light nuclei, is taken as : $\lambda_V = 1.0$ and $\lambda_W = 0.8$. In this case, data were found to be poorly reproduced. A better agreement was obtained with λ_V and λ_W left as free parameters (dashed line). However, even with these uncommon values, the elastic scattering at forward angles was still not well reproduced. The reaction framework was reconsidered. Since the ^{8}He(p,d) cross sections were found as large as the elastic ones in the angular range from 30 to $60^\circ_{c.m}$, it was a clear indication of strong coupled-channel (CC) effects[8] from the pick-up reaction (p,d) to the elastic scattering. Therefore, the data were analyzed using the coupled-reaction-channel (CRC) method and microscopic potentials, following the procedure detailed in Ref. 13.
From the coupled-channels Born approximation (CCBA) analysis,[9] the spectroscopic factor (S.F.) for neutron pick-up to the ^{7}He$_{gs}$ was C^2S =4.4 ±1.3; through the improved continuum discretized CC analysis- using microscopic potentials and explicit coupling to the (p,d) channel- the best fit gave C^2S =3.3[8] and both elastic and (p,d) data were reproduced.[8]

Within a CC analysis using microscopic potentials and explicit coupling to the

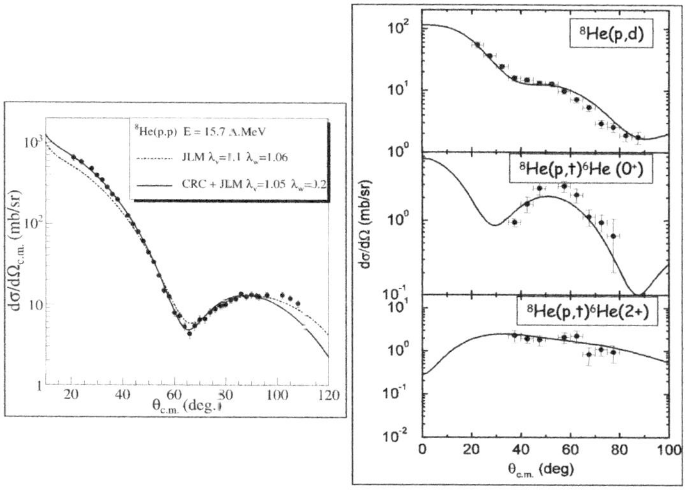

Fig. 1. Experimental and calculated cross sections for ^{8}He(p,p) and (p,d) at 15.7 A.MeV.[8,10] In the coupled reaction scheme, including the (p,d) and (p,t) transfer channels with the S.F. described in the text, the whole set of reactions are reproduced.[10]

(p,d) and also to (p,t) channels it was possible to reproduce both elastic and (p,d), (p,t) data, as shown by the full curve in Fig. 1 (left and upper panel of the right side). The S.F. of the $^8He_{g.s.}$ respected to $^7He_{3/2^-}$ g.s.[8] and ^{6}He(0^+) and ^{6}He(2^+) were deduced[10] : the C^2S values are 2.9, 1.0 and 0.014, respectively. These factors are in agreement with the results obtained from the analysis of the quasi-free scattering of ^{8}He measured at GSI:[14] they found the $^8He/^6He_{0^+}$ overlap of 1.3±0.1; and the S.F. for $^8He/^7He_{3/2^-}$ was of 3.3±0.3. The S.F. correspond to a structure for the ^{8}He ground state mainly built on the $^6He_{0^+} + 2n$ configuration. This is in contrast with the previously admitted structure for ^{8}He, described by the COSMA model and considered as a mixing between $^6He_{0^+} + 2n$ and $^6He_{2^+} + 2n$ configurations. In our analysis, we found that it was needed to describe the ^{8}He ground state as built on a mixing between the $(1p_{3/2})^4_\nu$ and the $(1p_{3/2})^2_\nu\ (1p_{1/2})^2_\nu$ configurations[10] to interpret our complete data set. This gave also a consistent analysis of the previous measurements, using the same S.F. inputs as for the SPIRAL data:

– through the CCBA the previous ^{8}He(p,d) data measured at 50 A.MeV[15] was reproduced;[9]

– the ^{8}He(p,t)^{6}He(0^+) and ^{8}He(p,t)^{6}He(2^+) reactions at 61.3 A.MeV[16] were reproduced by the CRC calculations.[10]

This analysis demonstrates how important the reaction framework is to draw correct conclusions on the S.F. and to deduce the microscopic structure. Moreover,

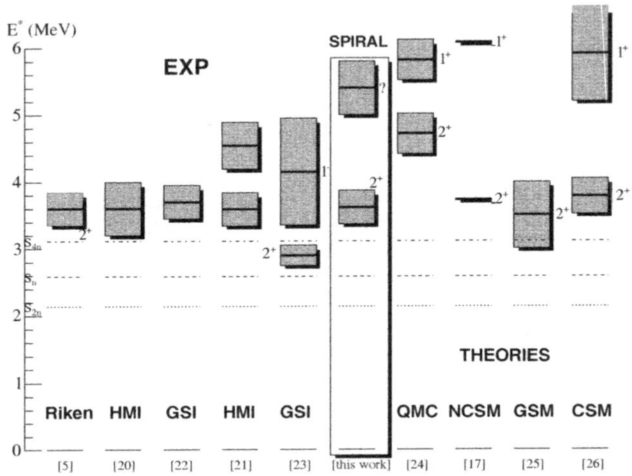

Fig. 2. Experimental and theoretical spectra of ^{8}He. The areas drawn for each resonance correspond to the FWHM widths.

from the analysis within the CRC framework, we have extracted the features of the density distributions of the ^{8}He ground state.[18] The calculations were done with the ^{8}He+p JLM potential including the ground state densities given either by the ab-initio NCSM[17] or by the COSMA model. The elastic data are better reproduced using the NCSM. The rms radii for the proton, neutron and matter distributions are respectively 2.00, 2.59 and 2.46 fm, to be compared to the corresponding rms radii given by the COSMA model: 1.69, 2.74 and 2.52 fm. This comparison shows that the ^{8}He is better modeled with a neutron-skin thickness of $\langle r^2\rangle_{\rm n}^{1/2} - \langle r^2\rangle_{\rm p}^{1/2} \simeq 0.6$ fm, rather than equal to 1 fm. It means that the proton-neutron correlations needed to describe the ^{8}He structure are different from the ones assumed in the COSMA model. In summary, the ground state is found to be a neutron-skin structure close to the one given by the NCSM,[17] but different from the COSMA model[2] which assumes an $\alpha + 4n$ configuration corresponding to a pure $(1p_{3/2})^4_\nu$ sub-shell closure. Preliminary CRC calculations of the (p,p') scattering were done, including only the (p,d) coupling. The microscopic potential was obtained from the JLM calculations using the NCSM ground state and transition densities.[17] It is found that the CRC calculations of the angular distributions overestimate the data with this set of densities. To test various profiles of the transition densities, and to deduce the quantitative characteristics of the transition densities, it is now required to take into account in the CRC calculations not only the (p,d) but also the (p,t) coupling effect on the (p,p') reaction. This work is in progress.

The ^{7}He and ^{8}He excitation energy spectra were extracted by missing mass method respectively from the ^{8}He(p,d) and (p,p') reactions. The analysis of the ^{7}He spectrum is detailed in Ref. 9.

From the ^{8}He excitation energy spectrum, the characteristics of the first two resonant excited states were obtained: the first 2^+ is obtained at 3.62 ±0.14 MeV (width 0.3 ±0.2 MeV), and a second state was found at 5.4 ±0.5 MeV (width 0.5 ±0.3 MeV).[18] The analysis will be detailed in a forthcoming article.[19] As shown in Fig. 2, the values for the first 2^+ state are consistent with the previous data obtained by (p,p'),[5] by multi-nucleon transfer[20,21] or break-up[22] reactions, and in contrast with the data in.[23] The resonances predicted by various theories, like the ab-initio calculations[24] or the ab-initio NCSM[17] overestimate our results. In the Gamow Shell Model (GSM),[25] through the consistent description of the bound states and the particle continuum, the ground state of the weakly-bound ^{6,8}He is found bound by the continuum-coupling correlations, and the position of the first excited state 2^+ is well predicted. In the Continuum Shell Model (CSM),[26] which includes also the particle continuum effects, the predictions for the first two resonant states are consistent with our data.

4. Conclusions and prospective

The reaction framework is crucial to obtain a consistent analysis of the reaction channels and to deduce the spectroscopic factors and the ground state structure. It is shown that the (p,d) coupling changes deeply the features of the entrance potential and strongly affects the extraction of the structure information. The spectroscopic factors of the ^{8}He$_{gs}$ respected to ^{7}He$_{gs}$, ^{6}He(0^+) and ^{6}He(2^+) were deduced.[10] The ^{8}He ground state is mainly built on the $^6\mathrm{He}_{0^+} + 2n$ configuration and there is a significant probability of finding the valence neutrons in other configuration than the full sub-shell $(1p_{3/2})^4$, such as $(1p_{3/2})^2(1p_{1/2})^2$. The ^{8}He$_{gs}$ structure is found well described by a neutron-skin density, similar to the one obtained by the NCSM.[17] The results obtained in the case of ^{8}He also demonstrate the need for a complete data set of direct reactions in order to understand the coupled-channel effects between elastic, inelastic and transfer reactions. This is a necessary step to be able to investigate the structure of weakly-bound nuclei. The same features can be expected throughout the nuclear chart, and particularly for neutron-rich nuclei with low particle threshold and continuum states close to the ground state.

To study the evolution of the neutron excitations in the neutron-rich isotopes, systematical studies of low-lying unbound excited states will be pursued through direct reactions on proton target. Dedicated tools have been developed to improve the efficiency, angular coverage and resolution of our present devices. Recently, the new Si-strip telescope array MUST2 has been built in a DAPNIA-IN$_2$P$_3$-GANIL collaboration.[27] As for ^{8}He, the low-lying spectroscopy of nuclei close or at the drip-line will be used as benchmarks for the most recent microscopic structure models developed to take into account explicitly the continuum-coupling effects. Through the confrontation between predictions and experimental results along the extended isotopic chains, the ingredients of the structure models will be checked. It is expected that the parameters of the nuclear interaction, isospin-dependent, spin-isospin dependent terms, and also the modelling of the spin-orbit coupling, pairing terms and continuum coupling effects could be extensively investigated and reevaluated. In the following years, using beams of other drip-line nuclei like ^{24}O, it should be possible to carry out a complete (p,p') analysis. The present RIBF facility at RIKEN, in the mid-term future accelerators like SPIRAL2 at GANIL or FAIR at GSI, and in the long-term future the European facility EURISOL, will offer new exotic beams and the opportunity to extend the systematics of the low-lying resonances in the neutron-rich nuclei.

References

1. A. Lagoyannis et al., *Phys. Lett.* **B518**, 27 (2001).
2. M.V. Zhukov, A.A. Korsheninnikov and M.H. Smelberg, *Phys. Rev.* **C50** (1994) R1.
3. J. A. Tostevin and J. S. Al-Khalili, *Nucl. Phys.* **A616**, 418c (1997).
4. J. S. Al-Khalili and J. A. Tostevin, *Phys. Rev.* **C57**, 1846 (1998).
5. A.A. Korsheninnikov et al., *Phys. Lett.* **B316**, 38 (1993).
6. S. Ottini et al., *Nucl. Inst. Meth.* **A431**, 476 (1999).
7. Y. Blumenfeld et al., the MUST collaboration, *Nucl. Inst. Meth.* **A421**, 471 (1999).
8. F. Skaza et al., *Phys. Lett* **B619**, 82 (2005).
9. F. Skaza et al., *Phys. Rev.* **C73**, 044301 (2006).
10. N. Keeley et al., *Phys. Lett.* **B646**, 222 (2007).
11. J. P. Jeukenne, A. Lejeune and C. Mahaux, *Phys. Rev.* **C16**, 80 (1977).
12. P. Navrátil and W. E. Ormand, *Phys. Rev.* **C57**, 3119 (1998).
13. N. Keeley, N. Alamanos, and V. Lapoux, *Phys. Rev.* **C69**, 064604 (2004).
14. L.V. Chulkov et al., *Nucl. Phys.* **A759**, 43 (2005).
15. A.A. Korsheninnikov et al., *Phys. Rev. Lett.* **82**, 3581 (1999)
16. A.A. Korsheninnikov et al., *Phys. Rev. Lett.* **90**, 082501 (2003).
17. P. Navrátil and W.E. Ormand, *Phys. Rev. Lett.* **88**, 152502 (2002); P.Navrátil, private communication.
18. F. Skaza et al., *Nucl. Phys.* **A788**, 260c (2007).
19. V. Lapoux et al., in preparation.
20. W. von Oertzen, *Nucl. Phys.* **A588**, 129c (1995).
21. H.G. Bohlen et al., *Prog. Part. Nucl. Phys.* **42**, 17 (1999).
22. T. Nilsson et al., *Nucl. Phys.* **A583**, 795 (1995).
23. K. Markenroth et al., *Nucl. Phys.* **A679**, 462 (2001).
24. S. C. Pieper, R. B. Wiringa, and J. Carlson, *Phys. Rev.* **C70**, 054325 (2004).
25. N. Michel, W. Nazarewicz, M. Ploszajczak, J. Okolowicz, *Phys. Rev.* **C67**, 054311 (2003).
26. A. Volya and V. Zelevinsky, *Phys. Rev. Lett.* **94**, 052501 (2005).
27. E.C. Pollacco et al., the MUST2 collaboration, *Eur. Phys. J.* **A25**, 287 (2005).

SHELL GAP BELOW ^{132}Sn BASED ON THE EXCITED STATES IN ^{130}Cd AND ^{131}In

M. GÓRSKA

GSI, D-64291 Darmstadt, Germany
E-mail: m.gorska@gsi.de

Nuclear structure studies in vicinity of the doubly magic nuclei located far off stability gained recently an increased interest with studies of shell evolution phenomena. In addition, for the most exotic nuclei shell quenching is expected as a consequence of a change of the nuclear potential from Woods-Saxon to harmonic oscillator like shape. A question was raised about a possible quenching $N = 82$ shell just below ^{132}Sn. The main goal of the experiment performed at GSI was the search for isomeric states in nuclei closest to the magic nucleus in order to extract information on sustainability of the $Z = 50$, $N = 82$ shell gap. The experiment was performed within the RISING Stopped Beam Campaign. Beams of ^{136}Xe and ^{238}U at 600-900 MeV/u impinged on a ^{9}Be fragmentation target. Experimental setup consisted of the FRagment Separator (FRS) which delivered the race isotope species, and a Ge array with the efficiency of 35% at 100 keV and 11% at 1.3 MeV γ-ray energy. The decay of the isomeric states which survived the flight time through the FRS was measured in the RISING array. Isomeric states were identified in ^{131}In and ^{130}Cd, which is analogous in structure to ^{98}Cd, its half-life and the energy of the 2^{+} state indicate no need for shell quenching below ^{132}Sn. The core-exited isomer observed in ^{131}In provides a direct measure of the shell gap when accounting for the residual interaction in an empirical way. The results are compared to shell-model calculations.

Manusccript not received.

HEAVY NEUTRON-RICH NUCLEI PRODUCED IN THE FRAGMENTATION OF A ^{208}Pb BEAM

Zs. PODOLYÁK[1], S. STEER[1], S. PIETRI[1], M. GÓRSKA[2], P.H. REGAN[1], A.B. GARNSWORTHY[1], D. RUDOLPH[3], E. WARNER-MALENTO[4], T. SHIZUMA[1], R.HOISCHEN[2,3], J. GERL[2], H.J. WOLLERSHEIM[2], G. FARRELLY[1], T. KURTUKIAN-NIETO[4], G. BENZONI[5], J. BENLLIURE[4], F. BECKER[2], P. BEDNARCZYK[2], L. CÁCERES[2], P. DOORNENBAL[2], H. GEISSEL[2], J. GRĘBOSZ[2], A. KELIC[2], I. KOJOUHAROV[2], R. KUMAR[2], N. KURZ[2], F. MONTES[2], W. PROKOPOWICZ[2], T. SAITO[2], H. SCHAFFNER[2], S. TACHENOV[2], A. HEINZ[6], M. PFÜTZNER[7], A. JUNGCLAUS[8], B.A. BROWN[9], H. GRAWE[2], H. MAIER[2] FOR THE RISING COLLABORATION

[1]*Department of Physics, University of Surrey, Guildford, GU2 7XH, UK*
[2]*GSI, Planckstrasse 1, D-64291, Darmstadt, Germany*
[3]*Department of Physics, Lund University, S-22100, Lund, Sweden*
[4]*Universidad de Santiago de Compostela, Santiago de Campostela, Spain*
[5]*Universitá degli Studi di Milano and INFN Milano, I-20133, Milano, Italy*
[6]*WNSL, Yale University, 272 Whitney Avenue, New Haven, CT, 06520, USA*
[7]*IEP, Warsaw University, Hoża 69, PL-00-681, Poland*
[8]*Departmento di Teórica, Universidad Autonoma de Madrid, Madrid, Spain*
[9]*NSCL, Michigan State University, East Lansing, MI 48824-1321, USA*

Gamma-rays de-exciting isomeric states in a number of heavy neutron-rich nuclei have been observed following relativistic projectile fragmentation of a 1 GeV per nucleon ^{208}Pb beam. New information on the excited states of the spherical N=126 ^{204}Pt and ^{205}Au, N=125 ^{203}Pt, and on the oblate ^{198}Os have been obtained.

Keywords: Fragmentation; Neutron-rich nuclei; ^{204}Pt; ^{205}Au; ^{198}Os.

1. Layout

The present paper presents selected highlights of the experimental results obtained from the fragmentation of a relativistic energy ^{208}Pb beam. The focus is on the *structure* of $N \approx 126$ neutron-rich nuclei.

Studies of semi-magic nuclei are of special importance since they allow direct tests of the purity of the model wave functions. Information on the single-particle energies and two nucleon residual interactions can be derived from such experimental observables as the energies of the excited states and transition probabilities. Information on the neutron-rich N=126 nuclei is very scarce. Below the doubly magic ^{208}Pb nucleus there is experimental information on the excited stated of only two isotones: ^{207}Tl[1] and ^{206}Hg.[2] In the case of the odd-Z ^{205}Au only the groundstate is known, with spin-parity $(3/2^{+})$.[3] In the present experiment information has been obtained on the ^{204}Pt and ^{205}Au N=126 closed neutron shell nuclei.

Moving further away from the double magic ^{208}Pb, the heavy Os-Pt region is characterised by the presence of nuclei with different shapes in their ground-states, such us prolate, oblate, triaxial and spherical. Consequently this region is considered to be a crucial testing ground for nuclear models. The lighter isotopes are prolate deformed, and by adding more and more neutrons the shape changes into oblate than in the vicinity of the N=126 closed shell it has to became spherical. In the prolate-oblate transition region the nuclei can be described by a potential with similar minima corresponding to prolate and oblate shapes. The transitional region in the case of the platinum nuclei starts at around mass A=192 and persist till A≈200. In the case of osmium nuclei there is sudden shape change. The exact place where the prolate to oblate shape change for the ground state of the osmium isotopes occurs is not clear. Experimental information suggest that up to mass A=194 the osmium isotopes are prolate.[4] In ^{196}Os the available experimental information is too scarce, only two excited states were observed,[5] to draw a definite conclusion. From the present experiment, information on ^{198}Os has been obtained.

The lack of information on nuclei 'below' ^{208}Pb is due to the difficulties in populating these neutron-rich nuclei. However, fragmentation has proved to be an

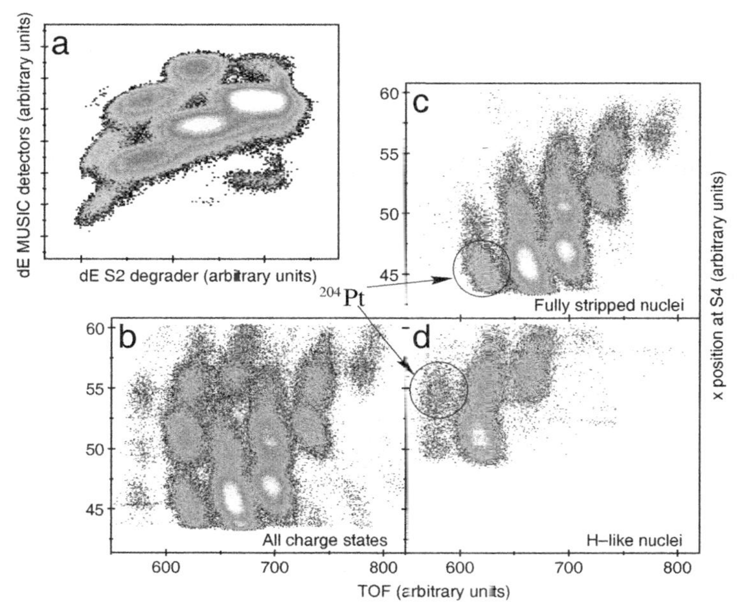

Fig. 1. Identification of the fragmentation products. (a) Energy loss at the final focal plane (S4) vs. energy loss at the middle focal plane (S2). It distinguishes between nuclei which do not change charge state at S2 from those which do. (b,c,d) energy loss vs. time-of-flight matrices without charge state condition (b), selecting fully-stripped nuclei (c) and H-like nuclei (d), respectively. The circles in (c) and (d) indicate the ^{204}Pt nuclei.

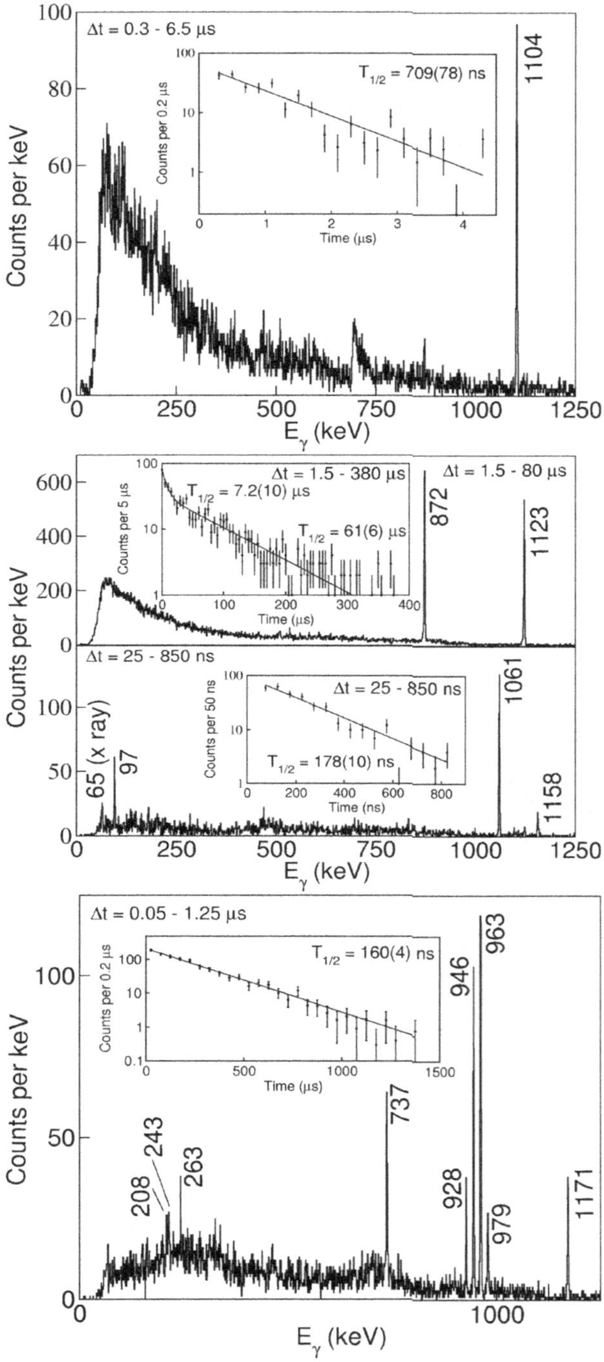

Fig. 2. Delayed gamma-ray spectra associated to ^{203}Pt *(top)*, ^{204}Pt *(middle)* and ^{205}Au *(bottom)*. All indicated half-lives are preliminary.

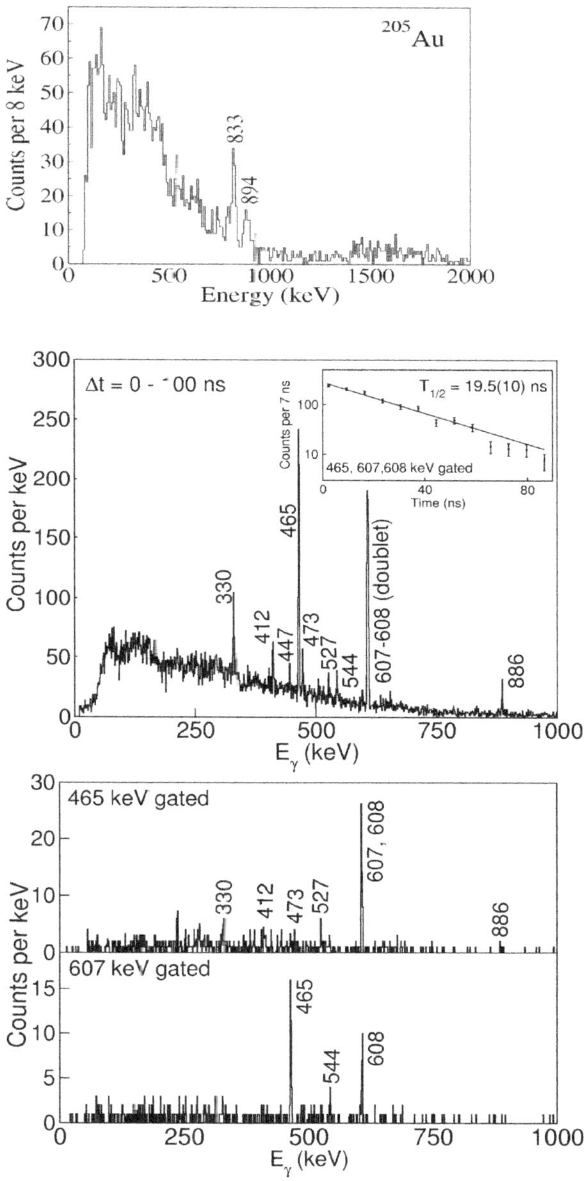

Fig. 3. Delayed conversion electron spectrum associated to ^{205}Au *(top)*. ^{198}Os singles spectrum taken within 100 ns after implantation *(middle)* and coincidence spectra obtained by gating on the 465 keV and 607 keV transitions *(bottom)*. All indicated half-lives are preliminary.

efficient tool to produce exotic nuclear species. When projectile fragmentation is combined with high sensitivity γ-detection arrays, structure information can be gained for otherwise inaccessible nuclei. The highest sensitivity is achieved with decay (both internal isomeric decay and beta decay) spectroscopy. In this technique the delayed gamma rays are correlated with the individually identified ion, therefore minimising the associated background radiation. Information on the excited states populated in this way can be obtained with only 1000 nuclei produced. In its simplest form the technique is sensitive to isomeric decays with lifetimes between 100 ns and 1 ms. The lower limit comes from the flight-time through the fragment separator (≈300 ns), the upper limit is determined by the necessity of correlating the delayed gamma rays with the implanted ion. However, shorter lifetimes could also be detected if the decay branch by electron conversion was hindered for specific charge states of the ion.[6] In the case of beta-decay, with the use of an active stopper the delayed gammas and beta can be correlated with the implantation of the incoming ion over more than 10 s.

2. Experimental setup

Heavy nuclear species were populated in relativistic energy projectile fragmentation. A beryllium target of thickness 2.5 g/cm^2 was bombarded with an E/A=1 GeV, ^{208}Pb beam provided by the SIS accelerator at GSI, Darmstadt, Germany.

The nuclei of interest were separated and identified using the FRagment Separator (FRS)[7] operated in standard achromatic mode with a wedge-shaped aluminium degrader in the intermediate focal plane of the separator. The identification of the fragments is based on the determined A/q (∼TOF), the energy loss in the ionisation chambers (≈ Z), and the longitudinal position of the nuclei at the intermediate (S2) and final (S4) focal planes of the FRS. For more details about the setup used in the present experiment see ref.[8] The identification process is illustrated in figure 1. for the case of the ^{204}Pt nucleus.

The transmitted nuclei were stopped in a catcher surrounded by the high-efficiency, high granularity Stopped RISING γ-ray spectrometer.[9] RISING consists of 15 cluster Ge detectors and has a full peak efficiency of 14% at 662 keV. Time-correlated gamma decays from individually identified nuclear species have been measured, allowing clean identification of isomeric decays.

3. Results

The results presented here were obtained from two different magnetic rigidity settings: one optimised to select fully stripped nuclei centred on the maximal transmission of ^{203}Ir (see fig.1.) and one centred on ^{199}Os. The detection of the previously identified γ rays following the de-excitation of isomeric states in ^{200}Pt,[10] ^{201,202}Pt ^{193}Re, ^{195}Os, ^{198}Ir[6] were used to confirm the calibration of the particle identification.

Delayed γ-ray spectra associated to ^{203,204}Pt, ^{205}Au and ^{198}Os nuclei are presented in figures 2 and 3. In addition, a conversion electron spectrum associated to ^{205}Au is also shown. This was taken in a separate experiment using an active, double sided Si strip detector, stcpper.

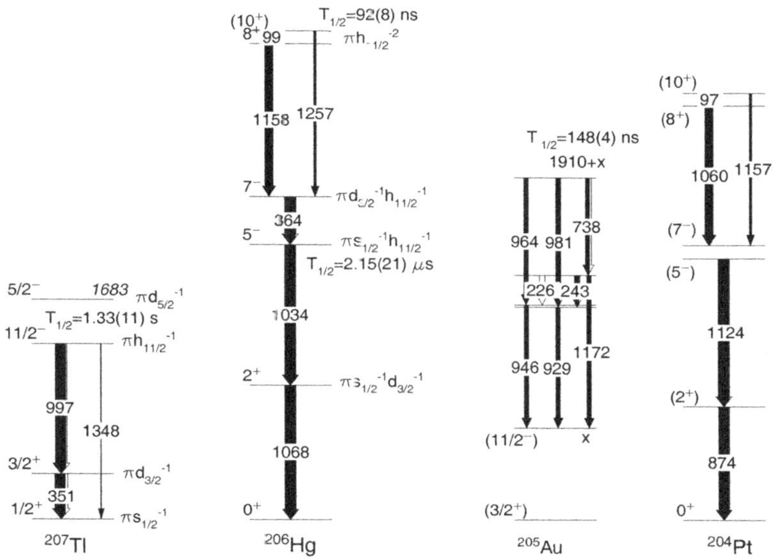

Fig. 4. Partial level schemes of N=126 nuclei: ^{207}Tl,[1] ^{206}Hg,[2] ^{205}Au and ^{204}Pt.

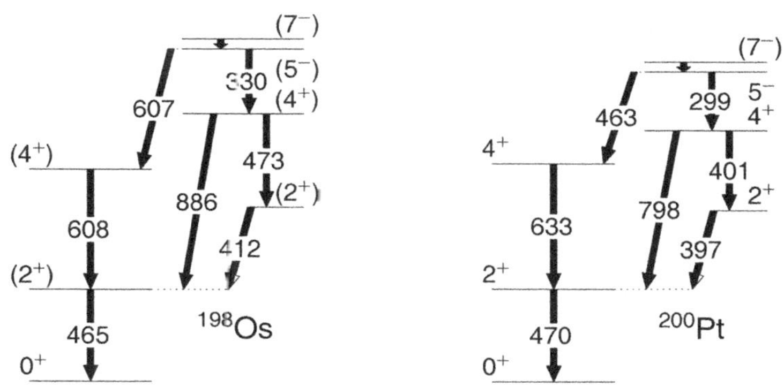

Fig. 5. Proposed level scheme of ^{198}Os. For comparison the partial level scheme of ^{200}Pt is also shown.[10]

4. Discussion

4.1. *N=126 nuclei*

The level schemes of ^{204}Pt and ^{205}Au have been obtained considering the systematics and comparison with shell-model calculations. Shell-model calculations have been performed using the OXBASH code.[11] The empirical interaction matrix elements are from[12] and are based on those of Kuo and Herling.[13] The proton-hole energies were taken from the experimental level scheme of ^{207}Tl. Partial level schemes of the ^{207}Tl, ^{206}Hg, ^{205}Au and ^{204}Pt N=126 nuclei are shown in figure 4. The energy of the $11/2^-$ $\pi h_{11/2}$ state in ^{205}Au is estimated from the conversion electron spectrum of figure 3. to be x≈910 keV.

4.2. ^{198}*Os*

The level scheme of ^{198}Os is shown in figure 5. The energies of the lowest excited states can be used to infer information about he character of the nuclei. Systematics related to the yrast 2^+ and 4^+ states as well the second 2^+ state (2_2^+) are presented in figure 6. In ^{198}Os, the ordering of the coincident 412 and 473 keV transitions cannot be determined from the present experiment. Here we considered the ordering shown in figure 5, resulting in an $E(2_2^+)/E(2^+)$ energy ratio of 1.89. By changing the order of the transitions the ratio is slightly modified to 2.02. The above assumption don't have any consequences on the discussion below. In addition, the spin-parity of the second excited state at 760(20) keV in ^{196}Os is not know.[5] As shown in figure 6, the platinum isotopes change slowly with atomic mass. The yrast $E(2^+)$ and $E(4^+)$ energies increase smoothly, but the $E(4^+)/E(2^+)$ energy ratio remains constant at about 2.5, the ratio expected for perfect O(6) γ-soft nuclei. The $E(2_2^+)/E(2^+)$ energy ratio is smoothly decreasing, and has a value of around ~2, which is lower than predicted for O(6) nuclei. These characteristics are consistent with the interpretation

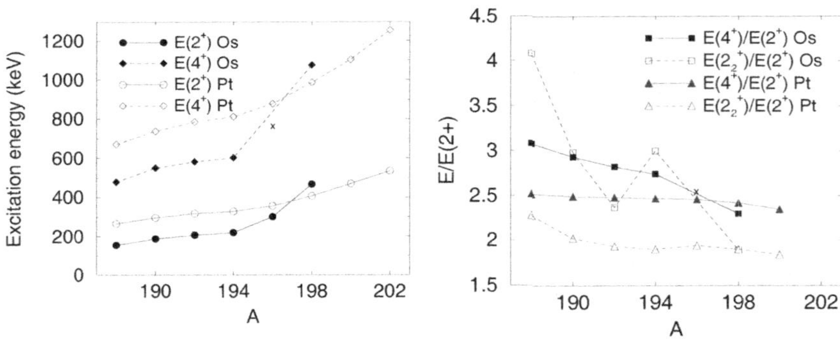

Fig. 6. Systematics of the $E(2^+)$, $E(4^+)$ energies and the $E(4^+)/E(2^+)$ and $E(2_2^+)/E(2^+)$ energy ratios in the even mass A=188–202 osmium and platinum isotopes. The values corresponding to the second excited state in ^{196}Os, which is either the yrast 4^+ or the second 2^+ state, are indicated with 'x'.

that these nuclei have low deformation oblate shape. In contrast to the platinum nuclei, in the osmium isotopes there is a sudden change at around A=196. The large $E(4^+)/E(2^+)$ energy ratio and the smooth change of the yrast $E(2^+)$ and $E(4^+)$ energies in the case of the A=188-194 osmium nuclei indicate that these isotopes are deformed. At A=196 there is an abrupt increase in the energy of the 2^+ state. This is even more obvious in the case of ^{198}Os, where both the $E(4^+)$ and $E(2^+)$ energies are much higher than in the lighter Os nuclei. Similarly, the $E(4^+)/E(2^+)$ and $E(2_2^+)/E(2^+)$ energy ratios drop to the lowest values for the osmium isotopes in the region. The low E(4)/E(2) is consistent with a low deformation and is less than the predicted 2.5 for gamma-soft O(6) nuclei by the Interacting Boson Model, while the low $E(2_2^+)/E(2^+)$ indicates the γ-softness of the system. We note that the energy ratios $E(4^+)/E(2^+)$=2.31 and $E(2_2^+)/E(2^+)\approx$2.0 are much lower than those expected for nuclei in the X(5) phase transition symmetry, 2.91 and >2.8, respectively.

5. Conclusions

Preliminary results about the structure of heavy neutron-rich nuclei ^{203,204}Pt, ^{205}Au and ^{198}Os have been presented. The interpretation of these results is in progress.

References

1. D. Eccleshall, M.J.L.Yates, *Phys. Lett.* **19**, 301 (1965).
2. B. Fornal et al., *Phys. Rev. Lett.* **87**, 212501 (2001).
3. Ch. Wennemann et al., *Z. Phys.* **A347**, 185 (1994); F.G. Kondev, *Nucl. Data Sheets* **101**, 521 (2004).
4. C. Wheldon et al., *Phys. Rev.* **C63**, 011304 (2000).
5. P.D. Bond, R.F. Casten, D.D. Warner, D. Horn, *Phys. Lett.* **B130**, 167 (1983).
6. M. Caamano et al, *Eur. Phys. J.* **A23**, 201 (2005).
7. H. Geissel *et al.*, *Nucl. Instrum. Methods Phys. Res. Sect.* **B70**, 286 (1992).
8. Zs. Podolyák *et al.*, *Eur. Phys. J. Special Topics* **150**, 165 (2007).
9. S. Pietri *et al.*, *Nucl. Instrum. Methods Phys. Res. Sect. B* 1079. P.H. Regan et al, *Nucl. Phys.* **A787**, 491c (2007).
10. S. Yates et al., *Phys. Rev.* **C37**, 1889 (1988).
11. B.A. Brown A. Etchegoyen, W.D M. Rae, The computer code OXBASH, MSU-NSCL report number 524.
12. L. Rydström *et al.*, *Nucl. Phys.* **A512**, 217 (1990).
13. T.T.S. Kuo and G.H. Herling, Naval Research Laboratory Report 2259 (Washington, DC, 1971).

BREAKUP AND INCOMPLETE FUSION IN REACTIONS OF WEAKLY-BOUND NUCLEI

D.J. HINDE*, A. DIAZ-TORRES, M. DASGUPTA and L.R. GASQUES

Department of Nuclear Physics,
Research School of Physical Sciences and Engineering,
Australian National University,
Canberra, ACT 0200, Australia
**E-mail: david.hinde@anu.edu.au*

J.A. TOSTEVIN

Department of Physics, University of Surrey,
Guildford, GU2 7XH, United Kingdom

Breakup and incomplete fusion are strongly related. Experimental evidence is first presented, then a new 3-D classical model with stochastic breakup introduced, which is able to consistently explain experimental observables such as complete and incomplete fusion cross-sections and sub-barrier breakup cross-sections.

Keywords: Weakly-bound Nuclei, Breakup, Incomplete Fusion, Isomer ratio

1. Introduction

Understanding the various phenomena occurring in the interactions of weakly bound nuclei is currently a major challenge. As discussed below, such interactions can be considered to highlight the limitations of current models, such as the continuum discretised coupled channels model, which is one of the main tools currently used to interpret a wide range of experimental measurements at near-barrier energies.

The process of incomplete fusion occurs when only part of the projectile fuses with the target nucleus. This can result from breakup of the projectile in the field of the target nucleus, which is expected to be significant in reactions of weakly-bound nuclei. Breakup can occur at large distances due to the Coulomb field (more likely the higher the beam energy), and due to the nuclear field at smaller distances. To model breakup, a number of approaches are used. The most common uses quantum mechanical scattering theory, based on the coupled-channels approach, in which the continuum above the breakup threshold is discretised in energy bins (called CDCC). Significant aspects of this approach are:

- the model provides a time-independent static solution
- breakup is not spatially localized
- flux can only be lost from the total wave-function ("total fusion")

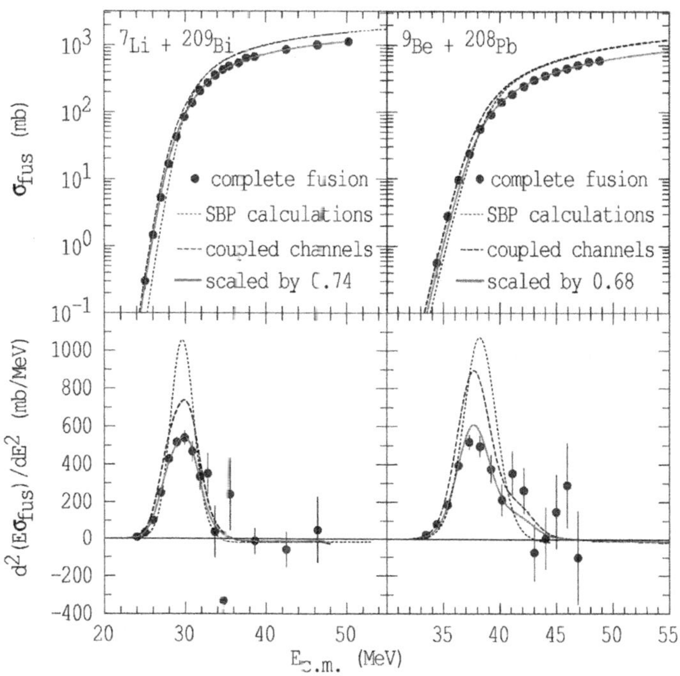

Fig. 1. Measured complete fusion cross sections (upper) and barrier distributions (lower) for reactions of weakly-bound stable nuclei. Determining the average fusion barrier energy allows the suppression factor for complete fusion to be determined, at around 0.7.

- fragment observables can only be determined where neither fragment is captured by the target, corresponding to no-capture break-up (NCBU).[1]

Thus although incomplete fusion (ICF) is closely related to breakup, it cannot currently be modeled using the CDCC approach, as the model cannot identify events resulting from the capture of one fragment but not the other. Only "total fusion" can be calculated,[2] corresponding to flux lost from the assumed 3-body model space.

Another theoretical approach, used up to now to model aspects of the reactions of halo nuclei, is to use the time-dependent wave-packet description.[3,4] Here the halo neutron wave-function, initially localized about the projectile core, can later be shared by the projectile and target nuclei. This picture is closer to our classical intuition, with breakup having some localization in time and space. A limited irreversibility has been included in the model, through an imaginary core-target potential, which means the model is currently limited to calculating total fusion.

Irreversibility would also be expected as a result of interactions of the halo neutron(s) with the constituent nucleons of the target nucleus, which in the calculation are once again beyond the assumed 3-body model space. Replacement of the core-target imaginary potential by a neutron-target imaginary potential would be of considerable interest in the case of peripheral collisions, as this should mock-up irreversible neutron-target interactions,[5] which occur at greater separation distances than core-target interactions.

A fully classical 2-D approach has also been developed,[6] to model incomplete fusion of nuclei breaking up into charged clusters. Key features of the model are:

- excitation to energies above the breakup threshold occurs through tidal forces
- breakup is localized in space and time
- classical trajectories of the breakup fragments allows calculation of all outcomes.

Although the breakup occurs at exactly the same point on the incident trajectory for the same initial conditions, which we would not necessarily expect in reality, the model showed some important results:[6]

- that capture of both fragments after breakup is significant, resulting to complete fusion (CF)
- that modeling the trajectories of the breakup fragments is very important in predicting the outcomes following breakup.

To fully understand the outcomes of interactions of weakly-bound nuclei, a realistic quantum-mechanical model is the ultimate goal. This should treat explicitly the irreversibility of the breakup and capture processes, which we feel is a key aspect in understanding nuclear reaction dynamics at near-barrier energies, where the timescale of the reaction is long. Currently, there are a few proposals on how this goal might be addressed, such as described in Ref.7, where a density matrix approach, with decoherence resulting from interactions outside the explicitly-included model space, is proposed as a possible solution. This and any other possible approaches need to be developed to allow quantitative calculations and understanding.

Experimentally too, the situation of incomplete fusion can be problematic, as the same reaction products as produced in ICF can sometimes be produced by complete fusion or transfer reactions, and these can be difficult or impossible to separate experimentally from incomplete fusion products.

In the case of a weakly-bound nucleus that breaks up into charged fragments, the clearest solution to this experimental problem is to measure the breakup process at an energy below the fusion barrier,[8] where the fragments cannot be captured, leading exclusively to no-capture break-up. Here, both CDCC and experiment should be unambiguous, and this could be a baseline for further developments in experiment and theory.

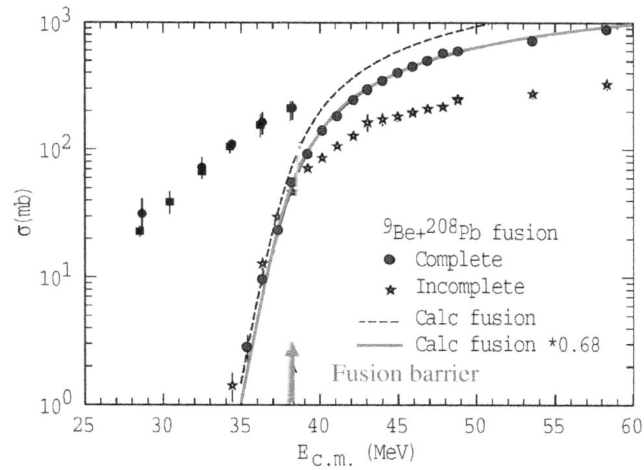

Fig. 2. The observation of substantial cross-sections for incomplete fusion above the barrier (stars) and for alpha-particle production below the barrier (black circles and squares at energies below the CF and ICF data) should be related through the dynamics of the breakup process.

2. Determination of suppression of complete fusion

The first definitive experimental determination of the suppression of complete fusion at above-barrier energies, associated with the interaction of a weakly bound nucleus, was for the ^{9}Be+^{208}Pb reaction.[9] Through precise cross section measurements, the average complete fusion barrier energy could be determined from the barrier distribution, essentially independent of the suppression factor.[9,10] In this case, complete fusion was defined as capture of all the charge of the projectile, as the capture or otherwise of the valence neutron that is not part of the alpha clusters could not be measured. Experimental results, along with those for ^{7}Li+^{209}Bi, are shown in Fig. 1, demonstrating the unambiguous determination of the suppression of complete fusion.

3. Breakup at sub-barrier energies

The relationship between the breakup process and incomplete fusion is hinted at in Fig. 2, where cross sections for complete and incomplete fusion, and alpha-particle production below the barrier are plotted. The large incomplete fusion yield is complemented by large alpha-particle production below-barrier (where there is no absorption). How can these two energy regions be related without a rigorous quantum-mechanical model including all the relevant physics?

One approach is illustrated in Fig. 3a, where the various processes producing α-

particles at sub-barrier energies have been disentangled,[8] and plotted as a function of the distance of closest approach on a Coulomb trajectory (a good approximation for sub-barrier energies). The prompt breakup shows a rapid increase in probability as the two nuclei get closer, which is well fitted by an exponential function (thick dark line). Extrapolation of this function to the fusion barrier radius ($\sim$ 11.5 fm) predicts a breakup probability that is almost 50%. Thus qualitatively at least, the prompt breakup probability is large enough to explain the fraction of incomplete fusion at above-barrier energies.

Can this empirical prompt breakup function be used to predict the properties of the complete and incomplete fusion observed at above-barrier energies? This question has led to the development of a new model, with classical trajectories in 3-D, and a stochastically sampled breakup function.[11] Quantitative predictions of above-barrier incomplete fusion observables can be made, and these, with comparisons to a range of data,[11,12] will now be described.

4. Classical stochastic model

This new approach[11] is based on the Monte Carlo method, in which reaction outcomes are determined for a large number of trajectories. The overall probability of breakup, the location of breakup on a given trajectory, and the conditions at breakup can be randomly sampled according to any physical scenario that is deemed suitable. It is not a model *of* breakup, but a model to *relate* breakup and incomplete fusion.

The effects of the Coulomb and nuclear two-body interactions that cause breakup are encoded in the local breakup probability $\mathcal{B}^L_{BU}(R)$, which is taken to be a function only of R, the projectile-target separation. $\mathcal{B}^L_{BU}(R)dR$ is the probability of breakup in the interval R to $R + dR$. Both experimental data[8] and CDCC calculations[11] show that the integral of this breakup probability (P_{BU}) along a given classical orbit is an exponential function of its distance of closest approach $R_{min}(E_0, L_0)$. Explicitly:

$$P_{BU}(R_{min}) = 2\int_{R_{min}}^{\infty} \mathcal{B}^L_{BU}(R)dR = A\,\exp(-\alpha R_{min}), \tag{1}$$

from which it follows uniquely that the local function has the same exponential form:

$$\mathcal{B}^L_{BU}(R) \propto \exp(-\alpha R). \tag{2}$$

The position of breakup on a given trajectory is chosen randomly, weighted according to the local breakup function $\mathcal{B}^L_{BU}(R)$. The exponential weighting will result in most breakup being in the vicinity of R_{min}, but all values of R are sampled. Following breakup, fragments F1 and F2 now interact with the target nucleus T, and with each other, via real central two-body potentials. The instantaneous dynamical variables of the excited projectile at breakup, its total internal energy ε_{12}, angular

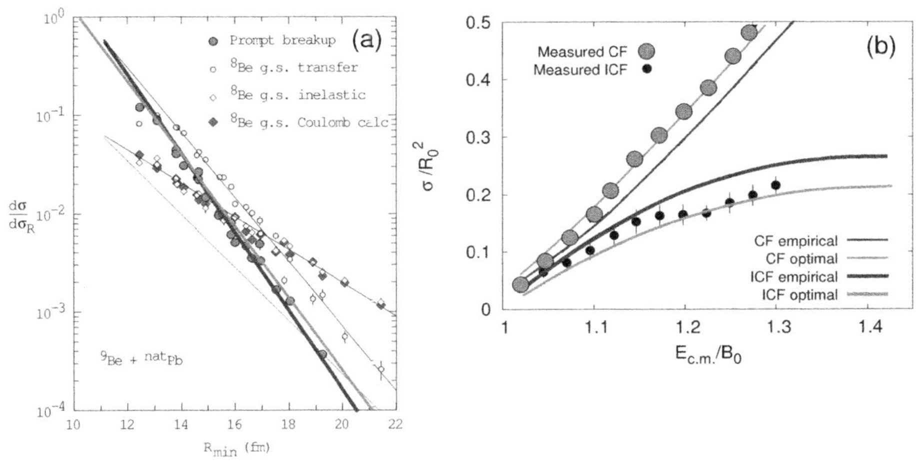

Fig. 3. (a) The probability of prompt breakup of ^{9}Be (filled circles) can be separated from sequential breakup of the ^{8}Be ground-state, and shows an exponential dependence on the distance of closest approach of a Coulomb trajectory. (b) Through the new dynamical model, sub-barrier breakup can be related to above-barrier incomplete fusion, and vice-versa, the "empirical" fit to the sub-barrier breakup resulting in complete and incomplete fusion cross sections close to those measured, whilst the "optimal" fit to the above-barrier data reproduces very satisfactorily the sub-barrier probabilities in (a).

momentum $\vec{\ell}_{12}$ and the separation of the fragments $\vec{d}_{12}$ are all Monte Carlo sampled.[11] Having fixed the positions and dynamical variables of the excited projectile fragments at the moment of breakup, the instantaneous velocities of F1, F2 and T are dictated by conservation of energy, linear momentum and angular momentum in the overall center of mass frame of the projectile and target system. Then the trajectories of F1, F2 and T are computed, which determine the number of ICF, CF and NCBU events, a fragment being assumed to be captured if the classical trajectory takes it inside the fragment-target barrier radius.

4.1. *Validation of the model against CDCC*

In order to validate the approach, NCBU observables,[1] which should be reliably calculated within the CDCC quantum mechanical approach, were compared with the classical model calculation for a simplified case[11] of the reaction of a pseudo-^{8}Be projectile (assuming a weakly-bound state of two α-particles) with a ^{208}Pb target. The CDCC and classical model calculations are matched as closely as possible. To allow comparison with CDCC, $P_{BU}(R_{min})$ was taken from the CDCC calculations to determine the local breakup function $\mathcal{B}^L_{BU}(R)$. Fig. 4 shows the good agreement

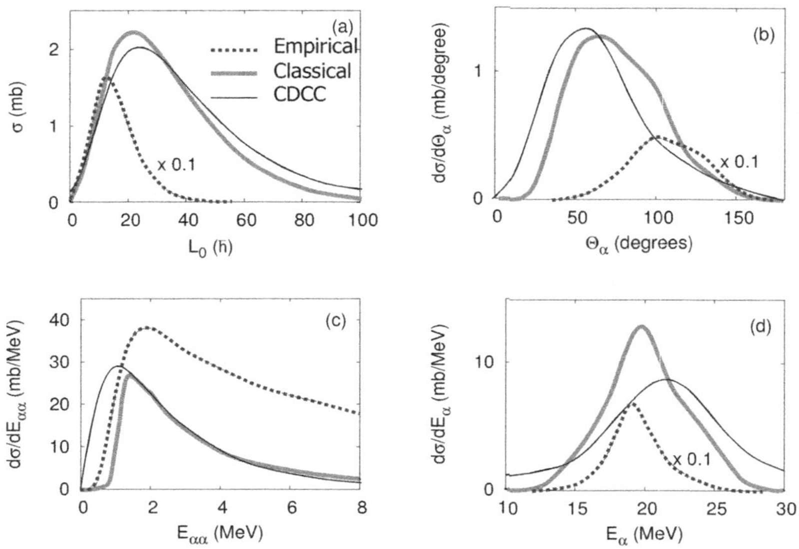

Fig. 4. Test calculations for 45 MeV ^{8}Be of break-up (no capture) cross sections (a) as a function of ^{8}Be orbital angular momentum (L_O), as a function of the opening angle (b) and relative energy (c) between the two alpha-particles, and as a function of the alpha-particle energy (d). Calculations are from CDCC, the classical model matched to CDCC, and the classical model fitting the empirical breakup function.

of observables between the CDCC calculations (thin black lines) and the classical model (thick lines).

The CDCC breakup function for the pseodo–^{8}Be has a much smaller slope then that found empirically for ^{9}Be. The sensitivity of the observables to changes in the breakup function are demonstrated in Fig. 4. The dotted lines result from using the empirical local breakup function $\mathcal{B}^L_{BU}(R)$ from the ^{9}Be+^{208}Pb experiment, as opposed to the CDCC breakup function (thick curves). The former results in breakup occurring on average much nearer to R_{min}, and thus to the target. The NCBU spin distribution (panel a) when using the empirical breakup function now peaks at smaller angular momenta with a corresponding shift in the peak of the α angular distribution to larger angles (panel b). The strong tidal forces experienced by the non-fused α-particles lead to an enhanced α-α yield of high relative energy (panel c) and smaller changes to the singles α energies (panel d).

4.2. *Comparison with experiment*

The model has been applied to experimental CF and ICF cross sections for the ^{9}Be+^{208}Pb reaction. Using $\mathcal{B}^L_{BU}(R)$ deduced from the empirical (measured) breakup function, the model predicts CF and ICF yields as shown by the dark lines in Fig.

3b. These are close to the experimental data (triangles and circles). Only a small change in $\mathcal{B}^L_{BU}(R)$ (and thus $P_{BU}(R_{min})$) is required to simultaneously give a very good reproduction of the data both below and above the fusion barrier (shaded lines in Figs. 3a and 3b).

This is the first quantitative result supporting the suggested linkage[8] between the breakup process observed below the barrier and suppression of complete fusion observed at energies above the barrier.

5. ICF product angular momentum distributions

The ICF process resulting from breakup can also be investigated through the heavy product angular momentum distributions. In the reactions of ^{6,7}Li and ^{9}Be with targets of ^{209}Bi and ^{208}Pb, the ICF products ^{212}At and ^{211}Po have high-spin isomers that decay by α-particle emission. Both isomeric and ground-state cross sections have been measured, allowing the determination of isomer ratios, which here are defined as the population of the isomeric state divided by the population of the ground-state. The isomer ratios are sensitive to the angular momentum distributions brought into the system,[13] which in turn are sensitive to the dynamics of the ICF process.

5.1. *Calibration reaction*

The isomer ratio is determined essentially by three factors:

- the angular momentum distribution of the compound nucleus
- the feeding from the compound states to the isomer and ground-state, which is independent of the reaction
- angular momentum fractionation of the various evaporation channels due to the increase in rotational energy with angular momentum.

Isomer ratio data are available for the α + ^{208}Pb[14] and α + ^{209}Bi[15] reactions. These lead to the same compound nuclei as are formed by ICF of ^{6,7}Li and ^{9}Be projectiles. Thus the angular momentum introduced in the ICF process can be calibrated against the α fusion reactions, which are well-understood.

The population of the isomeric state and the ground state have been determined empirically by splitting the angular momentum distribution of the compound nucleus into two regions, the lower feeding the ground-state, the upper feeding the isomer, as sketched in Fig. 5a. To account for the complex feeding pattern to the yrast line, the isomer ratio (R) was calculated using an effective angular momentum cutoff (J_{eff}) and a spreading parameter δ, assumed to be 0.5 (the results are quite insensitive to the latter value). Angular momentum fractionation was minimised by only considering the energy range where the product of interest was the dominant channel.[12] Thus for the α calibration reaction whose data are shown in Fig. 5b, J_{eff} was determined to be 10 by fitting the lower energies only.

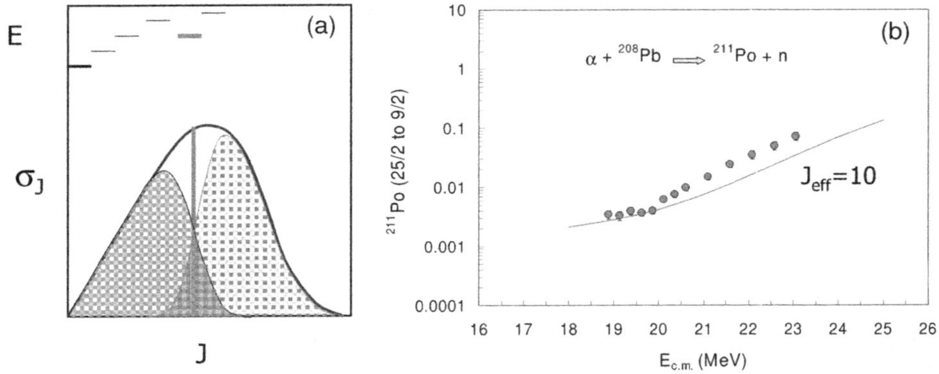

Fig. 5. (a) A high-spin isomer is populated by fusion with angular momentum above a critical effective value (close to the isomer spin), with some smoothing due to the preceding evaporation cascade. In a complete fusion reaction without breakup, the effective angular momentum can be determined - in this case to be 10, rather than the 25/2 of the state, due to nuclear structure effects.

5.2. *ICF isomer ratios*

Two extreme assumptions can be made regarding the angular momentum dependence of the ICF probability:[12]

- ICF occurs with equal probability for all partial waves
- only the highest angular momenta lead to ICF.

Knowing the fraction of flux resulting in ICF (~30%), and assuming that the angular momentum carried by the breakup fragment is proportional to its mass, isomer ratios can then be calculated under these two extreme assumptions. Comparing the experimental results shown in Fig. 6a with the two lower curves, neither the all-angular-momenta (full line) nor the high-angular-momenta (dashed line) assumptions come close to explaining the large measured ratios at the lower (near-barrier) energies for ICF in the ^{9}Be+^{208}Pb reaction (forming ^{211}Po after 2n evaporation). Clearly angular momenta are populated beyond those predicted by the simple picture described above.

Since the breakup fragments are charged, they will experience a mutual Coulomb repulsion that will result in a distribution of kinetic energies as well as modifying the point of interaction (impact parameter) with the target nucleus. Initially accounting for these effects in an empirical manner, the cross sections for each partial wave were folded with a single Gaussian distribution in angular momentum whose standard deviation was adjusted to best reproduce the experimental data. The upper curves

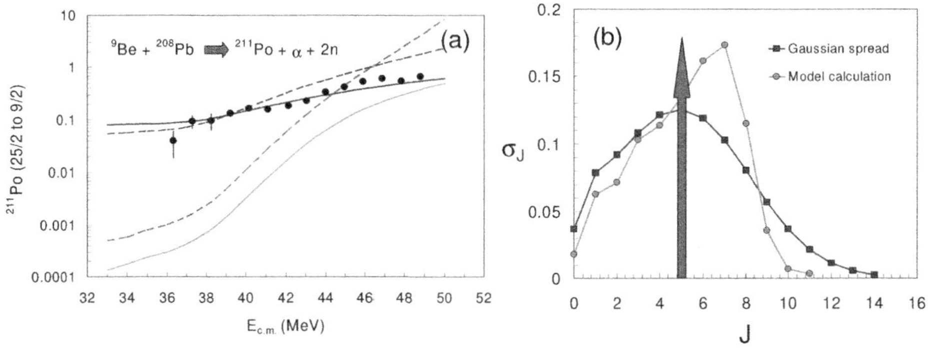

Fig. 6. The same isomer as in Fig.5 is populated in the incomplete fusion reaction. At near-barrier energies, the isomer ratio (panel(a)) is 100 times that of a simple model (lower full and dashed curves). Introducing a Gaussian angular momentum spread to each angular momentum (as shown in panel (b)) gives good agreement (upper curves in panel (a)). The classical breakup model predicts a similar spread in angular momentum as that required empirically. This is shown in (b) for ICF angular momentum $5\hbar$.

in Fig. 6a show the best fits for ICF at all angular momenta (full line), and ICF at high angular momenta (dashed line), requiring Gaussian spreading with FWHM 11 $\hbar$ and 7.5 $\hbar$ respectively.

Fig.6(b) shows the empirical angular momentum spread (Gaussian), and the distribution calculated with the 3-D classical trajectory model for a single ICF angular momentum at breakup of $5\hbar$. The two are in quite good agreement, indicating that future calculations with the classical trajectory model should be much closer to reproducing the measured isomer ratios than the initial calculations neglecting post-breakup trajectories.

6. Conclusions

Breakup and incomplete fusion have been quantitatively related through a 3-D classical trajectory model with breakup sampled stochastically according to an empirical exponential breakup function. Isomer ratios in the heavy ICF product are a new observable against which models of breakup and incomplete fusion can be tested.

Acknowledgments

The authors acknowledge support from an ARC Discovery grant, and JAT also acknowledges support also from the United Kingdom Science and Technology Facilities Research Council under Grant No. EP/D003628.

References

1. J.A. Tostevin et al., *Phys. Rev. C* **63**, 024617 (2001).
2. A. Diaz-Torres et al., *Phys. Rev. C* **65**, 024606 (2002); *Phys. Rev. C* **68**, 044607 (2003).
3. K. Yabana et al., *Prog. Theor. Phys.* **97**, 437 (1997), K. Yabana et al., *Nucl. Phys.* **A738**, 303 (2004).
4. M. Ito et al., *Phys. Lett.* **B637**, 53 (2006).
5. D.J. Hinde and M. Dasgupta, *Nature* **431**, 748 (2004).
6. M. Dasgupta et al., *Phys. Rev. C* **66**, 041602 (2002).
7. I.J. Thompson and A. Diaz-Torres, *P*rog. Theor. Phys. Suppl. **154**, 69 (2004).
8. D.J. Hinde et al., *P*hys. Rev. Lett. **89**, 272701 (2002).
9. M. Dasgupta et al., *Phys. Rev. Lett.* **82** (1999) 1395.
10. M. Dasgupta et al., *Phys. Rev. C* **70** (2004) 024606.
11. A. Diaz-Torres et al., *Phys. Rev. Lett.* **98** (2007) 152701.
12. L.R. Gasques et al., *Phys. Rev. C* **74** (2006) 064615.
13. R. Vandenbosch, *Annu. Rev. Nucl. Part. Sci.* **42**, 447 (1992).
14. A.R. Barnett and J. S. Lilley, *Phys. Rev. C* **9**, 2010 (1974).
15. W.J. Ramler et al., *Phys. Rev.* **114**, 154 (1959).

EXCITED STATES OF ^{13}B AND ^{8}He AND THEIR CLUSTER ASPECT

Y. KANADA-EN'YO

Yukawa Institute for Theoretical Physics, Kyoto University, Kyoto, Japan

Y. TANIGUCHI

Department of Physics/Yukawa Institute for Theoretical Physics, Kyoto University, Kyoto, Japan

M. KIMURA

Institute of Physics, University of Tsukuba, Tsukuba , Japan

Cluster states in light unstable nuclei have been investigated with antisymmetrized molecular dynamics. The results of ^{13}B suggest that largely deformed bands, $K^\pi = 3/2^-$ and $K^\pi = 1/2^+$, with intruder configurations appears in excited states around $E_x = 5 \sim 10$ MeV. The neutron structure of the $K^\pi = 3/2^-$ is analogous to the ground state of ^{12}Be. Structure of ^{8}He has been also investigated. It was found that the ground state of ^{8}He has both the j-j coupling feature($p_{3/2}$ closure) and the L-S coupling feature(^{4}He+$2n + 2n$). We theoretically predicted possible appearance of an excited 0^+ state with dineutron gas-like structure of ^{4}He+$2n$+$2n$.

1. Introduction

Owing to recent development of experimental and theoretical studies on unstable nuclei, it has been revealed that cluster structure is one of important features in light unstable nuclei as well as in stable nuclei. For instance, molecular-orbital structure,[1] which has been found in neutron-rich Be isotopes, is considered to be an origin of breaking of the $N = 8$ magic number in ^{11}Be and ^{12}Be. Recently, further exotic cluster states have been suggested to exist in excited states of light nuclei. In particular, a 3α gas-like state, which has been recently suggested in the 0_2^+ state of ^{12}C by Tohsaki et al.,[2] attracts great interests concerning with the BEC (Bose Einstein condensation) phenomena in nuclear system.

In this paper, we focused on molecular states and cluster gas-like states in light neutron-rich nuclei. Based on theoretical calculations with the antisymmetrized molecular dynamics, we discussed molecular states in ^{13}B, and dineutron structure in ^{8}He.

This paper is organized as follows. In the next section, we briefly explain the formulation of theoretical framework. The molecular orbital states are reported in section 3, and cluster gas-like states are discussed in section 4. Finally, we give a summary in section 5.

2. Formulation

We here briefly explain the formulation of the present calculations with the AMD method. The details of the AMD method for nuclear structure study are described in Refs. 3–6.

An AMD wave function is given by a Slater determinant of Gaussian wave packets;

$$\Phi_{\rm AMD}(\mathbf{Z}) = \frac{1}{\sqrt{A!}}\mathcal{A}\{\varphi_1, \varphi_2, ..., \varphi_A\}, \tag{1}$$

where the ith single-particle wave function is written by a product of spatial(ϕ), intrinsic spin(χ) and isospin(τ) wave functions as,

$$\varphi_i = \phi_{\mathbf{X}_i}\chi_i\tau_i, \tag{2}$$

$$\phi_{\mathbf{X}_i}(\mathbf{r}_j) \propto \exp\{-\nu(\mathbf{r}_j - \frac{\mathbf{X}_i}{\sqrt{\nu}})^2\}, \tag{3}$$

$$\chi_i = (\frac{1}{2}+\xi_i)\chi_\uparrow + (\frac{1}{2}-\xi_i)\chi_\downarrow. \tag{4}$$

$\phi_{\mathbf{X}_i}$ and χ_i are spatial and spin functions, and τ_i is isospin function which is fixed to be up(proton) or down(neutron). We used the optimum value of the width parameter ν for each system. Accordingly, an AMD wave function is expressed by a set of variational parameters, $\mathbf{Z} \equiv \{\mathbf{X}_1, \mathbf{X}_2, \cdots, \mathbf{X}_A, \xi_1, \xi_2, \cdots, \xi_A\}$.

In calculations of Be, B and C isotopes, we performed energy variation after spin parity projection(VAP) within the AMD model space in the same way as done in Refs.6,7. For He isotopes we applied the AMD+GCM. Based on the concept of GCM, we performed superposition of a number of AMD wave functions obtained by energy variation with constraints. The energy variation was performed for the parity-projected AMD wave function $\Phi^{\pm}_{\rm AMD}(\mathbf{Z})$ under constraints. In order to obtain basis wave functions, we adopted the total oscillator quanta and quadrupole deformation as the constraints.

3. Molecular orbital states in ^{13}B

As mentioned before, it is known that molecular orbital structure is favored in neutron-rich Be isotopes. In ^{12}Be, the intruder state drops down to be the ground state because of the well-developed molecular-orbital structure. This is consistent with the significant d-wave component in ^{12}Be, which was recently observed by neutron knock-out reactions.[8] On the other hand, a neighboring nucleus, ^{13}B, has the normal ground state. Therefore, it is challenging problem to search for molecular orbital states in the excited states of ^{13}B.

^{13}B is a nucleus with neutron magic number $N = 8$, and its ground state is the $3/2^-$ with normal configuration of neutron p-shell closure. Above the $3/2^-$ ground state of ^{13}B, many states are experimentally known to exist in the excitation energy $E_x \geq 3.5$ MeV region. Unfortunately, spins and parities of most of these states are unknown. Recently, the state at 4.83 MeV has been assigned to be a $1/2^+$ state by ^{4}He(^{12}Be,^{13}Bγ)X experiments.[9] Because of its strong production via proton-transfer to the ^{12}Be(0^+) state, this excited state is suggested to be a proton intruder state.

The calculated energy levels of the negative- and positive-parity states of ^{13}B are shown in Fig. 1. The effective interaction is same as that adopted in the previous calculations of ^{12}Be.[7] In addition to the ground $3/2_1^-$ state, we obtained many excited states with various J^π in the region $E_x \geq 4$ MeV. These states may correspond to the observed levels in this energy region. In the excited states, we found largely deformed bands, $K^\pi = 3/2^-$ and $K^\pi = 1/2^+$ (solid lines). These bands are composed of cluster structures. As for other excited states (disconnected filled circles), because their deformations are not so large, they are considered to be dominated by $0\hbar\omega$ or neutron $1\hbar\omega$ configurations.

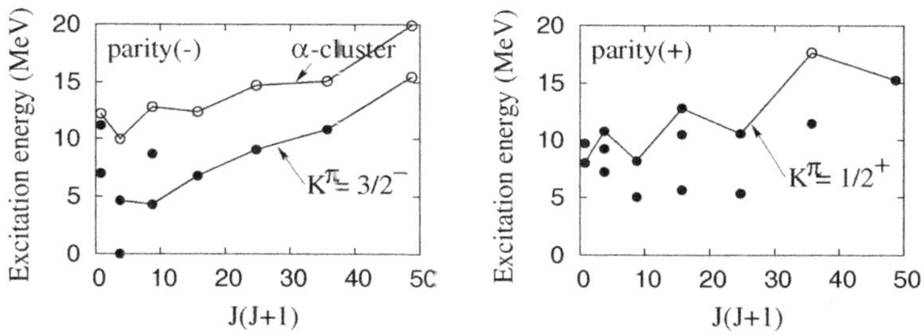

Fig. 1. Excitation energies of the negative-parity(left) and positive-parity(right) states of ^{13}B calculated by superposition of the basis wave functions. Filled circles are the energies of the J_n^π states, for which the VAP calculations were done. Open circles are the energies of the J_n^π states, which were obtained by diagonalization of Hamiltonian by superposing wave functions, but the VAP calculations were not performed for the corresponding J_n^π states.

In both the $K^\pi = 3/2^-$ and $K^\pi = 1/2^+$ bands, 2 α cores are formed. Let us look into intrinsic structures of these two bands, while focusing on the valence nucleons surrounding the cores.

In the $K^\pi = 3/2^-$ band, two valence neutrons occupy an approximately positive-parity orbital, and the other two neutrons and a proton are in orbitals with dominant negative-parity component. Since the positive-parity orbitals of the valence nucleons are associated with the sd-orbits, we can roughly regard this state as a neutron $2\hbar\omega$ excited state. The positive-parity orbital of the last two neutrons is largely deformed and has nodes along the 2α direction(longitudinal axis). This corresponds well to the so-called σ orbital in the molecular orbital picture.[1,10–14] It is interesting that the predicted $K^\pi = 3/2^-$ band has a quite similar neutron structure to that of the ^{12}Be ground state. It means that the $K^\pi = 3/2^-$ band consists of the neutron intruder states($2\hbar\omega$) and is interpreted as ^{12}Be(0_1^+)+p, where the ^{12}Be(0_1^+) has the intruder configuration and the additional proton strongly couples to the deformed core.

On the other hand, the band-head state $1/2^+_1$ of the $K^\pi = 1/2^+$ is the proton intruder state with a further large deformation. In this state, $1\hbar\omega$ excitation occurs in the proton shell. Namely, the last proton occupy a σ-like orbital, which is analogous to that of the highest neutron orbital in the $K^\pi = 3/2-$ band and also that in the ^{12}Be(0^+_1). This is extraordinary configuration because excitations are naively expected in the neutron side in case of neutron-rich nuclei. Such the proton excitation can be understood by the degeneration of the σ orbital with p orbits in the developed two-center structure. Recently, a state at 4.83 MeV has been assigned to be a $1/2^+$ state by ^{4}He(^{12}Be,^{13}Bγ)X experiments.[9] Because of strong production via the proton-transfer to the ^{12}Be(0^+) state and analysis of angular dependence, Ota *et al.* suggested this $1/2^+$ state to be a proton intruder state. The present prediction of the proton intruder configuration in the $1/2^+_1$ state is consistent with this observation though the theoretical excitation energy of the present calculation is slightly higher than the experimental value. Therefore, we assign the theoretical $1/2^+$ state to the experimental $1/2^+$(4.83 MeV).

4. Cluster gas-like states

Recently, Tohsaki *et al.* proposed a new type of cluster structure in the second 0^+ state ^{12}C, where 3 α clusters are weakly interacting.[2] This is a dilute gas state of α particles. Because of bosonic behaviour of the α particles in a s-orbit, this phenomena is called "alpha condensation" in association with Bose-Einstein Condensation(BEC) which has been suggested in dilute nuclear matter by Röpke et al.[15] It is a challenging subject to search for such cluster gas-like states with a dilute density in other nuclei. Many groups are now attacking this subject theoretically and experimentally. In analogy to the alpha condensation, dineutron condensation in neutron matter is a recent key issue in physics of unstable nuclei. Matsuo suggested that the dineutron correlation can be enhanced in dilute neutron matter.[16] Let us consider structure of ^{8}He from a point of view associated with the dineutron condensation, because two pairs of neutrons are possible in the system with four valence neutrons around the ^{4}He core. In analogy to ^{12}C, the ground state of ^{8}He may have a feature of the neutron $p_{3/2}$ closure or the SU(3)-limit p-shell configuration. Instead of the ground state, one can speculate appearance of the dineutron gas-like state with developed ^{4}He+$2n$+$2n$ structure in excited states.

We calculated the ground and excited states of He isotopes with the method of AMD+GCM. We used effective nuclear interactions consisting of the central force(Volkov[17] or MV1[18]), the spin-orbit force and Coulomb force. By taking care of energies of subsystems, we tuned the interaction parameters, w, b, h, m, for Wigner, Bartlett, Heisenberg and Majorana exchange terms in the the central force(Volkov or MV1), respectively. We used the spin-orbit force of the G3RS force[19] with a strength of u_{ls} = 2000 MeV. ^{6}He, ^{8}He and ^{10}He were calculated with AMD+GCM by using two cases of the central force. The adopted interaction parameters are explained in the caption of table 4. In the table, we also show the relative energies of subsystems and the nucleon-nucleon scattering lengths calculated with these interactions. We here estimate the energies of the ^{4}He, ^{4}He-n state with $J^\pi = 3/2^-$, and ^{4}He-^{4}He state with $J^\pi = 0^+$, by assuming the $(0s)^4$ state of ^{4}He and performing cluster-GCM calculations within the α-n and α-α cluster models for simplicity. The adopted v58 and m56 interactions systematically reproduce the energies of ^{4}He, ^{6}He and ^{8}He, though they overestimate the ^{10}He energy. The v58 interaction well describes the energies of subsystems but it has a disadvantage of too strong neutron-neutron interaction, while the m56 interaction reasonably reproduces the global features of the subsystem energies as well as the N-N phase shift.

Table 1. The theoretical values of scattering length $a_s(a_t)$ for singlet(triplet) even channel neutron separation energy of ^{5}He ($S_n(^5\text{He}) \equiv E(^4\text{He}) - E(^4\text{He-}n)$), 2α threshold energy of ^{8}Be, two-neutron separation energies of ^{6}He and ^{8}He ($S_{2n}(^6\text{He}) \equiv E(^4\text{He}) - E(^6\text{He})$ and $S_{2n}(^8\text{He}) \equiv E(^6\text{He}) - E(^8\text{He})$) calculated with the v58 interaction and the m56 interaction. The v58 interaction consist of Volkov No.2 central force with $w = 0.42$, $b = h = 0$, $m = 0.58$ and the G3RS-type spin-orbit force with the strength u_{ls}=2000 MeV. The m56 interaction consist of MV1(Modified Volkov No.1)-case(3) central force with $w = 0.44$, $b = h = 0.15$, $m = 0.56$ and the G3RS-type spin-orbit force with the strength u_{ls}=2000 MeV.

	exp.	v58	m56
a_t (fm)	5.42 (p-n)	9.7	4.2
a_s (fm)	−16.5 (n-n)	9.7	>100
$S_n(^5\text{He})$ (MeV)	−0.9	−0.7	−0.4
$2E(^4\text{He}) - E(^4\text{He-}^4\text{He})$ (MeV)	−0.1	0.6	−0.6
$S_{2n}(^6\text{He})$ (MeV)	1.0	1.3	1.1
$S_{2n}(^8\text{He})$ (MeV)	2.1	3.0	2.0

The calculated energy levels of ^{8}He are illustrated in Fig. 2(a). In both of the m56 and v58 results, the 2_1^+ state is the lowest excited state and the 0_2^+ state appears just above the 2_1^+ state. The 1_1^- and 3_1^- states are obtained in a higher energy region. In addition, in the present calculations with the m56 interaction, the 1_1^+, 0_1^- and 2_1^- states are obtained in almost the same energy region as the 1_1^- and 3_1^- states. In order to check the stability of the resonances against neutron decays, their properties should be carefully examined. In particular, the 1_1^-, 2_1^- and 0_1^- states can couple with $(0s)^2(0p)^3(1s)^1$ neutron configuration which has a valence $1s_{1/2}$ neutron with no centrifugal barrier.

Compared with the experimental data, the theoretical values of the 2_1^+ excitation energy are higher than the experimental data. However, it is important that the level structure for the excited states, 2_1^+, 0_2^+, 1_1^- and 3_1^-, is not sensitive to the adopted interaction though the relative position to the ground energy depends on the interaction. The 0_2^+ state is theoretically suggested to appear just above the 2_1^+ state. What is striking is that the 0_2^+ state has a remarkably large neutron radius

($r_{\text{r.m.s.r.}}$=3.1 fm in case m56) compared with the ground state($r_{\text{r.m.s.r}}$=2.6 fm in case m56) because of developed $^4\text{He} + 2n + 2n$ structure. In the obtained wave function of the 0_2^+ state, which is given by a superposition of the basis AMD wave functions, the amplitude is found to be widely distributed into the basis wave functions with various spatial configuration of $^4\text{He} + 2n + 2n$. This indicates a gas-like feature that the dineutrons are rather freely moving around the ^{4}He core. Therefore, we consider that the 0_2^+ state is the candidate of the cluster gas-like state with two dineutrons around the α core. In the experimental energy spectra, some excited states were observed above the 2_1^+ state. Spins and parities of these states are not definitely assigned yet.

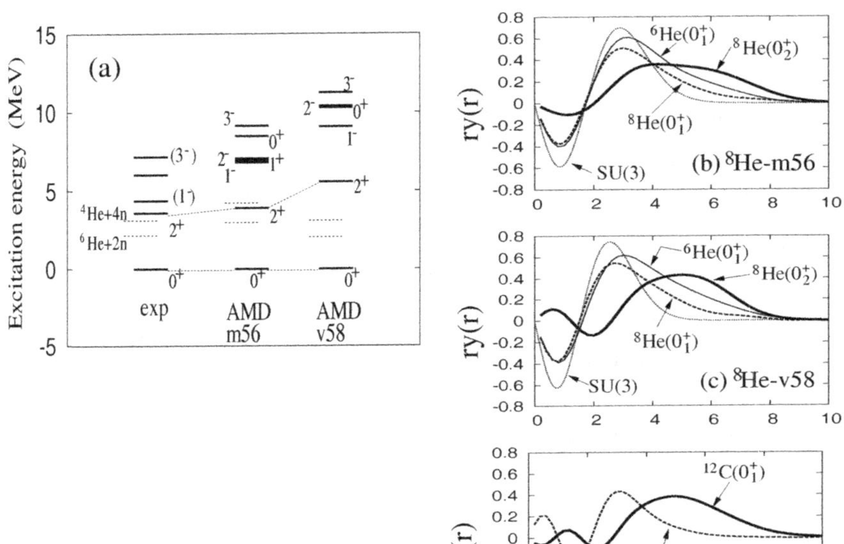

Fig. 2. (a)Energy levels of ^{8}He. The calculated results are those with the m56 and v58 interactions. (b),(c)Reduced width amplitudes $ry_{l=0}(r)$ for $^6\text{He}^{SU(3)}(0^+)$-^2n in the $^8\text{He}(0^+)$, and those for ^{4}He-^2n in the $^6\text{He}(0^+)$. The $^8\text{He}(0^+)$ and $^6\text{He}(0^+)$ wave functions are calculated by AMD+GCM with (b) the v58 and (c) the m56 interactions. The $^6\text{He}^{SU(3)}(0^+)$ is written by a SU(3)-limit ^{4}He-$2n$ cluster state. The ^{4}He cluster and the ^2n cluster are expressed by the $(0s)^4$ and $(0s)^2$ wave functions, respectively, where the size parameter for the $(0s)$ state is chosen to be the same value as the AMD+GCM wave functions; $b = 1/\sqrt{2\nu(^6\text{He}(0^+))}$ in the calculation of $ry(r)$ for $^6\text{He}(0^+)$ and $b = 1/\sqrt{2\nu(^8\text{He}(0^+))}$ in the calculation of $ry(r)$ for $^8\text{He}(0^+)$. The reduced width amplitudes for ^{4}He-^2n in the $^6\text{He}^{SU(3)}(0^+)$ are also shown. (d) The reduced width amplitudes for $^8\text{Be}^{SU(3)}(0^+)$-α in the $^{12}\text{C}(0^+)$ taken from.[20]

In order to investigate the features of dineutron cluster structure in the 0^+

states of ^{8}He, we calculated the reduced width amplitudes $ry(r)$ for the ^2n-cluster motion and the cluster probability $S^{\rm fac}$ by taking the overlap of the $^6{\rm He}^{SU(3)}(0^+)$-^2n cluster wave functions with the ^{8}He wave functions. In Fig. 2(b) and (c), we show the reduced width amplitudes in the $^8{\rm He}(0^+_1)$ and the $^8{\rm He}(0^+_2)$ wave functions obtained by the v58 and m56 interactions. Surprisingly, the ^2n-cluster motion in the $^8{\rm He}(0^+)$ states is quite similar to the α-cluster motion in ^{12}C. In particular, the $^8{\rm He}(0^+_2)$ state has the large amplitude of the dineutron cluster in the long distance region around $r = 4 - 6$ fm, which well corresponds to the peak position of the α-cluster motion in the $^{12}{\rm C}(0^+_2)$. The overlap between the $^8{\rm He}(0^+_2)$ and the $^6{\rm He}^{SU(3)}(0^+)$-^2n cluster wave functions is $S^{\rm fac} = 0.50$ and $S^{\rm fac} = 0.43$ in the v58 and the m56 results, respectively. Considering that the other ^2n cluster exists inside the $^6{\rm He}^{SU(3)}(0^+)$ core, it is regarded that the $^8{\rm He}(0^+_2)$ has the component of the developed ^{4}He+^2n+^2n clustering, where two dineutrons are moving in the $L = 0$ wave. From the analogy of the ^2n-cluster structure in the $^8{\rm He}(0^+_2)$ with the α-cluster structure in the $^{12}{\rm C}(0^+_2)$, the $^8{\rm He}(0^+_2)$ is considered to contain the dineutron gas-like structure.

In order to see more directly the ^2n-cluster gas-like nature, we assumed the ^2n condensate wave function in the $^4{\rm He} +^2 n +^2 n$ system and calculated the overlap with the obtained $^8{\rm He}(0^+)$ wave functions. The ^2n condensate wave function is given as follows,

$$\Psi_{\rm cond}(B) \equiv n_0 \int \prod_{i=1}^{k} \left\{ d^3\mathbf{S}_i \exp\left(-\frac{(\mathbf{S}_i - \mathbf{S}_C)^2}{B^2} \right) \right\} \Phi_{\rm Brink}(\mathbf{S}_C, \mathbf{S}_1, \mathbf{S}_2, \cdots \mathbf{S}_k), \quad (5)$$

where n_0 is the normalization factor and $\Phi_{\rm Brink}(\mathbf{S}_C, \mathbf{S}_1, \mathbf{S}_2, \cdots \mathbf{S}_k)$ is the Brink wave function for the $C + k(^2n)$-cluster system consisting the core(C) and k dineutrons(^2n) as,

$$\Phi_{\rm Brink}(\mathbf{S}_C, \mathbf{S}_1, \mathbf{S}_2, \cdots \mathbf{S}_k) \equiv \mathcal{A}\left\{ \varphi^{\rm C}(\mathbf{S}_C) \phi^{^2n}(\mathbf{S}_1) \phi^{^2n}(\mathbf{S}_2) \cdots \phi^{^2n}(\mathbf{S}_k) \right\}. \quad (6)$$

Here, the wave function of the ith ^2n, $\varphi^{^2n}(\mathbf{S}_i)$, is given by the $(0s)^2$ state with the size $b = 1/\sqrt{2\nu(^8{\rm He})}$ localized around $\mathbf{S}_i$. $\mathbf{S}_C$ is the mean position of the center of mass motion of the core, and is chosen to be $\mathbf{S}_C = -\frac{2}{A}(\mathbf{S}_1 + \mathbf{S}_2 + \cdots + \mathbf{S}_k)$. The $^8{\rm He}(0^+_2)$ has the maximum overlap as about 0.5 with the dineutron condensate wave function $\Psi_{\rm cond}(B)$ at $B = 4 \sim 5$. This is an strong indication of the dineutron gas-like component in the calculated $^8{\rm He}(0^+_2)$.

5. Summary

We theoretically investigated the molecular states and the cluster gas-like states in light nuclei based on calculations with antisymmetrized molecular dynamics. The theoretical results of ^{13}B suggest that the intruder states with cluster structures construct the rotational bands, $K^\pi = 3/2^-$ and $K^\pi = 1/2^+$, starting from 5 MeV and 8 MeV, respectively.

The ground and excited states of He isotopes were also investigated. In the ground state of ^{8}He, although the component of the $p_{3/2}$ sub-shell closure is dominant, the L-S coupling feature is also significantly contained because of the spin-zero dineutron correlation. The present results suggest that the 0_2^+ state may appear a few MeV above the 2_1^+ state. By analyzing dineutron structure, it was found that this state has a significant component of the developed ^{4}He+^2n+^2n structure where two dineutrons are moving around the ^{4}He core in S wave with a dilute density. We consider that the predicted 0_2^+ state is the candidate of the dineutron gas-like state, which is analogy to the α condensate state suggested in the ^{12}C(0_2^+).

Acknowledgments

The computational calculations of this work were performed by the Supercomputer Projects of High Energy Accelerator Research Organization(KEK) and also by super computers in YITP. This work was supported by Grant-in-Aid for Scientific Research from JSPS. It is also supported by the Grant-in-Aid for the 21st Century COE "Center for Diversity and Universality in Physics" from MEXT. Discussions in the workshops on cluster physics held at RCNP and those in the workshops YITP-W-06-17 and YITP-W-07-01 held at YITP were helpful to initiate and complete this work.

References

1. W. von Oertzen, M. Freer and Y. Kanada-En'yo, *Phys. Rep.* **432**, 43 (2006).
2. A. Tohsaki, H. Horiuchi, P. Schuck, and G. Röpke, *Phys. Rev. Lett.* **87**, 192501 (2001).
3. Y. Kanada-En'yo, H. Horiuchi and A. Ono, *Phys. Rev. C* **52**, 628 (1995); Y. Kanada-En'yo and H. Horiuchi, *Phys. Rev. C* **52**, 647 (1995).
4. Y. Kanada-En'yo and H. Horiuchi, *Prog. Theor. Phys. Suppl.***142**, 205 (2001).
5. Y. Kanada-En'yo, M. Kimura and H. Horiuchi, *Comptes rendus Physique Vol.4*, 497 (2003).
6. Y. Kanada-En'yo, *Phys. Rev. Lett.* **81**, 5291 (1998).
7. Y. Kanada-En'yo and H. Horiuchi, *Phys. Rev. C* **68**, 014319 (2003).
8. S. D. Pain *et al.*, *Phys. Rev. Lett.* **96**, 032502 (2006).
9. Ota *et al.*, talk in JPS meeting, March 2007, Tokyo, Japan.
10. S. Okabe, Y. Abe and H. Tanaka, *Prog. Theor. Phys.* **57**, 866 (1977).
11. M. Seya, M. Kohno, and S. Nagata, *Prog. Theor. Phys.* **65**, 204 (1981).
12. W. von Oertzen, *Z. Phys. A* **354**, 37 (1996); **357**, 355(1997).
13. N. Itagaki and S. Okabe, *Phys. Rev. C* **61**, 044306 (2000); N. Itagaki, S. Okabe and K. Ikeda, *Phys. Rev. C* **62**, 034301 (2000).
14. Y. Kanada-En'yo, H. Horiuchi and A. Doté, *Phys. Rev. C* **60**, 064304(1999); Y. Kanada-En'yo and H. Horiuchi, *Phys. Rev. C* **66**, 024305 (2002).
15. G. Röpke, A. Schnell, P. Schuck, and P. Nozieres, *Phys. Rev. Lett.* **80**, 3177 (1998).
16. M. Matsuo, *Phys. Rev. C* **73**, 044309 (2006).
17. A. B. Volkov, *Nucl. Phys.* **74**, 33 (1965).
18. T. Ando, K. Ikeda and A. Tohsaki, *Prog. Theory. Phys.* **64**, 1608 (1980).
19. N. Yamaguchi, T. Kasahara, S. Nagata and Y. Akaishi, *Prog. Theor. Phys.* **62**, 1018 (1979); R. Tamagaki, *Prog. Theor. Phys.* **39**, 91 (1968).
20. Y. Kanada-En'yo, *Prog. Theor. Phys.* **117**, 655 (2007).

NUCLEAR REACTIONS WITH WEAKLY-BOUND SYSTEMS: THE TREATMENT OF THE CONTINUUM

C.H. DASSO

Departamento de Física Atómica, Molecular y Nuclear, Facultad de Física
Universidad de Sevilla, Apartado 1065, 41080 Sevilla, Spain
E-mail: dasso@us.es

A. VITTURI

Dipartimento di Fisica "Galileo Galilei" and INFN, Padova, Italy
E-mail: vitturi@pd.infn.it

Different prescriptions for the treatment of continuum states in break-up processes involving weakly-bound nuclei are tested within a simple model. The model describes the evolution of a single particle wave function in one dimension, initially bound by a Woods-Saxon type potential and then perturbed by a time- and position-dependent external field. These processes generate inelastic excitation probabilities that – distributed over the bound and continuum states of the system – lead to either a partial or a total fragmentation of the final wave function. The numerical solution of this time-dependent Schroedinger equation is compared with the solution obtained within a coupled-channel scheme, where the total wave function is expanded over the single particle eigenstates of the Hamiltonian, and the continuum part of the spectrum is replaced by discretized states. We found that coupled-channel descriptions based on discretization of the continuum can only be accurate when proper choices are made regarding the number of discrete states, the energy mesh and the truncation of the space. This implies, even in simple cases, the use of a rather large (and often unpracticable) number of channels. The use of a more restricted number of channels may lead to misleading results.

1. Introduction

In spite of significant efforts invested over the past decade in the study of reactions with radioactive beams a number of important questions remain, disturbingly, either open or badly understood. A major standing topic in the case of weakly-bound systems is the role played in the reaction mechanism by the continuum states. Under debate it is, specifically, the proper way to incorporate continuum states into a standard coupled-channel reaction scheme.

A popular choice is provided by the so-called CDCC approximation scheme (Continuum Discretized Coupled Channels). In spite of the widespread application of this prescription, there is an actual need for examining the impact of the approximations made in the practical implementation of the formalism. Critical problems to be tested include the definition of continuum states with appropriate boundary conditions, the discretization procedure, the choice of size for the energy steps and truncation cutoff. Standard CDCC calculation are rather complex (and time consuming). They involve many channels, many angular momenta, complex potentials and interactions and so on. It is therefore not so easy to disentangle and check the validity of the different approximations involved. We propose here a simple model that is sufficiently rich to incorporate the basic physics of break-up processes in different situations. It can be (numerically) solved exactly, so that the different schemes can be easily tested.

2. The model

We set to describe the evolution of a single particle wave function initially bound in a one-dimentional Woods-Saxon potential $V_0(x)$ and perturbed by a time- and space-dependent interaction $V_{coup}(x,t)$. This coupling is for simplicity assumed to be of gaussian shape, $V_{coup}(x,t) = V_C \exp(-t^2/2\sigma_t^2)\exp(-(x-x_0)^2/2\sigma_x^2)$, but can be easily modified within the present implementation of the model as long as it stays separable. The potential $V_{coup}(x,t)$ is expected to simulate the main features of the excitation fields generated in a collision with a heavy partner. In particular, acting on the widths σ_t, σ_x and the asymmetry parameter x_0, one aims to describe different collision times (hence different bombarding energies), different ranges of the interaction and distances of closest approach (hence different impact parameters). The full evolution of the system is obtained by solving (numerically) the time-dependent Schrödinger equation *

$$i\hbar\partial\Psi(x,t)/\partial t = (H_0 + V_{coup}(x,t))\ \Psi(x,t)\ , \tag{1}$$

with

$$H_0 = -(\hbar^2/2\mu)\partial^2/\partial x^2 + V_0(x)\ . \tag{2}$$

For the initial condition (time large, negatixe) we take $\Psi(x, t \to -\infty) = \phi_N(x) \times \exp(iE_N t)$ where $\phi_N(x)$ is the spatial dependence of the chosen bound state of the potential $V_0(x)$.

Several examples of the time-dependent model calculations can be found in our ref.[1] We only report here two significant cases. Our first illustration refers to a case in which one starts with the particle in a deeply-bound level (N=3). Three snapshots of the evolution of the wave function at gradually increasing times are shown in Figure 2. One can immediately see that with the chosen interaction a large fraction of the evolving wave function escapes the potential well, populating unbound states in the continuum (note the change of scale in x). This can be interpreted as corresponding to final breakup events. As the perturbation was chosen asymmetric ($x_0 \neq 0$), breakup probabilities are different for one direction or the other. We can also see

*Work done within the time-dependent approach for weakly-bound nuclei can be found, e.g. in.[3,4] Specially close to our scheme is perhaps the earlier work of Bertsch and Bertulani.[5]

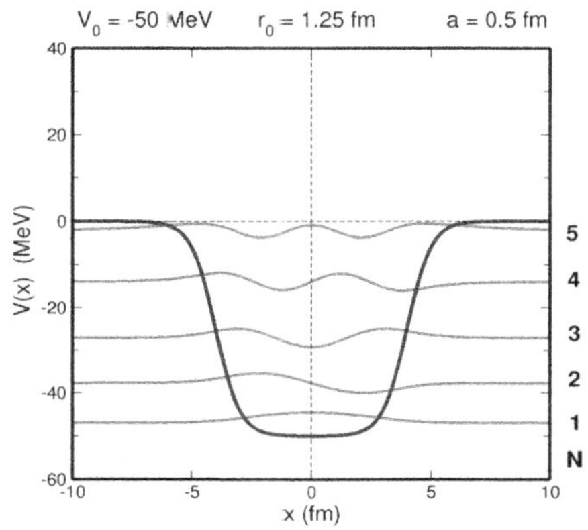

Fig. 1. Woods-Saxon potential and bound state wave functions used in the calculations.

that a fraction of the wave function remains localized, for very large times, in the potential region, revealing some finite probability of also populating bound states. Such process can therefore be interpreted as describing a partial breakup process. By looking at the time evolution of the probability current and the momentum distribution[1] one can indeed verify that the two external wave packets correspond to fragments moving in opposite directions, with momentum distributions with opposite signs. The parts of the wave function associated with the left fragment, the central part and the right fragment therefore yield the probability of break-up to the left, no break-up and break-up to the right, respectively. These quantities are directly related to "experimentally" measurable quantities.

The second case (illustrated in fig. 3) refers to a calculation in which one starts with the particle occupying the last bound orbital (N=5), a situation that one expects to simulate well conditions in the vicinity of the drip lines. No appreciable part of the final wave function remains in this case within the potential region, leading to a nearly complete breakup. Once again, this interpretation of the process can be confirmed by examining the current and momentum distributions.

The solution of the problem has not involved sofar any expansion of the wave functions into any basis whatsoever. In particular, it has not yet been necessary to introduce explicitly any scattering states, since all continuum effects are automatically taken into account. However, to yield information on the final Q-value distribution of the emerging particle, one needs to expand the final wave fuction in a set of basis energy eigenstates that include the continuum.

This raises the problem of defining the proper boundary conditions (two linearly independent solutions for each energy) with the necessary orthogonality condition. A possible line of action is to take the two solutions corresponding to scattering processes from the right and left sides. An alternative choice – equally valid – could be to take symmetric and antisymmetric combinations (stationary states, no current). The final results, obviously, will be independent of the chosen basis as long

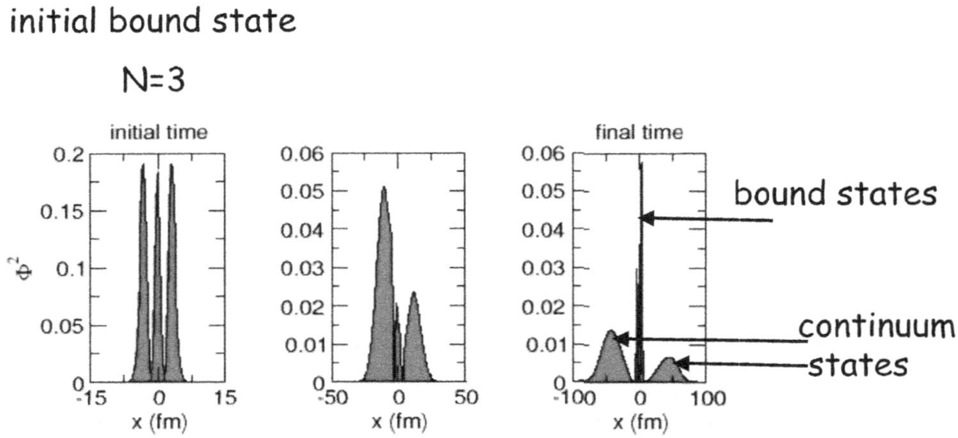

Fig. 2. Time evolution of the wave function probability. An asymmetric perturbation has been used, starting from the initial N=3 state.

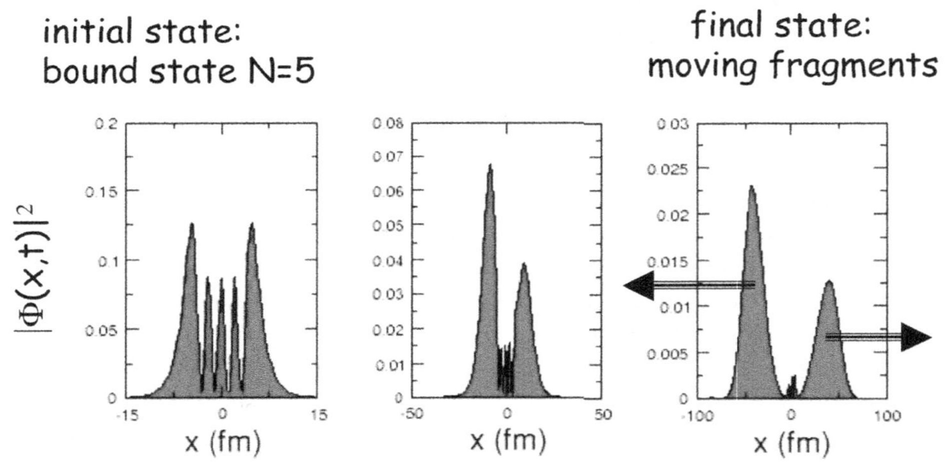

Fig. 3. Time evolution of the wave function probability. An asymmetric perturbation has been used, starting from the initial, weakly-bound, N=5 state.

as one takes two linearly independent solutions for each energy.

The total continuum Q-value distributions in the two considered cases are shown (shaded curves) in figs. 4 and 5. Note that in the case of a reaction starting from the weaky-bound orbital the energy distribution is more concentrated, as expected, at lower excitation energy.

Scattering problems as the ones discussed above are normally posed in terms of a

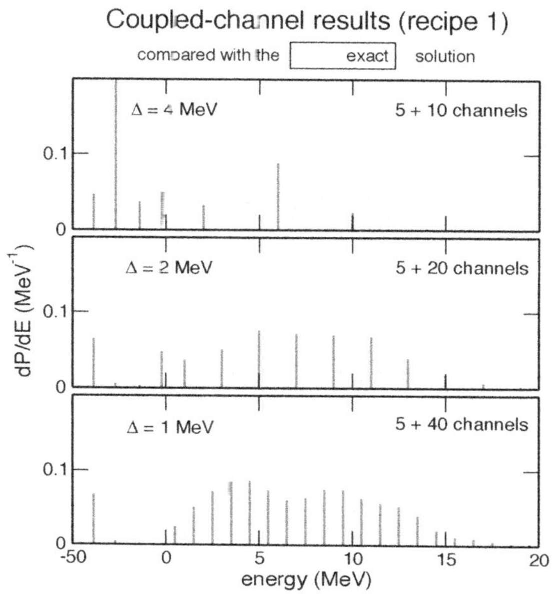

Fig. 4. Case of partial break-up starting from a well-bound orbital (N=3). Comparison of final Q-value distribution obtained in the exact calculation and with the discretized continuum (recipe I). In the later case, different energy meshes are used in the different frames. The energy cut-off was fixed at 20 MeV.

coupled-channel formalism, where the Schrödinger equation is solved by expanding the total wave function into a stationary basis. With this approach one constructs matrix elements of the perturbing interaction among the different states and solve (in time) the coupled set of equations associated with the expansion amplitudes. This is the preferred approach in situations where the direct numerical solution of the full time-dependent many-body wave function is too involved.

In those cases in which particles are promoted to positive energies one must necessarily enlarge the basis to include what is, in principle, an infinite number of continuum states. A procedure to discretize these continuum states is imperative and a number of different prescriptions have been advanced. In our case we have the "exact" solution of our problem and we can therefore use it as a reference to test these approximation schemes.

The general idea is that each energy slice corresponds to a single state. As a first recipe one can assume, for example, that each discrete state is represented by the true continuum state at the central energy, and take the matrix elements calculated precisely at those average energies. These states are normalized to a Dirac delta function and must be weigthed by a factor associated with the energy interval ΔE. The comparison to the reference calculation with of the coupled channel results for the two considered cases is also given in figs. 4 and 5, for different choices of the energy mesh (and consequently different number of basis states).

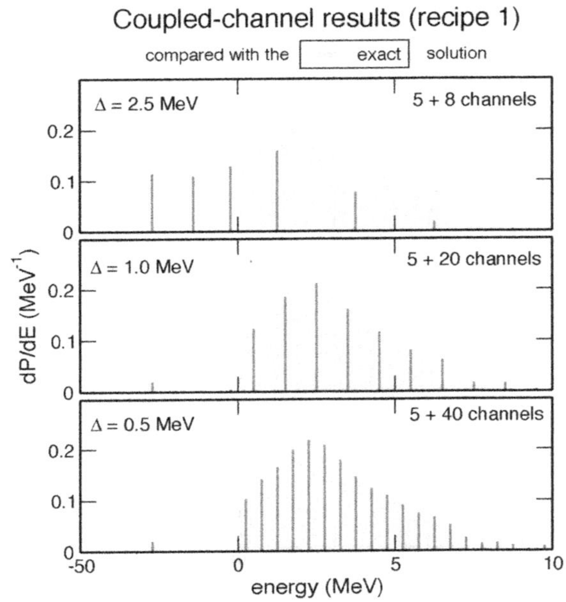

Fig. 5. Case of break-up starting from a weakly-bound orbital (N=5). Comparison of final Q-value distribution obtained in the exact calculation and with the discretized continuum (recipe I). In the later case, different energy meshes are used in the different frames. The energy cut-off was fixed at 10 MeV.

A second, more popular, recipe is again to associate each slice to a single state, but this time defined as an average over many continuum states in the interval ΔE. The states Φ_n(x) turn out (in the limit of an average over infinite equi-spaced states!) to be normalized to a Kroneker delta. Is this a necessary (or desired) requirement? The outcomes of this second recipe is given in fig. 6.

An essential parameter in the calculation, besides the value of the energy mesh ΔE, is the value of the maximum considered energy E_{cutoff}. The sensitivity of the results due to a poor choice of this parameter is illusterated in fig. 7.

Within the coupled-channels approach one can easily test different truncations for the coupling matrix elements. There has been speculation on the importance of continuum-continuum couplings. The importance of these couplings is illustrated in fig. 8. It is clear from the figure that calculations that exclude the couplings among the discretized continuum states are unable to provide a good description of the process.

In a similar way one can test the importance of higher-order scattering processes by solving just the first-order time-dependent coupled equations. The results clearly indicate the inherent multi-step character of the excitation mechanism.

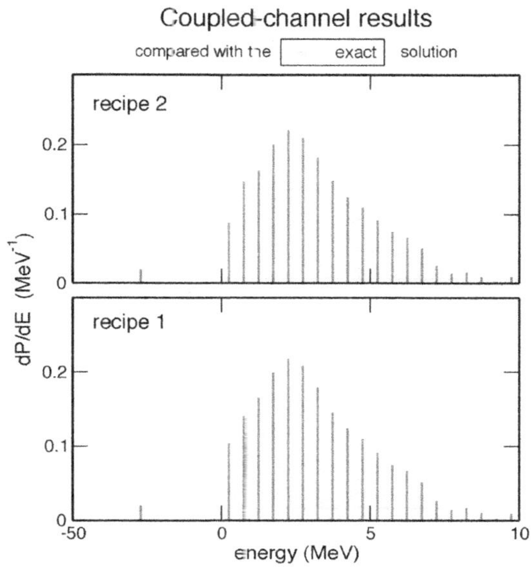

Fig. 6. Case of break-up starting from a weakly-bound orbital. The exact Q-value distribution is compared with those obtained with different prescriptions for the continuum discretization (recipe I and II). In both discretized calculations an energy mesh of 0.5 MeV and an energy cutoff of 10 MeV were used.

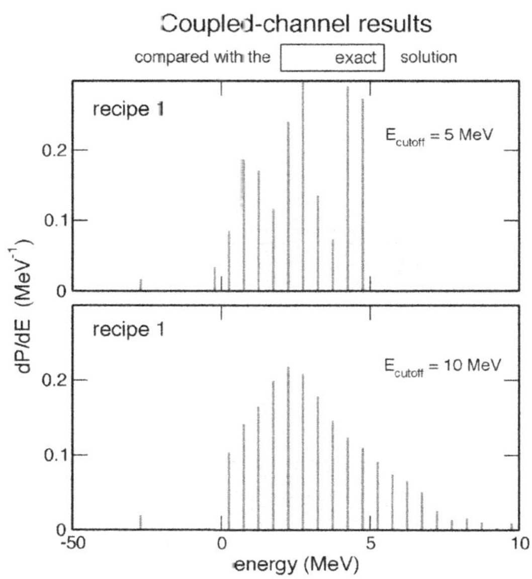

Fig. 7. Case of break-up starting from a weakly-bound orbital. Comparison of different energy cutoff for the continuum discretization.

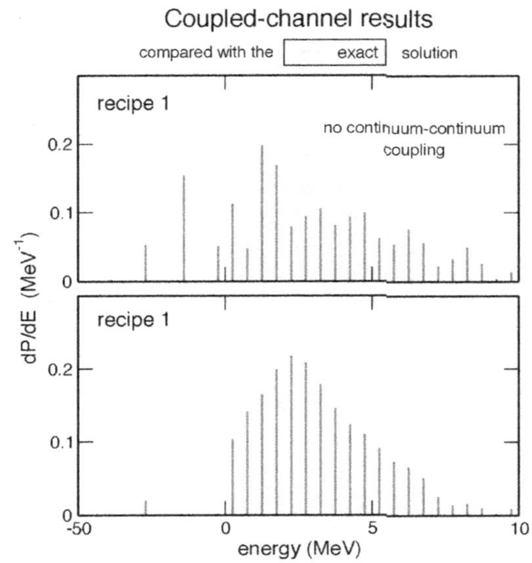

Fig. 8. Case of break-up starting from a weakly-bound orbital. Role of continuum-continuum couplings.

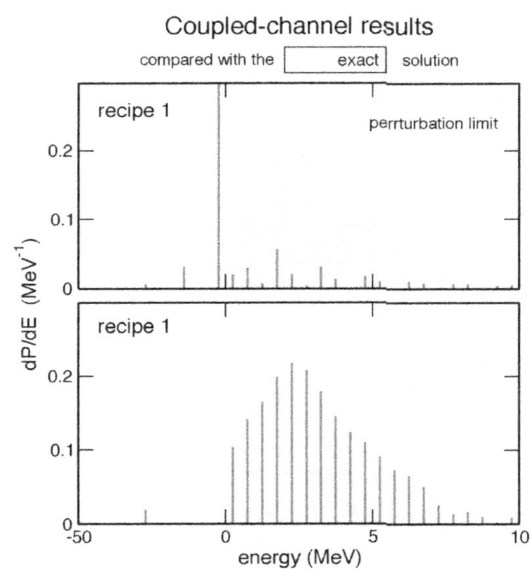

Fig. 9. Case of break-up starting from a weakly-bound orbital. Role of multistep processes.

3. Conclusions

The comparison with the exact calculations shows that coupled channel descriptions based on discretization of the continuum can be accurate when a proper choice is made of the number of discrete states, of the energy mesh and of the energy cutoff. This may imply, even in simplified cases, the use of a rather large (and unpracticable) number of channels. The use of a more restricted number of channels may lead to misleading results. The key role of multistep processes events in break-up reactions has been clearly confirmed.

This work has been partially supported by the Spanish Ministerio de Educación y Ciencia, by the European regional development fund (FEDER) under project numbers FIS2005-01105, FPA2005-04460 and by the exchange program INFN/MEC.

References

1. C.H. Dasso and A. Vitturi, *Nucl. Phys. A* **787**, 476 (2007).
2. J.Al-Khalili and F.Nunes, *J. Phys. G* **29**, R89 (2003).
3. T. Kido, K. Yabana, and Y. Suzuki, *Phys. Rev. C* **53**, 2296 (1996);
   M. Ito, K. Yabana, and T. Nakatsukasa, M. Ueda, *Phys. Lett. B* **637**, 53 (2006).
4. P. Capel, G. Goldstein and D. Baye, *Phys. Rev. C* **70**, 064605 (2004).
5. G.F. Bertsch and C.A. Bertulani, *Nucl. Phys. A* **556**, 136 (1993);
   C.A.Bertulani and G.F.Bertsch, *Phys. Rev. C* **49**, 2839 (1994).

DYNAMIC EVOLUTION OF THREE-BODY DECAYING RESONANCES

A.S. JENSEN*, D.V. FEDOROV, H.O.U. FYNBO, R. ÁLVAREZ-RODRÍGUEZ

Department of physics and Astronomy, University of Aarhus, DK-8000 Aarhus C, Denmark
**E-mail: asj@phys.au.dk*

E. GARRIDO

Instituto de Estructura de la Materia. CSIC, Serrano 123, E-28006 Madrid, Spain

We use complex scaling and the the hyperspherical adiabatic expansion method to compute resonance wave functions for several nuclear three-body systems. These wave functions often change character as the hyperradius increases. The properties at large hyperradii are decisive for the momentum distributions after decay into the three constituents. We demonstrate by several examples that this dynamic evolution is crucial to understand the relation between initial and final states in reactions and beta-delayed decays. For ^{11}Li(1^-) we see reminiscences of the Efimov effect in the continuum. For ^{6}Li(2^+) we see normal small isospin mixing at small distances and substantial mixing at large distances. For ^{12}C(0^+,1^+) the computed energy distributions determined at large distances are found to agree with recent measurements. The decay mechanisms depend on the resonance properties and can be either sequential or direct or a mxiture.

1. Motivation and basics

Three particles emerging from an initial state populated in beta-decay or in a reaction can all be accurately measured in coincidence.[1] Information about the process is contained in this three-body continuum final state which arises from some initial condition and evolve via non-observable intermediate configurations. A similar trivial example is α-decay where the discrete energy is given by total energy and momentum conservation, and angular distribution is determined by the angular momentum of the decaying state. For three particles in the final state the energies of the emerging particles are non-trivial and varying from discrete to smooth continuous distributions.[2]

The slightly simpler problem of a resonance, populated somehow, decaying into three particles is not yet theoretically understood. The reasons for this surprising lack of description become clear when we try to be specific and formulate a model. The difficulties are that the necessary three-body model at large distances (only three particles are in the final state) (i) must arise from a many-body small-distance structure with many, most likely different, degrees of freedom, (ii) the observables are determined at large distances where the wave functions must be very accurately computed, (iii) the final state is in the three-body continuum, (iv) both short-range and Coulomb interactions are important, (v) the asymptotic large-distance properties can be any combination of coherent two- and three-body discrete and continuum structures.

To solve the problem we compute three-body resonance wave functions by use of complex scaling and the hyperspherical expansion method.[3,4] We first choose the hyperspherical coordinates (ρ, Ω) based on the mass-scaled Jacobi coordinates $(\boldsymbol{x}, \boldsymbol{y})$ defined by

$$x^2 = \frac{m_i m_k}{m(m_i + m_k)} r_{ik}^2 \,, \quad y^2 = \frac{m_j(m_i + m_k)}{m(m_i + m_k + m_j)} r_{j,ik}^2 \,, \quad \boldsymbol{r_{ik}} = \boldsymbol{r_k} - \boldsymbol{r_i} \,, \tag{1}$$

$$\rho^2 = x^2 + y^2 \,, \quad \tan\alpha = \frac{x}{y} \,, \quad \Omega = \{\Omega_x, \Omega_y, \alpha\} \,, \tag{2}$$

where m is a reference mass. Then we select two-body interactions V_{ij} reproducing known low-energy scattering data, and we allow for a three-body interaction V_{3b}. The Faddeev equations are solved with the corresponding decomposition of the wave function Ψ, i.e.

$$\Psi(\rho, \Omega) = \sum_n f_n(\rho)\Phi_n(\rho, \Omega) = \sum_n f_n(\rho)\left(\phi_{12}^{(n)}(\rho, \Omega) + \phi_{13}^{(n)}(\rho, \Omega) + \phi_{23}^{(n)}(\rho, \Omega)\right) \,, \tag{3}$$

where f_n is the hyperradial wave function and $\phi_{ij}^{(n)}$ are the angular Faddeev components. The Fourier transform $\tilde{\Psi}$ of the coordinate-space resonance wave function Ψ has the same angular dependence as Ψ, see Ref. 2. The ρ-coordinate is exchanged with the momentum κ and a Breit-Wigner shape appears around the resonance value:

$$\frac{\tilde{\Psi}(\kappa_0, \Omega_k)}{\kappa^2 - \kappa_0^2} \,, \quad E_{res} = \frac{\hbar^2 \kappa_0^2}{2m} \,, \quad \kappa^2 = k_x^2 + k_y^2 \,, \tag{4}$$

where E_{res} is the resonance energy, and κ^2 is related to the partial momenta (k_x, k_y) as ρ^2 is related to x^2 and y^2. The final state energy distributions are found by absolute square and integration over unobserved momenta.

The large-distance boundary condition after complex scaling is exponential decay of each of the hyperradial wave functions f_n Ref. 4. The small-distance boundary condition is defined in three-body coordinates by this procedure where the many-body degrees of freedom contract to three particles at the nuclear surface. This is precisely as for α-emission where the energy is preselected and the possible preformation factor accounts for the probability of finding an α-particle at the nuclear surface. Here the α-particle is replaced by the three-body structure.

The crucial elements in the computations are the Faddeev decomposition, a large hyperharmonic basis for each $\phi_i^{(n)}(\rho, \Omega)$, accurate large-distances outgoing waves of radial wave functions f_n, and the correct energy of the three-body resonance obtained by tuning the three-body potential. Furthermore, the adiabatic expansion provides a mathematical division between intermediate states corresponding to sequential decay via two-body resonances or direct decay to the three-body continuum.

2. Efimov effect for ^{11}Li(1^-)

The Efimov effect occurs in a three-body system with at least two very large s-wave scattering lengths. To find two states is in practice impossible in nuclei where the neceesary vanishing of the Coulomb interaction uniquely point to two-neutron systems. Then two masses are necessarily small compared to the third mass and the radial extension is very large demanding extremely large scattering lengths.[5]

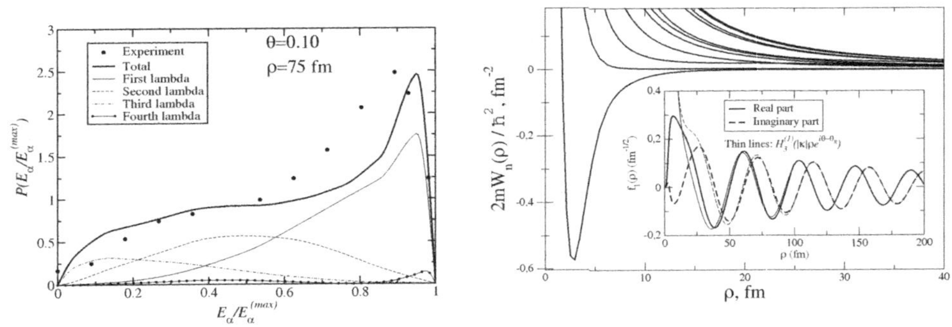

Fig. 1. The left side shows the energy distribution of the α-particle after decay of the 2^+-resonance in ^{6}He. The complex scaling angle is $\theta = 0.10$ and $\rho = 75$ fm where convergence is reached. The points are extracted from measurements.[6] Contributions from the lowest 4 adiabatic potentials are shown. The right side shows the lowest adiabatic potentials $W_n(\rho)$ for the ^{11}Li(1^-) nucleus in the three-body ^{9}Li+n+n model. The n-core and n-n scattering lengths are $a_{nc} \approx a_{nn} \approx$20 fm. The inset shows the lowest hyper-radial resonance function with its large distance asymptotics – the Hankel function. The complex scaling angle is $\theta = 0.15$, and the resonance angle $\theta_R = 0.12$ corresponds to the resonance energy of about $0.4 - 0.1i$ MeV.

The α-particle energy distribution is shown in fig. 1 for decay of the 2^+-resonance in ^{6}He. The rather pronounced peak at high energy corresponds to two neutrons moving together in the opposite direction of the α-particle. This is *not* the prominent n+^{5}He-structure at short distance provided by the dominating n-α $p_{3/2}$-attraction. The resonance structure must have changed from small to large distance where the energy distribution is determined. The reason is that s-waves are energetically preferred at large distances. The dynamic evolution is therefore a compromise between keeping the lowest energy and maintaining the structure. The reason is *not* that an intermediate quasi-stable state acts as a doorway state. Only s-wave attraction is responsible corresponding to a large neutron-neutron scattering length.

The Efimov effect requires at least one more large scattering length which can be found in the 1^- resonance in ^{11}Li (^{9}Li+$n+n$). Now all three scattering lengths are relatively large. The lowest adiababtic potentials are shown in fig. 1 together with the oscillating asymptotic hyperradial wave functions.[7] To see the clean effect we first compute the energy distributions shown in fig. 2 with only s-waves. The lowest adiabatic function is already very accurate and stable as seen by variation of ρ. The

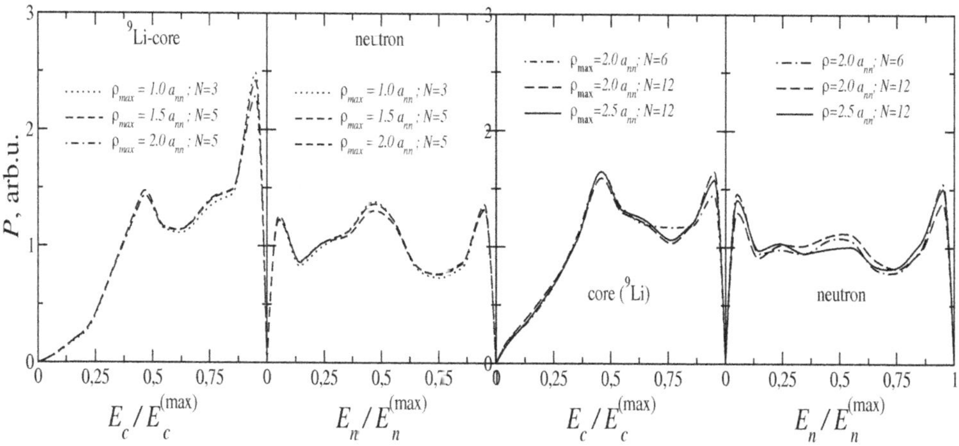

Fig. 2. Energy distributions of the ^{9}Li-core (left and center-right) and (center-left and right) the neutrons from the decay of ^{11}Li(1^-) in the three-body ^{9}Li + n + n model with realistic interactions (right) and (left) for s-wave neutron-neutron and n-core interactions (scattering length $a_{nc} \approx a_{nn} \approx 50$ fm). The curves have different ρ and numbers of adiabatic channels N.

^{9}Li-core energy distribution has two peaks, i.e. one at high energy corresponding to the two neutrons moving opposite to the core and one at intermediate energy corresponding to one ^{9}Li-neutron system moving opposite to the other neutron, i.e. neutron and core-emission.

The right side of fig. 2 shows the same energy distributions but now calculated with realistic interactions reproducing all other known ^{11}Li cluster properties. The results are remarkably similar with all peaks remaining, although less pronounced, from the schematic model. This can be viewed as the trace of the Efimov effect in the continuum. In any case the equally strong p-wave ^{9}Li-neutron attraction does not maintain the corresponding structure at large distance.

3. Isospin mixing for ^{6}Li(2^+)

The decay of the 2^+-resonance at 5.37 MeV in ^{6}Li, which is dominated by isospin 1, has experimentally given contradictory results, i.e. concluded to have either the normal small[8,9] or substantial isospin mixing.[10,11] These experiments measure different quantities and could therefore be consistent as demonstrated recently.[12] The method is again computation of the adiabatic potentials followed by solving the radial equation for outgoing boundary conditions with an appropriately adjusted three-body potential. Unusually high accuracy is needed because the bound deuteron state must fully decouple at large distance and because the isospin mixing arise from small couplings of near-lying adiabatic potentials.

We show in fig. 3 the variation with ρ of isospin content in a crucial adiabatic potential labeled as number 3. At around 15 fm the isospin mixing inceases strongly from being very small to about 40%. This is when the three particles just moved

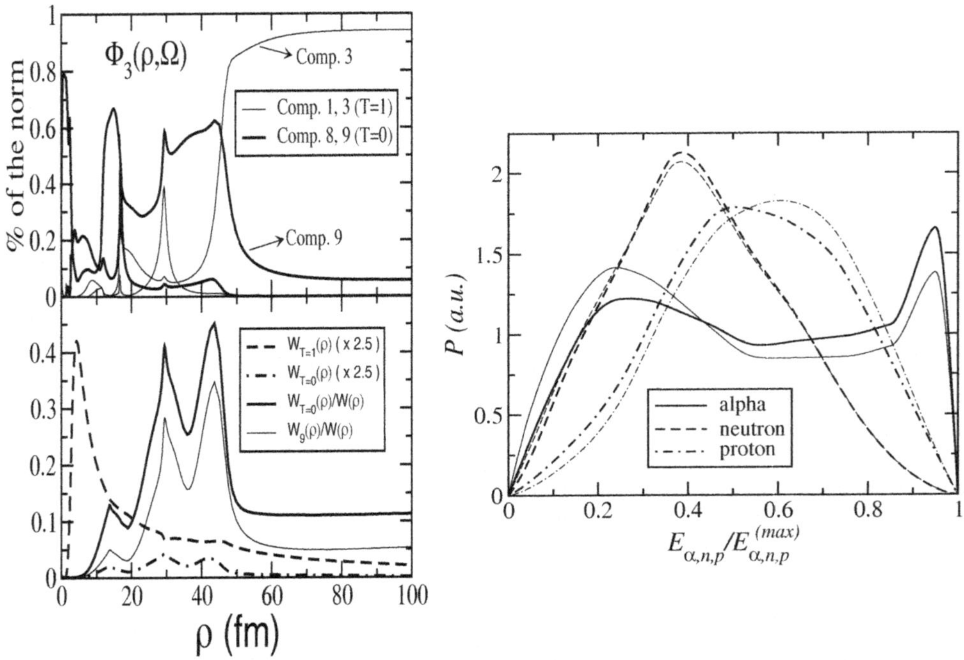

Fig. 3. The left side shows the partial wave decomposition of the third adiabatic potential for ^{6}Li(2^+). Strongly varying with distance, e.g.. component 3 and 9: $(\ell_x, \ell_y, L, s_x, S, T) = (1,1,1,1,1,1), (2,2,1,1,1,0)$. Bottom: Weight of the T=1 (dashed, large at short distance) and T=0 (dot-dashed, visible contribution at larger distances). Thick solid is relative weight of T=0, 40% between 20 − 50 fm. Larger ρ give about 10% of T=0. The right side shows the energy distributions after decay of ^{6}Li(2^+), α (solid), neutron (dashed), proton (dot-dashed). Thin is without T=0 components.

outside the range of the strong interaction leaving the Coulomb interaction to mix the symmetries. The resulting almost equal mixture of the two isospins 0 and 1 remains, apart from fluctuations, unchanged until $\rho \approx 50$ fm where the centrifugal force becomes decisive. The Coulomb interaction is the same independent of angular momentum of the states but the centrifugal barrier gives the advantage to the lowest angular momentum which happens to have isospin 1. The stable result is an isopsin mixture of about 10% at large distance which is where the observabe is determined. The experiment claiming small isospin admixture measured the fraction of deuterons emerging from the resonance decay. However, the neutron-proton combinations can appear in the continuum with both isospins. Again the dynamic evolution of the resonance from small to large distances is responsible for the measured values.

The observable energy distributions are also shown in fig. 3 both with and without the isospin 0 component which only changes the distributions marginally.

4. Energy distributions for $^{12}C(0^+,1^+)$

The energy distributions after three-body decay are determined by the structure of the wave functions at large distance. The predicted dynamic evolution can then be tested by comparison with experimental results. The beta-delayed decays measured for ^{12}C are almost ideal because of the precision and different decay modes.[1,13,14] We show in fig. 4 the lowest adiabatic potentials for 0^+ and 1^+ where we already added the three-body potentials. We get one 1^+-resonance and three 0^+-states where the bound ground state and one 0^+-resonance are dominated by components related to the lowest potential. The second 0^+-resonance arise mainly from the second potential.

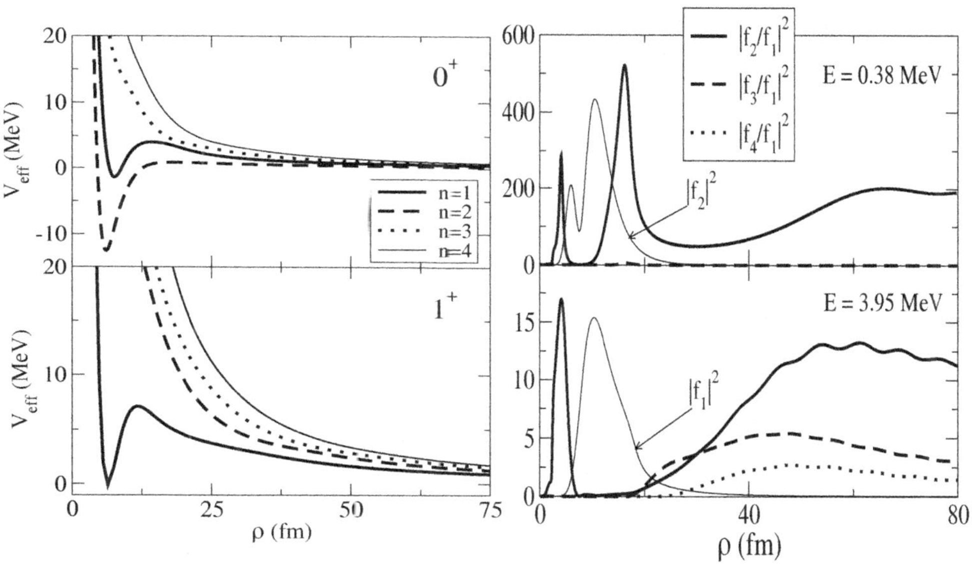

Fig. 4. The left side shows the real parts of the four lowest adiabatic potentials as functions of ρ for the 0^+ and 1^+ resonances of ^{12}C. The two-body interaction is from[16,17] and the three-body Gaussian potentials, $V_{3b} = S\exp(-\rho^2/b^2)$, have $b = 6$ fm, $-S = 20, 92$ MeV for 0^+ and 1^+, see Ref. 15. The right side shows the four lowest radial wavefunctions as functions of ρ for each of the two 0^+ resonances of ^{12}C at 0.38 MeV and 3.95 MeV above threshold.[14] Ratios and the small-distance dominating wavefunctions are given.

The structure changes dramatically with ρ for the second 0^+-resonance as shown in fig. 4. Both resonances are for large ρ dominated by the same component which describes $^8Be(0^+)$ plus one α far away, i.e. in both cases the same intermediate configuration in the sequential decay.

The energy distributions are shown in fig. 5 for the 1^+ and the second 0^+-resonance.[18] The experimental 1^+ distribution is remarkably well reproduced by inclusion of the two most important adiabatic potentials. The variation with ρ is also very stable indicating reliability and high accuracy. There seems to be no quasi-

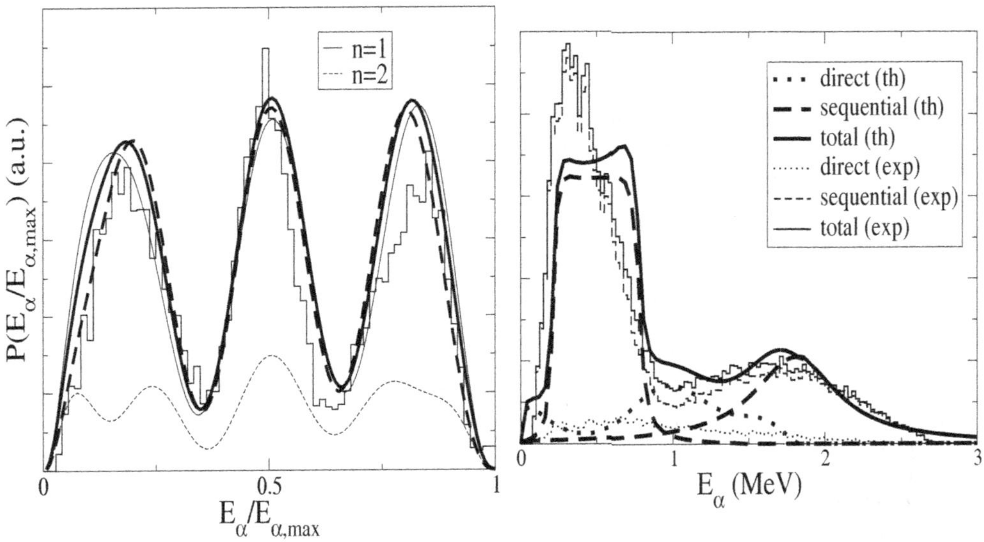

Fig. 5. The left side shows the α-particle energy distribution for the 1^+ resonance of ^{12}C at 5.43 MeV above threshold at an excitation of 12.71 MeV. The energy is measured in units of the maximum possible, i.e. $2 \times 5.43/3$ MeV. The thick solid and dashed curves are for coordinate space wavefunctions at $\rho = 70, 100$ fm. The thin curves are contributions from separate adiabatic potentials. The momentum space computation fall on top of the $\rho = 100$ fm curve (thick dashed). The histogram is the experimental distribution.[14] The right side shows the α-particle energy distribution for the second 0^+ resonance of ^{12}C at 4.3 MeV above threshold at an excitation of 11.2 MeV.[13,14] The interactions in fig. 1 give an energy of 3.95 MeV above threshold.[18] We use 2.8 MeV to account approximately for interference and beta-feeding distortion. The maximum energy for the most likely resonance position is then $2 \times 2.8/3 \approx 1.8$ MeV. The histograms are experimental direct (the small part), sequential, and total distribution.

stable intermediate configuration. The decay is better described as direct to the three-body continumm. The parameters in the computations are only related to the $\alpha-\alpha$ interaction and adjustment of the three-body energy to the experimental value. The agreement is also remarkable because the 1^+-resonance cannot be described as a cluster state. This implies that the boundary condition at the nuclear surface constructed by use of our three-body potential is sufficient, at least in this case, for computation of the energy distribution. This conclusion is far from being trivial.

In fig. 5 we also show the result for the second very broad 0^+-resonance. From fig. 4 we know that both 0^+-resonances preferentially decay sequentially via the ^{8}Be(0^+) ground state. Unfortunately this configuration is not accurately described at the computed relatively modest ρ-values of less that about 100 fm. On the other hand we know the asymptotic behavior and can therefore replace the corresponding adiababtic component by this expression. Since this is two consecutive two-body decays the resulting energy distribution is completely determined by well known two-body kinematics. To account for the energy dependence of the beta-feeding and for the interference between the two 0^+-resonances we shift the position of the second 0^+-resonance to 2.8 MeV. The result is in fair overall agreement with the experimental distribution.

5. Summary and conclusions

Reactions and beta-delayed decays often are interpreted as proceeding via population and subsequent decays of resonances. We investigate three-body decays of many-body resonances by use of a three-body cluster model. The large-distance asymptotic behavior is then explicitly part of the model. The short-distance boundary condition is mocked up by a three-body potential adjusted to reproduce the position of the decaying resonance. The analogy to α-emission is complete. We assume the decays are independent of the previous history of populating the resonance.

The resonance wave functions change sometimes substantially from small to large distances. This dynamic evolution is essential in undertanding details of the decays. We show that the decay of ^{11}Li(1^-) changes to s-wave character at large distance and exhibits reminiscences of the Efimov effect in the continuum. For ^{6}Li(2^+) we see normal small isospin mixing at small distances and substantial mixing at large distances. For ^{12}C(0^+,1^+) we find direct decay to the three-body continuum for the 1^+-resonance which cannot be described as a cluster state. We find preferentially sequential decay via the ^{8}Be ground state for the two lowest 0^+-resonances. All experimental data are remarkably well reproduced.

References

1. H.O.U. Fynbo et al., *Phys.Rev.Lett.* **91**, 082502 (2003).
2. E. Garrido et al., *Nucl. Phys. A* **781**, 387 (2007).
3. E. Nielsen, D.V. Fedorov, A.S. Jensen and E. Garrido, *Phys. Rep.* **347**, 373 (2001).
4. D.V. Fedorov, E. Garrido, and A.S. Jensen, *Few-body Systems* **33**, 153 (2003).
5. A.S. Jensen et al., *Rev. Mod. Phys.* **76**, 215 (2004).
6. B.V. Danilin et al., *Sov. J. Nucl. Phys.* **46**, 225 (1987).
7. E. Garrido, D.V. Fedorov and A.S. Jensen, *Phys. Rev. Lett.* **96**, 112501 (2006).
8. Y. Suzuki and K. Yabana. *Phys. Lett. B* **272**, 173 (1991).
9. K. Arai, Y. Suzuki, and K. Varga, *Phys. Rev. C* **51**, 2488 (1995).
10. N.O. Gaiser et al., *Phys. Rev. C* **38**, 1119 (1988).
11. P. Niessen et al., *Phys. Rev. C* **45**, 2570 (1992).
12. E. Garrido et al., *Phys. Lett.B* **648**, 274 (2007).
13. C. Aa. Diget et al., *Nucl. Phys. A* **760**, 3 (2005).
14. C. Aa. Diget, PhD-thesis, IFA, University of Aarhus, 2006.
15. R. Alvarez-Rodriguez et al., *Eur Phys. J. A* **31**, 303 (2007).
16. D. V. Fedorov and A. S. Jensen, *Phys. Lett. B* **389**, 631 (1996).
17. S. Ali and A.R. Bodmer, *Nucl. Phys.* **80**, 99 (1966).
18. R. Alvarez-Rodriguez et al., *Phys. Rev. Lett.* **99**, 072503 (2007).

PRERAINBOW OSCILLATIONS IN ^{3}He SCATTERING FROM THE HOYLE STATE OF ^{12}C AND ALPHA PARTICLE CONDENSATION

S. OHKUBO*

Department of Applied Science and Environment,
Kochi Women's University, Kochi 780-8515, Japan
**E-mail: shigeo@cc.kochi-wu.ac.jp*
www.cc.kochi-wu.ac.jp/~ shigeo/

Y. HIRABAYASHI

Information Initiative Center, Hokkaido University, Sapporo 060-0811, Japan
E-mail: hirabay@iic.hokudai.ac.jp

^{3}He+^{12}C scattering is studied in a coupled channel method by using a double folding model with microscopic wave functions of ^{12}C. Experimental angular distributions in elastic and inelastic scattering to the 2^+ (4.44 MeV), 0_2^+ (7.65 MeV) and 3^- (9.63 MeV) states of ^{12}C are well reproduced. It is found that the Airy minimum of the prerainbow oscillations for the Hoyle state is considerably shifted to a larger angle due to its dilute density distribution compared with that of the normal ground state in agreement with the idea of α particle condensation.

Keywords: ^{3}He+^{12}C scattering, nuclear rainbow, prerainbow, alpha particle condensation.

1. Introduction

Bose-Einstein condensation (BEC) has been well known in a dilute gas.[1] The macroscopic properties such as superconductivity and superfluidity in both ^{3}He and ^{4}He are understood in relation to BEC. Recently it has been speculated that the 0_2^+(7.65 MeV) state of ^{12}C, the Hoyle state, is a Bose-Einstein condensate of three α particles.[2] Uegaki *et al.*[3] and Kamimura and Fukushima[4] studied the α cluster structure of ^{12}C thoroughly in the microscopic cluster model and showed that the 0_2^+ state of ^{12}C, has a loosely coupled three α cluster structure with an $\alpha\otimes^8$Be configuration. Recently it has been shown that the wave functions of Uegaki *et al.*[3] and Kamimura and Fukushima[4] are almost completely equivalent to the wave function of an α particle condensate that the three α particles are sitting in the lowest 0s state in a dilute gas.

A macroscopic property that is peculiar to BEC such as superconductivity and superfluidity has not been observed in the case of α particle condensation. Recently Kokalova *et al.*[5] proposed a new experimental way of testing BEC of α particles in nuclei by directly observing the enhancement of α particle emission and the multiplicity partition of the possible emitted α particles. It is important to find a phenomenon which strongly reflects the properties of Bose-Einstein condensation of α particles.

We show that the dilute property of the matter density due to Bose-Einstein condensation can be seen in the nuclear refractive phenomena. Nuclear rainbow scattering has been powerful in the study of nucleus-nucleus interaction when absorption is incomplete.[6] It has been shown that rainbow scattering and the evolution of the Airy minimum can also be seen in inelastic scattering.[7] It is expected that the refractive effect becomes much larger and can be seen clearly at low incident energies. Recently a new concept of prerainbow has been proposed at the lower energy region.[8] The refractive index n is related to the optical potential V as follows:

$$n(r) = \sqrt{1 - \frac{V(r)}{E_{c.m.}}}. \tag{1}$$

There is no useful experimental data in inelastic α particle scattering to the Hoyle state in the low energy region. Fortunately, for ^{3}He scattering from ^{12}C there is an experimental angular distribution at 34.7 MeV measured by Fujisawa *et al.*,[9] which had been unnoticed for many years and to which no theoretical attention from the viewpoint of α particle condensation in the Hoyle state had been paid.

2. The double folding model

We study elastic and inelastic ^{3}He+^{12}C scattering in the microscopic coupled channel method by taking into account simultaneously the 0_1^+ (0.0 MeV), 2^+ (4.44 MeV), 0_2^+ (7.65 MeV), and 3^- (9.63 MeV) states of ^{12}C. The diagonal and coupling potentials for the ^{3}He+^{12}C system are calculated with the double folding model:

$$V_{ij}(\mathbf{R}) = \int \rho_{00}^{(^3\mathrm{He})}(\mathbf{r}_1)\, \rho_{ij}^{(^{12}\mathrm{C})}(\mathbf{r}_2)\, v_{\mathrm{NN}}(E, \rho, \mathbf{r}_1 + \mathbf{R} - \mathbf{r}_2)\, d\mathbf{r}_1 d\mathbf{r}_2\ , \tag{2}$$

where $\rho_{00}^{(^3\mathrm{He})}(\mathbf{r})$ is the ground state density of ^{3}He taken from Cook *et al.*,[10] while v_{NN} denotes the density-dependent M3Y effective interaction (DDM3Y).[11] $\rho_{ij}^{(^{12}\mathrm{C})}(\mathbf{r})$ represents the diagonal ($i = j$) or transition ($i \neq j$) nucleon density of ^{12}C calculated in the resonating group method by Kamimura *et al.*[4]

The folding potential is very sensitive to the wave functions used, which serves as a good test of the validity of the wave function. In the analysis we introduce the normalization factor N_R for the real part of the potential and phenomenological imaginary potentials with a Wood-Saxon form factor (volume absorption) and a derivative of the Wood-Saxon form factor (surface absorption) for each channel.

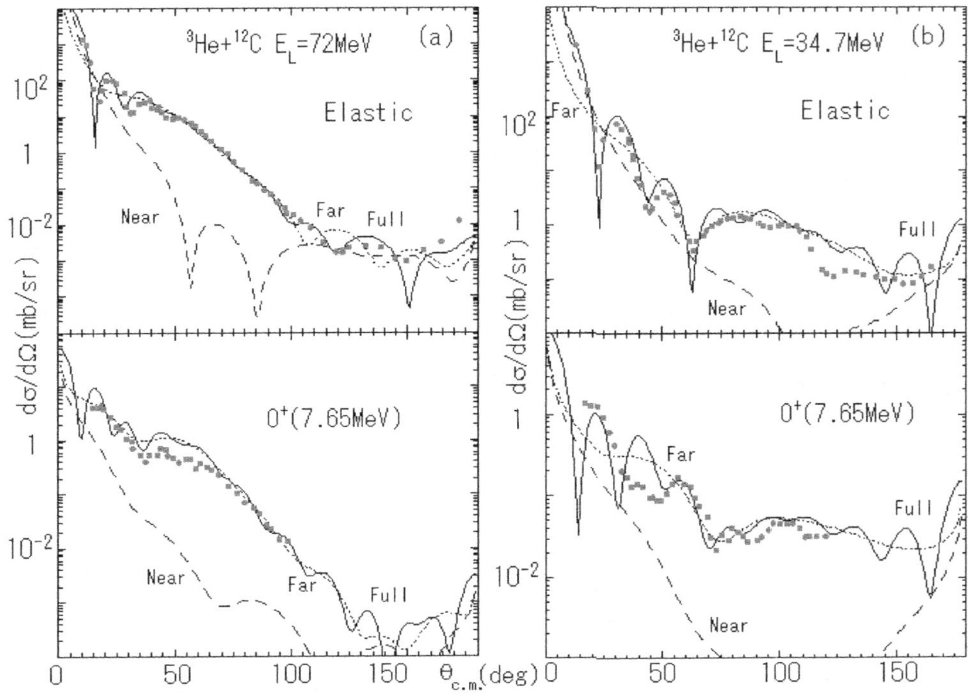

Fig. 1. The calculated angular distributions (solid lines) in ^{3}He+^{12}C scattering at E_L=34.7 and 72 MeV are decomposed into farside (dotted lines) and nearside (dashed lines) components and compared with the experimental data (points).[9,13]

3. Analysis of refractive ^{3}He+^{12}C scattering

In Fig. 1 angular distributions calculated using a coupled channel method at E_L=34.7 MeV and 72 MeV are compared with the experimental data. The calculation reproduces the experimental angular distributions for the ground state and the Hoyle state as well as the 2^+ and 3^- states.[14]

By decomposing the calculated scattering amplitude into farside and nearside contributions, the Airy minimum of the rainbow at 72 MeV and the prerainbow oscillations at 34.7 MeV for the Hoyle state is identified. At E_L=72 MeV the first Airy minimum $A1$ appears at 35° for the 0_2^+ state. For elastic scattering a clear minimum is not seen in the angular distribution of the farside cross sections and the Airy minimum in the experimental data is obscured by the interference between the farside and nearside amplitudes. On the other hand, the $A1$ minimum for the 0_2^+ state is clearly seen in the farside cross sections because the minimum is shifted to a larger angle where the nearside contribution is much smaller.

The situation is more clearly seen in the Airy structure at low the incident energy region where no typical rainbow falloff of the dark side appears. At E_L= 34.7 MeV

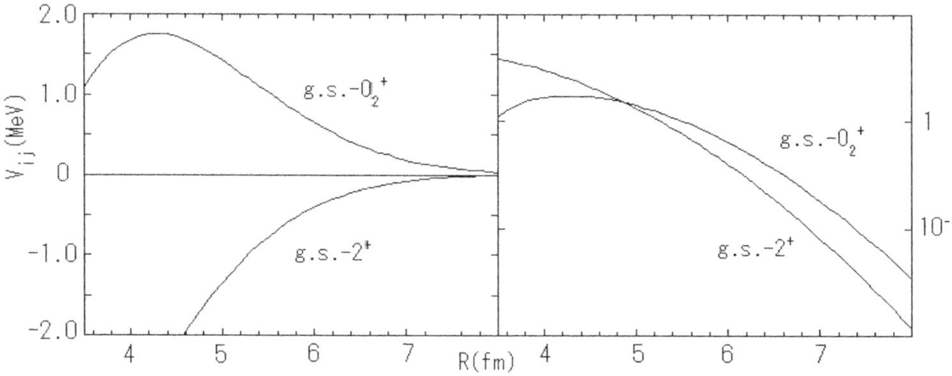

Fig. 2. Coupling potentials for the g.s.-2_1^+ and g.s.- 0_2^+ channels in ^{3}He+^{12}C scattering at E_L=34.7 MeV are shown on a linear scale (left) and their absolute values are shown on a logarithmic scale (right) to emphasize the difference at the surface region.

in Fig. 1 the Airy minimum A1 appears at 60° for elastic scattering and 75° for the 0_2^+ state. The latter is much shifted to a larger angle and the Airy minimum is not at all obscured by the nearside contributions. For the 0_2^+ state at E_L= 34.7 the nearside contributions are much smaller than the farside contributions compared with the elastic scattering case.

Thus the difference of the refraction between the ground state and the 0_2^+ state is much more clearly seen in the prerainbow oscillations at 34.7 MeV than in the rainbow at 72 MeV. This shows that the incident ^{3}He is strongly refracted in the Hoyle state in accordance with the previous finding in α+^{12}C scattering[15] that the state has a large lens composed of three α particles in a dilute density distribution.

In Fig. 2 the coupling potentials for the g.s.-2_1^+ and g.s.- 0_2^+ channels are compared. It is clear that the coupling potential for the g.s.-0_2^+ channel is considerably extended to the outer region, which enhances the inelastic scattering to the Hoyle state in the surface region.

In fact, in Fig. 3 we see that even at the low energy region where so many high partial waves are not involved, inelastic scattering to the Hoyle state occurs at large angular momenta (that is, large radius) compared with that to the normal g.s. and 2_1^+ states. This becomes clearer at the high energy region as was discussed in α particle scattering from ^{12}C at E_L=139 MeV.[16] These facts are in accordance with the previous finding that the Hoyle state has a large radius compared with the normal ground state.

4. Summary

We have shown that the prerainbow oscillation is useful for studying the dilute density distribution due to Bose-Einstein condensation of α particles. The present

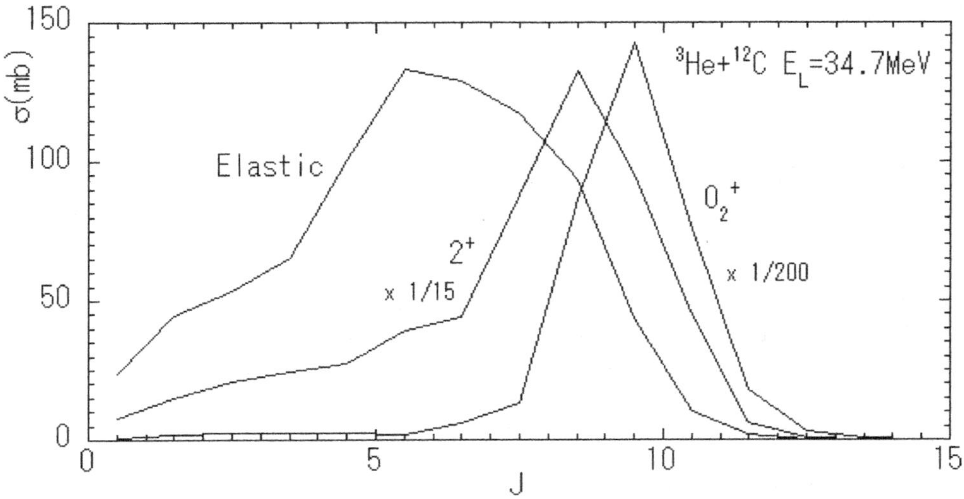

Fig. 3. Calculated partial cross sections for elastic and inelastic ^{3}He scattering to the 2_1^+ and 0_2^+ state of ^{12}C at E_L=34.7 MeV are shown as a function of total angular momentum.

approach is applicable not only to ^{3}He scattering but also to heavy ion rainbow scattering. The ^{16}O+^{12}C rainbow and prerainbow scattering will be useful to reconfirm the present conclusions because the refractive effect is very strong and clear Airy minima of higher order can be expected.

It has been also suggested that the $\frac{1}{2}^-$(8.86 MeV) state in ^{13}C , the 0^+(9.746 MeV) state in ^{14}C and the 0^+(~29 MeV) state in ^{16}C may be a state with one, two and four additional neutrons to the 0_2^+ state of ^{12}C. If the above states have a dilute density distribution, the prerainbow oscillations for these states would be considerably different from a state with a normal density distribution. This kind of experiment is highly desired.

Acknowledgments

One of the authors (S.O.) has been supported by a Grant-in-aid for Scientific Research of the Japan Society for Promotion of Science (No. 16540265).

References

1. A. J. Leggett, *Rev. Mod. Phys.* **73**, 307 (2001).
2. A. Tohsaki, H. Horiuchi, P. Schuck, and G. G. Röpke, *Phys. Rev. Lett.* **87**, 192501 (2001).
3. E. Uegaki, S. Okabe, Y. Abe, and H. Tanaka, *Prog. Theor. Phys.* **57**, 1262 (1977); E. Uegaki, Y. Abe, S. Okabe, and H. Tanaka, *ibid* **59**, 1031 (1978); **62**, 1621 (1979).

4. M. Kamimura, *Nucl. Phys.* **A351**, 456 (1981); Y. Fukushima and M. Kamimura, in *Proceedings of the International Conference on Nuclear Structure, Tokyo, 1977*, edited by T. Marumori *J. Phys. Soc. Japan Suppl.* **44**, 225 (1978).
5. Tz. Kokalova, N. Itagaki, W. von Oertzen, and C. Wheldon, *Phys. Rev. Lett.* **96**, 192502 (2006).
6. D. T. Khoa, W. von Oertzen, H. G. Bohlen, and S. Ohkubo, *J. Phys.* **G 34**, R111 (2007).
7. F. Michel and S. Ohkubo, *Phys. Rev.* **C 70**, 044609 (2004).
8. F. Michel, G. Reidemeister, and S. Ohkubo, *Phys. Rev. Lett.* **89**, 152701 (2002).
9. T. Fujisawa, S. Yamaji, K. Matsuda, S. Motonaga, F. Yoshida, H. Sakaguchi, and K. Masui, *J. Phys. Soc. Japan* **34**, 5 (1973).
10. J. Cook and R. J. Griffiths, *Nucl. Phys.* **A366**, 27 (1981).
11. A. M. Kobos, B. A. Brown, R. Lindsay, and G. R. Satchler, *Nucl. Phys.* **A425**, 205 (1984).
12. Y. Funaki, A. Tohsaki, H. Horiuchi, P. Schuck, and G. Röpke, *Phys. Rev.* **C 67**, 051306(R) (2003).
13. A. S. Dem'yanova, E. F. Svinareva, S. A. Goncharov, S. N. Ershov, F. A. Gareev, G. S. Kazacha, A. A. Ogloblin, and J. S. Vaagend, *Nucl. Phys.* **A542**, 208 (1992).
14. S. Ohkubo and Y. Hirabayashi, *Phys. Rev.* **C 75**, 044609 (2007).
15. S. Ohkubo and Y. Hirabayashi, *Phys. Rev.* **C 70**, 041602(R) (2004).
16. S. Ohkubo and Y. Hirabayashi, to appear in *AIP Proceedings of the 2nd International Conference on Frontiers in Nuclear Structure, Astrophysics and Reactions (FINUSTAR2)* (Aghios Nikolaos, Crete, Greece, 10-14 Sept., 2007).

ANGULAR DISPERSION BEHAVIOR IN HEAVY ION ELASTIC SCATTERING

Q. WANG*, J.L. HAN, Z.G. XIAO, H.S. XU, Z.Y. SUN, Z.G. HU, X.Y. ZHANG, H.W. WANG, R.S. MAO, X.H. YUAN, Z.G. XU, T.C. ZHAO, H.B. ZHANG, H.G. XU, H.R. QI, Y. WANG, F. JIA, L.J. WU, X.L. DING, Q. GAO, H. GAO, Z. BAI, Y.P. ZHANG, S.L. LI, J.Q. LI, G.Q. XIAO and G.M. JIN

Institute of Modern Physics, Chinese Academy of Sciences, Lanzhou 730000, China
** E-mail: wangqi@impcas.ac.cn*

Z.Z. REN

Department of Physics, Nanjing University, Nanjing 210008, China

S.G. ZHOU

Institute of Theoretical Physics, Chinese Academy of Sciences, Beijing 100080, China

S.Q. ZHANG, D.Y. PANG

Department of Technical Physics, Peking University, Beijing 100871, China

W. XU, G.T. FAN

Shanghai Institute of Applied Physics, Chinese Academy of Sciences, Shanghai 201800, China

S.Yu. KUN

Centro de Ciencias Físicas, National University of Mexico (UNAM), Cuernavaca, Mexico
Center for Nonlinear Physics, RSPhysSE,
The Australian National University, Canberra ACT 0200, Australia

The differential cross sections for elastic scattering products of ^{17}F and ^{17}O on ^{208}Pb have been measured at Radioactive Ion Beam Line at Lanzhou (RIBLL), China. The angular dispersion plots of $\ln(d\sigma/d\theta)$ versus θ^2 are obtained from the angular distribution of the elastic scattering differential cross sections. Systematical analysis on the angular dispersion for the available experimental data indicates that there is an angular dispersion turning angle at forward angular range beyond the grazing angle. This turning angle can be clarified as nuclear rainbow in classical deflection function. And the exotic behavior of the nuclear rainbow angle offers a new probe to investigate the halo and skin phenomena.

Keywords: Elastic scattering; Angular dispersion; Deflection function; Nuclear rainbow

1. Introduction

Nuclear halo state in ^{11}Li was reported in 1985 by I.Tanihata et al.[1], since then halo nuclei have been studied extensively both experimentally and theoretically. It is known that there are neutron halos, proton halos and many halo candidates for weakly bound nuclei[1–14]. It is believed that elastic scattering in heavy ion collisions can provide reliable information on the structure of weakly bound nuclei. And the classical deflection function is a basic and fruitful tool for describing elastic scattering in heavy ion collisions[15–17].

In this paper, the experimental data of elastic scattering differential cross section of weakly bound nucleus ^{17}F on ^{208}Pb target are presented and analyzed. We find that the fitting of the experimental angular dispersion plot $\ln(d\sigma/d\theta)$ versus θ^2 of elastic scattering products is out of line, there is a turning point with two different slopes at small angle. It is clarified as a nuclear rainbow angle in the classical deflection function. The exotic behavior of the nuclear rainbow angle provide a new probe to identify the abnormal structure of weakly bound nuclei.

2. Experimental Setup

The experiment was performed with the Radioactive Ion Beam Line in Lanzhou (RIBLL)[18–20] at Institute of Modern Physics, Chinese Academy of Sciences, China. Three position sensitive Parallel Plate Avalanche Counters (PPAC1-PPAC3) were arranged in the experimental measurement. The active area of each PPAC is 100×100 mm^2 and the position resolution is 0.5mm. PPAC1 and PPAC2 were set in front of the ^{208}Pb target to determine the position and the direction of the incident particle on the target. There are five pieces of silicon detectors (325μm in thickness) in a Si array that were used to detect the residual energy of the products with an energy resolution better than 1%. PPAC3 was placed in front of the Si detector array with an angle resolution of less than 0.2° to determine the scattering angle of the outgoing particle on the Si detectors. Due to the momentum dispersion ($\Delta P/P=\pm0.1\%$) of RIBLL, the first excited states both in ^{17}F (at 0.4953 keV) and in ^{208}Pb (at 2.614 MeV) can not be separated from elastic particles.

3. Experimental results and angular dispersion

The differential cross sections ($d\sigma/d\theta$) have been measured at forward angular range θ_{lab} from 6° to 20° for the elastic scattering products of ^{17}F and ^{17}O induced from the reactions of ^{17}F+^{208}Pb at 141MeV and ^{17}O+^{208}Pb at 128MeV, respectively. The angular dispersion plots of $\ln(d\sigma/d\theta)$ versus θ^2 are plotted in Fig.1 for ^{17}F and ^{17}O. Two different behaviors are obviously presented. In order to fit the experimental data, two straight lines with different slopes for the products ^{17}F are needed but only one straight line for the nucleus ^{17}O in the same angular range measured.

In heavy ion peripheral collisions an expression of angular dispersion has been suggested by Strutinsky[21] and it is widely used in the analysis of experimental data[22–28]. For a peak-forward angular distribution, the differential cross section ($d\sigma/d\theta$) at a small angular range can be reduced as

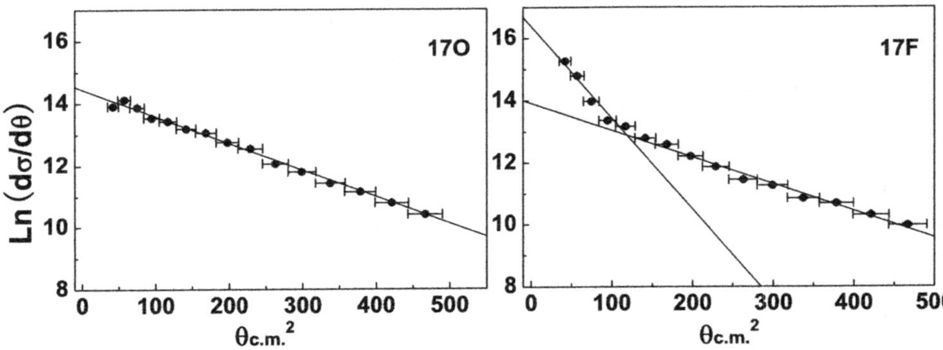

Fig. 1. Angular dispersion plots $ln(d\sigma/d\theta)$ vs θ^2 of elastic scattering $^{17}F/^{17}O$ on target ^{208}Pb

$$\frac{d\sigma}{d\theta} \propto e^{-(\theta/\xi)^2}. \tag{1}$$

In the expression Eq.1, the logarithm of the differential cross section $\ln(\mathrm{d}\sigma/\mathrm{d}\theta)$ is a linear function of the square of the scattering angle θ^2. The slope represents so-called angular dispersion parameter ξ. The angular dispersion ξ is determined both by the quantal dispersion coefficient ξ_q and by the dynamic quantity ξ_r,

$$\xi = \left(\xi_q^2 + \xi_r^2\right)^{1/2}. \tag{2}$$

For a pure Rutherford scattering, the nuclear interaction and then, the dynamic angular dispersion is negligible. However, there are both quantal angular dispersion and dynamic angular dispersion due to nuclear interaction in elastic scattering. This is a very important different behavior, there is a sole slope for the pure Rutherford scattering without the nuclear reaction, and there is another different slope for the elastic scattering including the nuclear reaction. As soon as there is an evolution from a pure Rutherford scattering to an elastic scattering, then there is a turning point of the two slopes in the angular dispersion plot of $\ln(\mathrm{d}\sigma/\mathrm{d}\theta)$ vs θ^2.

In order to test and verify our experimental results we collect the available original elastic scattering experimental data by other groups.[29–34] Their angular dispersion trend as shown in Fig.2 are reduced from the original experimental data of differential cross section angular distribution.

For all of the reaction systems shown in Fig.2, there are two straight lines to fit the experimental data with two different slopes in the angular dispersion plane of $\ln(\mathrm{d}\sigma/\mathrm{d}\theta)$ vs θ^2. Angular dispersion turning angle θ_{tu} corresponds to a turning point of two different slopes. It is clearly that the θ_{tu} is at forward angular range beyond the grazing angle θ_{gr} of the reaction system. The ratio between the θ_{tu} and the θ_{gr} are small about 0.20 for the halo nuclei ^{6}He and weakly bound nucleus ^{17}F and large about 0.44 for the stable nuclei ^{4}He, ^{6}Li and ^{16}O as listed in the table 1.

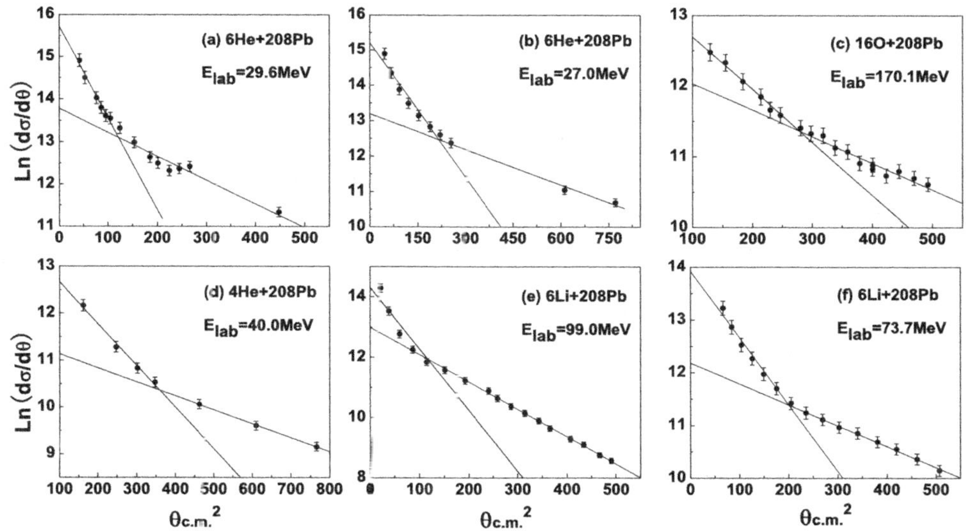

Fig. 2. Angular dispersion plots of $ln(d\sigma/d\theta)$ vs θ^2 in center of mass system for different elastic scattering channels of (a) ^{6}He, (b) ^{6}He, (c) ^{16}O, (d) ^{4}He, (e) ^{6}Li and (f) ^{6}Li

Table 1. The turning point angle θ_{tu} of the angular dispersion in the heavy ion elastic scattering

Reaction System	E_{lab} (MeV)	$E_{c.m.}/V_{CB}$ [1]	θ_{gr} [2]	Angular range Detected	θ_{tu}	θ_{tu}/θ_{gr}	Projectile Nucleus	Reference
^{17}F+^{208}Pb	141.0	1.5	59°	6 - 20°	11°	0.19	skin/halo	our work
^{17}O+^{208}Pb	128.0	1.5	57°	6 - 20°	>21°	>0.37	stable	our work
^{6}He+^{208}Pb	29.6	1.4	60°	>6°	11°	0.18	halo	Ref. 29
^{6}He+^{208}Pb	27.0	1.3	72°	>6°	15°	0.21	halo	Ref. 30
^{16}O+^{208}Pb	170.1	2.0	38°	>11°	17°	0.45	stable	Ref. 31
^{4}He+^{208}Pb	40.0	1.9	43°	>12°	19°	0.44	stable	Ref. 32
^{6}Li+^{208}Pb	99.0	1.9	41°	>9°	17°	0.41	stable	Ref. 33
^{6}Li+^{208}Pb	73.7	1.9	41°	>9°	17°	0.41	stable	Ref. 34

Note: [1] Coulomb barrier was calculated as discussed in Ref. 35.
[2] Grazing angle values are taken from the experimental data or calculated by non-relativistic kinematics, see Ref. 36.

4. Deflection function and discussions

We would like to present an understanding of the angular dispersion turning angle based on a classical picture of deflection function in heavy ion collisions. The classical deflection angle θ as a function of the impact parameter b can be wrote as

$$\Theta(b) = \pi - 2\int_0^{b/r_{\min}(b)} dw(1 - \frac{V[r(w)]}{E} - w^2)^{1/2} \qquad (3)$$

with

$$r(w) = b/w \tag{4}$$

The distance of closest approach $r_{\min}(b)$ can be obtained from energy and angular momentum conservation. The first important step to calculate the deflection function is to structure a proper effective heavy ion interaction potential $V_{eff}(r)$ quantitatively. Suppose $V_{eff}(r)$=$V_N(r)$+$V_{Clb}(r)$, in addition to the Coulomb potential $V_{Clb}(r)$, a nuclear potential $V_N(r)$ is usually assumed to be attractive. For all of the reaction system we discussed in this Letters, the incident energy E greater than the critical energy E_{crit}, there are both Coulomb rainbow and nuclear rainbow[15–17,38] in the trajectory of the classical deflection function as shown in Fig.3.

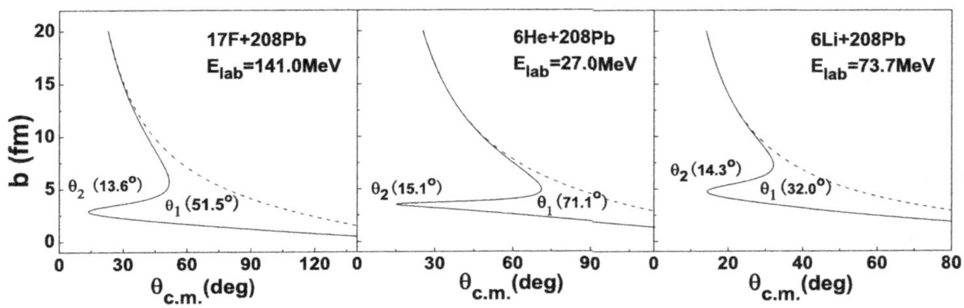

Fig. 3. Examples for classical deflection function calculations

In the view of the dimension of scattering angle in the deflection function, for those elastic scattering products emitted in very forward angles smaller than the nuclear rainbow angle θ_2 as shown in Fig.3, there is only the Coulomb interaction. As soon as the scattering angle increases to larger than θ_2, there are both two interaction potential, nuclear and Coulomb, so that the point at θ_2 is a turning point of different reaction mechanisms, a pure Coulomb scattering and an elastic scattering. It is consistent with the slope change due to the contribution from the dynamic angular dispersion with nuclear interaction in the angular dispersion plot as shown in Fig.1 and 2. This is to say that the angular dispersion tuning angle θ_{tu} corresponds to the nuclear rainbow angle θ_2.

Nevertheless, the angular dispersion of elastic scattering products of ^{17}O are approximately on a straight line as shown in Fig.1, there is no turning angle θ_{tu}. This is because of too limited angular range less than 20° measured in the experiment and the θ_{tu} is predicted about 25° by the systematic analysis listed in table 1.

In our calculations to the classical deflection function, a real Woods-Saxon type is taken as our nuclear potential. In order to eliminate the uncertainty of the potential parameters, three quantities are used as a set of criterion, the grazing angle θ_{gr} as the Coulomb rainbow angle,[37] the angular dispersion turning angle θ_{tu} as the nuclear rainbow angle and the critical angular momentum L_{cr} as another limitative condition. As examples, Fig.3 presents our calculation results of the classical deflection function, and the related real potential parameters are listed in Table 2. If these three limitative conditions are reasonable, then it should be reflected by a positive result to fit the differential cross sections of elastic scattering.

Table 2. Optical potential parameters for the different reaction systems

Reaction System	E_{lab} (MeV)	Potential Parameters						χ^2
			Real			Imaginary		
		V_0	R_0	a	W	Rw	Aw	
^{6}He+^{208}Pb	27.0	29.0	6.70	1.38	6.49	8.30	1.680	0.057
^{16}O+^{208}Pb	170.1	122.0	6.20	1.42	6.12	11.30	0.680	0.584
^{4}He+^{208}Pb	40.0	20.2	7.05	1.04	3.00	11.57	0.246	0.524
^{6}Li+^{208}Pb	99.0	36.5	6.40	1.06	6.61	11.24	0.644	0.636
^{6}Li+^{208}Pb	73.7	36.6	6.55	1.13	5.13	11.57	0.552	0.191

Elastic scattering differential cross sections have been fitted by FRESCO code.[39] As examples, Fig.4 reproduces very well the elastic scattering differential cross sections. Their optical potential parameters are listed in Table 2.

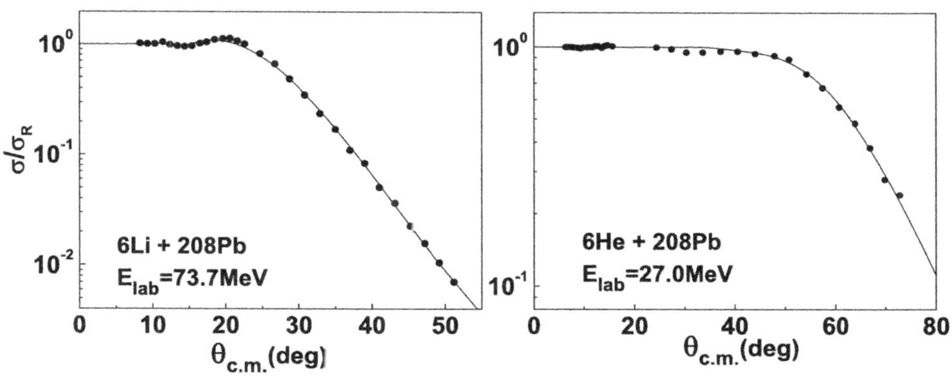

Fig. 4. Fit to the differential cross sections of elastic scattering products by FRESCO code

5. Summary

Differential cross sections for elastic scattering products of ^{17}F on ^{208}Pb at 141MeV has been measured. The angular dispersion plot of $\ln(d\sigma/d\theta)$ versus θ^2 is obtained. Systematical analysis indicate that (1) there is an angular dispersion turning angle at forward angular range beyond the grazing angle; (2) classical deflection function can be used to clarify the angular dispersion turning angle; (3) the nuclear rainbow angle is just the dispersion turning angle; (4) there is an exotic behavior, smaller angular dispersion turning angle and more diffuseness in the nuclear potential, for the weakly bound nuclei.

This work is supported by the National Natural Science Foundation of China (Grant No. 10675149 and 10535010), the Chinese Academy of Sciences (Grant No. 0730030YF0) and the Major State Basic Research and Developping of China (Grant No. 2007CB815000).

References

1. I. Tanihata *et al.*, *Phys. Rev. Lett.* **55**, 2676 (1985).
2. W. Mittig *et al.*, *Phys. Rev. Lett.* **59**, 1889 (1987).
3. M. G. Saint-Laurent *et al.*, *Z. Phys.* **A332**, 457 (1989).
4. Z. Z. Ren *et al.*, *Phys. Rev.* **C53**, R572 (1996).
5. B. A. Brown and P. G. Hansen, *Phys. Lett.* **B381**, 391 (1996).
6. R. Morlock *et al.*, *Phys. Rev. Lett.* **79**, 3837 (1997).
7. A. Navin *et al.*, *Phys. Rev. Lett.* **81**, 5089 (1998).
8. H. Y. Zhang *et al.*, *et al.*, *Nucl. Phys.* **A707**, 303 (2002).
9. Z. Z. Ren *et al.*, *Phys. Rev.* **C57**, 2752 (1998).
10. C. J. Lin *et al.*, *et al.*, *Phys. Rev.* **C66**, 067302 (2002).
11. R. E. Warner *et al.*, *Phys. Rev.* **C52**, R1166 (1995).
12. X. Z. Cai *et al.*, *Phys. Rev.* **C65**, 024610 (2002).
13. Z. H. Liu *et al.*, *Chin. Phys. Lett.* **21**, 1711 (2004).
14. A. S. Jensen *et al.*, *Rev.Mod.Phys.* **76**, 215 (2004).
15. W. Nörenberg, *Basic concept in the description of collisions between heavy nuclei,in Heavy ion collisions*, ed. Bock R, Vol.2 (North-Holland publishing company, Amsterdam, 1980) pp. 1-43.
16. E. Wilhelm, *Elastic Scattering, in Treatise on Heavy-Ion Science*, ed. D.Allan Bromley, Vol.1(Plenum press, New york and London,1984) pp.135-290.
17. J. M. Hu, *Nuclear Theory*, Vol.2 (Atonic Energy Press, 1996) pp.265.
18. Z. Sun *et al.*, *Nucl. Instr. and Meth.* **A503**, 496 (2003).
19. A. Ozawa *et al.*, *Nucl. Instr. and Meth.* **B247**, 155 (2006).
20. Z. H. Li *et al.*, *Phys. Rev.* **C72**, 064327 (2005).
21. V. M. Strutinsky, *Phys. Lett.* **B44**, 245 (1973).
22. A. Y. Abul-magd and M. H. Simbel, *Phys.Lett.* **B83**, 27 (1979).
23. A. Gobbi and W. Nörenberg, *Dissipative collisions, in Heavy Ion Collisions*, ed. Bock R,Vol.2(North-Holland Pub.Co.Amsterdam, 1980) pp.127-274.
24. A. De Rosa *et al.*, *Phys. Rev.* **C40**, 627 (1989).
25. A. De Rosa *et al.*, *Phys. Rev.* **C41**, 2062 (1990).
26. A. De Rosa *et al.*, *Phys. Rev.* **C44**, 747 (1991).
27. Q. Wang *et al.*, *Chin. Phys. Lett.* **21**, 1911 (2004).
28. Y. C. Dong *et al.*, *High Energy Phys. and Nucl. Phys.* **29**, 147 (2005).
29. K. Rusek *et al.*, *Phys. Rev* **C67**, 041604R (2003).
30. O. R. Kakuee *et al.*, *Nucl. Phys.* **A728**, 339 (2003).
31. S. D. Baker and J.A. Mcintyre, *Phys. Rev.* **161**, 1200 (1967).
32. R. Perry *et al.*, *Phys. Rev* **C24**, 1471 (1981).
33. P. Schwandt *et al.*, *Phys. Rev* **C24**, 1522 (1981).
34. R. Huffman *et al.*, *Phys. Rev* **C22**, 1522 (1980).
35. J. R. Huizenga, *USERDA Progress Reprot COO-3496-65*, 131 (1977).

36. J. B. Marion, *Nuclear Reaction Analysis* (North-Holland publishing Company, 1968).
37. R. Dasilveira, *Phys. Lett.* **B45**, 211 (1973).
38. D. T. Khoa *et al.*, *J. Phys.* **G34**, R111 (2007).
39. I. J. Thompson, *Comput. Phys Rep* **C7**, 167 (1988).

MICROSCOPIC OPTICAL POTENTIAL IN RELATIVISTIC APPROACH

ZHONG-YU MA*, JIAN RONG

China Institute of Atomic Energy,
Beijing 102413, P.R. of China
**E-mail: Mazy12@ciae.ac.cn*

YIN-QUN MA

Deaprtment of Physics, Taiyuan Normal University,
Taiyuan 030001, P.R. of China

The relativistic microscopic optical potential (RMOP) in the nucleon-nucleus scattering is studied in the framework of the Dirac Brueckner Hartree-Fock (DBHF) approach. The real part of the nucleon self-energy in asymmetric nuclear matter is calculated with the G-matrix, while the imaginary part is obtained from the polarization diagram. Nuclear optical potentials in finite nuclei are derived from the nucleon self-energies in asymmetric nuclear matter through a local density approximation. The RMOP is applied to study the nucleon scattering off stable nuclei. A satisfactory agreement with the experimental data is found. This is achieved without readjusting phenomenologically the RMOP derived from DBHF plus polarization diagram. The complex nucleon-nucleus optical potential is extended to microscopic optical potentials of nucleus-nucleus interaction by a folding method. The elastic scattering data of ^{6}He at 229.8MeV on ^{12}C target are analyzed within the standard optical model.

Keywords: Relativistic microscopic optical potential; Dirac Brueckner Hartree-Fock (DBHF) approach; Nucleus-nucleus scattering

1. Introduction

The optical model potential is an essential tool in the study of nuclear reactions. Recently, the relativistic mean field (RMF) approach has been of great success in describing not only the ground state properties of stable nuclei, but also those of exotic nuclei. There have been also various applications to investigate the nucleon-nucleus scattering. Studies of the relativistic microscopic optical potential (RMOP) in the past years were based on several models, such as the RMF,[1,2] relativistic Hartree-Fock (RHF),[3] and Dirac-Brueckner Hartree-Fock (DBHF)[4,5] approaches. The nucleon self-energies were calculated in symmetric nuclear matter and the isospin dependence was usually neglected. A reasonable description of the differential cross sections, analyzing powers and spin rotation functions of proton scattering off stable nuclei has been achieved in an energy range of 50-500 MeV without readjusting parameters.[3,5]

With the development of experiments on newly constructed radioactive beam facilities, the investigation of the isospin dependence of the optical potential becomes more and more important. Therefore we shall study the RMOP microscopically in the framework of the DBHF approach. Recently, a new decomposition of the Dirac structure of nuclear self-energy in the DBHF has been extended to the calculations in asymmetric nuclear matter.[6] The optical potential of a nucleon in the nuclear medium is identified with the nucleon self-energy. In the DBHF we investigate the proton and neutron self-energies in asymmetric nuclear matter, therefore the RMOP.[7] The real part of the nucleon self-energy in asymmetric nuclear matter is calculated with the G matrix in the Hartree-Fock approach and the imaginary part of the the nucleon self-energy is obtained by the G-matrix polarization diagram. The nucleon-nucleus optical potentials in finite nuclei are obtained by the local density approximation (LDA), in which the space distribution of the optical potential is directly connected with the density and asymmetry parameters of the nuclear self-energies in the asymmetric nuclear matter. The nucleon scattering off nucleus is analyzed in a Schrödinger-type equation, which is obtained by eliminating the lower component of the Dirac spinor in the Dirac equation. A satisfactory agreement with experimental data has been found.

The study of the nucleus-nucleus optical potentials is an object of importance not only for the description of elastic cross sections but also as an ingredient in the description of all the phenomena which occur when two nuclei collide.[8] The interest has been renewed recently due to the development of radioactive beam facilities in the world. In particular, it is very important to understand the complex optical potential for composite projectiles from a microscopic point of view not only to understand the relevant reaction dynamics involved but also to develop a practical tool for predicting optical potentials of colliding systems for which the elastic scattering measurement is absent or difficult, such as in the case of neutron-rich or proton-rich β-unstable nuclei. In this work we shall extend our method to investigate the microscopic optical potentials (MOP) of nucleus-nucleus scattering by a folding method. The elastic scattering data of ^{6}He at 229.8MeV on ^{12}C target are analyzed within the standard optical model.

2. Relativistic microscopic optical potential in nucleon scattering off nucleus

Starting from a bare NN interaction V the NN effective interaction in the nuclear medium is calculated by summing up all ladder diagrams in the DBHF approach. The DBHF G can be decomposed into the bare NN interaction and a correlation term: $G = V + \Delta G$.[6] The correlation term is parameterized by four vertices: scalar, vector and isoscalar, isovector. Due to the characteristic of the short range correlation, they can be described by infinite masses and finite ratios of strengths to the corresponding masses. The nucleon self-energies in nuclear matter are calculated with V and ΔG in the relativistic Hartree-Fock approach. The exchange term produces weak energy dependence of the nucleon self-energy, which takes account of the antisymmetry of the nucleon with all nucleons in the nuclear medium. The DBHF nucleon self-energy in asymmetric nuclear matter has the general form:

$$\Sigma^i(k, k_F, \beta) = \Sigma^i_s(k, k_F, \beta) - \gamma_0 \Sigma^i_0(k, k_F, \beta) + \boldsymbol{\gamma} \cdot \mathbf{k} \Sigma^i_v(k, k_F, \beta) \ , \qquad (1)$$

where i stands for proton or neutron. $\beta = (\rho_n - \rho_p)/(\rho_n + \rho_p)$, ρ_n, ρ_p are the asymmetry parameter, neutron and proton densities, respectively. Due to the isovector meson exchanges, the proton and neutron self-energies are distinguished. The imaginary part of the nucleon self-energy can be obtained by the G-matrix polarization diagram. An effective nucleon interaction was introduced in order to avoid difficulties caused by π-meson and simplify the calculation.[7] Four scalar and vector mesons with density dependent coupling constants were introduced to reproduce the saturation curves and nucleon self-energies at various densities and asymmetry parameters calculated with the DBHF G-matrix.

It is well known that the optical potential of a nucleon in the nuclear medium is equivalent to its self-energy. For finite nuclei the nucleon potential is obtained by means of the LDA, in which the space dependence of the relativistic microscopic optical potential is directly connected with the density of the target nucleus and asymmetry parameter in asymmetric nuclear matter. The Dirac equation of the projectile nucleon in the mean field of the target nucleus can be written as:

$$[\boldsymbol{\alpha} \cdot \mathbf{k} + \gamma_0 (M + U^i_s) + U^i_0 + V_c] \psi^i = E \psi^i, \qquad (2)$$

where

$$U^i_s = \frac{\Sigma^i_s - M\Sigma^i_v}{1 + \Sigma^i_v}, \qquad U^i_0 = \frac{-\Sigma^i_0 + E\Sigma^i_v}{1 + \Sigma^i_v} \ . \qquad (3)$$

In these expressions, M is the mass of nucleons, U^i_s, U^i_0 are Lorentz scalar potential and time like component of a four-vector potential, respectively. $E = \varepsilon + M$, and ε is the energy of the projectile in the center-of-mass system. V_c is the Coulomb potential.

By eliminating the lower component of the Dirac spinor in Eq. 2, a Schrödinger-type equation[7,9] is obtained for the upper component:

$$\left[\frac{p^2}{2E} + V^i_{eff}(r) + V_C(r) + V^i_{so}(r) \boldsymbol{\sigma} \cdot \mathbf{L} \right] \Phi^i(\mathbf{r}) = \frac{E^2 - M^2}{2E} \Phi^i(\mathbf{r}), \qquad (4)$$

where V^i_{eff} and V^i_{so} are the central and spin-orbit part of the Schrödinger-equivalent potential, respectively, which are known as the nucleon-nucleus optical potential in the non-relativistic approach. The explicit expressions for V^i_{eff} and V^i_{so} are

$$\begin{aligned} V^i_{eff} &= \frac{M}{E} U^i_s + U^i_0 + \frac{1}{2E} \left[(U^i_s)^2 - (U^i_0 + V_C)^2 \right] + V^i_D \ , \\ V^i_{so} &= -\frac{1}{2ErD^i(r)} \frac{dD^i(r)}{dr} \ , \end{aligned} \qquad (5)$$

where $D^i(r) = M + E + U^i_s - U^i_0 - V_c$ and a small Darwin term V^i_D is usually neglected. The Schrödinger equivalent potential yields exactly the same scattering phase shifts

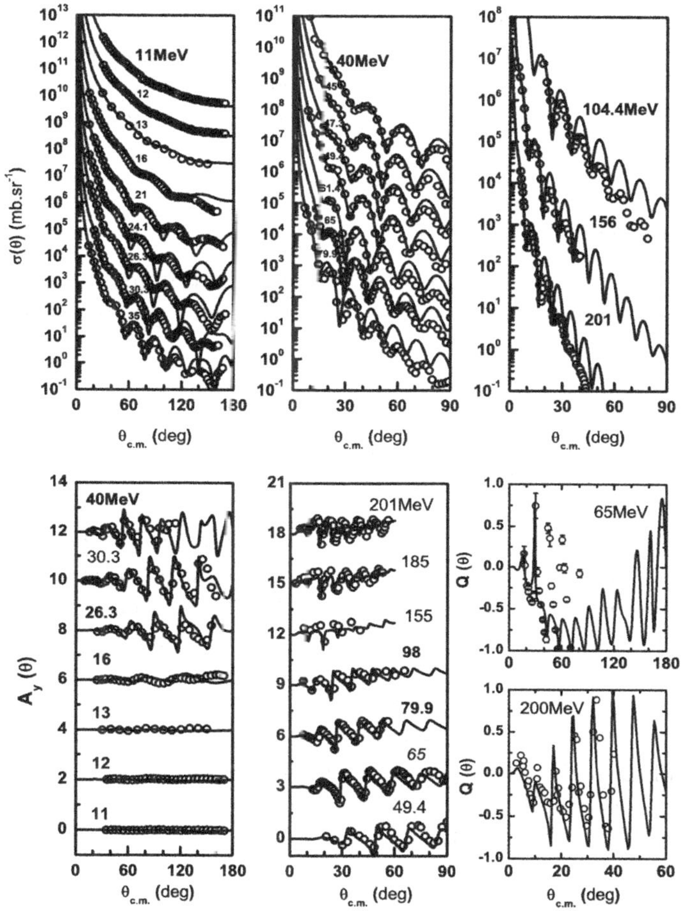

Fig. 1. Elastic scattering differential cross sections and spin observables in $p+^{208}$Pb at $E_p <$ 200MeV. Note that the differential cross sections are offset by powers of 10 (for $E_p <$ 100MeV) and 100 (for $E_p >$ 100MeV), while the analyzing powers are shifted by increments of 2 (for $E_p <$ 50MeV) and 3 (for $E_p >$ 50MeV), respectively.

as the original relativistic potential in the four-component Dirac equation. Therefore we have obtained the microscopic proton- and neutron-nucleus interaction in the framework of the DBHF approach, which are complex with both real and imaginary parts. Including the isovector meson exchanges the microscopic optical potential also depends on the neutron and proton density distributions of the target nucleus.[7]

We calculate the differential cross sections, analyzing powers and spin rotation functions of proton scattering off spherical nuclei at various energies of $E_p < 200MeV$ and compare with the experimental data. In Fig.1, we show the comparison of the theoretical results (solid curves) with the experimental data of the differential cross section $\sigma(\theta)$, analysing power $A_y(\theta)$ and spin rotation function

Q for $p+^{208}$Pb. It is found that the calculated cross sections as well as spin observables are in rather good agreement with the experimental data at low energies. As the energy increases, the theoretical differential cross sections are overestimated at large angles in comparison with the experimental data, but they keep the same diffraction patterns.

3. ^{6}He scattering off ^{12}C

In a simple practical approach to deal with a composite particle scattering, one considers the target just a scatterer, and the nucleus-nucleus optical potential can be obtained by a folding method. The proton- and neutron-nucleus optical potentials are folded with the corresponding proton and neutron density distributions in the projectile.

$$V_{FM}(\mathbf{R}) = \sum_{i=p,n} \int \rho_i(\mathbf{r}_i) V^i_{eff}(\mathbf{s}_i) d\mathbf{r}_i \ , \tag{6}$$

where $\mathbf{R}$ is the separation distance between two centers of the colliding nuclei. $\mathbf{r}_i$ is the coordinate of the proton (neutron) at the center of mass frame of the projectile, while $\mathbf{s}_i$ the vector between the proton (neutron) in the projectile and the center of mass of the target, $\mathbf{s}_i = \mathbf{R} - \mathbf{r}_i$. ρ_i is the density distribution of proton (neutron) at the projectile. In this method the overlap of two colliding nuclei has not taken into account, which would in general overestimate the real potentials and underestimate the imaginary ones, respectively. We shall discuss this effect late.

With the folding method we could obtain the nucleus-nucleus microscopic optical potential with the real and imaginary part, simultaneously, which appears in a one-body standard Schrödinger equation,

$$\left[-\frac{\hbar^2}{2\mu}\nabla^2 + U_{opt}(\mathbf{R}) + U_C(\mathbf{R}) \right] \chi(\mathbf{R}) = E_{c.m.}\chi(\mathbf{R}) \ . \tag{7}$$

Here, μ is the reduced mass of the pair, $E_{c.m.}$ is the center of mass energy of the relative motion. $U_C(\mathbf{R})$ is the Coulomb potential, which can be given as usual by

$$U_C(R) = \begin{cases} \frac{Z_P Z_T e^2}{R} & R > R_C \\ \frac{Z_P Z_T e^2}{2R_c}(3 - (\frac{R}{R_C})^2) & R \leq R_C \end{cases} , \tag{8}$$

where Z_P and Z_T are the charge numbers of the projectile and target, respectively. $R_C = r_0(A_P^{1/3} + A_T^{1/3})$ is the Coulomb radius of two charged spheres, $r_0 = 1.2$ fm, A_P, A_T are the mass numbers of the projectile and target, respectively.

Applying this method we analyze the recent experimental data on ^{6}He scattering off ^{12}C at the bombarding energy $E_{lab} = 229.8 MeV$ at GANIL.[10] ^{6}He is a halo nucleus with a dispersed neutron distribution. We adopt a halo-type density, which was obtained by calculations in a three-body model.[11]

$$\begin{aligned} \rho_p(r) &= 2e^{-(r/a)^2}/\pi^{3/2}a^3 \ , \\ \rho_n(r) &= 2e^{-(r/a)^2}/\pi^{3/2}a^3 + 4(e^{-(r/b)^2}/3\pi^{3/2}b^5)r^2 \ , \end{aligned} \tag{9}$$

with $a = 1.55$ fm, $b = 2.24$ fm. A two parameter Fermi function is used for the density distribution of ^{12}C,[10]

$$\rho(r) = \frac{\rho_0}{1 + exp((r - r_0)/a)}, \tag{10}$$

with r_0=2.1545 fm and a=0.425 fm. The density is normalized to the atomic number of ^{12}C, then ρ_0 is determined as 0.207 fm^{-3}. The real and imaginary parts of the optical potential $U_{opt}(R)$ for ^{6}He+^{12}C reaction at 229.8 MeV are calculated by folding the isospin dependent microscopic optical potentials in $p(n)$+^{12}C with the density of ^{6}He.

The elastic differential cross section for ^{6}He scattering off ^{12}C at 229.8MeV are calculated with the MOP. In order to reproduce the experimental data one introduces phenomenologically normalization factors on real and imaginary parts of the MOP. It is found that the imaginary part of the MOP obtained is too weak and an enhancing factor of $N_I = 3.0$ on the imaginary part of the optical potential is required when one keeps the real potential unchanged ($N_R = 1$). This can be understood that ^{6}He is a loosely bound nucleus, a breakup process is one of the main processes in the reaction, which is not considered in the microscopic optical potential of $p(n)$+^{12}C. The imaginary part of the $p(n)$+^{12}C optical potential was calculated from the polarization diagram of G matrix, which takes account only the processes of particle-hole type excitations.[7] Thus the calculated imaginary part of the optical potential in the reaction of ^{6}He+^{12}C is much too weak. In addition for a weakly bound or halo nucleus its particle threshold is close to the ground state, which implies a strong coupling to the continuum during the interaction of the nucleus with a target. A special treatment of the interaction potential to consider explicitly transitions to the low-lying excited states, to the resonances and breakup states may also required.

The differential cross section calculated directly by the folded potential with $N_I = 3.0$ is plotted in Fig.2 with a solid curve. The agreement between our theoretical calculation and the experimental data is impressive, which is rather encouraging without adjusting more free parameters. As mentioned above that the overlap effect between two colliding nuclei has not taken into account in our model. The results without normalization factor on the real part may illustrate that the overlap of two colliding nuclei is nonsignificant in the collision with a weakly bound nucleus.

In Fig.2 we also plot the comparison with the results based on various theoretical models. The dotted curve was calculated with a double folding model (DFM).[10] In this model the real part of the optical potential was obtained by double folding of the CDM3Y6 effective nucleon-nucleon interaction, and the imaginary part was based on a phenomenological Woods-Saxon type potential with including a dynamic polarization potential (DPP). A frozen density approximation was adopted, where the total density of the two overlapping nuclei was the sum of the two densities. In their calculation a best fit to the experimental data was obtained by adding a complex surface potential with a repulsive real part designed to simulate the polarization effects caused by the projectile breakup. The dash-dotted curve corresponds

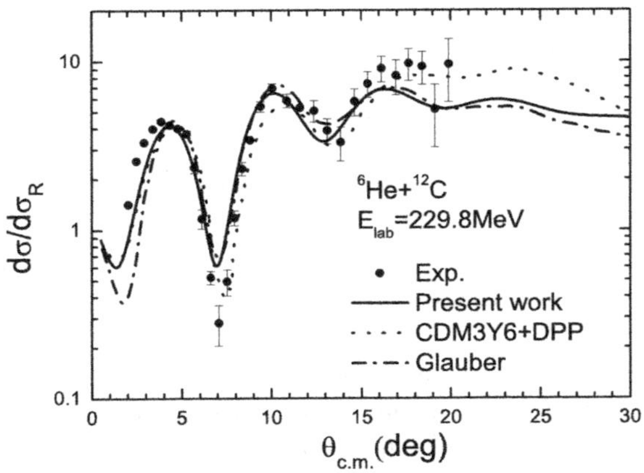

Fig. 2. Angular distribution of the elastic differential cross section for ^{6}He+^{12}C scattering at 229.8MeV. The solid curve shows the results of the optical model calculations with an enhancing factor 3 on the imaginary potential. The dotted, and dot-dashed curves show the results of calculations by the double folding model with CDM3Y6 + DPP and Glauber-model, respectively

to the calculation in a few-body Glauber model (FBGM),[12] where the ^{6}He wave function is factorized as a three-body wave function consisting of the α core and two neutrons. The target is considered just a scatterer and the effective phases in the Glauber model are calculated from a nucleon-target or a core-target optical potentials. It is found that all models could well reproduce the experimental data. At forward angle, our results agree with those of DFM+DPP, while the FBGM produces too deeper minimum. Our results well describe the position of the second minimum in the differential cross section, but the value is slightly weak, and so did FBGM. In general our results are closer to the results of DFM+DPP and the experimental data.

Actually, a fully microscopic investigation of the nucleus-nucleus scattering is much more complicated and difficult.[8] Inelastic and multi-reaction channels, absorption and breakup processes are involved in the nucleus-nucleus reactions. A proper treatment of the density overlap of two colliding nuclei is still not clear, which depends on various factors, such as the structure of two colliding nuclei, bombarding energy, etc. In our present investigations we have not attempted to deal microscopically with these dynamic reaction process and the density overlap between two colliding nuclei. In comparison with the experimental data, we investigate phenomenologically the effect of the dynamic processes and overlap density in the nucleus-nucleus scattering, especially for loosely bound nuclei.

4. Summary

We have investigated the RMOP and its isospin dependence in asymmetric nuclear matter in the framework of the DBHF approach. The nucleon self-energy is calculated with the DBHF G-matrix. The imaginary part of the nucleon self-energy with $E > 0$ is obtained by the G matrix polarization diagram. The nucleon-nucleus optical potentials in finite nuclei is related to the self-energy in nuclear matter by means of the local density approximation. With the optical potentials determined, we have calculated differential cross sections and spin observables of proton scattering off stable nuclei at $E_p < 200$ MeV. The agreement with the experimental data is reasonable. It should be emphasized that the RMOP obtained from the DBHF G matrix without free parameters could well describe the proton elastic scattering.

A microscopic optical model of nucleus-nucleus interaction is presented with a folding model of this nucleon-nucleus optical potential. The elastic scattering data of ^{6}He at 229.8 MeV on a ^{12}C target have been analyzed within the standard optical model. A factor of 3 on the imaginary potential to take account phenomenologically the breakup and complicated processes is introduced. The calculated ^{6}He+^{12}C elastic scattering differential cross section is in good agreement with the experimental data.

Acknowledgments

This work is supported by the National Natural Science Foundation of China under Grant Nos 10475116, 10535010, Major State Basic Research Development Program in China Under Contract Number 2007CB81500 and the European Community Project Asia-Europe Link in Nuclear Physics and Astrophysics, CN/ASIA-LINK/008(94791). ZYM gratefully acknowledges Nguyen Van Giai and Hermann Wolter for many stimulating discussions.

References

1. C. J. Horowitz and B. D. Serot, *Nucl. Phys. A* **399**, 529 (1983).
2. C. J. Horowitz, *Nucl. Phys. A* **412**, 228 (1984).
3. Ma Zhong-yu, Zhu ping, Gu ying-qi et al., *Nucl. Phys. A* **490**, 619 (1988).
4. Ma Zhong-yu and Chen Bao-qiu, *J. Phys. G* **18**, 1543 (1992).
5. B. Q. Chen and A.D. Mackellar, *Phys. Rev. C* **52**, 878 (1995).
6. E. Schiller and H. Müther, *Eur. Phys. J. A* **11**, 15 (2001).
7. J. Rong, Z. Y. Ma, and N. V. Giai, *Phys. Rev. C* **73**, 014614 (2006).
8. G. R. Satchler and W. G. Love, *Phys. Rep.* **55**, 183 (1979).
9. M. Jaminon, C. Mahaux, and P. Rochus, *Phys. Rev. C* **22**, 2027 (1980).
10. V. Lapoux, N. Alamanos, F. Auger et al., *Phys. Rev. C* **66**, 034608 (2002).
11. J. S. Al-Khalili, J. A. Tostevin, and I. J. Thompson, *Phys. Rev. C* **54**, 1843 (1996).
12. B. Abu-Ibrahim and Y. Suzuki, *Nucl. Phys. A* **728**, 118 (2003).

EXOTIC NUCLEI STUDIED IN DIRECT REACTIONS AT LOW MOMENTUM TRANSFER – RECENT RESULTS AND FUTURE PERSPECTIVES AT FAIR

PETER EGELHOF

Gesellschaft für Schwerionenforschung, GSI, Planckstr.1, D-64291 Darmstadt, Germany

for the

IKAR AND EXL COLLABORATIONS

Light-ion induced direct reactions, like for example elastic and inelastic scattering, transfer-, charge exchange-, or knock out-reactions, have been proved in the past, for the case of stable nuclei, to be powerful tools for obtaining nuclear structure information, and were also applied within the last decade for the investigation of light exotic nuclei with radioactive beams in inverse kinematics. In particular, it turned out that in many cases essential nuclear structure information is deduced from high-resolution measurements at low momentum transfer.

As an example an overview on most recent results on intermediate energy elastic proton scattering, a meanwhile well established method for the investigation of nuclear matter distributions of halo nuclei, is given. Differential cross section data for small angle proton scattering from ^{12,14}Be and ^{8}B at energies near 700 MeV/u were obtained and analyzed using the Glauber multiple scattering theory. Preliminary results on the nuclear matter radii and nuclear matter distributions are presented. In particular for ^{8}B, supposed to be a candidate for a proton halo, preliminary results confirm the existence of a pronounced proton halo structure. For the case of neutron-rich Li nuclei recent results from laser spectroscopy combined with results from proton scattering allowed to deduce the neutron radii and the neutron skin sizes of the ^{8,9,11}Li isotopes.

The experimental conditions at the future international facility FAIR will provide unique opportunities for nuclear structure studies on nuclei far off stability, and will allow to explore new regions in the chart of nuclides of high interest for nuclear structure and nuclear astrophysics. In particular, the predicted luminosities will allow for the investigation of direct reactions with stored and cooled radioactive beams at internal H, He, etc. targets of the new storage ring NESR. This technique enables high resolution measurements down to very low momentum transfer and provides a gain in luminosity from accumulation and recirculation of the radioactive beams. A brief overview on the research objectives and the technical realization of the EXL setup (EXotic nuclei studied in Light-ion induced reactions at the NESR storage ring), supposed to be a highly efficient, high resolution universal detection system, applicable to a wide class of reactions, is presented.

1. Introduction

One of the most powerful classical methods for obtaining spectroscopic information on the structure of nuclei is the investigation of light-ion induced direct reactions. The study of such reactions, as for example, elastic and inelastic proton-, deuteron-, alpha-scattering, one- and few-nucleon transfer reactions, or charge exchange and knock-out reactions, etc. contributed in the past substantially to our present knowledge about stable nuclei. In the recent past the use of good-quality secondary radioactive ion beams, now available at several major accelerator facilities world wide, enabled to extend such studies on exotic nuclei by using the method of inverse kinematics. Despite the experimental challenge in performing such experiments with relatively low secondary beam intensities, investigations with radioactive beams have opened new territories of nuclei in the nuclear chart to be explored. Within the last two decades a number of first generation experiments on direct reactions with radioactive beams have been performed. Such experiments were performed up to date only at external targets, and were mainly concentrated on light neutron-rich nuclei with mass number A < 20. Nevertheless, when reaching extreme neutron/proton ratios near the neutron drip line, allowed the discovery of new exciting phenomena of nuclear structure.

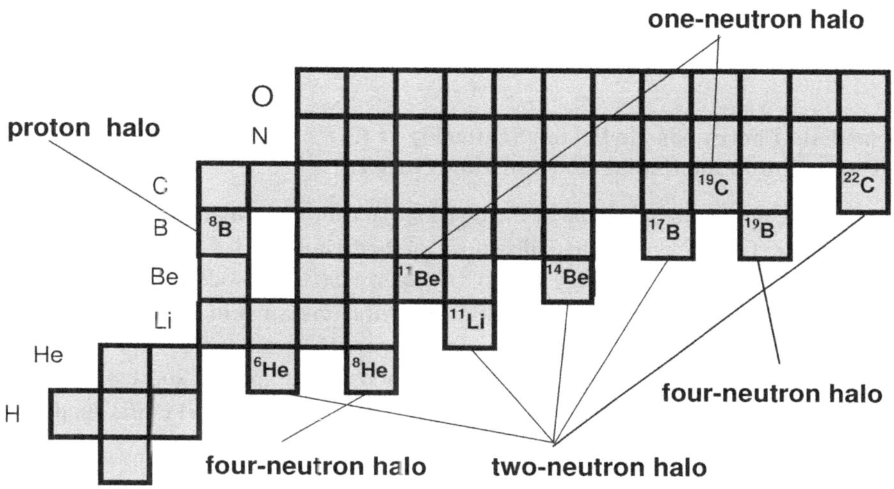

Figure 1. Chart of nuclides for light elements. Isotopes (with Z ≤ 8) for which a halo structure was discovered experimentally are indicated.

For example, one of the most outstanding discoveries was the finding that the nuclear matter may appear under certain conditions with a qualitatively new type of nuclear structure, so-called "halo"-structure. Compared to the stable nuclei and those close to stability, in which all the protons and neutrons are essentially distributed uniformly over the nuclear volume, it was found that some light neutron-rich nuclei located at or near the neutron drip line exhibit a widely extended low-density distribution of loosely bound valence neutrons (the "halo") surrounding a compact distribution of the majority of nucleons (the "core"). Meanwhile evidence for extended halo structures for a number of neutron-rich He, Li, Be, B and C, etc. isotopes, located at or close to the neutron dripline (see figure 1), has been obtained from various experimental investigations on total reaction cross sections, knockout reactions, breakup reactions, transfer reactions, and elastic and inelastic scattering (see [1] for an overview). Besides the determination of the size and shape of the nuclear matter distributions of such nuclei, the single-particle and few-body structure of these nuclei was investigated, and resonant states beyond the dripline were discovered.

The present article will concentrate on high resolution experiments in the region of very low momentum transfer, where in many cases essential nuclear structure information can be deduced, not accessible otherwise. Because such experiments can, due to the low recoil velocities of the target-like reaction products, not be performed with standard external targets, new experimental techniques are required for such investigations. In the following two examples along this direction, one concerning recent investigations on elastic proton scattering at low momentum transfer by using the technique of active targets, the other focusing on future perspectives for low momentum transfer experiments at the future facility FAIR (Facility for Antiproton and Ion Research [2-5]) within the EXL project [4,5,6], where direct reactions will be studied from the interaction of high quality radioactive beams, cooled to the lowest phase space densities, with internal targets of the New Experimental Storage Ring NESR.

2. Intermediate Energy Elastic Proton Scattering at Low Momentum Transfer – a Tool to Study the Radial Shape of Halo Nuclei

The size of nuclei and the radial shape of the distribution of nuclear matter and charge are fundamental properties of nuclei, and therefore of high interest for various fields in nuclear physics. Over the years a large variety of experimental methods were developed, using leptonic probes (as electrons, muons, etc.) for investigating nuclear charge distributions, and hadronic probes (as protons, α-particles, etc.) for exploring the distributions of nuclear matter. While all these methods were applied successfully for many years for the study of stable nuclei, the investigation of the size and radial shape of exotic nuclei has become a new and exciting field of research.

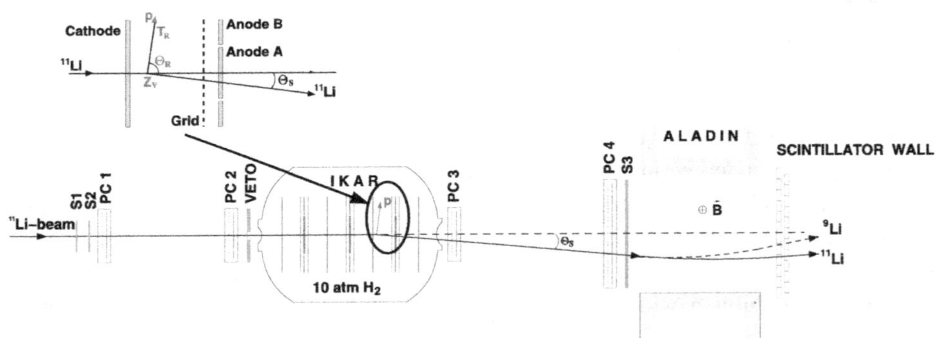

Figure 2. Schematic view of the experimental setup for small-angle elastic proton scattering on exotic nuclei in inverse kinematics. The central part shows the hydrogen-filled ionisation chamber IKAR which serves simultaneously as a gas target and a detector system for recoil protons. In 6 independent sections of this detector the recoil angle θ_R, the recoil energy T_R, and the vertex point Z_V are determined (see insert). The forward spectrometer consisting of four multiwire proportional chambers PC1 - PC4 determines the scattering angle θ_S of the projectile. The scintillation counters S1 - S3, VETO are used for beam identification and trigger. The ALADIN magnet with a position sensitive scintillator wall behind allows to identify the scattered beam particle and to discriminate break-up channels.

Proton-nucleus elastic scattering at intermediate energies around 700 - 1000 MeV is known to be a method well established for obtaining accurate nuclear matter distributions

[7]. This method was applied at GSI Darmstadt for the first time for the investigation of exotic nuclei by using the technique of inverse kinematics. The simulation [8,9] have clearly demonstrated that proton scattering in the region of small momentum transfer is particularly sensitive to the nuclear matter radius, and the halo structure of nuclei. Thus, besides determining precisely the nuclear matter radius, the shape of the nuclear matter distribution in halo nuclei can be explored already from data in the limited range of momentum transfer at small scattering angles.

Differential cross sections for elastic proton scattering at small momentum transfer were measured at GSI Darmstadt by the IKAR collaboration at energies around 700 MeV/u in inverse kinematics for the neutron-rich helium isotopes ^{6}He, ^{8}He [9,10], for the neutron-rich lithium isotopes ^{8}Li, ^{9}Li, and ^{11}Li [11,12], and most recently for the ^{12,14}Be and ^{8}B nuclei. The experimental method (see figure 2) is based on the high-pressure hydrogen-filled ionization chamber IKAR, which serves simultaneously as gas target and recoil proton detector. The use of this dedicated active-target technique enabled us to detect the low energy recoil protons with sufficient angular and energy resolution covering the full solid angle of interest (for details see [10]). In addition projectile scattering angles were measured with multiwire tracking detectors and the elastically scattered particles were identified by a magnetic rigidity analysis to allow for a safe separation of breakup channels.

3. Recent Results on Halo Structures in Exotic Li, Be and B Isotopes

The differential cross sections deduced [13] for the example of the most recently investigated isotopes ^{12}Be and ^{14}Be are displayed in figure 3. Note that the data shown are still preliminary. For establishing the nuclear matter density distributions from the measured cross sections, the Glauber multiple scattering theory was applied. For modelling the nuclear matter distributions, parametrical functions were used. The free parameters of these density distribution parametrizations were then determined by a least-square fit of the calculated cross sections to the experimental ones. A reasonably good description of the cross section data is obtained as may be seen in Fig. 3 (left part).

The present results (still preliminary!) on the nuclear matter densitiy distributions obtained for the ^{12,14}Be isotopes are displayed in figure 3 (right side). As compared to the nuclear matter distribution deduced for ^{12}Be, which exhibits a behaviour of "normal nuclei", a rather extended nuclear matter distribution is obtained for ^{14}Be, the matter density decreasing much slower with the radius parameter than the one for ^{12}Be, and therefore representing a clear signature for a pronounced neutron halo in ^{14}Be. This is also reflected in the deduced nuclear matter radii, where the result for ^{14}Be: $R_m = 3.78$ (37) fm has to be compared with $R_m = 2.88$ (7) fm for ^{12}Be. The close resemblance of the distribution deduced for the core of ^{14}Be (with $R_{core} = 2.76$ (4) fm) with the matter distribution of ^{12}Be supports the picture of a two-neutron halo structure in ^{14}Be as predicted from various theoretical calculations.

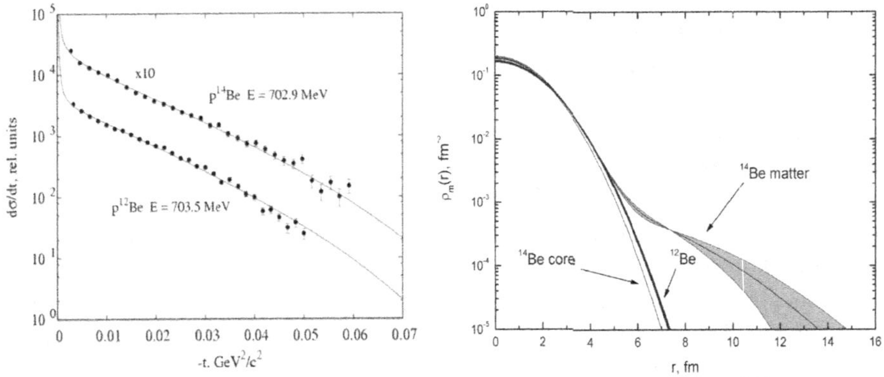

Figure 3. Differential cross sections versus the four-momentum transfer squared -t for $p^{12,14}Be$ elastic scattering (left) and the nuclear matter and nuclear core density distributions deduced from the experimental cross section data. Note that the data are still preliminary.

With ^{8}B for the first time an isotope supposed to be a candidate for a proton halo was investigated by the present method (see also fig. 1). From the preliminary results [13] it can be concluded that the halo structure of ^{8}B is confirmed. The deduced matter radius of ^{8}B amounts to $R_m = 2.88\ (13)$ fm. The extended nuclear matter distribution of ^{8}B, and in particular its long range tail, may also have a significant relevance for the astrophysical S (E) factor for $E \Rightarrow 0$ for the $^{7}Be\ (p,\gamma)\ ^{8}B$ reaction. This needs to be investigated in more detail in near future when final results from the present experiment are available.
For the $^{8,9,11}Li$ isotopes, the results on the nuclear matter distributions of which were already published in [11,12], the independent determination of the charge radii from laser spectroscopy experiments recently performed [14] allowed for the first time in a model independent way to deduce the neutron radii R_n and the neutron skin size δ_{np} of these nuclei by using the relations

$$\delta_{np} = R_n - R_p \qquad R_n^{\ 2} = \frac{1}{N_n} * \left(AR_m^{\ 2} - N_p R_p^{\ 2}\right)$$

The results are summarized in Table 1. Note that in the present contribution all nuclear matter distributions and radii are defined as point matter distributions and radii, and consequently the charge radii displayed in table 1 are point charge radii, deduced from the measured charge radii by unfolding from the charge radius of the proton. As seen from table 1 the results for ^{11}Li, in particular the close resemblance of the values for R_{core} and R_p, and the deduced values for R_n and δ_{np} support the existence of a pronounced neutron halo in ^{11}Li.

Table 1. Summary of nuclear radii of the $^{8,9,11}Li$ isotopes as deduced from experimental investigations on proton scattering [11,12] and laser spectroscopy [14]. Note that all values given denote point matter, and point charge radii, respectively. For details see text.

nucleus	R_m, fm	R_{core}, fm	R_{halo}, fm	R_p, fm	R_n, fm	δ_{np}, fm
^{8}Li	2.50 (6)	--	--	2.15 (3)	2.69 (9)	0.54 (10)
^{9}Li	2.44 (6)	--	--	2.06 (4)	2.61 (9)	0.55 (10)
^{11}Li	3.42 (11)	2.52 (2)	5.98 (32)	2.33 (4)	3.75 (15)	1.42 (16)

4. Perspectives at the Future Facility FAIR: The EXL Project

The final part of this contribution is intended to focus on the "long range" perspectives and possibilities which will be opened at the future international facility FAIR (Facility for Antiproton and Ion Research) with a particular emphasis on experiments with radioactive beams at storage rings. FAIR will provide the european and international science community with a technically novel and in many respects unique accelerator system for a multidisciplinary scientific program, including nuclear structure physics with radioactive ion beams, hadron physics with antiproton beams, physics of compressed nuclear matter with relativistic nuclear collisions, and many more (see [2,3,4,5] for an overview). The layout of the FAIR accelerator and storage ring complex is displayed in figure 4 (right side). It constitutes a substantial extension of the capabilities of the present GSI facility (left side of fig. 4) by delivering ion beams of much increased intensity, of higher energy, and of improved beam quality.

The expected performance of FAIR will, among others, provide unique opportunities for nuclear structure studies on nuclei far off stability, and will allow for exploring completely new and unknown regions in the chart of nuclides, of high interest for nuclear structure and astrophysics. The part of FAIR, which is foreseen for experiments with radioactive ion beams (see [4,5,15] for an overview), will include a dedicated storage ring system consisting of a Collector Ring (CR) and the New Experimental Storage Ring (NESR) for accumulation, storage and cooling of the large emittance fragment beams provided by the new superconducting large-acceptance fragment separator (Super-FRS). Its design aims for a total gain factor of about 10^4 in the intensity of secondary beams as compared to present facilities. Concerning the investigation of direct reactions with radioactive beams, this performance will provide sufficient luminosities in many cases to allow the application of new and promising experimental concepts, such as the use of internal ^{1}H, ^{2}H, ^{3}He, ^{4}He, etc. targets at the NESR for nuclear reaction studies, and the use of the intersecting electron ring (e^-A-Collider) for performing elastic and inelastic electron scattering experiments. Alternatively the secondary radioactive beams can be transferred to experimental areas foreseen for external target experiments [15]. Of course such new possibilities will demand the new and challenging experimental concepts.

The present article will focus on the EXL-project (Exotic nuclei studied in Light-ion induced reactions at the NESR storage ring), which is an integral part of FAIR. The objective of the EXL project is to capitalize on light-ion induced direct reactions in inverse kinematics by using novel storage ring techniques and a universal detector system (for details see [4,5,16]). Due to their spin-isospin selectivity, light-ion induced direct reactions at intermediate to high energies are an indispensable tool in nuclear structure investigations as evident from investigations of stable nuclei in the past.

The essential nuclear structure information is deduced from high-resolution measurements at low-momentum transfer. Because of the kinematical conditions of inverse kinematics in case of beams of unstable nuclei, low-momentum transfer measurements turn out to be an exclusive domain of storage ring experiments. Luminosities are superior by orders of magnitude compared to experiments with external targets. The EXL program thus utilizes one of the outstanding features of FAIR, i.e., the availability of a multi-storage-ring complex coupled to the superconducting fragment separator Super-FRS (see fig. 4).

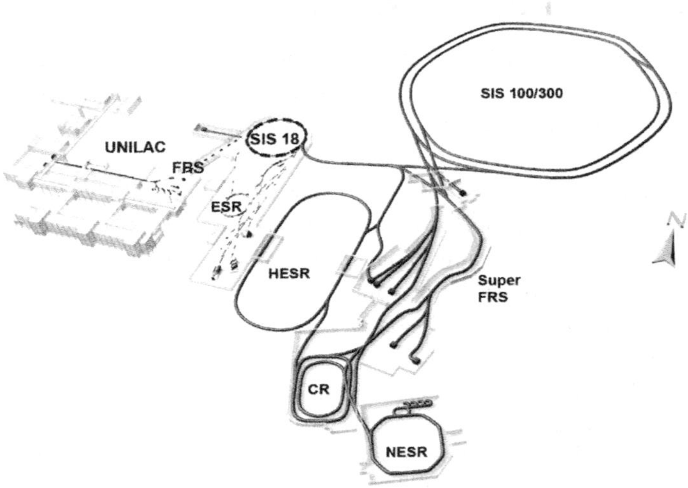

Figure 4. Schematic layout of the present GSI facility (left) and the FAIR facility (right). At FAIR the existing GSI accellerators UNILAC and SIS18 serve as an injector for the two large synchrotron rings SIS-100/300. The beams are distributed to various experimental areas or to the production targets for the anti-proton or radioactive-isotope beam facilities. For beam storage and manipulation several rings are foreseen including the high-energy storage ring HESR for anti-protons, the collector ring CR, and the experimental storage ring NESR for ions. The storage-ring complex provides possibilities of beam cooling as well as target areas for scattering experiments. The present fragment separator FRS and the new superconducting fragment separator Super-FRS for preparation of rare-isotope beams are indicated as well.

The key physics issues being covered within the EXL program address

- nuclear matter distributions near the drip lines (halo -, skin - structures, etc.),
- the isospin-dependence of the single-particle shell structures (magic numbers, shell gaps, spectroscopic factors),
- nucleon-nucleon correlations and cluster formation,
- new collective modes (different deformations for protons and neutrons; giant and pygmy resonances of various multipolarities),
- in-medium interactions in asymmetric and low-density nuclear matter,
- the astrophysical r- and rp-processes (Gamow-Teller transitions; dipole strength; neutron-capture).

A variety of light-ion induced direct reactions, such as elastic scattering of type *(p, p)*, *(α, α)*, inelastic scattering of type *(p, p')*, *(α, α')*, charge exchange reactions of type *(p, n)* , *(d,* ^{2}He), quasi-free scattering of type *(p, 2p)*, *(p, pn)*, *(p, p* + cluster), and transfer reactions of type *(p, t)*, *(p,* ^{3}He), *(p, d)*, *(d, p)*, serve in such investigations. The relevant physics observables (energy and momentum transfer) need to be extracted from the kinematical quantities (recoil momentum and kinetic energy). Since the domain of low momentum transfer is of interest, extremely thin targets are requested, resulting in too low luminosities if external targets would be used. Likewise, due to their production mechanism, a large momentum spread and large emittance are inherent to the secondary ion beams which would deteriorate a measurement of the target-recoil momenta and kinetic energies if not counteracted. These problems can be overcome using stored and cooled secondary beams of unstable nuclei interacting with thin internal gas-jet targets. This method provides

- high luminosities due to the continuous beam accumulation and beam recirculation;
- high resolution detection of low-energy recoil particles due to beam cooling, and thin targets;
- low-background conditions due to pure, windowless H, He, etc. targets;
- and the possibility to reduce beam components populating isomeric states.

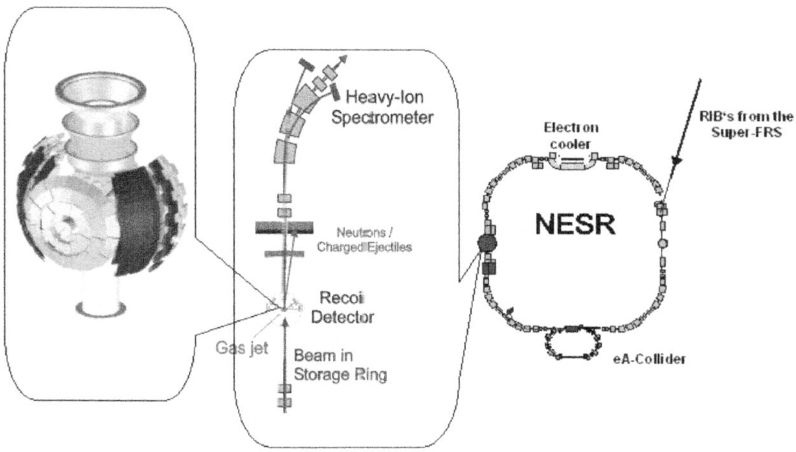

Figure 5. Layout of the EXL setup at the NESR storage ring. The insert on the left displays a view of the target recoil detector surrounding the internal target (for details see text).

For studies with the EXL apparatus, the FAIR storage rings are operated in the following scheme: Secondary beams of unstable nuclei are produced by fragmentation or fission reactions, are separated in the Super-FRS fragment separator, and then accumulated in the Collector Ring (CR). Finally the beams are transferred into the NESR ring where the measurement is performed. Within the EXL Technical Proposal, the design of a complex detection setup was investigated with the aim to provide a highly efficient, high-resolution universal detection system, applicable to a wide class of reactions (see figure 5). The apparatus is foreseen being installed at the internal target of the NESR storage cooler ring. Since a fully exclusive measurement is envisaged, the detection system includes

- a Si-strip and Si(Li) detector array for recoiling target-like reaction products; completed by slow-neutron detectors, and a scintillator array of high granularity for γ-rays and for the total-energy measurement of more energetic target recoils;
- detectors in forward direction for fast ejectiles from the excited projectiles, i.e., for neutrons and light charged particles;

- and heavy-ion detectors for the detection of beam-like reaction products.

All detector components will practically cover the full relevant solid angle and have detection efficiency close to unity. Very recently a feasibility study for the EXL project was performed at the present storage ring ESR at GSI, where prototype detectors of all major components of the final EXL detector setup were already subject to performance and background test, under realistic storage-ring operation conditions with stored stable beams. The results obtained [17,18] confirm the feasibility of the EXL setup, and are therefore considered as an important milestone towards the realisation of the EXL project.

References

1. B. Jonson, *Phys. Rep.* **389**, 1 (2004).
2. W.F. Henning, *Nucl. Phys. A* **734**, 654 (2004).
3. H.H. Gutbrod, *Nucl. Phys.* A **752**, 457c (2005).
4. Nuclear Physics News (ISSN: 1050-6896) 16 No. 1, 5-14 (2006).
5. Baseline Technical Report, Vol. 1 and 4, July 2006, http://www.gsi.de/fair/reports/btr.html.
6. http://ns.ph.liv.ac.uk/~mc/EXL/ns-instrum-exl.html.
7. G.D. Alkhazov et al., *Phys. Rep. C* **42**, 89 (1978).
8. P. Egelhof et al., *Prog. Part. Nucl. Phys.* **46**, 307 (2001).
9. G.D. Alkhazov et al., *Nucl. Phys. A* **712**, 269 (2002).
10. S.R. Neumaier et al., *Nucl. Phys. A* **712**, 247 (2002).
11. P. Egelhof et al., *Eur. Phys. J. A* **15**, 27 (2002).
12. A.V. Dobrovolsky et al., *Nucl. Phys. A* **766**, 1 (2006).
13. S. Ilieva, A. Inglessi et al., to be published.
14. R Sanchez et al., *Phys. Rev. Lett.* **96**, 033002 (2006).
15. T. Aumann, *Progr. Part. Nucl. Phys.* **59**, 3 (2007).
16. P. Egelhof et al., *Phys. Scripta* **T104**, 151 (2003).
17. S. Ilieva et al., Proc. of the 7th Int. Conf. on Radioactive Nuclear Beams, *Eur. Phys. J. ST* **150**, 357 (2007).
18. N. Kalantar-Nayestanaki et al., contribution to ISPUN07 Proceedings.

ISOTOPIC TEMPERATURES AND SYMMETRY ENERGY IN SPECTATOR FRAGMENTATION

M. DE NAPOLI[*,1)], C. SFIENTI[2)], S. BIANCHIN[2)], A. LE FÈVRE[2)], J. ŁUKASIK[2,8)], W. TRAUTMANN[2)], P. ADRICH[2)], T. AUMANN[2)], C.O. BACRI[3)], T. BARCZYK[4)], R. BASSINI[5)], , C. BOIANO[5)], A.S. BOTVINA[2)], A. BOUDARD[6)], J. BRZYCHCZYK[4)], A. CHBIHI[1)], J. CIBOR[8)], B. CZECH[8)], J.-E. DUCRET[6)], H. EMLING[2)], J. FRANKLAND[7)], M. HELLSTRÖM[2)], D. HENZLOVA[1)], K. KEZZAR[2)], G. IMMÉ[1)], I. IORI[5)], H. JOHANSSON[2)], A. LAFRIAKH[3)], , E. LE GENTIL[6)], Y. LEIFELS[2)], W.G. LYNCH[9)], J. LÜHNING[2)], , U. LYNEN[2)], Z. MAJKA[4)], M. MOCKO[9)], W.F.J. MÜLLER[2)], A. MYKULYAK[10)], H. ORTH[2)], A N. OTTE[2)], R. PALIT[2)], P. PAWŁOWSKI[8)], A. PULLIA[5)], G. RACITI[1)], E. RAPISARDA[1)], H. SANN[2)], C. SCHWARZ[2)], H. SIMON[2)], K. SÜMMERER[2)], M.B. TSANG[9)], C. VOLANT[6)], M. WALLACE[9)], H. WEICK[2)], J. WIECHULA[2)]. A. WIELOCH[4)] and B. ZWIEGLINSKI[10)]

[1)] *Dipartimento di Fisica, Universitá degli studi di Catania and LNS-INFN, 64, Via S. Sofia, I-95123 Catania, Italy*

[2)] *Gesellschaft für Schwerionenforchung, D-64291 Darmstadt, Germany*

[3)] *Institut de Physique Nucléaire, IN2P3-CNRS et Université, F-91406 Orsay, France*

[4)] *M. Smoluchowski Institute of Physics, Jagiellonian Univ., Pl-30059 Kraków, Poland*

[5)] *Istituto di Scienze Fisiche, Università degli Studi and INFN, I-20133 Milano, Italy*

[6)] *DAPNIA/SPhN CEA/Saclay, F-91191 Gif-sur-Yvette, France*

[7)] *GANIL, CEA et IN2P3-CNRS, F-14076 Caen, France;*

[8)] *H. Niewodniczański Institute of Nuclear Physics, Pl-31342 Kraków,Poland*

[9)] *Department of Physics and Astronomy and NSCL, MSU, East Lansing, MI 48824, USA;*

[10)] *A. Sołtan Institute for Nuclear Studies, Pl-00681 Warsaw, Poland*

* *E-mail. denapoli@lns.infn.it*

The influence of the isotopic composition in the breakup of projectile spectators at relativistic energies has been studied at the GSI laboratory with the ALADiN spectrometer. Besides a primary beam of ^{124}Sn, also secondary beams of ^{124}La and ^{107}Sn produced at the FRS fragment separator have been used in order to extend the range of isotopic compositions. The breakup temperatures obtained from the double ratios of isotopic yields have been extracted and compared with the limiting-temperature expectation. Moreover, from isoscaling analysis an apparent symmetry-energy term of the EOS has been extracted.

Keywords: Fragmentation; Isotopic temperatures; Isoscaling; Symmetry energy.

1. Introduction

Challenging motivations for isotopic studies in nuclear multifragmentation are derived from the importance of the density dependence of the symmetry-energy term of the nuclear equation of state for astrophysical applications and for effects linked to the manifestation of the nuclear liquid-gas phase transition. Different isotopic compositions are predicted for the coexisting liquid and gas phases, with the gas being more neutron rich than the liquid in asymmetric $N \neq Z$ matter.[1] This difference stems from the decrease in the symmetry energy in nuclear matter as the density is decreased. The expected magnitude of this density dependence, however, is model dependent and very poorly constrained by existing data.[2] If the phase transition for asymmetric nuclei still manifests itself as a plateau in the caloric curve[3] (i.e. the correlation between the temperature of the system and its excitation energy), this could be influenced by the degree of asymmetry as well as by the mass of the system undergoing the fragmentation. In particular it has been suggested[4] that for proton-rich systems the phenomenon of vanishing of the limiting temperature[5] should appear. On the other hand, SMM calculations predict nearly isospin-invariant temperatures for the coexistence region.[6] Therefore, in order to distinguish whether the breakup temperature is determined by the binding properties of the excited hot nuclear system or by the phase space accessible to it by fragmentation, a systematic investigation on projectile-spectator fragmentation has been undertaken at the ALADiN spectrometer at the GSI[7] (Fig. 1). Four different projectiles, ^{124}Sn, ^{197}Au, ^{124}La and ^{107}Sn, all with an incident energy of 600 AMeV on ^{116}Sn and ^{197}Au targets, have been studied. The two latter beams have been delivered by the FRagment Separator (FRS) of the GSI as products of the fragmentation of a primary ^{142}Nd beam at 890 AMeV on a ^{9}Be production target.

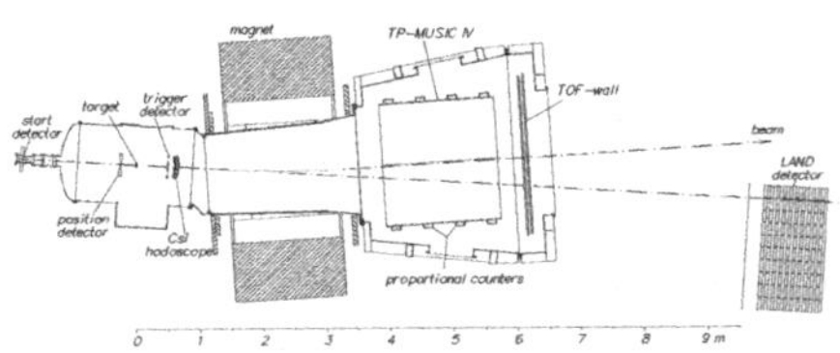

Fig. 1. Cross sectional view of the ALADIN setup.

The necessity of low beam intensities for the best operational condition of the ALADiN setup ($\cong$ 2000 particles/sec) and the possibility of using a thick target, in order to achieve high interaction rates, are indeed conditions compatible with radioactive-ion-beam experiments. Moreover, the inverse kinematics offers the possibility of a threshold-free detection of all heavy fragments and residues and thus gives a unique access to the breakup dynamics. The measurement of the charge and the momentum vector of all projectile fragments with $Z \geq 2$ has been performed, with high efficiency and high resolution, with the TP-MUSIC IV detector.[8] Using the reconstructed values for the rigidity and pathlength, the charge of the particle measured by the TP-MUSIC IV, and the time-of-flight given by the TOF-Wall, the velocity and the momentum vector can be calculated for each detected charged particle. The knowledge of velocity and momentum allows then the calculation of the particle's mass.

2. Isotopic effects in light-fragment production

The mass resolution obtained for projectile fragments entering into the acceptance of the ALADiN spectrometer is about 3% for fragments with Z=3 and decreases to 1.5% for Z≥6. Masses are thus individually resolved for fragments with atomic number Z≤10. The elements are resolved over the full range of atomic numbers up to the projectile Z with a resolution of ΔZ≤0.2 obtained with the TP-MUSIC IV detector.[7] The mean N/Z of the semi-peripheral isotope distributions of light-fragments in the range 2≤Z≤12 is presented in Fig. 2 The values obtained for

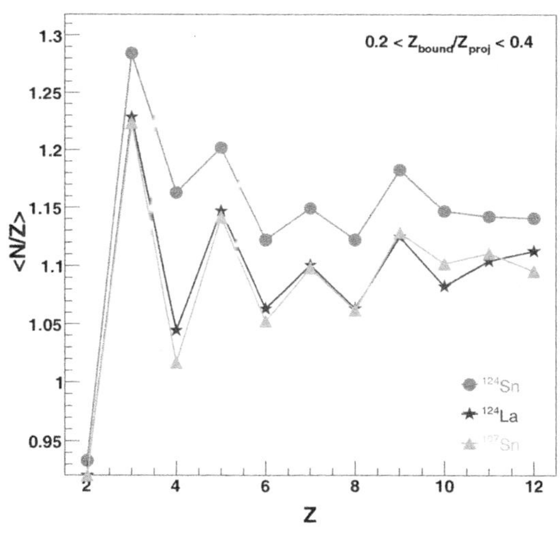

Fig. 2. Mean N/Z of the semi-peripheral light-fragments distribution.

^{124}Sn are larger than those of ^{124}La or ^{107}Sn as expected from the different N/Z of the original projectiles. Their odd-even staggering is more strongly pronounced for the neutron-poor ^{107}Sn and ^{124}La. Inclusive data obtained with the FRS fragment separator at GSI for ^{238}U and ^{56}Fe fragmentations on titanium targets at 1 AGeV bombarding energy confirm that the observed patterns are systematic,[9,10] exhibiting at the same time nuclear structure effects characteristic for the isotopes produced and significant memory effects of the isotopic composition of the excited system by which they are emitted. This has the consequence that, because of its strong variation with Z, the neutron-to-proton ratio <N>/Z is not a useful observable for studying nuclear matter properties. For this purpose, techniques, such as the isoscaling (Par. 4), will have to be used which cause the nuclear structure effects to cancel out.

3. Limiting temperature

The temperatures used in the caloric curve studies[3,5] were deduced from a double-ratio of isotopic yields. Under the assumptions of low density and chemical equilibrium a grand-canonical treatment[11] yields for the double ratio:

$$\frac{Y_{(A_1,Z_1)}/Y_{(A_2,Z_2)}}{Y_{(A_3,Z_3)}/Y_{(A_4,Z_4)}} = a\ exp(\frac{\Delta B}{T}). \tag{1}$$

with ΔB being the double difference of the binding energies of the chosen isotopes and a a statistical factor containing spin and mass terms. Furthermore, for best sensitivity the binding energy difference ΔB should be comparable or larger than the temperatures to be measured.[12] The thermometry with the ^{3,4}He isotope pairs used in the nuclear caloric curve[3] benefits from the largest difference ΔB = 20.6 MeV of the binding energies of these two nuclei but it is not the only choice. There are also other combinations of isotopes which can be expected to provide the necessary sensitivity for the measurement of temperatures in the MeV range. From the measured ratios, the isotopic temperatures have been extracted and an overall agreement with the previous systematic[13] have been obtained for the T_{HeLi}, T_{BeLi}, T_{CLi}, and T_{CC} thermometers. In Fig. 3 in particular, the apparent T_{HeLi} and T_{BeLi} temperatures for the ^{124}Sn and ^{124}La, ^{107}Sn spectator systems are reported as a function of the normalized Z_{bound}. This latter quantity, being the sum of the atomic numbers Z_i of all projectile fragments with $Z_i>1$, represents a measure of the size of the spectator and, in a geometrical picture, is monotously correlated with the impact parameter, as well as with the excitation energy. By comparing the two systems with the same mass but different isospin-content, an upper limit of about 0.7±0.1 MeV can be observed in the T_{HeLi} thermometer whereas almost no difference (hardly 0.1 MeV in average) between the two systems is observed in the case of the T_{BeLi} thermometer.

The small difference observed in the case of T_{HeLi} could be attributed to the dependence of the ^{6}Li/^{7}Li ratio only on the isotopic composition of the system.[14] In the case of T_{BeLi} both isotopic ratios ^{7,9}Be and ^{6,8}Li exhibit a dependence on the N/Z of the original projectiles, which cancels out in the double ratio. The invariance of the isotopic temperature with the isotopic composition of the system (Fig. 3) is inconsistent with the limiting temperature predictions of 2 MeV for the ^{124}Sn and ^{124}La systems and seems to favor a statistical interpretation.[6]

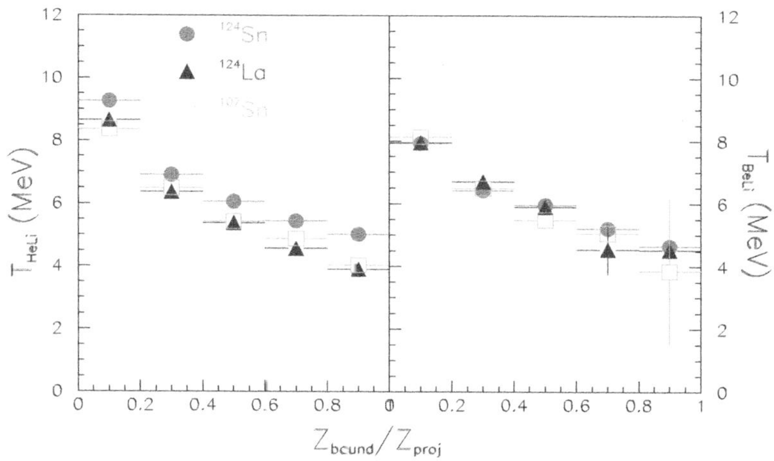

Fig. 3. Comparison between the isotopic temperatures T_{HeLi} and T_{BeLi} as a function of Z_{bound}/Z_{proj} for ^{124}Sn and ^{124}La, ^{107}Sn projectile spectators.

4. Isoscaling and Symmetry energy

Isotopic scaling, also called isoscaling, has been shown to be a phenomenon common to many different types of heavy-ion reactions.[15–18] It is observed by comparing products yields Y_i from reactions which differ only in the isotopic composition of the projectiles or targets or both. Isoscaling refers to an exponential dependence of the measured yield ratios $R_{21}(N,Z)$ on the neutron number N and the proton number Z of the detected products. The scaling expression:

$$R_{21}(N,Z) = Y_2(N,Z)/Y_1(N,Z) = C\ exp(N+Z) \tag{2}$$

describes rather well the measured ratios over a wide range of complex particles and light fragments.[18] For illustration, the yield ratios $R(^{124}\mathrm{Sn}/^{107}\mathrm{Sn})$ obtained for the fragmentation of the ^{124}Sn and ^{107}Sn projectiles in the $Z_{bound} = 20 \div 30$ interval, covering the region of maximum fragment production, are plotted in Fig. 4 as a function of the neutron (top panel) and proton (bottom panel) number. The lines are best fits to the data points with one common pair of values α and β for elements with $3 \leq Z \leq 10$. The slope parameter α decreases strongly with the centrality of the collision, extending from α=0.826 to α=0.320 for the most central event group.[19] In the grand-canonical approximation, assuming that the temperature T is about the same, the scaling parameters α and β are proportional to the difference of the neutron and proton chemical potentials for the two systems divided by the temperature, $\alpha = \mu_n/\mathrm{T}$ and $\beta = \mu_p/\mathrm{T}$. Of particular interest is their connection with the γ coefficient of the symmetry-energy term $E_{sym} = \gamma\ (\mathrm{A}-2\mathrm{Z})^2/\mathrm{A}$. It has been obtained from the statistical interpretation of isoscaling within the SMM[15] and expanding-emitting-source model[18] and confirmed by an analysis

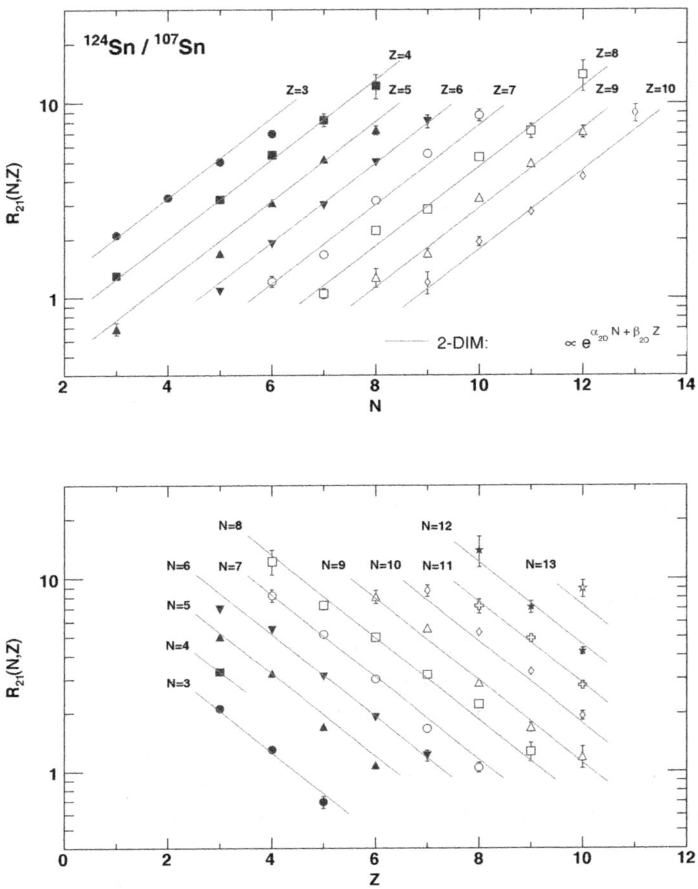

Fig. 4. Nuclear yield ratios for He to Ne isotopes from the fragmentation of ^{124}Sn and ^{107}Sn projectiles at 600 AMeV as a function of the neutron number N and proton number Z. The lines are the results of exponential fits.

of reaction dynamics.[20] The relation is

$$\alpha\, T = \Delta\mu_n = \mu_{n,2} - \mu_{n,1} \approx 4\,\gamma\,(\frac{Z_1^2}{A_1^2} - \frac{Z_2^2}{A_2^2}) = 4\,\gamma\,\Delta(Z^2/A^2) \tag{3}$$

where Z_1, A_1 and Z_2, A_2 are the charge and mass numbers of the neutron-poor and neutron-rich systems, respectively, at the breakup. The coefficient γ of the symmetry term can then be obtained from isoscaling knowing the temperature and the isotopic compositions.[21]

Starting from the Eq. 3, using the measured isotopic temperatures (Par. 3), the isoscaling α parameters and calculating the $\Delta(Z^2/A^2)$ from the isotopic composition of the incident projectiles, the γ term was extracted for different bins of Z_{bound}/Z_{proj} (Fig. 5). In the figure the current data are compared[22] with the one obtained in

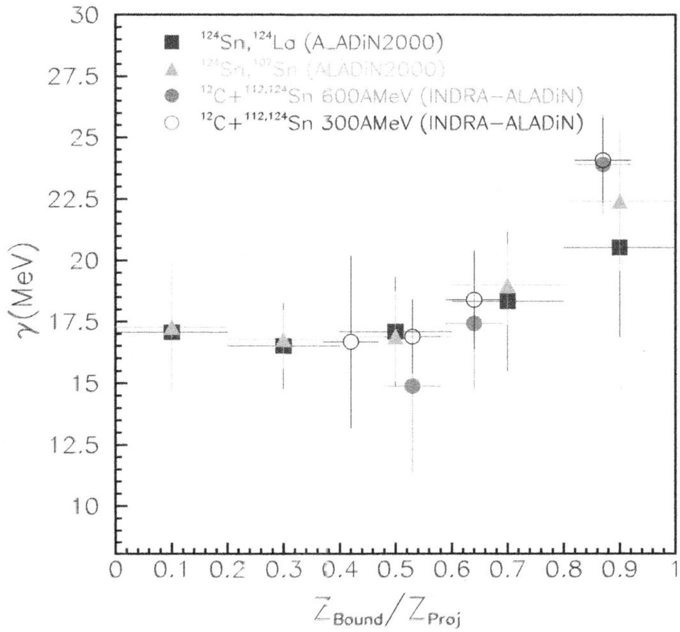

Fig. 5. γ factor obtained from the isoscaling coefficient α for the ^{107,124}Sn+Sn and ^{124}Sn,^{124}La,+Sn reactions at 600 AMeV as a function of the centrality of the collision. The present data are compared[22] with similar results from ref.[21]

ref[21] from the analysis of the target-spectator fragmentation at 600AMeV studied with the INDRA detector at the GSI. For very peripheral collisions (Fig. 5), the term reaches values near 25 MeV, that is its value at zero temperature and normal density. Going towards small Z_{bound}, the symmetry energy decreases saturating in a plateau when the gas phase is reached.

5. Conclusions

In the most recent ALADIN experiment isospin effects in the break-up of projectile spectators at relativistic energies have been studied. From the double ratios of Z≤4 isotopes, the isotopic temperatures have been determined. The nearly invariance of the isotopic temperatures with the isotopic composition in the entrance channel is inconsistent with the limiting temperature predictions. Clearly a more detailed analysis is needed to understand why the limiting temperature approximation, which nicely reproduces the dependence on the mass of the fragmenting system of the plateau of their caloric curves,[5] seems to fail when applied to the isospin degree of freedom. A better understanding of the relationship between the limiting temperature and the critical temperature for what concern the isospin degree of freedom and its interplay with finite-size effect and Coulomb interaction will be essential. Isoscaling has been observed for ^{107,124}Sn+Sn and ^{124}Sn,^{124}La,+Sn reactions at 600

AMeV. Of particular interest is the connection of the isoscaling parameters with the symmetry-term in the nuclear equation of state which has been consistently established with several methods. The coefficient γ is proportional to the isoscaling coefficient α, to the temperature of the systems and to its isotopic composition at the breakup. Assuming that the isotopic compositions are practically equal to those of the original systems, an apparent symmetry-term coefficient γ has been determined from the measurement of the isoscaling coefficient α and the isostopic temperature of the system. The results are found to be close to the normal-density cóefficient for peripheral collisions but drop to lower values at central impact parameters reaching a saturation in the gas phase. This saturation could reflect the fact that in the gas region, while the temperature still increases with the excitation energy, the density may reman constant. However, the present results depend, crucially, on the isotopic evolution of the multifragmenting system as it approaches the chemical freeze-out and, more qualitatively, on the treatment of sequential decay in the analysis. These questions deserve further attention.

References

1. H. Müller and B.D. Serot, *Phys. Rev. C* **52**, 2072 (1995).
2. for reviews see, e.g., C. Fuchs and H.H. Wolter, *Eur. Phys. J. A* **30**, 5 (2006).
3. J. Pochodzalla et al., *Phys. Rev. Lett.* **75**, 1040 (1995).
4. J. Besprovany and S. Levit, *Phys. Lett. B* **217**, 1 (1989).
5. J. Natowitz et al., *Phys. Rev. C* **65**, 034618 (2002).
6. A. S. Botvina et al., *Phys. Rev. C* **74**, 044609 (2006).
7. C. Sfienti et al., *Nucl. Phys. A* **749**, 83c (2005).
8. C. Sfienti et al., Proc. of the *XLI International Winter Meeting on Nuclear Physics*, Bormio, 26 Jan. 1 Feb. 2003, Ricerca Scientifica ed Educazione Permanente, Supp. N.120, p.323.
9. M. V. Ricciardi et al., *Nucl. Phys. A* **733**, 299 (2004).
10. P. Napolitani et al., *Phys. Rev. C* **70**, 054607 (2004).
11. S. Albergo et al., *Il Nuovo Cimento A* **89**, 1 (1985).
12. M.B. Tsang et al., *Phys. Rev. Lett* **78**, 3836 (1997).
13. W. Trautmann et al., to be published.
14. C. Sfienti et al., Proc. of the *XLV International Winter Meeting on Nuclear Physics*, Bormio, Jan. 14-21 2007, Ricerca Scientifica ed Educazione Permanente, Supp. N.124.
15. A.S. Botvina, O.V. Lozhkin and W. Trautmann, *Phys. Rev. C* **65**, 044610 (2002).
16. G.A. Souliotis et al., *Phys. Rev. C* **68**, 024605 (2003).
17. W.A Friedman, *Phys. Rev. C* **69**, 031601(R) (2004).
18. M.B. Tsang et al., *Phys. Rev. C* **64**, 054615 (2001).
19. S. Bianchin et al., Proc. of the *XLV Int. Winter Meeting on Nuclear Physics*, Bormio, Jan. 14-21 2007, Ricerca Scientifica ed Educazione Permanente, Supp. N.124.
20. A. Ono et al., *Phys. Rev. C* **68**, 051601(R) (2003).
21. A. Le Fèvre et al., *Phys. Rev. Lett.* **94**, 162701 (2005).
22. M. De Napoli, *Mass and Isospin Effects in Multifragmentation*, PhD Thesis (2005).

MULTI-CHANNEL ALGEBRAIC SCATTERING THEORY AND THE STRUCTURE OF EXOTIC COMPOUND NUCLEI

K. AMOS* and P. FRASER*

School of Physics, Melbourne University,
Victoria 3010, Australia
**E-mail: amos@physics.unimelb.edu.au; pfraser@physics.unimelb.edu.au*

D. VAN DER KNIJFF*

Advanced Computing Research, Information Division,
Melbourne University, Victoria 3010, Australia
**E-mail: dirk@unimelb.edu.au*

L. CANTON* and G. PISENT*

Istituto Nazionale di Fisica Nucleare, sezione di Padova,
e Dipartimento di Fisica dell'Università di Padova,
via Marzolo 8, Padova I-35131
**E-mail: luciano canton@pd.infn.it; gualtiero.pisent@pd.infn.it*

S. KARATAGLIDIS*

Department of Physics and Electronics, Rhodes University,
Grahamstown 6140, South Africa
**E-mail: S.Karataglidis@ru.ac.za*

J. P. SVENNE*

Department of Physics and Astronomy, University of Manitoba,
and Winnipeg Institute for Theoretical Physics,
Winnipeg, Manitoba, Canada R3T 2N2
**E-mail: svenne@physics.umanitoba.ca*

A Multi-Channel Algebraic Scattering (MCAS) theory is presented with which the properties of a compound nucleus are found from a coupled-channel problem. The method defines both the bound states and resonances of the compound nucleus, even if the compound nucleus is particle unstable. All resonances of the system are found no matter how weak and/or narrow. Spectra of mass-7 nuclei and of ^{15}F, and MCAS results for a radiative capture cross section are presented.

Keywords: Coupled-channel theory compound-nucleus spectra, capture cross sections

1. Introduction

The MCAS method[1,3] has been formed to solve coupled-channel Lippmann-Schwinger equations for a chosen two-body system. The theory is built upon sturmian expansions of whatever one chooses to be the matrix of interaction potential functions for the system. The set of sturmians that form the basis of the expansions are determined from the self-same interaction potentials and, if necessary, can be selected to ensure that the Pauli principle is not violated. The approach, which is most suited to deal with low energies, facilitates a systematic determination of all sub-threshold bound and resonance states of the compound nucleus within any energy range considered.

While the required starting matrix of potentials may be constructed from any model of nuclear structure, to date we have used just a simple collective rotational model to define those potentials with deformation taken to second order. Also, to date only low energy scattering from light mass nuclei have been considered so that just the ground, first, and second excited states of the target nucleus have been used to define the channel couplings.

Isospin symmetry in mirror scattering systems, save only for inclusion of the Coulomb interaction, has been used with MCAS to estimate the spectra of nuclei that are just outside of the proton drip line. ^{15}F is one such case.[4] The $A = 7$ system also has been considered and, with a unique nuclear potential, the level structures of ^{7}He, ^{7}Li, ^{7}Be, and ^{7}B have been obtained.[3] We have also used the scheme to consider low energy properties of ^{3}H-^{4}He and ^{3}He-^{4}He scattering to correlate different views of the spectra of ^{7}Li and ^{7}Be. Using the di-cluster structure of ^{7}Be found using MCAS, the astrophysical S-factor for the radiative capture of ^{3}He by ^{4}He has been evaluated.

2. The MCAS theory

In brief, the MCAS approach is based upon using sturmian functions as a basis set to expand the chosen interaction potentials. Each element in the interaction matrix then is a sum of separable interactions. The analytic properties of the S-matrix from a separable Schrödinger potential gives the means by which a full algebraic solution of the multichannel scattering problem can be realized. As all details of the MCAS theory have been published,[1] only salient features are repeated herein. Consider a coupled-channel system for each allowed scattering spin-parity J^π. With the MCAS method, one solves the Lippmann-Schwinger (LS) integral equations in momentum space, i.e.

$$T_{cc'}(p,q;E) = V_{cc'}(p,q) + \frac{2\mu}{\hbar^2}\left[\sum_{c''=1}^{\text{open}} \int_0^\infty V_{cc''}(p,x)\frac{x^2}{k_{c''}^2 - x^2 + i\epsilon} T_{c''c'}(x,q;E)\, dx \right.$$
$$\left. - \sum_{c''=1}^{\text{closed}} \int_0^\infty V_{cc''}(p,x)\frac{x^2}{h_{c''}^2 + x^2} T_{c''c'}(x,q;E)\, dx \right], \tag{1}$$

where the index c denotes the quantum numbers that identify each channel uniquely. Such requires specification of potential matrices $V_{cc'}^{(J^\pi)}(p,q)$. The open and closed channels have channel wave numbers k_c and h_c for $E > \epsilon_c$ and $E < \epsilon_c$ respectively, and μ is the reduced mass. Solutions of Eq. (1) are sought using expansions of the potential matrix elements in (finite) sums of energy-independent separable terms,

$$V_{cc'}(p,q) \sim \sum_{n=1}^{N} \chi_{cn}(p)\ \eta_n^{-1}\ \chi_{c'n}(q)\ . \tag{2}$$

The key to the method is the choice of the expansion form factors $\chi_{cn}(q)$. Optimal ones have been found from the sturmian functions that are determined from the actual (coordinate space) model interaction $V_{cc'}(r)$ initially chosen to describe the coupled-channel problem. The η_n are energy scales associated with the sturmian functions. Details are given elsewhere.[1]

Calculations of the multichannel T- and scattering (S-) matrices involve a Green's function matrix,

$$\begin{aligned}(G_0)_{nn'} = \frac{2\mu}{\hbar^2}\Bigg[&\sum_{c=1}^{\text{open}} \int_0^\infty \chi_{cn}(x)\frac{x^2}{k_c^2 - x^2 + i\epsilon}\chi_{cn'}(x)\ dx \\ &- \sum_{c=1}^{\text{closed}} \int_0^\infty \chi_{cn}(x)\frac{x^2}{h_c^2 + x^2}\chi_{cn'}(x)\ dx\Bigg]\,,\end{aligned} \tag{3}$$

The bound states of the compound system are defined by the zeros of matrix determinant for energy $E < 0$, namely the zeros of $\{|\boldsymbol{\eta} - \mathbf{G}_0|\}$ when all channels in Eq. (3) are closed. The $\boldsymbol{\eta}$ is a diagonal matrix with entries $(\eta)_{nn'} = \eta_n\ \delta_{nn'}$.

Elastic scattering observables are determined by the on-shell values ($k_1 = k_1' = k$) of the scattering matrices. For the elastic scattering of neutrons (spin $\frac{1}{2}$) from spin-zero targets $c = c' = 1$, and $S_{11} \equiv S_\ell^J = S_\ell^{(\pm)}$ are,[1]

$$S_{11} = 1 - i\pi\frac{2\mu k}{\hbar^2}\sum_{nn'=1}^{M} \chi_{1n}(k)\frac{1}{\sqrt{\eta_n}}\left[\left(\mathbf{1} - \boldsymbol{\eta}^{-\frac{1}{2}}\mathbf{G}_0\boldsymbol{\eta}^{-\frac{1}{2}}\right)^{-1}\right]_{nn'}\frac{1}{\sqrt{\eta_{n'}}}\chi_{1n'}(k). \tag{4}$$

Diagonalization of the complex-symmetric matrix,

$$\sum_{n'=1}^{N} \eta_n^{-\frac{1}{2}}\left[\mathbf{G}_0\right]_{nn'}\eta_{n'}^{-\frac{1}{2}}\tilde{Q}_{n'r} = \zeta_r\tilde{Q}_{nr}\,, \tag{5}$$

establishes the evolution of the complex eigenvalues ζ_r with respect to energy. Resonant behavior occurs when an eigenvalue, ζ_r, as a function of E, crosses the unit circle in the Gauss plane near the point (1,0). It is evident that the elastic channel S matrix has a pole structure at the corresponding energy where the modulus of one of these eigenvalues approach unity, since

$$\left[\left(\mathbf{1} - \boldsymbol{\eta}^{-\frac{1}{2}}\mathbf{G}_0\boldsymbol{\eta}^{-\frac{1}{2}}\right)^{-1}\right]_{nn'} = \sum_{r=1}^{N}\tilde{Q}_{nr}\frac{1}{1-\zeta_r}\tilde{Q}_{n'r}\,. \tag{6}$$

3. Results

3.1. *The spectrum of ^{15}F ($p+^{14}O$)*

Spectra, known and calculated (using MCAS[4]) for ^{15}C and ^{15}F, are shown in Fig. 1. The known properties of ^{15}C agree well with MCAS results for the $n+^{14}$C system.

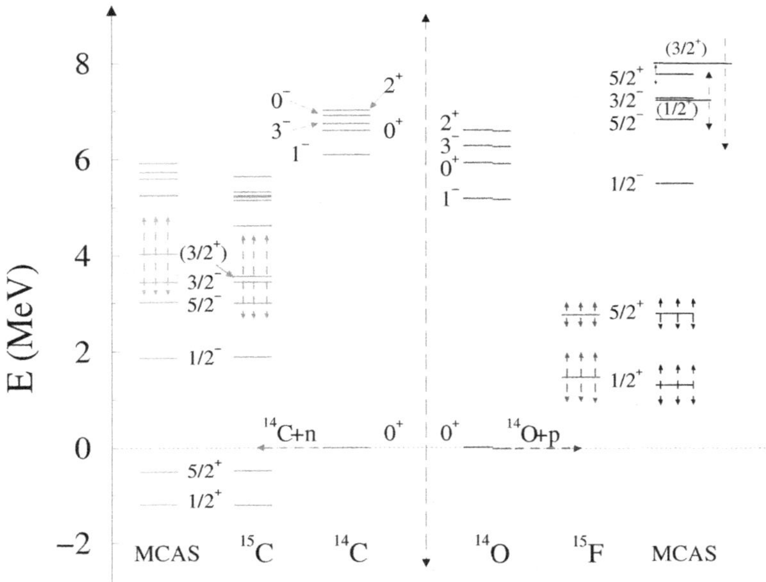

Fig. 1. Low energy spectra of ^{14,15}C and of ^{14,15}O, empirical[2] and from results of MCAS calculations. The zero of energy is set to that of the relevant mass-14 ground state.

Details have been specified fully in a recent paper.[4] The spectrum of ^{15}C has two bound states of spin-parities $\frac{1}{2}^+$ (ground) and $\frac{5}{2}^+$. They are described dominantly as a single *sd* shell neutron on the ^{14}C ground state. Then there are three quite narrow resonances, all having negative parity, which lie within the spread of a broad $\frac{3}{2}^+$ resonant state. That broad $\frac{3}{2}^+$ state was seen very clearly in the cross section from a measurement of ^{14}C(d,p). It is noteworthy that there are no other bound states; in particular none having negative parity. Such would occur if, in the $n+^{14}$C system, the $0p_{\frac{1}{2}}$ neutron orbit was not Pauli blocked. However, there are negative parity resonances, and to find them in our evaluations of ^{15}C, required that the neutron $0p_{\frac{1}{2}}$ orbit only be Pauli hindered.[4] With the nuclear interaction set, by including Coulomb interactions in MCAS evaluations, properties of the $p+^{14}$O system (giving particle unstable ^{15}F) were determined. Those results are shown on the right of Fig. 1.

The two lowest resonance states of ^{15}F have been observed in scattering cross-sections. Those data,[7,8] and our results, are shown in Fig. 2. In the top panel we show the cross sections from ^{14}O scattering from hydrogen. The MCAS result (solid

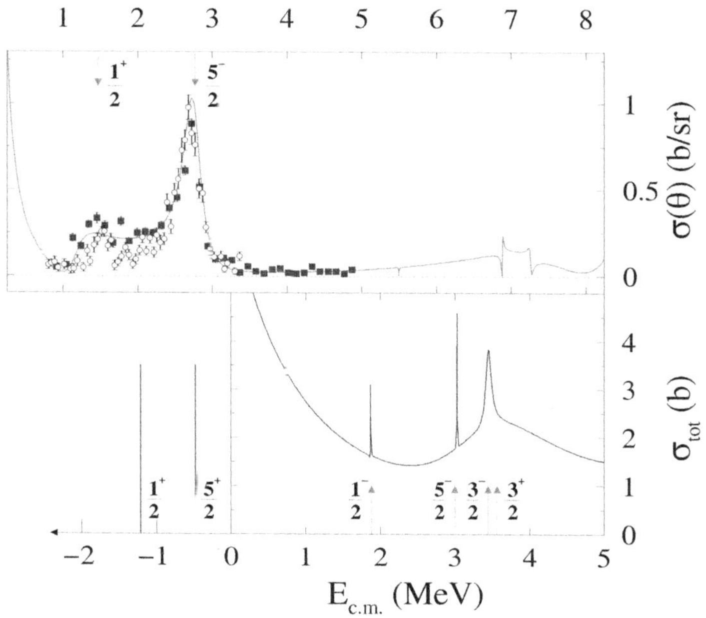

Fig. 2. The elastic cross sections from scattering of ^{14}O ions from hydrogen at $\theta_{c.m.} = 180°$ (top) and the predicted (MCAS) total cross section for n-^{14}C scattering (bottom).

curve) compares reasonably well with the data and it is as good as has been found with other analyses.[4] In the bottom panel of Fig. 2 is a prediction of the total scattering cross section of neutrons from ^{14}C for energies to 5 MeV. There are four obvious resonances; three quite narrow (of negative parity) and a very broad $\frac{3}{2}^+$ one. The three narrow resonances have partners in the p+^{14}O cross section. That is shown in the top panel, while the broad $\frac{3}{2}^+$ structure becomes more complex and overlaps with two other states. The zero of the energy scale has been placed to optimally match the $\frac{5}{2}^+$ bound state in ^{14}C to the centroid of the analogous resonance state in ^{15}F.

Analogues of the negative parity resonances in ^{15}C are predicted in ^{15}F. But the origin of these new, narrow negative-parity resonances in ^{15}F differ from those of the observed low-lying ones. They are compound resonances and are very affected by the Pauli-hindrance of the proton $0p_{\frac{1}{2}}$ orbit in the 0_2^+ and 2^+ excited states of ^{14}O.

3.2. *Spectra of ^{7}He, ^{7}Li, ^{7}Be, ^{7}B*

A single nuclear interaction matrix of potentials was found whose use with appropriate Coulomb terms in MCAS calculations gave spectra of these nuclei in good agreement with known ones. For ^{7}Li (p+^{6}He) and ^{7}Be (n+^{6}Be), the results are listed in Table 1. Therein the numbers in square brackets are the known experimen-

Table 1. Bound spectra of ^{7}Li and of ^{7}Be [a].

J^π	^{7}Li Exp.	p+^{6}He	^{3}H+^{4}He	^{7}Be Exp.	n+^{6}Be	^{3}He+^{4}He
$\frac{3}{2}^-$	−10.0	−10.0	−10.3	−10.7	−11.0	−10.1
$\frac{1}{2}^-$	−9.5	−9.5	−9.74	−10.2	−10.7	−9.88
$\frac{7}{2}^-$	−5.3 [0.06]	−5.3	−5.4 [0.08]	−6.1 [0.18]	−6.4	−6.0 [0.18]
$\frac{5}{2}^-$	−3.4 [0.92]	−3.4	−3.3 [0.83]	−4.0 [1.2]	−4.5	−4.0 [1.19]
$\frac{5}{2}^-$	−2.3 [0.08]	−0.3		−3.5 [0.44]	−1.6	
$\frac{3}{2}^-$	−1.2 [4.7]	−2.2		−−	−−	
$\frac{1}{2}^-$	−0.9 [2.8]	−0.9			−2.1	
$\frac{7}{2}^-$	−0.4 [0.44]	−0.4		−1.4 [?]	−1.7	
$\frac{3}{2}^-$	−−	−−		−0.8 [1.8]	−3.3	

Note: [a] All energies are in reference to nucleon + mass 6 nucleus thresholds

tal widths obtained from t+α (^{3}He +α) reactions. The nucleon emission thresholds lie higher and so no nucleon widths arise from the MCAS calculations. However, those evaluations produced 12 states to 15 MeV excitation in ^{7}Li with the lowest 9 matching known spin-parity states in the spectrum. The next three calculated levels are in the energy region in which two resonant states of undetermined spin-parity are known. The five lowest lying states in the known spectrum of ^{7}Be compare reasonably with the MCAS values. The calculations give more states than are known to date above an excitation energy of $\sim$ 8.5 MeV, and there are a few crossings.

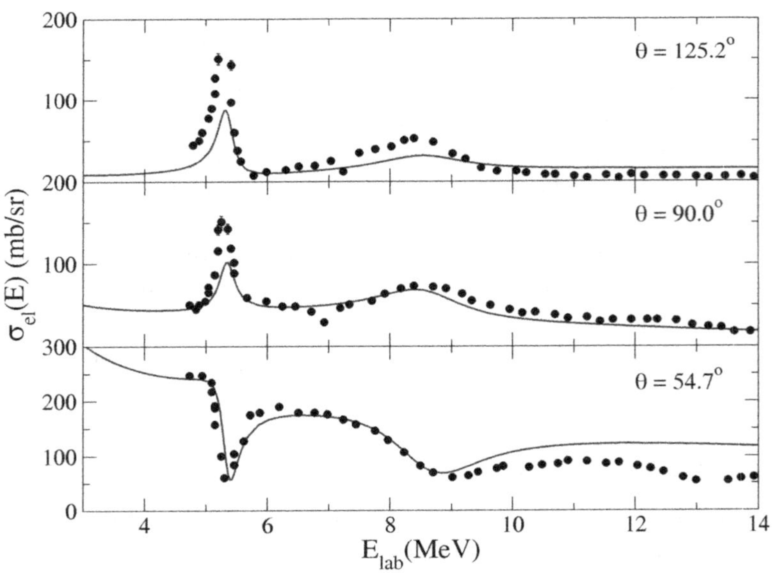

Fig. 3. Cross sections from α(^{3}He,^{3}He)α scattering.

To consider other open reaction channels, we have used MCAS to evaluate ^{3}H- and ^{3}He-α scattering using a potential model form of a cluster system. A nuclear interaction was found that gave the results listed in the fourth and seventh columns in Table 1. The calculated results agree well with the known values. Note that the energies are listed relative to the relevant nucleon+nucleus thresholds.

We used that interaction with MCAS to find elastic scattering cross sections at three center of mass scattering angles $(54.7°, 90.0°, 125.2°)$ at which data have been taken. The results of some of those calculations are shown by the solid curves in Fig. 3. The comparisons with data are very good, adding confirmation to our definition of the basic di-cluster-interaction potential.

3.3. *MCAS and capture cross sections*

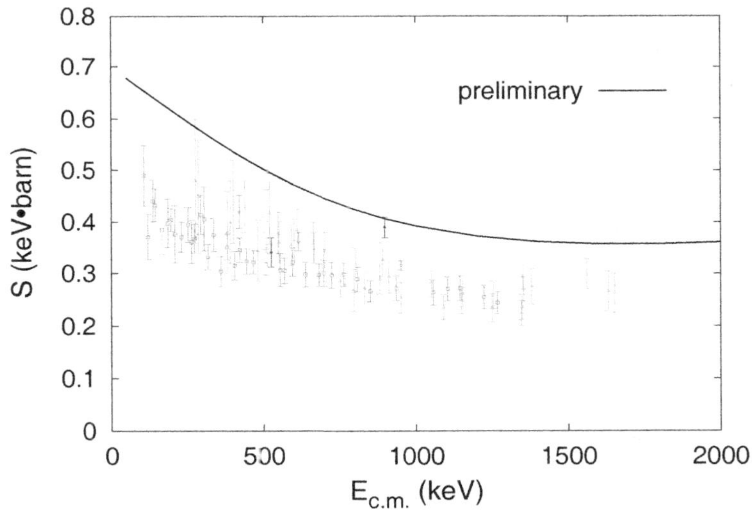

Fig. 4. The calculated astrophysical S-factor from ^{3}He-^{4}He capture compared with data.[6]

Recently, Canton, Levchuck, and Shebeko[5] used the MCAS di-cluster model wave function, though with an approximate scattering wave function (hence the results are preliminary), to evaluate the radiative capture cross section $\sigma_R(E)$ and from which the astrophysical S-factor is determined by

$$S(E) = \sigma_R(E)\, E \exp\{2\pi\eta(E)\}. \tag{7}$$

Therein $\eta(E)$ is the Sommerfeld parameter. The results of the calculated S-factor are compared in Fig. 4 with data.[5] The calculated S-factor overestimates the measured data but by only 20%. There has been no arbitrary scaling used, i.e. no asymptotic normalization or adjusted spectroscopic factor. Given that the bound state of ^{7}Be is not expected to be 100% a di-cluster, these results are most encouraging. Of note, both scattering and bound state functions can be given by MCAS so that there is no orthogonality problem as in past radiative capture calculations.

4. Conclusions

Applications of the MCAS theory have been made to discern the structure of the compound nuclei underlying various di-cluster nuclear systems of light mass. Despite the simplicity of the collective model potentials used, good agreement has been found with available data, even regarding a nucleus lying beyond the proton drip line. Use of MCAS information to evaluate radiative capture cross sections is a promising new development.

Acknowledgements

This research was supported by the Italian MIUR-PRIN Project "Fisica Teorica del Nucleo e dei Sistemi a Più Corpi", by the Natural Sciences and Engineering Research Council (NSERC), Canada, and by the National Research Foundation of South Africa.

References

1. K. Amos, L. Canton, G. Pisent, J.P. Svenne and D. van der Knijff, *Nucl. Phys.* **A728**, 65 (2003).
2. F. Ajzenberg-Selove, *Nucl. Phys.* **A523**, 1 (1991).
3. L. Canton, G. Pisent, K. Amos, S. Karataglidis, J. P. Svenne and D. van der Knijff, *Phys. Rev. C* **74**, 064605 (2006).
4. L. Canton, G. Pisent, J. P. Svenne, K. Amos and S. Karataglidis, *Phys. Rev. Lett.* **96**, 072502 (2006).
5. L. Canton, L. Levchuck and A. Shebeko, private communication.
6. D. Bemmerer *et al.*, *Phys. Rev. Lett.* **97**, 122502 (2006).
7. V.Z. Goldberg *et al.*, *Phys. Rev. C* **69**, 031302(R) (2004).
8. F.Q. Guo *et al.*, *Phys. Rev. C* **72**, 034312 (2005).

RESULTS FOR THE FIRST FEASIBILITY STUDY FOR THE EXL PROJECT AT THE EXPERIMENTAL STORAGE RING AT GSI

N. KALANTAR-NAYESTANAKI AND H. MOEINI

Kernfysisch Versneller Instituut (KVI), University of Groningen, Groningen The Netherlands
E-mail: nasser@kvi.nl

F. AKSOUH, K. BECKERT, P. BELLER, K. BORETZKY, A. CHATILLON, A. CORSI, P. EGELHOF, H. EMLING, G. ICKERT, S. ILIEVA, C. KOZHUHAROV, T. LE BLEIS, C. LE XUAN, Y. LITVINOV, K. MAHATA, J. P. MEIER, F. NOLDEN, U. POPP, H. SIMON, M. STECK, T. STÖHLKER, H. WEICK, A. ZALITE*

Gesellschaft für Schwerionenforschung Darmstadt, Germany

O KISELEV

Johannes Gutenberg Universität Mainz, Mainz, Germany

J. JOURDAN, D. ROHE, D. WERTHMÜLLER

Universität Basel, Basel, Switzerland

S. PASCHALIS

University of Liverpool, Liverpool, United Kingdom

For the EXL collaboration †

This contribution presents some results from the first feasibility measurement performed at GSI using a 350 MeV/nucleon ^{136}Xe beam and a Hydrogen gas-jet target. In this feasibility study, one element of every possible detection part of the future EXL detection system was investigated.

1. Introduction

Nuclear structure studies have received a boost in the last few years with the advent of radioactive ion beams. With these new beams, predictions of the models based on the knowledge obtained in experiments with stable isotopes will be put to a stringent test. Questions such as: how does the matter distribution look in neutron-rich nuclei near the neutron drip line can only be answered with the use

*Present address: Petersburg Nuclear Physics Institute, Gatchina, Russia
†http://ns.ph.liv.ac.uk/%7Emc/EXL/ns-instrum-exl.html

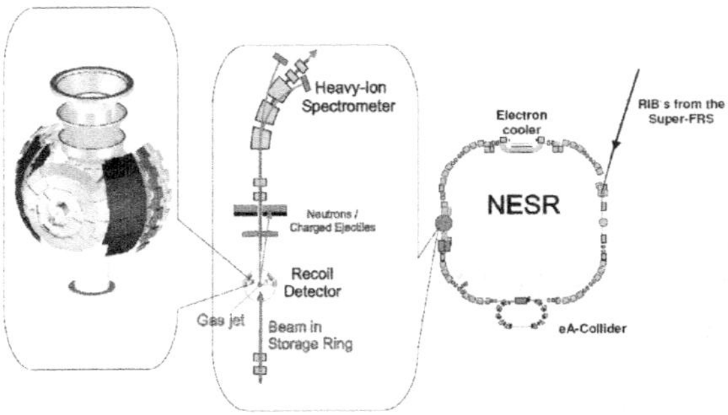

Fig. 1. Schematic drawing of the EXL setup at the New Experimental Storage Ring, NESR, and its central part.

of radioactive ions. One can study new phenomena such as the neutron halo and skins in these nuclei.[1] Many other aspects of nuclear structure such as the formation and the decay of giant resonances for these nuclei, deformation effects which could be different for protons and neutrons and their dependence on the N/Z ratio, in-medium interaction between protons and neutrons in very asymmetric nuclear matter, etc. can be studied with the help of radioactive ions.

In many of the studies, one should probe the periphery of the nucleus with light ions. For this, one would have to impinge a small momentum transfer to the system, namely the scattering angle of the light ejectile in the center of mass should be close to zero. In the experiments with the radioactive heavy ions, it is generally the beam which contains the radioactive ions leading to the so-called "inverse kinematics". In this kinematics, small momentum transfer corresponds to the light ion moving to large laboratory scattering angles with very small energies. This, in turn, leads to the situation in which one has to use very thin targets in order to allow the low-energy particle to leave the scattering area. The ring experiments are ideal for these type of measurements as one uses the same beam of ions several million times per second resulting in an acceptable luminosity.

For the future FAIR (Facility for Anti-proton and Ion Research) at GSI,[2] there is a plan to build a ring (New Experimental Storage Ring, NESR) in which the EXL (EXotic nuclei studied in Light-ion induced reactions) detector[3] will be placed. This detector should be capable of measuring the energy and the coordinates of light low-energy recoils, light high-energy ejectiles, heavy-ion ejectiles, and photons. In Fig. 1, a schematic drawing of the EXL detector setup and in particular its central part, namely Silicon strip detectors for the detection of low and high-energy light recoil ions is shown. This part consists of about 700 detection elements. In the present feasibility study, one Si-strip detector was mounted and tested. In addition, an element from each detection component was installed in the existing Experimental Storage Ring (ESR).[4] The experimental setup along with some results will be discussed here.

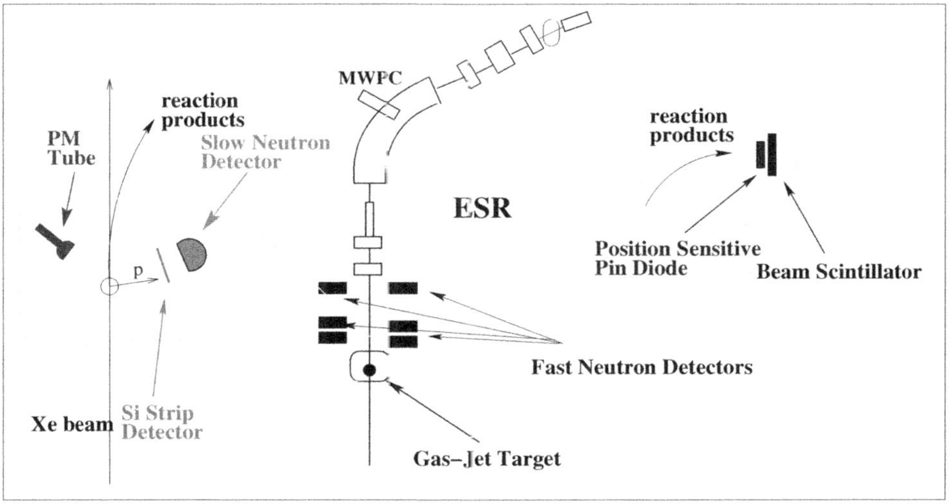

Fig. 2. The part of the ESR ring containing the target and the detectors used in the present measurement. For details see text.

2. Experimental setup

In order to examine the response of various detector elements in the ring environment, a test measurement was set up in the existing Experimental Storage Ring (ESR) at GSI in Darmstadt, Germany. Part of the ring where the target and the detector elements are placed is schematically shown in Fig. 2. The target used for this measurement is a hydrogen gas jet. A 4×4 cm^2 Silicon-strip detector of one millimeter thickness is placed in order to stop and measure the energy and the position of the slow recoil protons scattering to angles between 74° and 90° in the laboratory frame. This detector is placed in the ultra high vacuum of the ring and is also used to monitor and measure the relative and absolute luminosity through comparisons with the predicted elastic proton-scattering cross sections. Behind this detector and outside the vacuum chamber, a neutron detector (made out of liquid scintillation material) is placed to detect the low-energy neutrons. On the left side of the target, a photo-tube is placed to monitor the luminosity via the detection of the ultra-violet light emerging from the collision of the heavy ions with hydrogen. Downstream of the target, there were a number of plastic scintillators along with iron converters for the detection of fast neutrons and protons. Further downstream and around the first bend, a multi-wire proportional chamber is placed. This detector was also used as a luminosity monitor, detecting the Xe ions deflected out of the central orbit due to atomic charge-exchange reaction. Finally, as can be seen in Fig. 2, after the dipole and quadrupole magnets, a position-sensitive pin diode is placed. This detector is backed by another scintillator. The combination of these elements is used to look at projectile-like particles. The number of ^{136}Xe ions circulating in the ring was around 10^9. This, coupled to target density of around 2×10^{12}

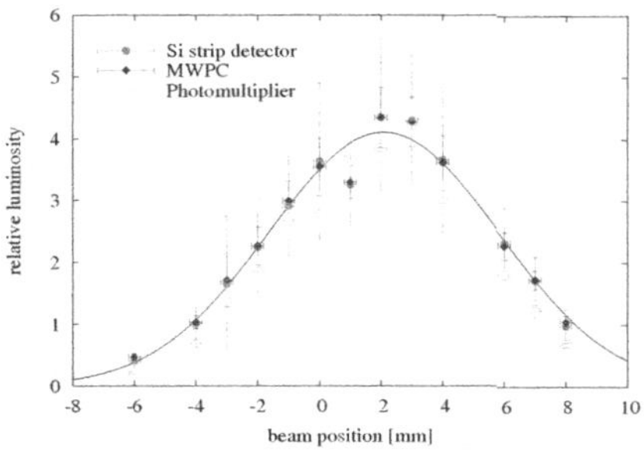

Fig. 3. The target profile as measured by three different detection systems.

atoms/cm^{-2} and a circulation frequency of about 2 MHz, results in a luminosity of about 4×10^{27} cm^{-2}s^{-1}. This luminosity is close to those planned for the future facility.

3. Selected results

In order to measure the size of the gas-jet target, the beam was moved across the target. This resulted in a measurement of the relative luminosity by the Si-strip detector, the MWPC and the photo-multiplier tube. As can be seen from Fig. 3, the results of the measurements by all three probes agree reasonably with each other. The size of the target obtained with this method is around 7.5 mm once one deconvolutes the beam size of around 5 mm.

The elastic proton-Xenon scattering can be measured by counting the number of low-energy protons at angles close to 90° in the laboratory frame which corresponds to an angle of 0° in the center of mass. The Silicon strip detector had 5 groups each with 8 strips. The first and the last strips of each group were read out; all strips of the same group were connected via resistors, representing resistive-chain read-out. The elastic scattering cross section, measured using the first group which is physically closest to 90°, are shown in Fig. 4. The data points which are here plotted as a function of the Mandelstam variable, $-t = 2mE$, are normalized to the theory at lower values of t. The normalization value used for this plot is in reasonable agreement with the nominal target thickness and the beam current cited above. This indicates that the nominal values are very close to the actual values. It is also clear that the measurements extend beyond the momentum transfer t if one assumes a point target. This is shown by the simulations in which the target size is changed from 0 to 7.5 mm. Note that the simulations, which use the theoretical predictions, are also normalized to the theoretical prediction at lower values of t. Up to a value of $t \approx 0.01$ (GeV/c)^2, the shape of the simulations agree reasonably well with the experimental results. However, beyond this value, there is an abrupt drop in the number of counts. This has to do with the operation of the Silicon strip detector in this experiment. The results of this measurement are consistent with predictions of the calculations including a matter radius close to known values in

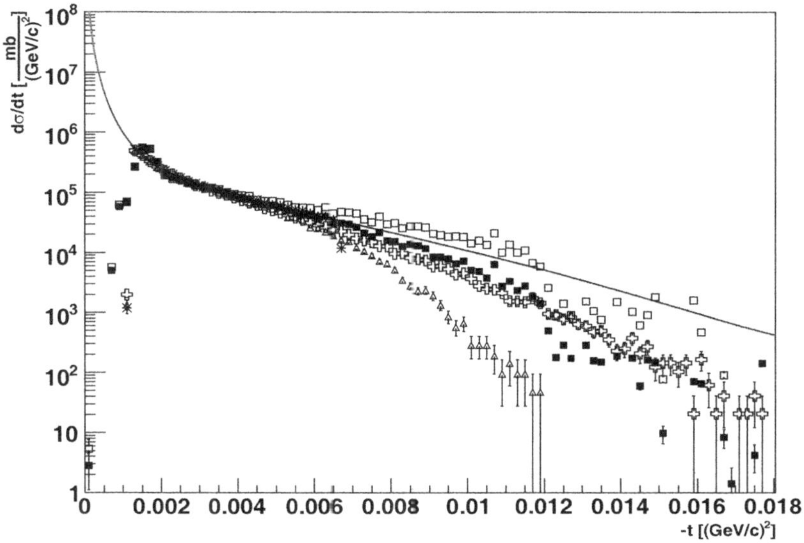

Fig. 4. The elastic proton-Xenon cross sections as a function of the Mandelstam variable, t. The solid squares are the experimental data and the line is the result of the theoretical prediction. The other symbols show the results of the simulations using the theoretical prediction with different FWHM for the target size (stars: point-like, triangles: 3 mm, and hollow crosses: 7.5 mm.) The hollow squares represent the data corrected for the target size of 7.5 mm. These can be compared with the theoretical prediction. All the histograms have been normalized to the theory at low values of t.

this region of stable nuclei. It is exactly this sort of measurements which will yield the value of the matter radius for the radioactive ions.

In the proton-Xenon interaction, there are many other channels which are present. For example, $p(^{136}\text{Xe}, pn)^{135}\text{Xe}$ and $p(^{136}\text{Xe}, 2pxn)^{132,133}\text{I}$ channels which are of physical interest have been identified in the present measurement. This is done with the help of coincidence measurement between the pin-diode which is used for the detection of the heavy ions and the detectors which are placed downstream of the target for the detection of fast nucleons (cf. Fig. 2). It is clear from Fig. 5, that without the coincidence measurement, the identification of these channels is not possible. The spectrum which is dominated by other channels becomes much cleaner once one requires that another (fast) proton or neutron has hit the small-angle detectors.

4. Summary

The first feasibility measurement for the EXL project with heavy ions in a ring impinging on a gas-jet target has successfully been performed. The results show that such experiments involving detectors in ultra high vacuum and close to the orbiting beam are feasible. The first results can already be compared to theoretical

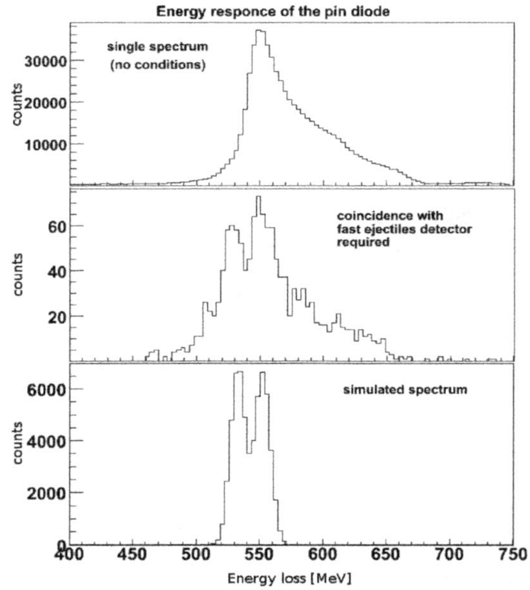

Fig. 5. The channel selection with the help of the coincidence measurements. The top panel shows the raw spectrum measured with the pin-diode detector. The middle panel shows the same spectrum once the coincidence condition is imposed. The bottom picture shows the results of the simulations which agree quite well with the data. The left and the right peaks correspond to I and Xe, respectively.

prediction. Improvements are being made in order to place more detectors in a new vacuum chamber which is being designed and built.

References

1. P. Egelhof et al., *Eur. Phys. J.* **A15**, 27 (2002)
2. http://www.gsi.de
3. http://www.gsi.de/fair/reports/btr.html (EXL technical report)
4. S. Ilieva et al., *Proceedings of the 7th international conference on Radioactive Nuclear Beams, Eur. Phys. J. ST* **150**, 357 (2007).

COULOMB EXCITATION OF ISOLDE NEUTRON-RICH BEAMS ALONG THE $Z = 28$ CHAIN

P. VAN DUPPEN

Instituut voor Kern- en Stralingsfysica
University of Leuven (K.U. Leuven), B-3001 Leuven Belgium
E-mail: piet.vanduppen@fys.kuleuven.be

Results from the recently commissioned REX-ISOLDE (CERN) post-accelerator facility are reported. Coulomb excitation with purified beams of neutron-rich zinc isotopes (including $N = 50$ ^{80}Zn) and with isomeric beams of copper isotopes were performed using the MINIBALL germanium array. The data are compared to large scale shell-model calculations and provide information on the fragility of the $N = 40$ sub-shell closure, stability of the $N = 50$ shell closure and the onset of deformation in this region of the nuclear chart. The specific nuclear structure around the $Z = 28$, $N = 40$ shells make Coulomb excitation an experimental tool for induced depopulation of a nuclear isomer. The presentation is concluded with a brief overview of the REX-ISOLDE physics program and with an outlook towards the intensity and energy upgrade of the ISOLDE complex (so-called HIE-ISOLDE).

Manuscript not received.

THE GAMMA DECAY OF THE PYGMY RESONANCE FAR FROM STABILITY AND THE GDR AT FINITE TEMPERATURE

G. BENZONI, A. BRACCO, F. CAMERA, A. CORSI, S. LEONI, B. MILLION AND O. WIELAND

INFN Sez. di Milano and Dipartimento di Fisica Università di Milano
via Celoria 16, 20133 Milano, Italy
E-mail: Bracco@mi.infn.it

The study of the gamma decay of the Giant Dipole Resonance is a powerful tool to investigate nuclear structure at extreme conditions of temperature, angular momentum and isospin. Here we present and discuss the results of two different measurements. The first experiment focuses on the search for pygmy strength in the neutron rich ^{68}Ni nucleus using Coulomb excitation of the radioactive ^{68}Ni beam at 600 MeV/u. Evidence for strength exceeding the standard tail of the GDR is found and results are in rather good agreement with theory. The second measurement is focusing on the study of the dynamic dipole which is formed in heavy ions collisions before equilibration of the compound nucleus. In this experiment the gamma decay from the system ^{16}O+^{116}Sn at 8 and 15 MeV/u has been measured together with the recoiling nuclei and the emitted charged particles. The experimental findings are rather well reproduced by the Bremsstrahlung calculations obtained within the reaction dynamic model based on the BNV approach.

1. Introduction

The measurement of the gamma decay of the Giant Dipole Resonance (GDR) allows for the investigation of different aspects of nuclei at extreme conditions, going from the nuclear response at external excitations to the properties of reaction dynamics of heavy ion collisions.

In the case of nuclei far from stability, the electric dipole (E1) response at energy around the particle separation energy is expected to have an enhanced strength as compared with the standard tail of the Lorentzian centered at $\approx$ 15 MeV. There is presently interest in studying this strength around the particle binding energy not only to test mean field theories but also because it affects reaction rates in astrophysical scenarios. The accumulation of E1 strength around the particle separation energy is commonly denoted as "pygmy" dipole resonance due to the smaller size of this strength in comparison with the "giant" dipole resonance dominating the E1 response and exhausting the Thomas-Reiche-Kuhn oscillator sum rule. The question of the evolution of the E1 strength in neutron nuclei far from stability is presently one of the most interesting topics since it provides information on the properties of neutron skins and consequently on the symmetry term relevant for the study of neutron stars. So far data, obtained using neutron break-up reactions, are only available for neutron rich O and Sn isotopes [1,2]. The unstable ^{68}Ni nucleus represents a good case to search for pygmy structures being this nucleus located in the middle of the long isotopic Ni chain having at the extremes the doubly magic ^{48}Ni and ^{78}Ni, while being experimentally accessible with the present radioactive beam facilities. Preliminary experimental results on ^{68}Ni are presented in section 2.

The second topic addressed in this proceeding is the gamma decay in heavy-ion reactions where the projectile and the target have different charge asymmetries. Gamma rays are emitted from a dynamical electric dipole which is formed in fusion reactions during the charge equilibration process in addition to the statistical decay from the thermally equilibrated compound nucleus. The origin of the dynamic dipole is related to the fact that, in dissipative collisions, energy and angular momentum are quickly distributed among all single particle degrees of freedom, while charge equilibration takes place on larger time scales. Consequently for charge asymmetric entrance channels one expects pre-equilibrium photon emission from the dipole oscillation in the isospin transfer dynamics at the time of CN formation. A good way to test the models describing the pre-equilibrium emission of gamma rays as due to the dynamic dipole is to measure reactions characterized not only by a dipole moment in the entrance channel but also by the presence of sizable pre-equilibrium emission, the latter evidenced by charged particle emission. In section 3 this problem is discussed in connection with the measurement of the reaction ^{16}O + ^{116}Sn (characterized by $(N/Z)_{proj} = 1$ and $(N/Z)_{targ} = 1.32$) at bombarding energies of 130 and 250 MeV.

2. Search for the pygmy resonance in ^{68}Ni

The search for the pygmy dipole resonance presented in this paper is performed by measuring the gamma decay following Coulomb excitation of a ^{68}Ni beam at 600 MeV/u. At this energy the dipole excitation dominates over other excitation modes [3]. This is the first experiment to measure the dipole response through its gamma decay at such high energies.

The radioactive ^{68}Ni beam was produced by fragmentation of the primary ^{86}Kr beam delivered, at 900 MeV/u, by the SIS synchrotron at GSI, with an intensity of $\approx 10^{10}$ particles per spill. The spill was approximately 6 seconds long with a period of 10 seconds. The primary beam was fragmented on a ^{9}Be target with a thickness of 4 g/cm^2. The ^{68}Ni ions were selected using the FRagment Separator (FRS) which was set in order to select and transport a cocktail beam of which ^{68}Ni nuclei accounted approximately for 30%, as shown in figure 1.

The transported ions were made to react on a secondary Au target. The particle identification after the target was performed by the CATE calorimeter [4,5] which, in the present experiment, consisted of nine position sensitive Si detectors coupled to four 6-cm thick CsI scintillators, arranged to equally share the intensity of the incident beam. A total of approximately $3 \cdot 10^7$ ^{68}Ni events were collected. The gamma-ray emission from Coulomb-excited ^{68}Ni was measured using the RISING array [5] located at the final focus of the FRS spectrometer. The gamma detection array consisted of 15 HPGe cluster detectors of the Euroball array [6], each formed by 7 germanium crystals, located in rings of 16°, 33° and 36°, of 7 HPGe segmented clusters from the Miniball array [7] (at 90°), and of 8 BaF_2 detectors from the HECTOR array [8] at 88° and at 142°. Only events with gamma multiplicity equal to one in which the ions were stopped in the CATE calorimeter were further analysed. The peak in the energy spectrum measured with CATE corresponding to ^{68}Ni has a mass resolution of approximately 1%, which is sufficient to discriminate between different masses.

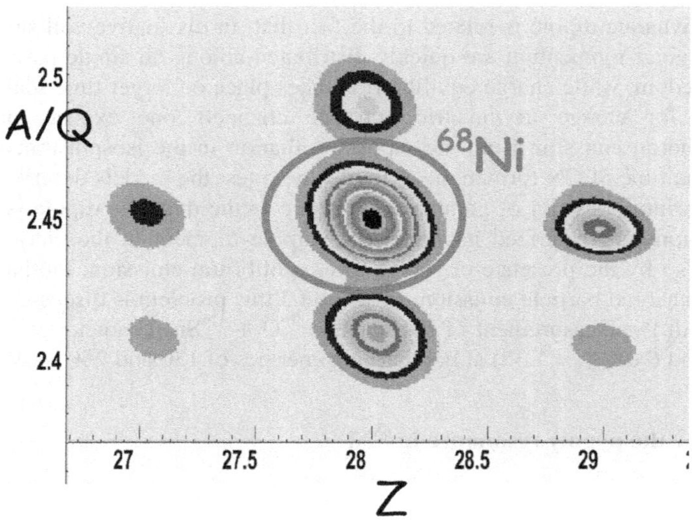

Figure 1. The figure shows the charge (horizontal axis) and the A/Q values (vertical axis) of the ions produced by fragmentation of ^{86}Kr at 900 MeV/u and transported through the FRS to finally imping on the secondary Au target.

In order to disentagle the contribution of the pygmy dipole resonance a series of conditions were applied to the gamma-ray energy spectrum. The main constraint is the detection of ^{68}Ni, before and after the interaction on the Au target. A condition on the time spectrum measured with the different types of detectors was also requested. The gamma-ray energy spectrum was constructed after applying a Doppler correction on an event-by-event base. The trajectory of the incident secondary beam produced by fragmentation are reconstructed by tracing the particles using the beam tracking detectors of the FRS system. A peak structure centered between 10-11 MeV is visible in the gamma-ray energy spectrum measured by BaF_2 detectors, as shown in figure 2. The width of this peak is much larger than the intrinsic energy resolution of the detector. A simulation was made for the detector response for gamma rays of 10.5 MeV using the GEANT libraries. We included in the simulation the experimental conditions which contribute to the broadening of lines corresponding to a given gamma-ray transition. The experimental broadening is dominated by the opening of the detectors and by the straggling in the target, the first giving rise to a distribution of energies due to the variation of incident angles ($\Delta\theta$) and the second a variation of the velocity and hence of the v/c ($\Delta\beta$). The result of the GEANT simulation is shown in figure 2 with a line superimposed to the measured peak. In the inset of the figure the peak structure is shown after subtracting a linear background. A peak structure at the same energy is observed also in the gamma-ray energy spectrum measured with the HPGe detectors of RISING at forward angles. It should be noted that model predictions for the E1 strength for this nucleus also show an enhanced strength in this energy region. This is true for both predictions obtained within the RPA approach [9] and the ones using the relativistic mean field theory [10].

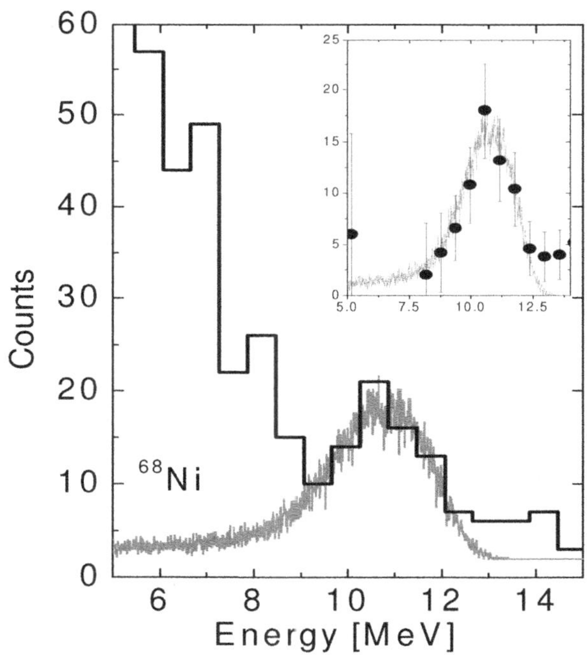

Figure 2. Energy spectrum measured with the BaF_2 detectors at ~90°, after Doppler correction. The line is a simultation of the detector response calculated with GEANT. The inset shows the peak after background subtraction.

3. Excitation of the dynamical dipole in the charge-asymmetric reaction $^{16}O + ^{116}Sn$

The experiment aiming at the investigation of the dynamical dipole formed in the initial phase of the collision dynamics in heavy ion reactions was performed at LNL using beams from the Tandem-Alpi accelerator complex. The experimental setup consisted of a combination of three detector systems: the first is the GARFIELD array [11] for the measurement of charged particles, the second is the HECTOR setup made of large volume BaF_2 detectors [8] for high-energy γ rays, and the third is a set of two Position Sensitive Parallel Plate Avalanche Counter telescopes (PSPPAC) to detect heavy ions.
The GARFIELD apparatus is a large acceptance system covering an angular range from 29^0 to 82^0. The GARFIELD drift chamber, filled with CF_4 gas is azimuthally divided into 24 sectors each with 8 ΔE-E telescopes. More specifically the ΔE signal is given by a drift chamber, while the E signal by CsI(Tl) scintillation detectors.

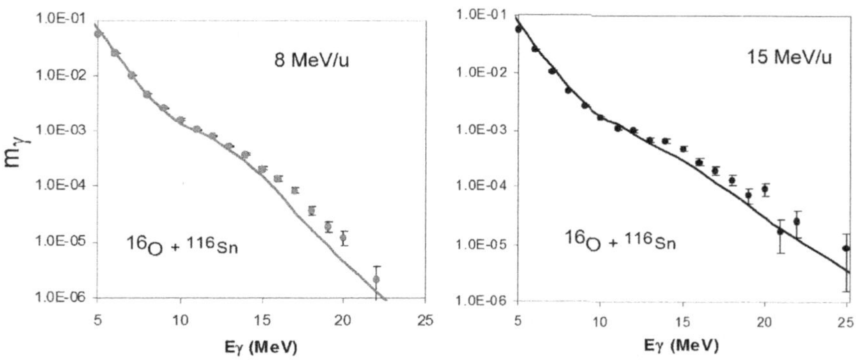

Figure 3. The energy spectra of gamma rays measured with BaF_2 detectors at two different bombarding energies. The measured spectra are compared with statistical model calculations (shown with the continuous lines) using the parameters described in the text.

The recoiling heavy nuclei produced in the reactions were detected using two pairs of Position Sensitive Parallel Plate Avalanche Counters (PSPPACs). Each pair consists of two PSPPACs positioned one behind the other at 150 cm from the target in order to cover the angular range from 4^0 to 12^0. The TOF measurement of the recoiling nuclei provided a precise discrimination between the fusion events and other reaction channels corresponding to reaction products with higher velocities.
The high-energy gamma rays were measured with the HECTOR system consisting of 8 large (14.5 x 17 cm) BaF_2 scintillators. The BaF_2 detectors were calibrated using the 15.1 MeV resonant state populated by the reaction

$d(^{11}B,n\gamma)^{12}C$ at beam energy of 19.1 MeV. In order to monitor the beam intensity we used a Faraday cup, located downstream approximately 2 meters from the target center. The measurement of the charged particles was important in this study to determine the presence of pre-equilibrium emission. In particular the ^{16}O-induced reaction has a sizable contribution from pre-equilibrium emission of alpha particles [12]. A detailed study of the alpha and the proton spectra is presented in reference [13].

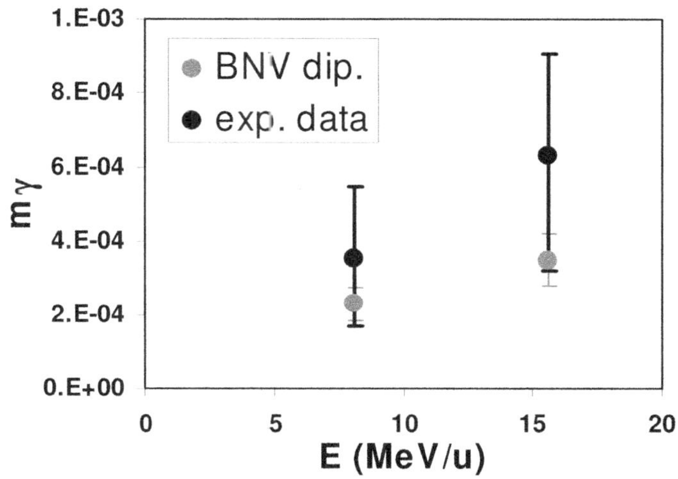

Figure 4. Multiplicity of the gamma rays corresponding to pre-equilibrium emission. This represents the excess yield as compared to the statistical model. The error bars of the data include the statistical error and the uncertainty related to the energy involved in the pre-equilibrium emission. This uncertainty is mainly due to the emission of neutrons, that is deduced from models and not directly measured. The predictions based on calculations within the collisional model using the BNV approach are also shown.

The multiplicities of the high-energy gamma rays measured in coincidence with evaporation residues are shown in figure 3. In the same figure the statistical model calculations made at the excitation energy $E_1 = E_{CN} - E_{PE}$ (where E_{CN} is the CN excitation energy obtained from kinematics and E_{PE} is the energy removed by pre-equilibrium particle emission) are shown. We used as parameters of the GDR the values of E_{GDR} and Γ_{GDR} deduced from the measurement of $^{132}Ce^*$ made with the same experimental array, produced by the $^{64}Ni+^{68}Zn$ reaction leading to a fully thermalized compound nucleus [12]. The response function of the detectors, calculated with the GEANT libraries was also folded to the calculations. One can note that, starting from a transition energy of 12 MeV, there is an excess yield as compared to the emission of the thermalized CN at excitation energy of 95 and 165 MeV.

The multiplicity of the measured excess yield, compared with the statistical model, is integrated from 8 to 20 MeV and shown in figure 4. Both points correspond to the emission from a dynamical dipole with moment D = 8.6 fm as obtained from the expression: $D = r_0(A_p^{1/3} + A_t^{1/3})/A * Z_pZ_t (N_t/Z_t - N_p/Z_p)$. The error bars of the data in figure 4 include the statistical error and the uncertainty related to the energy lost in the pre-equilibrium emission. This uncertainty is mainly related to the emission of neutrons, which is not measured but deduced from systematics (see [13]).
In the same figure predictions obtained within the BNV transport model are shown [14,15]. It should be noted that, in spite of the rather good agreement between experiment and theory, the calculated spectral distribution is peaking 3-4 MeV lower than the measured one.

4. Conclusion

In this contribution we have presented results of two measurements of the gamma decay from the giant dipole resonance.
The first is the search of the pygmy resonance in the exotic nucleus ^{68}Ni using the Coulomb excitation technique. The measured gamma ray spectrum shows a peak centered around 10-11 MeV, corresponding to an enhanced strength as compared to that of the tail of the standard GDR lorentzian function. Such excess is also predicted by calculations of two different types, one based on the RPA approach and the other obtained using relativistic mean field theories. This first evidence of the pygmy resonance in a measurement using the virtual photon scattering technique opens possibilities for more systematics studies in the future.
The second topic here discussed is that of the emission from a dynamic dipole in heavy ion reactions. The present study of the high-energy gamma rays produced in the asymmetric reaction ^{16}O + ^{116}Sn at two different bombarding energies has shown the existence of a pre-equilibrium gamma-ray emission due to a dynamic dipole moment, which is formed before full equilibration is reached. The intensity of the gamma-ray pre-equilibrium emission is rather well reproduced by a model based on the BNV model. If, on one hand, the excitation mechanism seems to be understood, the spectral distribution of the gamma rays is still debated: in fact, the average energy of the predicted spectra of the Bremsstrahlung radiation is always $\approx$ 3-4 MeV lower than the measured one.
Precise tests with stable beams are important before this problem can be attacked better in more asymmetric configurations as those that could be reached using the future generation of radioactive beams.

Acknowledgments

The authors acknowledge all the collaborators of the RISING experiment (see the author list of reference [16]) and of the GARFIELD experiment (see the author list of reference [12]) whose work has been essential for the success of the experiments. We thank the LNS-Catania theoretical group and M. Di Toro for providing the BNV-Code and fruitful discussions.

References

1. A. Leistenschneider et al., *Phys. Rev. Lett.* **86**, 5442 (2001) and E. Tryggestad et al., *Phys. Lett. B* **541**, 32 (2002).
2. P. Adrich et al., *Phys. Rev. Lett.* **95**, 132501 (2005).
3. T. Glashmacher, *Ann. Rev. Nucl. Part. Sci.* **48**, 1.(1998).
4. R. Lozeva et al., *Nucl. Instrum. Methods A* **562**, 298 (2006).
5. H.J. Wollersheim et al., *Nucl. Instrum. Methods A* **537**, 637 (2005).
6. G. de Angelis, A. Bracco and D. Curien, *Eur. News* **34**, 181 (2003) and J. Simpson, *Z. Phys. A* **358**, 139 (1997).
7. P. Reiter et al., *Nucl. Phys. A* **701**, 209c (2002).
8. M. Kmiecik et al., *Phys. Rev. C* **70**, 064317 (2004).
9. D. Sarchi et al., *Phys. Lett. B* **601**, 27 (2004) and G. Colo' private communications.
10. D. Vretner et al., *Nucl. Phys. A* **692**, 496 (2001) and J. Liang et al. *Phys. Rev. C* **75**, 054320 (2007).
11. F. Gramegna et al., *Nucl. Inst. Meth. A* **389**, 474 (1997).
12. O. Wieland et al., *Phys. Rev. Lett.* **97**, 012501(2006) .
13. S. Barlini et al., *submitted to Phys. Rev. C* (2007).
14. V. Baran et al., *Nucl. Phys. A* **679**, 373 (2001).
15. V. Baran et al., *Phys. Rev. Lett.* **87**, 182501 (2001).
16. A. Bracco et al., *Acta Physica Polonica B* **38**, 1229 (2007).

THERMAL PAIRING IN NUCLEI

NGUYEN DINH DANG

Nishina Center for Accelerator-Based Science, RIKEN
Hirosawa 2-1, Wako city, Saitama 351-0198, Japan
and
Institute for Nuclear Science & Technique,
P.O. Box 5T-160, Nghia Do, Hanoi, Vietnam.
E-mail: dang@riken.jp

The modified Hartree-Fock-Bogoliubov (MHFB) theory at finite temperature is derived for finite nuclei.[1] In the limit of constant pairing parameter, the MHFB theory yields the modified BCS (MBCS) theory.[2] These are the microscopic theories that can describe the crossover region at temperature T around the critical value T_c of the BCS superfluid-normal (SN) phase transition. By requiring the unitarity conservation of the particle-density matrix, the derivation of these theories is achieved by constructing a modified quasiparticle density matrix, where the fluctuation of the quasiparticle number is microscopically built in. This matrix can be directly obtained from the usual quasiparticle-density matrix by applying the secondary Bogoliubov transformation, which includes the quasiparticle occupation number.

The calculations of the thermal pairing gap, total energy, heat capacity, quasiparticle and pairing correlation functions were carried out within MBCS theory for the Richardson model[3] as well as realistic single-particle spectra. The Richardson model under consideration has varying Ω equidistant levels and N particles with a level distant equal to 1 MeV. It is shown that the limitation of the configuration space sets a limiting temperature T_M up to which the MBCS theory can be applied. Enlarging the space in the half-filled case ($\Omega = N$) by one valence level ($\Omega = N + 1$) extends T_M to a much higher temperature so that the predictions by the MBCS theory can be compared directly with the exact results up to $T \sim 4$ - 5 MeV even for small N. The MBCS gap does not collapse, but decreases monotonously with increasing T. The total energy and heat capacity predicted by the MBCS theory are closer to the exact results than those predicted by the BCS theory, especially in the region of the SN phase transition predicted within the BCS theory. The discontinuity in the BCS heat capacity at the critical temperature T_c is smoothed out within the MBCS theory, especially for small N, showing the disappearance of SN phase transition in very light systems. With increasing N the peak at T_c in the heat capacity becomes more pronounced, showing a phase-transition-like behavior in heavy systems. The effect of approximated particle-number projection using the Lipkin-Nogami method is also discussed.

An application of the MBCS theory to the description of the damping of giant dipole resonances (GDR) in hot nuclei shows that, because of the existence of the pseudo gap, the GDR width remains nearly constant at temperatures up to around 1 MeV in tin isotopes in good agreement with the recent experimental systematic.[4]

References

1. N.D. Dang and A. Arima, *Phys. Rev. C* **68**, 014318 (2003).

2. N.D. Dang and V. Zelevinsky, *Phys. Rev. C* **64**, 064319 (2001).
3. N.D. Dang, *Nucl. Phys. A* **784**, 147 (2007).
4. N.D. Dang and A. Arima, *Phys Rev. C* **68**, 044303 (2003).

Manuscript not received.

MOLECULAR-ORBITAL AND DI-NUCLEI STATES IN Ne AND F ISOTOPES

MASAAKI KIMURA

University of Tsukuba, Tsukuba 305-8571, Japan
E-mail: masaahi@nucl.ph.tsububa.ac.jp

In this talk, the molecular-orbital states and di-nuclei states in Ne and F isotopes are reported and discussed based on the theoretical study by the AMD+GCM.[1] The successful description of Be isotopes by assuming the molecular-orbital structure suggests a new type of clustering and the binding mechanism that is peculiar to $N \neq Z$ nuclei. Therefore, it is of importance to investigate whether the molecular-orbital structure and other new clustering aspects exist in heavier system. Ne and F isotopes are promising candidates, since the core nucleus ^{20}Ne and ^{19}F have the prominent cluster structure. In all of Ne and F isotopes, we have found various kinds of the cluster structure in the excited states. In the case of ^{22}Ne, the cluster structure is classified into two types. The first one is the "molecular-orbital bands" that have an α+^{16}O cluster core and two valence neutrons orbiting around the core. The $K^\pi = 0_2^+, 0_1^-$ and 1^- bands belong to this type and have one or two valence neutrons occupying σ molecular-orbital. Especially, the $K^\pi = 0_2^+$ and 0_1^- bands are regarded as parity doublet. These bands are built a few MeV below the α+^{18}O threshold energy and correspond to the observed states with large α reduced widths.[2] The second one is the "di-nuclei" bands that have the α+^{18}O cluster structure. The $K^\pi = 0_3^+, 0_2^-$ bands belong to this type. In these bands, two valence neutrons orbit only around the ^{16}O cluster. They are built a few MeV above the threshold energy and correspond to the observed α+^{18}O molecular resonances.[3] Thus, the α clustering systematic is much more complicated and rich in $N \neq Z$ nuclei. Among the obtained states, the molecular-orbital states of F isotopes show an interesting behaviour. As neutron number increases, the excitation energy of the molecular orbital band becomes drastically small. It is related to the breaking of the neutron magic number $N = 20$. When the α+^{15}N cluster core develops, the system is strongly deformed. This large deformation lowers the energy of the σ molecular-orbital and hence the energy of the molecular-orbital states. One must note that the σ molecular orbital has a large overlap with the Nilson [3,0,0,1/2] orbital that intrudes into the *sd*-orbital in the neutron rich $N = 20$ isotopes. These states can be corresponds to the observed low-lying states in ^{25,27}F.[4]

References

1. M. Kimura, *Phys. Rev. C* **75**, 034312 (2007).
2. W. Scholz et al., *Phys. Rev. C* **6**, 893 (1972).
3. V. Z. Goldberg et al., *Phys. Rev. C* **69**, 024602 (2004).
4. Z. Elekes, et al., *Phys. Lett. B* **599**, 17 (2004).

Manuscript not received.

LOW-MOMENTUM INTERACTIONS FOR NUCLEI

ACHIM SCHWENK

TRIUMF, 4004 Wesbrook Mall, Vancouver, BC, Canada, V6T 2A
E-mail: schwenk@triumf.ca
Research supported by NSERC and the National Research Council of Canada

Nuclear interactions with variable momentum cutoffs, known generically as "$V_{\text{low }k}$," show great promise for few- and many-body calculations.[1–3] Changing the cutoff leaves observables unchanged by construction, but shifts contributions between the potential and the sums over intermediate states in loop integrals. These shifts can weaken or largely eliminate sources of nonperturbative behavior such as strong short-range repulsion and the tensor force. An additional advantage is that the corresponding three-nucleon (3N) interactions become perturbative at lower cutoffs and are thus tractable.[2] As a result, few- and many-body calculations can be greatly simplified or converge more rapidly by lowering the cutoff, and we are able to perform the first calculations of intermediate-mass nuclei with 3N interactions. We find improved convergence for nuclei with low-momentum interactions with smooth cutoffs[4] (which have technical advantages in oscillator spaces) and when chiral EFT interactions are evolved to low momentum. Finally, the cutoff variation can provide estimates for theoretical uncertainties, which is important for the extrapolation to the extremes of isospin, density and temperature.

There are a wide range of on-going developments in few- and many-body systems, where we will highlight 3N interactions as a current frontier in nuclear physics. Recent studies of the coupled-cluster method with $V_{\text{low }k}$ show direct convergence and that low-momentum 3N interactions are accurately treated as effective zero-, one- and two-body terms.[5] This will enable us to investigate the contributions of 3N interactions to valence-shell interactions. Studies of light nuclei with $V_{\text{low }k}$ in the no-core shell model approach are in progress. For many-body systems, the developments include improved calculations of neutron[6] and nuclear matter, a new power counting for nuclear matter, and density functional calculations based on low-momentum interactions.

References

1. S.K. Bogner, T.T.S. Kuo, A. Schwenk, *Phys. Rep.* **386**, 1 (2003).
2. A. Nogga, S.K. Bogner, A. Schwenk, *Phys. Rev. C* **70**, 061002(R) (2004).
3. S.K. Bogner, A. Schwenk, R.J. Furnstahl, A. Nogga, *Nucl. Phys. A* **763**, 59 (2005).
4. S.K. Bogner, R.J. Furnstahl, S. Ramanan, A. Schwenk, *Nucl. Phys. A* **784**, 79 (2007).
5. G. Hagen, T. Papenbrock, D.J. Dean, A. Schwenk, A. Nogga, M. Włoch, P. Piecuch, arXiv:0704.2854 [nucl-th].
6. B. Friman, K. Hebeler, A. Schwenk, L. Tolos, arXiv:0704.1404 [nucl-th].

Manuscript not received.

NONRELATIVISTIC NUCLEAR ENERGY FUNCTIONALS INCLUDING THE TENSOR FORCE

GIANLUCA COLÒ*, PIER FRANCESCO BORTIGNON

Dipartimento di Fisica, Università degli Studi and INFN, Sezione di Milano, 20133 Milano, Italy
** E-mail: gianluca.colo@mi.infn.it*

HIROYUKI SAGAWA

Center for Mathematical Sciences, University of Aizu, Aizu-Wakamatsu, Fukushima 965-8560, Japan

In the present volume, several contributions are devoted to the implementations of the Density Functional Theory (DFT) in atomic nuclei, either in the nonrelativistic or covariant formalism. By restricting ourselves to the nonrelativistic case, in this work we deal with some recent improvement of the theory. We argue that tensor correlations are important to be included in the functionals; in particular, we study the role of the tensor terms of the Skyrme effective interaction on the spin-orbit splittings in the N=82 isotones and Z=50 isotopes. We conclude that the experimental isospin dependence of these splittings cannot be described by calculations employing the usual Skyrme parametrizations, but is very well accounted for when the tensor interaction is introduced. The necessity to include other kind of correlations, namely those associated with the particle-vibration coupling, is discussed by using the specific example of ^{132}Sn.

Keywords: Density Functional Theory; Hartree-Fock theory; tensor force; single-particle states; particle-vibration coupling

1. Introduction

The atomic nucleus is a strongly correlated many-body system. The so-called *ab-initio* methods, which start from the bare nucleon-nucleon (NN) interaction and generate highly correlated wavefunctions, seem to be, at present, difficult to be applied above mass number A≈12. For heavier nuclear systems, one is obliged to resort to alternative models, in which the complexity of the wavefunction is reduced. This is possible if an "effective" interaction is employed, which takes implicitly into account the effect of correlations. We define effective interaction every interaction which is meant to be used in connection with a given many-body approximation scheme. One of the most widely used approaches in nuclear physics is the so-called self-consistent mean-field (SCMF) method based on an effective interaction $V_{\rm eff}$.[1] Starting from the effective Hamiltonian $H_{\rm eff} = T + V_{\rm eff}$, where T is the kinetic energy, one calculates the total energy as the expectation value of this Hamiltonian over the most general Slater determinant $|\Phi\rangle$. If this approximation of the exact energy functional is a good one the model, although formally mean field, can effectively include a large class of many-body correlations.

In nuclear structure, both zero-range interactions like the one proposed by Skyrme at the end of the fifties,[2] and parametrized since the seventies by the Orsay group,[3] or finite-range interactions like the Gogny force,[4] are quite popular and lead to satisfactory descriptions of many nuclear properties. Nuclear masses can be obtained with errors which are below $\approx$1 MeV, and r.m.s. radii and deformed shapes can be described in an overall satisfactory way along the periodic table. In fact, this is an indication that the self-consistent mean field based on these interactions is, among existing approaches, the one which is closest to a realization of the Density Functional Theory (DFT) for nuclei.

Recently, within many papers appeared in the literature, attention has been paid to the importance of including the tensor correlations in the existing energy functionals, with the aim of explaining the behavior of the single-particle states in stable and exotic systems. In fact, the original interaction proposed by T.H.R. Skyrme[2] includes a tensor component. Afterwards, with a few exceptions,[5–7] there has not been a clear assessment of the role played by the tensor component of the Skyrme force: in particular, not so much has been said about how the single-particle states are affected by tensor correlations.

The new efforts in setting a relationship between the tensor correlations and the evolution of the shell structure of atomic nuclei, have been stimulated by the present (and future) availability of experimental data in exotic nuclei, thanks to the new radioactive beams facilities. Theoretical studies have elucidated the specific effects of the proton-neutron tensor force within the framework of the shell-model calculations. In Ref.[8] it has been pointed out that the tensor force provides a specific attraction (repulsion) between protons and neutrons lying in $j_>$ and $j_<$ ($j_>$ and $j_>$, or $j_<$ and $j_<$) orbitals (cf. also[9]). In the framework of mean-field calculations employing effective interactions, the issue has been addressed in Ref.[10] using the Gogny interaction and in Ref.[11] using the Skyrme interaction. In the latter work, it has been shown that the effects of the tensor force can be understood in a very transparent way by inspecting the Hartree-Fock (HF), or HF plus Bardeen-Cooper-Schrieffer (HF plus BCS), formulas. Recent works about the relevance of the tensor force when one considers the evolution of single-particle states have been carried out by other groups as well.[12–14]

The outline of the paper is as follows. Our theoretical framework is described briefly in Sec. 2. In the following part, namely in Sec. 3, we present the detailed study of the Sn isotopes and N=82 isotones, as an evidence of how the tensor correlations improves the description of single-particle states. Finally, we discuss together the inclusion of the tensor force and the inclusion of correlations beyond mean-field in the specific example of ^{132}Sn (Sec. 4). We draw our conclusions in Sec. 5.

2. Theoretical framework

In the present paper, we employ a Skyrme parameter set plus the triplet-even and triplet-odd tensor zero-range tensor terms, which read

$$\begin{aligned} V_{\text{eff}}^{(tensor)} &= \frac{T}{2}\{[(\sigma_1 \cdot k')(\sigma_2 \cdot k') - \frac{1}{3}(\sigma_1 \cdot \sigma_2)k'^2]\delta(r_1 - r_2) \\ &+ \delta(r_1 - r_2)[(\sigma_1 \cdot k)(\sigma_2 \cdot k) - \frac{1}{3}(\sigma_1 \cdot \sigma_2)k^2]\} \\ &+ U\{(\sigma_1 \cdot k')\delta(r_1 - r_2)(\sigma_2 \cdot k) - \frac{1}{3}(\sigma_1 \cdot \sigma_2)[k' \cdot \delta(r_2 - r_2)k]\}. \end{aligned} \quad (1)$$

In the above expression, the operator $k = (\nabla_1 - \nabla_2)/2i$ acts on the right and $k' = -(\nabla_1 - \nabla_2)/2i$ acts on the left. The coupling constants T and U denote the strength of the triplet-even and triplet-odd tensor interactions, respectively; we treat these coupling constants as free parameters in the following study. The tensor interactions (1) give contributions both to the binding energy and to the spin-orbit splitting, which are, respectively, quadratic and linear in the proton and neutron spin-orbit densities,

$$J_q(r) = \frac{1}{4\pi r^3} \sum_i v_i^2(2j_i + 1) \left[j_i(j_i + 1) - l_i(l_i + 1) - \frac{3}{4} \right] R_i^2(r). \quad (2)$$

In this expression $q = 0(1)$ labels neutrons (protons), while where $i = n, l, j$ runs over all states having the given q. The v_i^2 is the BCS occupation probability of each orbital and $R_i(r)$ is the radial part of the wavefunction. It should be noticed that the exchange part of the central Skyrme interaction gives the same kind of contributions to the total energy density and spin-orbit splitting. The spin-orbit potential is given by

$$U_{s.o.}^{(q)} = \frac{W_0}{2r}\left(2\frac{d\rho_q}{dr} + \frac{d\rho_{q'}}{dr}\right) + \left(\alpha\frac{J_q}{r} + \beta\frac{J_{q'}}{r}\right), \quad (3)$$

where the first term on the r.h.s comes from the Skyrme spin-orbit interaction whereas the second term includes both the central exchange and the tensor contributions, that is, $\alpha = \alpha_C + \alpha_T$ and $\beta = \beta_C + \beta_T$. The central exchange contributions are written in terms of the usual Skyrme parameters,

$$\begin{aligned} \alpha_C &= \frac{1}{8}(t_1 - t_2) - \frac{1}{8}(t_1 x_1 + t_2 x_2) \\ \beta_C &= -\frac{1}{8}(t_1 x_1 + t_2 x_2). \end{aligned} \quad (4)$$

Basic definitions of all quantities derived from the Skyrme parameters can be found in Refs.[3,15] The tensor contributions are expressed as

$$\begin{aligned} \alpha_T &= \frac{5}{12}U \\ \beta_T &= \frac{5}{24}(T + U). \end{aligned} \quad (5)$$

The central exchange contributions (4) have been neglected when fitting most of the Skyrme parameter sets, and when performing most of the previous HF calculations. In this work, we employ the SLy5 parameter set[15] which has been fitted with the same protocol of the more widely used SLy4 set and should consequently have similar quality. In the case of SLy5, the central exchange contributions are included in the fit and we take them into account here.

Except for the double-magic systems, we perform HF-BCS in order to take into account the pairing correlations. Our pairing force is a zero-range, density-dependent one, namely

$$V = V_0 \left(1 - \left(\frac{\rho\left(\frac{\vec{r}_1+\vec{r}_2}{2}\right)}{\rho_0}\right)^{\gamma}\right) \cdot \delta(\vec{r}_1 - \vec{r}_2). \tag{6}$$

The parameters of this force (that is, V_0=680 MeV·fm^3, γ=1 and ρ_0=0.16 fm^{-3}) have been fixed along the Z=50 isotopic chain in connection with the SLy4 set in Ref.[16] Therefore, we employ the same force here both for the neutron and the proton pairing interactions in the 50-82 shell, neglecting small readjustments which could be made to account for the Coulomb anti-pairing effect in the case of protons.

One should remark that since the Skyrme spin-orbit contribution (proportional to W_0) gives a value of the spin-orbit splitting which is linear in the derivatives of the proton and neutron densities, the associated mass number and isospin dependence is very moderate in heavy nuclei. On the other hand, the second term in Eq. (3) depends on the spin density J_q which has a more peculiar behavior. J_q gives essentially no contribution in the spin-saturated cases, but it increases linearly with the number of particles if only one of the spin-orbit partners is filled. The sign of the J_q will change depending upon the quantum numbers of the orbitals which are progressively filled: that is, the orbital with $j_>$ gives a positive contribution to J_q while the orbital with $j_<$ gives a negative contribution to J_q. This must be kept in mind to understand the results which are discussed below.

3. Results for the Z=50 isotopes and N=82 isotones

In Fig. 1, the energy differences of the proton single-particle states, $\varepsilon(h_{11/2})-\varepsilon(g_{7/2})$, along the Z=50 isotopes are shown as a function of the neutron excess N-Z. The original SLy5 interaction fails to reproduce the experimental trend both qualitatively and quantitatively. On the other hand, we can see a substantial improvement due to the introduction of the tensor interaction with (α_T, β_T)=(-170,100) MeV·fm^5. This choice gives a very nice agreement with the experimental data in the range $20 \leq$ (N-Z) ≤ 32, both quantitatively and qualitatively. The HF+tensor results can be qualitatively understood by simple arguments, looking at Eq. (3). In the Z=50 core, only the proton $g_{9/2}$ orbital dominates the proton spin density J_p (cf. Eq. (2)); consequently, with a negative value of α_T, the spin-orbit potential (3) is enlarged in absolute value (notice that W_0 is positive and the radial derivatives of the densities are negative), the values of the proton spin-orbit splittings are increased, and

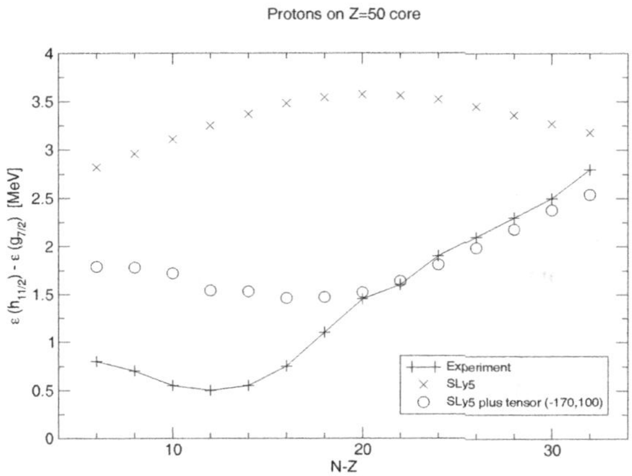

Fig. 1. Energy differences between the $1h_{11/2}$ and $1g_{7/2}$ single-proton states along the Z=50 isotopes. The calculations are performed without (crosses) and with (circles) the tensor term in the spin-orbit potential (3), on top of SLy5 (which includes the central exchange, or J^2, terms). The experimental data are taken from ref.[17] See the text for details.

the energy difference $\varepsilon(h_{11/2}) - \varepsilon(g_{7/2})$ is reduced with respect to HF-BCS without tensor. This reduction is seen better around N-Z≃20: in fact, ^{120}Sn is, to a good extent, spin-saturated as far as the neutrons are concerned so that one gets no contribution from J_n. It should be noticed that the term in α does not give any isospin dependence to the spin-orbit potential for a fixed proton number, but only the term in β can be responsible for the isospin dependence. In a pure HF description, from N-Z=6 to 14, the $g_{7/2}$ neutron orbit is gradually filled and J_n is reduced. Then, the positive value of β_T enlarges in absolute value the spin-orbit potential and increases the spin-orbit splitting, so that the energy difference $\varepsilon(h_{11/2}) - \varepsilon(g_{7/2})$ becomes smaller. Because of pairing, this decrease is not so pronounced in the results of Fig. 1. Moreover, from N-Z=14 to 20, the $s_{1/2}$ and $d_{3/2}$ neutron orbits are occupied and in this region the spin density is not so much changed since the $s_{1/2}$ orbital does not provide any contribution. Instead, for N-Z=20 to 32, the $h_{11/2}$ orbital is gradually filled. This gives a positive contribution to the spin-orbit potential (3) and the the spin-orbit splitting becomes smaller. $\varepsilon(h_{11/2}) - \varepsilon(g_{7/2})$ consequently increases, and this effect is well pronounced in our theoretical results.

In Fig. 2, the energy difference $\varepsilon(i_{13/2}) - \varepsilon(h_{9/2})$ for neutrons outside the closed N=82 core is plotted as a function of the neutron excess. The notation is the same of the previous figure. Essentially, the same arguments already made in the previous paragraph can be applied in order to understand the results; simply, we should remind that the proton number is increasing from the right (where the last nucleus displayed is ^{132}Sn) to the left (where the first isotope plotted is ^{150}Er). The $1g_{7/2}$ and $2d_{5/2}$ orbitals are rather close in energy, above the last occupied proton state $1g_{9/2}$

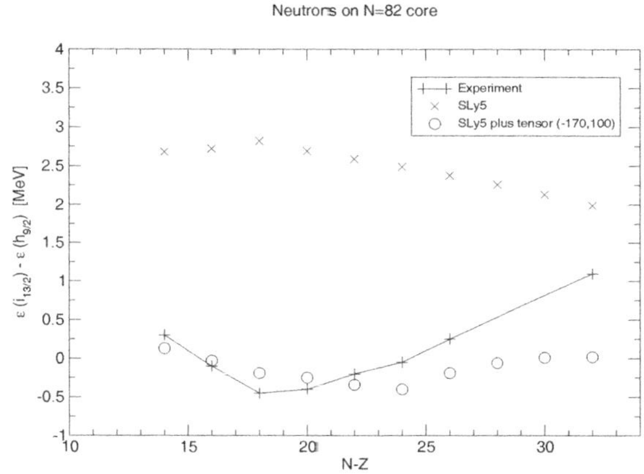

Fig. 2. Energy differences between the $1i_{13/2}$ and $1h_{9/2}$ single-neutron states along the N=82 isotones. The calculations are performed without (crosses) and with (circles) the tensor term in the spin-orbit potential (3), on top of SLy5 (which includes the central exchange, or J^2, terms). The experimental data are taken from ref.[17] See the text for details.

of the Z=50 core, and their occupations are affected by the pairing correlations introduced by the BCS approximation. These two proton orbitals have opposite effect on the spin orbit potential (3). Because of its larger value of j, the $1g_{7/2}$ orbital turns out to play a more important role on the spin-orbit potential, when the tensor interaction is included, in the nuclei with N-Z decreasing from 32 to 24. Accordingly, with positive β_T the neutron spin-orbit potential is enlarged in absolute value: the spin-orbit splitting is made larger for these isotones, so that the $i_{13/2}$ orbital is pushed down and the $h_{9/2}$ is pushed up. These changes make the energy gap $\varepsilon(i_{13/2}) - \varepsilon(h_{9/2})$ smaller for the nuclei from N-Z=32 (^{132}Sn) to N-Z=24 (^{140}Ce). Then, the occupation of the $2d_{5/2}$ orbital reverses the trend around N-Z=22 (^{142}Nd). The theoretical trend remains the same until N-Z=14, since the effect of the $2d_{3/2}$ occupation is counterbalanced by the occupation of the $1h_{11/2}$ which is not much higher and enters the active BCS space.

4. The effect of particle-vibration coupling

As discussed at length in Ref. [18] the coupling of single-particle states to vibrational states has the net effect of increasing the level density around the Fermi surface by about 30%, by shifting occupied and unoccupied states in opposite directions. In our work, we have taken into account this coupling in the case of the nucleus ^{132}Sn (i.e., without pairing correlations), by using the formalism which is described in detail in Ref.[19] We have calculated the single-particle states first within simple Skyrme-HF, and then by including the tensor terms with the parameters mentioned above. We have finally added perturbatively the energy shift due to the coupling with phonons calculated by means of Random Phase Approximation (RPA). The results are depicted in Fig. 3, where we display the valence neutron hole states.

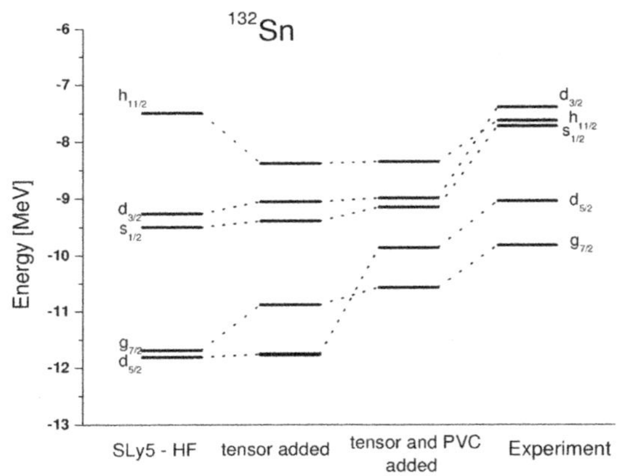

Fig. 3. Results for the neutron hole states in the nucleus ^{132}Sn. The four columns, from left to right, correspond respectively to: simple HF, HF including the tensor terms, HF including the tensor terms plus the energy shift due to coupling with vibrations and experimental energies.

One can notice that the tensor force is quite important for the shift it can gives to the $h_{11/2}$ state. Both tensor correlations and the particle-vibration coupling cooperate in giving the correct ordering and energy separation between the $g_{7/2}$ and $d_{5/2}$ states. Moreover, the large compression of levels provided by the particle-vibration coupling is clearly evident in Fig. 3. The final result, as compared with the experimental finding, is extremely better than pure HF, in keeping with the fact that it provides the correct magnitude of the energy range of the valence shell. The only feature which is still not reproduced is the ordering of the two last occupied levels, $d_{3/2}$ and $h_{11/2}$. Apparently, no mean field model can provide the correct ordering of these two levels.[1] Although still work has to be done to explain this, as well as other, systematics of single-particle levels, it is clear from Fig. 3 that including particle-vibration coupling on top of a complete Skyrme energy functional which includes the tensor terms, looks like a promising strategy.

5. Conclusions

The calculations based on effective interactions are, at present, the only scheme which can be applied in the case of medium-heavy nuclei. They can be viewed as an approximate realization of the DFT in nuclei. Recently, there has been debate on whether the available functionals include all the necessary terms. In particular, much attention has been focused in the tensor correlations.

In this contribution, we show that by including the tensor terms on top of existing Skyrme functionals, it is possible to explain the isotope (or isotone) trend of single-particle states. We have worked out in detail the case of the Z=50 isotopes and N=82 isotones. We have fitted optimal parameters. In another contribution included in the present volume,[20] it is shown that the same parameters provide good results also for lighter systems.

By combining together the mean field results which include the tensor correlations, and the coupling of single-particle states with nuclear shape vibrations, we have tried to reproduce in detail the spectrum of ^{132}Sn. Our results are preliminary, and more work along this line is in progress. However, it looks quite promising to try to extend this strategy to a number of different systems, in keeping with what has been shown in Fig. 3.

References

1. See M. Bender, P.-H. Heenen, P.-G. Reinhard, *Rev. Mod. Phys.* **75**, 121 (2003) for a recent review.
2. T.H.R. Skyrme, *Nucl. Phys.* **9**, 615 (1959).
3. D. Vautherin, D.M. Brink, *Phys. Rev. C* **5**, 626 (1972); M. Beiner, H. Flocard, N. Van Giai, Ph. Quentin, *Nucl. Phys. A* **238**, 29 (1975).
4. J. Dechargé, D. Gogny, *Phys. Rev. C* **21**, 1568 (1980); J.F. Berger, M. Girod, D. Gogny, *Comp. Phys. Comm.* **63**, 365 (1991).
5. Fl. Stancu, D.M. Brink and H. Flocard, *Phys. Lett. B* **68**, 108 (1977).
6. F. Tondeur, *Phys. Lett. B* **123**, 139 (1983).
7. K.F. Liu *et al.*, *Nucl. Phys. A* **534**, 1, 25 and 48 (1991).
8. T. Otsuka, T. Suzuki, R. Fujimoto, H. Grawe and Y. Akaishi, *Phys. Rev. Lett.* **95**, 232502 (2005).
9. Cf. the contribution by T. Otsuka in this volume.
10. T. Otsuka, T. Matsuo and D. Abe, *Phys. Rev. Lett.* **97**, 162501 (2006).
11. G. Colò, H. Sagawa, S. Fracasso, P.F. Bortignon, *Phys. Lett. B* **646**, 227 (2007).
12. D.M. Brink and Fl. Stancu, *Phys. Rev. C* **75**, 064311 (2007).
13. J. Dobaczewski, in: T. Duguet, H. Esbensen, K. M. Nollett, and C. D. Roberts (eds.), *Proceedings of the Third ANL/MSU/JINA/INT RIA Workshop*, (World Scientific, Singapore, 2006).
14. T. Lesinski *et al.*, *Phys. Rev C* **76**, 014312 (2007).
15. E. Chabanat *et al.*, *Nucl. Phys. A* **635**, 231 (1998).
16. S. Fracasso and G. Colò, *Phys. Rev. C* **72**, 064310 (2005).
17. J. P. Schiffer *et al.*, *Phys. Rev. Lett.* **92**, 162501 (2004).
18. C. Mahaux, P. F. Bortignon, R. A. Broglia, C. H. Dasso, *Phys. Rep.* **120**, 1 (1985).
19. G. Colò, T. Suzuki and H. Sagawa, *Nucl. Phys. A* **695**, 167 (2001).
20. Cf. the contribution of Wei Zou *et al.* in this volume.

NEW ASPECTS ON DYNAMICS IN NUCLEI DESCRIBED BY COVARIANT DENSITY FUNCTIONAL THEORY

P. RING* and D. PENA

Physics Department, Technical University Munich,
Garching, D85748, Germany
**E-mail: ab_ring@ph.tum.de*

The description of nuclear dynamics in the framework of Covariant Density Functional Theory (CDFT) is based on the Relativistic Random Phase Approximation (RRPA) or, in superfluid systems, on the Relativistic Quasi-particle RPA (RQRPA). This method has been used in the past extensively for the calculation of many kinds of resonances in spherical systems. Here we report on an extensions of this method: a new relativistic QRPA code has been developed for the description of deformed systems and first applications for the description of E1 pygmy modes and M1 scissor-like excitations are presented.

1. Introduction

New experimental facilities with radioactive nuclear beams make it possible to investigate the nuclear chart not only along the narrow line of stable isotopes but also in areas of large neutron- and proton excess far from the valley of β-stability. Low-lying collective dipole excitations are expected to play an important role in astrophysical applications in neutron rich systems, where the presence of a low-lying resonant component of the E1 strength has a strong influence on the radiative neutron-capture rates in the r-process.[1] This situation has stimulated enhanced efforts on the theoretical side to understand the dynamics of the nuclear many-body problem by microscopic methods. For the large majority of medium heavy and heavy nuclei, a quantitative microscopic description is only possible by density functional theory. Although this theory can, in principle, provide an exact description of the many-body dynamics[2] if the exact density functional is known, in nuclear physics one is far from a microscopic derivation of this functional. The most successful schemes use a phenomenological ansatz incorporating as many symmetries of the system as possible and adjust the parameters of these functionals to ground state properties of characteristic nuclei all over the periodic table.

Of particular interest are covariant density functionals[3,4] because they are based on Lorentz invariance. This symmetry not only allows to describe the spin-orbit coupling in a consistent way which has an essential influence on the underlying shell, but it also put stringent restrictions on the number of parameters in the corresponding functionals without reducing the quality of the agreement with experimental data. A very successful example is the Relativistic Hartree-Bogoliubov model.[5] It combines a density dependence with pairing correlations based on an effective interaction of finite range. A large variety of nuclear phenomena have been described over the years within this model. the equation of state in symmetric nuclear matter, ground state properties of finite spherical and deformed nuclei all over the periodic table.[6] Recently this model has been also applied very successfully for the description of excited states, such as rotational bands in normal and super-deformed nuclei[7] and collective vibrations[8] in the framework relativistic RPA of QPRA.

So far, all the relativistic RPA and quasiparticle RPA calculations were restricted to spherical nuclei. On the other hand it is well known, that only semi-magic nuclei have spherical shape and that most of the other nuclei in the nuclear chart are deformed. Important astrophysical processes, such as the r-process or the s-process pass though large areas os nuclei with deformation. The description of nuclear ground states in deformed regions is well established since many years[6,9] and very successful. In order to investigate also collective excitations in deformed nuclei a new deformed quasiparticle RPA code has been developed. We start from a relativistic density functional with non-linear meson exchange (NL-ME) or with density dependent nucleon-meson couplings (DD-ME) and solve in a first step the relativistic mean (RMF) field equations for the deformed ground state. Pairing correlations are taken into account by a monopole pairing force in the BCS approximation. In the second step the relativistic QRPA-equations

$$\begin{pmatrix} A & B \\ B^* & A^* \end{pmatrix} \begin{pmatrix} X \\ Y \end{pmatrix}_\mu = \Omega_\mu \begin{pmatrix} 1 & 0 \\ 0 & -1 \end{pmatrix} \begin{pmatrix} X \\ Y \end{pmatrix}_\mu \tag{1}$$

are solved in the basis of the deformed Dirac wave functions of the ground state. Because of the deformation rotational symmetry is broken and angular momentum is no longer a good quantum number. Since we work in an axial basis, the only good quantum numbers are the projection of the angular momentum onto the symmetry axis K and the parity π. The effective interaction used in the QRPA-equations 1 is obtained as the second derivative with respect to the density from the same energy functional which is used for the calculation of the ground state. The matrix elements are calculated in momentum space using Fourier Bessel techniques in the axial basis.[10] They are multiplied with the usual BCS occupation factors.[11] No new parameters are needed and as usual in RPA and QRPA theory the Goldstone modes connected with the various broken symmetries are found at zero energy and can be fully separated from the other physical modes. This provides a very sensitive test for the accuracy of the solution of the relativistic QRPA Eqs. (1).

There are five Goldstone modes, two of them with the quantum numbers $K^\pi = 0^-$ and $K^\pi = 1^-$ correspond to the violation of translational symmetry in the direction along the symmetry axis and perpendicular to the symmetry axis, one of them with the quantum number $K^\pi = 1^+$ corresponds to the violation of rotational symmetry around an axis perpendicular to the symmetry axis and, in superfluid systems, two further modes with the quantum number $K^\pi = 0^+$ correspond to the

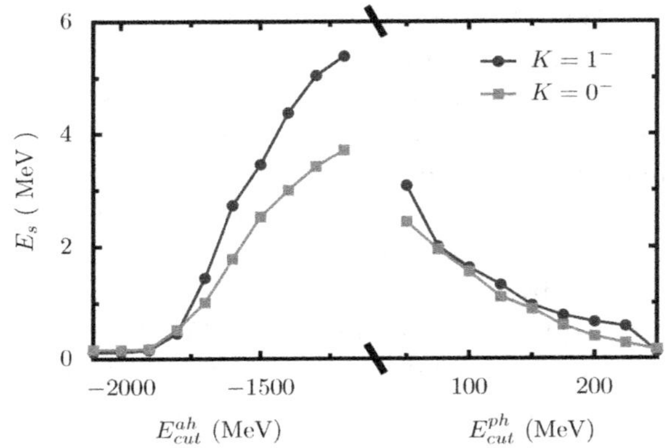

Fig. 1. The translational Goldstone modes and their dependence on the size of the configuration for *particle-hole*-configurations and for *antiparticle-hole* configurations.

violation of particle number for neutrons and protons.

Fig. 1 shows the dependence of the translational Goldstone modes on the size of the configuration space for the nucleus ^{20}Ne calculated with the parameter set NL3.[12] Two curves are plotted, one for the $K^\pi = 0^-$ mode and one for the $K^\pi = 1^-$ mode. We see that one has to include nearly all possible $2qp$-pairs in the configuration space in order to bring the Goldstone mode close to zero. In ^{20}Ne, that amounts to roughly five thousand pairs. The situation improves greatly in heavier nuclei, where usually 5% percent of all possible $2qp$-pairs are enough to decouple the spurious modes at energies around 0.5MeV.

2. The electric dipole response in Ne-isotopes

In recent years, much theoretical and experimental efforts have been devoted to the study of the low lying collective soft modes in nuclei far from stability (for a recent review see[13]). They correspond to oscillations of the mantle composed by excess neutrons against a core formed from the rest of the neutrons and protons. The structure of such modes is deeply linked with the single particle structure near the Fermi surface, and thus its study can provide valuable information regarding the performance of the different functionals. So far, most of these experimental and theoretical investigations have been devoted to spherical nuclei.

Recent experiments at GANIL[14] found in ^{26}Ne an enhancement of the isovector dipole strength in the low energy region. Tentatively, it has been identified as a possible pygmy mode. In Fig. 2 we show results of a theoretical QRPA study of the dipole response in the chain of Ne isotopes, from ^{20}Ne to ^{28}Ne. Two different relativistic density functionals are used, the parameter set NL3[12] with a non-linear meson coupling and the density dependent set DD-ME2.[15] We show excitations along the symmetry axis with the quantum numbers $K^\pi = 0^-$ as well as those perpendicular to the symmetry axis with $K^\pi = 1^-$. Since these nuclei have prolate

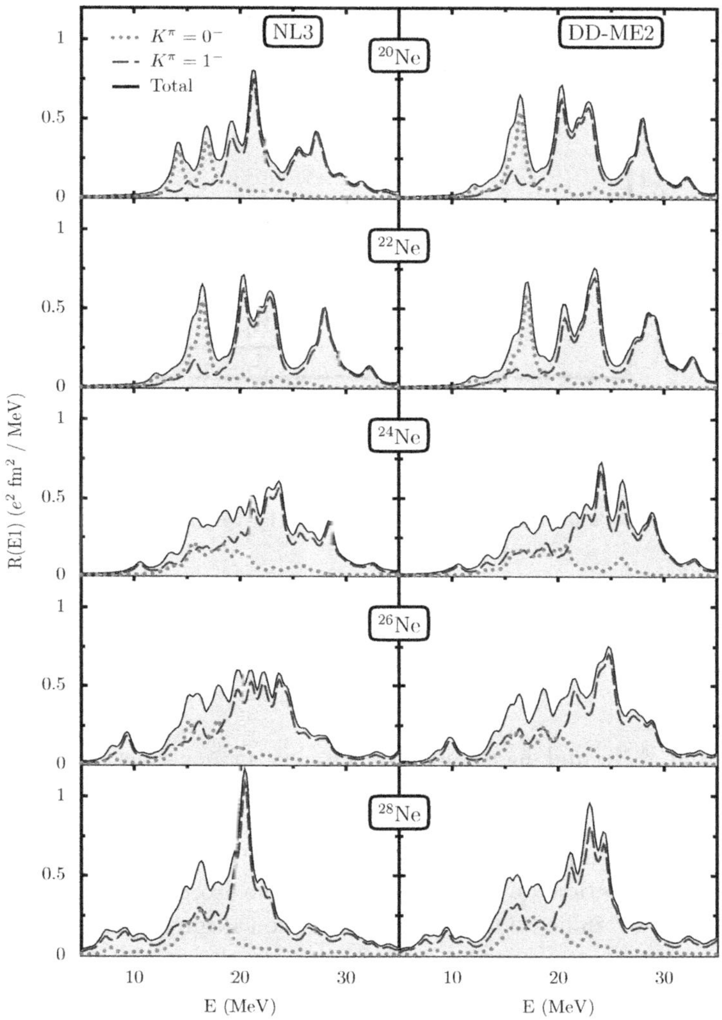

Fig. 2. Evolution of the E1 excitation strength from ^{20}Ne to ^{28}Ne for two different relativistic functionals.

shapes the strength due to the $K^\pi = 0^-$ mode should lie at lower energies compared to the $K^\pi = 1^-$ mode, because for prolate shapes the nuclear potential is flatter along the symmetry axis than perpendicular to it. Therefore it is possible to relate the nuclear deformation with the energy separation of the two modes.[16,17]

In Fig. 2 the response in the energy region between 15 MeV and 25 MeV corresponds to the IVGDR. Its strength is heavily fragmented in several peaks in an energy interval of about 3-4 MeV for both excitation modes. The high collectivity of these peaks indicate a very coherent excitation pattern that fits into the properties

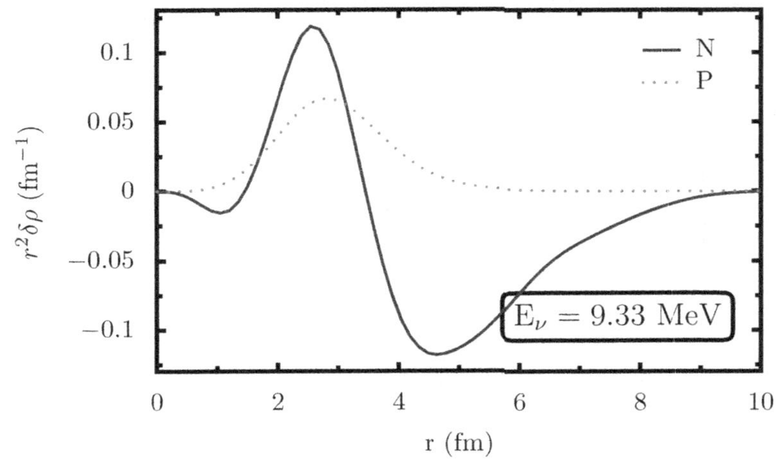

Fig. 3. Transition densities for the low lying peaks in the E1 response of ^{26}Ne, calculated with the parameter set NL3.

of a giant resonance.

Starting from ^{20}Ne and increasing the number of neutrons two main effects are observed: a) an increased fragmentation of the dipole strength, and b) the appearance of low lying strength below 10 MeV. These results are similar to those obtained for oxygen isotopes within the spherical RHB+RQRPA framework.[18]

We focus our attention on the properties of the isotope ^{26}Ne, where recent experimental results[14] indicate the presence of a non negligible dipole response in the 8-12 MeV energy region. Neutron contributions to the isovector dipole strength in ^{26}Ne dominate in the low energy region. The strength function below 12 MeV is fragmented into several peaks, with major contributions at 7 MeV and 9 MeV for $K^\pi = 1^-$ and $K^\pi = 0^-$ transitions, respectively.

In Fig. 3 we consider, as an example, the peak at 9.33 MeV and display the transition density in the laboratory system, which is obtained from the transition density in the intrinsic frame by projection onto good angular momentum. In contrast to the well known radial dependence of the IVGDR transition densities (proton and neutron oscillate with opposite phases), the proton and neutron transition densities are mostly in phase in the nuclear interior. Furthermore, there is almost no contribution from the protons in the surface region. The strong neutron transition density displays a long tail in the radial coordinate. The different radial dependence of the transition densities that correspond to the soft low-energy states as compared to those of the giant resonances are due to the contribution of the loosely bound excess neutrons.

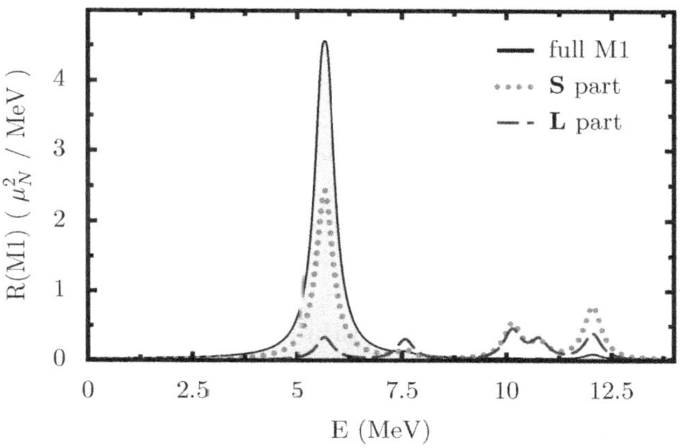

Fig. 4. M1 Excitation strength for ^{20}Ne, using the NL3 parameter set. A very well developed peak can be seen around 5.8 MeV Its structure is composed mostly of spin flip transitions.

3. The magnetic dipole response in ^{20}Ne

Since the eighties it is known[19] that the magnetic dipole response in deformed nuclei contains scissor modes. The search for such modes was stimulated by the prediction[20–23] of a collective M1 mode, where the deformed proton core oscillates in a motion of rotational character against the deformed neutron core, the so called Two Rotor Model (TMR).

This mode has been detected in most of the deformed nuclei ranging from the fp-shell to Rare-Earth and Actinide regions. The mode has been well characterized, and it is well established that it is fragmented into several closely packed M1 transitions. A byproduct of the systematic study of the "scissors mode" was the discovery of spin excitations. Inelastic proton scattering experiments on ^{154}Sm and other deformed nuclei[24] found a sizable and strongly fragmented M1 spin strength distributed over an energy range of 4 MeV to 12 MeV. The experimental discovery stimulated the proliferation of theoretical investigations, were most of the microscopic studies[25] were carried out in the RPA or TDA approximation.

As a first application of the RHB+RQRPA framework we investigated the low-lying magnetic dipole excitations in ^{20}Ne. This nucleus offers several advantages. Its ground state is well deformed and exhibits a prolate shape in the RHB model, with a quadrupole deformation parameter $\beta \approx 0.5$. Another advantage is the reduced number of nucleons to take into account in the calculations, which translates in fast running times and thus in the possibility of detailed analysis. With the optimal number of oscillator shells for a full precision ground state calculation, the number of qp pairs never exceeds five thousand. These reasons make ^{20}Ne the ideal theoretical playground where to introduce the concepts that can later be used in the study of more complex systems.

Fig. 4 shows the magnetic dipole response in ^{20}Ne for the parameter set NL3. The number of quasiparticle pairs is around five thousand. The rotational spurious mode is well separated, situated for both functionals below 0.1 MeV, i.e. no admixture with the vibrational response is observed. In Fig. 4 we show the M1-response as well as the orbital and spin contributions

Only one prominent peak is found at 5.7 MeV. Calculations performed with other reliable parameter sets (such as NL3, DD-ME2 and PC-F1) show good agreement, with an energy dispersion of less than half an MeV in the position of the main peak. Regrettably, there is no experimental data available for the magnetic response in this nucleus.

Regarding the contributions from the orbital and spin components of the M1 operator to the total response strength, we find in Fig. 4 two differentiated energy regions. Around the main excitation peak at 6 MeV there is an enhancement of the response due to the addition of the orbital and spin contributions. In contrast, we find the opposite in the energy region above 6.5 MeV, both contributions cancel. This feature of the M1 strength distribution is much more evident in heavier nuclei. The main contribution to the total response strength comes from spin excitations. The supposed orbital character of the low lying spectra in the M1 transitions is eclipsed by the preponderance of spin flip strength, three times larger than the orbital response.

4. Conclusions and outlook

Time dependent density functional theory is an important tool to study excitations of vibrational character in nuclei far from stability. The self-consistent solution of the relativistic QRPA equations is now also possible in deformed systems. First applications for the electric dipole response in the chain of Ne-isotopes are presented. Besides the well known IVGDR resonance low-lying E1-strength is found in deformed neutron rich Ne-isotopes. These results are in agreement with recent experiments at GANIL and they supports the interpretation of this mode as a pygmy dipole resonance, where the excess neutrons oscillate against the isospin-saturates neutron-proton core. In the magnetic dipole response in ^{20}Ne a sharp peak is found at around 6 MeV. It is too a large extend a spin-mode. However it also has considerable admixtures of orbital motion, which is interpreted as a scissor mode in this nucleus.

This method can now be applied for a large number of investigations in medium heavy and heavy nuclei, in a first step for a systematic study of low-lying electric and magnetic dipole strength over large regions of the periodic table. The conventional giant multipole resonances can be investigated in heavy deformed nuclei. Replacing the quasiparticle pairs of like particles by proton-neutron pairs deformed proton-neutron RPA and QRPA can be formulated and this opens the door for the wide area of nuclear spin-isospin excitations, in particular the Isobaric Analog Resonance (IAR) and the Gamov-Teller Resonance (GTR), but also for many types of weak processes as the as the β-decay and neutrino-reactions in deformed nuclei.

Acknowledgements

This work has been supported in part the European Community project Asia-Europe Link in Nuclear Physics and Astrophysics, CN/ASIA-LINK/008(94791), by the Alexander von Humboldt Stiftung and by the DFG cluster of excellence "Origin and Structure of the Universe" (www.universe-cluster.de).

References

1. S. Goriely, *Phys. Lett. B* **436**, 10 (1998).
2. W. Kohn and L. J. Sham, *Phys. Rev. C* **137**, 1697 (1965).
3. P. Ring, *Prog. Part. Nucl. Phys.* **37**, 193 (1996).
4. D. Vretenar, A. V. Afanasjev, G. A. Lalazissis, and P. Ring, *Phys. Rep.* **409**, 101 (2005).
5. T. Gonzales-Llarena, J. L. Egido, G. A. Lalazissis, and P. Ring, *Phys. Lett.* **B379**, 13 (1996).
6. Y. K. Gambhir, P. Ring, and A. Thimet, *Ann. Phys.* (N.Y.) **198**, 132 (1990).
7. A. V. Afanasjev, P. Ring, and J. König, *Nucl. Phys.* **A676**, 196 (2000).
8. Z.-Y. Ma, A. Wandelt, N. Van Giai, D. Vretenar, P. Ring, and L.-G. Cao, *Nucl. Phys.* **A703**, 222 (2002).
9. W. Pannert, P. Ring, and J. Boguta, *Phys. Rev. Lett.* **59**, 2420 (1987).
10. D. Pena, Phd thesis, Technische Universität München (unpublished), 2007.
11. P. Ring and P. Schuck, *The nuclear many-body problem* (Springer, Heidelberg, 1980).
12. G. A. Lalazissis, J. König, and P. Ring, *Phys. Rev. C* **55**, 540 (1997).
13. N. Paar, D. Vretenar, E. Khan, and G. Colò, *Rep. Prog. Phys.* **70**, 691 (2007).
14. Gibelin, Phd thesis, Université Paris XI Orsay (unpublished), 2006.
15. G. A. Lalazissis, T. Nikšić, D. Vretenar, and P. Ring, *Phys. Rev. C* **71**, 024312 (2005).
16. M. Danos, *Nucl. Phys.* **5**, 23 (1958).
17. K. Okamoto, *Phys. Rev.* **110**, 143 (1958).
18. N. Paar, P. Ring, T. Nikšić, and D. Vretenar, *Phys. Rev. C* **67**, 034312 (2003).
19. D. Bohle, A. Richter, W. Stephen, A. E. L. Dieperink, N. lo Iudice, F. Palumbo, and O. Scholten, *Phys. Lett. B* **137**, 27 (1984).
20. N. lo Iudice and F. Palumbo, *Nucl. Phys.* **A289**, 461 (1977).
21. R. R. Hilton, *Ann. Phys.* (N.Y.) **214**, 258 (1992).
22. T. Suzuki and D. J. Rowe, *Nucl. Phys.* **A286**, 307 (1977).
23. F. Iachello, *Nucl. Phys.* **A358**, 89c (1981).
24. D. Frekers, H. Wortche, and A. Richter, *et al.*, *Phys. Lett. B* **244**, 170 (1990).
25. R. Hilton, W. Höhenberger, and P. Ring, *Eur. Phys. J* **A1**, 257 (1998).

THEORETICAL STUDIES ON GROUND-STATE PROPERTIES OF SUPERHEAVY NUCLEI *

ZHONGZHOU REN, CHANG XU and QIJUN ZHI

Department of Physics, Nanjing University,
Hankou Road, Nanjing 210093, China
E-mail: zren@nju.edu.cn

The ground-state properties of superheavy nuclei with proton number 109−117 are investigated by both the macroscopic-microscopic model (MM) and the relativistic mean-field model (RMF). The alpha-decay energies and half-lives of these exotic nuclei by different theoretical models are analyzed in detail. The spontaneous fission half-lives of superheavy nuclei (Z=104−114) are also discussed and some possible candidates of superheavy nuclei are suggested for future experiments.

1. Introduction

The synthesis of superheavy nuclei has received particular attention in nuclear physics in recent years.[1,2] Since 1995, the superheavy elements Z=113-116,118 have been successfully synthesized at GSI, Dubna, RIKEN and Lawrence Berkeley National Laboratory of USA due to the rapid development of modern accelerators and detectors.[3–10] Many theoretical researches were also performed to reproduce the experimental data and to make predictions for future experiments.[11–15] Experimentally, the superheavy nuclei are usually identified by alpha-particle emission or spontaneous fission. Therefore, alpha-decay and spontaneous fission are two important nuclear decay modes for superheavy nuclei. Study on these decay modes can provide detailed information on the nuclear structure of superheavy nuclei.

The first purpose of this paper is to calculate the ground state properties of some superheavy nuclei by the macroscopic-microscopic (MM) model and the relativistic mean-field model (RMF). Then alpha decay energies of these superheavy nuclei are extracted from corresponding ground state binding energies . With the help of the calculated decay energies, the alpha-decay half-lives are predicted for the superheavy nuclei of unknown mass region by using the Viola-Seaborg formula with new parameters. Finally, we will analyze the spontaneous fission half-lives of superheavy nuclei with proton number Z=104−114 and the systematical law of spontaneous fission half-lives of these nuclei is discussed.

*This work is supported by the National Natural Science Foundation of China (Grant Nos. 10535010, 10735010, 10775068), by the 973 National Major State Basic Research and Development of China (Grant No. 2007CB815004).

2. The formalism of the macroscopic-microscopic model

In the macroscopic-microscopic approach, the total potential energy is the sum of a macroscopic term and a microscopic term representing the shell plus pairing correction.[12] It can be written as:

$$E_{tot} = E_{mac}(Z, N, \bar{\epsilon}) + E_{mic}(Z, N, \bar{\epsilon}) \tag{1}$$

where E_{mic} is :

$$E_{mic} = E_{shell}(Z, N, \bar{\epsilon}) + E_{pair}(Z, N, \bar{\epsilon}) \tag{2}$$

Here the macroscopic energy is given by:[12]

$$\begin{aligned} E_{LD} &= -a_v(1 - K_v(\frac{N-Z}{A})^2)A \\ &+ \frac{3}{5}\frac{e^2 Z^2}{R_c}(B_c(\bar{\epsilon}) - \frac{5\pi^2}{6}(\frac{d}{R_c})^2) \\ &+ a_s(1 - K_s(\frac{N-Z}{A})^2)A^{2/3}B_s(\bar{\epsilon}) \\ &+ \begin{cases} -12/\sqrt{A} & \text{odd-odd nuclei} \\ 0 & \text{odd-even nuclei} \\ -12/\sqrt{A} & \text{even-even nuclei} \end{cases} \end{aligned} \tag{3}$$

where $\bar{\epsilon}$=($\varepsilon_2, \gamma, \varepsilon_4$), and ε_2, γ, ε_4 are the quadrupole and hexadecapole deformation degrees of freedom. The liquid-drop parameters are chosen in the following way[12]

$$a_v = 15.4941 \text{ MeV}, \quad a_s = 17.9439 \text{ MeV},$$
$$K_v = K_s = 1.7826, \quad d = 0.546 \text{ fm}, \; R_c = 1.2249A^{1/3} \text{ fm},$$

We adopt the modified single-particle oscillator potential as the microscopic potential and only consider ε_2 and ε_4 deformation. The modified single-particle oscillator potential used here is of the form[12]

$$h = h_{ho}(\varepsilon_2, \gamma) + 2\hbar\omega_0\rho^2\sqrt{\frac{4\pi}{9}}\varepsilon_4 V_4(\gamma) + V' - \omega j_x \tag{4}$$

Because the nuclear spin of ground state is zero for even-even nuclei, so we neglect the last term ωj_x in our calculation. The term $h_{ho}(\varepsilon_2, \gamma)$ is the anisotropic harmonic oscillator potential[12]

$$h_{ho}(\varepsilon, \gamma) = \frac{p^2}{2m} + \frac{1}{2}m(\omega_x^2 x^2 + \omega_y^2 y^2 + \omega_z^2 z^2), \tag{5}$$

ω_x, ω_y, ω_z are expressed in quadrupole deformation parameters. $V_4(\gamma)$ is the hexadecapole potential, it is of the form:

$$V_4 = a_{40}Y_4^0 + a_{42}(Y_4^2 + Y_4^{-2}) + a_{44}(Y_4^4 + Y_4^{-4}) \tag{6}$$

where a_{4i} are chosen as:

$$a_{40} = \frac{1}{6}(5\cos^2\gamma + 1) \; , \; a_{42} = -\frac{1}{12}\sqrt{30}\sin 2\gamma \; , \tag{7}$$

$$a_{44} = \frac{1}{12}\sqrt{70}\sin^2\gamma, \tag{8}$$

The term V' is introduced to reproduce the level sequence as observed in nuclei. It is of the form[12]

$$V' = -\kappa(N)\hbar\overset{o}{\omega}_o\{2l_t \cdot s + \mu(N)(l_t^2 - \langle l_t^2\rangle_N)\} \quad (9)$$

the oscillator strength $\overset{o}{\omega}_o$ is chosen as

$$\hbar\overset{o}{\omega}_{0p/n} = \frac{41}{A^{1/3}}(1 \mp \frac{N-Z}{3A}) \quad (\text{MeV}) \quad (10)$$

for protons and neutrons respectively. The parameters κ and μ are dependent on the main oscillator quantum number.[12]

We use the Strutinsky's shell correction method to calculate the shell energy and the BCS method is used to calculate the pairing energy[12]

$$E_{shell} = E_{sp} - \widetilde{E}_{sp} \quad (11)$$

where E_{sp} is the total single-particle energy, $\widetilde{E}_{sp}$ is the smoothed total single-particle energy[12]

$$\widetilde{E}_{sp} = \int_{-\infty}^{\varepsilon_f} \varepsilon\, \widetilde{g}(\varepsilon) d\varepsilon \quad (12)$$

and

$$\widetilde{g}(\varepsilon) = \frac{1}{\gamma\sqrt{\pi}} \sum_i n_i \exp\{-(\varepsilon - \varepsilon_i)^2/\gamma^2\} \quad (13)$$

We also perform systematic calculations in the framework of the relativistic mean-field model. The detailed formalism of the relativistic mean-field model can be found in Refs.[13-15] and we do not repeat them here.

3. Numerical results and discussions

The detailed theoretical results of the α-decay energies and half-lives for the superheavy elements are given in Table 1. In Table 1, the first column denotes the nuclei. The alpha-decay energies from the macroscopic-microscopic model are given in the second column. The third and fourth columns are the alpha-decay energies from the relativistic mean-field model with TMA and NLZ2 force parameters, respectively. The fifth columns are the alpha-decay energies taken from Ref.[5] (denoted by *) and Ref.[16] (denoted by #). The calculated half-lives by the Viola-Seaborg formula with new parameters[12] are listed in column 6-8. The experimental half-lives[16] are given in the last column of Table 1.

One can see from Table 1 that the calculated α-decay energies from the macroscopic-microscopic model are in reasonable agreement with the experimental data. The maximum difference between the calculated results and the experimental data is approximately within 0.5 MeV. The calculated results of the macroscopic-microscopic model are also close to the results of relativistic mean-field model. For the alpha decay half-lives, the calculated and experimented values are generally within a factor of 10^2. We note that the theoretical half-lives are very sensitive to

Table 1. The calculated α-decay energies and half-lives of superheavy elements are compared with the results from RMF models.

	Q_α MM	Q_α TMA	Q_α NLZ2	Q_α Exp	T(ms) MM	T(ms) TMA	T(ms) NLZ2	T(ms) Exp
$^{275}109$	10.05	9.71	10.43	10.48*	1.59E+02	1.36E+03	1.64E+01	1.23E+01
$^{274}109$	10.33	9.74	10.97	10.5#	6.84E+01	2.61E+03	1.85E+00	2.54E+01#
$^{273}109$	10.39	9.79	10.89	10.82#	2.07E+01	1.80E+00	1.23E+00	1.80E+00#
$^{272}109$	10.70	9.89	10.4	10.6#	8.15E+00	1.43E+01	4.54E+01	1.43E+01#
$^{271}109$	10.44	9.99	10.1	10.14#	1.55E+01	9.19E+01	1.17E+02	9.19E+01#
$^{279}111$	10.41	10.45	9.96	10.52*	7.83E+01	6.17E+01	1.25E+03	4.09E+01
$^{278}111$	10.64	10.5	10.26	10.72#	4.73E+01	1.07E+02	4.48E+02	2.99E+01#
$^{277}111$	10.75	10.43	10.85	11.18#	1.08E+01	6.95E+01	6.19E+00	1.02E+00#
$^{276}111$	10.90	10.75	11.39	11.32#	1.10E+01	2.52E+01	7.82E-01	1.13E+00#
$^{275}111$	11.10	10.89	11.62	11.55#	1.56E+00	4.95E+00	1.04E-01	1.48E-01#
$^{283}113$	10.25	11.28	9.76	10.23*	8.37E+02	2.32E+00	2.08E+04	8.34E+02
$^{282}113$	10.16	11.36	10.01		3.62E+03	3.52E+00	9.38E+03	
$^{281}113$	11.46	11.28	10.33		8.92E-01	2.32E+00	5.42E+02	
$^{280}113$	11.58	11.2	10.76		1.11E+00	8.31E+00	9.75E+01	
$^{279}113$	12.65	11.06	11.17		2.71E-03	7.71E+00	4.21E+00	
$^{293}117$	11.47	11.31	10.42		1.23E+01	2.95E+01	5.64E+03	
$^{292}117$	11.47	11.00	10.58	11.6#	2.35E+01	3.97E+02	4.86E+03	1.41E+01#
$^{291}117$	11.67	10.64	11.22	11.9#	4.19E+00	1.45E+03	4.88E+01	1.26E+00#
$^{290}117$	11.93	10.57	11.52		2.51E+00	5.16E+03	2.17E+01	
$^{289}117$	12.31	11.09	11.75		1.51E-01	1.02E+02	2.75E+00	
$^{288}117$	12.63	12.27	11.95		8.10E-02	4.56E-01	2.27E+00	
$^{287}117$	12.76	13.25	12.15		1.90E-02	2.09E-03	3.56E-01	

the alpha decay energies and a small variation of the decay energy may result in a change of several orders of the magnitude of alpha-decay half-life. From Table 1, we can also see that the calculated half-lives of many isotopes of the Z=113 and Z=117 chains are within the range of current experimental capabilities. It will be very interesting to compare the above theoretical predictions with the experimental data in future.

Besides the alpha-decay mode, spontaneous fission is another limiting factor that may prevent the formation of long-lived superheavy elements.[17,18] From the theoretical point of view, the alpha-decay and spontaneous fission share the same underlying quantum tunnelling mechanism in physics. However, the spontaneous fission process of heavy and superheavy nuclei is much more complex as compared with alpha decay process. Microscopic calculation of spontaneous fission half-lives is very difficult due to both the complexity of fission process and the uncertainty of the height and shape of fission barrier. In previous studies, we have analyzed the systematical law of the spontaneous fission half-lives of heavy nuclei and a long lifetime line $\mathbf{N} - Z - 52$ is discussed in detail.[17,18] Here we extend this long lifetime line to heavier mass region with Z=104−114 and the spontaneous fission half-lives in this region can be approximately described by.

Table 2. Comparison of experimental and calculated spontaneous fission half-lives for isotopes of Z=104−114.

Decay	N−Z−52	T_{sf}^{exp}(s)	T_{sf}^{cal}(s)	Ref.
^{254}Rf	-6	2.3×10^{-5}	3.0×10^{-4}	[19]
^{256}Rf	-4	6.2×10^{-3}	3.8×10^{-3}	[19]
^{258}Rf	-2	1.4×10^{-2}	2.5×10^{-2}	[19]
^{260}Rf	0	2.0×10^{-2}	8.6×10^{-2}	[19]
^{262}Rf	2	2.1×10^{0}	1.6×10^{-1}	[19]
^{258}Sg	-6	2.9×10^{-3}	3.0×10^{-4}	[19]
^{260}Sg	-4	7.0×10^{-3}	3.8×10^{-3}	[19]
^{262}Sg	-2	1.5×10^{-2}	2.5×10^{-2}	[20]
^{264}Sg	0	3.7×10^{-2}	8.6×10^{-2}	[20]
^{266}Sg	2	4.4×10^{-1}	1.6×10^{-1}	[21]
^{264}Hs	-4	2.0×10^{-3}	3.8×10^{-3}	[19]
^{270}Ds	-2	8.0×10^{-2}	2.5×10^{-2}	[16]
^{282}112	6	8.2×10^{-4}	7.5×10^{-2}	[6]
^{284}112	8	1.0×10^{-1}	2.0×10^{-2}	[4]
^{286}114	6	2.6×10^{-1}	7.5×10^{-2}	[6]

$$\mathrm{Log}_{10}\mathrm{T}_{\mathrm{sf}}(s) = c_1 + c_2(\mathbf{N} - \mathrm{Z} - 52) + c_3(\mathbf{N} - \mathrm{Z} - 52)^2 \tag{14}$$

where the values of the parameters are c_1=−1.063; c_2=0.200; c_3=−0.0351; We list the detailed results from this simple law in Table 2. Surprisingly, we can see from Table 2 that the calculated spontaneous fission half-lives agree remarkably well with the experimental ones. The mean deviation for 15 even-even nuclei is 0.642, which implies that the experimental spontaneous fission half-lives are generally reproduced within a factor of 5. This level of agreement is very good because one usually considers a deviation such as 10^{4-5} is still acceptable for spontaneous fission half-lives. To show the agreement more clearly, we give the comparison of experimental and calculated spontaneous fission half-lives for Rf and Sg isotopes in Fig.1. It is seen from Fig.1 that the theoretical curve follows the experimental points very well. For the odd-A and odd-odd nuclei, the spontaneous fission half-lives are much longer than those of their neighborhood even-even nuclei. The systematical law can also yield reasonable agreement between experiment and theory by taking the block effect into account.

4. Summary

To conclude, we have systematically investigated the ground-state properties of superheavy nuclei (Z=109−117) by the macroscopic-microscopic model (MM) and the relativistic mean-field model (RMF). The theoretical alpha-decay energies and half-lives are predicted for some superheavy nuclei to be synthesized in experiments. The spontaneous fission half-lives of superheavy nuclei (Z=104−114) are also analyzed and a systematical law is proposed. The systematical law of spontaneous

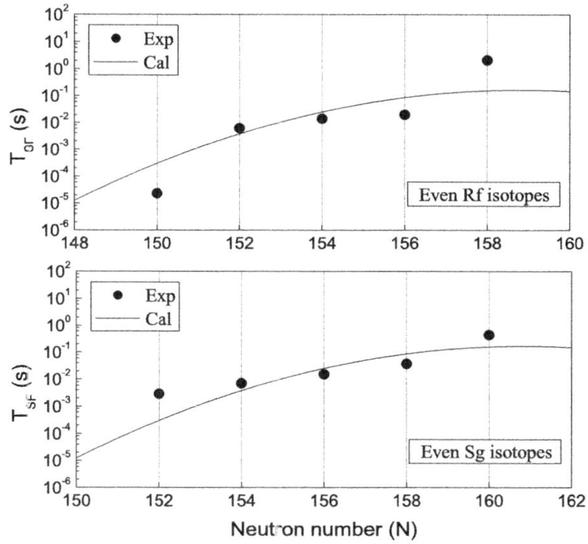

Fig. 1. The comparison of experimental and calculated spontaneous fission half-lives of Rf and Sg isotopes.

fission half-lives reproduces the experimental data very well and may be used to make predictions for unknown mass range in future.

References

1. S. Hofmann and G. Münzenberg, *Rev Mod. Phys.* **72**, 733 (2000).
2. Yu. Ts. Oganessian *et al.*, *Nature* **400**, 242 (1999).
3. Yu. Ts. Oganessian *et al.*, *Phys. Rev.* **C63**, 011301 (2001).
4. Yu. Ts. Oganessian *et al.*, *Phys. Rev.* **C70**, 064609 (2004).
5. Yu. Ts. Oganessian *et al.*, *Phys. Rev.* **C72**, 034611 (2005).
6. Yu. Ts. Oganessian *et al.*, *Phys. Rev.* **C74**, 044602 (2006).
7. T. N. Ginter *et al.*, *Phys. Rev.* **C67**, 064609 (2003).
8. K. Morita *et al.*, *Nucl. Phys.* **A734**, 101 (2004).
9. K. Morita *et al.*, *J. Phys. Soc. Jpan.* **73**, 2593 (2004).
10. C. M. Folden III *et al.*, *Phys. Rev. Lett.* **93**, 212702 (2004).
11. P. Möller *et al.*, *At. Data and Nucl. Data tables.* textbf66, 131 (1997).
12. Qijun Zhi, Yinchen Mao, Zhong-zhou Ren *Frontiers of Physics in China* **1**, 143 (2006).
13. Zhong-Zhou Ren *et al.*, *Phys. Rev.* **C67**, 064302 (2003).
14. Zhong-Zhou Ren *et al.*, *Phys. Rev.* **C66**: 064306 (2002).
15. Zhong-Zhou Ren, *Phys. Rev.* **C65**: 051304(R) (2002).
16. G. Audi *et al.*, *Nucl. Phys.* **A729**, 337 (2003).
17. Chang Xu and Zhong-Zhou Ren, *Phys. Rev.* **C71**, 014309 (2005).
18. Zhong-Zhou Ren and Chang Xu, *Nucl. Phys.* **A759**, 64 (2005).
19. N. E. Holden and D. C. Hoffman, Pure. Appl. Chem. **72**, 1525 (2000).
20. K. E. Gregorich, J. M. Gates, Ch. E. Düllmann, R. Sudowe, S. L. Nelson, M. A. Garcia,

I. Dragojević, C. M. Folden III, S. H. Neumann, D. C. Hoffman, and H. Nitsche, Phys. Rev. C **74**, 044611 (2006).

21. J. Dvorak, W. Brüchle, M. Chelnokov, R. Dressler, Ch. E. Düllmann, K. Eberhardt, V. Gorshkov, E. Jäger, R. Krücken, A. Kuznetsov, Y. Nagame, F. Nebel, Z. Novackova, Z. Qin, M. Schädel, B. Schausten, E. Schimpf, A. Semchenkov, P. Thörle,7 A. Türler, M. Wegrzecki, B. Wierczinski, A. Yakushev, and A. Yeremin, *Phys. Rev. Lett.* **97**, 242501 (2006).

NEW RESULTS IN THE STUDY OF SUPERFLUID NUCLEI: MANY-BODY EFFECTS, SPECTROSCOPIC FACTORS

P.F. BORTIGNON*, L. MICENE and E. VIGEZZI

Dipartimento di Fisica, University of Milan,and INFN
Sezione di Milano, I-20133 Milano, Italy
** E-mail: pierfrancesco.bortignon@mi.infn.it, enrico.vigezzi@mi.infn.it*

F. BARRANCO

Escuela Superior de Ingenieros Industriales,Universidad de Sevilla
41092 Sevilla, Spain
E-mail: barranco@us.es

R.A. BROGLIA

Dipartimento di Fisica, University of Milan,and INFN
Sezione di Milano, I-20133 Milano, Italy, and
The Niels Bohr Institute, University of Copenhagen
D-2100 Copenhagen, Denmark
E-mail: ricardo.broglia@mi.infn.it

A quantitative calculation of the state-dependent pairing gap requires to go beyond the quasiparticle approximation, treating in detail the breaking of the single-particle strength (self-energy processes), and the induced interaction arising from the exchange of (QRPA) collective vibrations between pairs of nucleons, not to mention the vertex-corrections. Different Skyrme forces have been used for the HF basis to obtain different level densities at the Fermi energy, and special attention has been paid to the calculation of spectroscopic factors with the comparison with the (rather old) existing data. The agreement is encouraging, but the need of new data will be discussed.
Mention will be done of two other important works, discussing the relation of the typical induced-interaction matrix elements with those of the Gogny force and of the v_{low-k} nucleon-nucleon potential, and pointing-out also in finite nuclei the (relatively small) screening effect of the exchange of spin modes in the induced interaction.

Keywords: bare and induced pairing interaction, self-energy effects, spectroscopic factors

1. Introduction

In the revival of nuclear structure studies produced by the availability of radioactive beams, the problem of nuclear pairing, introduced in the realm of nuclear physics 50 years ago,[1] is again in the forefront of the theoretical interest.[2,3] This is because, for drip-line nuclei with very small separation energy, the pairing gap Δ is of the same order of the Fermi energy E_F, and the pairing component of the effective interaction plays a basic role in the stability of the systems. In the most common approach the

HF+BCS or HFB mean-field equations are solved using effective interactions of zero- or finite-range, eventually density dependent,[4] with parameters fitted mainly to the ground-state (pairing) properties of chosen nuclei.

However, a quantitative calculation of the state-dependent pairing gap requires to go beyond the quasiparticle approximation, treating in detail the breaking of the single-particle strength (self-energy processes), and the induced interaction arising from the exchange of (QRPA) collective vibrations between pairs of nucleons, not to mention the vertex-corrections. Therefore, we have followed in a number of recent papers[5–7] a different strategy, starting from an HF calculations with different Skyrme interactions (to have different effective masses, that is different level density). Then, we solve the pairing problem with a bare interaction in the $^1S_0$ channel. Finally, we add correlations beyond mean-field, as shown in Fig. 1. It is stressed, that the order of magnitude of the induced pairing interaction is of the same, well known order of the self-energy contribution,[8] and eventually considerably larger than the bare force from the exchange of low-frequency modes.[6,9]

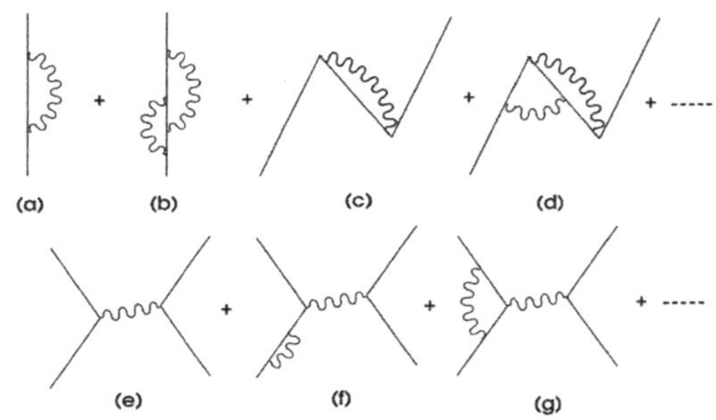

Fig. 1. Renormalization processes arising from the coupling between quasiparticles (straight lines) and phonons (wavy lines). Processes (a-d) renormalize the self-energy of the quasiparticles, while processes (e-g) involve the exchange of a vibration, and are the source of the induced pairing interaction

2. Formalism

The basic ingredients in the processes of Fig. 1 are the particle-vibration vertices

$$V(a, b\lambda) = h(a, b\lambda)(u_a u_b - v_a v_b)$$
$$W(a, b\lambda) = -h(a, b\lambda)(u_a v_b + v_a u_b) \tag{1}$$

where u e v are the usual BCS amplitudes and

$$h(a, b\lambda) = (-1)^{(j_a+j_b)}(2j_a+1)^{-1/2}(2\lambda+1)^{-1/2} < j_b||k_\lambda \beta_\lambda Y_\lambda||j_a >, \tag{2}$$

with $k_\lambda \beta_\lambda$ the phonon transition potential, calculated according to the collective model of Ref.[9]

The appropriate framework to perform self-consistent calculations with the couplings described above, to obtain a renormalized, fragmented quasiparticle a, is the solution of the Dyson (Nambu-Gor'kov) equation with the normal and anomalous density Green function.[10–12]

We follow the equivalent treatment of Ref.,[5,13] diagonalizing the matrix

$$\begin{pmatrix} E_a & V(a,b\lambda) & 0 & W(a,b\lambda) \\ V(a,b\lambda) & (E_b+E_\lambda) & W(a,b\lambda) & 0 \\ 0 & W(a,b\lambda) & -E_a & -V(a,b\lambda) \\ W(a,b\lambda) & 0 & -V(a,b\lambda) & -(E_b+E_\lambda) \end{pmatrix} \begin{pmatrix} x_a \\ C_{b\lambda} \\ y_a \\ D_{b\lambda} \end{pmatrix} = E \begin{pmatrix} x_a \\ C_{b\lambda} \\ y_a \\ D_{b\lambda} \end{pmatrix} \tag{3}$$

and obtaining for every eigenvalue $E_a^i > 0$ new occupation factors

$$\begin{aligned} U_a^i &= x_a^i u_a - y_a^i v_a \\ V_a^i &= x_a^i v_a + y_a^i u_a \end{aligned} \tag{4}$$

for the fragmented eigenstate (a,i).

Introducing the BCS pairing field in the matrix projected on the quasiparticle space, we obtain the new rather transparent matrix

$$\left[\begin{pmatrix} E_a & \Delta_a \\ \Delta_a & -E_a \end{pmatrix} + \begin{pmatrix} \Sigma_{11}(E_a^i) & \Sigma_{12}(E_a^i) \\ \Sigma_{12}(E_a^i) & \Sigma_{22}(E_a^i) \end{pmatrix}\right] \begin{pmatrix} x_a^i \\ y_a^i \end{pmatrix} = E_a^i \begin{pmatrix} x_a^i \\ y_a^i \end{pmatrix} \tag{5}$$

being E_a è equal to $|e_a - \lambda_F|$ and

$$\begin{aligned} \Sigma_{11}(E) &= \sum_{b\lambda} \left(\frac{|V(a,b\lambda)|^2}{E-(E_b+E_\lambda)} + \frac{|W(a,b\lambda)|^2}{E+(E_b+E_\lambda)} \right) \\ \Sigma_{22}(E) &= \sum_{b\lambda} \left(\frac{|W(a,b\lambda)|^2}{E-(E_b+E_\lambda)} + \frac{|V(a,b\lambda)|^2}{E+(E_b+E_\lambda)} \right) \\ \Sigma_{12}(E) &= \sum_{b\lambda} V(a,b\lambda)W(a,b\lambda) \left(\frac{1}{E-(E_b+E_\lambda)} - \frac{1}{E+(E_b+E_\lambda)} \right) \end{aligned} \tag{6}$$

Note that Σ_{12} is vanishing in normal systems.

The total gap for each eigenstate is given by[12]

$$\Delta_a^* = (\Sigma_{12} + \Delta_a)/Z \tag{7}$$

with

$$Z = 1 - (\Sigma_{11} + \Sigma_{22})/2E. \tag{8}$$

3. Results

In the very recent work of Ref.,[14] see also Ref.,[5] we used for our calculation of the pairing properties of the atomic nucleus ^{120}Sn, three Skyrme interactions, SgII, SkM* and SLy4 (see Ref.[15] and references therein) to obtain the HF mean-fields. They are characterized by the HF effective masses[8] m^* shown in Fig. 2.

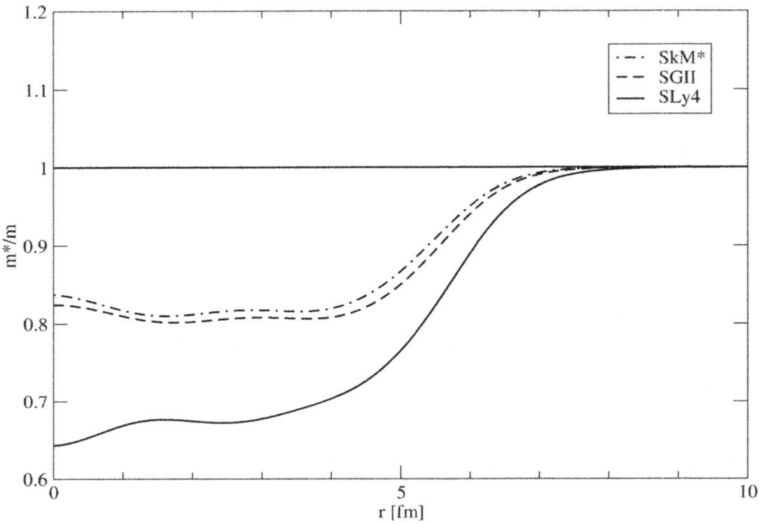

Fig. 2. Spatial dependence of the effective mass calculated in ^{120}Sn and associated with the three Skyrme interactions discussed in the text: SgII (dash-dotted line), the SLy4 (solid line) and SkM* (dashed line).

Then, extended BCS calculations (including pairs of particles with different number of nodes) were performed using the bare nucleon-nucleon v_{14} Argonne potential. The results are shown in Fig. 3(a).

The relative trend of the gap values obtained with the different interactions is understood, having in mind Fig. 2 and that[9]

$$\Delta \propto \exp\left(-\frac{1}{G\rho}\right) \tag{9}$$

where ρ is the level density, and G stays for the constant matrix elements of a simplified pairing interaction, and with

$$\rho(E^*) \propto \exp\left(2\sqrt{aE^*}\right) \tag{10}$$

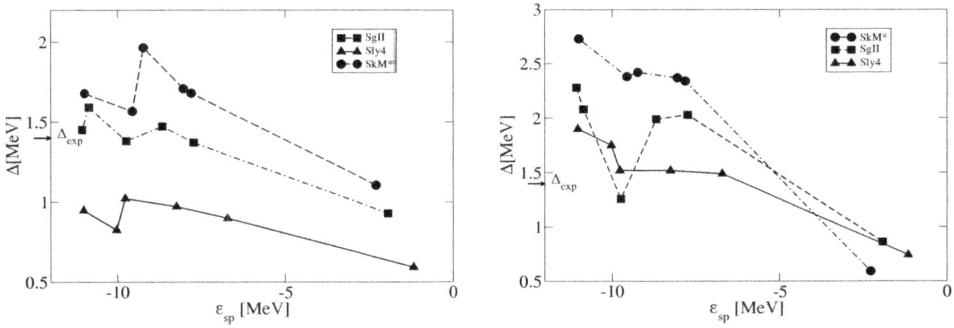

Fig. 3. (a) State-dependent pairing gaps obtained solving the gap equation for neutrons in ^{120}Sn with the Argonne interaction, using different mean fields, obtained with the SgII (squares), the SLy4 (triangles) and the SkM* interactions (dots). The symbols indicate the values of the pairing gaps associated with the quasi particle levels close to the Fermi energy, which have , in order of increasing energy, the quantum numbers $J^\pi = 5/2^+, 1/2^+, 7/2^+, 3/2^+, 11/2^-$ and $7/2^-$. (b) State-dependent pairing gaps obtained taking into account the renormalization processes discussed in the text.

$$a \propto \frac{A}{\epsilon_F} \propto m^*(\epsilon_F) \tag{11}$$

In Fig. 3(b), we display the state-dependent pairing gaps obtained taking into account the renormalization processes of Fig. 1. For the vibrations we couple to, the collective low-lying (surface) phonons are used, with their experimental properties. The attractive character of the contribution (e) produces the antiscreening effect, increasing substantially the gap values.

From the square of the U_a^i and V_a^i quantities, the distributions of the quasiparticle strength can be obtained and compared with the spectroscopic factors from transfer reactions like ^{120}Sn(d,p)^{121}Sn[16] or ^{120}Sn(^{3}He,α)^{119}Sn.[17] This is done in Figs. 4–6. The agreement is encouraging. Theoretical spectra appear to be too fragmented (an unusual feature[18]), but one should also consider that the (rather old) data have low resolution and cover a small interval in excitation energy.

In Fig. 7, from Ref.,[6] we show the relation of the typical matrix elements of the calculated induced interaction with those of the Gogny force (fitted to large extent also to the pairing properties of selected nuclei) and of the v_{low-k} nucleon-nucleon potential.

In Ref.,[7] the (relatively) small screening effect produced in finite nuclei from the exchange of spin modes is discussed, being the most important mechanism in neutron matter, see e.g.[19,20]

4. Perspectives

For the next future, we need to be self-consistent in the p-h channel, but also we need new data, as discussed by Betty Tsang in her talk.

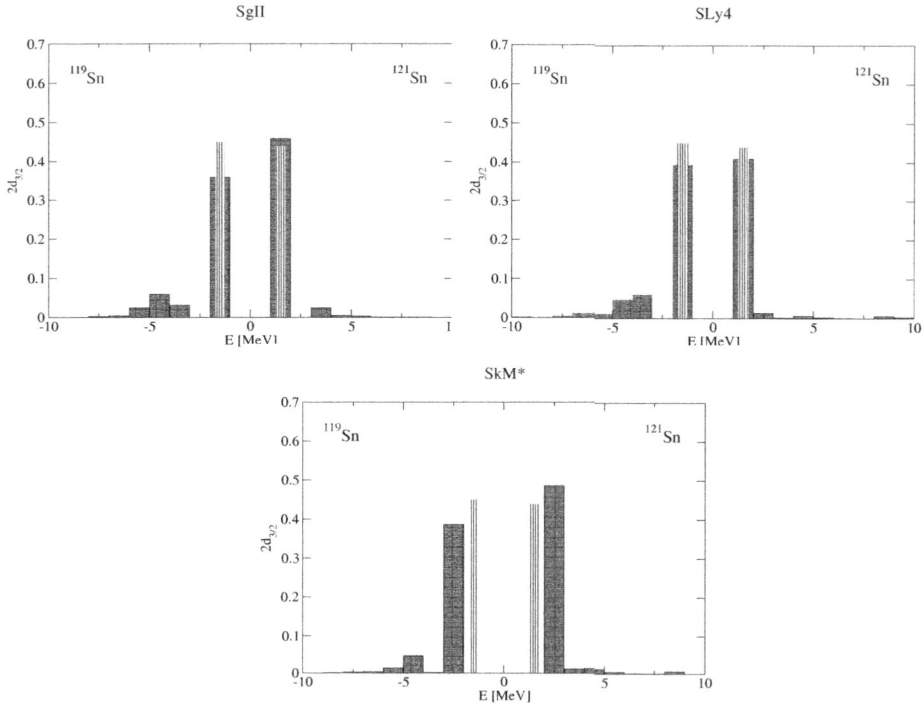

Fig. 4. The thick bars show the calculated spectroscopic factors associated with one-neutron pick-up and stripping reactions for the $3/2^+$ quasiparticle in ^{120}Sn, obtained integrating the single-particle strength in bins of 1 MeV width, using the interactions SgII (a), SLy4 (b), or SkM* (c) to produce the Hartree-Fock mean field. The thin bars show the experimental spectroscopic factors obtained from transfer reactions in two different experiments, as discussed in the text.

References

1. A. Bohr, B.R. Mottelson and D. Pines, *Phys. Rev.* **110**, 936 (1958).
2. D.J. Dean and M. Hjorth-Jensen, *Rev. Mod. Phys.* **75**, 607 (2003).
3. D.M. Brink and R.A. Broglia, *Nuclear Superfluidity. Pairing in Finite Systems* (Cambridge Univ. Press, Cambridge, 2005)
4. G. F. Bertsch and H. Espensen, *Ann. Phys.* (NY) **209**, 227 (1991); N. Sandulescu, P. Schuck and X. Viñas, *Phys. Rev.* **C71**, 054303 (2005).
5. F. Barranco et al., *Eur. Phys. J.* **A21**, 57 (2004).
6. F. Barranco, P.F. Bortignon, R.A. Broglia, G. Colò, P. Schuck and X. Viñas, *Phys. Rev.* **C72**, 054314 (2005).
7. G. Gori et al, *Phys. Rev.* **C72**, 011302(R) (2005).
8. C. Mahaux, P.F. Bortignon, R.A. Broglia and C.H. Dasso, *Phys. Rep.* **120**, 1 (1985).
9. A. Bohr and B.R. Mottelson, *Nuclear Structure* Vol. **II** (Benjamin, Reading 1975).
10. J.R. Schrieffer, *Theory of Superconductivity* (Addison Wesley, Redwood, 1964).
11. J. Terasaki et al., *Nucl. Phys.* **A697**, 126 (2002).
12. J. Terasaki et al., *Progr. Theor. Phys.* **108**, 495 (2002).
13. V. Van der Sluys et al., *Nucl. Phys.* **A551**, 210 (1993).
14. L. Micene, *Master Thesis* (unpublished, 2007).

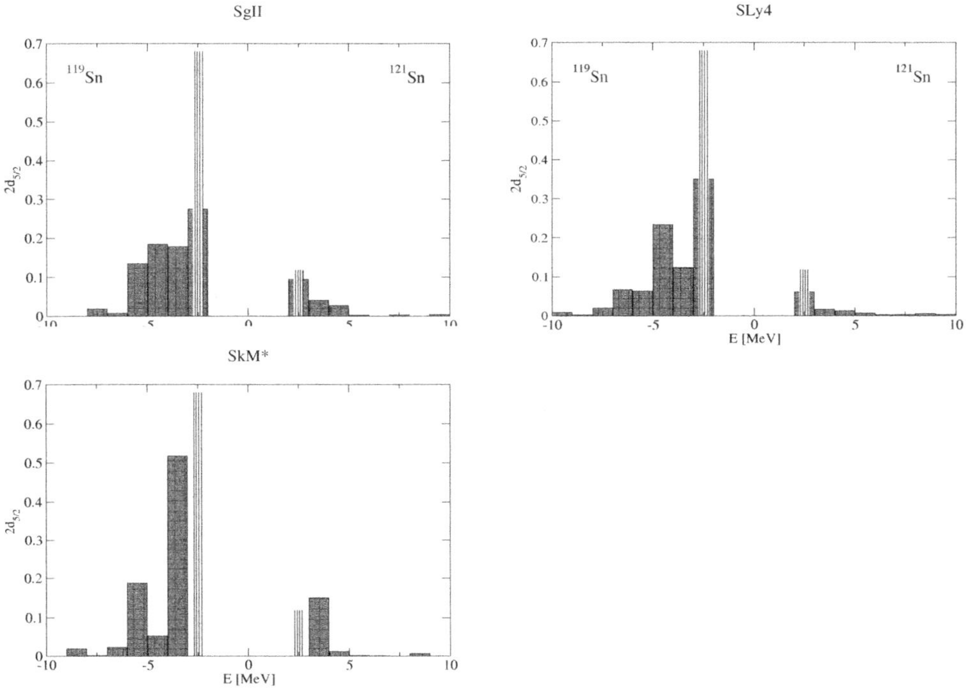

Fig. 5. The same as Fig. 4 but for the $5/2^+$ quasiparticle.

15. E. Chabanat et al., *Nucl. Phys.* **A627**, 710 (1997).
16. J. Bechara and O. Dieztch, *Phys. Rev.* **C12**, 90 (1975).
17. E. Gerlic et al., *Phys. Rev.* **C21**, 124 (1980).
18. G.F. Bertsch, P.F. Bortignon and R.A. Broglia, *Rev. Mod. Phys.* **55**, 287 (1983).
19. L.G. Cao, U. Lombardo and P. Schuck *Phys. Rev.* **C74**, 064301 (2006).
20. Several Talks at the INT Workshop (June 2007) on *The Neutron Star Crust and Surface: Observations and Models* (http://www.int.washington.edu)

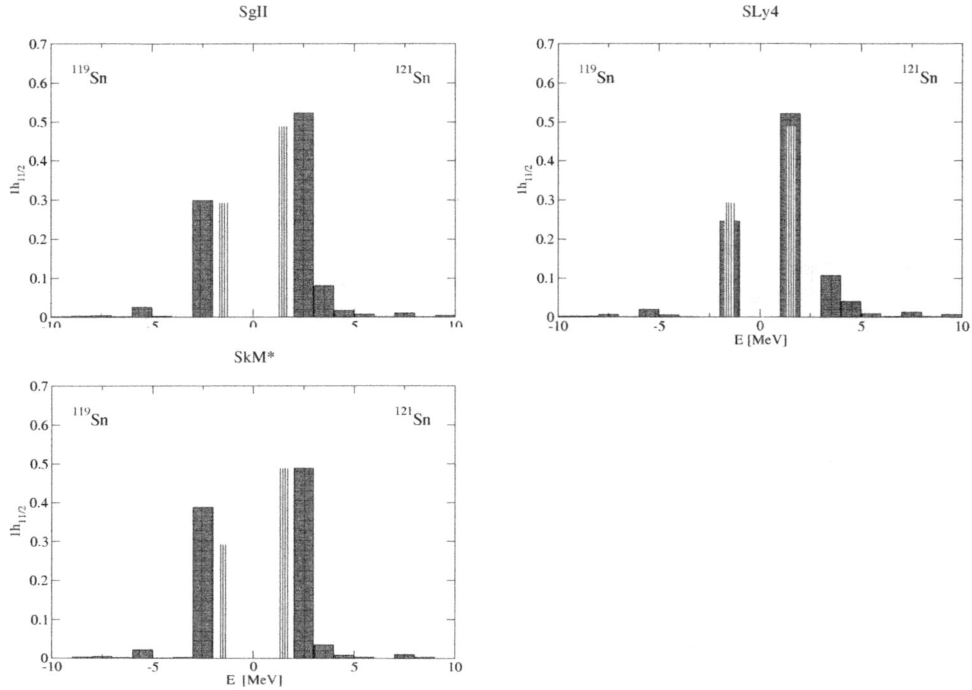

Fig. 6. The same as Fig. 4 but for the $11/2^-$ quasiparticle.

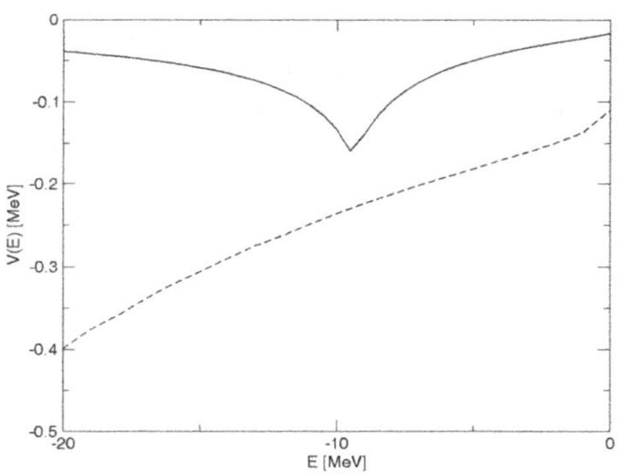

Fig. 7. Diagonal pairing matrix elements of the induced interaction (dash-dotted curve), obtained in ^{120}Sn as a function of the single-particle energy using a semiclassical formalism (from Ref.[6]). They are compared with the matrix elements of the Gogny force (solid curve) and with the matrix elements of the v_{low-k} potential (dashed curve). The mean field is a Woods-Saxon with effective mass equal to the bare mass.

NEW EFFECTIVE NUCLEON-NUCLEON INTERACTION FOR THE MEAN-FIELD APPROXIMATION

V. KIM AU and S. SHLOMO

Cyclotron Institute, Texas A&M University,
College Station, Texas 77840, USA
E-mail: auvuong@comp.tamu.edu, shlomo@comp.tamu.edu
http://cyclotron.tamu.edu

B. K. AGRAWAL

Theory Division, Saha Institute of Nuclear Physics
Kolkata, India

We first present a new and more realistic Skyrme type interaction, named KDE0, obtained by fitting Hartree-Fock results to an extensive experimental data under some constraints and using the simulating annealing method. We then discuss the issue of deducing the nuclear matter incompressibility coefficient from data on the energies of the breathing mode within relativistic and non-relativistic models.

Keywords: Hartree-Fock; Skyrme interaction; Nuclear matter compressibility; Simulated annealing method.

1. Introduction

Since the early work of Brink and Vautherin,[1] significant efforts have been made to refit the parameters of the Skyrme type effective nucleon-nucleon (NN) interaction to better reproduce experimental data. We note that the parameters of the Skyrme interaction available in the literature were obtained by fitting of the Hartree-Fock (HF) results to experimental data on properties of a few stable closed-shell nuclei. Recently, we have implemented, for the first time,[2] the simulated annealing method (SAM)[3] to determine the values of the Skyrme parameters. For obtaining a more realistic parameterization of the Skyrme interaction, we apply the SAM to fit the HF results to an extensive set of experimental data, including some constraints. In the following we present the ingredients used in determining the new KDE0 Skyrme interaction.[2] We also show the results of fully self-consistent HF based random phase approximation (RPA) calculation for the centroid energy, E_0, of the isoscalar giant monopole resonance (ISGMR) obtained using the KDE0 and other Skyrme interactions. We compare these results with those obtained within a relativistic models and discuss the current status of the nuclear matter incompressibility coefficient, $K_{\text{n.m.}}$, which is important in the study of properties of nuclei, heavy ion collisions, and in Nuclear Astrophysics.

2. Hartree-Fock with Skyrme interaction

We adopt the following standard form for the Skyrme type effective NN interaction:[1,4]

$$\begin{aligned}V_{12} &= t_0\,(1 + x_0 P_{12}^{\sigma})\,\delta(\mathbf{r}_1 - \mathbf{r}_2)\\ &+\frac{1}{2}t_1\,(1 + x_1 P_{12}^{\sigma}) \times \left[\overleftarrow{k}_{12}^{2}\delta(\mathbf{r}_1 - \mathbf{r}_2) + \delta(\mathbf{r}_1 - \mathbf{r}_2)\,\overrightarrow{k}_{12}^{2}\right]\\ &+t_2\,(1 + x_2 P_{12}^{\sigma})\,\overleftarrow{k}_{12}\delta(\mathbf{r}_1 - \mathbf{r}_2)\,\overrightarrow{k}_{12}\\ &+\frac{1}{6}t_3\,(1 + x_3 P_{12}^{\sigma})\,\rho^{\alpha}\left(\frac{\mathbf{r}_1 + \mathbf{r}_2}{2}\right)\delta(\mathbf{r}_1 - \mathbf{r}_2)\\ &+iW_0\,\overleftarrow{k}_{12}\delta(\mathbf{r}_1 - \mathbf{r}_2)(\overrightarrow{\sigma_1} + \overrightarrow{\sigma_2}) \times \overrightarrow{k}_{12},\end{aligned} \tag{1}$$

where t_i, x_i, α and W_0 are the parameters of the interaction. In our HF calculations, we use all the components of this interaction, Eq. (1), employing the following approximations:

(i) We adopt the direct term and neglect the exchange term of the Coulomb energy to compensate for the missing contributions due to charge symmetry breaking (CSB) and correlations.

(ii) For the center of mass correction to the binding energy we use the harmonic oscillator value: $E_{CM}^{HO} = \frac{3}{4}\hbar\omega$ with $\hbar\omega = \frac{\hbar^2}{mA\langle r^2\rangle}\sum_i[\mathcal{N}_i + \frac{3}{2}]$.

(iii) The charge mean-square radius, fitted to the experimental data, is obtained as

$$\langle r_{ch}^2\rangle = \langle r_p^2\rangle_{\rm HF} - \frac{3}{2\nu A} + \langle r^2\rangle_p + \frac{N}{Z}\langle r^2\rangle_n + \frac{1}{Z}\left(\frac{\hbar}{mc}\right)\sum_{nlj\tau}(2j+1)\mu_\tau\langle\vec{\sigma}\cdot\vec{l}\rangle_{lj}, \tag{2}$$

with $\langle r^2\rangle_n = -0.12$ fm^2 and $\langle r^2\rangle_p = 0.801$ fm^2. The second term on the RHS of (2) is due to the center of mass motion and the last term is due to spin-orbit effect.

3. Determining the Skyrme interaction using the SAM

We use the simulated annealing method (SAM) to determine the parameters of the new Skyrme type effective NN interaction by searching for the global minimum in the hyper-surface of the χ^2 function,

$$\chi^2 = \frac{1}{N_d - N_p}\sum_{i=1}^{N_d}\left(\frac{M_i^{\rm exp.} - M_i^{\rm th.}}{\sigma_i}\right)^2. \tag{3}$$

Here N_d is the number of experimental data points, N_p is the number of fitted parameters, σ_i is the theoretical uncertainty and $M_i^{\rm exp.}$ and $M_i^{\rm th.}$ are the experimental and the corresponding theoretical values, respectively, which are associated with a certain observable. Note that the values of χ^2 depends on the Skyrme parameters, since, the $M_i^{\rm th.}$ in Eq. (3) is calculated by using the HF approach with the Skyrme interaction.

3.1. *Implementing the SAM to search for the minimum of χ^2*

We now describe the steps employed in the SAM for determining the parameters of the Skyrme interaction.[2]

(1) The Skyrme parameters t_i, x_i and α for a fixed value of W_0 can be expressed in terms of the quantities B/A, $K_{\text{n.m.}}$, $\rho_{\text{n.m.}}$, m^*/m, E_s, J, L, κ, and G_0' denoting the binding energy per nucleon, incompressibility coefficient, nuclear matter density, effective mass, surface energy, symmetry energy coefficient, the quantity which is related to the slope of the symmetry energy coefficient ($L = 3\rho dJ/d\rho$), the enhancement factor of the energy weighted sum rule (EWSR) of the isovector giant dipole resonance (IVGDR) and Landau parameter, respectively.
(2) Define a vector **v**

$$\mathbf{v} \equiv (B/A, K_{\text{n.m.}}, \rho_{\text{n.m.}}, m^*/m, E_s, J, L, \kappa, G_0', W_0). \tag{4}$$

(3) Calculate χ^2_{old} for a given set of experimental data and the corresponding HF results (using an initial guess for the Skyrme parameters).
(4) Determine a new set of Skyrme parameters by the following steps:
   (a) use a uniform random number to select a component v_r of the vector **v**,
   (b) use another random number η $(-1 < \eta < 1)$ and the displacement d_r to get a new value for v_r

$$v_r \to v_r + \eta d_r, \tag{5}$$

   (c) use this modified **v** to generate a new set of Skyrme parameters
(5) Go back to HF to calculate χ^2_{new}
(6) The new set of Skyrme parameters is accepted only if

$$\mathcal{P}(\chi^2) = e^{(\chi^2_{\text{old}} - \chi^2_{\text{new}})/T} > \beta, \tag{6}$$

   where β $(0 < \beta < 1)$ is another random number,
(7) Start with an initial value of the effective temperature $T = T_i$, and repeat steps 4 - 6 for a large number of loops.
(8) Reduce the effective temperature T as $T = \frac{T_i}{k}$ and repeat steps 1 - 7, with $T_i = 1.25$ and $k = 1, 2, 3, ...$
(9) Keep doing this until hopefully reaching the global minimum of χ^2.

3.2. *Experimental data and some constraints*

We have fitted the values of the Skyrme parameters to an extensive experimental data together with a few additional constraints (see Table 1). The data set consists of the binding energies for 14 nuclei ranging from normal to exotic (proton- or neutron-rich) ones, charge root mean square (rms) radii for 7 nuclei, spin-orbit splittings for the $2p$ proton and neutron (bare) single particle states of the ^{56}Ni nucleus, and the rms radii for $1d_{5/2}$ and $1f_{7/2}$ valence neutron orbits in the ^{17}O and ^{41}Ca nuclei,

Table 1. Selected experimental data for the binding energy B, charge rms radius $r_{ch.}$, rms radii of valence neutron orbits r_v, spin-orbit splitting S-O, breathing mode constrained energy E_0 and critical density $\rho_{cr.}$ used in the fit to determine the parameters of the Skyrme interaction.

Properties	Nuclei
B	^{16,24}O, ^{34}Si, ^{40,48}Ca, ^{48,56,68,78}Ni, ^{88}Sr, ^{90}Zr, ^{100,132}Sn, ^{208}Pb
$r_{ch.}$	^{16}O, ^{40,48}Ca, ^{56}Ni, ^{88}Sr, ^{90}Zr, ^{208}Pb
$r_v(\nu 1d_{5/2})$	^{17}O
$r_v(\nu 1f_{7/2})$	^{41}Ca
S-O	$2p$ orbits in ^{56}Ni
$\rho_{cr.}$	nuclear matter
E_0	^{90}Zr, ^{116}Sn, ^{144}Sm, ^{208}Pb

respectively. Also included in the fit are the critical density ρ_{cr}, determined from the stability conditions for the Landau parameters, to be close to 2.5 times the nuclear matter saturation density ρ_0. Additional constraints imposed on the Skyrme parameters are (i) the quantity $P = 3\rho(dS/d\rho)$, which is directly related to the slope of the symmetry energy S, must be positive for matter densities up to $3\rho_0$, a condition imposed by neutron star models ; (ii) the enhancement factor κ, associated with the EWSR for the IVDGR, should lie in the range of 0.1 - 0.5; and (iii) the Landau parameter G_0', important for the spin properties of finite nuclei and nuclear matter, should be positive at $\rho = \rho_0$. Also included in the fit are the experimental values of the breathing-mode constrained energy of ^{90}Zr, ^{116}Sn, ^{144}Sm and ^{208}Pb. Using these experimental data together with the additional constrains, we have carried out two different fits named as KDE0, where the Coulomb exchange term was neglected, and KDE (see Table 2). Figure 1 demonstrates the rapid convergence of the SAM method. The values of the parameters of the resulting KDE0 and KDE Skyrme interactions are given in Table 2 and compared with those of the Lyon[5] SLy7 Skyrme interaction. We emphasize that the parametrization KDE0 provides a better fit to the experimental data. The selection of the experimental data and the additional constraints ensures that the KDE0 interaction can be used to study the bulk ground-state properties of nuclei beyond the stability lines, as well as the properties of neutron stars.

4. Relativistic and Non-relativistic models

The importance of carrying out fully self-consistent calculations was demonstrated recently in Refs.[11,12] In the Table 3, we present the results of fully self-consistent HF-RPA calculation[12] for E_0 obtained using various Skyrme interactions and compare them with the relativistic mean-field based RPA (RRPA) results.[13] It is seen from Table 3 that the parameter set SK255 of the Skyrme interaction, deduced in Ref.,[7] yields for E_0 values which are quite close to the RRPA results with the NL3 interaction. Moreover, for the SK255 parameter set, one finds a good agreement with experimental data for E_0 for all nuclei considered, provided that the corresponding excitation energy ranges used in determining E_0 are the same as those

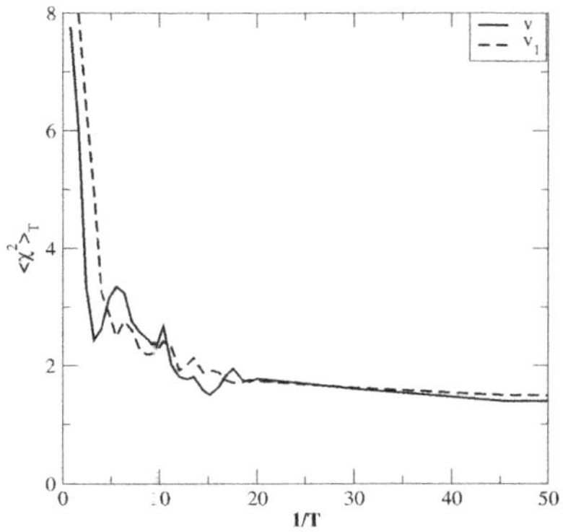

Fig. 1. Variation of the average value of $\langle\chi^2\rangle_T$ as a function of the inverse of the control parameter T for the KDE0 interaction for the two different choices of the starting parameters.

Table 2. The values of the Skyrme parameters for KDE0 and KDE interactions obtained by minimizing the χ^2. For the sake of comparison we have also listed the values of the parameters for the SLy7 interaction. The values in parenthesis are the standard deviations for the corresponding Skyrme parameters.

Parameter	KDE0	KDE	SLy7
t_0(MeV·fm^3)	-2526.5110 (140.6256)	-2532.8842 (115.3165)	-2482.41
t_1 (MeV·fm^5)	430.9418 (16.6729)	403.7285 (27.6336)	457.97
t_2 (MeV·fm^5)	-398.3775 (27.3099)	-394.5578 (14.2610)	-419.85
t_3(MeV·fm$^{3(1+\alpha)}$)	14235.5193 (680.7344)	14575.0234 (641.9932)	13677.0
x_0	0.7583 (0.0655)	0.7707 (0.0579)	0.8460
x_1	-0.3087 (0.0165)	-0.5229 (0.0298)	-0.5110
x_2	-0.9495 (0.0179)	-0.8956 (0.0270)	-1.0000
x_3	1.1445 (0.0862)	1.1716 (0.0767)	1.3910
W_0 (MeV·fm^5)	128.9649 (3.3258)	128.0572 (4.3943)	126.00
α	0.1676 (0.0163)	0.1690 (0.0144)	0.1667

used in obtaining the experimental data. Therefore, by demonstrating that it is possible to build bona fide Skyrme forces with associated values of K close to that obtained in relativistic models, we have resolved the discrepancy in the extracted value of K within relativistic and non-relativistic models.[7] It is also seen from Table 3 that the value of K increases with the value adopted for the symmetry energy coefficient J (and its slope).

This work was supported in part by the US National Science Foundation under Grant No. PHY-0355200 and the US Department of Energy under the Grant No. DOE-FG03-93ER40773. V. Kim Au was a Nishina Memorial Fellow at RIKEN.

References

1. D. Vautherin and D. M. Brink, *Phys. Rev. C* **5**, 626 (1972).
2. B.K. Agrawal, S. Shlomo and V. Kim Au, *Phys. Rev. C* **72**, 014310 (2005).

Table 3. Fully self-consistent HF-RPA results for the ISGMR centroid energy $E_0 = m_1/m_0$ (in MeV) obtained using the interactions SGII,[6] KDE0,[2] SK255[7] and compared with relativistic RPA results obtained with the NL3 interaction[8] (the energy range $\omega_1 - \omega_2$ (MeV) and the experimental data is taken from Refs.[9,10]). The incompressibility ($K_{n.m.}$) and symmetry energy (J) coefficients are given in units of MeV.

Nucleus	$\omega_1 - \omega_2$	Expt.	NL3	SK255	SGII	KDE0
^{90}Zr	$0 - 60$		18.7	18.90	17.89	18.03
	$10 - 35$	17.81±0.30		18.85	17.87	17.98
^{116}Sn	$0 - 60$		17.1	17.31	16.36	16.58
	$10 - 35$	15.85±0.20		17.33	16.38	16.61
^{144}Sm	$0 - 60$		16.1	16.21	15.26	15.46
	$10 - 35$	15.40±0.40		16.19	15.22	15.44
^{208}Pb	$0 - 60$		14.2	14.34	13.57	13.79
	$10 - 35$	13.96±0.20		14.38	13.58	13.84
$K_{n.m.}$			272	255	215	229
J			37.4	37.4	26.8	33.0

3. S. Kirkpatrik, C.D. Gelatt Jr. and M.P. Vecchi, *Science* **220**, 671 (1983).
4. E. Chabanat, P. Bonche, P. Haensel, J. Meyer and R. Schaeffer, *Nucl. Phys. A* **627**, 710 (1997).
5. E. Chabanat, P. Bonche, P. Haensel, J. Meyer and R. Schaeffer, *Nucl. Phys. A* **635**, 231 (1998).
6. N.V. Giai and H. Sagawa, *Phys. Lett. B* **106**, 379 (1981).
7. B.K. Agrawal, S. Shlomo and V. Kim Au, *Phys. Rev. C* **68**, 031304(R) (2003).
8. G.A. Lalazissis, J. Konig and P. Ring, *Phys. Rev. C* **55**, 540 (1997).
9. D.H. Youngblood, Y.W. Lui, H.L. Clark, B. John, Y. Tokimoto and X. Chen, *Phys. Rev. C* **69**, 034315 (2004).
10. D.H. Youngblood, Y.W. Lui, B. John, Y. Tokimoto, H. L. Clark and X. Chen, *Phys. Rev. C* **69**, 054312 (2004).
11. B.K. Agrawal and S. Shlomo, *Phys. Rev. C* **70**, 014308 (2004).
12. T. Sil, S. Shlomo, B.K. Agrawal and P.G. Reinhard, *Phys. Rev. C* **73**, 034316 (2006).
13. D. Vretenar, T. Niksic and P. Ring, *Phys. Rev. C* **68**, 024310 (2003).

LINEAR RESPONSE CALCULATIONS WITH THE TIME-DEPENDENT SKYRME DENSITY FUNCTIONAL

T. NAKATSUKASA* and K. YABANA

Center for Computational Sciences and Institute of Physics, University of Tsukuba, Tsukuba 305-8571, Japan

**Present address: Theoretical Nuclear Physics Laboratory, RIKEN Nishina Center, Wako 351-0198, Japan*

T. INAKURA

Institute of Physics, University of Tsukuba, Tsukuba 305-8571, Japan

Nuclear giant dipole resonances are investigated with the time-dependent Skyrme density-functional theory in the three-dimensional coordinate space. Numerical calculations are performed using two different methods: Real-time method and finite amplitude method. We investigate electric dipole responses in even-even Ne isotopes. and show that the low-lying pygmy peaks appear in deformed $^{26-30}$Ne, but not seen in $^{20-24}$Ne.

Keywords: Nuclear Structure; Giant Dipole Resonance; Time-Dependent Hartree-Fock Method; Random-Phase Approximation.

1. Introduction

Atomic nuclei exhibit a variety of collective modes of excitation. In particular, the giant resonances always have been of central interest in nuclear structure and reaction studies. The giant resonances correspond to the most fundamental oscillations in nuclei, in the sense that they exhaust major portions of the energy-weighted sum-rule values and that the nucleus strongly absorbs energy from an external field, acting as a whole. Indeed, the giant resonances can be qualitatively described in terms of semiclassical models of the Fermi liquid.[1]

Among many kinds of giant resonances, the isovector giant dipole resonance (GDR) is the most famous and exhausts almost 100 % of the sum-rule value. It is well explained by ordinary hydrodynamical models. This is rather exceptional because quantitative description of other modes requires a treatment as a Fermi liquid.[1] The hydrodynamical models have a close connection to the time-dependent density-functional theory (TDDFT). In the limit of $\hbar \to 0$, the TDDFT equation goes over into the Vlasov equation. Therefore, the TDDFT should provide a microscopic description of an appropriate hydrodynamical model. We have proposed the real-time TDDFT method combined with the absorbing-boundary condition for a linear response function in the continuum.[2,3] This may be regarded as an extension of the continuum random-phase approximation (RPA), made applicable to a deformed system.

The real-time method is simple and efficient, but has a problem with the spurious zero modes.[3] The presence of the zero modes often leads to a violation of the linear (small-amplitude) approximation. Recently, we have proposed a new computational approach to the RPA calculation, "finite amplitude method" (FAM),[4] to overcome this difficulty yet keeping simplicity of the real-time method. This method does not require explicit evaluation of complex residual interactions of the Skyrme functional. Instead, it resorts to numerical differentiation of the single-particle Hamiltonian with respect to the single-particle ket and bra states. We have developed a computer code to calculate response functions with the Skyrme functional in the fully self-consistent manner.

In this paper, we show real-time calculation of the TDDFT in the continuum for ^{24}Mg and the FAM calculation for Ne isotopes. We show how well the Skyrme functional describes the GDR resonances in known nuclei, then discuss pygmy resonances in neutron-rich nuclei.

2. Real-time method

The TDDFT is simply expressed by the time-dependent Kohn-Sham equations,

$$i\hbar\frac{\partial}{\partial t}\phi_i(\vec{r},t) = \left\{h[\rho(t)](\vec{r},t) + V_{\text{ext}}(\vec{r},t)\right\}\phi_i(\vec{r},t), \tag{1}$$

$$\rho(\vec{r},t) = \sum_{i=1}^{A}\phi_i(\vec{r},t)\phi_i^*(\vec{r},t), \tag{2}$$

where $h[\rho]$ is the Kohn-Sham Hamiltonian which is a functional of the time-dependent density $\rho(\vec{r},t)$ and their derivatives. Although the proof given in Ref. 5 guarantees that the exact many-body dynamics can be described by Eq. (1), the density functional depends on the initial state. Normally, one uses the same functional as that used in the static calculation for the ground state, which is called "adiabatic approximation". We adopt this approximation in the present work. We use the Skyrme energy functional of Ref. 6. The functional contains densities which are odd with respect to the time reversal; the spin density $\vec{s}(\vec{r})$, the current density $\vec{j}(\vec{r})$, etc. These time-odd densities are irrelevant in the ground state, however, some of them are important in the time-dependent phenomena[3] and necessary to keep the local Galilean invariance.[7]

$$h[\rho(t)]\phi_i(\vec{r},\sigma,\tau,t) = \sum_{\sigma'}\Big\{-\vec{\nabla}\frac{\hbar^2}{2m_\tau^*(\vec{r})}\vec{\nabla}\delta_{\sigma\sigma'} + U_\tau(\vec{r})\delta_{\sigma\sigma'} + \vec{V}_\tau(\vec{r})\cdot\langle\sigma|\vec{\sigma}|\sigma'\rangle$$
$$+ i\vec{C}_\tau(\vec{r})\cdot\vec{\nabla}\delta_{\sigma\sigma'} + i\vec{W}_\tau(\vec{r})\cdot(\langle\sigma|\vec{\sigma}|\sigma'\rangle\times\vec{\nabla})\Big\}\phi_i(\vec{r},\sigma',\tau,t), \tag{3}$$

where the indexes σ and τ indicates the intrinsic spin ($=\pm 1$) and isospin (n and p), respectively. m_τ^* is an effective mass for nucleons ($\tau = n, p$). In finite nuclei, the effective mass also depends on the coordinates. The spin-independent part has two components, $U_\tau(\vec{r})$ and $\vec{C}_\tau(\vec{r})$. The former is even and the latter is odd with respect to the time reversal. The spin-dependent part also has a time-even term, $\vec{W}(\vec{r})$, and a time-odd term, $\vec{V}(\vec{r})$.

Each Kohn-Sham orbital is evolved in time as

$$\phi_i(t) = \exp\Big(-\frac{i}{\hbar}\int_{t_0}^{t} h[\rho(t')]dt'\Big)\phi_i(t_0), \tag{4}$$

where we omit the external field V_{ext} for simplicity. This is a self-consistent equation: The Kohn-Sham Hamiltonian depends on the time-dependent density $\rho(t)$. The time evolution for a small time step Δt can be approximated as

$$\begin{aligned}\phi_i(t+\Delta t) &\approx \exp\Big(-\frac{i}{\hbar}h[\rho(t+\Delta t/2)]\Delta t\Big)\phi_i(t),\\ &\approx \sum_{n=0}^{M}\frac{1}{n!}\Big(-\frac{i}{\hbar}h[\rho(t+\Delta t/2)]\Delta t\Big)^n\phi_i(t).\end{aligned} \tag{5}$$

In order to justify the Taylor expansion in Eq. (5), the quantity, $\Delta t\epsilon_{\text{max}}/\hbar$, should be smaller than unity where ϵ_{max} is the maximum energy of the Kohn-Sham orbitals in the adopted model space.

In order to treat the single-particle continuum, we use the absorbing boundary condition (ABC).[2,3,8–10] A nice feature of the ABC method is its simplicity. First, we divide the coordinate space into the inner (relevant) and outer (irrelevant) spaces. The center-of-mass coordinate of the ground state is at the origin. The entire model space is the three-dimensional space of $r < R_0$, while the inner interacting region is defined by $r < R_c$ ($R_c < R_0$). Then, the time-dependent Kohn-Sham Hamiltonian is modified as

$$h[\rho(\vec{r},t)] \to h[\rho(\vec{r},t)] - i\eta(\vec{r}). \tag{6}$$

Here, $\eta(\vec{r})$ is active only in the outer space ($R_c < r < R_0$) and zero in the interacting region. The suitable asymptotic behavior is obtained by a suitable choice of this absorbing potential.[2,3,9,10] Now, we may use the vanishing boundary condition at the edges of the model space, $\phi_i(r \geq R_0, t) = 0$ for any value of t. Thus, even for the continuum states, we may use the same algorithm as that for the bound states.

Let us show results calculated with the SGII functional for the isovector dipole response in ^{24}Mg. The calculated ground state of ^{24}Mg is prolate. We take the external field as

$$V_{\text{ext}}(\vec{r}) = \epsilon d_K(\vec{r}) = \epsilon \sum_{t=n,p} e_t r Y_{1K}(\hat{r}), \tag{7}$$

where the recoil charge is defined by $e_n = (1-\tau_3)Z/2A$ and $e_p = (1+\tau_3)N/2A$. The parameter ϵ should be small enough to validate the linear (small-amplitude) approximation. After calculating the time evolution of the $E1$ moment $D_K(t) \equiv \sum_i\langle\phi_i(t)|d_K^\dagger|\phi_i(t)\rangle$, the Fourier transform leads to the $E1$ strength distribution. See Ref. 3 for details of the calculation.

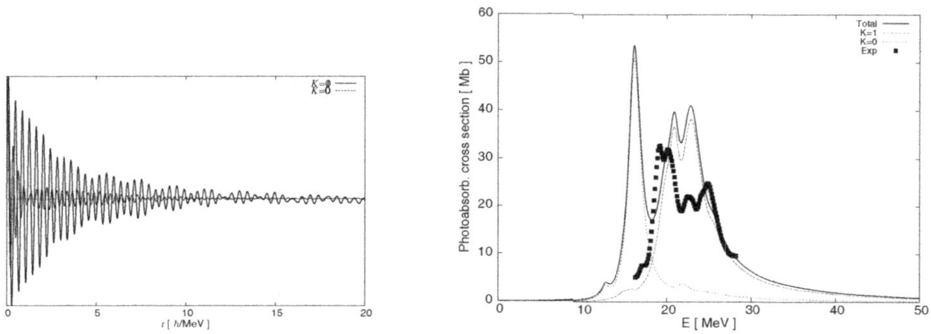

Fig. 1. Left: Time evolution of calculated $E1$ moment $D_0(t)$ for ^{24}Mg.
Right: Calculated photoabsorption cross section in ^{24}Mg as a function of photon energy. It is given by a sum of $K = 0$ and $K = 1$ components (dashed and dotted lines). Squares indicate the experimental data.[11]

The calculated $D_K(t)$ is shown in the left panel of Fig. 1 for $K = 0$. We perform the same calculation with the $K = 1$ external field, though it is not shown in Fig. 1. Then, performing the Fourier transform, we obtain the photoabsorption cross section shown in the right panel of Fig. 1. There is a clear deformation splitting. The calculated $K = 0$ (lower) peak is at $E = 16.2$ MeV. The experimental data suggest that this is around $E = 19$ MeV. The $K = 1$ (higher) peak is calculated at $E = 20 \sim 24$ MeV. This position is also slightly lower than the experimental data. However, we may say that the overall feature of the GDR is qualitatively reproduced in the TDDFT calculation with ABC.

3. Finite amplitude method

In this section, we recapitulate the finite amplitude method (FAM). See Ref. 4 for more details. The method is formulated in the frequency representation. The linearized TDDFT equation for the transitional density $\delta\rho(\omega)$ is given by

$$\omega\ \delta\rho(\omega) = [h_0, \delta\rho(\omega)] + [V_{\rm ext}(\omega) + \delta h(\omega), \rho_0], \tag{8}$$

where $h_0 = h[\rho_0]$ and ρ_0 is the density in the ground state. Since the KS Hamiltonian $h[\rho]$ can be regarded as a functional of bra and ket KS orbitals, we denote it as $h[\langle\phi|, |\phi\rangle]$ for a later purpose. We express the transition density as

$$\delta\rho(\omega) = \sum_i \left\{ |X_i(\omega)\rangle\langle\phi_i^0| + |\phi_i^0\rangle\langle Y_i(\omega)| \right\}, \tag{9}$$

introducing the forward and backward amplitudes. Here, $|\phi_i^0\rangle$ indicate the single-particle states in the ground state. This is equivalent to the Fourier decomposition of the time-dependent single-particle orbitals,[12]

$$|\delta\phi_i(t)\rangle = \sum_\omega \left\{ |X_i(\omega)\rangle e^{-i\omega t} + |Y_i(\omega)\rangle e^{i\omega t} \right\}. \tag{10}$$

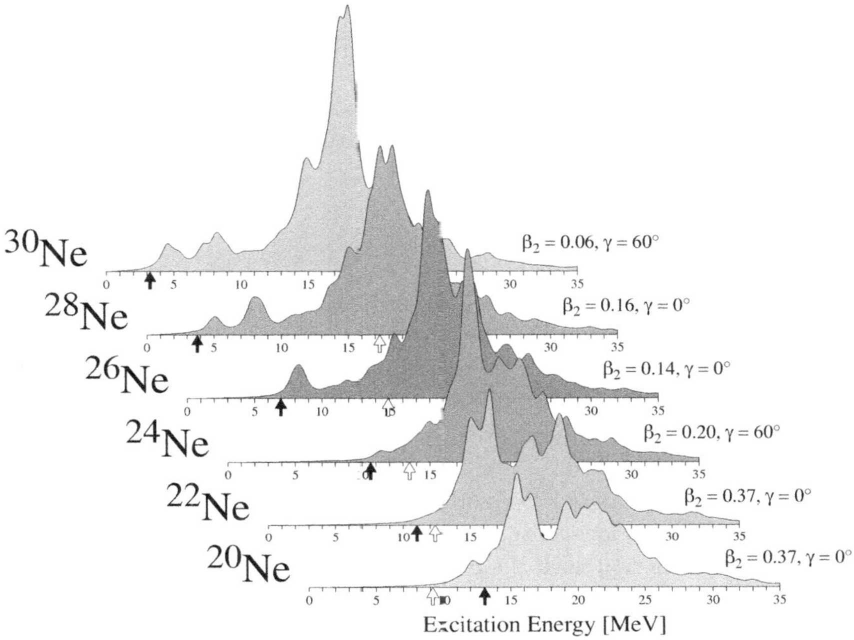

Fig. 2. $E1$ strength distribution for even-even Ne isotopes calculated with the SkM* functional. The filled (open) arrows indicate neutron (proton) separation energy.

Multiplying both sides of Eq. (8) with a ket of hole states $|\phi_i^0\rangle$, we have

$$\omega|X_i(\omega)\rangle = (h_0 - \epsilon_i)|X_i(\omega)\rangle + \hat{Q}\left\{V_{\rm ext}(\omega) + \delta h(\omega)\right\}|\phi_i^0\rangle, \tag{11}$$

where $\hat{Q}$ is a projection operator onto the particle space, $\hat{Q} = 1 - \sum_j |\phi_j^0\rangle\langle\phi_j^0|$. Another equation can be derived by multiplying a bra state $\langle\phi_i^0|$ with Eq. (8).

$$\omega\langle Y_i(\omega)| = -\langle Y_i(\omega)|(h_0 - \epsilon_i) - \langle\phi_i^0|\left\{V_{\rm ext}(\omega) + \delta h(\omega)\right\}\hat{Q}. \tag{12}$$

The essential idea of the FAM is that one can evaluate $\delta h(\omega)$ in Eqs. (11) and (12) utilizing the following numerical linearization.

$$\delta h(\omega) = \lim_{\eta\to 0}\frac{1}{\eta}\left(h\left[\langle\psi'|, |\psi\rangle\right] - h_0\right), \tag{13}$$

where $\langle\psi_i'| = \langle\phi_i^0| + \eta\langle Y_i(\omega)|$ and $|\psi_i\rangle = |\phi_i^0\rangle + \eta|X_i(\omega)\rangle$. In the linear order in η, Eq. (13) is exact. In practice, the parameter η is taken finite but small enough to validate the linearization. In this manner, we can calculate the induced residual field without explicit calculation of residual $\delta h/\delta\rho$. All we need is the calculation of the KS Hamiltonian h with different bra and ket states. Therefore, this is achieved by a minor modification of the static DFT computer code.

The accuracy and usefulness of the FAM is demonstrated in Ref. 4. Here, we show fully self-consistent calculation of the $E1$ strength distribution in Ne isotopes. Figure 2 shows those in even-even Ne isotopes from $A = 20$ to 30. The prominent deformation splitting is seen in ^{20,22}Ne. At low energy below 10 MeV, a pygmy strength is seen in $^{26-30}$Ne. In ^{26}Ne, the low-energy strength is a single peak, while it becomes double-peak structure in ^{28,30}Ne. There is an experimental indication for the $E1$ peak at $E \sim 8$ MeV. Our calculation suggests that this peak consists of both $K = 0$ and $K = 1$ components and is strongly coupled to the octupole mode. The detailed analysis is under progress.

4. Conclusion

We have presented the linear response calculations based on the time-dependent density functional theory using the Skyrme density functional. Using the real-time method, we calculate the photoabsorption cross section for ^{24}Mg and compare it with experimental data. The calculated GDR peak position is higher by $2 \sim 3$ MeV, but its qualitative feature is reproduced. Next, we calculate $E1$ strength distribution in even-even Ne isotopes ($A = 20 \sim 30$), using the finite amplitude method. We predict a single low-energy peak in ^{26}Ne and double peaks for ^{28,30}Ne. It will be of significant interest to observe low-energy $E1$ strength in these neutron-rich nuclei, to check the TDDFT prediction with the Skyrme functional.

Acknowledgments

This work is supported by the Grant-in-Aid for Scientific Research in Japan (Nos. 17540231, 18540366, 18036002), by the PACS-CS project of Center for Computational Sciences, University of Tsukuba, and by the Large Scale Simulation Program Nos. 06-14 (FY2006) and 07-20 (FY2007) of High Energy Accelerator Research Organization (KEK).

References

1. P. Ring and P. Schuck, *The nuclear many-body problems*, Texts and monographs in physics (Springer-Verlag, New York, 1980).
2. T. Nakatsukasa and K. Yabana, *J. Chem. Phys.* **114**, 2550 (2001).
3. T. Nakatsukasa and K. Yabana, *Phys. Rev. C* **71**, p. 024301 (2005).
4. T. Nakatsukasa, T. Inakura and K. Yabana, *Phys. Rev. C* **76**, p. 024318 (2007).
5. E. Runge and E. K. U. Gross, *Phys. Rev. Lett.* **52**, 997 (1984).
6. P. Bonche, H. Flocard and P. H. Heenen, *Nucl. Phys. A* **467**, 115 (1987).
7. J. Dobaczewski and J. Dudek, *Phys. Rev. C* **52**, 1827 (1995).
8. R. Kosloff and D. Kosloff, *J. Comp. Phys.* **63**, 363 (1986).
9. D. Neuhasuer and M. Baer, *J. Chem. Phys.* **90**, 4351 (1989).
10. M. Child, *Mol. Phys.* **72**, 89 (1991).
11. National Nuclear Data Center, http://www.nndc.bnl.gov/.
12. G. Bertsch and S. F. Tsai, *Phys. Rep.* **18**, 125 (1975).

DISSIPATIVE DYNAMICS WITH EXOTIC BEAMS

M. DI TORO *, M. COLONNA, V. GRECO, G. FERINI, C. RIZZO, J. RIZZO

Laboratori Nazionali del Sud INFN, I-95123 Catania, Italy, and Physics-Astronomy Dept., University of Catania
* *E-mail: ditoro@lns.infn.it*

V. BARAN

Dept.of Theoretical Physics, Bucharest Univ., and NIPNE-HH, Magurele, Bucharest, Romania

H. H. WOLTER

Dept. für Physik, Universität München, D-85748 Garching, Germany

M. ZIELINSKA-PFABE

Smith College, Northampton, Mass. USA

Heavy Ion Collisions (HIC) represent a unique tool to probe the in-medium nuclear interaction in regions away from saturation and at high nucleon momenta. In this report we present a selection of reaction observables particularly sensitive to the isovector part of the interaction, i.e. to the symmetry term of the nuclear Equation of State (EoS) At low and Fermi energies the behavior of the symmetry energy around saturation influences dissipation and fragment production mechanisms. Predictions are shown for fusion, deep-inelastic and fragmentation collisions induced by neutron rich projectiles. At all energies the isospin transport data are supplying valuable information on value and slope of the symmetry term below saturation. The importance of studying violent collisions with radioactive beams in this energy range is finally stressed.

Keywords: Charge Equilibration, Symmetry Energy, Isospin Transport, Isospin Flows

1. Introduction

The symmetry energy E_{sym} appears in the energy density $\epsilon(\rho,\rho_3) \equiv \epsilon(\rho) + \rho E_{sym}(\rho_3/\rho)^2 + O(\rho_3/\rho)^4 + ..$, expressed in terms of total ($\rho = \rho_p + \rho_n$) and isospin ($\rho_3 = \rho_p - \rho_n$) densities. The symmetry term gets a kinetic contribution directly from basic Pauli correlations and a potential part from the highly controversial isospin dependence of the effective interactions.[1] Both at sub-saturation and supra-saturation densities, predictions based of the existing many-body techniques diverge rather widely, see.[2] We take advantage of new opportunities in theory (development of rather reliable microscopic transport codes for HIC) and in experiments (availability of very asymmetric radioactive beams, improved possibility of measuring event-by-event correlations) to present results that are severely constraining the existing effective interaction models. We will discuss dissipative collisions in the low energy range, from just above the Coulomb barrier up to about hundred $AMeV$. In this way we can probe in detail the symmetry energy in dilute matter. The transport codes are based on mean field theories, with correlations included via hard nucleon-nucleon collisions and via stochastic forces, self-consistently evaluated from the mean phase-space trajectory, see.[1,3–5] Stochasticity is essential in order to get distributions as well as to allow the growth of dynamical instabilities. The isovector part of the EoS has been tested systematically by using two different behaviors of the symmetry energy below saturation: one ($Asysoft$) where it is a smooth decreasing function towards low densities, and another one ($Asystiff$) where we have a rapid decrease,[1,6]

2. The Prompt Dipole γ-Ray Emission

The possibility of an entrance channel bremsstrahlung dipole radiation due to an initial different N/Z distribution was suggested at the beginning of the nineties.[7,8] After several experimental evidences, in fusion as well as in deep-inelastic reactions, see[9] and refs. therein, we have now a good understanding of the process and stimulating new perspectives from the use of radioactive beams.

During the charge equilibration process taking place in the first stages of dissipative reactions between colliding ions with different N/Z ratios, a large amplitude dipole collective motion develops in the composite dinuclear system, the so-called dynamical dipole mode. This collective dipole gives rise to a prompt γ-ray emission which depends: i) on the absolute value of the initial dipole moment

$$D(t=0) = \frac{NZ}{A}\,|R_Z(t=0) - R_N(t=0)| =$$
$$\frac{R_P + R_T}{A} Z_P Z_T \left| (\frac{N}{Z})_T - (\frac{N}{Z})_P \right|, \tag{1}$$

being $R_Z = \frac{\Sigma_i x_i(p)}{Z}$ and $R_N = \frac{\Sigma_i x_i(n)}{N}$ the center of mass of protons and of neutrons respectively, while R_P and R_T are the projectile and target radii; ii) on the fusion/deep-inelastic dynamics; iii) on the symmetry term, below saturation, that is acting as a restoring force.

A detailed description is obtained in a microscopic approach based on semiclassical transport equations, of Landau-Vlasov type,[11] where mean field and two-body collisions are treated in a self-consistent way, see details in.[12] Realistic effective Skyrme-like interactions are used. The numerical accuracy of the transport code has been largely improved in order to have reliable results also at low energies, just above the threshold for fusion reactions.[13,14] The resulting physical picture is in good agreement with quantum Time-Dependent-Hartree-Fock calculation.[15]

We present here some results for the $^{32}S + ^{100}Mo$ (N/Z asymmetric) reaction at 6 *and* 9 $AMeV$, recently studied vs. the "symmetric" $^{36}S + ^{96}Mo$ counterpart in ref.[9] In Fig.1 (left columns) we draw the time evolution for $b = 2fm$ of the dipole moment in the r-space (solid lines), $D(t) = \frac{NZ}{A}(R_Z - R_N)$ and in p−space

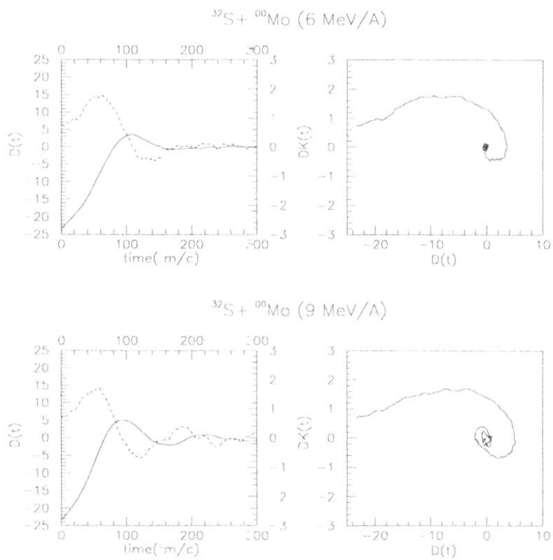

Fig. 1. The time evolution of the dipole mode in $r-$space $D(t)$ (solid lines) and $p-$space $DK(t)$ (dashed lines, in fm^{-1}) and the correlation $DK(t) - D(t)$ at incident energy of $6AMeV$ and $9AMeV$ for $b = 2fm$.

(dashed lines), $DK(t) = (\frac{P_p}{Z} - \frac{P_n}{N})$, with P_p (P_n) center of mass in momentum space for protons (neutrons), is just the canonically conjugate momentum of the $D(t)$ coordinate, i.e. as operators $[D(t), DK(t)] = i\hbar$ see.[14–16] On the right hand side columns we show the corresponding correlation $DK(t) - D(t)$ in the phase space. We choose the origin of time at the beginning of the *dinuclear* phase. The nice "spiral-correlation" clearly denotes the collective nature of the mode. From Fig.1 we note that the "spiral-correlation" starts when the initial dipole moment $D(t = 0)$, the geometrical value at the touching point, is already partially quenched. This is the reason why the dinucleus dipole yield is not simply given by the "static" estimation but the reaction dynamics has a large influence on it.

A clear energy dependence of the dynamical dipole mode is evidenced with a net increase when we pass from $6AMeV$ to $9AMeV$. At lower energy, just above the Coulomb barrier, a longer transition to a dinuclear configuration is required which hinders the isovector collective response. When the collective dipole response sets in the charge is already partially equilibrated via random nucleon exchange.

From the dipole evolution given from the Landau-Vlasov transport we can directly apply a bremsstrahlung *("bremss")* approach[16] to estimate the "direct" photon emission probability ($E_\gamma = \hbar\omega$):

$$\frac{dP}{dE_\gamma} = \frac{2e^2}{3\pi\hbar c^3 E_\gamma}|D''(\omega)|^2, \tag{2}$$

where $D''(\omega)$ is the Fourier transform of the dipole acceleration $D''(t)$. We remark that in this way it is possible to evaluate, in *absolute* values, the corresponding pre-equilibrium photon emission.

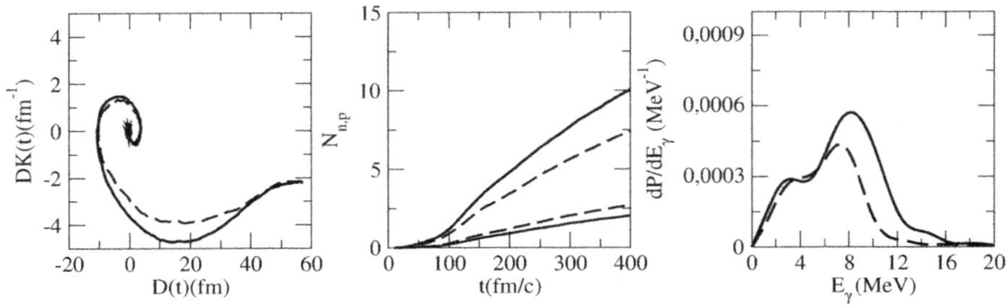

Fig. 2. ^{132}Sn $+^{58}$Ni system ($E = 10AMeV$, $b = 4fm$). Left Panel: DK-D Spirals. Central panel: neutron (upper) and proton (lower) emissions. Right panel: γ spectrum. Asy-soft (solid lines) and asy-stiff (dashed lines) symmetry energies.

We add a couple of comments of interest for the experimental selection of the Dynamical Dipole: i) The centroid is always shifted to lower energies (large deformation of the dinucleus); ii) A clear angular anisotropy should be present since the prompt mode has a definite axis of oscillation (on the reaction plane) at variance with the statistical *GDR*. In a very recent experiment the prompt dipole radiation has been investigated with a 4π gamma detector. A strong dipole-like photon angular distribution $(\theta_\gamma) = W_0[1 - a_2 P_2(cos\theta_\gamma)]$, θ_γ being the angle between the emitted photon and the beam axis, has been observed, with the a_2 parameter close to -1, see.[10]

The prompt dipole radiation also represents a nice cooling mechanism on the fusion path. It could be a way to pass from a *warm* to a *cold* fusion in the synthesis of heavy elements with a noticeable increase of the *survival* probability,[17]

$$\frac{P_{surv,dipole}}{P_{surv}} = \frac{P_\gamma P_{surv}(E^* - E_\gamma)}{P_{surv}(E^*)} + (1 - P_\gamma) > 1, \frac{P_{surv}(E^* - E_\gamma)}{P_{surv}(E^*)} > 1 \quad (3)$$

The use of unstable neutron rich projectiles would largely increase the effect, due to the possibility of larger entrance channel asymmetries. In order to suggest proposals for the new *RIB* facility *Spiral* 2,[18] we have studied fusion events in the reaction $^{132}Sn +^{64} Ni$ at $10AMeV$,[19] We expect a *Monster* Dynamical Dipole, the initial dipole moment $D(t = 0)$ being of the order of 50fm, about two times the larger values probed so far, allowing a detailed study of the symmetry potential, below saturation, responsible of the restoring force of the dipole oscillation and even of the damping, via the fast neutron emission. In the Fig.2 we present some preliminary very promising results. The larger value of the symmetry energy for the *Asysoft* choice at low densities, where the prompt dipole oscillation takes place, leads to some clear observable effects: i) Larger Yields, as we see from the larger amplitude of the "Spiral" (left panel) and finally in the spectra (right panel); ii)

Larger mean gamma energies, shift of the centroid to higher values in the spectral distribution (right panel); iii) Larger width of the "resonance" (right panel) due to

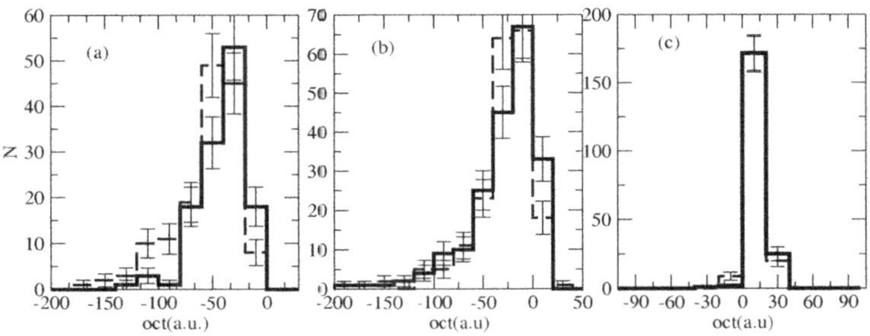

Fig. 3. Distribution of the octupole moment of primary fragments for the $^{132}Sn +^{64} Ni$ reaction at 10 $AMeV$ (impact parameters (a):$b = 6fm$, (b):$7fm$, (c):$8fm$). Solid lines: asysoft. Dashed lines: asystiff

the larger fast neutron emission (central panel). We note the opposite effect of the Asy-stiffness on neutron vs proton emissions. The latter point is important even for the possibility of an independent test just measuring the N/Z of the pre-equilibrium nucleon emission.

3. Isospin effects on Deep-Inelastic Collisions

Dissipative semi-peripheral collisions at low energies, including binary and three-body breakings, offer a good opportunity to study phenomena occurring in nuclear matter under extreme conditions with respect to shape, excitation energy, spin and N/Z ratio (isospin). In some cases, due to a combined Coulomb and angular momentum (deformation) effect, some instabilities can show up.[20] This can lead to 3-body breakings, where a light cluster is emitted from the neck region. Three body processes in collisions with exotic beams will allow to investigate how the development of surface (neck-like) instabilities, that would help ternary breakings, is sensitive to the structure of the symmetry term around (below) saturation,[21,22] In order to suggest proposals for the new RIB facilities, we have studied again the reaction $^{132}Sn +^{64} Ni$ at 10 $AMeV$ in semicentral events, impact parameters $b = 6, 7, 8fm$, where one observes mostly binary exit channels, but still in presence of large dissipation.

The Wilczynski plots, kinetic energy loss vs. deflection angle, show slightly more dissipative events in the $Asystiff$ case, consistent with the point that in the interaction at lower densities in very neutron-rich matter (the neck region) we have a less repulsive symmetry term. In fact the neck dynamics is rather different in the two cases, as it can be well evidenced looking at the deformation of the PLF/TLF residues. The distribution of the octupole moment over the considered ensemble of events is shown in Fig.3 for the three considered impact parameters.

Except for the most peripheral events, larger deformations, strongly suggesting a final 3-body outcome, are seen in the $Asystiff$ case. Now, due to the lower value of the symmetry energy, the neutron-rich neck connecting the two systems survives a longer time leading to very deformed primary fragments, from which eventually small clusters will be dynamically emitted. Finally we expect to see effects of the different interaction times on the charge equilibration mechanism, probed starting from entrance channels with large N/Z asymmetries, like $^{132}Sn(N/Z = 1.64) +^{58} Ni(N/Z = 1.07)$. Moreover the equilibration mechanism is also directly driven by the strength of the symmetry term.

4. Isospin Distillation with Radial Flow

In central collisions at 30-50 MeV/A, where the full disassembly of the system into many fragments is observed, one can study specifically properties of liquid-gas phase transitions occurring in asymmetric matter.[1,5,23,24] For instance, in neutron-rich matter, phase co-existence leads to a different asymmetry in the liquid and gaseous phase: fragments (liquid) appear more symmetric with respect to the initial matter, while light particles (gas) are more neutron-rich. The amplitude of this effect depends on specific properties of the isovector part of the nuclear interaction, namely on the value and the derivative of the symmetry energy at low density.

This investigation is interesting in a more general context: in heavy ion collisions the dilute phase appears during the expansion of the interacting matter. Thus we study effects of the coupling of expansion, fragmentation and distillation in a two-component (neutron-proton) system. This leads us to suggest new correlation observables to probe the symmetry term at sub-saturation density.

We will focus on central collisions, $b = 2\ fm$, at $50AMeV$ considering symmetric reactions between systems having three different initial asymmetries: $^{112}Sn +^{112} Sn,^{124} Sn +^{124} Sn,^{132} Sn +^{132} Sn$, with $(N/Z)_{in} = 1.24, 1.48, 1.64$, respectively. The average N/Z of emitted nucleons (gas phase) and Intermediate Mass Fragments (IMF) is presented in Fig.4 as a function of the initial $(N/Z)_{in}$ of the three colliding systems. One observes that, generally, the gas phase (right panel) is more neutron-rich in the asysoft case, while IMF's (left panel) are more symmetric. This is due to the larger value of the symmetry energy at low density for the asysoft parametrization.[25]

Now we investigate in more detail correlations between fragment isotopic content and kinematical properties. The idea is that fragmentation originates from the break-up of a composite source that expands with a given velocity field. Since neutrons and protons experience different forces, one may expect a different radial flow for the two species. This trend should be reflected in the fragment asymmetries.

As a measure of the isotopic composition of the IMF's, we will consider the

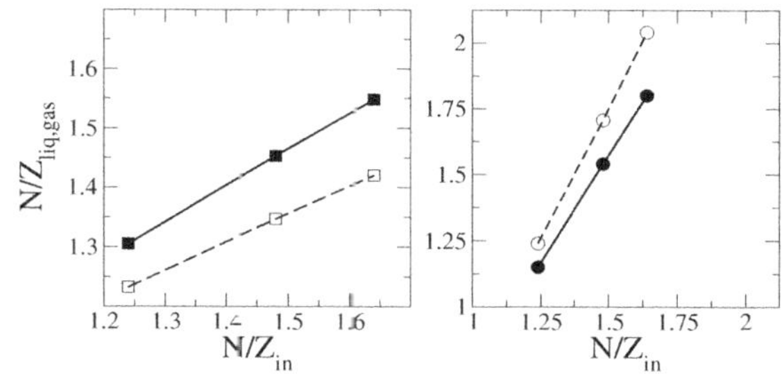

Fig. 4. The N/Z of the liquid (left) and of the gas (right) phase is displayed as a function of the system initial N/Z. Full lines and symbols refer to the asystiff parameterization. Dashed lines and open symbols are for asysoft.

sum of neutrons, $N = \sum_i N_i$, and protons, $Z = \sum_i Z_i$, of all IMF's in a given kinetic energy range, in each event. Then we take the ratio N/Z and we consider the average over the ensemble of events.

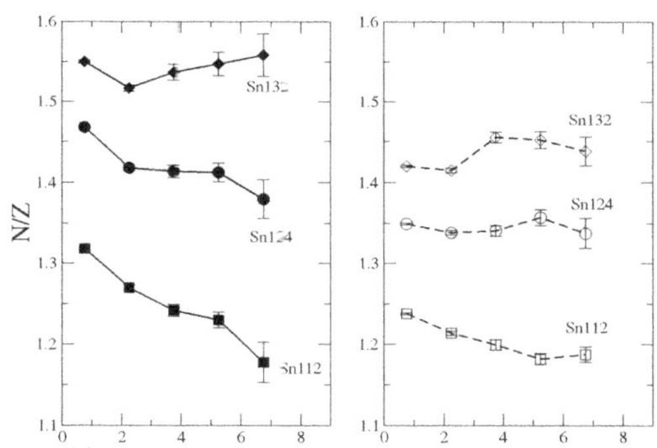

Fig. 5. The fragment N/Z (see text) as a function of the kinetic energy. Left panel:asystiff; Rigth panel: asysoft.

This observable is plotted in Fig.5 for the three reactions. The behaviour observed is rather sensitive to the iso-EoS. For the proton-rich system, the N/Z decreases with the fragment kinetic energy, especially in the asystiff case, where the symmetry energy is relatively low at low density.[25] In this case, the Coulomb repulsion pushes the protons towards the surface of the system. Hence, more symmetric fragments acquire larger velocity. The decreasing trend is less pronounced in the asysoft case (right panel) because Coulomb effects on protons are counterbalanced by the larger attraction of the symmetry potential. In systems with higher initial asymmetry, the decreasing trend is reversed, due to the larger neutron repulsion in neutron-rich systems. Larger slopes are always observed in the asysoft case.

In conclusion, this analysis reveals the existence of significant, EoS-dependent correlations between the N/Z and the kinetic energy of IMF's. This correlation is linked to the different forces experienced by neutrons and protons along the fragmentation path, that in turn depend on the detail of the isovector part of the nuclear interaction.

5. Isospin Transport in Peripheral Collisions at the Fermi Energies

Now we investigate peripheral collisions between similar systems with different isospin, specifically collisions of different combinations of Sn isotopes. The isospin is used here both as a tracer of the reaction mechanism, as well as an observable of interest with respect to the iso-EoS. In the previous section we have seen the isospin dynamics in instability regions and fragment formation, here we study more directly the isospin transport in binary events.

The isospin transport can be discussed in a compact way by means of the chemical potentials for protons and neutrons as a function of density ρ and isospin I.[26] From this the p/n currents can be expressed as

$$\mathbf{j}_{p/n} = D^{\rho}_{p/n} \nabla \rho - D^{I}_{p/n} \nabla I \tag{4}$$

with $D^{\rho}_{p/n}$ the drift, and $D^{I}_{p/n}$ the diffusion coefficients for transport, which are given explicitly in ref.[26] Of interest here are the differences of currents between protons and neutrons which have a simple relation to the density dependence of the symmetry energy

$$\begin{aligned} D^{\rho}_{n} - D^{\rho}_{p} &\propto 4I \frac{\partial E_{sym}}{\partial \rho} \,, \\ D^{I}_{n} - D^{I}_{p} &\propto 4\rho E_{sym} \,. \end{aligned} \tag{5}$$

Thus the isospin transport due to density gradients, i.e. isospin migration, depends on the slope of the symmetry energy, or the symmetry pressure, while the transport due to isospin concentration gradients, i.e. isospin diffusion, depends on the absolute value of the symmetry energy. In peripheral collisions discussed here, residues of about normal density are in contact with the neck region of density below saturation density. In this region of density a stiff iso-EoS has a smaller value but a larger slope compared to a soft iso-EoS. Correspondingly we expect opposite effects of these models on the migration and diffusion of isospin.

We will in particular discuss the so-called Imbalance Ratio (also called Rami- or transport-ratio), which is defined as[27]

$$R^{x}_{P,T} = \frac{2x^{M} - (x^{H} + x^{L})}{(x^{H} - x^{L})} \,. \tag{6}$$

Here, x is an isospin sensitive quantity that is to be investigated with respect to equilibration, as e.g. the asymmetry $I = (N-Z)/(N+Z)$ which is considered here primarily, but also other quantities, such as isoscaling coefficients, ratios of production of light fragments, etc. The indices H and L refer to the symmetric reaction between the heavy (n-rich) and the light (n-poor) systems, while M refers to the mixed reaction. P, T denote the rapidity region, in which this quantity is measured, in particular the projectile and target rapidity regions. Clearly, this ratio is ± 1 in the projectile and target region, respectively, for complete transparency, and oppositely for complete rebound, while it is zero for complete equilibration.

The imbalance ratio is related to the interacting time and so is also varying with the Momentum Dependence (MD) of the effective forces.[26,28,29] In fact in an analysis in terms of impact parameters EoS and MD effects are mixed. A good, measurable, correlation appears to be to the dissipation, i.e. to the kinetic energy loss per particle

$$E_{loss} = E_{cm} - \frac{E_{kin} + E_{pot}^{Coul}}{A_{PLF} + A_{TLF}}, \; E_{cm} = \frac{E_{lab}}{A_P} \frac{A_P A_T}{(A_P + A_T)^2}, \tag{7}$$

where $A_P, A_T, A_{PLF}, A_{TLF}$ are the masses of the initial projectile and target, and of the final projectile-like and target-like fragments, respectively. Here E_{cm} is the initial energy available in the cm system. E_{kin} and E_{pot}^{Coul} are the final kinetic energies of the fragments in the cm system, and the Coulomb potential energy, respectively.

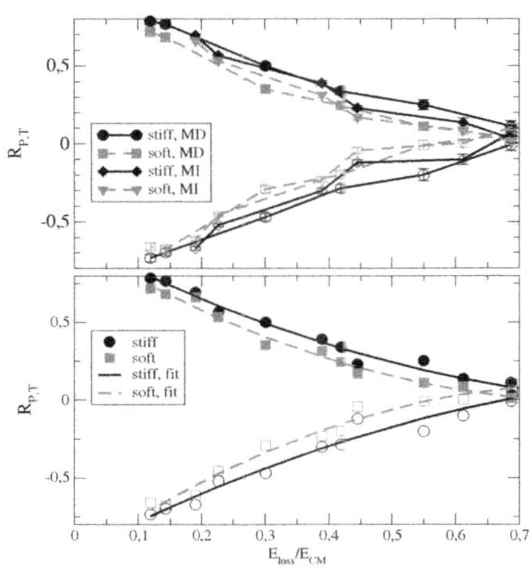

Fig. 6. Imbalance ratio as a function of relative energy loss: (upper panel) Separately for stiff (solid) and soft (dashed) iso-EoS, and for MD (circles and diamonds) and MI (squares and triangles)interactions, in the projectile region (full symbols) and the target region (open symbols). (lower panel) Quadratic fit to all points for the stiff (solid), resp. soft (dashed) iso-EoS.

We have performed detailed calculations for Sn isotopes at various beam energies. In Fig.6 we present the imbalance ratio as a function of the relative energy loss,

In the upper panel we separate the results for MD and MI interactions and stiff and soft iso-EoS, but in each case the results for 35 and 50 MeV collected together and connected by lines. We see that the curves for the soft *EoS* (dashed) are generally lower in the projectile region (and oppositely for the target region), i.e. show more equilibration, that those for the stiff *EoS*. In order to emphasize this trend we have, in the lower panel of the figure, collected together the values for the stiff (circles, solid) and the soft (squares, dashed) iso-EoS, and fitted them by a quadratic curve. The difference between the curves for the stiff and soft iso-EoS in the lower panel then isolates the effect of the iso-EoS from kinematical effects depending on the interaction time. It is seen, that there is a systematic effect of the symmetry energy of the order of about 20 percent, which should be measurable.

6. Perspectives

We have shown that *violent* collisions of n-rich heavy ions from low to Fermi energies can bring new information on the isovector part of the in-medium interaction, qualitatively different from equilibrium *EoS* properties, in particular in dilute nuclear matter. We have presented quantitative results in a wide range of beam energies, suggesting several isospin-sensitive observables. At low energies we have seen isospin effects on the rather exciting new prompt dipole radiation and on the dissipation in deep inelastic collisions, at Fermi energies the Iso-EoS sensitivity of the isospin transport in fragment reactions.

In conclusion the results presented here appear very promising for the possibility of exciting new results from dissipative collisions with radioactive beams.

References

1. V. Baran, M. Colonna, V. Greco, M. Di Toro, *Phys. Rep.* **410**, 335 (2005).
2. C. Fuchs, H.H. Wolter, *Eur. Phys. Jour. A* **30**, 5 (2006).
3. A. Guarnera, M. Colonna, P. Chomaz, *Phys. Lett. B* **373**, 267 (1996).
4. M. Colonna et al., *Nucl. Phys. A* **642**, 449 (1998).
5. P.Chomaz, M.Colonna, J.Randrup, *Phys. Rep.* **389**, 263 (2004).
6. M.Colonna, M. Di Toro, G. Fabbri, S. Maccarone, *Phys. Rev. C* **57**, 1410 (1998).
7. P. Chomaz, M. Di Toro, A. Smerzi, *Nucl. Phys. A* **563**, 509 (1993).
8. P.F. Bortignon et al., *Nucl. Phys. A* **583**, 101c (1995).
9. D. Pierroutsakou et al., *Phys. Rev. C* **71**, 054605 (2005).
10. D. Pierroutsakou et al. (Medea Collab.), *Prompt Dipole Radiation in Fusion Reactions, Phys. Rev. Lett.* 2007, submitted.
11. D.M. Brink, M. Di Toro, *Nucl. Phys. A* **372**, 151 (1981).
12. V. Baran et al., *Nucl. Phys. A* **600**, 111 (1996).
13. M. Cabibbo et al., *Nucl. Phys. A* **637**, 374 (1998).
14. V. Baran et al., *Nucl. Phys. A* **679**, 373 (2001).
15. C. Simenel et al., *Phys. Rev. Lett.* **86**, 2971 (2000).
16. V. Baran, D. M. Brink, M. Colonna, M. Di Toro, *Phys. Rev. Lett.* **87**, 182501 (2001).
17. L. Bonanno, *Effetti di radiazione diretta dipolare sulla sintesi degli elementi superpesanti*, Master Thesis, Catania Univ. 2004.

18. M. Lewitovicz, *Challenges of the SPIRAL 2 Project*, NN06 Conf., Rio de Janeiro; http://ganinfo.in2p3.fr/research/developments/spiral2
19. Letter of Intent for the new SPIRAL2 Facility at GANIL, 2007.
20. M. Colonna, M. Di Toro, A. Guarnera, *Nucl. Phys. A* **589**, 160 (1995).
21. V. Baran, M. Colonna, M. Di Toro, *Nucl. Phys. A* **730**, 329 (2004).
22. M. Di Toro, A. Olmi, R. Roy, *Eur. Phys. Jour. A* **30**, 65 (2006).
23. H. Mueller and B.D. Serot, *Phys. Rev. C* **52**, 2072 (1995)
24. B.A. Li and C.M. Ko, *Nucl. Phys. A* **618**, 498 (1997)
25. V. Baran *et al.*, *Nucl. Phys. A* **703**, 603 (2002).
26. V. Baran *et al.*, *Phys. Rev. C* **72**, 064620 (2005).
27. F. Rami *et al.*, *Phys. Rev. Lett.* **84**, 1120 (2000).
28. B.A. Li, L.W. Chen, *Phys. Rev. C* **72**, 064611 (2005).
29. J. Rizzo et al, *Investigation of the Isovector Equation of State in Peripheral Heavy Ion Collisions at Fermi Energy*, *Nucl. Phys. A* 2007, submitted.

EXPLORING THE SYMMETRY ENERGY OF ASYMMETRIC NUCLEAR MATTER WITH HEAVY ION REACTIONS

M.B. TSANG

National Superconducting Cyclotron Laboratory
Michigan State University, East Lansing, Michigan 48824, USA

The nuclear symmetry energy governs many properties of neutron-rich nuclei and of neutron stars. Neutron star radii, moments of inertia, phase transitions in the stellar interior and the cooling of proto-neutron stars are among the stellar properties that are most strongly affected. A number of interesting experimental observables have been identified that provide opportunities to probe the density dependence of the symmetry energy via nucleus-nucleus collisions. In this talk, we discuss measurements and simulations of the isospin diffusion and neutron to proton emission observable over a wide incident energy range. This work is supported by the National Science Foundation under Grant Nos. PHY-01-10253, PHY-0606007.

Manuscript not received.

INVARIANT MASS SPECTROSCOPY OF HALO NUCLEI

T. NAKAMURA*

Department of Physics, Tokyo Institute of Technology,
2-12-1 Meguro, Tokyo, 152-8551, Japan
**E-mail: nakamura@phys.titech.ac.jp*

Y. KONDO and T. SUGIMOTO

RIKEN Nishina Center for Accelerator-Based Science, Hirosawa 2-1, Wako, Saitama 351-0198, Japan

Invariant mass spectroscopy is a powerful tool to study low-lying excitation of the halo nuclei. We have applied the invariant mass spectroscopy to the E1 excitation of ^{11}Li and to the 2^+ transition of ^{14}Be. The former study used the Coulomb breakup of ^{11}Li on Pb at 70 MeV/nucleon, which revealed a strong E1 transition at very low excitation energies. Such a strong soft E1 excitation indicated dineutron-like spatial correlation in ^{11}Li. The latter study used inelastic scattering of ^{14}Be on carbon and proton targets at about 70 MeV/nucleon. The first 2^+ state was observed in energy spectra for both targets. The extracted deformation parameters were smaller than those of the neighboring ^{12}Be nucleus, indicating the hindered collectivity at the drip-line nucleus ^{14}Be.

Keywords: Neutron Halo, Coulomb breakup, Collectivity

1. Introduction

One of highlights of physics with unstable nuclear beam is the study of loosely bound nuclei near and on the drip-line, where nuclear halo often appears. Weak binding in such nuclei implies that continuum is very close to the ground state. In such cases, invariant mass spectroscopy, which is used to investigate structures above the particle decay threshold, becomes a powerful method, as has successfully been applied to the spectroscopy of halo nuclei.

The low-lying strong E1 transitions (soft E1 excitation) has been intensively studied as a unique feature of halo nuclei by using the invariant mass method.[1–11] It should be noted that the soft E1 excitation is in sharp contrast to the ordinary nuclei whose E1 strength is exhausted by the giant dipole resonance (GDR). The invariant mass method also provides a useful tool to study the discrete resonant states above the particle-decay threshold. For instance, the first 2^+ states of drip-line nuclei are such cases. We present here recent achievements with invariant mass spectroscopy: 1) Coulomb breakup of the two-neutron halo nucleus ^{11}Li, and 2) excitation of the first 2^+ state of ^{14}Be by inelastic scattering with proton and carbon targets.

2. Invariant mass method

For the reaction of halo nuclei, a target nucleus plays a role as a probe to excite a halo nucleus which is provided as a radioactive beam. One can then choose a spectroscopic method depending on whether an excited state to be studied is located in the bound region or in the unbound continuum region. For bound excited states, in-beam γ ray spectroscopy can be applied as demonstrated successfully for the low-lying states of radioactive nuclei as in ^{12}Be.[12–14] For the latter case, the invariant mass spectroscopy is a suitable method.

In the invariant mass spectroscopy, momentum vectors of breakup particles after excitation are measured in coincidence, to reconstruct the invariant mass M^* of the excited state, namely,

$$M^* = \sqrt{\sum_i E_i^2 - \sum_i \vec{P}_i^2}, \tag{1}$$

where the E_i and $\vec{P}_i$ respectively denote the total energy and momentum of the decay particle 'i'. Then the relative energy and the excitation energy is given as

$$E_{\rm rel} = M^* - \sum_i M_i \tag{2}$$

$$E_{\rm x} = E_{\rm rel} + S, \tag{3}$$

where M_i and $E_{\rm rel}$ are the rest mass and relative energy of breakup particles, respectively. $E_{\rm x}$ and S respectively stand for the excitation energy, and separation energy of the projectile. For instance, in Coulomb breakup of ^{11}Li, a ^{11}Li projectile excited by a virtual photon from a high-Z target is broken up into ^{9}Li and two neutrons. The excitation energy is then reconstructed from a measurement of ^{9}Li and two neutrons in coincidence.

An advantage of the invariant mass method using radioactive nuclear beam is that this scheme can realize rather good energy resolution since the invariant mass is determined solely by the momenta of outgoing particles and thus the resolution of M^* and hence $E_{\rm rel}$ is not affected by the poor energy resolution of the secondary beam. Furthermore, the energy resolution $\Delta E_{\rm rel}$ is scaled by the incident energy per nucleon and the relative energy, not by the total energy of the beam and ejectile. For instance, for the incident energy of 70 MeV/nucleon and the relative energy of 1 MeV, the energy resolution of about 200 keV can be achieved for a measurement with about 1% momentum resolution.

3. Coulomb Breakup of ^{11}Li

We show results of recent kinematically complete measurement of Coulomb breakup of a two-neutron halo nucleus ^{11}Li using the invariant mass method.[15] Soft E1 excitation has so far been studied in particular for the case of one-neutron halo nucleus, such as ^{11}Be[1,4,5] and ^{19}C.[2] For ^{11}Be Coulomb breakup, $B(E1)$ strength distribution peaking at about $E_{\rm x}$ ~0.8 MeV with the strength of about 3 W.u. has successfully been explained by the direct breakup mechanism.[1,4,5] In this mechanism, large $B(E1)$ strength can be interpreted as an amplified image of the large amplitude of the halo radial wave function since it is described as a Fourier-Bessel transformation of the halo wave function. Selectivity to the s-wave halo component of the ground state in direct breakup mechanism has thus been used to extract the spectroscopic factor for the halo configuration.

On the other hand, nature of E1 response of two-neutron halo nuclei such as ^{11}Li has not been understood well. The difficulty is due mainly to controversial situation in the previous three Coulomb-breakup experiments of ^{11}Li.[6–8] The $B(E1)$ distributions obtained in these three experiments were inconsistent as compared in Fig. 3 of Ref.[15] This discrepancy may be attributed to difficulties in detecting two neutrons unambiguously. As theoretical work on ^{11}Li has demonstrated ,[16–27] a precise determination of the soft E1 excitation spectrum would place significant constraints on the reaction mechanism, two-neutron correlation, core-n correlation, and the relative role of $(s_{1/2})^2$ and $(p_{1/2})^2$ configurations in the ^{11}Li ground state. Hence, an improved measurement of the soft E1 strength distribution should be highly desirable to clarify the exotic nature of the two-neutron halo nucleus.

In order to obtain data with much higher statistics and much less ambiguity caused by cross-talk events in detecting two neutrons, Coulomb breakup of ^{11}Li on a Pb target was measured at 70 MeV/nucleon at RIKEN. A secondary beam of ^{11}Li was produced by fragmentation of a 100 MeV/nucleon primary ^{18}O beam in a thick Be production target and separated using the RIPS radioactive beam line at RIKEN.[28] The ^{11}Li beam bombarded a Pb target of thickness 346 mg/cm^2 with an average energy of 70 MeV/nucleon. The momentum of the beam ion was obtained by tracing the trajectory with two PPACs and by measuring the time of flight (TOF) using a thin plastic scintillator and an RF signal. The outgoing particles, a ^{9}Li ion and two neutrons, were emitted in a narrow kinematical cone at forward angles. The trajectory of the ^{9}Li ion was bent by a large-gap dipole magnet, tracked with a drift chamber, and then the ^{9}Li ion hit a hodoscope. The momentum of the ^{9}Li ion was thus obtained using the tracking information in combination with the TOF. The momentum of neutrons was obtained using the TOF and position information at the neutron detector arrays, which were composed of 54 rods of plastic scintillators arranged into two layers. Two layer arrangement was used to disentangle the cross talk events by use of the causality analysis. The momentum vectors of outgoing three particles were then combined to extract the invariant mass of the excited ^{11}Li nuclei. The detail of the experiment and this technique of exclusion of cross-talk events is described in Ref.[15]

Figure 2 shows the deduced $B(E1)$ distribution of ^{11}Li (solid circles) as a function of the relative energy ($E_{\rm rel}$) of ^{9}Li+n+n. A strong peak at $E_{\rm rel} \sim 300$ keV ($E_{\rm x}$=$E_{\rm rel}$+$S_{2n} \sim 600$ keV) is clearly seen, which is in sharp contrast to those from the previous three experiments that peaked at $E_{\rm rel}$=600–1000 keV. This difference can be due to the insensitivity at low relative energies in the previous experiments as discussed in Ref.[15] As is compared in Fig. 2, the current $B(E1)$ spectrum is in good agreement with a theoretical three-body calculation (solid curve),[17] where strong two-neutron correlation and final state interactions are taken into account.

Such neutron-neutron correlation in the ground state of ^{11}Li can be examined by using the non-energy weighted E1 cluster sum rule written by,

$$B(E1) = \int_{-\infty}^{+\infty} \frac{dB(E1)}{dE_{\rm rel}} dE_{\rm rel} = \frac{3}{4\pi}\left(\frac{Ze}{A}\right)^2 \langle r_1^2 + r_2^2 + 2\vec{r_1}\cdot\vec{r_2}\rangle, \quad (4)$$

where r_1, r_2 are the coordinates of two valence neutrons relative to the core. The term $\vec{r_1}\cdot\vec{r_2}$ contains the opening angle θ_{12} of two valence neutrons, which is a measure of the two-neutron correlation. The integrated $B(E1)$ strength obtained in the current experiment is 1.42±0.18 e^{2}fm^2 (4.5±0.6 Weisskopf units) for $0 \leq E_{\rm rel} \leq 3$ MeV. According to the theory, this value can be extrapolated to the cluster sum of 1.78±0.22 e^{2}fm^2, which corresponds to the mean opening angle of $\langle\theta_{12}\rangle = 48^{+14}_{-18}$ degrees. This value is smaller than that for the model with independently-moving two neutrons in the halo, *i.e.*, $\langle\theta_{12}\rangle$=90 degrees. More detailed analysis on two neutron correlation as well as n-^{9}Li correlation is now in progress for further clarification of the nature of the two-neutron halo nucleus.

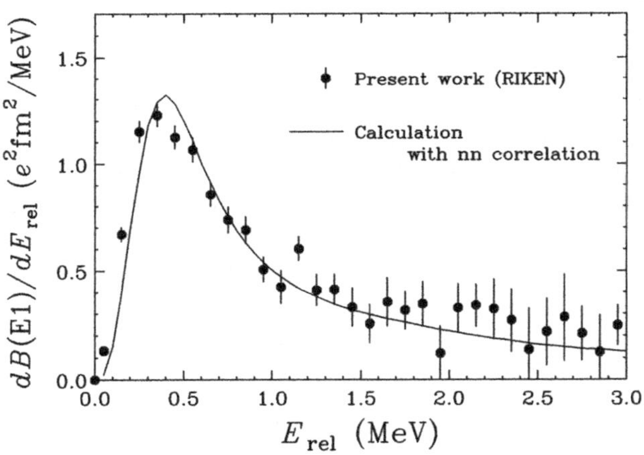

Fig. 1. The $B(E1)$ distribution of ^{11}Li obtained in the current Coulomb breakup experiment (solid circles). The data are compared with the three-body theory with neutron-neutron correlation and final state interactions,[17] as shown by the solid curve.

4. Inelastic scattering of ^{14}Be

The nucleus ^{14}Be is the heaviest bound beryllium isotope, and is known to have a two-neutron halo structure due to the weak binding (S_{2n}=1.26(13) MeV). The neighboring nucleus ^{12}Be has recently drawn much interest due to the melting of N=8 magicity at this nucleus. Hence, it is very interesting to study how the shell structure changes from ^{12}Be to ^{14}Be as one reaches the drip line.

The first 2^+ state (2_1^+) of an even-even nucleus can be a benchmark of the shell closure. However, that of ^{14}Be has not yet been fixed. Bohlen *et al.*[29] reported a candidate for the 2_1^+ state at E_x=1.59(13) MeV using the ^{14}C(^{14}C,^{14}O)^{14}Be reaction. Korsheninnikov *et al.*[30] reported observation of the state at E_x=1.6(1) MeV, using inelastic scattering on a proton target, consistent with the value in Ref.[29] On the other hand, experiments using the invariant mass method with the inelastic scattering of ^{14}Be could not find such a state.[11,31,32] There were also no experiments that extracted the transition probability between the ground state and the 2_1^+ state to obtain the information of collectivity.

To determine the energy of 2_1^+ state ($E_x(2_1^+)$) and extract its transition probability, we measured the inelastic scattering of ^{14}Be on carbon and proton targets using the invariant mass method. The different sensitivity between carbon and proton targets to the neutron- and proton excitation may enable us to extract the collectivity of neutron- and proton part in ^{14}Be independently. A secondary beam of ^{14}Be bombarded a carbon target (377 mg/cm^2) at 68 MeV/nucleon and a liquid hydrogen target (227 mg/cm^2) at 69 MeV/nucleon. Experimental setup is nearly identical to the one in ^{11}Li. The detail of the experiment for the carbon inelastic scattering is seen in Ref.[33]

Figure 2 shows the relative energy spectrum of ^{12}Be+n+n in the excitation of ^{14}Be by the carbon target. A clear peak appears around $E_{\rm rel}$=0.27(1) MeV above the neutron decay threshold. We also installed the γ ray detectors surrounding the target. Since no γ ray was observed in coincidence with the peak, we can thus determine unambiguously the excitation energy of this state as $E_x(=E_{\rm rel}+S_{2n})$=1.54(13) MeV.

The inset of the figure shows the angular distribution obtained by gating this peak. The experimental angular distribution is compared with the DWBA calculations shown by curves, which were obtained for 4 sets of optical potential parameters: Set A was taken from the ^{11}Be+^{12}C reaction at 49.3 MeV/nucleon,[34] Set B from the quasi-elastic ^{14}Be + ^{12}C reaction,[35] and Sets C and D from the elastic scattering of ^{12}C+^{12}C at 84.6 MeV/nucleon, and ^{12}C+^{16}O at 93.9 MeV/nucleon,[36] respectively. We find that the experimental data are consistent with the angular distributions calculated for ΔL=2, providing firm evidence that the state observed at E_x=1.54(13) MeV is the first 2^+ state of ^{14}Be. Among the four calculations, we note that those with optical potential parameters A,C, and D are in good agreement with the data, with reduced χ^2 values (χ^2/N) of 1.14, 1.52, and 1.54, respectively. The calculation with parameter set B fails to reproduce the data with χ^2/N = 3.96, probably due to the fact that the parameters in Ref.[35] are influenced by the quasi-elastic channel. Hence, we adopt here the parameter sets A,C, and D in extracting the deformation length (δ). The value is δ=1.18±0.13 fm, obtained as the average value of the three cases: δ=1.26(12) fm, 1.17(11) fm, and 1.12(10) fm for Sets A, C, and D, respectively.

The case of proton target also showed a peak corresponding to the first 2^+ state together with an angular distribution which is well explained by 2^+ transition. The

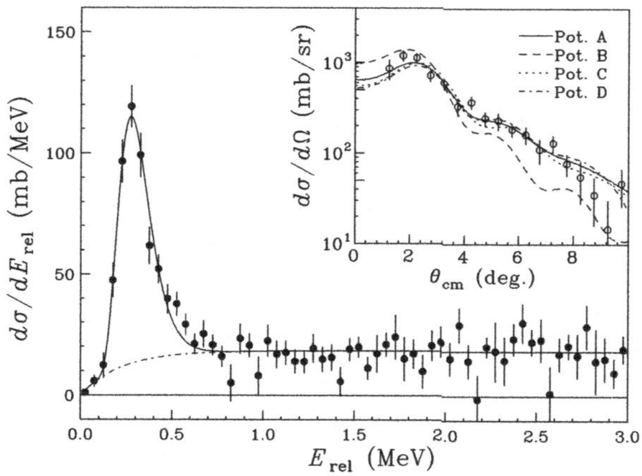

Fig. 2. Relative-energy spectrum of ^{12}Be+n+n in the excitation of ^{14}Be by the carbon target. A strong peak is observed around E_{rel}=0.3 MeV. Spectrum is fitted with a single peak on an arbitrary background function as shown by the lines. Inset: Differential angular cross section for the peak observed in the relative energy spectrum. The angular pattern is found well fitted with the calculation with the L=2 transition.

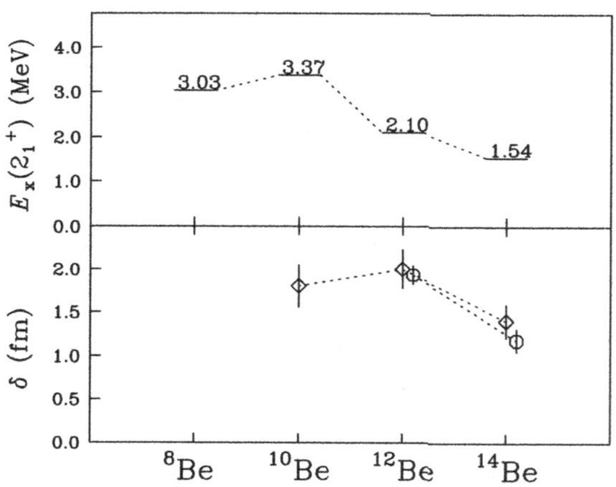

Fig. 3. Comparison of excitation energies of the first 2^+ state (upper panel) and the deformation lengths (lower panel) for the even-even Be isotopes. Deformation lengths obtained with carbon inelastic scattering are shown by open circles, while those with proton inelastic scattering are shown by open diamonds. Smaller E_x and deformation lengths in ^{14}Be relative to other Be isotopes are clearly seen.

extracted deformation length for the proton inelastic scattering was δ=1.40(19) fm. The excitation energy and the deformation lengths of ^{14}Be is compared with those of other Be nuclei in Fig. 3. In a naive picture, as in Raman's systematics,[37] a smaller $E_x(2_1^+)$ value indicates a more deformed state and larger transition probability. However, contrary to this simple picture, the deformation lengths extracted for

^{14}Be are significantly smaller than those for ^{12}Be, suggesting a smaller quadrupole collectivity in ^{14}Be. This may suggest the quenching of the effective charges in ^{14}Be.[33] Preliminary analysis combining proton and carbon results shows that such quenching is in particular strong for excitation in the proton part. Further analysis using microscopic calculations is now in progress.

5. Summary

In summary, we have demonstrated the powerfulness of the invariant mass spectroscopy by showing the recent experimental results of Coulomb breakup of ^{11}Li and the 2^+ excitation of ^{14}Be by carbon and proton targets. In the Coulomb breakup of ^{11}Li, we observed strong E1 transitions peaking around $E_{\rm rel}$=0.3 MeV in the energy spectrum. The deduced $B(E1)$ distribution was in excellent agreement with a three-body model incorporating the two-neutron correlations. This may suggest a spatial di-neutron like correlation of two halo neutrons. Comparison with the non-energy weighted E1 cluster sum rule provided an additional indication for the significant spatial correlation of the two valence neutrons in the ^{11}Li ground state.

We also showed the results of 2^+ excitation in ^{14}Be with proton and carbon inelastic scattering. We observed the first 2^+ state in the energy spectrum. The spin-parity assignment was confirmed by the angular distribution. The deformation lengths extracted for ^{14}Be were smaller than those for other Be isotopes, which suggested the quenching of the quadrupole collectivity at the drip-line of Be isotopes.

Acknowledgments

The experiments of the Coulomb breakup of ^{11}Li and the inelastic scattering of ^{14}Be with a carbon target (R301n experiment) have been done in collaboration with A.M. Vinodkumar, T. Sugimoto, N. Aoi, H. Baba, D. Bazin, N. Fukuda, T. Gomi, H. Hasegawa, N. Imai, M. Ishihara, T. Kobayashi, Y. Kondo, T. Kubo, M. Miura, T. Motobayashi, H. Otsu, A. Saito, H. Sakurai, S. Shimoura, K. Watanabe, Y.X. Watanabe, T. Yakushiji, Y. Yanagisawa and Y.Yoneda. The inelastic scattering with a proton target (R364n experiment) was done in collaboration with those in R301n experiment and with Y.Satou, N.Matsui, T.Okumura, Y.Hashimoto, T.Nakabayashi, M.Shinohara, N.Endo, S.Kawai, M.Kiyama, Y.Matsuda, T.Onishi, J.Ong, S.Takeuchi, N.Tamaki, Y.Togano. The present work was supported in part by a Grant-in-Aid for Scientific Research (No.15540257) from the Ministry of Education, Culture, Sports, Science and Technology (MEXT, Japan).

References

1. T. Nakamura *et al.*, *Phys. Lett. B* **331**, 296 (1994).
2. T. Nakamura *et al.*, *Phys. Rev. Lett.* **83**, 1112 (1999).
3. U. Datta Pramanik *et al.*, *Phys. Lett. B* **551**, 63 (2003).
4. R. Palit *et al.*, *Phys. Rev. C* **68**, 034318 (2003).
5. N. Fukuda *et al.*, *Phys. Rev. C* **70**, 054606 (2004).

6. K. Ieki *et al.*, *Phys. Rev. Lett.* **70**, 730 (1993); D. Sackett *et al.*, *Phys. Rev. C* **48**, 118 (1993).
7. S. Shimoura *et al.*, Phys. Lett. B **348**, 29 (1995).
8. M. Zinser *et al.*, *Nucl. Phys. A* **619**, 151 (1997).
9. T. Aumann *et al.*, *Phys. Rev. C* **59**, 1252 (1999).
10. J. Wang *et al.*, *Phys. Rev. C* **65**, 034306 (2002).
11. M. Labiche *et al.*, *Phys. Rev. Lett.* **86**, 600 (2001).
12. H. Iwasaki *et al.*, *Phys. Lett. B* **481** 7 (2000).
13. H. Iwasaki *et al.*, *Phys. Lett. B* **491** 8 (2000).
14. S. Shimoura *et al.*, *Phys. Lett. B* **560** 31 (2003).
15. T. Nakamura *et al.*, *Phys. Rev. Lett.* **96** 252502 (2006).
16. H. Esbensen, *Phys. Rev. C* **44**, 440 (1991).
17. H. Esbensen and G.F. Bertsch, *Nucl. Phys. A* **542**, 310 (1992); H. Esbensen, private communication.
18. H. Esbensen, G.F. Bertsch, and C.A. Bertulani, *Nucl. Phys. A* **581**, 107 (1995).
19. A. Cobis, D.V. Fedorov, and A.S. Jensen, *Phys. Lett. B* **424**, 1 (1998); E. Garrido, D.V. Fedorov, and A.S. Jensen, *Nucl. Phys. A* **708**, 277 (2002).
20. J.M. Bang and RNBT collaboration, *Phys. Rep.* **264**, 27 (1996).
21. A. Pushkin, B. Jonson, and M.V. Zhukov, *J. Phys. G* **22**, L95 (1996).
22. I.J. Thompson *et al.*, *J. Phys. G.* **24**, 1505 (1998).
23. A. Bonaccorso and N. Vinh Mau, *Nucl. Phys. A* **615**, 245 (1997).
24. C. Forssén, V.D. Efros, and M.V.Zhukov, *Nucl. Phys. A* **706**, 48 (2002).
25. T. Myo, S. Aoyama, K. Kato, and K. Ikeda, *Phys. Lett. B* **576**, 281 (2003).
26. Yu.A. Lurie and A.M. Shirokov, *Ann. of Phys.* **312**, 284 (2004).
27. K. Hagino and H. Sagawa, *Phys. Rev. C* **72**, 044321 (2005).
28. T. Kubo *et al.*, *Nucl. Instrum. and Methods Phys. Res. B* **70**, 309 (1992).
29. H.G. Bohlen *et al.*, *Nucl. Phys. A* **583**, 775 (1995).
30. A.A. Korsheninnikov *et al.*, *Nucl. Phys. A* **616**, 189c (1997).
31. B. Jonson *Phys. Rep.* **389**, 1 (2004).
32. H. Simon *et al.*, *Nucl. Phys. A* **791**, 267 (2007).
33. T. Sugimoto *et al.*, *Phys. Lett. B* **654**, 160 (2007).
34. P. Roussel-Chomaz, private communication; M. Takashina, private communication; R.C. Johnson, J.S. Al-Khalili, and J.A. Tostevin, *Phys. Rev. Lett.* **79**, 2771, (1997); The optical potential parameters are shown in Table II of Ref..[5]
35. M. Zahar *et al.*, *Phys. Rev. C* **48**, R1484 (1993).
36. M.-E. Brandan, *Phys. Rev. Lett.* **60**, 784 (1988).
37. S. Raman *et al.*, *At. Data Nucl. Data Tables* **78**, 1 (2001).

CORE ⊗ $(sd)^3$ STRUCTURES IN ^{17}C, ^{16}C AND ^{15}C UP TO HIGH EXCITATION ENERGIES

H. G. BOHLEN*, W. VON OERTZEN, CH. SCHULZ, TZ. KOKALOVA and C. WHELDON

Hahn-Meitner-Institut Berlin
Glienocker Str. 100, 14109 Berlin, Germany
**E-mail: bohlen@hmi.de*
www.hmi.de

R. KALPAKCHIEVA

Flerov Laboratory for Nuclear Reactions, JINR
RU-141980 Dubna, Russia
and
Institute for Nuclear Research and Nuclear Energy, BAS,
Blvd. Tsarigradsko Shosse 72, BG-1784 Sofia, Bulgaria

T. N. MASSEY

Department of Physics and Astronomy, Ohio University
Athens, Ohio 45701-2979, USA

M. MILIN

Ruđer Bošković Institute,
Bijenicka 54, HR-10002 Zagreb, Croatia

Structures of ^{17}C of the type ^{14}C$\otimes\nu$(sd)^3 have been studied up to high excitation energies using the (^{12}C,^{9}C) three-neutron transfer reaction on ^{14}C at 231 MeV incident energy. In this reaction (3p2h)-configurations are directly populated selectively. The large negative Q-value (Q_0 = -46.93 MeV) supports this selectivity for three-neutron configurations and high-spin states by dynamical matching conditions in the reaction mechanism, and higher-order transitions are suppressed. Tentative assignments could be made in this way. The experimental results are compared to shell-model calculations. The same structures have been also populated in ^{16}C and ^{15}C using the (^{12}C,^{9}C) reaction on ^{12}C and ^{13}C targets. Nine states in the three carbon isotopes ^{17}C, ^{16}C and ^{15}C show strong similarities in their properties. This concerns the excitation energies, which agree within $\pm$0.23 MeV except for a constant shift of excitation energies in ^{16}C by 5.82 MeV and in ^{15}C by 6.65 MeV with respect to ^{17}C, and these states have also similar widths and cross-section ratios. Therefore tentative assignments are discussed in the same way for ^{16}C and ^{15}C. Furthermore, the levels of the three isotopes have been investigated for band structures using the tentative assignments and shell-model calculations. Two bands could be identified in each isotope.

Keywords: Transfer reactions A = 15, 16, 17; Three-neutron configurations, Shell-model calculations; Tentative assignments.

1. Introduction

States in ^{17}C with three-neutron (sd)^3 configurations, which are coupled to the core of ^{14}C, are expected to exist in a broad excitation energy range. These structures are characterised as (3p2h) states. We have studied these configurations recently[1] using the direct three-neutron transfer reaction ^{14}C(^{12}C,^{9}C)^{17}C. Previously three states of even parity were known, all of them below the neutron threshold (S_n= 0.729 MeV): the ground state with J^π=3/2$^+$, and two excited states at 0.21 MeV and 0.33 MeV,[2,3] with probable assignments 1/2$^+$ and 5/2$^+$,[1] respectively. Before only two-neutron transfer reactions have been measured on ^{14}C, for example the (t,p) reaction.[4–6] In this case states of ^{16}C of ^{14}C$\otimes$(sd)^2 configurations with (2p2h) character and even parity are populated. Core excitations would lead to (3p3h) odd parity states, but no such states could be definitely identified experimentally, only a state at 6.11 MeV received the tentative assignments (2$^+$, 3$^-$, 4$^+$), including an odd parity.

Odd parity states were identified in ^{16}C at excitation energies above 7 MeV in our measurement of the ^{13}C(^{12}C,^{9}C)^{16}C three-neutron transfer reaction.[7] The (3p3h) configurations are populated easily in this case, because the ^{13}C target nucleus serves as a core with a built-in neutron 1p1/2-hole, therefore the odd parity is normal in *this* reaction. In Ref. 7 these structures could be localised in ^{16}C up to 10.4 MeV excitation energy.

For the correct description of the (3pxh) structures (x=2, 3) in theoretical calculations it is important to treat all *three* neutrons of the (2s1d)^3 configuration as valence particles. Two of the three neutrons may couple not only to 0$^+$ (giving rise also to single-neutron configurations in ^{17}C based on the ^{16}C ground state), but also to 2$^+$ or 4$^+$, which increases the number of configurations tremendously. Strong configuration mixing and large deformations are expected.

The (^{12}C,^{9}C) three-neutron transfer reaction, which was used to investigate the level structure of ^{17}C, has a very large negative Q-value of −46.93 MeV, which requires a high incident energy. In this case both, the high projectile velocity and the large negative Q-value will lead to a suppression of higher order contributions like target particle-hole excitations in combination with the three-neutron transfer.

2. Measurements and spectra analysis

The measurements of the (^{12}C,^{9}C) reactions have been performed at the Q3D magnetic spectrograph at ISL, Hahn-Meitner-Institut Berlin. The ^{12}C beam of 231.3 MeV incident energy was delivered by the isochron-cyclotron at ISL. The target had a content of 384 μg/cm^2 ^{14}C and 50 μg/cm^2 ^{12}C. For this reason a spectrum of the ^{12}C(^{12}C,^{9}C)^{15}C reaction has been measured on a separate ^{12}C target (thickness 200 μg/cm^2) for background subtraction. Particles were identified in the focal plane by the measurement of the energy-loss (ΔE), the rest-energy (E) and the time-of-flight (ToF) through the spectrograph, further details about the detector setup can be found in Ref. 1. Figure 1 shows the correlation between the ΔE

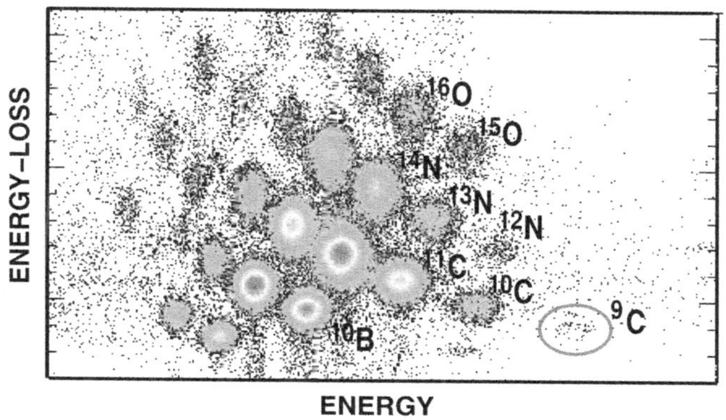

Fig. 1. Identification plot for particles arriving in the focal plane of the magnetic spectrograph.

versus the E signals, the ^{9}C events are encircled. They are further gated in a plot of ΔE versus ToF and then projected on to the position coordinate X measured in the focal plane. An overall energy resolution of 220 keV was obtained for the ^{12}C target, and 250 keV for the ^{14}C target due to its larger thickness.

The calibrated spectra are shown in Fig. 2 for ^{17}C (panel (a)), ^{16}C (panel (c)), and in the case of ^{15}C for two different magnetic field settings, one for higher excitation energies of ^{15}C (panel (b)) and the other one for lower ones (panel (d)). In all cases Breit-Wigner line-shapes have been used for resonances. Broad three-body distributions exist in all spectra, which are described by the sequential decay of outgoing highly excited ^{10}C*, which decay in-flight into detected ^{9}C and a neutron.

3. Population of core⊗(sd)^3 structures

The actual three-neutron transfer reactions have extremely large negative ground-state Q-values, ^{15}C: $Q_0=-38.79$ MeV; ^{16}C: $Q_0=-39.48$ MeV; ^{17}C, $Q_0=-46.93$ MeV. The total Q-values even increase up to about −60 MeV with increasing excitation energies. Therefore the grazing angular momenta in the entrance and exit channels, L_i and L_f, respectively, are different by about $\Delta L \approx 9\hbar$ or more at the incident beam energy of 231 MeV. This large difference ΔL in angular momentum is called dynamical mismatch. In such a situation the population of high-spin states is favored, i. e., stretched angular momentum configurations of the three neutrons coupled to high-spin states can bridge this difference, at least partially. Contrary to this, three-neutron configurations with strong 2s1/2-occupancies will be disfavored by the dynamical mismatch due to the $\ell=0$ neutron orbit.

The effect of dynamical matching on the population of states is seen in the ^{15}C-spectrum, Fig. 2 (d). For the ground state of ^{15}C $(1/2^+)$ no count is observed, and only very few for the first excited state $(5/2^+)$. Both states have single-particle character with *one* neutron in the 2s1/2 and 1d5/2 shell, respectively. Both other

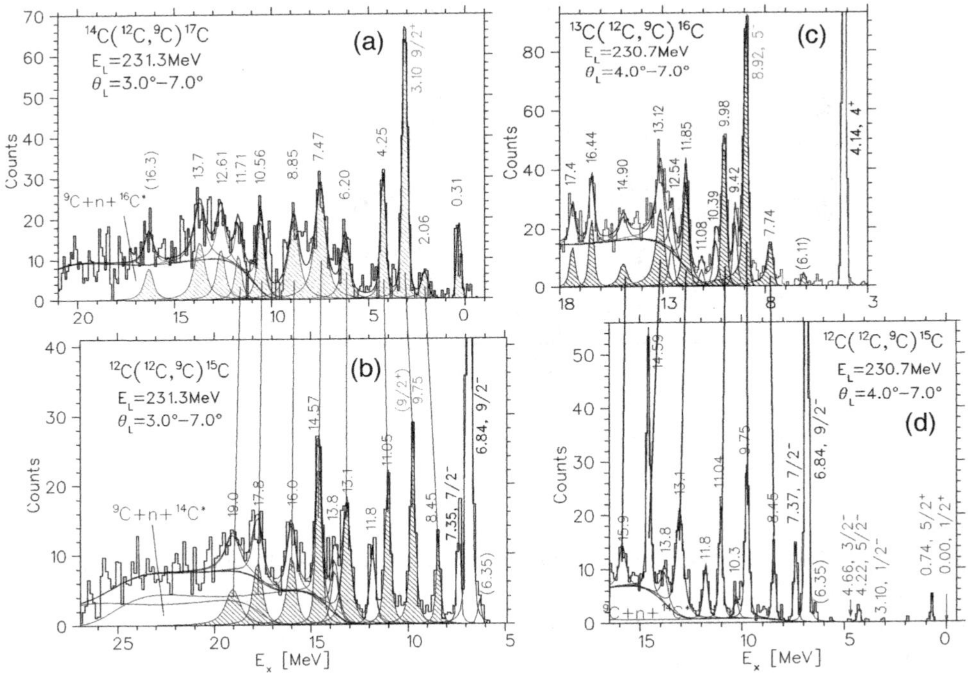

Fig. 2. Spectra of the (^{12}C,^{9}C) reaction on ^{14}C (a), ^{12}C (b) and (d), and on ^{13}C (c) for the spectroscopy of ^{17}C, ^{15}C, and ^{16}C, respectively. In panel (a) the background contributions resulting from the ^{12}C content of the ^{14}C target are already subtracted (see details in Ref. 1). The lines connect corresponding states.

transferred neutrons just fill the holes in the 1p1/2 shell and couple to 0^+. Also the next three excited states are only weakly populated. These states have known configurations of *two* neutrons in the (sd)-shell and only one neutron in the 1p1/2 shell. The $(sd)^2$ neutron pair can couple to the spins 0^+, 2^+ and 4^+ and the following final spins are obtained: in the first case (0^+) a single $1/2^-$ state (3.10 MeV), in the second case a doublet $3/2^-$, $5/2^-$ (4.66 MeV, 4.22 MeV) and in the third case a doublet $7/2^-$, $9/2^-$ (7.37 MeV, 6.84 MeV). From all five states only the $9/2^-$ state is strongly populated, only this has a stretched configuration. When all *three* neutrons are transferred to the (sd)-shell, we distinguish the low excitation energy region, where contributions from the 1d5/2 and 2s1/2 shells are dominant and from the 1d3/2 shell weak, and the region starting at about 4 MeV higher energies, where about one neutron occupies in most cases the 1d3/2 shell. In the first case stretched configurations couple to a maximum of spin of $9/2^+$, either from two neutrons in the 1d5/2 shell and one in the 2s1/2 shell, or from all three neutrons in the 1d5/2 shell. According to the selective reaction mechanism and matching rules, $J^\pi = 9/2^+$ must be the assignment for the next strongest observed state of ^{15}C at 9.75 MeV.

We use the same arguments for assignments of excited states in ^{17}C. Here the neutron (1p)-shell is closed, and the lowest-lying states are represented by $(2s1d)^3$ configurations. As before, the strongest low-lying state, observed at 3.10 MeV, must have a stretched $9/2^+$ configuration. The 1d3/2-strength should become important at excitation energies about 4.0 - 4.4 MeV higher (deduced from the 1d3/2-strength relative to 1d5/2 in ^{15}C and ^{13}C[3,9]). A strong peak with the stretched configuration $[(5/2^+)^2_{4+}]\otimes(3/2^+) = 11/2^+$ is expected. Taking the 3.10 MeV state of ^{17}C as reference, this $11/2^+$ state should be localised at about 7.3 MeV. Indeed, the next strongest peak is observed at 7.47 MeV (see Fig. 2 (a)), it is assigned $(11/2^+)$.

When the spectra of ^{17}C and ^{15}C are plotted with a relative energy shift of 6.65 MeV (this is the difference of excitation energies for the stretched $9/2^+$ states at 9.75 MeV (^{15}C) and 3.10 MeV (^{17}C), respectively) as shown in Fig. 2 (a) and (b), a strong similarity becomes obvious. Most observed states of ^{15}C are then aligned with states of ^{17}C, as indicated by the straight lines. But also the populated strength is very similar. The states at 3.10 MeV and 9.75 MeV in ^{17}C and ^{15}C, respectively, form a pair with the same structure (except for the number of holes in the (1p)-shell), and this is also true for the pair at 7.47 MeV and 14.57 MeV.

This observation can be extended to ^{16}C, when (i) the spins $9/2^+$ from the three neutrons in the (sd)-shell and $1/2^-$ from the neutron 1p1/2-hole are coupled to the stretched configuration 5^-, and (ii) an energy shift of 5.82 MeV is used (see Fig. 2 (c), (d)). The latter value results from the difference between 8.92 MeV (^{16}C) and 3.10 MeV (^{17}C)). The members of a pair are called *corresponding* states according to the very similar structure. Such pairs of states in ^{16}C and ^{15}C are (8.92 MeV, 9.75 MeV) for $J^\pi = 9/2^+$, and (13.12 MeV, 14.59 MeV) for $11/2^+$.

However, many other corresponding states exist as indicated by the lines in Fig. 2 and as discussed in Ref. 1. These excitation energies are listed in Table 1, rows 2, 3, and 4, with already subtracted energy shifts (the energies of group 3 have therefore all the same value 3.10 MeV). For all corresponding states of ^{17}C and ^{16}C, the same values within ±0.23 MeV are obtained, and maximum deviations of ±0.63 MeV for ^{17}C and ^{15}C, and this even over a range of almost 12 MeV excitation energy. This is a very strong hint for a common structure of the members of a pair. Further support of this conclusion is obtained from the comparison of the widths and cross-section ratios given in rows 5 – 9 of Table 1. Here the agreement is not as perfect, but it is still significant in many cases. The observed corresponding states of the three carbon isotopes ^{17}C, ^{16}C, and ^{15}C must have very similar three-neutron $(sd)^3$ configurations coupled to the cores of ^{14}C, ^{13}C, and ^{12}C, respectively.

4. Shell model calculations and possible band structures

We have performed shell model (SM) calculations to understand more about the structure of low-lying states of ^{17}C and the three-neutron configurations. The shell model code OXBASH[10] and the 'psdmwkpn' interaction[11] have been used. The results are compared to the experimental data in Table 1 and for ^{17}C also in Fig. 3.

In Table 1 four rows are used for each isotope. The first row shows the mea-

Table 1. Comparison of experimental excitation energies E_x of ^{17}C, ^{16}C and ^{15}C, the widths Γ and ratios of cross sections. Tentative or definite spin-parity assignments are discussed in the text. Shell model (SM) calculations for ^{17}C, ^{16}C and ^{15}C are compared to the experimental results.

Corresponding states, No.	1	2	3	4	5	6	7	8	9		
$E_x(^{17}C)_{exp}$ [MeV]	0.31	2.06	**3.10**	4.25	6.20	**7.47**	8.85	10.56	11.71		
$E_x(^{16}C)_{exp}$ **−5.82** [MeV]	0.29	1.92	**3.10**	4.16	6.03	**7.30**	9.08	10.62	11.58		
$E_x(^{15}C)_{exp}$**−6.65** [MeV]	(-0.30)	1.80	**3.10**	4.40	6.45	**7.92**	9.35	11.15	12.35		
$\Gamma(^{17}C)$ [MeV]	–	0.25	0.10	0.14	0.35	0.58	0.66	0.30	0.30	0.45	0.6
$\Gamma(^{16}C)$ [MeV]	<0.03	0.20	≤0.10	0.12	0.22	0.40	0.30	0.15	0.20		
$\Gamma(^{15}C)$ [MeV]	–	0.08	0.10	0.10	0.30	0.13	0.30	0.50	(1.0)		
$d\sigma(^{16}C)/d\sigma(^{17}C)$	–	2.9	2.1	2.7	3.2	1.5	0.6	1.3	1.6		
$d\sigma(^{15}C)/d\sigma(^{16}C)$	–	2.6	1.4	1.6	2.3	1.6	2.3	2.7	–		
$E_x(^{17}C)_{exp}$ [MeV]	0.31	2.06	**3.10**	4.25	6.20	**7.47**	8.85	10.56	11.71	12.6	13.7
$J^\pi(^{17}C)_{exp}$	$5/2^+$	$(7/2, 3/2)^+$	$\mathbf{9/2^+}$	$(5/2\text{-}9/2)^+$	$(7/2, 5/2)^+$	$\mathbf{(11/2^+)}$	$(9/2^+)$	$(13/2^+)$	$(11/2^+)$		
$E_x(^{17}C)_{SM}$ [MeV]	0.02	2.59	3.00	4.11	4.97	6.84	8.61	10.17	11.13		
$J^\pi(^{17}C)_{SM}$ (3p2h)	$5/2^+_1$	$7/2^+_1$	$9/2^+_1$	$5/2^+_3$	$7/2^+_2$	$11/2^+_1$	$9/2^+$	$13/2^+_1$	$11/2^+_3$		
$E_x(^{16}C)_{exp}$ [MeV]	(6.11)[a]	7.74	**8.92**	9.98	11.85	**13.12**	14.90	16.44	17.4		
$J^\pi(^{16}C)_{exp}$	(3^-)	$(4 - 1)^-$	$\mathbf{5^-}$	$(3 - 5)^-$	$(4 - 3)^-$	$\mathbf{(6 - 5)^-}$	(5^-)	(7^-)	(6^-)		
$E_x(^{16}C)_{SM}$ [MeV]	5.71	7.71	8.72	9.48	12.55	12.37	14.94				
$J^\pi(^{16}C)_{SM}$ (3p3h)	3^-_1	4^-_1	5^-_1	3^-_3	4^-	6^-_1	5^-				
$E_x(^{15}C)_{exp}$ [MeV]	(6.35)	8.45	**9.75**	11.05	13.1	**14.57**	16.0	17.8	19.0		
$J^\pi(^{15}C)_{exp}$	$(5/2^+)$	$(7/2, 3/2)^+$	$\mathbf{9/2^+}$	$(5/2\text{-}9/2)^+$	$(7/2, 5/2)^+$	$\mathbf{(11/2^+)}$	$(9/2^+)$	$(13/2^+)$	$(11/2^+)$		
$E_x(^{15}C)_{SM}$ [MeV]	6.49	8.71	11.15	10.98	13.64	13.73	16.18				
$J^\pi(^{15}C)_{SM}$ (3p4h)	$5/2^+_2$	$7/2^+_2$	$9/2^+_3$	$5/2^+$	$7/2^+$	$11/2^+_2$	$9/2^+$				

[a] The ^{16}C state at 6.11 MeV is also observed in ^{14}C(t,p) reactions.[5,6]

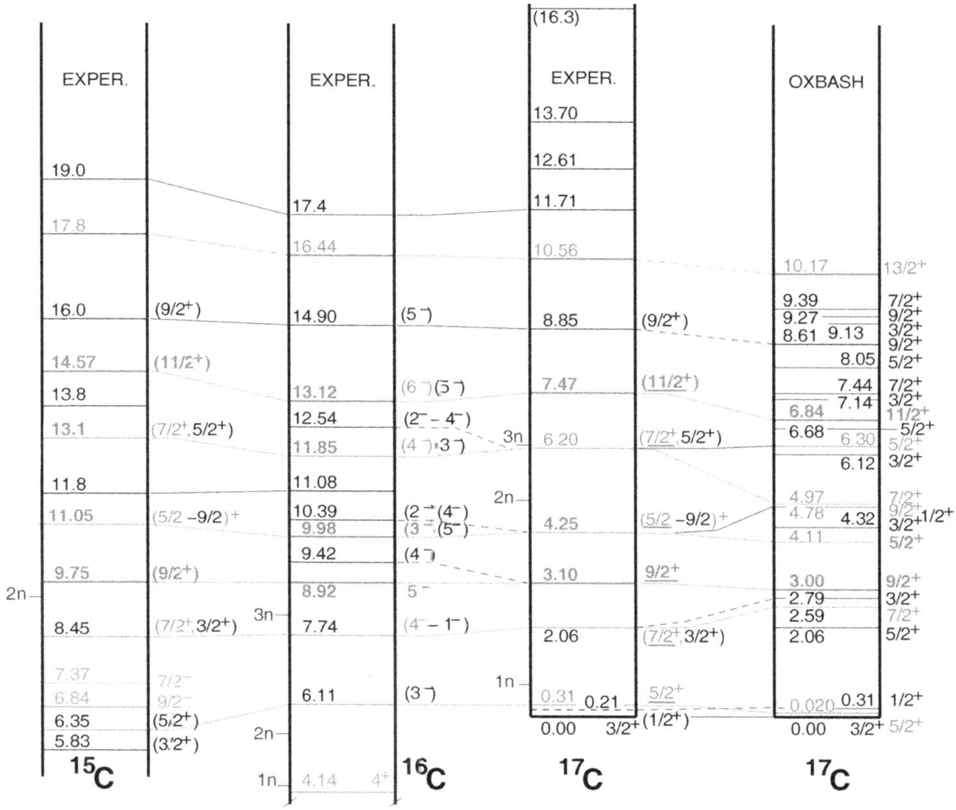

Fig. 3. Comparison of level schemes of ^{15}C, ^{16}C, and ^{17}C for states populated in the three-neutron transfer reactions and for states with 3n-configurations obtained in shell model calculations for ^{17}C.

sured excitation energies, the second one definite or tentative assignments using all available information including also decay properties of the states and the widths,[1] and in the third and fourth rows excitation energies and spin-parities from the SM calculations taking the arguments from matching conditions in the reaction and high spin values into account. In the calculations for ^{17}C the lowest-lying SM-states with J^{π} in the range from $5/2^+$ to $13/2^+$ have in all cases an experimental counterpart with the same J^{π}, which was assigned already in many cases tentatively from different arguments, and with an excitation energy nearby (see Fig. 3).

The SM-calculations for ^{17}C, ^{16}C, ^{15}C (Table 1) confirm that almost the same $(sd)^3$ configurations exist for the corresponding states. In Fig. 3 these states are connected by lines between the experimental level schemes.

With these tentative spin-parity assignments the data have been inspected for possible band structures. For ^{17}C we use the $5/2^+_1$ state at 0.31 MeV as the band head, because this state is well populated in the three-neutron transfer. From the

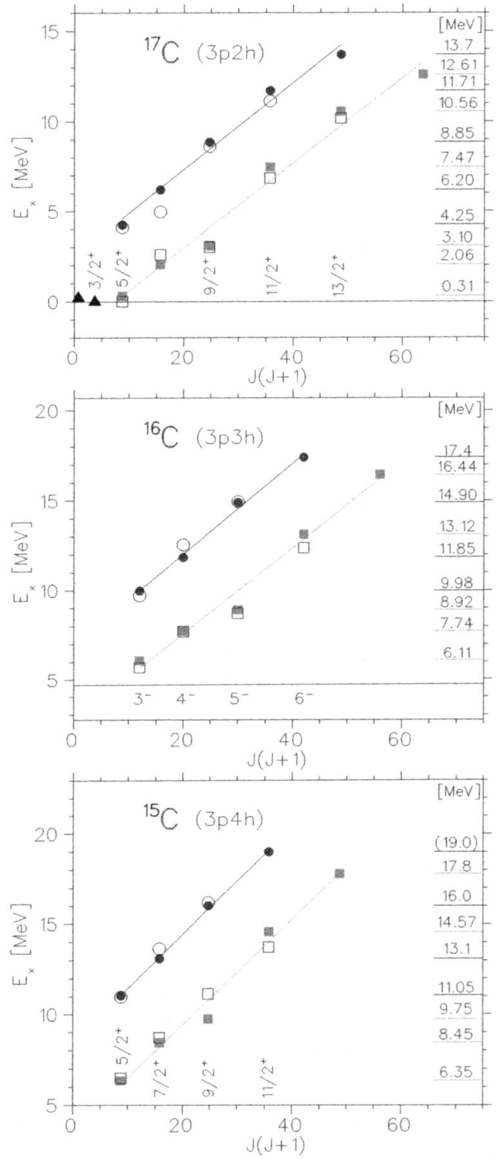

Fig. 4. Band structures of ^{17}C (upper panel), ^{16}C (central panel) and ^{15}C (lower panel) observed in the (^{12}C,^{9}C) three-neutron transfer reaction. The filled plot symbols correspond to excited states populated in this experiment and the open symbols to shell model calculations. Experimental excitation energies of the states are given in each panel on the right.

level schemes (Fig. 3) the following members with increasing spin are identified: 0.31 MeV, $5/2_1^+$, ; 2.06 MeV, $(7/2_1^+)$; 3.10 MeV, $9/2_1^+$; 7.47 MeV, $(11/2_1^+)$; 10.56 MeV, $(13/2_1^+)$. A second band may be localised at 4.25 MeV starting with the tentatively assigned band head ($5/2^+$) and the further members: 6.20 MeV, ($7/2^+$); 8.85 MeV, ($9/2^+$); 11.71 MeV, ($11/2^+$). These bands are shown in Fig. 4 (upper panel) together with the results of SM-calculations.

The bands found in ^{17}C have *corresponding* bands in ^{16}C and ^{15}C as expected from the structure of corresponding states. In Fig. 3 the members of these bands are connected by lines, and these bands are plotted in Fig. 4 (central and lower panels, respectively).

5. Summary

The level schemes of the neutron-rich carbon isotopes ^{17}C, ^{16}C, and ^{15}C have been studied using the three-neutron transfer reaction (^{12}C,^{9}C) on ^{14}C, ^{13}C, and ^{12}C targets. The large negative Q-values up to about −60 MeV impose rather restrictive matching conditions for the transfer cross sections, with strong selectivity for stretched configurations in the final state. Therefore preferentially high-spin states in the three isotopes with the same corresponding structures are populated. Most of these states can be ordered into two bands in each isotope with core⊗(sd)^3 structure. The experimental results are reproduced by shell-model calculations.

References

1. H. G. Bohlen *et al.*, *Eur. Phys. J. A* **31**, 279 (2007).
2. M. Stanoiu *et al.*, *Eur. Phys. J. A* **20**, 95 (2004).
3. Z. Elekes *et al.*, *Phys. Lett. B* **614**, 174 (2005).
4. D. P. Balamuth *et al.*, *Nucl. Phys. A* **290**, 65 (1977).
5. H. T. Fortune, M. E. Cobern *et al.*, *Phys. Rev. Lett.* **40**, 1236 (1978).
6. R. R. Sercely, R. J. Peterson and E. R. Flynn, *Phys. Rev. C* **17**, 1919 (1978).
7. H. G. Bohlen, R. Kalpakchieva, *et al.*, *Phys. Rev. C* **68**, 054606 (2003).
8. S. E. Darden, G. Murillo and S. Sen, *Phys. Rev. C* **32**, 1764 (1985).
9. F. Ajzenberg-Selove, *Nucl. Phys. A* **523**, 1 (1991).
10. W. Rae *et al.*, *Code OXBASH, Technical Report 524* (MSU Cycl. Lab., 1985).
11. E. K. Warburton, B. A. Brown, *Phys. Rev. C* **46**, 923 (1992).

LIGHT NEUTRON-RICH NUCLEI STUDIED BY ALPHA-INDUCED REACTIONS

S. SHIMOURA*

Center for Nuclear Study (CNS), University of Tokyo,
Wako, Saitama 351-0198, Japan
**E-mail: shimoura@cns.s.u-tokyo.ac.jp*

The structure of light neutron-rich nuclei, especially for vicinities of ^{12}Be and ^{22}O nuclei, has been studied by the α induced reactions in inverse kinematics at intermediate energies.

Alpha inelastic scattering to highly excited states in ^{12}Be were studied by using invariant mass spectroscopy in inverse kinematics. A new multipole decomposition analysis taking into account the angular correlation of decaying process originated from the alignments of the residual nucleus was applied to extract transferred angular momenta.

The proton transfer reactions (α,t) were performed for spectroscopy of proton single-particle states characterized by mean field in neutron-rich nuclei.

Keywords: Alpha induced reactions; gamma-ray spectroscopy; Invariant mass spectroscopy; Single particle states; Cluster structure.

1. Intdoduction

Nuclei apart from the stability line attract much attention, since provided rich information on nuclear structures as a function of the proton and the neutron numbers where exotic natures, such as a neutron halo, soft collective excitations, and disappearance of ordinary magic numbers, have been observed. One of the powerful experimental approaches for investigating exotic nuclei is the measurements of the decays of the excited states populated by the direct reactions of the RI beams at intermediate energies. In this report, recent experimental results on α induced direct reactions on light neutron-rich nuclei by the inverse kinematics combined with invariant-mass/γ-ray spectroscopy.

2. Inverse Kinematics

One of the most important techniques in experiments using intermediate-energy unstable nuclear beams is inverse kinematics, where the beam or beam-like nuclei are to be investigated (physical targets) through interactions with the target nuclei (physical probes). For measurements of α induced direct reaction with the inverse kinematics, we developed a liquid helium target system.[1] The beam-like nucleus is produced by interacting with target nucleus through a direct reaction and subsequently decays by emitting particle(s) and/or photon(s). Experimentally, instead of detecting the recoils of the targets, all the emitted particles and/or photons are detected to determine their excited energies by reconstructing the invariant masses or decay energies.

Because of the kinematic focusing in intermediate-energy reactions, velocities of all the emitted particles are almost same as the beam velocity, which enables us to detect the decay particles with a large acceptance by using detectors placed at a forward direction. The extracted momentum vectors are used to reconstruct the decay energy, which is defined as the difference between the invariant mass and the threshold energy. The excitation energy is deduced by adding the separation energy for the decay channel to the decay energy.

When the decays are accompanied by emissions of γ-ray(s), it is required to correct Doppler shifts from fast-moving emitters. Since the magnitudes of the shifts amount to ± 30 % for $\beta = 0.3$, a segmented array of γ-ray detectors should be used to measure the emission angle of the γ-ray as well as its energy. It is noted that the Doppler-shift correction of the γ-ray is the same as the deduction of the decay energy from the invariant mass of the γ and the emitting particle.

By using tracking information of secondary beams and position information of particle and γ detectors, we also deduce angular distributions of the primary reaction and angular correlations of the decaying process which relate to the alignment of the excited states.

3. Alpha Inelastic Scatterings

Alpha inelastic scatterings at intermediate energies are useful for exciting the isoscalar states. For stable nuclei the (α,α') reactions have been extensively studied at 60 A MeV to extract isoscalar E0, E1, and E2 strengths in excited states.[2–5] It is interesting to extend such studies for unstable nuclei having exotic nature in the ground states. Firstly, I report measurements of the α inelastic scatterings to the low-lying 2^+ and 1^- states in the neutron-rich nucleus ^{12}Be at 60 A MeV. It is noted that several experimental evidences of the breaking of the $N = 8$ shell closure in the ^{12}Be have been reported.[6–9]

Figures 1 and 2 respectively show the angular distributions of the differential cross section of the α inelastic scatterings and the angular distribution of the γ decays for the 2_1^+ (closed symbols) and 1_1^- (open symbols) states in the exit channel. The lines denote the predictions of DWBA calculations with a folding model, where the transition potentials are calculated by folding the collective transition densities.[10,11] As shown in the figures the DWBA calculations predict experimental data very well not only for the angular distribution but also for the alignments of the excited states.

The same distorting and transition potentials were used for a new type of multipole decomposition analysis (MDA) for the ^{4}He(^{12}Be,^{12}Be$^* \rightarrow ^6$He+^{6}He) reaction, where the angular distributions of the decaying ^{6}He particles depending on the align-

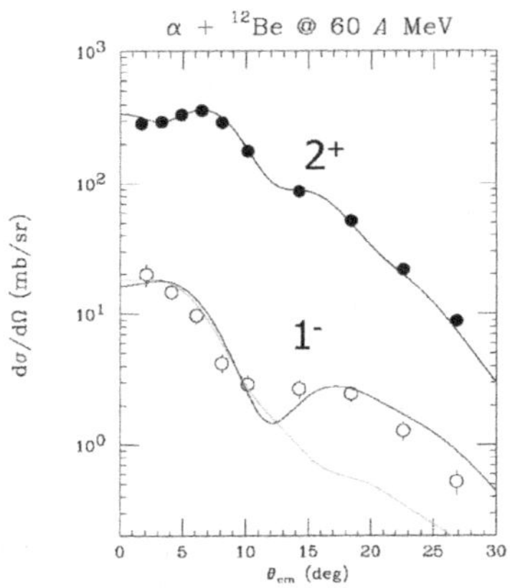

Fig. 1. Angular distributions of α inelastic scatterings ^{4}He(^{12}Be,^{12}Be*) at 60 A MeV to the 2_1^+ (closed symbols) and the 1_1^- (open symbols) states in the ^{12}Be nucleus. Solid lines denote the predictions of DWBA calculations.

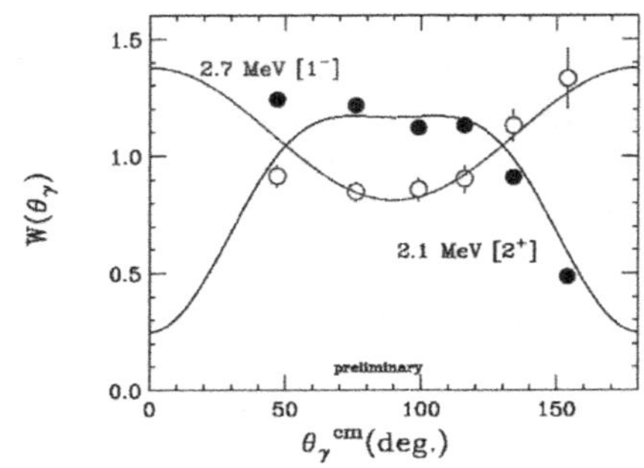

Fig. 2. Angular distributions of γ rays from the 2_1^+ (closed symbols) and the 1_1^- (open symbols) states in the ^{12}Be nucleus populated by α inelastic scatterings. Solid and dashed lines denote the predictions for the 2_1^+ and the 1_1^- states, respectively, by using statistical tensors calculated by DWBA.

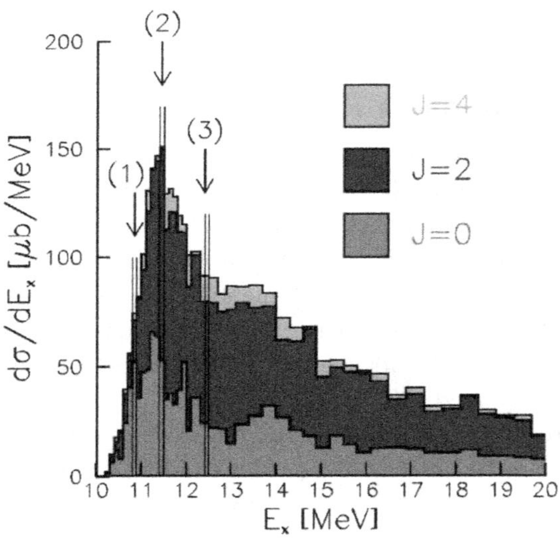

Fig. 3. Excitation energy spectrum of the ^{4}He(^{12}Be,^{12}Be*) at 60 A MeV deduced by the decay energy of the ^{12}Be* $\rightarrow$ ^{6}He+^{6}He process.

ments of the transferred angular momenta were included simultaneously.[12] Figure 3 shows an excitation energy spectrum deduced by the invariant mass of the final ^{6}He+^{6}He system, which consists of several multipoles (ℓ=0, 2, and 4).

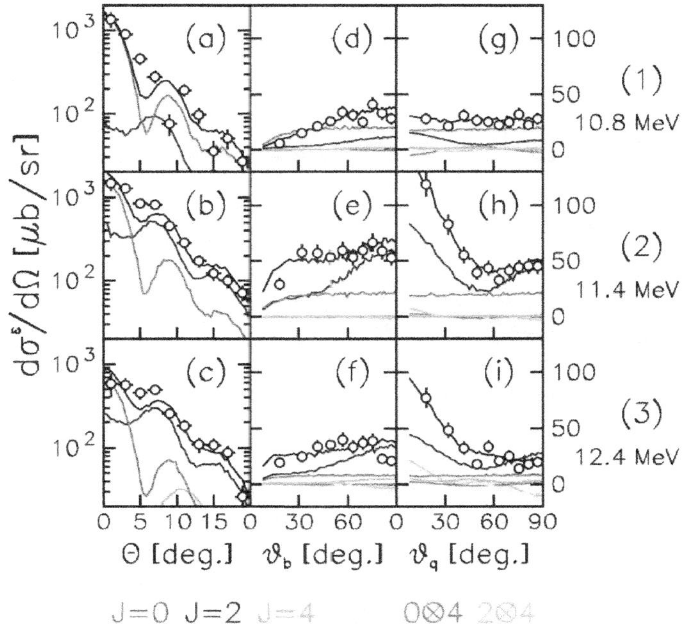

Fig. 4. Angular distributions of the ^{4}He(^{12}Be,^{12}Be*) reaction and angular correlations of the ^{12}Be* → ^{6}He+^{6}He process for E_x=10.8, 11.4 and 12.4 MeV. Lines denotes the components of the MDA and their sums (thick).

The present MDA was performed by using DWBA angular distribution of each multipole and the decay amplitudes obtained by the statistical tensors calculated by DWBA where amplitudes of different multipoles interfere each other. As shown in Fig. 4, the MDA reproduces both the angular distribution and the angular correlations well and provides the excitation energy spectrum for each multipole (Fig. 5). It is noted that the angular correlations exhibit clearly the characteristics of the multipole: almost isotropic for 10.8 MeV region indicates the dominance of ℓ=0 and the other two show large contributions from ℓ=2.

Each spectrum was fitted by a sum of several peaks and a background, which suggests existence of resonances with ℓ=0 and 2. Assuming that the peak positions correspond to the resonances, an yrast diagram is constructed (Fig. 6) together with the previously reported states.[13,14] The parallel three lines in the figure correspond to the moment of inertia of the ^{6}He+^{6}He system where the two ^{6}He touch each other at their surfaces. The present results may be members of molecular bands which may be connected to the results of ref.[14] and qualitatively consistent with recent theoretical predictions of molecular bands.[15,16]

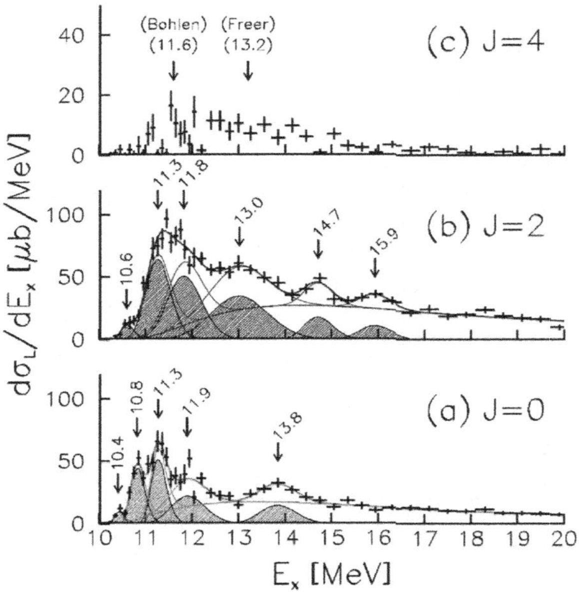

Fig. 5. Excitation energy spectra for ℓ=0, 2, and 4. Lines denote the results of fit assuming the sum of several peaks and backgrounds.

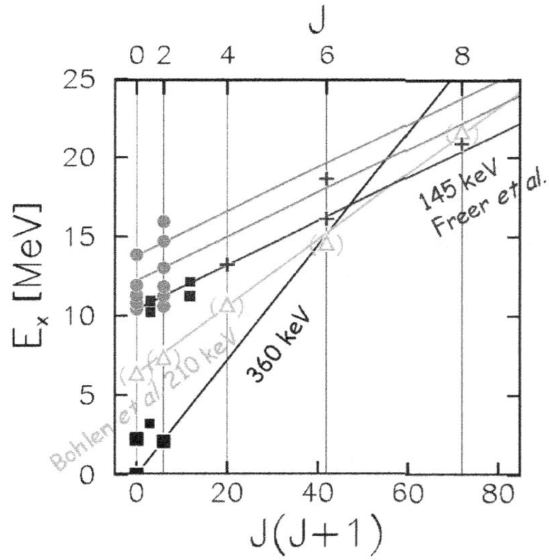

Fig. 6. Yrast diagram of the ^{12}Be nucleus. Closed circles denote the present results and the other symbols are the previously reported results.

4. Proton Transfer Reaction (α,t)

In order to investigate the shell evolution in the nuclei far from the stability line, excited states with single-particle nature are interested. The single-particle states are populated by nucleon transfer reactions such as (d,p) and (d,n) reactions, which are effective below 10 A MeV region, because of the matching condition in the momentum. In secondary-beam experiments with the inverse kinematics this fact requires a use of relatively thinner target or results a small cross section at higher incident energies. However, in the nucleon transfer from the α particle the restriction from the matching condition becomes weaker for higher incident energies, because of the larger separation energy and the larger Fermi momentum than the deuteron.[17] Based on this consideration, we performed measurement of the (α,t) reaction on the ^{12}Be, which populated excited states in ^{13}B.[18]

Here, we used the GRAPE which consists of six pairs of segmented Ge detectors[19,20] for resolving excited states with close energies in the odd A nucleus ^{13}B.

Figure 4 demonstrates the γ-ray spectrum taken by the GRAPE after the correction of the Doppler shifts. There are several peaks corresponding to the transitions of the known excited states to the ground states (3.48, 3.53, 3.68, 3.71, 4.13 and 4.83 MeV). The yields of γ-rays are deduced by fitting the spectrum with a sum of the response functions of the corresponding energies. The 4.83-MeV state having the largest yield could have a proton single-particle nature, comparing to the reported yields to this state in other reactions.[21–24]

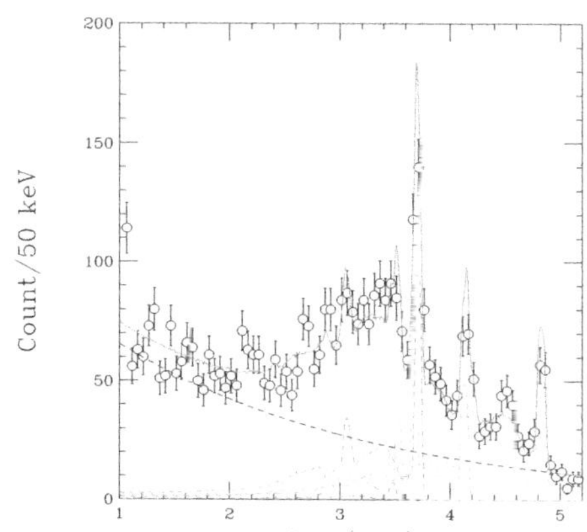

Fig. 7. Doppler-corrected γ-ray spectrum obtained in coincidence with ^{13}B (preliminary).

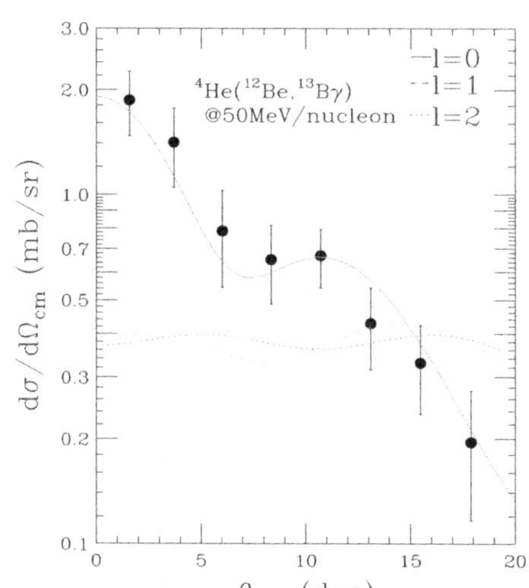

Fig. 8. Angular distribution of the nucleon transfer reaction. Solid, dashed and dotted lines denotes preliminary predictions of DWBA calculation with ℓ = 0, 1, and 2, respectively.

In order to determine the transferred angular momentum (ℓ) to the 4.83-MeV state, the angular distribution of the transfer reaction is examined (Fig. 4). As shown in the figure, a forward peaking angular distribution is deduced, which suggests ℓ=0 corresponding to J^π=1/2$^+$. The ℓ=0 transfer is supported by the comparison of the DWBA predictions assuming ℓ=0, 1, and 2. The present result indicates that the 4.83-MeV state is a proton intruder state.

Although the ground state of the ^{13}B nucleus is known to be spherical, the excited state may be deformed because of a large deformation of the ^{12}Be nucleus.[9,25] If the proton is bound by the deformed ^{12}Be core, the proton intruder orbit of [220]1/2 come down to the p-shell region. The present 1/2$^+$ state may be this deformed intruder state, which may be a kind of shape coexistence in the ^{13}B nucleus.

5. Summary

Alpha induced reactions (inelastic scatterings and nucleon transfers) combined with invariant-mass / γ spectroscopy become new tools for the spectroscopy of excited states in exotic nuclei. Although the qualities (emittance, intensity, etc.) of the secondary beams worse, coincident measurements make it possible to perform detailed spectroscopy of the excited states by the combination of the analyses of the angular distributions of the primary process and the decay process. New information on changes in the nuclear structure as a function of proton and neutron numbers such as shell evolution and change of nuclear shapes will be investigated extensively by such detailed measurements, especially in the RIBF at RIKEN, and extensive analysis.

Acknowledgments

The author would like to thank the members of the CNS-Rikkyo-RIKEN-Tokyo-Kyoto collaboration for the present works and projects.

The present work was supported by the Grant-in-Aid for Scientific Research (Nos. 11440081 and 15204018) by the Ministry of Education, Culture, Sports, Science and Technology.

References

1. H. Ryuto et al., *Nucl. Instr. Meth. Phys. Res. A* **555**, 1 (2005).
2. D. Youngblood, Y.-W. Lui and H. Clark, *Phys. Rev. C* **60**, 014304 (1999).
3. D. Youngblood, Y.-W. Lui and H. Clark, *Phys. Rev. C* **64**, 067301 (2001).
4. D. Youngblood, Y.-W. Lui and H. Clark, *Phys. Rev. C* **65**, 034302 (2002).
5. B. John et al., *Phys. Rev. C* **68**, 014305 (2003).
6. H. Iwasaki et al., *Phys. Lett. B* **491**, 8 (2000).
7. H. Iwasaki et al., *Phys. Lett. B* **481**, 7 (2000).
8. S. Shimoura et al., *Phys. Lett. B* **560**, 31 (2003).
9. S. Shimoura et al., *Phys. Lett. B* **654**, 87 (2007).
10. G.R. Satchler, *Nucl. Phys. A* **472**, 215 (1987).
11. M. Harakeh and A.E.L. Dieperink, *Phys. Rev. C* **23**, 2329 (1981).

12. A. Saito, PhD thesis, Rikkyo University, (Tokyo, Japan, 2006), pp. ix + 119.
13. H. Bohlen et al., in *Proc. Int. Symposium on Exotic Nuclei 2001 (EXON2001)*, 2002.
14. M. Freer et al., *Phys. Rev. Lett.* **82**, 1383 (1999).
15. Y. Kanada-Enyo and H. Horiuchi, *Phys. Rev. C* **68**, 014319 (2003).
16. M. Ito and Y. Sakuragi, *Phys. Rev. C* **62**, 064310 (2000).
17. S. Shimoura, *J. Phys. G* **31**, S1759 (2005).
18. S. Ota et al., submitted to *Phys. Lett. B*.
19. M. Kurokawa et al., *IEEE Trans. Nucl. Sci.* **50**, 1309 (2003).
20. S. Shimoura, *Nucl. Instr. Meth. Phys. Res. A* **525**, 188 (2004).
21. F. Ajzenberg-Selove et al., *Phys. Rev. C* **17**, 1283 (1978).
22. Y. Guimarães et al., *Phys. Rev. C* **61**, 064609 (2000).
23. N. Aoi et al., *Phys. Rev. C* **66**, 014301 (2002).
24. R. Kalpakchieva et al., *Eur. Phys. Jour. A* **7**, 451 (2000).
25. I. Hamamoto and S. Shimoura, to be published in *J. Phys. G*.

FUSION AND DIRECT REACTIONS AROUND THE COULOMB BARRIER FOR THE SYSTEM ^{6}He + ^{64}Zn

V. SCUDERI[*,1,2], A. DI PIETRO[2], P. FIGUERA[2], F. AMORINI[1,2], C. ANGULO[3], G. CARDELLA[2], E. CASEREJOS[3], M. LATTUADA[1,2], A. MUSUMARRA[4], M. PAPA[2], M.G. PELLEGRITI[5], R. RAABE[6], F. RIZZO[1,2]

[1] *Dipartimento di Fisica e Astronomia, Universitá degli studi di Catania, 62, Via S. Sofia, I-95123 Catania, Italy*

[2] *INFN-Laboratori Nazionali del Sud and Sezione di Catania, 64, Via S. Sofia, I-95123 Catania, Italy*

[3] *CRC Louvain la Neuve, Belgium*

[4] *Dipartimento di Metodologie Fisiche e Chimiche per l'Ingegneria, Universitá di Catania, Catania, Italy*

[5] *GANIL, Caen, France*

[6] *Instituut voor Kern-en Stralingsfysica Katholieke Universiteit Leuven, Leuven, Belgium*

[*] *E-mail: scuderiv@lns.infn.it*

Fusion and direct processes have been studied for the ^{6}He+^{64}Zn reaction at energies around the Coulomb barrier, in order to investigate the effects of the projectile halo structure on the reaction mechanisms. As comparison, the ^{4}He+^{64}Zn reaction was also studied. No effect has been observed on the fusion cross section for the ^{6}He when compared with the reaction induced by the stable well bound isotope, whereas a large yield for direct reactions due to transfer and break-up events has been evidenced.

Keywords: Reactions induced by unstable nuclei; Fusion reactions; direct reactions

1. Introduction

In the last years a lot of work has been performed in order to understand which effects one should expect on reaction mechanisms at energies around the Coulomb barrier, when using halo nuclei as projectiles. In the case of reaction induced by halo, but also weakly bound projectiles, due to the particular structure of these nuclei, the system in question is able to exploit a multitude of other available degrees of freedom. In particular, in the reaction dynamics the halo structure implies static effects due to the long range of the radial wave-functions associated with the valence orbitals and a reduction in the barrier height which may lead to a larger sub-barrier fusion cross-section. Besides the purely static effects, one also expects other effects of dynamical nature associated to the low break-up energy threshold. Dynamic effects cause a strong coupling of the break-up channels with the fusion one.

Theoretically, the role of the break-up process of weakly bound projectiles in fusion and also in scattering mechanisms have been extensively investigated via Coupled reaction Channel (CC) calculations. Recently, in order to proper describe the reaction dynamics in collisions involving weakly bound nuclei, since the ground state of the nucleus lies close to the continuum, Continuum-Discretised Coupled-Channel (CDCC) calculations [1,2] have been developed. In these calculations the coupling not only to bound but also to continuum states (break-up states) is included. CDCC calculations predict that dynamical effects (coupling) enhance the sub barrier total fusion (complete + incomplete) cross-section with respect to the no-coupling case. In addition, due to the break-up, a suppression of the total fusion cross section is predicted by CDCC calculations at energies above the barrier.[2] However, the calculated cross-section and hence the amount of sub-barrier enhancement, depends strongly on the treatment of the continuum, as explained

Opposite results have been obtained with a different approach, developed by M. Ito in[3] et al.,[4] based on the time dependent wave packet formalism which uses a three body model (core, halo and target). According to this model, the cross-section in the neutron halo case is slightly suppressed than in the non-halo case. However according to the authors of [4] the discrepancy with the CDCC calculations of [1,2] is generated by the smaller value adopted for the maximum halo-core relative-angular-momentum used in the CDCC calculation and values of lmax $\geq$ 30 must be used. Also from the experimental point of view there are many open issue. Due to low intensity of radioactive beams, which are several order of magnitude smaller than stable beams intensity, is extremely difficult to measure at sub barrier energies. Moreover, due to the low measured yield, fusion excitation function do not extend down to deep sub barrier energies. To compensate for the low intensity of the beams, large efficiency detector arrays have been developed e.g.[5] and activation experiments have been performed e.g.[6,7]

From an analysis of existing data, when fusion excitation function for halo nuclei are compared with those for stable isotopes, no strong effect on fusion cross section has been found,[7–9] with the exception of two systems: ^{6}He on ^{209}Bi [10–13] and on ^{206}Pb,[14] in which an enhancement of the fusion cross section around and below the barrier is evidenced by the author.

We have, recently, extended the study of the elastic scattering, fusion and transfer + break-up channels for the ^{6}He+^{64}Zn reaction, we had already measured at energies around the barrier,[7] up to energies above the Coulomb barrier (about two times V_{CB}). As comparison, the ^{4}He+^{64}Zn reaction was also studied in the same center of mass energy. In this paper we will discuss the results for the fusion and transfer + break-up processes for the reaction ^{6}He+^{64}Zn.

2. The experiment

Using the ^{6}He beam available at Louvain la Neuve, we have measured the fusion excitation function in the energy range 7.5 MeV$\leq$Ecm$\leq$16.5 MeV. The fusion cross-

section has been measured, by using an activation technique detecting the atomic X-rays emitted from the radioactive evaporation residues (E.R.) after decaying by Electron Capture (E.C), as discussed in.[7]
In the new measurement a set-up similar to the previous experiment was used. The experimental set-up is sketched in fig. 1. An array of Si strip detectors surrounding a thin (500 $\mu g/cm^2$) ^{64}Zn target was used to detect the light particles (^{6}He, α and p) produced in the reaction. The thin ^{64}Zn target was angled at 45° in order to allow detection of particles emitted at 90°. The angular range covered was $5°\leq\theta\leq120°$. He isotopes were discriminated from H using the time of flight (TOF) information. Downstream the thin target a stack of thick ^{64}Zn targets ($\sim$2 $\mu g/cm^2$), each followed by a Nb catcher ($\sim$3.5 $\mu g/cm^2$), was irradiated by the same beam while measuring at the same time with the thin target the elastic scattering and other reaction channels. Short and long activation runs were performed to optimize the production of E.R. with different half-lives. The catcher thickness was chosen in order to stop the E.R. produced in the previous ^{64}Zn target and to slow down the beam. The target activity was measured off-line using Pb shielded large area Si(Li) detectors whose intrinsic efficiency is 100% for X-ray energies of our interest (around 10 keV). The advantage of measuring X-ray is the very low background that can be obtained with proper passive shields which allowed us to measure with X-ray counting rates as low as 0.3 counts/h. To extract the activity of the long lived residues, the irradiated targets were measured at Laboratori Nazionali del Sud for few months.

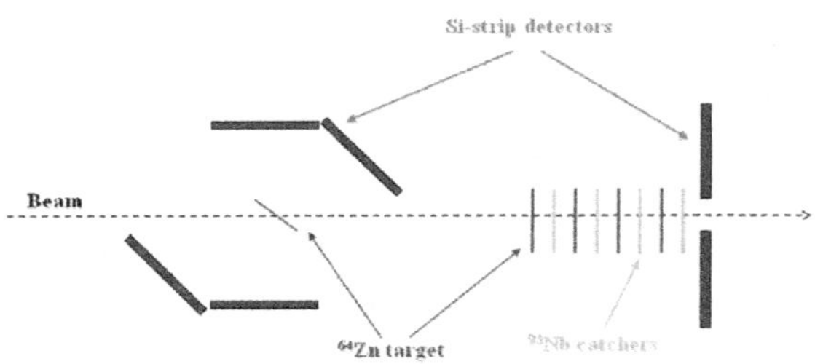

Fig. 1. Sketch of the experimental setup.

3. Experimental results

As widely discussed in,[7,15] one can clearly identify the E.R. by atomic number from the energy of the X-ray lines. Following the activity curves of each element as function of time it is also possible to discriminate different isotopes by their half-lives. Detail of the experimental procedure used to extract the production cross-section of the various heavy residues can be found in.[7,15]
The excitation functions obtained by summing up at each energy the contribution

of all radioactive heavy fragments produced in the two reactions ^{4,6}He+^{64}Zn are shown in fig. 2. An apparent enhancement of the fusion cross-section seems to be

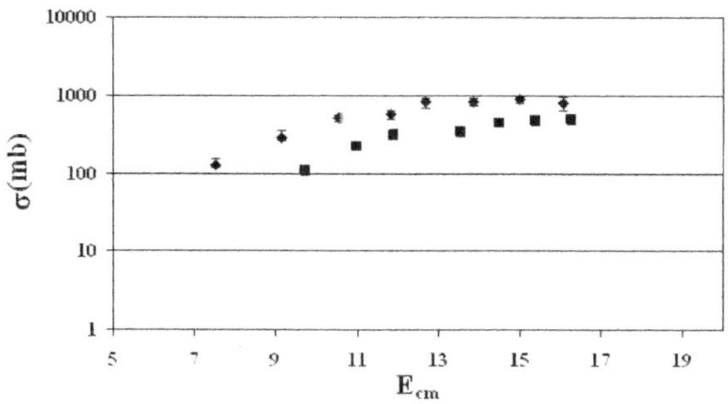

Fig. 2. Heavy fragments excitation function for the reactions ^{6}He+^{64}Zn (diamonds) and ^{4}He+^{64}Zn (squares).

present in the reaction induced by the ^{6}He beam with respect to the ^{4}He one, in all the measured energy range. However, the comparison between the measured yield for the different evaporation channels and statistical model calculations,[7] performed with code CASCADE,[16] showed a large discrepancy for the ^{65}Zn. The excess in the yield measured for this particular channel could be due to the contribution of the 1 and 2 neutron transfer which produces the radioactive ^{65}Zn nucleus also formed in alpha-neutron evaporation process.

Indeed, indication of the presence of a strong transfer component comes from the analysis of the charged particle coincidences. In fig. 3 is shown, as histogram, the light particle energy spectrum corresponding to α-α and α-p coincidences, measured at backward angles (θ=90°) at Elab=10 MeV. In the figure the experimental spectrum is compared with the corresponding spectrum (symbols) of α-α and α-p calculated using a Monte Carlo version of CASCADE code where all the experimental constraints have been included. As can be seen the experimental spectrum cannot be reproduced by fusion-evaporation processes. The energy and position of the observed peak corresponds to α particles emitted in 2n transfer reactions.

Thus the apparent enhancement of the fusion cross section was imputed to a large contribution due to transfer processes in agreement with what observed by different authors for collisions induced by the ^{6}He on different targets ^{209}Bi,[12] ^{238}U [8] and ^{63,65}Cu.[17] In order to obtain the fusion excitation function we subtracted the contribution due to the transfer, using the same procedure used in,[7] i.e. by replacing the measure value for ^{65}Zn with the one calculated using the statistical model code CASCADE. The results are shown in figure 4. It seems that no effect on the fusion cross section due to the halo structure of the ^{6}He nucleus is present, at least in the energy range where these data have been measured.

In order to take the different system size and Coulomb barrier height into consider-

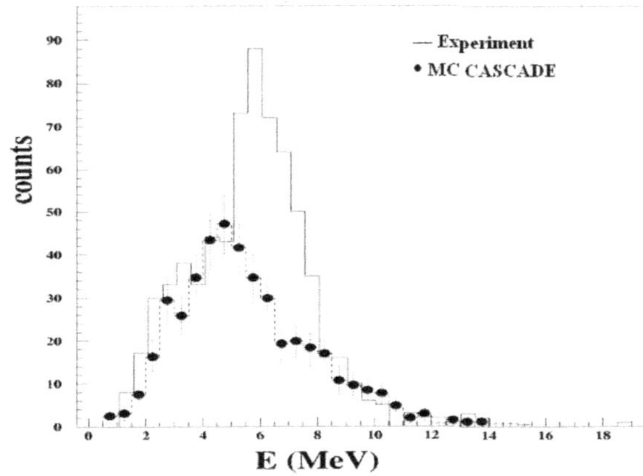

Fig. 3. Energy spectrum of events with charged particle multiplicity 2 at backward angles for the $^{6}He+^{64}Zn$ reaction at Elab=10 MeV (histogram). The symbols represent the contribution of fusion evaporation processes calculated using a Monte Carlo statistical code.

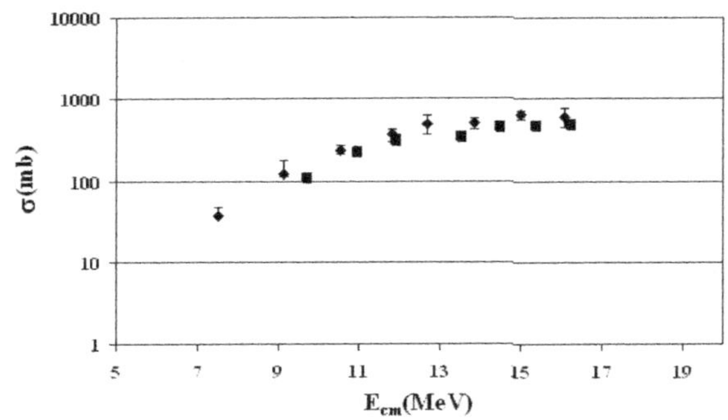

Fig. 4. $^{6}He+^{64}Zn$ (diamonds) excitation function, obtained by substituting the measured ^{65}Zn contribution with the one calculated by CASCADE, and $^{4}He+^{64}Zn$ (squares) excitation function.

ation, without washing out effects due to the long tail of the halo nucleons density distribution or to the low binding energy, we reduce the data as suggested in.[18] As one can see from fig. 5, no effect is observed in the $^{6}He+^{64}Zn$ fusion excitation function.

Recently we have extended the measurement of the fusion excitation function for the ^{4}He + ^{64}Zn system at lower energies using the same activation technique in order to compare the two excitation functions up to the lowest energy data point measured for the ^{6}He induced reaction. The results concerning these new data are not shown in fig. 5 since the analysis is still in progress.

As above mentioned, in this experiment other channels were also analysed. In particular the elastic scattering angular distributions and the α particle angular distri-

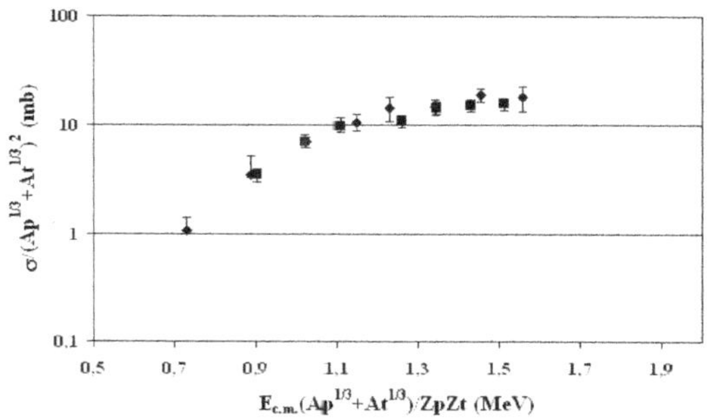

Fig. 5. ^{6}He+^{64}Zn (diamonds) and ^{4}He+^{64}Zn (squares) fusion excitation function plotted after reducing σ and VC as suggested in.[13]

butions have been measured at four beam energies Elab=10, 13.6, 15, 18 MeV. The last two energies concern this new experiment. As said, in this paper we will report only on direct reactions.

The alpha particle angular distributions for the ^{6}He + ^{64}Zn at Elab= 13.6 MeV is shown in fig. 6. From the behaviour of the α-particle angular distribution, which

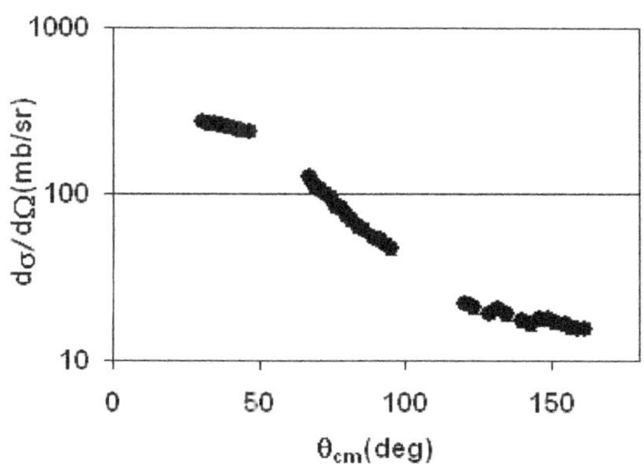

Fig. 6. α-particle angular distribution for ^{6}He+^{64}Zn at Elab=13.6 MeV.

is peaked and has a maximum that moves to smaller angles as the beam energy increases, the origin of such alphas is attributed to direct processes: transfer and break-up. This represents a further confirm of the strong contribution of the direct component to the total reaction cross section at sub barrier energies.

4. Conclusions

In the last years more and more papers have been published on reactions, in particular fusion, involving weakly bound and halo nuclei. Measuring fusion cross-sections with low intensity radioactive beams is not an easy task and most of the available data are not really exploring the region below the barrier. The reactions induced by the ^{6}He nucleus have shown that the dominant process at energies below the barrier is of direct type, mainly 2n and 1n transfer and break-up. However, different experimental results concerning the possible enhancement of the sub-barrier fusion in halo nuclei induced reactions still remain. In fact, the enhancement of fusion around and below the barrier evidenced in ^{6}He on ^{209}Bi and ^{206}Pb is not seen (within the error bars and energy range explored) in the other systems studied although, as said, few data exist below the barrier. In order to further clarify this topic new data in collisions induced by different nuclei and extending below the barrier are necessary.

References

1. K. Hagino et al., *Phys. Rev. C* **61**, 037602 (2000).
2. A. Diaz-Torres et al., *Phys. Rev. C* **65**, 024606 (2002).
3. L.F. Canto et al., *Phys. Rep.* **424**, 1 (2006).
4. M. Ito et al., *Phys. Lett. B* **37**, 53 (2006).
5. T. Davinson et al., *Nucl. Instr. and Meth. A* **454**, 350 (2000).
6. C. Signorini et al., *Eur. Phys. Jour. A* **2**, 227 (1998).
7. A. Di Pietro et al., *Phys. Rev. C* **69**, 044613 (2004).
8. R. Raabe et al., *Nature* **431**, 823 (2004).
9. C. Signorini et al., *Nucl. Phys. A* **735**, 329 (2004).
10. J.J. Kolata et al., *Phys. Rev. Lett.* **81**, 4580 (1998).
11. J.J. Kolata et al., *Eur. Phys. J. A* **13**, 117 (2002).
12. P.A. De Young et al., *Phys. Rev. C* **71**, 051601 (2005).
13. N. Alamanos et al., *Phys. Rev. C* **65**, 054606 (2002).
14. Yu.E. Penionzhkevich et al., *Phys. Rev. Lett* **96**, 162701 (2006).
15. A. Di Pietro et al., *Phys. Atomic Nuclei* **69**, 1366 (2006).
16. F. Puhlhofer et al., *Nucl. Phys. A* **280**, 267 (1977).
17. A. Navin et al., *Phys. Rev. C* **70**, 044601 (2004).
18. P.R.S. Gomes et al., *Phys. Rev. C* **71**, 017601 (2005).

ANALYZING POWER MEASUREMENT FOR PROTON ELASTIC SCATTERING ON ^{6}He

S. SAKAGUCHI*, T. UESAKA, T. KAWABATA and Y. SASAMOTO

Center for Nuclear Study, University of Tokyo,
Hirosawa 2-1, Wako, Saitama 351-0198, Japan
**E-mail: satoshi@cns.s.u-tokyo.ac.jp*

T. WAKUI and M. ITOH

Cyclotron and Radioisotope Center, Tohoku University,
6-3 Aramaki, Aoba, Sendai, Miyagi 980-8578, Japan

T. KAWAHARA

Department of Physics, Toho University,
Miyama 2-2-1, Funabashi, Chiba 274-0072, Japan

Y. ICHIKAWA, H. IWASAKI, H. KUBOKI, T. NAKAO, H. SAKAI,
M. SASANO, D. SUZUKI, Y. TAKAHASHI and K. YAKO

Department of Physics, University of Tokyo,
Hongo 7-3-1, Bunkyo, Tokyo 113-0033, Japan

H. OKAMURA, K. SUDA and A. TAMII

Research Center for Nuclear Physics, Osaka University,
Mihogaoka 10-1, Ibaraki, Osaka 567-0047, Japan

Y. MAEDA

Department of Physics, Kyushu University,
Hakozaki 6-10-1, Higashi, Fukukoka, Fukuoka 812-8581, Japan

K. ITOH

Department of Physics, Saitama University,
Shimoookubo 255, Sakura, Saitama 338-8570, Japan

N. AOI, N. SAKAMOTO, K. SEKIGUCHI and M. YAMAGUCHI

RIKEN (The Institute of Physical and Chemical Research),
Hirosawa 2-1, Wako, Saitama 351-0198, Japan

Y. HASHIMOTO, Y. SATOU and M. SHINOHARA

Department of Physics, Tokyo Institute of Technology,
Ohokayama 2-12-1, Meguro, Tokyo 152-8551, Japan

M. ICHIKAWA and R. MATSUO

Department of Physics, Tohoku University,
Aoba 6-3, Aramaki, Aoba, Sendai, Miyagi 980-8578, Japan

Spin-dependent asymmetry was measured for the $\vec{p}$–^{6}He elastic scattering at 71 MeV/nucleon making use of a solid polarized proton target specially designed for radioactive ion beam experiments. Absolute values of the analyzing power were deduced for the first time from the measured asymmetry and from the target polarization determined via the $\vec{p}$–^{4}He elastic scattering at 80 MeV/nucleon. Large discrepancy was found between obtained analyzing power data and predictions of g-matrix folding model.

Keywords: helium isotope, polarized proton, elastic scattering, analyzing power, spin-orbit interaction, neutron halo

1. Introduction

Scattering experiments making use of radioactive ion (RI) beams have expanded the scope of nuclear physics to nuclei far from the β-stability line. A plenty of new phenomena such as neutron skin/halo and changes of the shell structure have been discovered. On the other hand, direct reactions induced by polarized protons have revealed various aspects of nuclear structure and reaction mechanisms. Spin parities of single-particle states and neutron matter distributions are good examples of them. It is naturally expected that polarized protons play a central role also in the study of unstable nuclei and provide valuable information through polarization observables.

For the study of unstable nuclei with polarized protons, we have constructed a solid polarized proton target that is specially designed for RI beam experiments.[1,2] This target has a unique capability of operating under a low magnetic field of 0.1 T and at high temperature of 100 K. These modest conditions allow us to detect low energy (~10 MeV) recoiled protons which are essential for event identification. Measurement of polarization observable in RI beam experiments was realized for the first time by the solid polarized proton target.

Making use of this target, we measured analyzing power for the $\vec{p}$–^{6}He elastic scattering at 71 MeV/nucleon. Aim of the measurement is to obtain direct information on the spin-orbit potential between a proton and a ^{6}He particle. Since spin-orbit interaction is a surface effect, spin-orbit potential reflects surface structure of a nucleus. Polarization study of ^{6}He could provide us with new information on the extended neutron distribution of ^{6}He particle.

2. Experiment

The experiment was carried out at RIKEN Accelerator Research Facility (RARF) using RIKEN Projectile-fragment Separator (RIPS). A ^{6}He beam was produced through projectile fragmentation of a ^{12}C beam with an energy of 92 MeV/nucleon bombarded on to a Be target. The energy and the typical intensity of the ^{6}He beam were 71 MeV/nucleon and 250 kcps, respectively. The purity of the beam was 95%. The spot size on the secondary target was 10 mm FWHM in diameter. The angular spread of the beam was as small as 20 mrad FWHM.

Experimental setup is schematically illustrated in Fig. 1. A solid polarized proton target was used for a secondary target. Target material was a single crystal of naphthalene doped with small amount (0.005 mol%) of pentacene. The shape of the crystal was a thin disk of which diameter and thickness were 14 mmϕ and 1 mm, respectively. The number of protons in the target was $4.29 \pm 0.17 \times 10^{21}$ cm^{-2}. Protons were polarized in a low magnetic field of B =88 mT and at high temperature of T =100 K. Figure 2 shows the target polarization as a function of time. For the elimination of spurious asymmetry, the direction of polarization axis was reversed three times during the measurement by 180° pulse NMR method.[7] Depolarization effect due to the beam irradiation was sufficiently small. Absolute value of the target polarization was determined with measured spin-dependent asymmetry of the $\vec{p}$–^{4}He elastic scattering at 80 MeV/nucleon and with analyzing power data obtained by Togawa *et al.*[3] Average and the maximum polarization were 13.8% and 20.4%, respectively.

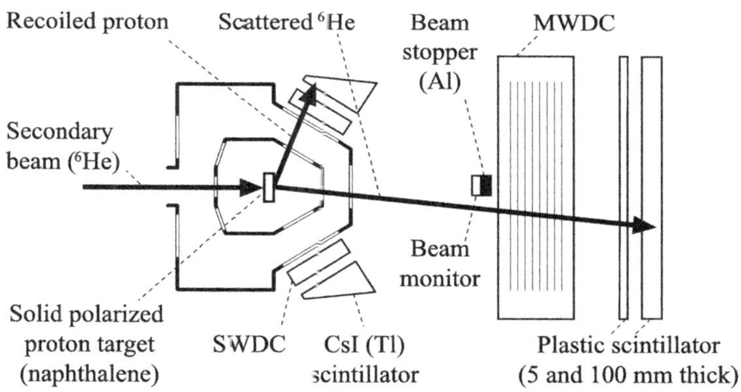

Fig. 1. Experimental setup is illustrated. A solid polarized proton target is bombarded by a ^{6}He beam produced by RIPS. Leading particles were detected by an MWDC and plastic scintillators. Recoiled particles were detected by SWDCs and CsI (Tl) scintillators.

Detection of recoiled protons is essential for event identification, because the solid proton target is a compound of protons and carbons. Recoiled particles were detected by a pair of counter telescopes which consisted of a single-wire drift chamber (SWDC) and a CsI(Tl) scintillation detector. Distance from the target to the SWDCs and CsI(Tl) scintillators were 140 and 175 mm, respectively. Sensitive area of the telescope was 85×60 mm^2, which covers the proton scattering angle of from 40° to 70°. Scattering angles of recoiled particles were determined by the SWDCs. CsI(Tl) scintillators provided information on total kinetic energies of the particles.

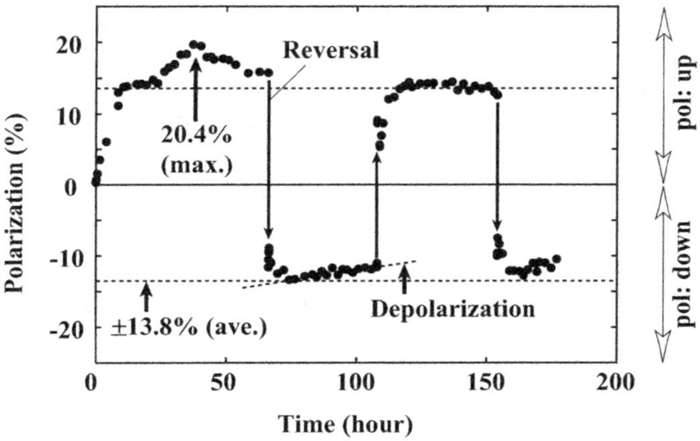

Fig. 2. Time development of target polarization. See text for details.

Leading particles were detected by a multi-wire drift chamber (MWDC) and plastic scintillator hodoscope placed 1 m downstream of the target. Sensitive area of the MWDC was 640×160 mm^2. Scattering angles of leading particles were determined by the MWDC. The energy deposit ΔE and the total energy E were measured with two layers of plastic scintillators whose thicknesses are 5 and 100 mm.

3. Results

^{6}He particles were identified by the standard $\Delta E - E$ method. The $\vec{p}$–^{6}He elastic scattering events were selected by requiring kinematical consistency to protons and identified ^{6}He particles. Here, difference between the azimuthal angle of a leading particle ϕ_{lead} and that of a recoiled particle ϕ_{rec} was checked. Since elastic scattering is the only channel whose final state is a two-body system, elastic scattering events can be selected by requiring a condition: $|\phi_{lead} - \phi_{rec}| \approx 0$. Polar angle correlation after this selection is shown in Fig. 3 (a). Clear loci of the elastic scattering events are identified on the solid curves which represent the result of kinematics calculation. Figure 3 (b) shows a histogram of θ_{lead} where θ_{rec} is gated from 57.5° to 60.0°. Yields were obtained by integrating the peaks corresponding to the elastic scattering events.

The analyzing power A_y was derived, cancelling spurious asymmetry, by

$$A_y(\theta) = \frac{1}{\overline{P}_y} \frac{\sqrt{Y(\theta)} - 1}{\sqrt{Y(\theta)} + 1} \tag{1}$$

$$Y(\theta) = \frac{N_L^{\uparrow}(\theta) \cdot N_R^{\downarrow}(\theta)}{N_L^{\downarrow}(\theta) \cdot N_R^{\uparrow}(\theta)} \tag{2}$$

Here, $\overline{P}_y$ is the average polarization of the target. N_i^j is yield where the subscript

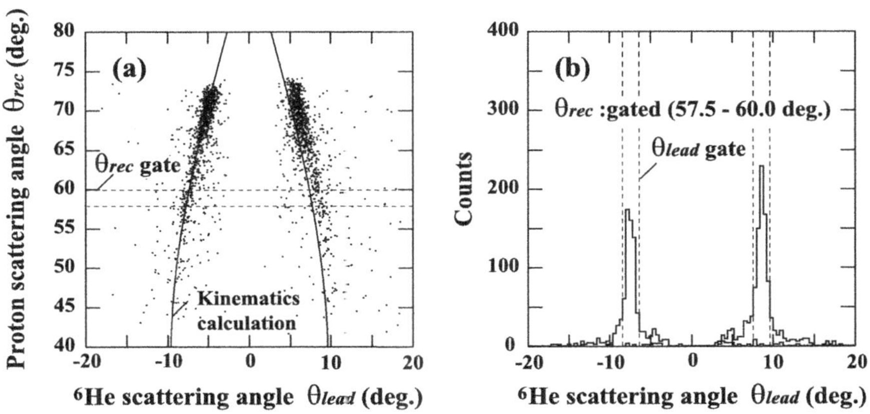

Fig. 3. (a): Polar angle correlation of protons and ^{6}He particles. (b): Histogram of θ_{lead} where θ_{rec} is gated from 57.5° to 60.0°.

$(i = L/R)$ and the superscript $(j = \uparrow / \downarrow)$ denote scattering direction of ^{6}He particles and polarizing direction of the target, respectively.

Global feature of the analyzing power was consistent with previous data measured by Hatano *et al.*[4] More detailed angular dependence was revealed due to higher statistics. Furthermore, absolute values of the analyzing power were determined for the first time using the target polarization measured via $\vec{p}$–^{4}He elastic scattering. Obtained data were compared with standard optical model calculations[5,6] based on a g-matrix folding model. While the differential cross section data were well reproduced, large discrepancy was found between the analyzing power data and optical model calculations.

References

1. T. Wakui, M. Hatano, H. Sakai, T. Uesaka and A. Tamii, *Nucl. Instrum. and Methods. A* **550**, (2005) 521.
2. T. Uesaka, M. Hatano, T. Wakui, H. Sakai and A. Tamii, *Nucl. Instrum. and Methods. A* **526**, (2004) 186.
3. H. Togawa *et al.*, *RCNP Annual Report*, (1987) 1.
4. M. Hatano, H. Sakai, T. Wakui, T. Uesaka *et al.*, *Eur. Phys. J.* A **25**, (2005) 255.
5. S.P. Weppner, O. Garcia and Ch. Elster, *Phys. Rev. C* **61**, (2000) 044601.
6. D. Gupta, C. Samanta and R. Kanungo, *Nucl. Phys. A* **674**, (2000) 77.
7. T. Wakui *et al.*, in *11th Internation Workshop on Polarized Sources and Targets*, November 2005, Tokyo, Japan

KNOCKOUT REACTION SPECTROSCOPY OF EXOTIC NUCLEI *

J.A. TOSTEVIN

Department of Physics, Faculty of Engineering and Physical Sciences, University of Surrey, Guildford, Surrey GU2 7XH, UK.
E-mail: j.tostevin@surrey.ac.uk

Direct reaction studies to probe the evolution of nuclear single-particle properties, and their implications for the structures of the very rarest short-lived nuclei, are a major topical interest. They pose some of the most precise and challenging tests of extensions of nuclear structure model predictions into mass and charge regions of large neutron excess or of neutron deficiency. Untangling the structural changes relies to a very large extent upon combining timely developments of novel spectroscopic methods for theoretical interpretation with a systematic programme of challenging measurements of direct nuclear reactions. Among such theoretical and experimental developments there has been major progress in the use of intermediate energy direct one- and in particular two-nucleon knockout reactions. Following the pioneering experiment of[1] the required theoretical developments can be found in.[2] This talk reviews the recent theoretical progress, the results of experiments to validate these.[3] The most recent measurements of both partial cross sections and of residue momentum distributions[4] aer interpreted together with the insights that are being gained on both nuclear structures[4] and confirmation of the associated reaction dynamics.[5,6]

References

1. D. Bazin *et al.*, *Phys. Rev. Lett.* **91**, 012501 (2003).
2. J.A. Tostevin *et al.*, *Phys. Rev. C* **70**, 064602 (2004).
   J.A. Tostevin, B.A. Brown, *Phys. Rev. C* **74**, 064604 (2006).
3. A. Gade *et al.*, *Phys. Rev. C* **74**, 021302(R) (2006).
   K. Yoneda *et al.*, *Phys. Rev. C* **74**, 021303(R) (2006).
4. A. Gade *et al.*, *Spectroscopy of ^{36}Mg: The interplay of normal and intruder configurations at the neutron-rich boundary of the Island of Inversion*, in the press.
5. J. A. Tostevin, *Reaction spectroscopy at fragmentation beam energies - recent advances in studies of two-nucleon removal, Eur. Phys. J. A* in the press.
6. J.A. Tostevin, *Acta Physica Polonica B* **38**, 1195 (2007).

Manuscript not received.

*This work is supported by the United Kingdom Engineering and Physical Sciences Research Council (EPSRC) under Grant No. EP/D003628.

EXOTIC NUCLEI, QUANTUM PHASE TRANSITIONS, AND THE EVOLUTION OF STRUCTURE

R.F. CASTEN

A.W. Wright Nuclear Structure Laboratory, Yale University, New Haven, Connecticut, 06520-8124, USA

E-mail: rick@riviera.physics.yale.edu

This paper focuses on the opportunities for nuclear structure physics provided by new generation facilities for exotic beams of nuclei far from the valley of stability. These nuclei offer the possibility to study the physics of weakly bound, strongly interacting quantal systems, as well as the evolution of structure across wide swaths of the nuclear chart. This talk will focus on the latter area, with particular emphasis on the link between microscopic and macroscopic perspectives on nuclear structure. In particular, the evolution of collective behavior, and especially the occurrence of quantal phase transitions in the equilibrium shapes of nuclei, will be discussed both from the perspective of nuclei as geometric objects and from that of the microscopic interactions that drive configuration mixing, changes in shell structure, and the development of collectivity and deformation in nuclei. Important for the former discussion will be the new concept of critical point symmetries and the plethora of related new models it has spawned. The latter discussion will focus on the effects of the valence proton-neutron interaction, including ways of extracting those interaction strengths empirically from nuclear masses as well as new theoretical calculations of them.

1. Introduction

Exotic nuclei provide the venue for a rebirth in interest in the science of nuclei that has occurred in the last decade and which will accelerate in the next, as new generations of facilities for the study of nuclei far from stability come on line. This development has been well documented and acknowledged, very recently for example, by a National Academy study[1] in the U.S. which stated that "Nuclear science is entering a new era of discovery in understanding how nature works at the most basic level and in applying that knowledge in useful ways." This new era is driven by access to literally thousands of new nuclei, heretofore either inaccessible or accessible with intensities orders of magnitude weaker than will be available within the next ten years.

The physics of exotic nuclei is critical to a comprehensive understanding of nuclear systems because, already, studies with existing facilities have shown that some of the most fundamental concepts underlying our understanding of nuclei do not apply to nuclei far off stability. The best known of these concepts is that of the magic numbers that define the locus of shell closures: in light nuclei, it has become apparent that these change away from stability,[2–7] due either to accumulating effects of residual interactions, to changes in those residual interactions in weakly bound nuclei, or to more basic changes in the mean field potential itself induced by the different spatial distributions of the weakly bound nucleons that generate the mean field. Even as basic an idea as the viability of the ansatz of individual particle motion may alter near the drip lines, as pairing, which is known to be density-dependent, may rise in importance as scattering into the continuum becomes important.

The chart of the nuclides reveals four frontiers of this new science. Three are well known: proton-rich nuclei, neutron-rich nuclei, and the heaviest nuclei. The fourth frontier is less often talked about but may ultimately be the most important. It is the internal one of the evolution of structure within the expanded boundaries of known nuclei. For example, studies of the Ni isotopes will eventually span four major shells and allow an unprecedented opportunity to see how structure evolves as a function of (in this case) neutron number and neutron binding.

In this paper we will focus on this last frontier and will approach it from the twin perspectives of macroscopic collectivity and its microscopic underpinnings. These perspectives reflect the two great challenges of much of modern science—complexity and simplicity. That is, expressed in the context of atomic nuclei: how can we understand these complex objects in terms of their simple ingredients (nucleons), their interactions, quantum mechanics, and conservation laws; and how is it that these complex objects can display such remarkably simple patterns. The latter can be expressed as follows. Nuclei consist of up to hundreds of nucleons, interacting primarily with the strong and Coulomb forces, and occupying roughly half the nuclear volume. One would expect a highly chaotic situation and yet, somehow, these nucleons orbit the nucleus on the order of 10^{21} times per second in nearly undisturbed paths with large mean free paths and the nucleus is able to generate extremely simple patterns of excitation energies and systematics with N and Z. Of course, we know the basics of the answer, lying primarily in the effects of the Pauli Principle, but new aspects of structure that appear far from stability will give us significant new insights into the origins of simple patterns in nuclei.

We will first briefly discuss some of the newest facets of simplicity in nuclear structural evolution and the emergence of collectivity and then turn to one key aspect of the microscopic underpinnings of this collectivity in the valence proton-neutron $(p-n)$ interaction.

2. Quantum phase transitional behavior

It has been known for half a century that some nuclei are spherical and others ellipsoidally deformed and that the transition between such shapes as a function of nucleon number can be very rapid. However, recent studies have shown that these shapes changes can have many of the properties of quantum phase transitions (QPT's). These are phase transitions that occur, not as a function of temperature, for example, but in the ground state equilibrium structure as a function of nucleon number. Of course, this concept must be approximate since nuclei are finite systems and since the control parameter, nucleon number, is not a continuous variable but, in fact, is limited to discrete integer values. However, this issue has been dealt with[8,9] and, while worth additional study, will not concern us further here.

Nuclei in shape transitional regions have long been among the most difficult to understand due to the competition of different degrees of freedom. However, a few years ago, Iachello developed[10,11] the concept of Critical Point Symmetries (CPS's) which attempt to describe nuclei at the phase transitional point in terms of an utterly simple ansatz, namely that the geometric potential as a function of beta can be described by a square well. The idea is illustrated for the case of a first order phase transition in Fig. 1. Here there is a competition between spherical and deformed shapes. For nuclei near closed shells the former dominates and $V(\beta)$ has a minimum at β=0. When there are many valence nucleons, of both types, nuclei take on deformed equilibrium shapes with a minimum in $V(\beta)$ at some finite value of β. The evolution of structure in a first order phase transition can be envisioned in terms of an equilibrium spherical shape, with a deformed minimum at higher energy and the decrease in energy of that deformed minimum with increasing valence nucleon number. The phase transition occurs when these two minima are degenerate: this situation is illustrated by the potential energy curve in Fig. 1.

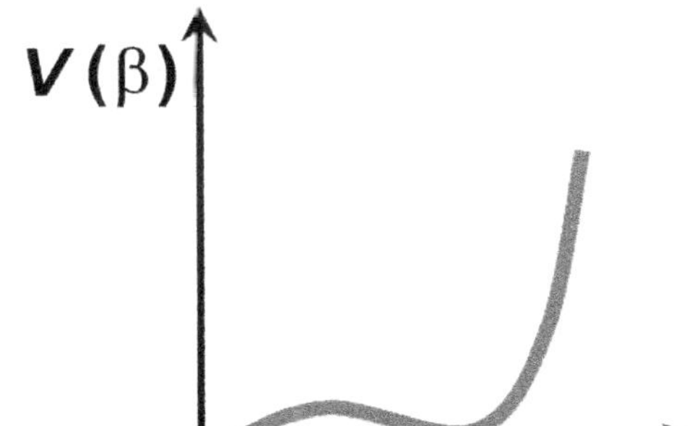

Fig. 1. Nuclear potential energy as a function of β at the critical point of a first order phase transition.

Iachello took the bold step of approximating this curve (itself a gross approximation of the real many-body situation) by a simple square well—that is, ignoring the small barrier in the middle of the potential and approximating the sharp rise in the potential at large β by a vertical wall. For a second order phase transition, the relevant description is labelled E(5) while for the first order case the CPS is denoted X(5). (Note that these descriptions, especially X(5), are not necessarily "symmetries" in the strict group theoretical sense of the word and also that they are approximate solutions to the Bohr Hamiltonian and are not directly connected with the IBA model.)

Immediately upon the development of these CPS's, empirical manifestations, in ^{134}Ba and ^{152}Sm were discovered.[12,13] We show the comparison of ^{152}Sm with X(5) in Fig. 2. Subsequently, another example was discovered in ^{150}Nd[14]. In Refs. [15,16], systematic searches for further cases were carried out. One of these, ^{130}Ce, is currently being studied with β decay.[17] Such studies are valuable as they begin to explore how CPS's and QPT's may be dependent on shell, on proton-neutron asymmetry, and on weak binding.

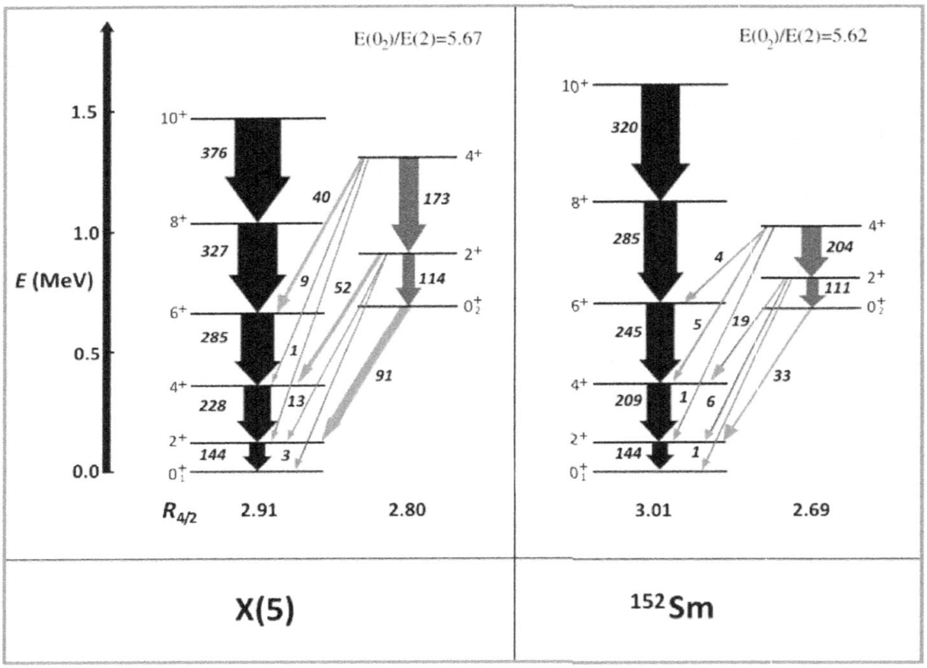

Fig. 2. Comparison of X(5) with the data on ^{152}Sm. Based on Ref. [13].

The agreement in Fig. 2, overall, is excellent. We stress that X(5) is completely parameter-free except for scale and that both intrinsic and quasi-rotational excitations are predicted on the same basis. Thus, for example, unique predictions such as the ratio of the energy of the first excited 0^+ state to the first 2^+ state and the yrast 6^+ state, predicted in X(5) to be 5.67 and 1.04, respectively, are inherent to X(5). Two obvious discrepancies, the expanded scale of energies of the 0_2^+-based sequence compared to experiment and the stronger inter-sequence $B(E2)$ values, have been much discussed. The former is actually common to many collective models and, in the case of X(5), probably relates[18] to the extreme assumption of a rigid wall potential at large beta. Calculations with a sloped wall correct this discrepancy. The experimental $R_{4/2}$ value in ^{152}Sm is 3.01 which is slightly to the rotor side of X(5) which predicts 2.91. As the rotor scheme is approached, interband transitions become much weaker and we are probably seeing the beginnings of this trend in ^{152}Sm. Note, however, the important point that X(5) excellently reproduces both the relative energy spacings within the 0_2^+-based sequence and the relative interband $B(E2)$ values.

Two neutron separation energies give further evidence about shape transitional regions. This is illustrated in Fig. 3. Overall, there is a general decrease with the addition of neutrons as these particles occupy successively less bound orbits. Major discontinuities occur just after magic numbers where the next neutrons occupy orbits at larger radii.

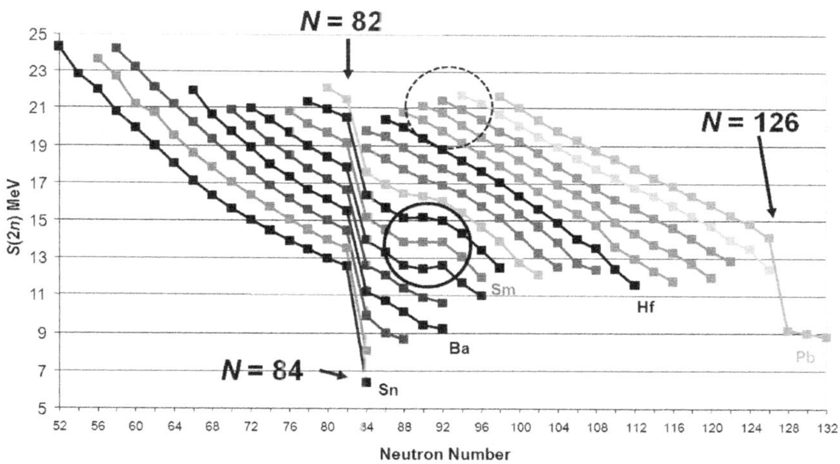

Fig. 3. Experimental two neutron separation energies in heavy nuclei. Constructed using masses from Ref. [19].

For our discussion, however, the most interesting feature is the smaller, but clearly visible, flattening (see circled nuclei) near the center of the figure near $N=90$ in the Nd, Sm, and Gd nuclei. This is precisely where the spherical-deformed QPT that we discussed above occurs. Here, nuclei suddenly energetically prefer deformed shapes and hence the separation energy halts its normal decline with neutron number. This can be an important observable, as seen in the dashed circle above that at $N \sim 90$. Here nuclei such as ^{168}W have some spectroscopic characteristics (*e.g.*, $R_{4/2}$) of X(5) but there is clearly no discontinuity in the separation energies. Since $R_{4/2}$ values near 2.9 can occur for a variety of structures, the combination of spectroscopic and mass data can provide refined structural insights.

The success of the concept of CPS's has spawned a large number of related geometrical models,[20–25] all solutions to the Bohr Hamiltonian. These include potentials of the form β^{2n}, where n=1,2,3,. and the Davidson potential. For the former group, the solutions range from a type of anharmonic vibrator for n=1 to an X(5)-like potential for large n. The Davidson potential can actually span essentially the gamut from pre-deformed, through a transition region (albeit different from X(5) since the Davidson potential does not have a flat bottom), to well-deformed structures.

3. Microscopic underpinnings to the development of emergent collectivity in the many-body nuclear system

Having discussed the emergence of collectivity, shape transitions, and deformation from a macroscopic, collective model approach, it is crucial to try to understand the origins of these phenomena from a microscopic perspective. It has long been known that structure evolves largely because of a competition between the pairing force among identical nucleons, which favors spherical shapes and the $p-n$ interaction which fosters configuration mixing which is essential to the development of coherence, collectivity and, ultimately deformation. In this section we will focus on recent studies of the valence $p-n$ interaction, and, in particular, on empirical extractions of average $p-n$ interactions of the last two neutrons with the last two protons and their comparison with theory.[26–30]

Figure 4 shows a plot of these interactions, called δV_{pn}, in the $Z-N$ plane where the strength of the interaction is indicated by the darkness of the boxes for each nucleus. Despite the fact that this region covers entire major shell regions in both protons and neutrons, where we still do not have the ability to do realistic microscopic calculations of excitation spectra because of the huge Hilbert spaces involved, the patterns in Fig. 4 are simple to understand.

The residual $p-n$ interaction is primarily short range and attractive. Therefore, proton and neutron orbits with high spatial overlap should have the largest $p-n$ interactions. This simple idea is elegantly illustrated by the data in Fig. 4. Shell structure (for normal parity orbits) in heavy nuclei has a generic pattern more or less independent of shell, namely ranging from high j-low n early in a shell to low-j, high n at shell end. Thus, if the protons and neutrons are filling their respective shells to the same fractional degree, they should be in highly overlapping orbits. This occurs along the diagonal in Fig. 4 where, indeed, the largest δV_{pn} values occur.[28] Moreover, larger δV_{pn} values would be expected in the lower left and upper right quadrants, where protons and neutrons are either both particles or both holes than in the upper left quadrant where protons are filling the second half of their shell while neutrons are filling the first half of theirs. This, in fact, is what is observed in Fig. 4. Clearly the figure also shows the real need for new mass measurements that will give δV_{pn} values for the lower right quadrant where small values would be expected.

These $p-n$ interactions are the key to inducing collectivity and deformation in nuclei. The interactions lead to configuration mixing, which, in turn, gives non-

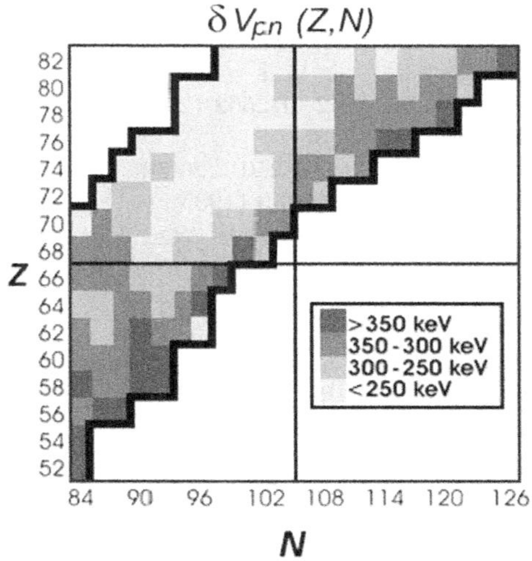

Fig. 4. δV_{pn} (Z, N) values extracted from measured masses in the major shell regions for Z=50-82 and N=82-126. Based on Ref. [28].

uniform occupation probabilities for different magnetic substrates, and hence favors non-spherical shapes. The monopole component of the same interactions[31] also leads to shifts in single-particle energies which can alter shell structure. This effect was suggested very early on[32] as the explanation for the onset of deformation at N=90 in the rare earth region and has received considerable confirmation recently[2–7] in studies of changing shell structure in light nuclei.

Clearly the future for studies such as these is very bright as next generation exotic beam facilities will give access to new regions of nuclei far from stability where structure and structural evolution is likely to be considerably different as the proton-neutron number asymmetry grows, as new proton and neutron orbit combinations are occupied, and as effects of weak binding and the continuum grow in importance. It will be fascinating to see how this plays out in the next decade.

Acknowledgments

I am grateful to my principal collaborators and colleagues with whom I have discussed these topics, especially F. Iachello, Victor Zamfir, E.A. McCutchan, R.B. Cakirli, and W. Nazarewicz. Work supported in part by the US DOE under grant number DE-FG02-91ER-40609.

References

1. Rare Isotope Science Assessment Committee, Board on Physics and Astronomy, Division on Engineering and Physical Sciences, National Research Council of the National Academies, *Scientific Opportunities with a Rare-Isotope Facility in the United States*, (The National Academies Press, Washington, DC, 2007).

2. R. Ibbotson *et al.*, *Phys. Rev. Lett.* **80**, 2081 (1998).
3. A. Ozawa, T. Kobayashi, T. Suzuki, K. Yoshida and I. Tanihata, *Phys. Rev. Lett.* **84**, 5493 (2000).
4. S.N. Liddick *et al.*, *Phys. Rev. Lett.* **92**, 072502 (2004).
5. R.V.F. Janssens, *Nature* **435**, 897 (2005).
6. V. Tripathi *et al.*, *Phys. Rev. Lett.* **94**, 162501 (2005).
7. T. Motobayashi *et al.*, *Phys. Lett. B* **346**, 9 (1995).
8. F. Iachello, N.V. Zamfir and R.F. Casten, *Phys. Rev. Lett.* **81**, 1191 (1998).
9. R.F. Casten, Dimitri Kusnezov and N.V. Zamfir, *Phys. Rev. Lett.* **82**, 5000 (1999).
10. F. Iachello, *Phys. Rev. Lett.* **85**, 3580 (2000).
11. F. Iachello, *Phys. Rev. Lett.* **87**, 052502 (2001).
12. R.F. Casten and N.V. Zamfir, *Phys. Rev. Lett.* **85**, 3584 (2000).
13. R.F. Casten and N.V. Zamfir, *Phys. Rev. Lett.* **87**, 052503 (2001).
14. R. Krücken *et al.*, *Phys. Rev. Lett.* **88**, 232501 (2002).
15. D.S. Brenner, *Mapping the Triangle*, eds. A. Aprahamian, J.A. Cizewski, S. Pittel and N.V. Zamfir, (AIP Conf. Proc. **638**, New York, 2002), p. 223.
16. R.M. Clark *et al.*, *Phys. Rev.* **C 68**, 037301 (2003).
17. A. Mertz *et al.*, private communication and to be published.
18. M.A. Caprio, *Phys. Rev.* **C 69**, 044307 (2004).
19. G. Audi, A.H. Wapstra and C. Thibault, *Nucl. Phys. A* **729**, 337 (2003).
20. D. Bonatsos, D. Lenis, N. Minkov, P.P. Raychev and P.A. Terziev, *Phys. Rev.* **C 69**, 014302 (2004).
21. D. Bonatsos, D. Lenis, N. Minkov, D. Petrellis, P.P. Raychev and P.A. Terziev, *Phys. Rev.* **C 70**, 024305 (2004).
22. D. Bonatsos, D. Lenis, D. Petrellis, P.A. Terziev and I. Yigitoglu, *Phys. Lett. B* **632**, 238 (2006).
23. D. Bonatsos, D. Lenis, N. Pietralla and P.A. Terziev, *Phys. Rev.* **C 74**, 044306 (2006).
24. D. Bonatsos, D. Lenis, E.A. McCutchan, D. Petrellis and I. Yigitoglu, *Phys. Lett. B* **649**, 394 (2007).
25. L. Fortunato, *Eur. Phys. J. A* **26**, 1 (2005).
26. R.B. Cakirli, D.S. Brenner, R.F. Casten and E.A. Millman, *Phys. Rev. Lett.* **94**, 092501 (2005).
27. D.S. Brenner, R.B. Cakirli and R.F. Casten, *Phys. Rev.* **C 73**, 034315 (2006).
28. R.B. Cakirli and R.F. Casten, *Phys. Rev. Lett.* **96**, 132501 (2006).
29. M. Stoitsov, R.B. Cakirli, R.F. Casten, W. Nazarewicz and W. Satula, *Phys. Rev. Lett.* **98**, 132502 (2007).
30. Y. Oktem, R.B. Cakirli, R.F. Casten, R.J. Casperson and D.S. Brenner, *Phys. Rev.* **C 74**, 027304 (2006).
31. K. Heyde, P. Van Isacker, R.F. Casten and J.L. Wood, *Phys. Lett.* **155B**, 303 (1985).
32. R.F. Casten, D.D. Warner, D.S. Brenner and R.L. Gill, *Phys. Rev. Lett.* **47**, 1433 (1981).

STRUCTURE OF EXOTIC NUCLEI IN THE MEDIUM MASS REGION

TAKAHARU OTSUKA

Department of Physics and Center for Nuclear Study,
University of Tokyo, Hongo, Bunkyo-ku, Tokyo 113-0033, Japan,
E-mail: otsuka@phys.s.u-tokoyo.ac.jp

Nishina Center, RIKEN,
Hirosawa, Wako-shi, Saitama, 351-0198, Japan

National Superconducting Cyclotron Laboratory, Michigan State University,
East Lansing, Michigan, 48824, USA

In this talk, I will overview recent studies on the evolution of the shell structure in stable and exotic nuclei, and will show results of related shell-model calculations. This shell evolution is primarily due to the tensor force. The robust mechanism and some examples will be presented. Such examples include the disappearance of existing magic numbers and the appearance of new ones. The nuclear magic numbers have been believed, since Mayer and Jensen, to be constants as 2, 8, 20, 28, 50, ... This turned out to be changed, once we entered the regime of exotic nuclei. This shell evolution develops at many places on the nuclear chart in various forms. For instance, the Island of Inversion picture has been changed considerably.

Keywords: exotic nuclei, shell evolution, tensor force, pion, island of inversion

1. Trends of Physics of Exotic Nuclei

Exotic (or unstable) nuclei have faced nuclear physics with new phenomena, features, facets and ideas. Although exotic nuclei have been objects of nuclear physics since its very beginning, systematic studies have started after the discovery of extraordinarily large radius of ^{11}Li by Tanihata *et al.* in 1985.[1] This experiment was made possible by the radioactive ion (rare isotope) beam (RI-beam) technology,[1] and in fact many experiments have been performed since then, changing the landscape of nuclear structure physics.

The nucleus ^{11}Li is known for its the neutron halo, which is a direct consequence of the loose binding of last neutrons. This nucleus can be reached by adding four neutrons to the stable nucleus ^{7}Li. In fact, in the 90's, light exotic nuclei around ^{11}Li on the nuclear chart, for which the neutron drip line is not very far, have been studied intensively, and the motion of weakly bound neutrons

In the 21st century, besides light ones, heavier exotic nuclei have become objects of systematic extensive studies, partly due to further developments of RI-beam experiments. Let us imagine the nuclear chart, particularly its left-lower corner. The stability line and the drip line drift apart from each other as we move to higher atomic number, Z. In fact, contrary to Li isotopes, the neutron number, N, increases by more than ten units from the stability line to the neutron drip line for Mg isotopes. One can then raise a question what happens between the stability and drip lines on the nuclear chart. The area between the two lines becomes wider and wider as Z becomes higher, while the nuclei in this area are still well bound except on and very near the drip line.

In this talk, I shall state how the structure of many nuclei between the stability and drip lines looks like, what is the underlying mechanism if some changes occur, and what one can expect in exotic nuclei to be explored.

2. Island of Inversion - Revised View -

Back to the year 1990, the so-called Island of Inversion was proposed[2] for the nine nuclei such as ^{30,31,32}Ne, ^{31,32,33}Na, and ^{32,33,34}Mg, incorporating earlier studies, *e.g.*, Refs.[3,4] These nine nuclei were predicted to have abnormal ground states with 2p2h excitation from the *sd* shell across the N=20 magic gap, which is supposed not to change much. This Island of Inversion is an isolated area in the nuclear chart. However, the constant N=20 gap may not correspond to reality. Figure 1 shows the conventional constant gap and the varying gap. The latter is calculated from the sdpf-M interaction[5,6] which has been very successful for the shell model description of exotic Ne, Na, Mg, Al, Si isotopes, as discussed in *e.g.* Refs.[7–11] A small gap for Z smaller in Fig. 1 produces deformation for nuclei outside the Island of Inversion of,[2] by bringing intruder dominant configurations into ground and low-lying states. Thus, the Island of Inversion is not like the one proposed by Warburton *et al*,[2] and the "western" boundary of the Island should be moved and may not be sharp. Other boundaries may change from the original ones too.

We now move to the mechanism varying the gap this way.

3. Evolution of the shell

The structure of bound nuclei depends much on their shell structure. Not only single particle properties but also deformation can be strongly influenced by the change of shell structure. The shell structure has been believed to be basically common not only among stable nuclei but also between stable and exotic (unstable) nuclei. The shell structure of stable nuclei has been proposed by Mayer and Jensen in 1949 with magic numbers, 2, 8, 20, 28, 50 ,...[12] These magic numbers can be conceived from three basic features; (i) short-range attraction of nucleon-nucleon (NN) interaction, (ii) density saturation, (iii) spin-orbit splitting. Note that for (i), an isotropic attraction is in mind like the central potential. From (i) and (ii), the Harmonic Oscillator potential can be derived,[13,14] and the combination with (iii) produces the magic numbers of Mayer and Jensen. Thus, the magic numbers of

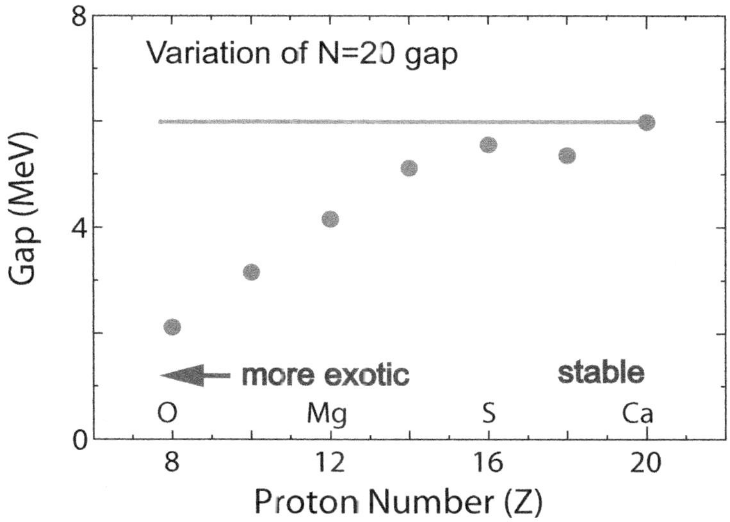

Fig. 1. Variation of N=20 gap as a function of Z. The dots are calculated from the shell model interaction sdpf-M containing appropriate amount of the tensor force. The horizontal bar is a conventional expectation of a constant gap.

Mayer and Jensen are quite robust, as they originate in very basic properties.

The key to this answer has been found recently to be the tensor force.[15] The tensor force is not among the three features mentioned above. While the tensor force itself has been known and studied over half a century, not everything has been clarified. Figure 2 illustrates intuitively how the tensor force works for two interacting nucleons in two orbitals. Here, we use notations

$$j_> = l + 1/2 \quad \text{and} \quad j_< = l - 1/2 \,, \tag{1}$$

where l is the orbital angular momentum. Namely, in the former case, the orbital angular momentum and spin are parallel, whereas they are opposite in the latter. The tensor force works only if spins of two nucleons are parallel coupled to the total S=1. So spins can be fixed as being "up", but the orbital motion can be in either way (See Fig. 2). If two nucleons are in orbits $j_<$ and $j'_>$, they are moving in opposite directions as in Fig. 2(a). The relative momentum is high at "collision", and the spatial wave function of the relative motion is suppressed along the direction of collision. Thus, the wave function is stretched along the spin S, and the tensor force works attractively. If two nucleons are in orbits $j_>$ and $j'_>$, they are moving together as in Fig. 2(b), and the tensor force works repulsively. Thus, the tensor force changes single-particle energy depending on the numbers of nucleons in other orbitals, following the rule proposed in Fig 2 (See also Fig. 2 of[15]).

Many phenomena in the structure of exotic nuclei have been understood in terms of the tensor force, and the mechanism shown in Fig. 2 always gives us the correct picture immediately.

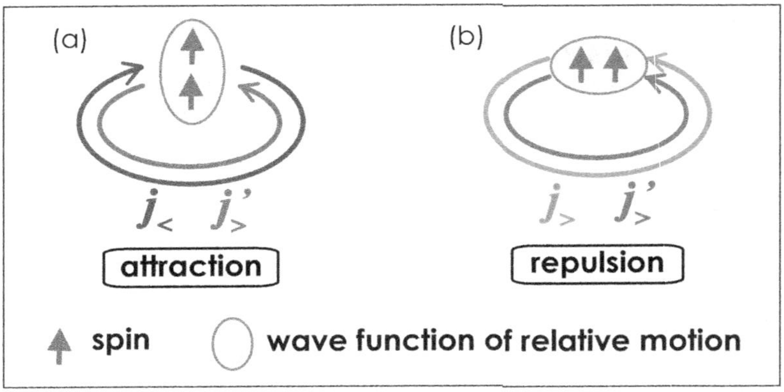

Fig. 2. Intuitive illustration of the tensor force acting two nucleons on orbits j and j'. Original colored figure is Fig. 2 of Ref.[15]

For instance, the gap change discussed in the previous section is due to the proton occupancy into the $1d_{5/2}$ orbit and the strong tensor attraction between $1d_{5/2}$ and $1d_{3/2}$. For oxygen (Z=8), N=16 gap is larger and dominant over the N=20 gap, but the tensor-force effect mentioned just above shifts $1d_{3/2}$ downwards, while shifts $1f_{7/2}$ upwards, ending up with the "normal" N=20 gap for "normal" and stable nuclei like ^{40}Ca. Thus, the magic number changes between N=16 and 20 in going from oxygen (Z=8) to silicon (Z=14), as discussed first in Ref.[16] in terms of the spin-isospin flip coupling which turned out to be a result of the tensor force later.[15,17]

Figure 3 shows the energy splitting between $1h_{11/2}$ and $1g_{7/2}$ orbits of protons in Sb (Z=51) isotopes as a function of the neutron number. The splitting is plotted for the variation from the splitting at N=64. Schiffer *et al.* reported experimental values[18] as shown in Fig. 3. It was not possible to explain the gradual increase of the splitting within existing mean field models. However, once we calculate this splitting by including the tensor force, the theoretical result showed a remarkable agreement first by a shell-model[15] with the $\pi + \rho$ meson exchange potential,[19] and later by a mean-field model.[17] This is the precious case where the tensor force effect has been examined with experimental data over a long chain of isotopes.

In Fig. 3, two mean-field models are used. One is Gogny model[20] with D1S interaction.[21] This model has been very successful, but does not contain the tensor force, similarly to usual calculations by Skyrme model.[22,23] The other model, GT2, has recently been introduced in[17] based on the Gogny model with explicit inclusion of the tensor force, giving rise to a good agreement to experiment as shown in Fig. 3.

Figure 3 contains another important aspect. The theoretical prediction goes up to N=104, where the splitting decreases. The D1S result meets the GT2 result at N=104, because the tensor force effects are canceled for the spin saturation. The small increase of the splitting is due to the 2-body spin-orbit force.[17] We emphasize that the effects of the tensor force become evident once the systematic behavior over long isotope chain is obtained.

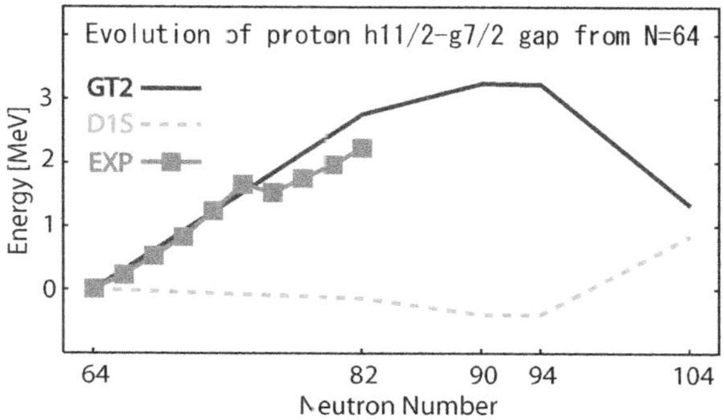

Fig. 3. Evolution of the proton $1h_{11/2}$-$1g_{7/2}$ splitting measured from the value at N=64. Original colored figure is Fig. 4 of.[17]

At this point, we mention other theoretical works incorporating the tensor force into mean field calculations. Stancu, Brink and Flocard[24] introduced tensor-like zero-range interaction in 1977. Although this work has been quoted at a modest rate, after the publication of[17] many papers based on this interaction have been written and submitted.[25–29] Relativistic approaches are also being made.[30,31] For the understanding of the shell evolution, it seems that the inclusion of the tensor force in mean field calculations become a standard process due to the impact of recent studies,[15,17] while the shell evolution has been studied also in the shell model calculation (see for example Refs.[32,33]).

4. A new shell-model approach

From recent analysis of shell-model interaction, the tensor-force component in the shell-model effective interaction is very close to the π + ρ meson exchange potential.[19] This seems to be somewhat related to the Chiral Perturbation idea of Weinberg.[34] Along this line, we have obtained a new shell model interaction, and its results for exotic Si isotopes are in a good agreement with experiments.[35–37] The potential energy surface, where the tensor force is switched on or off in the inter shell channel in one calculation, appears to be very instructive. The tensor force produces the pronounced oblate minimum, whereas the other calculation gives a spherical minimum. Thus, one sees the crucial role of the tensor force also for the deformation. The nucleus ^{42}Si has been the object of a debate on its magic nature.[37,38] The present result seem to explain both small cross section and large deformation, but we should wait until the calculation of the cross section. There are many other models and calculations leading to different conclusions.[37,38]

5. Summary

In summary, we have made manifest that the tensor force changes magic numbers and shell structure of exotic nuclei from those of stable nuclei. Its effect can be seen at many places on the nuclear chart. The tensor force can be crucial for deformation, and can affect the weak process transition strength. The variation of shell structure can be seen experimentally by considerably increasing the neutron number. Thus, we need RI-beam experiments to establish the properties of the tensor force. As the tensor force is characteristic to meson exchange processes, we can now see that Yukawa's meson exchange forces create intriguing and exciting structures in exotic nuclei, and affect the nucleosynthesis through r-process etc. The structure of exotic nuclei may be more directly linked to the nucleon-nucleon interaction, because the stability and restoring force of the mean field become weaker. We mention that the tensor force has, in some cases, effects beyond single-particle energies even for low-energy spectrum[39] and neutrino reactions.[40]

Acknowledgments

The author thanks Drs. Y. Utsuno, N. Shimizu, R. Fujimoto, Professors M. Honma, T. Mizusaki, T. Suzuki, H. Grawe, and Y. Akaishi for their fruitful collaborations in theoretical works. Many experimentalists are acknowledged for their valuable collaborations. This work has been supported in part by the RIKEN-CNS large-scale nuclear structure calculation project, and by the JSPS core-to-core project, Int. Res. Network on Exotic Femto Systems (EFES).

References

1. I. Tanihata *et al.*, *Phys. Rev. Lett.* **55** 2676 (1985).
2. E.K. Warburton, J.A. Becker and B.A. Brown, *Phys. Rev. C* **41**, 1147 (1990).
3. C. Thibault *et al.*, *Phys. Rev. C* **12**, 644 (1975); G. Huber *et al.*, *Phys. Rev. C* **18**, 2342 (1978).
4. D. Guillemaud-Mueller *et al.*, *Nucl. Phys. A* **426**, 37 (1984).
5. Y. Utsuno, T. Otsuka, T. Mizusaki, and M. Honma, *Phys. Rev. C* **60**, 054315 (1999).
6. Y. Utsuno *et al.*, *Phys. Rev. C* **70**, 044307 (2004).
7. M. Belleguic, *et al.*, *Phys. Rev. C* **72**, 054316 (2005).
8. Zs. Dombradi *et al.*, *Phys. Rev. Lett.* **96**, 182501 (2006).
9. V. Tripathi *et al.*, *Phys. Rev. Lett.* **94**, 162501 (2005).
10. V. Tripathi *et al.*, *Phys. Rev. C* **73**, 054303 (2006).
11. G. Neyens *et al.*, *Phys. Rev. Lett.* **94**, 022501 (2005).
12. M.G. Mayer, *Phys. Rev.* **75** 1969 (1949); O. Haxel, J.H.D. Jensen and H.E. Suess, *Phys. Rev.* **75** 1766 (1949).
13. A. Bohr and B.R. Mottelson: *Nuclear Structure*, Vol. 1, (Benjamin, New York 1969).
14. T. Otsuka, in *Euroschool Lectures in Physics*, Vol. **III**, ed. by E. Roeckl and J. Al-Khalili (Springer, 2008).
15. T. Otsuka, T. Suzuki, R. Fujimoto, H. Grawe and Y. Akaishi, *Phys. Rev. Lett.* **95**, 232502 (2005).
16. T. Otsuka, *et al.*, *Phys. Rev. Lett.* **87**, 082502 (2001).
17. T. Otsuka, T. Matsuo and D. Abe, *Phys. Rev. Lett.* **97**, 162501 (2006).

18. J.P. Schiffer *et al.*, *Phys. Rev. Lett.* **92**, 162501 (2004).
19. F. Osterfeld, *Rev. Mod. Phys.* **64**, 491(1992).
20. J. Decharge and D. Gogny, *Phys. Rev. C* **21**, 1568 (1980).
21. J.F. Berger, M. Girod and D. Gogny, *Nucl. Phys. A* **428**, 23c (1984); *Comput. Phys. Commun.* **63**, 365 (1991),
22. T.H.R. Skyrme, *Nucl. Phys.* **9**, 615 (1959).
23. D. Vautherin and D.M. Brink, *Phys. Rev. C* **5**, 626 (1972).
24. Fl. Stancu, D.M. Brink and H. Flocard, *Phys. Lett.* **68B**, 108 (1977).
25. B.A. Brown *et al.*, *Phys. Rev. C* **74**, 061303 (2006).
26. G. Colo *et al.*, *Phys. Lett. B* **646**, 227 (2007).
27. D.M. Brink and F. Stancu, *Phys. Rev. C* **75**, 064311 (2007).
28. M. Zalewski, W. Satula and J. Dobaczewski, private communication.
29. T. Lensinki *et al.*, private communication.
30. G. Lalazissis *et al.*, private communication.
31. W.H. Long *et al.*, private communication.
32. N.A. Smirnova, A.De Maesschalck, A. Van Dyck and K. Heyde, *Phys. Rev. C* **69**, 044306 (2004).
33. E. Caurier *et al.*, *Rev. Mod. Phys.* **77** 427 (2005).
34. S. Weinberg, *Phys. Lett. B* **251**, 288 (1990).
35. R.W. Ibbotson *et al.*, *Phys. Rev. Lett.* **80**, 2018 (1998).
36. C.M. Campbell *et al.*, *Phys. Rev. Lett.* **97**, 112501 (2006).
37. B. Bastin *et al.*, *Phys. Rev. Lett.* **99**, 022503 (2007).
38. J. Fridmann *et al.*, *Nature* **435**, 922 (2005).
39. T. Otsuka, M. Honma and D. Abe, *Nucl. Phys. A* **788**, 3 (2007).
40. T. Suzuki, *et al.*, *Phys. Rev. C* **74**, 034307 (2006).

PAIRING CORRELATIONS IN HALO NUCLEI

H. SAGAWA

Center for Mathematical Sciences, University of Aizu, Ikki-machi, Aizu-Wakamatsu, Fukushima 965-8580, Japan

K. HAGINO

Department of Physics, Tohoku University, Sendai, 980-8578, Japan

Paring correlations in weakly bound halo nuclei ^{6}He and ^{11}Li are studied by using a three-body model with a density-dependent contact interaction. We investigate the spatial structure of two-neutron wave function in a Borromean nucleus ^{11}Li. The behavior of the neutron pair at different densities is simulated by calculating the two-neutron wave function at several distances between the core nucleus ^{9}Li and the center of mass of the two neutrons. With this representation, a strong concentration of the neutron pair on the nuclear surface is quantitatively established for neutron-rich nuclei. Dipole excitations in ^{6}He and ^{11}Li are also studied within the same three-body model and compared with experimental data. The small open angles between the two neutrons from the core are extracted empirically by the B(E1) sum rule together with the rms mass radii, indicating the strong di-neutron correlation in the halo nuclei.

Keywords: three-body model; di-neutron correlation ;coulomb break-up.

1. Introduction

Pairing correlations play a crucial role in many Fermion systems, such as liquid ^{3}He, atomic nuclei, and ultracold atomic gases. When an attractive interaction between two Fermions is weak, the pairing correlations can be understood in terms of the well-known BCS mechanism, that shows a strong correlation in the momentum space. If the interaction is sufficiently strong, on the other hand, one expects that two Fermions form a Bosonic bound state and condense in the ground state of many-body system.[1] The transition from the BCS-type pairing correlation to the Bose-Einstein condensation (BEC) takes place continuously as a function of the strength of attractive interaction. This feature is referred to as the BCS-BEC crossover.

It has been feasible by now to study the structure of nuclei on the edge of neutron drip line. Such nuclei are characterized by a dilute neutron density around the nuclear surface so that one can investigate the pairing correlations at several densities,[2] ranging from the normal density in the center of nucleus to a diluted density at the surface. The pairing correlations are predicted to be strong at the surface, but rather weak at the normal density and also far outside of the core. Thus, the weakly bound nuclei will provide an ideal environment to study the dynamics of pairing correlations in relation with the BCS-BEC crossover phenomenon.

In this talk, we discuss the manifestation of the BCS-BEC crossover phenomenon in *finite* neutron-rich nuclei. We particularly study the ground state wave function of a two-neutron halo nuclei, ^{6}He and ^{11}Li. These nuclei are known to be well described as a three-body system consisting of two valence neutrons and the core nucleus(^{4}He or ^{9}Li).[3–5] A strong di-neutron correlation as a consequence of pairing interaction between the valence neutrons has been shown theoretically in ^{11}Li.[5,6] We take this nucleus to study BCS-BEC crossover features in connection to the strong two-neutron correlation, which has recently been observed experimentally in low-lying dipole strength in ^{11}Li.[7]

2. Three-body model and di-neutron correlation

In order to study the pair wave function in ^{11}Li, we solve the following three-body Hamiltonian,[5,6]

$$H = \hat{h}_{nC}(1) + \hat{h}_{nC}(2) + V_{nn} + \frac{\vec{p}_1 \cdot \vec{p}_2}{A_c m}, \tag{1}$$

where m and A_c are the nucleon mass and the mass number of the inert core nucleus, respectively. $\hat{h}_{nC}$ is the single-particle Hamiltonian for a valence neutron interacting with the core. We use a Woods-Saxon potential for the interaction in $\hat{h}_{nC}$. The diagonal component of the recoil kinetic energy of the core nucleus is included in $\hat{h}_{nC}$, whereas the off-diagonal part is taken into account in the last term in the Hamiltonian (1). The interaction between the valence neutrons V_{nn} is taken as a delta interaction whose strength depends on the density of the core nucleus. Assuming that the core density is described by a Fermi function, it reads

$$V_{nn}(\vec{r}_1, \vec{r}_2) = \delta(\vec{r}_1 - \vec{r}_2)\left(v_0 + \frac{v_\rho}{1 + \exp[(R - R_\rho)/a_\rho]}\right), \tag{2}$$

where $R = |(\vec{r}_1 + \vec{r}_2)/2|$. We use the same value for the parameters as in Refs.[5,6]

The two-particle wave function $\Psi(\vec{r}_1, \vec{r}_2)$ is obtained by diagonalizing the three-body Hamiltonian (1) with a large model space which is consistent with the nn interaction, V_{nn}. To this end, we expand the wave function $\Psi(\vec{r}_1, \vec{r}_2)$ with the eigenfunction $\phi_{nljj_z}(\vec{r}_i)$ of the single-particle Hamiltonian $\hat{h}_{nC}$. In the expansion, we explicitly exclude those states which are occupied by the core nucleus.

The ground state wave function is obtained as the state with the total angular momentum $J = J_z = 0$. We transform it to the coordinate system with the relative and center of mass (cm) motions for the valence neutrons, $\vec{r} = \vec{r}_1 - \vec{r}_2$ and $\vec{R} = (\vec{r}_1 + \vec{r}_2)/2$.[8,9] The wave function is first decomposed into the total spin S=0 and S=1 components. Then, the coordinate transformation is performed for the S=0 component, which is relevant to the pairing correlation:

$$\Psi^{S=0}(\vec{r}_1, \vec{r}_2) = \sum_L f_L(r, R)\,[Y_L(\hat{\vec{r}})Y_L(\hat{\vec{R}})]^{(00)}\,|\chi_{S=0}\rangle, \tag{3}$$

where $|\chi_{S=0}\rangle$ is the spin wave function. The two-particle wave function is plotted for ^{11}Li in two different coordinates in Fig. 1. Fig. 1(a) is plotted as a function of $r \equiv r_1 = r_2$ and the angle between the valence neutrons, while the radial coordinates r and R are adopted for Fig. 1(c). The $S = 0$ component is only taken for Fig. 1(b). One observes two peaks in both the figures. The peak at smaller angle θ_{12} in Fig. 1(a) is referred to as "di-neutron configuration" , while the large angle is called "cigar-like configuration". One can see the di-neutron configuration has a long tail as a typical feature of halo wave function. Figure 1 (c) shows the square of two-particle wave function for the $L = 0$ component. One can clearly recognize the two peaked structure in the plot, corresponding to the di-neutron and the cigar-like configurations.

The $L = 0$ wave functions of ^{11}Li for different values of R are plotted in Fig. 2. Since we consider the density-dependent contact interaction, Eq. (2), this is effectively equivalent to probing the wave function at different densities. At $R = 0.5$

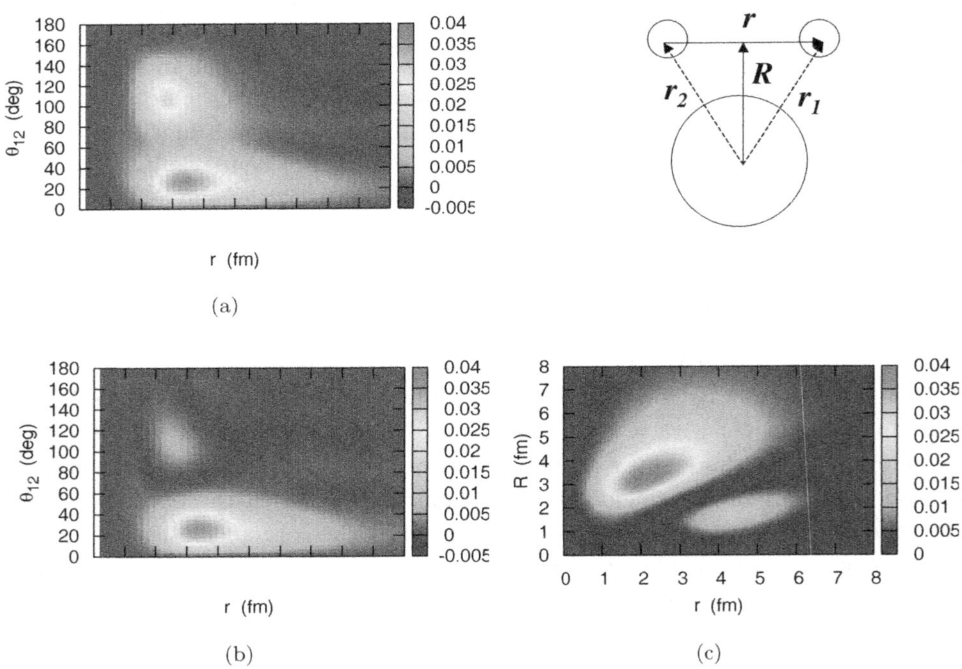

Fig. 1. A two dimensional plot of the square of the ground state two-particle wave function for ^{11}Li; (a) the total density, (b) the $S = 0$ component as a function of the radial coordinate $r \equiv r_1 = r_2$ and the angle between the valence neutrons θ_{12}. (c) $r^2R^2|f_{L=0}(r,R)|^2$ in Eq. (3), as a function of the relative distance r and the center of mass coordinate R for the valence neutrons as denoted in the upper right inset.

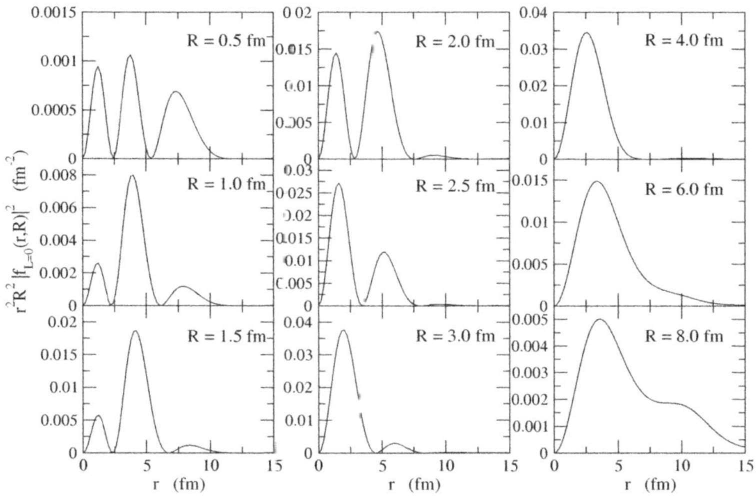

Fig. 2. The ground state two-particle wave functions, $r^2R^2|f_{L=0}(r,R)|^2$ of ^{11}Li as a function of the relative distance between the neutrons, r, at several center of mass distances R as indicated in the inset. Notice the different scales on the ordinate in the various panels.

fm, where the density is close to the normal density ρ_0, the two particle wave function is spatially extended and oscillates inside the nuclear interior. This oscillatory behavior is typical for a Cooper pair wave function in the BCS approximation, and has in fact been found in nuclear and neutron matters at normal density ρ_0.[10,11] As R increases, the density ρ decreases. The two-particle wave function then gradually deviates from the BCS-like behavior. At $R = 3$ fm, the oscillatory behavior almost disappears and the wave function is largely concentrated inside the first node at $r \sim 4.5$ fm. The wave function is compact in shape, indicating the strong di-neutron correlation, typical for BEC where many such pairs are present. At R larger than 3 fm, the squared wave function has essentially only one node, and the width of the peak gradually increases as a function of R. This behavior is qualitatively similar to the pair wave function in infinite matter.[10] We have confirmed using the same three-body model that this scenario also holds for another Borromean nucleus ^{6}He as well as for non-Borromean neutron-rich nuclei ^{16}C and ^{24}O.

The transition from the BCS-type pairing to the BEC-type di-neutron correlation can also clearly be seen in the root mean square (rms) distance of the two neutron system. We plot this quantity in Fig. 3(a) as a function of R. In order to compare it with the rms distance in nuclear matter, we relate the cm distance R with the density ρ using the Fermi-type functional form $\rho(R)/\rho_0 = [1+\exp((R-R_\rho)/a_\rho)]^{-1}$, as used in the nn interaction in Eq. (2). Fig. 3(b) shows the rms distance as a function of density ρ thus obtained. The rms distance shows a distinct minimum at $\rho \sim 0.4\rho_0$ ($R \sim 3.2$ fm). This indicates that the strong di-neutron correlation grows in ^{11}Li around this density. Notice that the probability to find the two-neutron pair

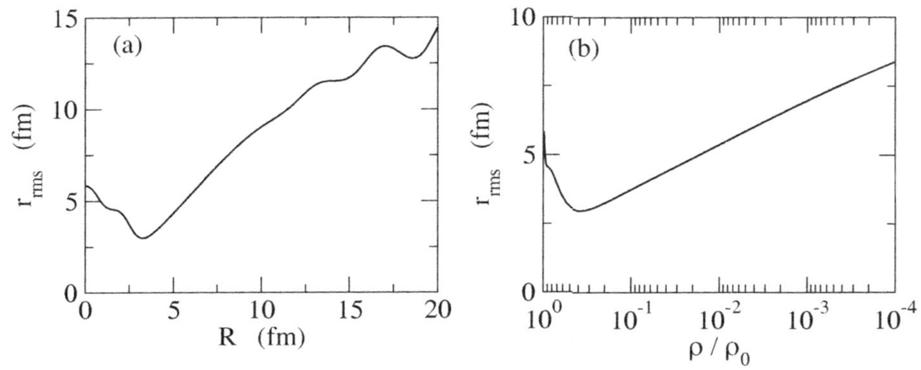

Fig. 3. The root mean square distance $r_{\rm rms}$ for the neutron pair.

is maximal around this region (see Fig. 1 (c)). The behavior of rms distance as a function of density ρ qualitatively well agrees with that in infinite matter (see Fig. 3 in Ref.[10]), although the absolute value of the rms distance is much smaller in the finite nucleus.

3. Dipole excitations and correlation angles

The rms distance $\sqrt{\langle r^2_{c-2n}\rangle}$ has an intimate relation to the B(E1) strength as,[3,4,12]

$$B(E1) = \frac{3}{\pi}\left(\frac{Ze}{A}\right)^2 \langle r^2_{c-2n}\rangle. \tag{4}$$

This relation is obtained with closure, which includes unphysical Pauli forbidden transitions to the states with negative excitation energies. Although the effect of Pauli forbidden transitions is not large, it leads to a non-negligible correction. In Ref.,[12,13] it has been proposed to estimate the experimental value for $\langle r^2_{c-2n}\rangle$ using the relation,

$$\langle r^2_{c-2n}\rangle_{\rm exp} = \frac{B(E1; E \le E_{\rm max})_{\rm exp}}{B(E1; E \le E_{\rm max})_{\rm cal}} \cdot \langle r^2_{c-2n}\rangle_{\rm cal}. \tag{5}$$

The dipole strength distributions for the ^{6}He and ^{11}Li nuclei obtained with the three model are shown in Fig. 4. Also shown by the solid curves are the B(E1) distributions smeared with the Lorentzian function with the width of $\Gamma = 0.2$ MeV. For the ^{6}He nucleus, we obtain the total B(E1) strength of 0.660 e^{2}fm^2 up to $E \le 5$ MeV and 1.053 e^{2}fm^2 up to $E \le 10$ MeV. These are in good agreement with the experimental values, B(E1; $E \le 5$ MeV)=0.59 $\pm$ 0.12 e^{2}fm^2 and B(E1; $E \le 10$ MeV)=1.2 $\pm$ 0.2 e^{2}fm^2.[15] For the ^{11}Li nucleus, we obtain the total B(E1) strength of 1.405 e^{2}fm^2 up to $E_{\rm rel} = E - S_{2n} \le 3$ MeV, which is compared to the experimental value, B(E1; $E_{\rm rel} \le 3$ MeV)=1.42 $\pm$ 0.18 e^{2}fm^2.[7] Again, the experimental data is well reproduced within the present model. From the calculated values for $\langle r^2_{c-2n}\rangle_{\rm cal}$, that is, 13.2 and

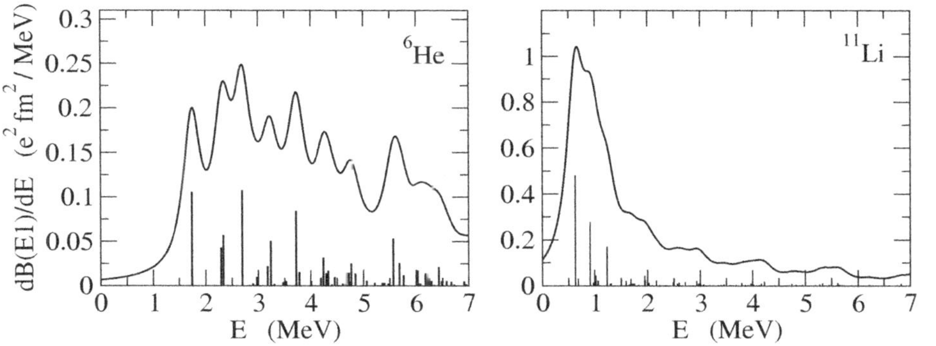

Fig. 4. The B(E1) distribution for the ^{6}He and ^{11}Li nuclei. The solid curve is obtained by a smearing procedure with a Lorentzian weighting factor of the width Γ=0.2 MeV.

26.3 fm^2 for ^{6}He and ^{11}Li, respectively, we thus obtain $\sqrt{\langle r_{c-2n}^2 \rangle_{\rm exp}} = 3.878 \pm 0.324$ fm and 5.15 ± 0.327 fm for ^{6}He and ^{11}Li, respectively. Notice that the value for the ^{6}He nucleus is somewhat larger than the one estimated in Ref.,[15] that is, 3.36 ± 0.39 fm.

We next evaluate the Coulomb breakup cross sections based on the relativistic Coulomb excitation theory.[14] These are obtained by multiplying the virtual photon number $N_{\rm E1}(E)$ to the B(E1) distribution shown in Fig. 4. The solid line in Figs. 5 (a) and (b) shows the Coulomb breakup cross sections thus obtained for ^{6}He+Pb reaction at 240 MeV/nucleon[15] and ^{11}Li+Pb reaction at 70 MeV/nucleon,[7] respectively. In order to facilitate the comparison with the experimental data, we smear the discretized cross sections with the Lorentzian function with an energy depen-

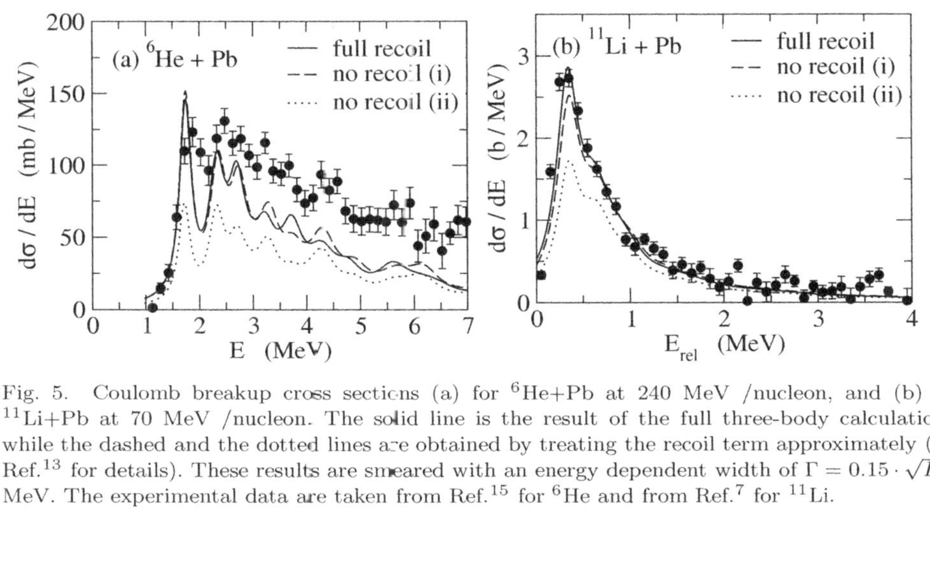

Fig. 5. Coulomb breakup cross sections (a) for ^{6}He+Pb at 240 MeV /nucleon, and (b) for ^{11}Li+Pb at 70 MeV /nucleon. The solid line is the result of the full three-body calculations, while the dashed and the dotted lines are obtained by treating the recoil term approximately (see Ref.[13] for details). These results are smeared with an energy dependent width of $\Gamma = 0.15 \cdot \sqrt{E_{\rm rel}}$ MeV. The experimental data are taken from Ref.[15] for ^{6}He and from Ref.[7] for ^{11}Li.

dent width, $\Gamma = \alpha \cdot \sqrt{E_{\rm rel}}$. We take $\alpha = 0.15\text{MeV}^{1/2}$ and 0.25 $\text{MeV}^{1/2}$ for ^{6}He and ^{11}Li, respectively. We see that the experimental breakup cross sections are reproduced remarkably well within the present three-body model, especially for the ^{11}Li nucleus.

Let us now discuss the geometry of the ^{6}He and ^{11}Li nuclei. Using the experimental value for $\langle r^2_{c-2n}\rangle$ obtained from the B(E1) distribution, one can extract the mean opening angle between the valence neutrons once an additional information is available. The mean opening angle can be extracted directly when the rms distance between the valence neutrons, $\langle r^2_{nn}\rangle$, is available. This quantity is related to the matter radius and $\langle r^2_{c-2n}\rangle$ in the three-body model,[3,6,12]

$$\langle r^2_m\rangle = \frac{A_c}{A}\langle r^2_m\rangle_{A_c} + \frac{2A_c}{A^2}\langle r^2_{c-2n}\rangle + \frac{1}{2A}\langle r^2_{nn}\rangle, \tag{6}$$

where $A_c = A - 2$ is the mass number of the core nucleus. The matter radii $\langle r^2_m\rangle$ can be estimated from interaction cross sections. Employing the Glauber theory in the optical limit, Tanihata *et al.* have obtained $\sqrt{\langle r^2_m\rangle} = 1.57 \pm 0.04$, 2.48 ± 0.03, 2.32 ± 0.02, and 3.12 ± 0.16 fm for ^{4}He, ^{6}He, ^{9}Li, and ^{11}Li, respectively.[16] Using these values, we obtain the rms neutron-neutron distance of $\sqrt{\langle r^2_{nn}\rangle} = 3.75 \pm 0.93$ and 5.50 ± 2.24 fm for ^{6}He and ^{11}Li, respectively. Combining these values with the rms core-di-neutron distance, $\sqrt{\langle r^2_{c-2n}\rangle}$, obtained with Eq. (5), we obtain the mean opening angle of $\langle\theta_{nn}\rangle = 51.56^{+11.2}_{-12.4}$ and $56.2^{+17.8}_{-21.3}$ degrees for ^{6}He and ^{11}Li, respectively. These values are close to the result of the three-body model calculation, $\langle\theta_{nn}\rangle$=66.33 and 65.29 degree for ^{6}He and ^{11}Li, respectively,[5] although the experimental values are somewhat smaller. An alternative way to extract the value $\sqrt{\langle r^2_{nn}\rangle}$ was reported by the three-body correlation study in the dissociation of two neutrons in halo nuclei.[17,18] The two neutron correlation function provides the experimental values for $\sqrt{\langle r^2_{nn}\rangle}$ to be 5.9 ± 1.2 and 6.6 ± 1.5 fm for ^{6}He, ^{11}Li, respectively.[17] When one adopts the presently obtained value for $\sqrt{\langle r^2_{c-2n}\rangle}$ with Eq. (5) instead of those in Refs.,[7,15] one obtains $\langle\theta_{nn}\rangle$=74.5 $^{+11.2}_{-13.1}$ and 65.2 $^{+11.4}_{-13.0}$ for ^{6}He and ^{11}Li, respectively. Notice that these values are in good agreement with the results of the three-body calculation[5] as is seen in Table 1.

Table 1. The geometry of the ^{6}He and ^{11}Li nuclei extracted from various experimental data. The mean opening angles calculated by the three-body model are also given in the last line for each nucleus in the table.

nucleus	$\sqrt{\langle r^2_{c-2n}\rangle}$ (fm)	$\sqrt{\langle r^2_{nn}\rangle}$ (fm)	method	$\langle\theta_{nn}\rangle$ (deg.)
^{6}He	3.88±0.32	3.75 ± 0.93	(matter radii)	51.56 $^{+11.21}_{-12.37}$
		5.9± 1.2	(neutron correlations)	74.5 $^{+11.2}_{-13.1}$
				66.33[5]
^{11}Li	5.15±0.33	5.50 ± 2.24	(matter radii)	56.2 $^{+17.8}_{-21.3}$
		6.6± 1.5	(neutron correlations)	65.2 $^{+11.4}_{-13.0}$
				65.29[5]

4. Summary

We studied the two-neutron (2n) wave function in the Borromean nuclei ^{6}He and ^{11}Li by using the three-body model with the density-dependent pairing force. We explored the spatial distributions of 2n wave function as a function of the cm distance R from the core nucleus and found that the structure of the 2n wave function alters drastically as R is varied. We also showed that the relative distance between the two neutrons scales consistently to that in the infinite matter as a function of density. These features are in close analogue to the characteristics of the BCS-BEC crossover phenomenon found in the infinite nuclear and neutron matters. We have used the same three-body model to analyze the B(E1) distribution as well as the Coulomb breakup cross section of the ^{6}He and ^{11}Li nuclei. We have shown that the strong concentration of the B(E1) strength near the continuum threshold can be well reproduced with the present model for both the nuclei. Using the calculated B(E1) strength, we extracted the experimental value for the rms distance between the core and di-neutron, which was then converted to the mean opening angle of the two valence neutrons. We have found that the mean opening angles thus obtained are in good agreement of the results of the three-body model calculation.

Acknowledgements

We thank H. Esbensen, J. Carbonell, and P. Schuck for fruitful collaborations which made this presentation possible. We thank also T. Aumann and K. Nakamura for valuable experimental information and discussions.

References

1. D.M. Eagles, *Phys. Rev.* **186**, 456 (1969); A.J. Leggett, *J. Phys.* **C41**, 7 (1980); P. Nozières and S. Schmitt-Rink, *J. Lowe Temp. Phys.* **59**, 195 (1985).
2. M. Matsuo, K. Mizuyama, and Y. Serizawa, *Phys. Rev.* **C71**, 064326 (2005).
3. G.F. Bertsch and H. Esbensen, *Ann. Phys. (N.Y.)* **209**, 327 (1991).
4. H. Esbensen and G.F. Bertsch, *Nucl. Phys.* **A542**, 310 (1992).
5. K. Hagino and H. Sagawa, *Phys. Rev.* **C72**, 044321 (2005).
6. H. Esbensen, G.F. Bertsch and K. Hencken, *Phys. Rev.* **C56**, 3054 (1999).
7. T. Nakamura *et al.*, *Phys. Rev. Lett.* **96**, 252502 (2006).
8. B.F. Bayman and A. Kallio, *Phys. Rev* **156**, 1121 (1967).
9. K. Hagino, H. Sagawa, J. Carbonell, and P. Schuck, *Phys. Rev. Lett.* **99**, 022506 (2007).
10. M. Matsuo, *Phys. Rev.* **C73**, 044309 (2006).
11. M. Baldo, U. Lombardo, and P. Schuck, *Phys. Rev.* **C52**, 975 (1995).
12. H. Esbensen, K. Hagino, P. Mueller, and H. Sagawa, *Phys. Rev.* **C76**, 024302 (2007).
13. K. Hagino and H. Sagawa, *Phys. Rev.* **C**, in press (2007).
14. C.A. Bertulani and G. Baur, *Phys. Rep.* **163**, 299 (1988).
15. T. Aumann *et al.*, *Phys. Rev.* **C59**, 1252 (1999).
16. I. Tanihata *et al.*, *Phys. Lett.* **B206**, 592 (1988); A. Ozawa *et al.*, *Nucl. Phys.* **A693**, 32 (2001).
17. F.M. Marques *et al.*, *Phys. Lett.* **B476**, 219 (2000).
18. C.A. Bertulani and M.S. Hussein, *arXiv* :0705.3998.

EXPERIMENTAL APPROACH TO HIGH-TEMPERATURE STELLAR REACTIONS WITH LOW-ENERGY RI BEAMS

S. KUBONO, H. YAMAGUCHI, G. AMADIO, S. HAYAKAWA, Y. WAKABAYASHI, Y. KURIHARA, J.J. HE[†], A. SAITO, H. FUJIKAWA, LE HONG KHIEM[††], M. NIIKURA

Center for Nuclear Study (CNS), University of Tokyo, Wako Branch at RIKEN
2-1 Hirosawa, Wako, Saitama, 351-0198 Japan

T. TERANISHI

Department of Physics, Kyushu University
6-10-1 Hakozaki, Higashi-ku, Fukuoka, 812-8581 Japan

S. NISHIMURA

Nishina Accelerator Research Center, RIKEN
2-1 Hirosawa, Wako, Saitama, 351-0198 Japan

S. KATO

Department of Physics, Yamagata University, Kojirakawa, Yamagata, Japan

C.S. LEE, Y.K. KWON

Department of Physics, Chung-Ang University
Seoul 156-756, Korea

I.S. HAHN, A. KIM

Department of Physics, Ewha Womans University
Seoul 120-750, Korea

The experimental efforts for the stellar reactions under high-temperature and high densities have been made as the major program using the RI beams from the CNS low-energy in-flight RI beam separator (CRIB) of University of Tokyo, in order to understand the evolution of the universe as well as various stellar phenomena. Specifically, two subjects of hydrogen burning are discussed here. One is a reaction study of the pp-chain and the second is of the explosive hydrogen burning, the rp-process. Some s-wave resonances have been identified by the thick target method in the crucial reaction processes in the hydrogen burning. The resonant scattering with the thick target method also succeeded in identifying inelastic resonant scattering, giving proton widths for the first excited state of the target nucleus. This provides very efficiently the reaction rate estimate for the process under high-temperature equilibrium condition. Possibilities of the CRIB facility in near future are also briefly discussed.

[†] Present address; School of Physics, University of Edinburgh, EH9 3JZ, United Kingdom
[††] JSPS Invitation Fellow.

[†††] A low-energy RI beam separator of CNS, University of Tokyo.

1 Introduction

1.1. *Hydrogen Burning under High Temperature and High-Density Conditions*

As is well known, nuclear physics plays a key role in the evolution of the universe as well as in various stellar phenomena in terms of synthesis of elements and energy generation. Specifically, the investigation of the nuclear phenomena under high temperatures and high densities has attracted nuclear physicists because nuclear physics of unstable nuclei can be a key element for understanding various phenomena such as supernovae and gamma-ray burst [1].

Since hydrogen is the most abundant element, hydrogen burning often plays an important role in the evolution of stars as well as various stellar events. Hydrogen burns quiescently in the early stage of stellar evolution, typically in the main sequence stage, whereas it burns explosively in the late stage of the star evolution, like in novae and supernovae. Thus, we have been focusing our effort to investigate experimentally the nuclear processes, especially those under high temperature and high density conditions of hydrogen. The pp-chain plays a major role in the main sequence stage in low-mass stars, and the CNO cycle in the massive stars. Even these quiescent burning nucleosynthesis stages involve unstable nuclei of relatively long half-lives such as ^{7}Be. The hot-pp chain, which would take place under higher temperature conditions, was proposed for investigating the evolution of very metal poor stars [2]. This process could provide another pathway for synthesis of Carbon and heavier elements through the proton-rich nuclear region in addition to the triple-alpha process. Astronomers are also approaching to the first generation stars. Recently, they reported extremely metal poor stars that have large overabundance for the light elements such as C, N, and O, but have extremely small amount of irons. However, most nuclear processes of the hot-pp chain are not known well yet. The hot-CNO cycle and the rp-process involve many reactions which are not known yet. Most of them involve proton-rich unstable nuclei. These processes will be related partly to the νp-process which has been recognized in the recent years and would play a key role in the very early epoch of type II supernovae because of the neutrino processes. This process is very much like high-temperature rp-process under a proton-rich environment [3,4].

It should be noted that the nucleosynthesis flows under high temperature conditions run close to the proton drip-line in the light mass region, and thus relevant level densities are very small, indicating statistical models are not applicable. This fact clearly indicates an importance of experimental efforts. It is of great interest to study the reactions directly with low-energy unstable nuclear beams.

1.2. *Low-Energy In-Flight RI Beam Facility CRIB for Nuclear Astrophysics*

Low-energy RI beams are definitely needed for investigating nuclear reactions in hydrogen burning process under high temperature and high density conditions. The low-energy RI beam in-flight separator, CRIB [5,6], was constructed year 2000 by the Center for Nuclear Study (CNS), University of Tokyo as the first joint program with RIKEN, and started RI beam productions in 2001. Figure 1 shows the plane view of CRIB. Although the target thickness is small for RI beam production with heavy ion beams at low-energies, the primary beam currents are nowadays are very high with ECR sources, and thus the final RI beam intensities may be not so small. To realize such RI beam facility, a project, named AVF upgrade project, started to establish a low-energy RI beam facility, involving development of ion sources, the AVF cyclotron, the beam line and the separator CRIB. The detail of the separator CRIB is described in refs. 5, 6. It consists of a double achromatic magnetic system, very much like ordinary fragment separators, together with a Wien Filter to purify further the RI beams. The first momentum dispersive focal plane is called F1, and the double achromatic focal plane F2. One may use a degrader at F1 for a better beam separation. One can do experiment at F2. Then, the secondary beam particles will be transferred to the focal plane F3 through the Wien Filter which adds another particle separation capability by velocity.

This filter is very powerful for light element separation. But of course, difficulties arise when the mass number increases. The primary beams are provided by the AVF cyclotron of RIKEN whose K number is now improved to 79. Thus, the RI beams produced have roughly 10 MeV/u or less, where one uses light ion reactions such as (p,n), (d,n), etc. in inverse kinematics. The nuclear species to be produced also are limited within a few nucleons away from the line of stability. The most favorable production reaction is the (p,n) reactions because the cross sections usually are large and the phase space to accept by CRIB is also very large, giving the final beam intensities of $10^{6\sim7}$ pps with a primary beam of 1 pμA. This intensity can be further increased by developing the production target technology as well as the ion source technology. High intensity primary beams more that 1 pμA now are available for many elements. The AVF cyclotron has a

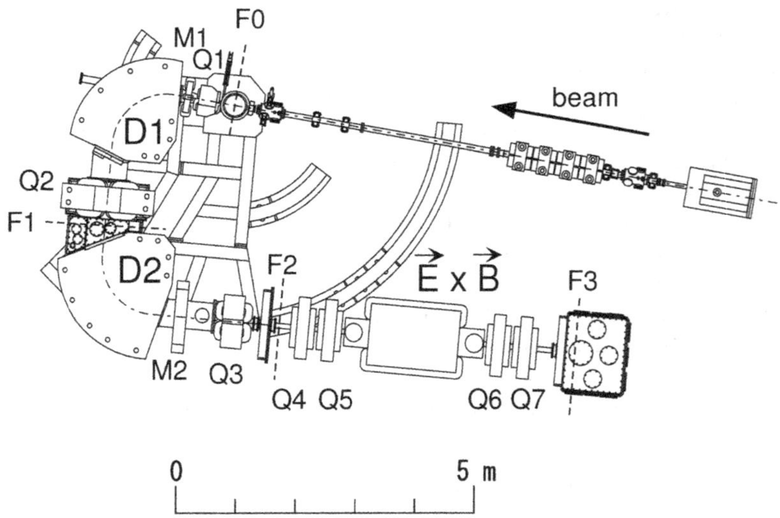

Fig. 1 Plane view of the low-energy in-flight RI beam separator CRIB, which has been installed in the E7 experimental hall of the RIKEN accelerator facility.

beam transmission of a few ~ 20 % at present. The redesigning the central region of the AVF cyclotron is in progress for higher transmission efficiency as well as for other Harmonic mode operations. Our recent effort for the production target is discussed in the last section. A flat-top acceleration mode has been successfully installed to the cyclotron, giving much better quality for heavy ion beams with better extraction efficiency.

2 Investigation of Stellar Reactions in the pp-Chain

The pp-chain is the main source for energy generation in low mass stars at the main sequence stage, where hydrogen burns under quiescent conditions. The pp-chain also involves relatively long-lived unstable nuclei at higher density and temperature conditions. The burning process in our sun has been long investigated both by the solar neutrino observations and laboratory studies of the nuclear reactions relevant. The main problem of the neutrino flux deficit has been explained by the neutrino oscillation [7]. However, the investigation of the solar model still needs nuclear data at very low energies with high precision for the reaction rates of the pp-chain. The nuclear data are not available for most reactions at the energies of interest. In addition, the data available at higher energies have large uncertainties for the cross sections. Among the reaction chain that leads to production of the high-energy neutrino emitter ^{8}B, the sensitivity of each astrophysical S-factor is predicted to be [8];

$$\Phi(^8B) = S_{11}^{-2.6}\, S_{33}^{-0.43}\, S_{34}^{0.81}\, S_{17}^{1.0}\, S_{e-7}^{-1.0}$$

The ^{3}He(^{3}He,2p)^{4}He reaction has been investigated many years at Gran Sasso. Although the experiment has reached the Gamow window region, it is still unclear if there is any resonance in the region of interest because of the experimental uncertainties. One possible way is to study a possible resonance by a ^{3}He transfer reaction on ^{3}He. Such investigation is in progress at CRIB.

One of the most crucial reactions among the reaction sequence above is ^{7}Be(p,γ)^{8}B. There are many extensive works reported, but they deviate considerably each other yet. The experimental data with high precision are demanded. Another problem is the way to extrapolate the data to the very low-energy, the Gamow window region. Barker made a precise analysis of the S_{17} data and derived an extrapolation curve, where he needed to have an s-wave 2^- resonance with a width more than 4 MeV at higher energy region in ^{8}B [9]. Rogachev and his collaborators made an experiment of ^{7}Be+p resonant scattering experiment for a limited excitation energy range, suggesting a 2^- s-wave resonance at around 3.5 MeV with $\Gamma = 4 - 8$ MeV [10].

An extensive experiment of ^{7}Be+p was recently performed using the thick target method at CRIB [5,6] for much wider excitation energy region. The s-wave resonances of 2^- and 1^- have been clearly identified together with the 3^+ known resonance at E_{cm} = 2.32 MeV [11]. The width of the 2^- s-wave resonance was carefully studied by analyzing the excitation functions at three different angles. The result is at Ex = 3.1 MeV Γ=2.8 MeV which is much smaller than requested to fit the low energy behavior by Barker. Thus, there remains a problem in extrapolation of the S_{17} value to the Gamow energy region in the middle of the sun, although the new resonance improves the fit at higher energy region.

The hydrogen burning in the mass region below mass 12 has a potential importance for investigating the evolution of the first generation stars, as discussed in the introduction. The hot pp-chain [12] might play a crucial role for synthesis of C and heavier elements under extremely metal poor conditions. Most reactions, however, await experimental efforts.

3 Investigation of Stellar Reactions in the rp-Process

3.1. *Nuclear Reactions in the Extended CNO Cycle*

In this mass region under hydrogen-rich environment, we have typically the CNO cycle and the extended CNO cycle. These processes play a basic role in evolution of stars. On the other hand, there are many observational data of elemental abundance relevant, and thus this mass region has been extensively investigated in the past years using both stable and unstable nuclear beams as well as by theories. However, many reactions are still very uncertain and needed to be investigated to clarify the burning process. The breakout process from the hot-CNO cycle is also an interesting subject in this region for studying the nature of explosive stellar events. There are some reactions being investigated with using low-energy RI beams at CRIB. The $^{18}F(p,\alpha)^{15}O$ reaction [13,14] is related to the observation of positron-electron annihilation gamma rays emitted in the occasion of novae, because ^{18}F is considered to be one of the main contributors for the annihilation gamma rays in the events. The (p,α) reaction destroys the nucleus ^{18}F. Now, the INTEGRAL project is reporting an observation of such annihilation gamma rays [15]. Here, a quantitative estimate of the annihilation gammas is really awaited for in astronomy. The reaction cross sections of the (p,α) reaction can be obtained by a direct method and also possibly by the Trojan Horse method.

Another interesting, and crucial subject for this mass region is the breakout process from the CNO cycle. There are two breakout processes suggested; one starting from ^{14}O and the other one from ^{15}O. Under a high temperature and high density condition just before the breakout, the most abundant nuclides are ^{14}O and ^{15}O together with abundant α particles, and thus the α induced reactions on ^{14}O and ^{15}O can be the ignition reactions. Especially, the process $^{14}O(\alpha,p)^{17}F(p,\gamma)^{18}Ne(\alpha,p)^{21}Na$ is considered to be the key process under high temperature rp-process like in X-ray bursts. No reaction in the chain is well understood at the stellar temperatures.

The crucial $^{14}O(\alpha,p)^{17}F$ reaction [16] was investigated some time ago using the low energy RI beam of ^{14}O at CRIB [5,6]. The reaction cross section was measured by a thick target method of a He gas target cooled down to about 30K that reduced the target length roughly by a factor of ten. The thick target method was applied [17]. The excitation function of the (α,p) reaction shows several peaks which correspond well to the known excited states in ^{18}Ne. Some yields around Ex = 6.2 MeV in ^{18}Ne were observed, which is consistent with the previous predictions. It should be of great interest to investigate further to determine definitely the cross section, because the measurement was made only at zero degree and has some uncertainties of other excited state contributions.

The first step of the hot-CNO cycle is the $^{13}N(p,\gamma)^{14}O$ reaction which was the first stellar reaction that involved a short-lived nucleus and successfully studied by the direct method. But, the contribution only from the first excited state above theproton threshold

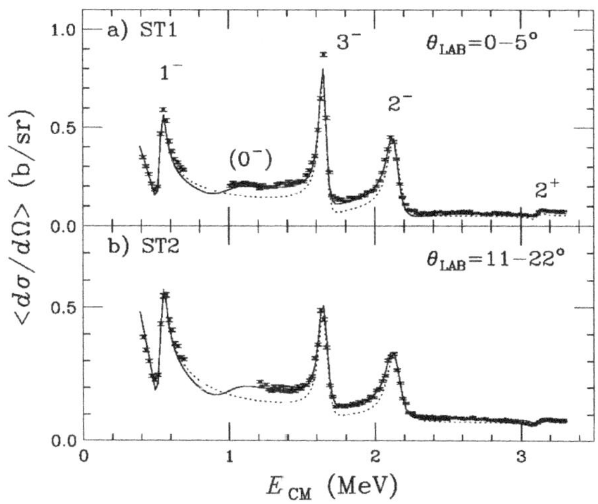

Fig. 2 Excitation functions of ^{13}N+p elastic scattering. The spin parities are for the resonances in ^{14}O.

was investigated. The low-lying levels in ^{14}O are considered to be of single particle nature, but the there were still some controversies for the structure. There should be an s-wave 0^- resonance, but there was no experimental evidence reported before. The spin parity of the 6.8 MeV state in ^{14}O was not known. If it is a d-wave resonance, as expected from the analog state, we should see a peak prominently in a resonant elastic scattering. Our recent experiment has confirmed this prediction by the experiment with a ^{13}N beam obtained by CRIB [18]. Figure 2 shows the excitation function of ^{13}N+p measured. The peak was well fitted by the R-matrix calculation of d-wave resonance with a large reduced width for the peak at Ecm = 2.1 MeV. In addition, the 0^- s-wave resonance was identified for the first time at 5.71 MeV as the second excited state. The effect to the nucleosynthesis was found to be small.

3.2. *Early Stage of the rp-Process*

There are many works on nucleosynthesis in explosive hydrogen burning in the past years in the mass region of A = 20 – 32 both by direct and indirect methods. We performed a series of experiments previously [19] using indirect methods for a possible pathway just after the breakout, $^{19}Ne(p,\gamma)^{20}Na(p,\gamma)^{21}Mg(\beta^+)^{21}Na(p,\gamma)^{22}Mg(\beta^+)^{22}Na(p,\gamma)$ $^{23}Mg(p,\gamma)^{24}Al(p,\gamma)^{25}Si$. Many new resonances just above the proton thresholds were discovered by using particle transfer reactions such as (^3He,^6He) [20]. However, resonance parameters are needed to determine the reaction rate of their contributions. The resonance parameters can be better investigated directly using the proton-rich unstable nuclear beams. Resonant proton elastic scattering provides information on the incident channel of the proton capture reactions of the rp-process. A series of such experimental studies are in progress using low-energy proton-rich RI beams at CRIB.

An interesting new result was obtained for the proton resonant scattering of ^{22}Mg + p, one of the results obtained with CRIB [21]. This is the incident channel of the ^{22}Mg(p,γ)^{23}Al stellar reaction which is important for the astronomical nuclear gamma ray observation of ^{22}Na as well as for the NeE problem of the isotopic anomalies in meteorites.

The experiment was performed using a primary beam of ^{20}Ne at 8.11 MeV/u. A ^{22}Mg beam was separated by CRIB at F2 focal plane. A thick target method was used to measure the excitation function of ^{22}Mg + p elastic scattering. The data clearly indicates inelastic resonant scattering of ^{22}Mg(1st)+p, together with elastic scattering of ^{22}Mg(g.s.)+p. Three resonances were observed at around 3 MeV in ^{23}Al, and one of them decay to the ground state only, but the other two decay mainly to the first excited state in ^{22}Mg. The decay widths for the excited state in ^{22}Mg ($\Gamma_{p'}$) were obtained from the R-matrix analysis, which are consistent with the OXBASH shell model calculations. The present method demonstrates to be useful for deducing the decay width for the excited states, although the method is very simple. It should be compared with a direct particle transfer reactions such as the (d,p) reaction which is sensitive only to the decay width to the ground state. This information is very useful to evaluate the reaction rate under high-temperature equilibrium conditions, where ^{22}Mg can be in the first excited state with a certain fraction, defined by partition [22]. For instance, about 4 % of ^{22}Mg would be in the first excited state at the equilibrium of $3\text{x}10^9$ K. The nuclear reactions under such condition will have a Gamow window based on the first excited state in ^{22}Mg (not ^{22}Mg(g.s.)), and thus resonances at higher energies in ^{23}Al would contribute for the reaction rate of ^{22}Mg(p,γ)^{23}Al.

4 Outlook

Low-energy RI beams can be available relatively easily by the in-flight method, as I demonstrated here. The RI beam intensities, however, are limited due to the effective target thickness available as well as by the primary heavy ion beam intensities at low energies. The latter point will be improved by developing the ion source technology and the accelerator properties. The keen problem is the durability of the production target against the primary beams. Since low-energy heavy ions deposit high energy densities to the target, the targets can be easily deteriorated. To overcome this problem, we made a cryogenic production target that uses liquid nitrogen [23]. The target cell can be directly cooled by the liquid nitrogen and the circulating target gas can be also cooled down in the liquid nitrogen tank. This cryogenic target was tested with high intensity heavy ion beams. One typical result is $2\text{x}10^8$ pps ^{7}Be by the ^{7}Li(p,n)^{7}Be reaction in inverse kinematics with a 5.6 MeV/u ^{7}Li beam of 2.8 eμA, where the production target was cooled down to about 85 – 90 K. To ensure the intensities of 10^8 pps for wider nuclear species, further development is needed from the ion source, the AVF cyclotron, and the separator. One may investigate radiative capture reactions of astrophysical interest such as (p,γ) and (α,γ).

A total development of an RI beam facility will enable one to produce reasonably high intensities of RI beams. This will be encouraging for a small accelerator facility who aims at developing low energy RI beam sciences.

In addition, the world-top, third-generation RI beam facility RIBF in RIKEN just has started the operation this year, providing a wide range of capabilities for nuclear astrophysics especially for the study of the r-process. A variety of nuclear astrophysics programs will be available in the RIKEN campus.

Acknowledgments

This work is partially supported by JSPS Joint Research Project under the Japan-Korea Basic Scientific Cooperation Program (2007-2009).

References

1. Proc. Origin of Matter and Evolution of Galaxies, eds. S. Kubono et al., *AIP Conf. Proc.* **847** (2006)
2. W. Aoki, Nuclei in the Cosmos VIII, eds. L. Buchmann et al., *Nucl. Phys. A* **758**, 209 (2005)
3. C. Fröhlich, et al., *Phys. Rev. Lett.* **96**, 142502 (2006)
4. S. Wanajo, *Astrophys. J.* **647,** 1323 (2006)
5. Y. Yanagisawa, et al., *Nucl. Instr. Meth. A* **539**, 74 (2005)
6. S. Kubono, *Eur. Phys. J.* **A13,** 217 (2002)
7. S.N. Ahmed, et al., *Phys. Rev. Lett.* **92**, 181301 (2004)
8. J.N. Bahcall and M.H. Pinsonneault, *Rev. Mod. Phys.* **64**, 885 (1992), and J.N. Bahcall, Solar Neutrinos, in Unsolved Problems in Astrophysics, Princeton University Press, (1997)
9. F.C. Barker and A.M. Mukhamedzhanov, *Nucl. Phys. A* **673**, 526 (2000)
10. G.V. Rogachev, J.J. Kolata, F.D. Becchetti, P.A. DeYoung, M. Hencheck, K. Helland, J.D. Hinnefeld, B. Hughey, P.L. Jolivette, L.M. Kiessel, H.-Y. Lee, M.Y. Lee, T.W. ODonnell, G.F. Peaslee, D. Peterson, D.A. Roberts, P. Santi, S.A. Shaheen, *Phys. Rev. C* **64**, 061601 (2001)
11. H. Yamaguchi, et al., Origin of Matter and Evolution of Galaxies, eds. S. Kubono et al., *AIP Conf. Proc.* **847**, 275 (2006)
12. M. Wiescher, J. Görres, S. Graff, L. Buchmann, and F. -K. Thielemann, *Astrophys.* J. **343**, 352 (1989)
13. N. de Sereville, et al., *Phys. Rev. C* **67**, 052801 (2003)
14. R.L. Kozub, et al., *Phys. Rev. C* **71**, 032801(R) (2005)
15. R. Diehl, Origin of Matter and Evolution of Galaxies, eds. S. Kubono et al., *AIP Conf. Proc.* **847**, 289 (2006)
16. M. Notani, et al., *Nucl. Phys. A* **746**, 113 (2004)
17. S. Kubono, *Nucl. Phys. A* **693**, 221 (2001)
18. T. Teranishi, et al., *Phys. Lett. B* **650**, 127 (2007)
19. S. Kubono, *Prog. Theor. Phys.* **96**, 275 (1996)

20. S. Kubono, et al., *Phys. Rev. C* **43**, 1821 (1991)
21. J.J. He, et al., to be published in *Phys. Rev. C*
22. C.E. Rolfs and W.S. Rodney, Cauldrons in the Cosmos, The University of Chicago Press (1988)
23. H. Yamaguchi, et al., to be published in *Nucl. Instr. Meth. A*

TRANSITION TO QUARK MATTER IN NEUTRON STARS *

G. X. PENG†

Institute of High Energy Physics, CAS, P.O. Box 918-4, Beijing 100049, China
Phys. Depart. of Graduate University, CAS, Yuquan Rd. 19A, Beijing 100049, China
† *E-mail: gxpeng@ihep.ac.cn*

A. LI

Dipartimento di Fisica e Astronomia, viale A. Doria 6, 95125 Catania, Italy
Department of Physics, Xiamen University, Xiamen 361005, China

U. LOMBARDO

Laboratori Nazionali del Sud, INFN, Via S. Sofia 62, 95123 Catania, Italy
Dipartimento di Fisica e Astronomia, viale A. Doria 6, 95125 Catania, Italy

The transition of nuclear matter to quark matter is studied under the condition of beta equilibrium. It is found that the mixed hadron-quark phase can occur near the saturation density, and can undergo a phase transition to pure quark matter at about 5—6 times the saturation. The onset of mixed and quark phase in neutron stars is compatible with the observed class of low mass ($1.4M_\odot$) neutron stars, whereas it hinders the occurrence of kaon condensation.

1. Introduction

So far the study of neutron stars has been mainly focussed on the relationship between the equation of state (EoS) of nuclear matter and the observed maximum mass. The connection has been achieved by solving the hydrostatic equilibrium equations based on general relativity. The first generation observed masses exhibited an average value around 1.5 solar masses. This value required a softening of the EoS which can be easily reached by the introduction of new degrees of freedom like hyperons, kaons or quarks accompanied or not by a phase transition. Sometime the softening was so large that the neutron star is predicted to collapse into a black hole, as for the SN1987A.[1] In the new generation of observations the masses are distributed within a large range, up to more than 2 solar masses (e.g. PSR J751+1807 : $M = 2.1 \pm 0.2\ M_\odot$) that requires a stiff EoS, i.e. hadronic matter without new degrees of freedom. Since it is hard to imagine pure hadronic matter to sustain the high pressure predicted in the inner core, new scenarios have to be

*talk given by U. Lombardo

advanced to explain the coexistence, in the phenomenology of neutron stars, of low and high mass spectrum.

Recently[2] it has been argued that the two observed classes of neutron stars might correspond to two different evolutionary scenarios of neutron stars. In one case, the hot and dense remnant of the supernova explosion rapidly evolves into a hybrid star, where the transition to a quark phase softens the nuclear matter so that $M \approx 1.5 M_{\odot}$; in the other case a slow evolution leads the neutron star to a large mass via a mass accreting from the coupling with a white dwarf. From this point of view the destiny of the remnant is strongly affected by the initial conditions,i.e. density, temperature, leptonization degree etc. For instance, if the mass of the remnant is below the mass threshold for quark nucleation the transition to the quark phase is forbidden.[3] If the mass is slowly accreting the transition is allowed. The role of temperature or other parameters defining the initial state of a new born neutron star has not yet studied.

In this contribution the transition from hadron phase (HP) to strange quark phase (SQP) in the inner core of a neutron star is investigated under the condition of beta-equilibrium. Several quark models have been adopted to study the quark phase, including the conventional MIT bag model. In our model the confinement is achieved by means of a density-dependent quark mass[4,5] derived from in-medium chiral condensates.[6] In the hadron sector we adopt the equation of state from Brueckner-Bethe-Goldstone (BBG) approach with three-body force (TBF).[7,8] This theory, being a completely microscopic approach, can easily incorporate degrees of freedom such as nucleon resonances ($\Delta(1232)$ or $N^*(1440)$), which are expected to appear at higher hadron density. It is found that the mixed hadron-to-quark phase can occur, for reasonable values of the confinement parameter, a little above the normal saturation density, and can undergo the transition to pure quark matter at about 5-6 times the saturation, which is quite different from the previous results from Nambu-Jona Laşinio (NJL) model in which the mixed quark phase can not appear at neutron-star densities.[9] Afterwards, the influence of the mixed and quark phase on the structure of compact stars is discussed by solving the Tolman-Oppenheimer-Volkov (TOV) equation and extracting the mass-radius plots for neutron stars. Finally, it is shown that the transition to the deconfined phase turns out to be incompatible with the onset of kaon condensation.

2. Phase diagram

Let us assume that the nuclear matter, consisting of nucleons and electrons, is in equilibrium with a gas of u, d, s quarks and electrons. We study the possibility of the occurrence of a phase transition from pure hadron matter to pure quark matter with an intermediate mixed phase. According to Glendenning[10] the condition of total charge neutrality has to be imposed with positive hadron charge compensated by negative quark charge. Dealing with a two component system we have to impose

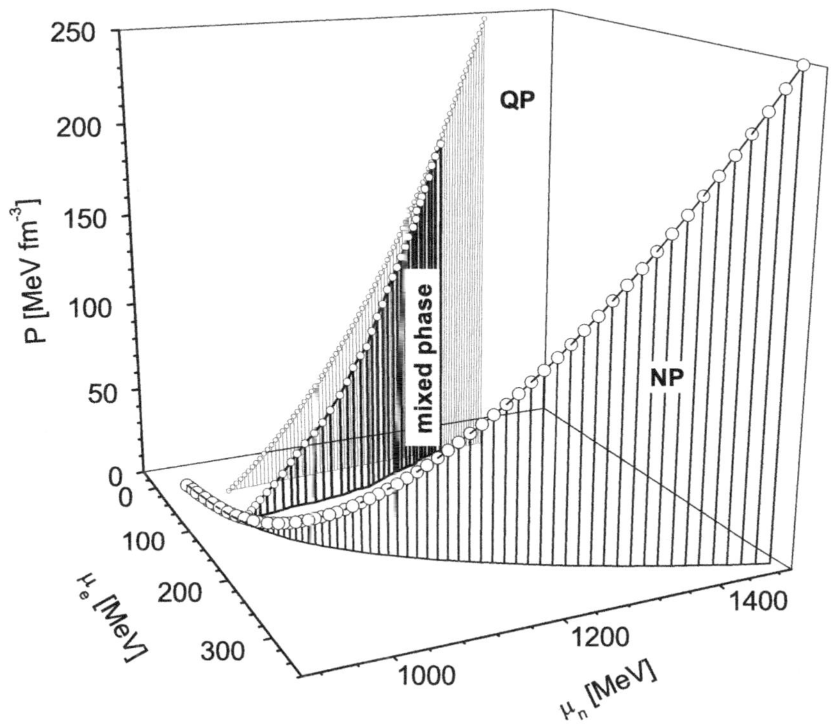

Fig. 1. Gibbs phase construction of hadron-to-quark transition.

the Gibbs equilibrium conditions between the two phases:

$$T_H = T_Q \quad (thermal) \tag{1}$$

$$\mu_H = \mu_Q \quad (chemical) \tag{2}$$

$$p_H = p_Q \quad (mechanical)\cdot \tag{3}$$

In our case the thermal equilibrium is automatically satisfied since $T = 0$. The chemical equilibrium condition leaves us with only two independent chemical potentials, e.g. μ_n and μ_e. As a consequence, expressing the hadron and quark pressures in terms of the chosen variables, the phase equilibrium will be marked by the intersect of the two surfaces $P_H(\mu_n, \mu_e)$ and $P_H(\mu_n, \mu_e)$. Fig. 1 shows what happens when the mixed phase occurs. In the lower crossing point the transition from pure hadron phase to mixed phase occurs, in the upper crossing point the transition from mixed phase to quark phase occurs.

The conservations of total baryonic density ρ_t and total charge $Q_t = 0$ are expressed by the equations

$$\rho_t = (1-\chi)\rho + \chi\rho_q \tag{4}$$

$$0 = (1-\chi)Q + \chi Q_q \tag{5}$$

where ρ,Q (ρ_q,Q_q) are the baryon (quark) density and charge, respectively. $\chi = V_q/V$ is the quark fraction expressed in terms of the quark volume-to-total volume ratio.

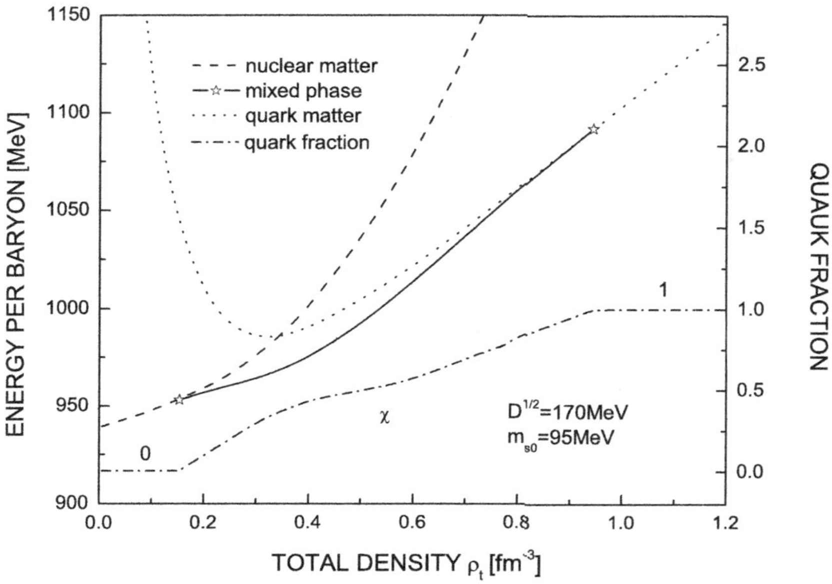

Fig. 2. The energy per baryon as a function of the total density in pure nuclear, mixed, and quark phase. The quark fraction χ in also given. In this calculation $D = (170\text{MeV})^2$.

In Fig. 2, we display the density dependence of the energy per baryon in pure nuclear matter, in pure quark matter, and in the mixture of nuclear and quark matter. It is seen that the nuclear matter is the most favorite phase at lower densities, and the quark matter is the most stable phase at higher densities, while at intermediate densities, the mixed phase has the lowest energy. The quark density n and nuclear density ρ are also plotted on the right axis. The density dependence of the pressure is given in Fig. 3.

3. Properties of hybrid stars

The structure of neutron stars is expected to be very much affected by the occurrence of the transition to quark matter,either in the presence of a mixed phase or for direct transition to pure quark phase. The study of the structure of hybrid stars passes through the solution of the Tolman-Oppenheimer-Volkov (TOV) equation

$$\frac{\mathrm{d}P}{\mathrm{d}r} = -\frac{GmE}{r^2}\frac{(1+P/E)(1+4\pi r^3 P/m)}{1-2Gm/r}, \tag{6}$$

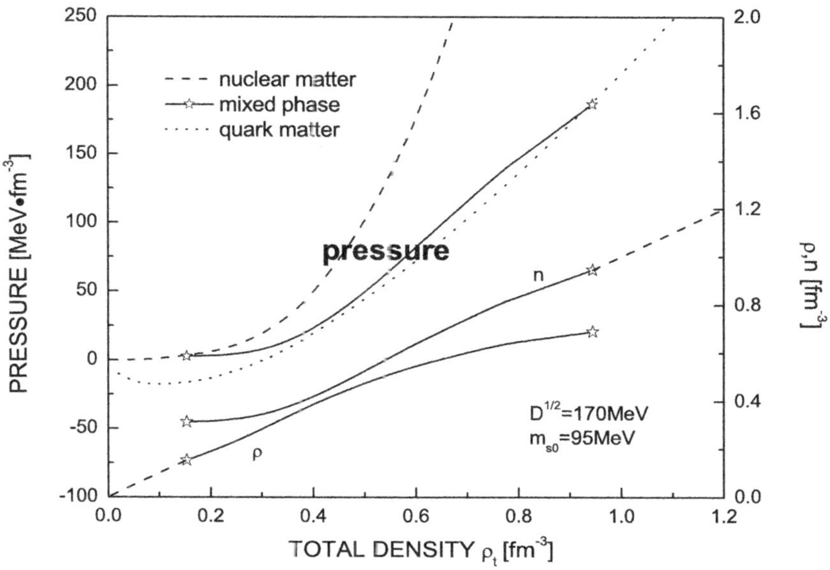

Fig. 3. The pressure is shown as a function of the total density in pure nuclear, mixed, and quark phase. The nuclear density ρ and the quark density n are also given. In this calculation $D = (170\text{MeV})^2$.

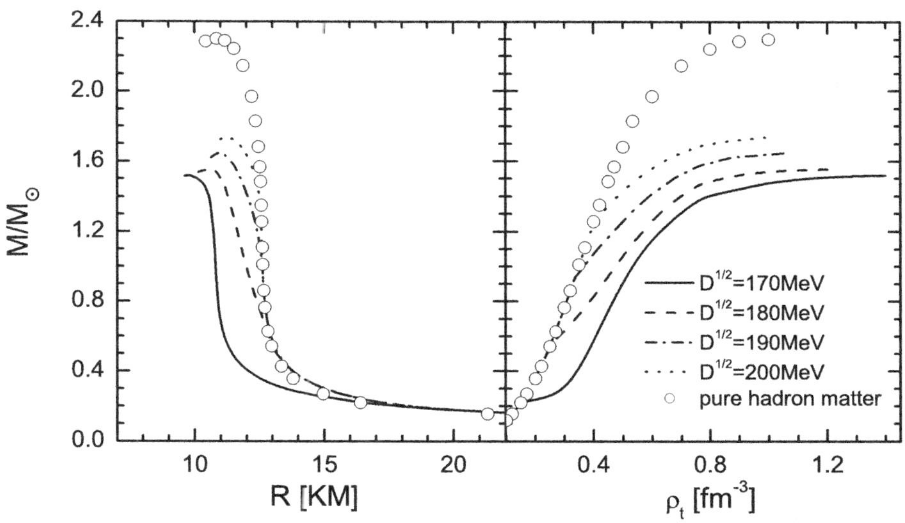

Fig. 4. Mass-radius relation of hybrid stars at $m_{s0} = 95$ MeV for four values of the confinement parameter $D^{1/2}$ of 170, 180, 190 200 MeV .

where $G = 6.707 \times 10^{-45}$ MeV^{-2} is the gravitational constant, r is the distance from the center of the star, $E = E(r)$ and $P = P(r)$ are, respectively, the energy density and pressure at the radius r. The subsidiary condition is

$$\mathrm{d}m/\mathrm{d}r = 4\pi r^2 E \tag{7}$$

with $m = m(r)$ being the mass within the radius r.

At variance with pure nuclear or quark stars, a hybrid star contains pure quark matter in the core, pure nuclear matter near the outer part, and a mixed phase between the quark and nuclear matter.

The resulting gravitational mass for the hybrid star is plotted in Fig.4, as a function of both radius and central density, for four values of the confinement parameter $D^{1/2}$ of 170, 180, 190, 200 MeV. We observe a decrease of the maximum mass for smaller D values of the constant D, from 1.74 $M_\odot$ at $D^{1/2} = 200$ MeV down to 1.6, 1.55, 1.52 $M_\odot$ corresponding to $D^{1/2} = 190, 180, 170$ MeV. This can be easily understood in this way: since quark phase sets in earlier at the lower D case, sometime even less than the nuclear saturation density (for example only 0.15 fm^{-3} for $D^{1/2} = 170$ MeV), then the quark population of the star would be more numerous. As a consequence, the stronger softening of the equation of state may only support less gravitational mass. In particular, the most favorable case of $D^{1/2}$ = 180 and 190 MeV, when the quark confinement appears at around $2\rho_0$ consistent with the experiments with heavy ion collisions, the predicted maximum mass can match very well the S branch of neutron stars mentioned by P. Haensel et. al[2] with a error less than 5%.

4. Kaon condensation

The condensation of K^- mesons in dense matter is originated by the reaction

$$e^- \to K^- + \nu\,. \tag{8}$$

if the effective mass of the K^- drops below the chemical potential of the electron, this reaction becomes possible in dense matter, indicating the presence of kaon condensation. Since almost all the studies of the medium effects on kaons[11] suggest the consistent picture that, since nuclear matter interacts attractively with kaons, the K^- mass drops down at higher density, so that the threshold condition for the onset of K^- condensation $\mu_e = m_K^*$, which follows from Eq. (8), could be very probably fulfilled in the center of the neutron stars[11] at $\rho \gtrsim 3\rho_0$. However, the deconfinement phase transition from the hadronic phase to the quark phase occurs also at rather low density, which leads to the appearance of mixed phase in neutron stars. Therefore, it is very interesting to explore how the quark deconfinement affects the threshold of K^- condensation.

We take the antikaon dispersion relation constrained by the heavy-ion data as empirical indication of an attractive antikaon potential in dense matter,[11] and combine this with the BHF model of nuclear matter together with the above quark model without bag constant, to calculate the effective kaon mass in hybrid stars. The result is illustrated in Fig. 5., where the effective kaon mass and the electron

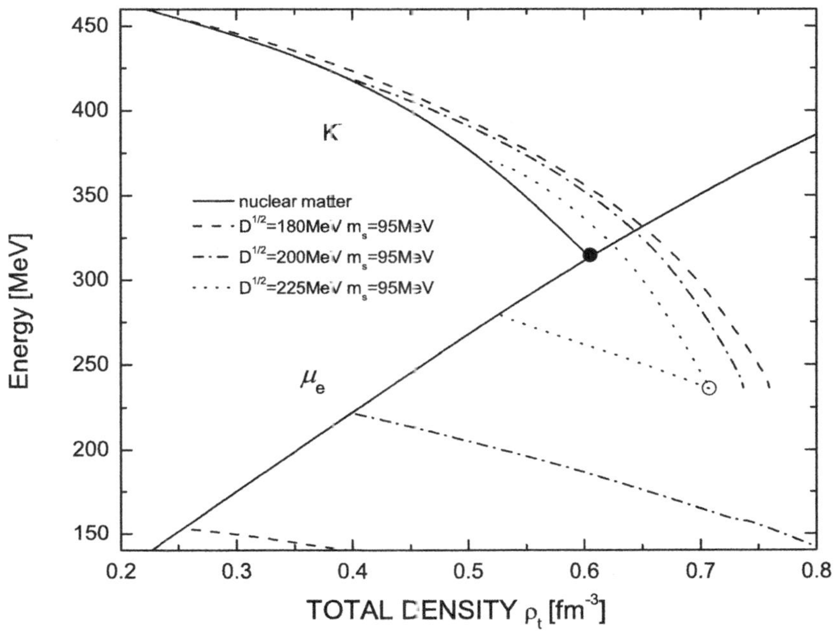

Fig. 5. The effective kaon mass and the electron chemical potential are shown in normal nuclear matter and also in mixed phase at $m_{s0} = 95$ MeV for three selected values of the confinement parameter $D^{1/2}$ of 180, 200, 225 MeV.

chemical potential are shown in normal nuclear matter and especially in mixed phase for three selected values of the confinement parameter D. One sees that in normal nuclear matter, K^- medium mass keeps decreasing with density and meets the value of electron chemical potential at ~ 0.6 fm^{-3} (which is indicated with a bullet), so the kaon condensation would be present for sure in this case; However, once the quark phase sets in at the total density of 0.26, 0.4, 0.53 fm^{-3} respectively for $D^{1/2}$ = 180, 200, 225 MeV, the K^- mass decrease from the matter seems to slow down a little bit because of the exact quarks; In addition, more conclusively, the electron population begin to drop instead, for the reason that the electric charge neutrality can be achieved more efficiently through the charged quarks themselves. As a result the threshold condition of K^- condensation is much more difficult to satisfy, unless the confinement parameter D is chosen to be extremely high (at least $D^{1/2}$=225 MeV) when the presence of quark phase is pushed to very high densities, then one may expect that m_K^* can finally equal to μ_e (which is indicated with a circle). Those high values of D may even not be realistic, therefore we'd rather conclude that the inclusion of quark phase may make the kaon condensation impossible in neutron stars, or at least hinder it very strongly.

Acknowledgments

G.X. Peng would like to acknowledge the support from NSFC (10675137, 10375074, 90203004) and KJCX3-SYW-N2. The work was performed within the Asia-Link project CN/ASIA-LINK/008 (94-791) of the European Commission.

References

1. H.A. Bethe and G.E. Brown, *ApJ L* **129**, 445 (1995).
2. P. Haensel, M. Bejger and J.L. Zdunik, arXiv:0705.4594v1 [astro-ph] 30 May 2007, submitted to A&A.
3. I. Bombaci, G. Lugones, I. Vidana, *Astron. Astrophys.* **462**, 1017 (2007).
4. S. Chakrabarty, S. Raha, and B. Sinha, *Phys. Lett. B* **229**, 112 (1989); S. Chakrabarty, *Phys. Rev. D* **43**, 627 (1991); **48**, 1409 (1993); **54**, 1306 (1996).
5. G. X. Peng, H. C. Chiang, J. J. Yang, L. Li, and B. Liu, *Phys. Rev. C* **61**, 015201 (2000).
6. G. X. Peng, *Nucl. Phys. A* **747**, 75 (2005).
7. P. Grangé, A. Lejeune, M. Martzolff, and J.-F. Mathiot, *Phys. Rev. C* **40**, 1040 (1989).
8. W. Zuo, A. Lejeune, U. Lombardo, and J.-F. Mathiot, *Nucl. Phys. A* **706**, 418 (2002); *Eur. Phys. J. A* **14**, 469 (2002).
9. M. Baldo, G. F. Burgio, P. Castorina, S. Plumari, D. Zappala', *Phys. Rev. C* **75**, 035804 (2007) and references quoted therein.
10. N. K. Glendenning, *Phys. Rev. D* **46**, 1274 (1992).
11. G. Q. Li, C.-H. Lee G. E. Brown, Phys. Rev. Lett. **79**, 5214 (1997); F. Weber, *Prog. Part. Nucl. Phys.* **54**, 193 (2005) and quoted therein.

RESEARCH AT VATLY: MAIN THEMES AND RECENT RESULTS

P.N. DIEP, P.N. DONG, P.T.T. NHUNG, P. DARRIULAT*,
N.T. THAO, D.Q. THIEU and V.V. THUAN

*VATLY, Institute for Nuclear Science and Technique,
5T-160 Hoang Quoc Viet, Nghia Do, Cau Giay, Hanoi, Vietnam
E-mail: darriulat@mail.vaec.gov.vn

A brief description of the research topics addressed by the Hanoi cosmic ray laboratory, VATLY, is given in the form of a progress report with respect to the situation in 2002 which had been presented at the ISPUN02 conference in Halong Bay.

Keywords: Cosmic rays; Air showers.

1. Introduction

VATLY, the Vietnam Auger Training LaboratorY, is a cosmic ray laboratory located in Hanoi[a] and associated with the Pierre Auger Observatory (PAO) in Argentina.[1] Its aim is to establish in Vietnam a research team of international stature and to promote astrophysics in the country.

The first measurements,[2-9] of the atmospheric muon spectrum in Hanoi, contributed useful data to the understanding of atmospheric neutrino oscillation experiments.[10] They show a strong east-west asymmetry due to the effect of the Earth magnetic field which is particularly important in Hanoi where the rigidity cutoff reaches 17 GeV.

Since then, VATLY has worked in close connection with the PAO, using data collected both there and at the Institute from equipment installed on the roof of the laboratory. These studies are summarized below, following a brief presentation of the PAO and of its physics.

2. The PAO

High energy cosmic rays are observed from the extensive air showers that they produce when entering the Earth atmosphere. One method consists in sampling the particle density on ground. another method consists in detecting the fluorescence light produced on nitrogen molecules along the shower axis. In both cases timing gives the direction of the shower induced by the primary cosmic ray and intensity

---

[a]Information and documentation concerning VATLY is available on http://www.inst.gov.vn/inst/English/About/VATLY/Vatly.htm

gives its energy, however with very different sources of systematic errors. The PAO[1] is the first large hybrid detector ever built: it combines the strengths of surface detector arrays and of air fluorescence detectors. It measures accurately ultra high energy (UHE) cosmic rays at the very end of their energy spectrum, from 10^{18} eV up. It aims at understanding the nature of the primaries – protons, heavier nuclei or else – at identifying possible sources – the product of their size by their magnetic field must be exceedingly large[11] to contain the primaries in the acceleration phase – and to observe the effect of photoproduction on the cosmic microwave background – the GZK effect.[12] Recent Auger publications deal with the anisotropy of UHE sources and their correlation with nearby active galactic nuclei[13,14] and with the GZK cut-off.[15]

The giant surface detector array of the PAO is made of water Cherenkov counters and covers 3000 km^2 with a triangular grid having a 1.5 km mesh size (1600 detectors of 10 m^2 area each). A 10^{20} eV shower involves 15 to 20 such detectors. Typical angular and energy resolutions are 1.5^o and 20%. The energy measurement relies on the dependence of the measured signals on the distance to the shower axis. The installation will be completed by the end of the year.

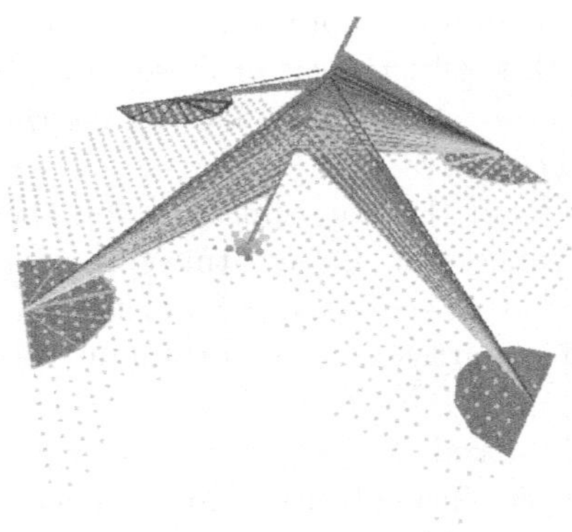

Fig. 1. The first event seen by all eyes and by the surface array was recorded on May 21st, 2007. The segments of the shower seen by each of the eyes and the water Cherenkov detectors hit are indicated.

The air fluorescence detector includes four stations of six eyes each, each eye covering a field of view of $30^o \times 28^o$ with a mirror focusing on an array of 22×20 pixels (photomultiplier tubes), each having 1.5^o aperture. They measure the induced fluorescence of nitrogen molecules in the near UV. Figure 1 shows the first event seen by all four eyes together with its trace in the surface detector array.

3. Studies using laboratory equipment

Studies using laboratory equipment aim at getting familiar with the behavior of the PAO detectors. They are centered on the operation, upgrading and study of a replica of one of its Cherenkov counters which was constructed on the roof of the laboratory (Figure 2).

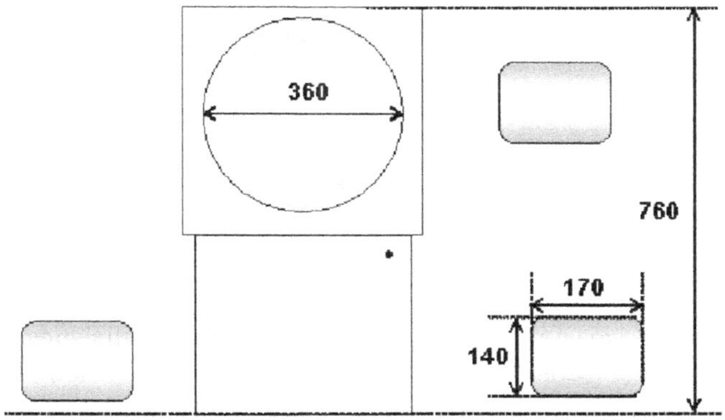

Fig. 2. Plan view of the set of Cherenkov counters installed on the roof of VATLY. The larger counter is a replica of a PAO counter. The smaller counters are horizontal cylinders. Their coincidence provides an unbiased trigger for the larger counter.

A study of its response to muons was performed[16,18] using a movable scintillator hodoscope located below it in the laboratory. The signals measured in each phototube were observed to be essentially proportional to the length of the muon track in water. Their distributions for vertical traversing muons was however wider, and its average value lower, than in the PAO (by a factor 4 in terms of light intensity). We have now improved[19] the water purity (new water filtering system) and the quality of the wall surface (aluminized mylar coating) in order to reach this goal. Moreover PAO photomultiplier tubes[b] are now used with a larger photocathode having a significantly better efficiency. Data collected with an independent trigger on extensive air showers have given evidence for a major improvement[20,21] of the optical quality of the counter.

Three additional smaller water tanks (3000 liters each), each seen by two photomultiplier tubes, have been installed[19] around the PAO Cherenkov counter in order to detect extensive air showers. The system is now operational and provides a trigger on vertical showers having energies in the 200 GeV range with a triple coincidence rate of 0.1 Hz and a core particle density of some 3 to 4 m^{-2}.[20,21] The signals in the trigger counters and in the larger counter are strongly correlated (Figure 3).

[b]Manufactured by Photonis, Brives-la-Gaillarde, France; we are indebted to Dr Pascal Lavoute (Photonis) and to the Auger Collaboration for a gift of 12 such photomultiplier tubes.

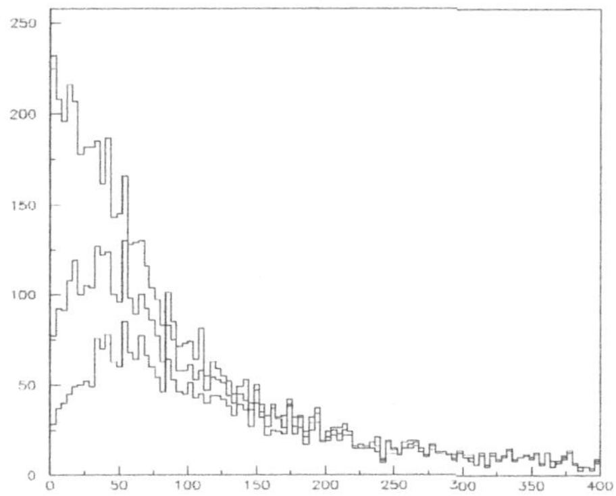

Fig. 3. Distributions of the signals recorded in the larger counter (ADC counts) for increasing signals in the trigger counters. They increase by 38% when the trigger rate is halved.

Finally, laboratory measurements of the photocathode efficiency of the PAO Photonis photomultiplier tubes are made in Hanoi in collaboration with a French PAO team (IPN Orsay) with the aim to understand its detailed dependence on impact point and other parameters of relevance.

4. Studies using PAO data

A simulation program has been written in order to understand in simple terms specific features of the detection of UHE showers in a fluorescence detector. The PAO was used as an illustration. In addition to the difficulty of positioning accurately the shower in the shower-detector plane in the case of monocular observation (pointing to the importance of making binocular, or better hybrid observations) other effects were considered: the compatibility between fluorescence and surface detections, very good in the PAO design; the necessity to extrapolate the observed shower profile outside the field of view, implying a variety of requirements for a good energy measurement; the use of parameters such as the spans in time and in atmosphere thickness covered by the field of view as indicators of the quality of the measurements performed.[22,23]

Detailed studies of the PAO surface detector signals are being made. These are stored in flash analog to digital converters (FADC, 25 ns bin size). Their time distributions are sums of muon and electron-photon signals which should be disentangled in order to learn about the nature of the primaries. Each particle produces Cherenkov light in the counter starting abruptly and decaying exponentially with a decay time of some 75 ns (typically 14 diffusions of the light). A Monte Carlo simulation of the optical properties of the tank reproduces well this general behavior.[24] Moreover, light collected in the first 25 ns is not equally shared between the three photomultiplier tubes but displays a strong asymmetry correlated with the azimuth of the incoming particles and with their impact points (Figure 4a). This feature can

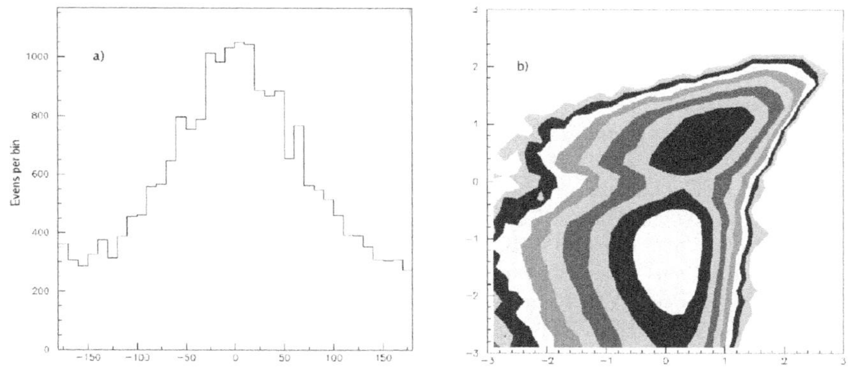

Fig. 4. a) Distribution of the angular difference (in degrees) between the azimuth of the shower and that of the PMT having the largest intensity in the first time bin of the FADC trace. b) The FADC trace, after deconvolution of the exponential decay of the light signals, is analysed for bins having a large content and a large asymmetry. To this effect two χ^2's are defined. One (abscissa) measures the significance of the asymmetry, the other (ordinate) the presence of a signal above noise. The lower ellipse corresponds to empty bins, the upper one to real signals, for which a clear correlation is visible.

be used to tag the start of individual signals in the FADC traces to help disentangling their pattern. A strategy along these lines is currently being elaborated and gives encouraging results (Figure 4b)

5. Teaching and training

VATLY has been actively contributing to the teaching of particle physics and astrophysics in the Ha Noi universities (national and pedagogic) and the Institute of Physics.

Modern astrophysics lectures were delivered for the first time to Vietnamese university students in 2004. VATLY is providing supervision for graduation, master and PhD theses in collaboration with the Hanoi Universities and with the Institute of Physics and Electronics.[25] The PhD program is under joint supervision with western universities yielding to a PhD degree in both Vietnam and the foreign country.

Acknowledgments

We are indebted to CERN, RIKEN, the World Laboratory, the EU Asia-Link Program CN/ASIA-LINK/008 (94791) the French CNRS and the Rencontres du Vietnam for invaluable material and financial contributions. We deeply acknowledge major support from the basic research program of the Ministry of Science and Technology of Vietnam. We are particularly grateful to the Pierre Auger Collaboration for their constant interest and support in the development of our group.

References

1. The Auger Collaboration, J. Abraham *et al.*, *Nucl. Inst. Meth. A* **523**, 50 (2004). A. Watson, *Highlights from the Pierre Auger Observatory*, presented at the 30th ICRC, Merida, Mexico, 2007.
2. P.N. Dinh *et al.*, *Nucl. Phys. B* **627**, 29 (2002).
3. P.N. Dinh *et al.*, *Nucl. Phys. B* **661**, 3 (2003).
4. P.N. Diep *et al.*, *Com. Phys. Vietnam* **14**, 56 (2003).
5. P.N. Dinh *et al.*, *Nucl. Phys. A* **722**, 439 (2003).
6. P.N. Diep *et al.*, *Nucl. Phys. B* **678**, 3 (2004).
7. P.N. Diep *et al.*, *Com. Phys. Vietnam* **15**, 55 (2005).
8. D.Q. Thieu *et al.*, Proc. Vth Renc. Vietnam (2004), The Gioi Pub., 2005, p.225.
9. P.N. Diep *et al.*, Proc. IXth Asia Pacific Phys. Conf., Hanoi (2004), The Gioi Pub., 2005, p.87.
10. SuperKamiokande Collaboration, *Phys. Rev. Lett.* **85**, 3999 (2000). M. Honda *et al.*, Proc. 2003 ICRC, Tsukuba, Japan, p.1415.
11. A.M. Hillas, *Ann. Rev. Astron. and Astrophys.*, **22**, 425 (1984).
12. K. Greisen, *Phys. Rev. Lett.* **16**, 748 (1966). G.T. Zatsepin and V.A. Kuzmin, *JETP Lett.* **4**, 78 (1966).
13. Pierre Auger Collaboration, J. Abraham *et al.*, *Astropart. Phys.* **27**, 244 (2007).
14. Pierre Auger Collaboration, to appear in Science, 2007.
15. T. Yamamoto *et al.*, *The UHECR spectrum measured at the Pierre Auger Observatory and its astrophysical implications*, presented at the 30th ICRC, Merida, Mexico, 2007.
16. P.T.T. Nhung, *Performance studies of water Cherenkov counters*, master thesis, Vietnam University of Sciences, Hanoi, 2006.
17. P.N. Diep *et al.*, Frontier of Basic Science, H. Takabe *et al. ed.*, Proc. Osaka-Hanoi Forum (2005), Osaka Univ. Press, 2006.
18. P.N. Diep *et al.*, Proc. VIth Renc. Vietnam, 2006, to be published.
19. P.N. Dong, *The Cherenkov counters of the VATLY laboratory*, master thesis, Hanoi University of Technology, 2006.
20. N.T. Thao, *The detection of extensive air showers in Hanoi*, master thesis, Vietnam University of Sciences, Hanoi, 2007.
21. P.N. Diep *et al.*, *Extensive air showers in Hanoi*, subm. for pub. in Comm. Phys. Vietnam, 2007.
22. P.N. Diep *et al.*, *Com. Phys. Vietnam*, **16**, 129 (2006).
23. P.N. Diep, *Detection and study of ultra high energy cosmic rays*, VAST, Institute of Physics and Electronics, master thesis, Hanoi, 2006.
24. D.T. The, *Optical properties of a Cherenkov counter*, Hanoi National University of Education, Hanoi, 2007.
25. In addition to the theses quoted above, one may mention:
D.Q. Thieu, *Thi Nghiem Do Phan Bo Goc Thong Luong Muon Khi Quyen Tai Ha Noi*, PhD thesis, Vietnam Atomic Energy Commission, 2006.
N.H.. Duong, *Khao Sat Do Bat Doi Xung Dong Tay Cua Thong Luong Muon Tai Ha Noi*, master thesis, Vietnam University of Sciences, Ho Chi Minh city, 2004.
K.T. Phuong, *The black hole in the centre of the Milky Way*, Vietnam University of Sciences, Hanoi, 2006.

STUDY OF THE ASTROPHYSICAL REACTION $^{13}C(\alpha,n)^{16}O$ BY THE TRANSFER REACTION $^{13}C(^{7}Li,t)^{17}O$

F. HAMMACHE*, M.G. PELLEGRITI, P. ROUSSEL, L. AUDOUIN, D. BEAUMEL, S. FORTIER

IPN-Orsay, IN2P3-CNRS, Université Paris XI, Orsay, 91406, France
** E-mail: hammache@ipno.in2p3.fr*

J. KIENER, A. LEFEBVRE-SCHUHL, V. TATISCHEFF

CSNSM, IN2P3-CNRS, Université Paris XI, Orsay, 91405, France

F. DESCOUVEMONT

Physique Nucléaire Théorique et Physique Mathématique, ULB CP229, B-1050 Brussels, Belgium

L. GAUDEFROY

GANIL, IN2P3-CNRS, Caen, France

M. STANOUI

GSI, Postfach 110552, D-64220 Darmstadt, Germany

The reaction $^{13}C(\alpha,n)^{16}O$ is considered as the main neutron source for s-process in low-mass asymptotic giant branch stars. At low energies of astrophysical interest (190 keV corresponding to temperature of 10^8 K), the contribution of the subthreshold state 6.356 MeV of ^{17}O to the $^{13}C(\alpha,n)^{16}O$ cross section should be taken into account. However, the results of previous studies of this contribution lead to different conclusions.

Hence, to further investigate the effect of this resonance, we performed a new precise measurement of the alpha spectroscopic factor, S_α, of the 6.356 MeV state using the transfer reaction $^{13}C(^{7}Li,t)^{17}O$ at two different incident energies. A brief description of the experiment will be given. The measured angular distributions and the obtained spectroscopic factors will be presented as well as their impact on $^{13}C(\alpha,n)^{16}O$ reaction rate.

1. Introduction

Nearly half of the heavy elements observed in the Universe are produced by a slow sequence of neutron capture reactions, the so-called s-process. For this astrophysical process, two neutron sources are used in all evolution models: the reactions $^{13}C(\alpha,n)^{16}O$ and $^{22}Ne(\alpha,n)^{25}Mg$. For the AGB stars of 1-3 solar masses at low temperatures, only the $^{13}C(\alpha,n)^{16}O$ reaction is considered as the main neutron source for the s-process[1–3] . In these stars, ^{13}C is synthesized in the He intershell of the star by the CNO cycle $^{12}C(p,\gamma)^{13}N(\beta^+)^{13}C$. This leads to the formation of the so-called ^{13}C pocket on the top of the intershell region, where the neutrons necessary for the s-process, are produced through the burning reaction $^{13}C(\alpha,n)$. Hence, all the models describing the s-process nucleosynthesis in these AGB stars depend critically on the neutron flux from the $^{13}C(\alpha,n)^{16}O$ reaction and so on the cross section of this reaction which occurs in these stars at temperatures around 10^8 K, i.e around the Gamow peak E_{cm} ~190 keV.

As the cross section of $^{13}C(\alpha,n)^{16}O$ reaction decreases drastically when the incident α energy decreases, direct measurement of this reaction at this energy is extremely difficult. Direct measurements[4] have only been performed down to 270 keV (see Fig. 1) too far away from the energy range of interest around 190 keV. The estimation of the cross section of $^{13}C(\alpha,n)^{16}O$ reaction at the energy of astrophysical interest depends critically on the way extrapolations are made. First one has to extrapolate the measured cross sections from high energies and consider in the same time the likely effect of the subthreshold state at 6.356 MeV of the compound nucleus ^{17}O in this energy region. R-matrix extrapolations[5] , which consider a large value of 0.7 for the α spectroscopic factor (S_α) of the subthreshold state at 6.356 MeV lead to a rapid increase of the S-factor[a] when the energy decreases to zero (see curve A in Fig. 1), while they lead to constant extrapolated S-factor at low energies when they use a null S_α (see curve B in Fig.1).

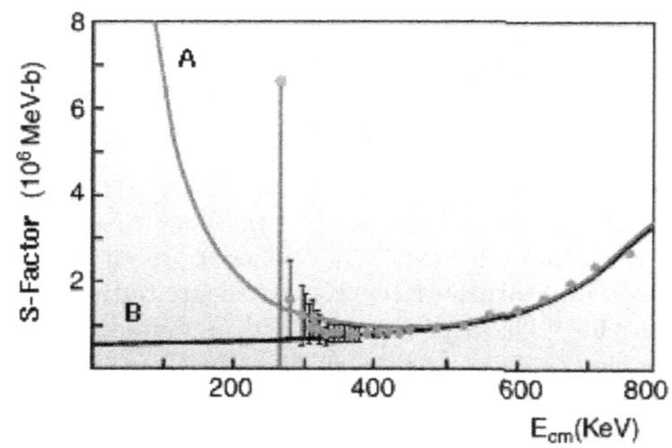

Fig. 1. $^{13}C(\alpha,n)^{16}O$ astrophysical S-factors as a function of center of mass energy. The data points are the direct measurements of Drotleff et al.[4] Curves A and B inferred from R-matrix calculations using an α spectroscopic factor for the subthreshold state at 6.356 MeV of about 0.7 and 0 respectively.

One has to note that all the astrophysical s-models use astrophysical S-factors based on α spectroscopic factors S_α ranging from 0.3 to 0.7. With these theoretically

[a]Note that because of the drop of the cross section at low energy due to the coulomb barrier, the astrophysical S-factor $S(E)=\sigma(E)\ E\ \exp(2\pi\eta)$ is more often used in nuclear astrophysics than the cross section itself for data representation(see Fig. 1).

predicted[6,7] values, ^{13}C(α,n)^{16}O reaction produce enough neutron flux to induce the s-process nucleosynthesis.

If the experimental points in Fig.1 suggest a rise of the S-factor when the energy decreases as expected with the usually used spectroscopic factors in the s-models ($S_\alpha \sim$ 0.3-0.7(curve A)), the error bars are too large to well discriminate curves A and B.

Kubono et al.[8] then suggested that a better way to determine the S-factor of ^{13}C(α,n)^{16}O at lower energies is via a measurement of the S_α of the subthreshold state at 6.356 MeV by an α transfer measurement to this state. They used ^{13}C(^{6}Li,d)^{17}O reaction at a laboratory energy of 60 MeV to determine the S_α values. Transfer angular distributions leading to different states of ^{17}O were precisely measured and the spectroscopic factor S_α of each state was determined from a finite-range DWBA analysis of the data. Their analysis gave a small value of $S_\alpha \sim 0.011$ for the 6.356 MeV $1/2^+$ state of ^{17}O which leads to a very small contribution of this subthreshold state to the cross section of ^{13}C(α,n)^{16}O at energies of astrophysical interest and thus to the S-factor. This result has dramatic effects on the neutron flux and subsequently on the onset of the s-process nucleosynthesis. However, a reanalysis of the same data by Keeley et al.[9] ends up to a drastically different conclusion since they found a value of S_α of about 0.4, forty times higher than Kubono's value and consistent with the values ($S_\alpha \sim$ 0.3-0.7) usually used in astrophysical calculations. In addition to these controversial results, the new determination of the contribution of the $1/2^+$ state to the total ^{13}C(α,n)^{16}O astrophysical S-factor, by the recent ^{13}C(^{6}Li,d)^{17}O ANC measurement,[10] was found 10 times smaller than Nacre[5] adopted value (based on S_α=0.1-0.7) but 5 times larger than the result of Kubono et al. at the energy of 190 keV.

Therefore, it appeared highly desirable to perform a new precise determination of this α spectroscopic factor through another transfer reaction ^{13}C(^{7}Li,t)^{17}O and an improved DWBA analysis. The use of ^{7}Li nuclei instead of ^{6}Li (Kubono et al.) to populate the various excited states of ^{17}O has some advantages:
(i) possible multi-step effects are less marked for the (^{7}Li,t) reaction than for (^{6}Li,d) one[11] .
(ii) transfer cross sections to low spin states are enhanced because of the non-zero α angular momentum in ^{7}Li leading to a better momentum matching. This was deduced from the study of the transfer reactions (^{6}Li,d) and (^{7}Li,t) on ^{12}C[12,13] .

Besides, it is useful to populate the same final states in ^{17}O using different transfer reactions. They involve different parameters such as internal spin, angular momenta and optical potentials and they allow to improve the systematic errors estimation introduced in DWBA analysis.

2. Experiment description

The experiment was performed using a ^{7}Li^{3+} beam provided by the Orsay TANDEM. Two self-supporting enriched ^{13}C targets, with a ^{13}C thickness of 0.072 $\pm$

0.004 mg/cm^2 and 0.133 $\pm$ 0.007 mg/cm^2 and an initial purity of about 90% were used. A ^{12}C target of 0.080 $\pm$0.004 mg/cm^2 was also used for calibration purposes and background subtraction. The absolute amount of ^{13}C and the final amount of ^{12}C in the ^{13}C target were deduced, respectively, from an α-energy loss measurement and by comparing the measured ratio between 28 MeV elastically scattered particles at 21° from the ^{12}C nuclei in ^{12}C and ^{13}C targets. The reaction products were analyzed with an Enge Split-pole magnetic spectrometer and detected at the focal plane by a 50 cm long position-sensitive gas chamber and a ΔE proportional gas-counter. The particle identification was made unambiguously using ΔE versus position measurements. The tritons were detected at angles ranging from 0 to 31 degrees corresponding to angles up to 43 degrees in the center of mass frame. Measurements at θ=0° were performed by stopping the beam in a thick absorber located inside the spectrometer. The overall energy resolution was about 70 keV in laboratory. The beam and ^{12}C amount were continuously monitored with a telescope of silicon detectors mounted inside the scattering chamber at θ_{lab}=35°.

Due to the presence of ^{12}C impurities, spectra coming from the ^{13}C(^{7}Li,t)^{17}O reaction were contaminated by the excited levels of ^{16}O, produced via the reaction ^{12}C(^{7}Li,t)^{16}O. So the (^{7}Li,t) reaction was measured on both ^{13}C and ^{12}C targets at each angle with the same setup.

3. Results

The experimental ^{13}C(^{7}Li,t)^{17}O differential cross sections measured for the 6.356, 3.055, 4.55 and 7.38 MeV, at the two incident energies of 34 and 28 MeV, are displayed in Fig.2a and Fig.2b, respectively. The data points displayed for 3.055 MeV in the 34 MeV left-column are Clark's measurements[14] at 35.5 MeV. The accuracy assigned to our measured cross sections includes the uncertainties on the peak yield, the number of target atoms, the solid angle and the integrated charge except for the zero degree run (no charge measurement) where the number of counts in the silicon monitor detector placed at 35° was used.

Calculations with finite-range DWBA method, using the FRESCO code[15] , were performed. For the triton channel, the optical model potentials used were taken from Garrett et al.[16] (POT I-f7/2 shell). Concerning the ^{7}Li channel, for the measurements at 34 MeV, several ^{7}Li optical potentials model were investigated within those given by Schumacher et al.[17] (Pot III) deemed as giving the best fit to the data without ad hoc parameter adjustment. Concerning the data at 28 MeV, we used both those from Schumacher et al. (Pot III, same as at 34MeV) as well as those provided from fitting our elastic scattering cross sections. The optical potential parameters finally selected are those giving the best fit for all the studied transitions in the (^{7}Li,t) reaction.

The adopted radius R=3.6 fm and diffuseness a=0.76 fm values provide a good fit of the angular distributions measured for the four studied states at both incident energies. The depth was adjusted to reproduce the binding energy of each

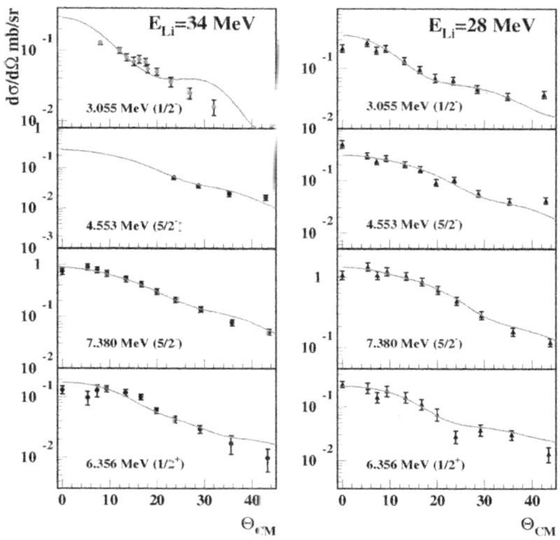

Fig. 2. Experimental differential cross sections of the ^{13}C(^{7}Li,t)^{17}O reaction obtained at 28 and 34 MeV, compared with finite-range DWBA calculations normalized to the data.

considered ^{17}O states and the number of radial nodes N were fixed by the oscillator energy conservation relation. The calculated angular distributions normalized to the data are shown in Figure 2. For both incident energies, the calculated curves agree fairly well with all the measured angular distributions of the different populated states. The α-spectroscopic factors were extracted from the normalization of the finite-range DWBA curves to the experimental data, $S_\alpha = \frac{\sigma_{exp}}{\sigma_{DW}}$. The spectroscopic factor for the overlap between α+t and ^{7}Li was taken to be 1.0. With this value, the α-spectroscopic factor of the reference state at 3.055 MeV is found to be S_α=0.26±0.06, similar to that obtained in nuclear model calculations S_α=0.25[18] . A close value was obtained by Keeley et al.[9]

The good agreement between the DWBA calculations and the measured differential cross sections of the different excited states of ^{17}O at the two bombarding energies of 28 MeV and 34 MeV respectively, gives strong evidence of the direct nature of the (^{7}Li,t) reaction populating these levels and confidence in our DWBA analysis. An S_α mean value of 0.29±0.11 is deduced for the state of interest at 6.356 MeV of ^{17}O, which is in good agreement with that obtained by Keeley et al.[9] and those used earlier (S_α ≈0.3-0.7) in the s-process models. The uncertainty on the extracted α spectroscopic factor for our state of interest (6.356 MeV) was evaluated from the dispersion of the deduced S_α values at the two incident energies and using different sets of optical potentials in the entrance[17] and exit channels[16] and different α-^{13}C well geometry parameters.

The α-reduced width γ_α^2 of about 13.5±4.3 keV for the 6.356 MeV state was obtained by using the expression $\gamma_\alpha^2 = \frac{\hbar^2 R}{2\mu} S_\alpha |\phi(R)|^2$[12] , where μ is the reduced mass and $\phi(R)$ is the radial part of the α-^{13}C cluster wave function calculated at R=7.5 fm. This radius was chosen in order to reach the coulomb asymptotic behavior of the radial part of the α-^{13}C wave function. The contribution of the $1/2^+$ state to the astrophysical S-factor when using our deduced γ_α^2 is shown in dashed curve in Figure 3.

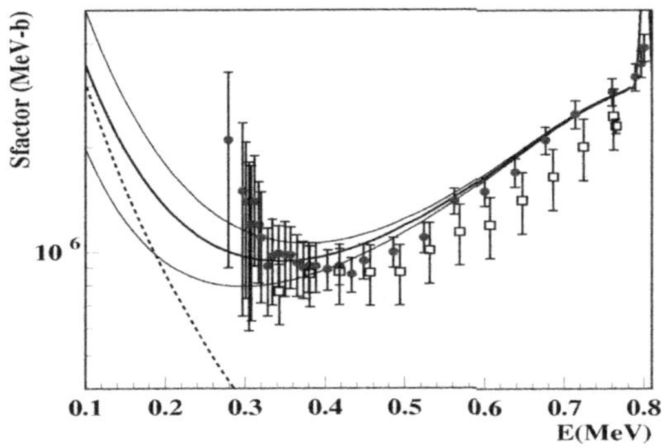

Fig. 3. Astrophysical S-factor for the ^{13}C(α,n)^{16} reaction with R-matrix calculations. The data are taken from Refs.[4,20] The contribution of the 6.356 MeV state is shown as the dashed curve. The thick curve correspond to our recommended γ_α^2 values, and the thin curves to the lower and upper limits.

Its value at 190 keV is 9.6×10^5 MeV-b, which is five times larger than in reference[10] . The present value for γ_α^2 (6.356 MeV) has been used to evaluate the ^{13}C(α,n)^{16}O S-factor at astrophysical energies. All ^{17}O resonances up to 7.38 MeV have been included in an R-matrix calculation. As the resonant structure near the $3/2^+$ resonance at 0.84 MeV is not well reproduced, we have readjusted the partial widths of this resonance assuming a non-resonant contribution. We find $\Gamma_\alpha = 0.09$ keV, and $\Gamma_n = 400$ keV. These results are slightly different from the recommended values[19] ($\Gamma_\alpha = 0.07$ keV, $\Gamma_n = 280$ keV), but are consistent with a ^{16}O+n measurement of Lister et al.[21] The resulting S-factor is shown in Fig.3 as a solid curve, with the uncertainty associated with the error bar on γ_α^2 (6.356 MeV). At the energy of astrophysical interest, E_{cm}=0.19 MeV, the contribution of the $1/2^+$ subthreshold state to the total S-factor is dominant ($\approx$ 70%).

Our calculated ^{13}C(α,n)^{16}O reaction rate, at temperature T=0.09 GK important for the s-process in low mass AGB stars, is found to be 1.3 times less than adopted in NACRE compilation; $5.41^{+1.8}_{-1.8}$ and $6.99^{+1.46}_{-4.96}$ $\times10^{-15}$cm^3 mol^{-1} s^{-1} respectively.

4. Conclusion

In summary, the reaction ^{13}C(α,n)^{16}O was investigated through the direct α transfer reaction (^{7}Li,t) at 28 and 34 MeV incident energies. The spectroscopic factor and hence the α-width of the ^{17}O subthreshold state at 6.356 MeV was deduced from a

DWBA analysis using different sets of selected optical and well parameters. These parameters were selected in order to fit the measured angular distributions of the various populated states at both incident energies. This procedure leads to a good evaluation of the uncertainties on S_α and α-reduced width. The α-reduced width was used in the calculation of the astrophysical S-factor and the $^{13}C(\alpha,n)^{16}O$ reaction rate at stellar temperatures. The latter was found 1.3 times lower than the NACRE adopted value but with a lower uncertainty. With our precise measurement the contribution of the $1/2^+$ state is found dominant at astrophysical energies and the error coming from it is no longer dominating the uncertainties in the total reaction rate.

References

1. R. Gallino et al., *Astrophys. J* **497**, 388 (1998)
2. M. Busso et al., *Annu . Rev. Astron. Astrophys* **37**, 239 (1999)
3. W. Aoki et al., *Astrophys. J. Lett* **536**, 197 (2000)
4. H. W. Drotleff et al., *Astrophys. J* **414**, 735 (1993)
5. C. Angulo et al., *Nucl. Phys. A* **656** 1 (1999)
6. G. M. Hale, *Nuclear physics A* **621**, 177 (1997)
7. P. Descouvemont, *Phys. Rev. C* **36**, 2206 (1987)
8. S. Kubono et al., *Phys. Rev. lett* **90**. 062501 (2003)
9. N. Keeley, K.W. Kemper, D T. Khoa, *Nuclear physics A* **726**, 159 (2003)
10. E. D. Johnson et al., *Phys. Rev. Lett* **97**, 192701 (2006)
11. K. Kemper et al., *Aust. J. Phys* **33**, 197 (1980)
12. F. D. Becchetti et al., *Nucl. Phys A* **305**, 293 (1978)
13. P. T. Debevec et al., *Phys. Rev. C* **9**, 2451 (1974)
14. M. E. Clark, K. W. Kemper, and J. D. Fox *Phys. Rev. C* **18**, 1262 (1978)
15. I.J. Thomson et al., *Comp. Phys. Rep* **7**, 167 (1988)
16. J. D. Garrett et al., *Nucl. Phys. A* **212**, 600 (1973)
17. P. Schumacher et al., *Nucl. Phys. A* **212**, 573 (1973)
18. H. Furutani et al., *Prog. Theor. Phys. Suppl* **68**, 193 (1980)
19. D. R. Tilley et al., *Nucl. Phys. A* **564**, 1 (1993)
20. C. R. Brune, I. Licot, R. W. Kavanagh, *Phys. Rev. C* **48**, 3119 (1993)
21. D. Lister et al., *Phys. Rev* **143**, 746 (1966)

SPIRAL2 AT GANIL: A WORLD LEADING ISOL FACILITY FOR THE PHYSICS OF EXOTIC NUCLEI

S.GALES

GANIL (DSM-CEA/IN2P3-CNRS)

Blvd. Henri Becquerel, F-14076 Caen cedex, France

During the last two decades, RIB has allowed the investigation of a new territory of nuclei with extreme N/Z called "terra incognita". Due to technical limitations, existing facilities were able to cover some part of the light mass region of this "terra incognita". The main goal of SPIRAL2 is clearly to extend our knowledge of the limit of existence and the structure of nuclei deeply in the medium and heavy mass region (A=60 to 140) which is to day an almost unexplored continent.

SPIRAL 2 is based on a high power, CW, superconducting driver LINAC, delivering 5 mA of deuteron beams at 40MeV (200KW) directed on a C converter+ Uranium target and producing therefore more 10^{13} fissions/s.

The expected radioactive beams intensities for exotic species in the mass range from A=60 to A=140, of the order of 10^6 to 10^{10}pps will surpass by two order of magnitude any existing facilities in the world. These unstable atoms will be available at energies between few KeV/n to 15 MeV/n. The same driver will accelerate high intensity (100μA to 1 mA), heavier ions up to Ar at 14 MeV/n producing also proton rich exotic nuclei.

In applied areas SPIRAL2 is considered as a powerful variable energy neutron source, a must to study the impact of nuclear fission and fusion on materials. The intensities of these unstable species are excellent opportunities for new tracers and diagnostics either for solid state, material or for radiobiological science and medicine.

The "Go" decision has been taken in May 2005. The investments and personnel costs amount to 190 M€, for the construction period 2006-2012. The project group has been completed in 2006, the construction of the accelerator started in the beginning of 2007 whereas detail design of the RIB production processes are underway. Construction of the SPIRAL2 facility is shared by ten French laboratories and a network of international partners. Under the 7FP program of European Union called "Preparatory phase for the construction of new facilities ", the SPIRAL2 project has been granted a budget of about 4M€ to build up an international consortium around this new venture.

Regarding the future physics program a call for Letter of intents has been launched in Oct 2006 .A very positive and enthusiastic response is the basis of new large international collaborations. Proposals for innovative new instrumentation and methods for the for SPIRAL2 facility are being examined by an International. Scientific Advisory Committee. The status of the construction of SPIRAL2 accelerator and technical R&D programs for physics instrumentation (detectors, spectrometers) in collaboration with EU and International partners will be presented.

1 Introduction

The search for a better understanding of nuclei and interactions between them as well as the way that matter is synthesized in the Universe, all depend crucially on our knowledge of the physics of the nucleus. In particular, the modern basic nuclear physics research is aiming to study exotic nuclei and the underlying physics, beyond the realm of the stable nuclei, even reaching the 'drip-lines' at the very edge of the nuclear chart. The physicists are seeking today to explore the entire chart.

New exotic shapes and excitation modes, halo-like and molecular structures, new modes of nuclear decay have been recently observed, while tests of fundamental symmetries, testing and refinement of the Standard Model of fundamental interactions, and exploration of the 'magic' numbers of protons and nuclei in very exotic nuclei are all enticing avenues of discovery. The study of radioactive nuclei, involved as they are in nucleosynthesis in the stars, leading to the creation of nuclei, atoms, molecules and all other complex structures our world is made of, has until now been strongly restricted by their short lifetimes and the limited production yields. By using new technologies, we are now able to produce accelerators and ingenious systems for producing beams of very exotic ions in quantities, which will permit their properties to be measured and understood. The quest for Rare Isotope Beams (RIB), which are orders of magnitude more intense than those currently available, is the main motivation behind the SPIRAL 2 project [1-6].

The importance of the RIB's has been underlined by NuPECC (Nuclear Physics European Collaboration Committee - an expert committee of the European Science Foundation) which recommend the construction of two new-generation complementary RIB facilities in Europe. One based on the in-flight fragmentation (IFF) as proposed for the FAIR facility at GSI (Darmstadt, Germany) [7] and the other on isotope-separation on-line (ISOL) method. In ISOL method, a thick target is bombarded with a primary light or heavy-ion beam or with a secondary neutron beam in case of fission targets. The produced radioactive nuclei are post-accelerated to energies from keV to tens of MeV. The ISOL method provides high-intensity beams of high optical quality comparable to stable beams.

The French Ministry of Research took a decision on the construction of SPIRAL 2 in the end of May 2005. Its construction cost, estimated to 190 M€ (including personnel and contingency), will be shared by French founding agencies CNRS/IN2P3 and CEA/DSM, the Region of Basse Normandie and the European partners. Funding from EU 7th framework and from others partnership countries are expected to contribute for about 20% to this budget. The construction will last about seven years (2006-2012) and operation of the facility will cost 8.5 M€ per year.

In October 2006, ESFRI (European Strategy Forum on Research Infrastructures) published a list of selected 34 Research Infrastructure projects corresponding to major needs of the European scientific community (all fields of science included) in the coming years. The FAIR and SPIRAL 2 are among the selected projects and were recommended as Research Infrastructures necessary to maintain Europe's position at the cutting-edge of world research.

As a result of the ESFRI selection process , under the 7FP program of European Union called "Preparatory phase for the construction of new facilities ", the SPIRAL2 project has been granted a budget of about 4M€ to build up an international consortium around this new venture.

The first beams are expected by the end of 2011, beginning of 2012. As an added bonus, the facility will also produce over 10^{15} neutrons per second, making it the world's most powerful source of fast neutrons for several years. SPIRAL 2 will reinforce

European leadership in the field of exotic nuclei and will serve a community of about 1000 scientists.

2 Description of the project

The quest for Rare Isotope Beams (RIB), which are orders of magnitude more intense than those currently available, is the main motivation behind the SPIRAL 2 project.

The new SPIRAL2 facility is a major improvement of the SPIRAL1 facility. More than 20 laboratories from 10 countries from 2001 to 2005 have developed the conceptual and the technical design of SPIRAL2. More than 200 scientists and engineers have contributed to the scientific case and the design of the facility [1,2].

The SPIRAL 2 facility (figure 1) is based on a high power, superconducting driver LINAC, which will deliver a high intensity, 40 MeV deuteron beam as well as a variety of heavy-ion beams with mass over charge ratio equals to 3 and energy up to 14.5 MeV/nucleon. Using a carbon converter, the 5 mA deuteron beam and a uranium carbide target, a fast-neutron induced fission is expected to reach a rate of up to 10^{14} fissions/s. The RIB intensities in the mass range from A=60 to A=140 will surpass by one or two order of magnitude any existing facilities in the world. A direct irradiation of the UC_x target with beams of deuterons, $^{3,4}He$, $^{6,7}Li$, or ^{12}C can be used if higher excitation energy leads to higher production rate for a nucleus of interest.

The neutron-rich fission RIB could be complemented by beams of nuclei near the proton drip line provided by fusion-evaporation or transfer reactions. For example, a production of up to $8x10^4$ atoms of 80Zr per second using a 200 μA $^{24}Mg^{8+}$ beam on a ^{58}Ni target should be possible. Similarly, the heavy- and light-ion beams from LINAG on different production targets can be used to produce high-intensity light RIB with ISOL technique.

The extracted RIB will be subsequently accelerated to energies of up to 20 MeV/nucleon (typically 6-8 MeV/nucleon for fission fragments) by the existing CIME cyclotron. Thus using different production mechanisms and techniques SPIRAL 2 would allow performing experiments in a wide range of neutron- and proton-rich nuclei far from the line of stability (figure 2).

Recent papers describing the technical and instrumentation developments related to the SPIRAL2 facility can be found in refs 8-11.

SPIRAL2 has also a remarkable potential for neutron-based research both for fundamental physics and various applications. In particular, in the neutron energy range from a few MeV to about 35 MeV this research would have a leading position for the next 10-15 years if compared to other neutron facilities in operation or under construction worldwide.

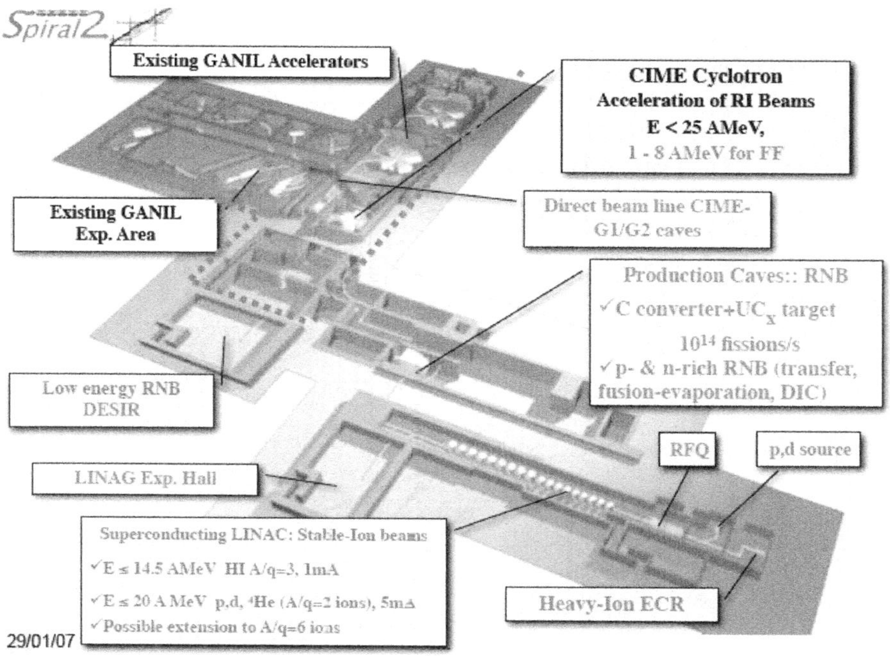

Figure1. Schematic layout of the SPIRAL 2 facility at GANIL.

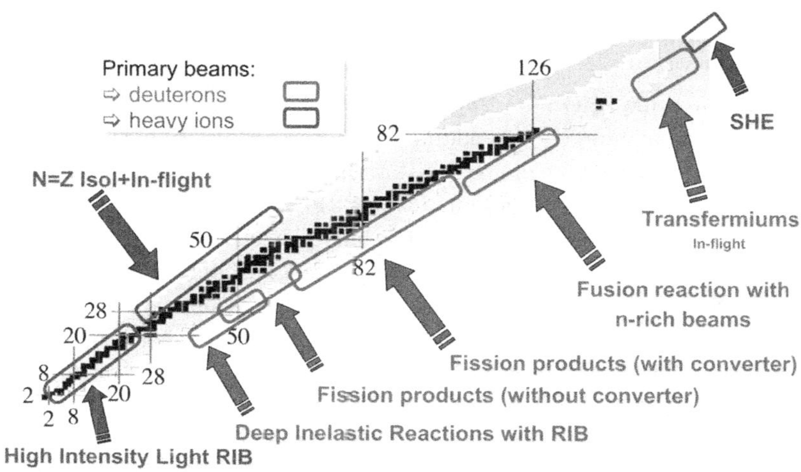

Figure 2. Regions of the chart of nuclei accessible for research of nuclei far from stability at SPIRAL2

3 Performances of the SPIRAL2 facility

3.1. *Intense stable beams from LINAG*

Very intense stable heavy ions beams will be available at the exit of the LINAG accelerator. These heavy ions (from He to Ca in a first step, then up to U), will reach intensities from 1pμA up to 1pmA using new PHOENIX ion sources in combination with A/Q=1/3 RFQ injector and later A/Q=1/6 for heavier ions (Kr to U) as illustrated in Fig.3

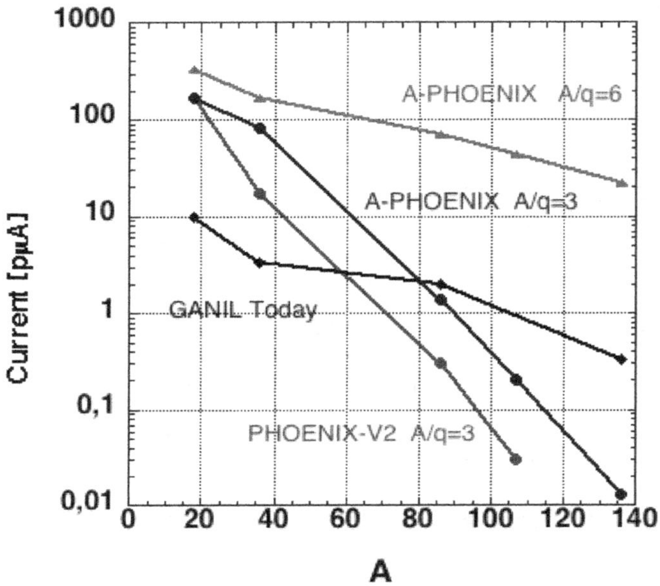

Figure 3. Expected intensities of stable heavy ion beams at the exit of the LINAG accelerator for various versions of PHOENIX ion sources.

The construction of the PHOENIX A/q=3 ion source has started at Grenoble (LPSC Grenoble-GANIL collaboration) [12-14)]. In addition a recent MoU has been signed between Argonne National Laboratory, Accelerator Division and GANIL-SPIRAL2 project. Under the agreed terms ANL has started the design of a new RFQ injector with A/Q= 6, devoted to the production of intense stable beams with mass higher than A= 40.

Reaching such beams intensities will open a wide range of physics experiments not accessible to day. Super Heavy Elements synthesis and spectroscopy, spectroscopy at or beyond the proton drip line, multinuclear transfer and deep inelastic scattering, study of mechanisms and reaction product distributions, production and study of isomers, ground state properties of rare nuclei.

A new device, S3 (for Super Separator Spectrometer) has been proposed for experiments with the very high intensity stable beams of SPIRAL2 coming from the LINAG accelerator [15-18].

Most of the physics cases listed above can be seen as either the study of rare processes or as the study of a secondary reaction with exotic ions produced in a first step reaction.

S3 shall be designed to answer most experimental needs of these physics cases. The crucial point of the device is to cope with the very high intensity beams provided by the LINAG accelerator. That means:

- the targets should be able to cope with very high power dissipation,
- the reaction products should be separated from the beam ions before they can be detected,
- the full optics of the spectrometer should reduce the background coming from scattered ions and activation products,

Meanwhile, the vast array of different physics that could be possible with S3 implies that it should be compatible with various detection settings .The basic design considered could be the following (see figure 4):

The primary beam will impinges on a primary target which should be made resistant to the high fluxes (rotating cooled targets, gas or liquid targets...)

The separation of the beam can proceed in two steps, the first being a high rejection device and an appropriate beam stopper.

After this first step, ions of interest could, if necessary, react on a secondary target. As they deal with moderate fluxes, this secondary target point could have gamma or particle prompt detection.

Next to this point, a secondary spectrometer helps with separation if very high purity is needed. It has some mass/momentum resolution and could rotate for reactions products at large angles.

The final focal plane can be equipped with the relevant detection system.

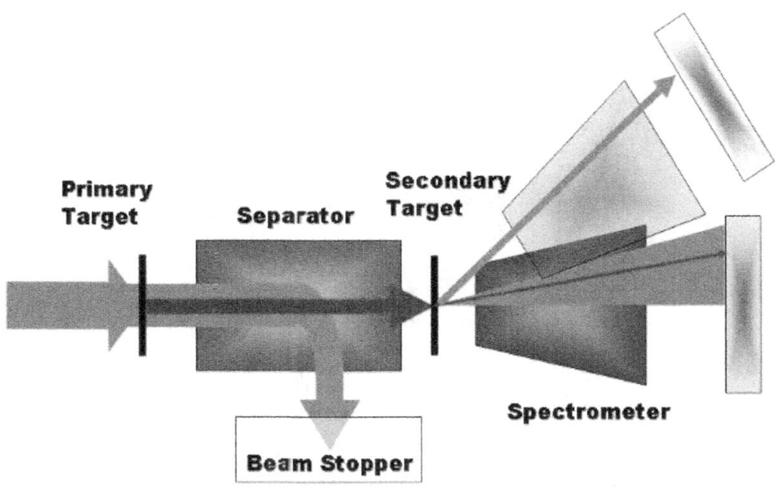

Figure. 4: Basic scheme for S3 design

3.2. *The radioactive ion production system*

The SPIRAL2 facility at GANIL will produce radioactive isotopes ranging from the lightest to very heavy elements beyond uranium. Different production mechanisms will be utilised. A lay out of the radioactive production vaults is shown in fig 5.

The radioactive neutron rich beams will be mainly produced via the fission process[19], with the aim of 5.10^{13} fissions/s .The retained solution is the fast neutrons induced fission in a depleted UCx target. The neutrons are generated by the beak-up of 40 MeV deuterons in a thick C target (converter). For the aimed high fission rate the converter should subtend a beam power of 200kW (see fig.1). With a length of 80mm and at about 40mm from the entrance of the converter, 5.10^{13} fissions/s can be reached in the target volume with an UC depleted target of $11 g/cm^{3}$. The resulting fission products accumulate in the target, diffuse from the surface from which they evaporate towards an ion source.

Few target-ion sources systems are being developed, often in the framework of international collaborations. Let us mention, an ECR 1+ "MONOBOB" for gaseous elements in test at GANIL[20], a FEBIAD,"IRENA" ion source well suited for short lifetimes at IPN Orsay[12], a Laser Ion source, particularly efficient for its selectivity and for metallic beams. An international collaboration around the Laser Ion Source will be built under the 7th European FP.

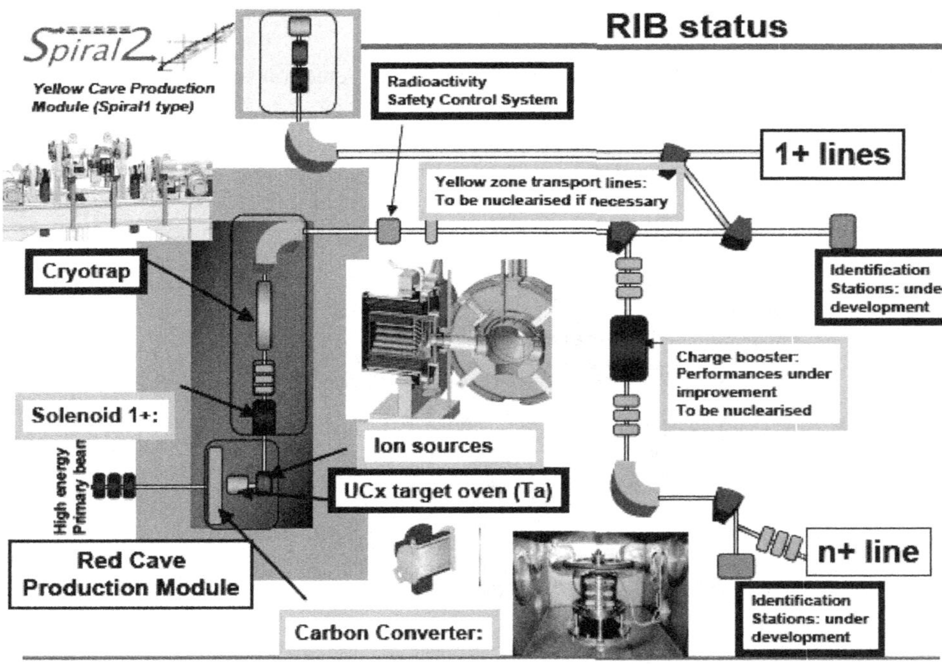

Fig5. Lay out of the radioactive production halls for the SPIRAL2 facility

For neutron rich fission products the whole area will be confined in a red zone (see fig 5) with only remote controls and maintenance. At the exit of the target ion source, Cryotraps are being developed in order to catch volatile gaseous atoms which are not ionized. Then, the 1+ ions are separated and charge bred in a ECR ion source in order to be accelerated by the existing CIME cyclotron up to 5-9 MeV/nucleon for masses A=70 to A=140.

The expected intensities (to be validated by measurements) of typical fission fragment beams of Kr and Sn isotopes from SPIRAL2 are compared to existing or planned RIB facilities in fig.6

For SPIRAL2 the RIB intensities in the mass range from A=60 to A=140 will be of the order of 10^6 to 10^{11} part/s. For example, the intensities should reach, 10^{9-10} part. /s for ^{132}Sn and 10^{11} pps for ^{92}Kr surpassing by two order of magnitude any existing facilities in the world (see fig.6)

Production of rare isotopes are also possible using a second target station (yellow cave production see fig.5) similar to SPIRAL1 cave where intense heavy-ion beams from LINAG can induce fusion evaporation reactions leading to rare isotopes for medium-mass, neutron-deficient nuclei. In addition, using transfer and deep-inelastic reactions, SPIRAL2 facility will produce intense secondary beams of light to medium mass nuclei closer to the line of stability. Examples of reactions leading to the production of light neutron rich and N=Z exotic nuclei are listed in Table.1 below.

4 Selected examples of the scientific opportunities at SPIRAL2

As discussed in previous section, the multi-beam driver LINAC will deliver intense heavy ion stable beams and may be dedicated to the synthesis of very heavy and Super-heavy nuclei. A new experimental Hall (AEL) will host the new S^3 "Super-Separator-Spectrometer" and a wide range of physics experiments not accessible today will be at reach.

In the same experimental hall a neutron –TOF facility is planned, thanks to the high energy and high intensity neutron flux available at SPIRAL2, the facility will offer a unique opportunity for material irradiations and cross-section measurements both for fission (notably accelerator driven systems (ADS) and Gen-IV fast reactors) and fusion related research, tests of various detection systems and of resistance of electronics components to irradiations, etc.

The DESIR collaboration[21)], formed after the SPIRAL2 workshop on low-energy physics at GANIL in July 2005, proposes the construction of an experimental facility to exploit the low-energy "exotic" beams from SPIRAL2. Experiments performed at other ISOL facilities (e.g., ISOLDE or ISAC) show that a high degree of purity is necessary for the ions beams. The most efficient and universal way of achieving isotopic ally pure beams is a high-resolution mass separator.

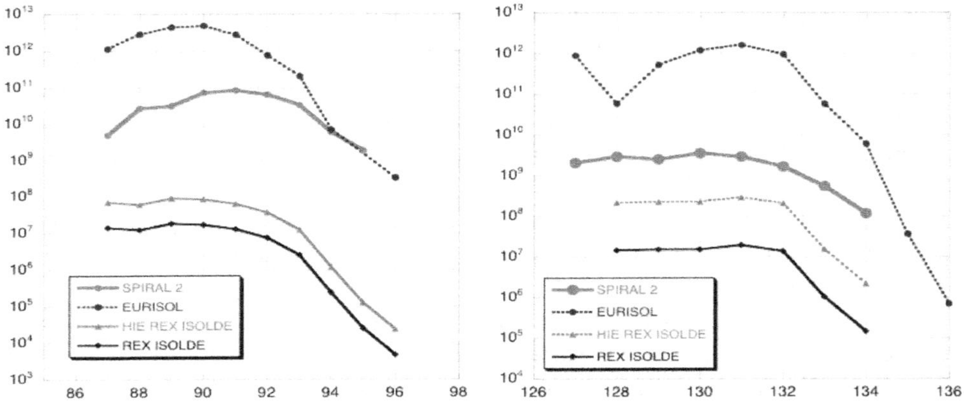

Fig.6 Expected intensities of Kr and Sn isotopes SPIRAL2 after acceleration compared to REX-ISOLDE present and future (HE-REX ISOLDE) and to EURISOL predictions.

Table.1 Calculated production yields for light and N=Z exotic nuclei using transfer or fusion evaporation reactions

RI Beam	Reaction	Production method	Yield (min. - max.) pps
6He	9Be(n,α)6He	ISOL	5x107 - 1012
11C	14N(p,α)11C	ISOL	107 - 3x1011
15O	15N(d,2n)15O	ISOL	3x107 - 1010
18Ne	19F(p,2n)18Ne	ISOL	6x106 - 7x109
34Ar	35Cl(p,2n)34Ar	ISOL	2x106 - 2x108
56Ni	58Ni(p,p2n)56Ni	Batch mode	2x104 - 108
58Cu	58Ni(p,n)58Cu	Batch mode	104 - 108
80Zn	24Mg+58Ni	In-flight	< 3x104

At SPIRAL2 and its low-energy beam facility DESIR, there is an opportunity of building three kinds of setup to try to cover the whole nuclear chart. A ß-NMR setup will be suitable to study precisely the moments of key cases and, in combination with collinear hyperfine spectroscopy, get unambiguous spin assignments. With a collinear spectroscopy setup, we will have the possibility to study series of isotopes in the intermediate and heavy mass regions. For the heavy elements, a double laser + RF spectroscopy in a Paul trap will be used to study the hyperfine anomaly and higher-order moments up to very high precision.

Shell structure of neutron-rich isotopes near the N=20 and 28 shell are to day the subject of intensive research all around the world. Shell quenching, weakening of the spin-orbit interaction, tensor component in the n-n force are investigated using the rich and powerful tools of nuclear reactions from nucleon transfer[22], in beam gamma spectroscopy [23], Coulomb excitation, fusion –evaporation and deep inelastic processes.

SPIRAL2 will be well suited to address these fundamental questions, thanks to the expected large coverage of RIB species, very high intensities and energy range covered by the facility. Two main regions of the mass chart in the vicinity of the N=50, and 82 shell closures are of special interest

a) The N=50 regions:

One important issue is to determine how thick can be the neutron skin for nuclei far from stability? From the measurement of the charge radius, one has access to the neutron radius. A key element would be germanium which is produced in large quantities in neutron stars. So the spectroscopy of this element and its neighbours (Zn, As, Ga, Se) over the magic shell N=50 using collinear laser spectroscopy and reactions will answer this question.

Another typical example of an element that has to be studied is of course nickel as close as possible to ^{78}Ni. Is this doubly-magic nucleus still spherical?

b) The N=82 regions:

A lot of nuclear models predict strong evolution of shapes and/or change of shell structure around neutron number N=60 and mass number A=110. Neutron skin is predicted to occur for the very neutron-rich Sn isotopes. Studying tin isotopes very far from stability will be a very stringent test for these nuclear models.

Nuclei were created during major events that made the evolution of the universe and are the witness of the origin of matter: while the lightest elements were created in the first minutes following the big bang, heavier ones are synthesized in stars. Understanding our world, its origin and its evolution is one goal of nuclear physics. However, these various nucleosynthesis processes involve radioactive nuclei. SPIRAL 2 will abundantly produce some of them opening new investigation field.

The future SPIRAL2 facility will contribute prominently to several areas of active research in nuclear astrophysics, such as explosive hydrogen burning, s-process and r-process nucleosynthesis, which are linked to astrophysical observations (novae, X-ray bursts, type 2 supernovae, etc...). For example, radiative capture cross-sections can be determined from spectroscopic factors (or Asymptotic Normalization Coefficients) obtained in transfer reaction or elastic resonant scattering.

The (d, p) reaction can be used to simulate the (n,γ) capture on medium-mass nuclei at SPIRAL2 near the r-process possible paths.

The way the energy of theses s or p states evolve from Z=50 down to Z=40 determines how neutron capture cross sections will evolve along the N=82 shell closure and how the r process synthesis of nuclei will be blocked at this shell closure. The neutron rich isotopes of Sn and Cd nuclei are excellent test cases. The neutron-capture through the soft Pigmy resonance could also play an important role if the energy of the soft pigmy vibration mode is peaked around the S_n value. The way this energy is shifted when moving from Sn to Cd isotopic is certainly an important issue to consider.

The scientific opportunities discussed above are just a few examples among many exciting avenues which can be investigated with SPIRAL2. New exotic shapes and excitation modes, halo-like and molecular structures, single particle properties and new magic numbers, new modes of nuclear decay have been recently observed, while studies of fundamental symmetries, testing and refinement of the Standard Model and of fundamental interactions in very exotic nuclei have all a huge potential of discovery.

5 Construction of the facility and International Collaborations

Since September 2005, a project construction group has been formed and the whole project has been organized around three main sections

- The driver accelerator
- The secondary radioactive beams (production and transport)
- The infrastructure (buildings, vacuum and cryogenics, power supplies…)

The project has been divided in each section into work packages.

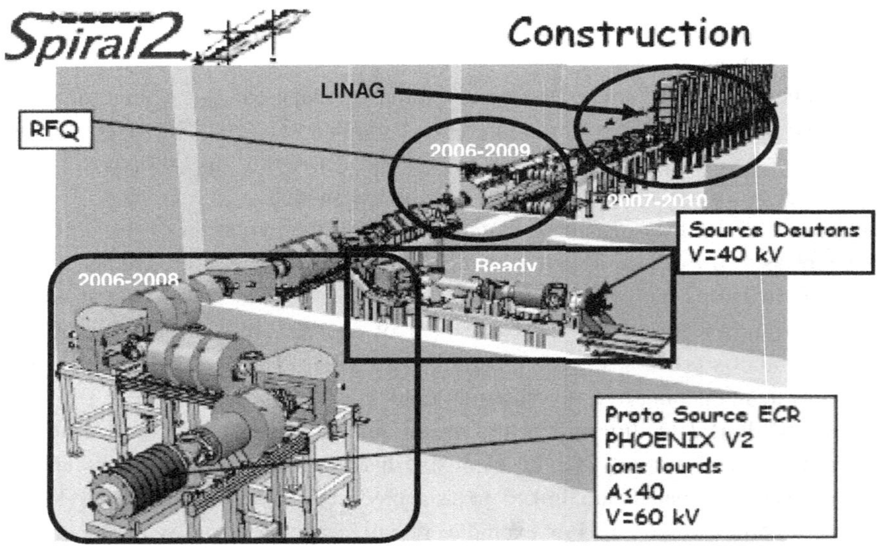

Figure.7 SPIRAL2 driver accelerator divided into main components

For the driver accelerator, the work packages are under the responsibility of IPN Orsay and DAPNIA/DSM/Saclay laboratories[24-27]. Figure.7 show the driver accelerator divided into main components and the associated target dates for completion. The R&D on prototypes' for the RFQ, SC cavities and RF couplers are completed and order of all driver components will be completed at the end of 2007.

A SPIRAL2-France consortium had been set to coordinate the efforts of 10 French nuclear physics laboratories from CNRS/IN2P3 and CEA. Moreover SPIRAL2 facility will be built by a large European and International collaboration. Several MoU have been already signed or are in final negotiation stages with JINR Dubna, Israel Romania, Isolde (CERN), Italy, UK ,Germany, Poland, Canada, Japan and US ,both for accelerator components (control system, beam loss monitors, beam lines) and/or construction of components of the RIB production units (neutron converter , beam lines, target-ions sources, laser ion sources, high-resolution magnetic separator ,....).

At the EU research organisation level, SPIRAL2 as well as FAIR have been selected by the ESFRI Committee and a proposal for the construction phase in the framework of the 7th FPRD has been recently granted 4M€ to coordinate the contributions of more than 20 Laboratories from EU, USA, Japan and India.

The process of formation of international collaborations around the SPIRAL 2 research programme has started in 2004 and pursued the next years with the organization of 5 specialised workshops held in 2004-2006 (Neutrons for science at SPIRAL 2, 13-14/12/2004,Physics with separated low energy beams at SPIRAL2, 4-5/07/2005,Future prospects for high resolution gamma spectroscopy at GANIL, 4-6/10/2005,Nuclear Astrophysics with SPIRAL2, 17-18/10/2005,SPIRAL2 Reactions, 19-21/10/2005).

In the beginning 2006 ,under the umbrella of SPIRAL2 Scientific Advisory Committee (SAC) a Call for letters of intent has been launched in March 2006 . A very positive and enthusiastic response from the International community has been obtained. 19 Letters of Intents (see refs. 1, 18), coming from 600 physicists and 34 countries have been received and evaluated by the SAC in October 2006.

After this evaluation, the formation of about 9 large international collaborations around major instruments is under progress .Next important step is the call for Technical proposals around the spring of 2008.

In terms of schedule for the new facility, the ambition is to start operation of the driver accelerator with intense stable beams before the end of 2011 and to complete the whole project including RIB production and post acceleration of fission fragments before the beginning of 2013.

6 Conclusion

Nuclear physics has been revolutionized by the recent development of the ability to produce accelerated beams of radioactive nuclei. For the first time it will be possible to study reactions to produce the 6000 to 7000 nuclei we believe exist rather than the 300 stable ones that nature provides. At the larger scale one wants to understand the limits of nuclear stability by producing exotic nuclei with vastly different numbers of neutrons and protons.

SPIRAL2 is major expansion of the SPIRAL facility at GANIL, which will help maintain European leadership in ISOL (Isotope separation on line) development, aiming at two orders of magnitude increase of the secondary beams available for nuclear physics studies.

The technical challenges of the acceleration, targetry and experimental equipments will provide essential knowledge on the road toward EURISOL[28)], the most advanced nuclear physics research facility presently imaginable and based on the ISOL principle. It is expected, that the realization of SPIRAL2 will substantially increase the know-how of technical solutions to be applied not only for EURISOL but also in a number of other European/ world projects. The scientific program, proposes the investigation of the most challenging nuclear and astrophysics questions aiming at the deeper understanding of the nature of matter.

SPIRAL2 will contribute to the physics of nuclear fission and fusion based on the collection of unprecedented detailed basic nuclear data, to the production of rare radioisotopes for medicine, to radiobiology and to material science.

References

1. www.ganil.fr/research/developments/spiral2/
2. "The scientific objectives of the SPIRAL2 project" *rapport WB_SP2_Final-1.pdf* June 2006. http://www.ganil.fr/research/developments/spiral2/index.html and Letter of Intention therein.
3. W. MITTIG et al., "LINAG Phase1", *internal report GANIL* (2002).
4. M.-G. SAINT-LAURENT et al. "SPIRAL phase II, European RTT" R0103 *internal report GANIL* (2001).
5. M.-H. MOSCATELLO, "SPIRAL2 at GANIL" *Seventeenth International Conference on the cyclotrons and their applications*, Tokyo, 2004 October 18-22, p. 518, Particle Accelerator Society of Japan (2005).
6. T. JUNQUERA et al., "High Intensity Linac Driver for the SPIRAL-2 Project : Design of Superconducting 88 MHz Quarter Wave Resonators (beta 0.12), Power Couplers and Cryomodules" *9th European Particle Accelerator Conference (EPAC 2004)* Luzern, July 2004, TUPLT058, p. 1285, European Physical Society Accelerator Group (EPS-AG) (2004).
7. http://www.gsi.de/fair/reports/btr.html FAIR Baseline Technical report (Sept 2006)
8. M. LEWITOWICZ, "The SPIRAL 2 project" *Tours Symposium on Nuclear Physics VI*, Tours (FR) 5th-8th September 2006, Vol. 891, p. 91 AIP Conference Proceedings (2007).
9. M. JACQUEMET, E. PETIT et al., "Status report on the SPIRAL2 project", *XVth International Conference on Electromagnetic Isotopes Separators and Techniques related to their applications EMIS 2007*, Deauville, 24^{t}h-29th June 2007, to be published.
10. S. GALES, "SPIRAL2 at GANIL : a world leading ISOL facility at the dawn of the next decade" *International Workshop on Nuclear Physics 28th Course radioactive Beams, Nuclear Dynamics and Astrophysics*, Erice (IT), September 16th-24th 2006, Vol. 59, p. 22, Progress in Particle and Nuclear Physics (2007).
11. M.-G. SAINT –LAURENT , " Future opportunities with SPIRAL2 at GANIL " to be published

12. M. CHEIK MHAMED et al., "Development of a plasma ion source for next-generation facilities" *Review of Scientific Instruments* **77**, 03A702 (2006).
13. T. LAMY et al "Status of charge breeding with electron cyclotron resonance in sources" *Review of Scientific Instruments* **77**, 03B101 (2006).
14. T. THULLIER et al., "A-Phoenix, an electron cyclotron resonance ion source or the SPIRAL2 facility" Caen, 2005 September 12-16, Vol. 77, p. 03A323 *Review of Scientific Instruments* (2006).
15. A.C.C VILLARI et al., "S3: the Super Separator Spectrometer for LINAG" *International Symposium on Exotic Nuclei* Khanty-Mansiysk (RU) July 17-22 j 2006, AIP Conference Proceedings **912**, p. 436 (2007).
16. A. DROUART et al., "Design Study of a Pre-Separator for the LINAG Super Separator Spectrometer" *XVth International Conference on Electromagnetic Isotope Separators and Techniques related to their Applications (EMIS2007)*, France (2007, June 24th-27th), to be published.
17. G. SAVARD et al., "Radioactive beams from gas catchers: RIBS with and without driver", *XVth International Conference on Electromagnetic Isotopes Separators and Techniques related to their applications.* Deauville, 24[t]h-29th June 2007, to be published.
18. S^3 Collaboration LOI-SPI2-03 , http://www.ganil.fr/research/developments/spiral2/loi_texts.html
19. M. FADIL et al., "About the production rates and the activation of the uranium carbide target in SPIRAL2", *XVth International Conference on Electromagnetic Isotopes Separators and Techniques related to their applications EMIS 2007*, Deauville, 24[t]h-29th June 2007, to be published.
20. C. HUET-EQUIBEC et al., *Nuclear Instruments and Methods in Physics Research Section B*, Volume **240**, Issue 3, p. 752 (2005).
21. DESIRCollaborationLOI-SPI2 ,see ref 18)
22. L.GAUDEFROY et al., *"Reduction of the spin-orbit splittings at the N=28 shell closure"* Phys. Rev. Lett. **97**, 092501 (2006)
23. B.BASTIN et al., *"Collapse of the N=28 shell closure in ^{42}Si"*, Phys. Rev. Lett. **99**, 022503 (2007)
24. R. FERDINAND et al., "SPIRAL2 RFQ prototype - First results" *10th European Particle Accelerator Conference (EPAC'06),* Edinburgh, 26-30 June 2006, MOPCH103 European Physical Society Accelerator Group (EPS-AG) (2006).
25. P.E.BERNAUDIN et al., "Design of the Low-Beta, quarter-wave resonator and its cryomodule for the for the SPIRAL 2 project " *9th European Particle Accelerator Conference (EPAC 2004)* Luzern, July 5-9 2004, TUPLT054, p 1276, European Physical Society Accelerator Group (EPS-AG) (2004).
26. G. OLRY et al., "Status of the beta 0.12 superconducting cryomodule development for SPIRAL2 project" *Proceedings of EPAC 2006, Edinburgh*, 26-30 June 2006, MOPCH146, European Physical Society Accelerator Group (EPS-AG) (2006).
27. Y. GOMEZ MARTINEZ et al., "Theoretical study and experimental result of the RF coupler prototypes of Spiral2" *Proceedings of EPAC 2006, Edinburgh*, THPCH160, European Physical Society Accelerator Group (EPS-AG) (2006).
28. http://www.eurisol.org/site01/index.php

MAGNETIC PROPERTIES OF LIGHT NEUTRON-RICH NUCLEI AND SHELL EVOLUTION

TOSHIO SUZUKI[1,2]

[1] *Department of Physics, College of Humanities and Sciences, Nihon University, Sakurajosui 3-25-40, Setagaya-ku, Tokyo 156-8550, Japan*
[2] *Center for Nuclear Study, University of Tokyo, Hirosawa, Wako-shi, Saitama 351-0198, Japan*
E-mail: suzuki@chs.nihon-u.ac.jp

TAKAHARU OTSUKA

Department of Physics and Center for Nuclear Study, University of Tokyo, Hongo, Bunkyo-ku, Tokyo 113-0033, Japan, and RIKEN, Hirosawa, Wako-shi, Saitama 351-0198, Japan

A new shell model Hamiltonian for p-shell nuclei which properly takes into account important roles of spin-isospin interactions, especially the tensor force, is shown to significantly improve the description of magnetic moments and Gamow-Teller transitions of p-shell nuclei as well as neutrino induced reactions on ^{12}C. The Hamiltonian is further modified to take full account of the roles of tensor and two-body spin-orbit interactions, and applied to study shell evolutions toward drip-lines and magnetic and spin properties of light neutron-rich nuclei, where the effects of the spin-isospin interactions become essential and more pronounced. Anomalous behavior of magnetic dipole transitions in ^{17}C and electric quadrupole transitions in ^{16}C and ^{12}Be recently observed are found to be rather well explained.

Keywords: shell model, tensor and spin-orbit interactions, neutron-rich nuclei, Gamow-Teller and electromagnetic transitions, neutrino-nucleus reactions, nucleosynthesis

1. Introduction

We have modified a shell model Hamiltonian for $p - sd$ shell by taking into account important roles of spin-isospin interactions. Magnetic and spin properties of p-shell nuclei have been significantly improved.[1] A general important role played by the tensor force for the proper shell evolution is pointed out.[2] Monopole terms of the tensor components of the new Hamiltonian, SFO,[1] have proper signs, that is, attractive (repulsive) between $j_{>}=\ell+1/2$ and $j_{<}=\ell$-$1/2$ ($j_{>}$ and $j_{>}$, or $j_{<}$ and $j_{<}$) orbits, though their magnitude is small compared with that of the $\pi + \rho$ meson exchange potential (see Fig. 1). We will further modify the SFO Hamiltonian later by taking full account of the tensor interaction (see sect. 2). Gamow-Teller transition strength in ^{12}C is enhanced compared with that obtained by the Millener-Kurath Hamiltonian, PSDMK2,[3] and comes closer to the experimental value.[4] We show calculated Gamow-Teller strength for ^{14}N $\rightarrow$ ^{14}C as one of the examples of the improvements realized by the SFO Hamiltonian in Fig. 1 (right figure). The almost vanishing $B(GT)$ value of the transition to the ground state of ^{14}C is nicely reproduced. The splitting of the strength into the three 2^+ states of ^{14}C observed at KVI and RCNP[6] is also reproduced in contrast to the case of ref.[7]

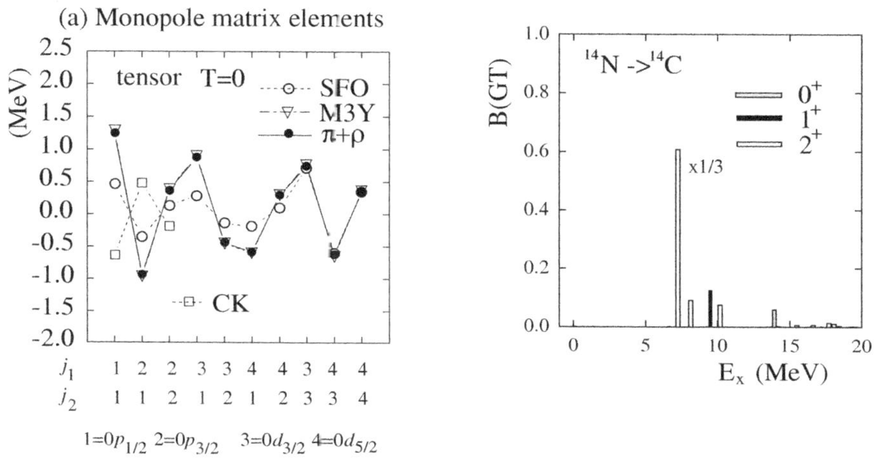

Fig. 1. (a) Monopole terms of the tensor components of the SFO and Cohen-Kurath interactions as well as the $\pi + \rho$ exchange potential and the M3Y[5] interaction. (b) Gamow-Teller strength for $^{14}\mathrm{N} \to {}^{14}\mathrm{C}$ obtained by the SFO Hamiltonian (right figure).

Neutrino induced reactions on ^{12}C and ^{4}He have been studied based on the new shell model Hamiltonians.[8] The reaction cross sections are found to be enhanced compared with those obtained by the conventional Hamiltonians and previous calculations[9] (see Fig. 2 for supernova neutrinos). The Gamow-Teller and spin-dipole transitions are the dominant contributions in ^{12}C while in ^{4}He most of the contributions are due to the spin-dipole transitions. A large quenching of the axial-vector coupling constant g_A is taken for the excitations of 2^- states in ^{12}C. Quenching factors for g_A used are 0.7 and 0.9 for 2^- and other multipoles, respectively, in ^{12}C, which are taken to be consistent with electron scattering experiments.[10,11] The longitudinal part of the 1^- form factor is also quenched by multiplying a factor of 0.7.[12]

The enhancement of the cross sections leads to that of the production yields of ^{7}Li and ^{11}B during supernova explosions[8] through the nucleosynthesis path shown in Fig.3 (see Fig. 4 also). We point out that in case of ν oscillations the yield ratio of ^{7}Li over ^{11}B is found to be sensitive to the ν mixing angle θ_{13} for the normal mass hierarchy, which can be used to determine the lower limit of θ_{13} as well as the neutrino mass hierarchy whether it is normal or inverted.[13]

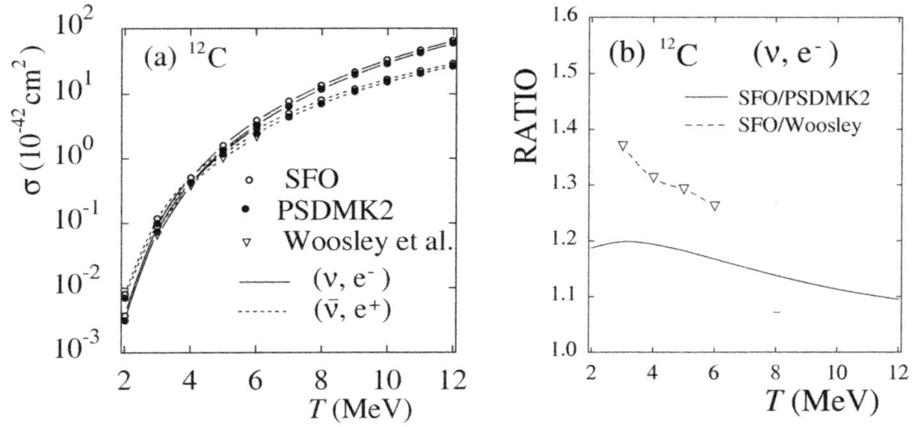

Fig. 2. Comparison of calculated charge-exchange ν-^{12}C reaction cross sections induced by supernova neutrinos with temperature T. (a) Results for the SFO, PSDMK2 and previous one[9] are shown. (b) Ratios of the cross sections are also shown.

2. Properties of Neutron-Rich Nuclei

2.1. *Effects of Two-Body Spin-Orbit Interaction*

We study nuclei near drip lines, where the effects of tensor interaction on shell evolutions become more pronounced. We first point out the important role of the two-body spin-orbit interaction in particular orbits, that is, between the p and s orbits. The monopole matrix elements of the two-body spin-orbit force are shown in Fig. 5 for the M3Y interaction,[5] that is due to the exchanges of σ, ω and ρ mesons, and the Millener-Kurath (MK) interaction.[3] The enhancement of the monopole term in the $0p$ and $1s$ orbits is found for the M3Y interaction (see Fig. 5). The enhancement is caused by the shorter range nature of the interaction compared with the MK interaction, and is important for lowering the $1/2^+$ state relative to the $1/2^-$ state, for example, in ^{11}Be.

2.2. *Full Inclusion of Tensor and Spin-Orbit Components in the Hamiltonian*

Now, we improve the SFO Hamiltonian further by taking full account of the important roles of the tensor and the two-body spin-orbit forces. The tensor and the spin-orbit components of the interaction are replaced by those of the $\pi + \rho$ meson exchange potential and the M3Y interaction, respectively, while the central components of the interaction are kept as they are.

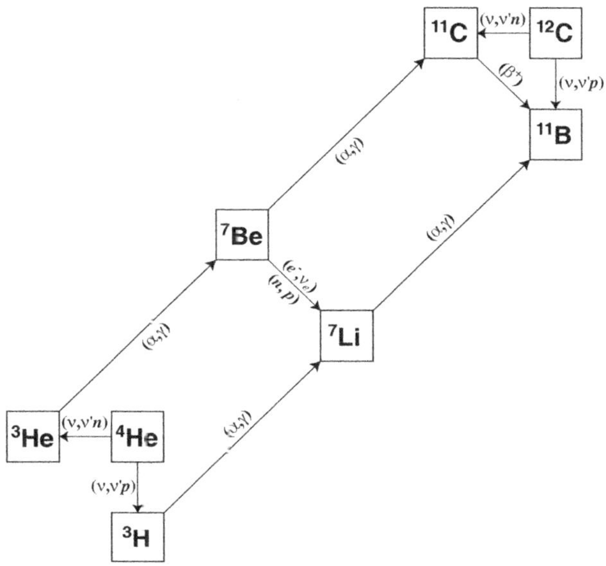

Fig. 3. Nucleosynthesis path of light elements, ^{7}Li and ^{11}B, during supernova explosions.

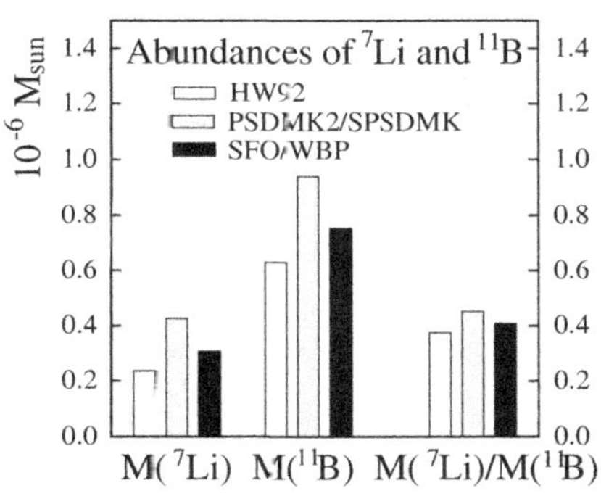

Fig. 4. Enhancement of production yields of ^{7}Li and ^{11}B and their ratios in a supernova explosion model.

$$V = V_C(SFO) + f * \{V_T(\pi + \rho) + V_{LS}(M3Y : \sigma + \omega + \rho)\} \tag{1}$$

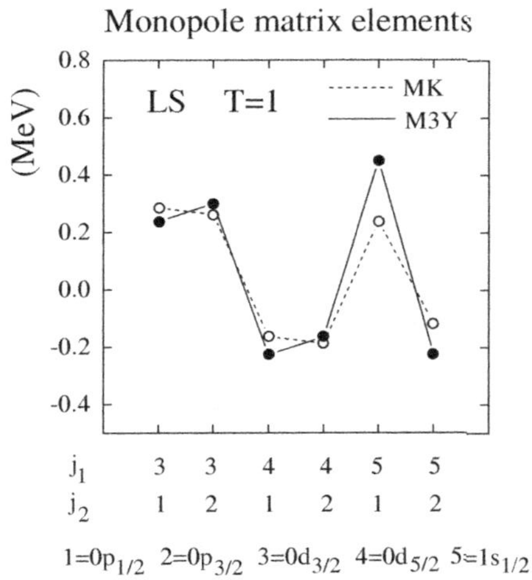

Fig. 5. Monopole terms of the two-body spin-orbit interactions, SFO and MK, in the triplet odd (T=1) channel.

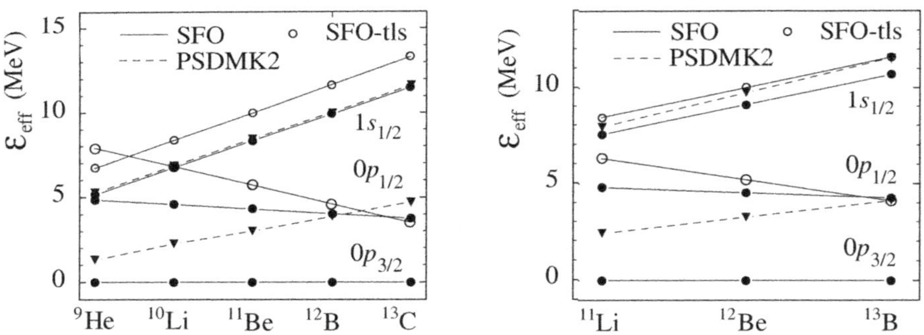

Fig. 6. Effective neutron single particle energies relative to that of the $0p_{3/2}$ orbit for N=7 and N=8 isotones.

Renormalization factors of f= 0.9 (0.7) are multiplied for the $p^2 - p^2$ ($p^2 - (sd)^2$) two-body tensor and spin-orbit interactions, and the shell gap between the $0p_{1/2}$ and $0p_{3/2}$ orbits is enlarged. Note that the central part, $V_C(SFO)$, includes the renormalization effects already. Effective single particle energies for isotones with N (neutron number) =7 and 8 are shown in Fig. 6. The new Hamiltonian, referred to SFO-tls from now on, gives more admixtures of the p-shell and sd-shell components toward the neutron-rich region. We see that for N=7 effective single particle energy of the $1s_{1/2}$ orbit comes below that of the $0p_{1/2}$ orbit in ^{9}He. The $1/2^+$ state is found to be the ground state in ^{9}He and the excitation energy of the $1/2^-$ state is calculated to be 1.5 MeV and 0.60 MeV for the SFO-tls and the SFO Hamiltonian, respectively. In case of ^{10}Li, 2^- is found to be the ground state, and the first excited state is 2^+ (E_x= 1.00 MeV) and 1^+ (E_x= 0.15 MeV) for the SFO-tls and the SFO Hamiltonian, respectively. This reflects the lowering of the $1s_{1/2}$ orbit toward the drip lines.

The Hamiltonian gives a good account of GT transitions in ^{12}C as well as in N=8 isotones, ^{11}Li and ^{12}Be, as shown in Fig. 7. (See ref.[1] for OFU*. The SFO-LS denotes a Hamiltonian with the spin-orbit component replaced by that of the M3Y interaction.) It also explains the E2 transition rate for the transition $0_2^+ \rightarrow 2_1^+$ in ^{12}Be. Calculated $B(E2)$ value is 8.1 $e^2 fm^4$ with the use of effective charges, e_p=1.2 and e_n=0.3. This value is consistent with the observed value of 7.0±0.6 $e^2 fm^4$.[14] Previous Hamiltonian, SFO',[1] gives 13.2 $e^2 fm^4$ while the PSDMK2 leads to 0.19 $e^2 fm^4$.

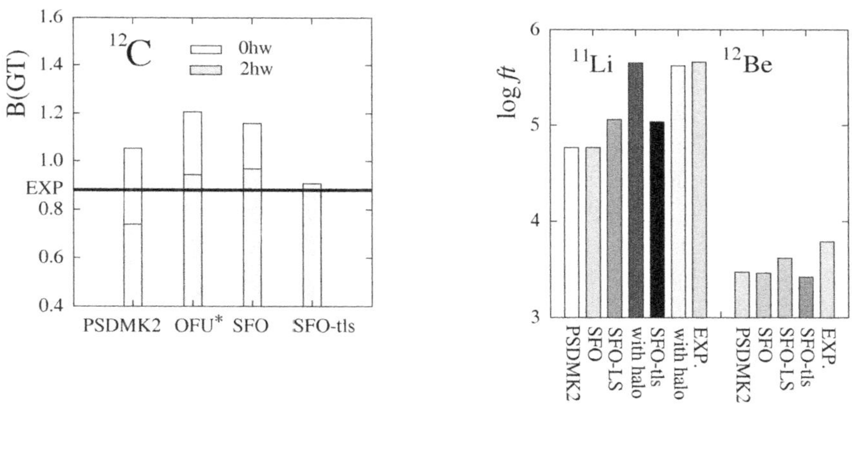

Fig. 7. (a) Calculated and experimental $B(GT)$ values for $^{12}\mathrm{C} \rightarrow {}^{12}\mathrm{N}$ (left figure). (b) Comparison of calculated and experimental $B(GT)$ values in ^{11}Li and ^{12}Be (right figure)

2.3. *Electromagnetic Transitions in C Isotopes*

We next discuss on electromagnetic transitions in neutron-rich carbon isotopes. Anomalously small $B(E2)$ has been observed in ^{16}C.[15] A small value of $B(E2)$ is also reported in ^{18}C.[16] Calculated and observed values of the $B(E2)$ values in C isotopes are shown in Fig. 8. Effective charges that decrease as neutron excess

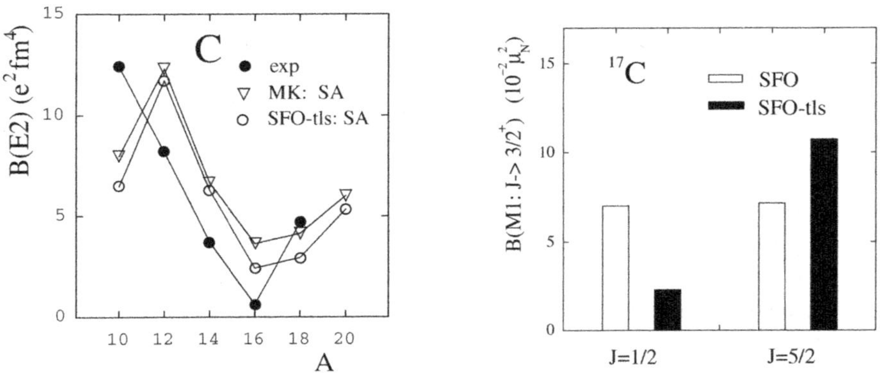

Fig. 8. (a) Comparison of calculated and observed $B(E2)$ values in C isotopes (left figure). (b) Calculated $B(M1)$ values for $1/2^+ \rightarrow 3/2^{3/2}$ and $5/2^+ \rightarrow 3/2^{3/2}$ transitions in ^{17}C (right figure).

increases are used.[17] The new Hamiltonian, SFO-tls, leads to smaller $B(E2)$ values and improves the agreement with experiments.

Finally, we discuss on magnetic dipole transitions in ^{17}C. There is an indication that the $B(M1)$ value for the $1/2_1^+ \rightarrow 3/2_{g.s.}^+$ transition is considerably suppressed, while that for the $5/2_1^+ \rightarrow 3/2_{g.s.}$ transition is not suppressed and consistent with conventional shell model calculations.[18] Note that within the $0d_{5/2}1s_{1/2}$ configuration space, the $B(M1)$ value for the $1/2^+ \rightarrow 3/2^+$ transition vanishes exactly as $1/2^+$ state has a single configuration, $\mid 1/2^+\rangle = \mid 0d_{5/2}^2(0^+) \times 1s_{1/2}\rangle$, since three $d_{5/2}$ neutrons can couple only to $3/2^+$, $5/2^+$ and $9/2^+$ states. This state can not be connected with the $3/2^+$ state with configurations, $\mid 0d_{5/2}^2(2^+) \times 1s_{1/2}\rangle$ and $\mid 0d_{5/2}^2(2^+, 4^+) \times 0d_{5/2}\rangle$, by the M1 transition. The new Hamiltonian gives a quite small value for the transition rate, $B(M1; 1/2^+ \rightarrow 3/2^+) = 0.023\ \mu_N^2$, while the SFO gives $B(M1) = 0.070\ \mu_N^2$ (see Fig. 8). The $B(M1)$ values for the $5/2^+ \rightarrow 3/2^+$ transition are calculated to be 0.107 μ_N^2 and 0.072 μ_N^2 for the SFO-tls and SFO Hamiltonians, respectively. A large quenching only in the $1/2^+ \rightarrow 3/2^+$ transition is realized by the new Hamiltonian, which can not be reproduced by conventional shell model Hamiltonians. The tensor interaction plays an important role to realize the magicity at N=16. Attractive (repulsive) tensor force between the neutron $0d_{5/2}$ orbit and the proton $0p_{1/2}$ ($0p_{3/2}$) orbit reduces the shell gap between the proton $0p_{1/2}$ and $0p_{3/2}$ orbits, thus leading to more particles in proton $0p_{1/2}$ shell. Protons in the $0p_{1/2}$ push up the neutron $0d_{3/2}$ orbit due to the repulsive nature of the tensor force. In this way, the neutron $0d_{3/2}$ orbit gets effectively inactive in the neutron configurations.

In summary, we have extended our previous work[1] to take full account of tensor and spin-orbit forces due to meson exchanges, and applied to study structures of light neutron-rich nuclei. The significant role of the tensor force shown here as a direct consequence of the π and ρ meson exchanges is consistent with the chiral perturbation idea of Weinberg[19]

Acknowledgements

The author would like to express their thanks to T. Yoshida, S. Chiba and T. Kajino for a collaboration of the studies on astrophysical subjects and neutrino-nucleus reactions. They are also grateful to M. Ishihara, H. Sakurai, and H. J. Ong for informing them of experimental data on C isotopes prior to publication.

References

1. T. Suzuki R. Fujimoto and T. Otsuka, *Phys. Rev. C* **67**, 044302 (2003).
2. T. Otsuka, T. Suzuki, R. Fujimoto, R. Grawe and Y. Akaishi, *Phys. Rev. Lett.* **95**, 232502 (2005).
3. D. J. Millener and D. Kurath, *Nucl. Phys. A* **255**, 315 (1975).
4. R. E. McDonald, J. A. Becker, R. A. Chalmers, and D. H. Wilkinson, *Phys. Rev. C* **10**, 333 (1974).
5. G. Bertsch, J. Borysowicz, H. McManus and W. G. Love, *Nucl. Phys. A* **284**, 399 (1977).
6. A. Negret et al., *Phys. Rev. Lett.* **97**, 062502 (2006).
7. S. Aroua et al., *Nucl. Phys. A* **720**, 71 92003).
8. T. Suzuki, S. Chiba, T. Yoshida, T. Kajino and T. Otsuka, *Phys. Rev. C* **74**, 034307 (2006).
9. S. E. Woosley, D. H. Hartmann, R. D. Hoffman and W. C. Haxton, *Astrophys. J.* **356**, 272 (1990).
10. C. Gaarde et al, *Nucl. Phys. A* **422**, 189 (1984).
11. T. E. Drake, E. L. Tomusiak and H. S. Caplan, *Nucl. Phys. A* **118**, 138 (1968); A. Yamaguchi, T. Terasawa, K. Nakahara amd Y. Torizuka, *Phys. Rev. C* **3**, 1750 (1971).
12. T. Suzuki, H. Sagawa and K. Hagino, *Phys. Rev. C* **68**, 014317 (2003).
13. T. Yoshida et al., *Phys. Rev. Lett.* **96**, 091101 (2006) and **94**, 231101 (2005); *Astrophys. J.* **649**, 319 (2006).
14. S. Shimoura et al., *CNS-REP-71* (2007), and *Phys. Lett. B* **560**, 31 (2003).
15. N. Imai et al., *Phys. Rev. Lett.* **92**, 062501 (2004).
16. H. J. Ong et al., *Doctor Thesis* (University of Tokyo, 2007).
17. H. Sagawa and K. Asahi, *Phys. Rev. C* **63**, 064310 (2001).
18. M. Ishihara and H. Sakurai, private communications;
    D. Suzuki, *Master Thesis* (University of Tokyo, 2006).
19. S. Weinberg, *Phys. Lett. B* **251** , 288 (1990).

MULTIPLE SCATTERING EFFECTS IN ELASTIC AND QUASI FREE PROTON SCATTERING FROM HALO NUCLEI

R. CRESPO*

Centro de Física Nuclear, Universidade de Lisboa,
Av. Prof. Gama Pinto 2, 1649-003 Lisboa, Portugal
Departamento de Física, Instituto Superior Técnico, Taguspark,
Av. Prof. Cavaco Silva, Taguspark, 2780-990 Porto Salvo, Oeiras, Portugal
**E-mail: raquel.crespo@tagus.ist.utl.pt*

A. DELTUVA, E. CRAVO, M. RODRÍGUEZ-GALLARDO, A.C. FONSECA

Centro de Física Nuclear, Universidade de Lisboa,
Av. Prof. Gama Pinto 2, 1649-003 Lisboa, Portugal

Multiple scattering higher order effects are studied in the elastic and breakup of ^{11}Be on a proton target at 150 MeV using the Faddeev/AGS scattering framework. The validity of using truncated multiple scattering frameworks is assessed.

Keywords: Nuclear reactions; Halo nuclei

1. Introduction

It is by now well established that the correct treatment of the few-body nature of halo nuclei ($n < A$) must be taken into account in the reaction formalism when studying the scattering from stable nuclei. We are interested here in the scattering of a proton from the two-body ($n = 2$) halo nucleus ^{11}Be which corresponds to the solution of a three-body problem. In the intermediate energy regime, this can be conveniently addressed by multiple scattering frameworks such as those developed by Watson,[1,2] Glauber,[3,4] and Faddeev/Alt, Grassberger, and Sandhas (AGS).[5,6]

It is fair to say that rigorous numerical techniques have been developed in the past twenty years for the solution of the Faddeev/AGS equations using the partial wave decomposition. However, only recently has the inclusion of the long range Coulomb interaction been made possible leading to numerically converged calculations[6] for elastic, transfer and breakup reactions using the screening and renormalization approach. Although most applications of few-body scattering have been confined to the three-nucleon system below π-production threshold where the number of partial waves is manageable, we have been able recently to solve Faddeev/AGS equations for nuclear reactions at intermediate energies using standard nuclear interactions.[7] With this achievement a new era has opened for the application of the Faddeev/AGS equations to reactions involving the scattering of protons from heavier nuclei, such as halo structures like ^{11}Be or ^{19}C assumed to be well described by a core and one valence neutron in the low and intermediate energy regime.

It has been shown[8] that the forward approximation made in the Glauber scattering framework naturally truncates the elastic scattering multiple scattering expansion to n order. That is, for example, for the elastic scattering of ^{11}Be from a proton target, the series is summed only up to the double scattering term. The validity of using a truncated multiple scattering expansion needs to be investigated if one wants to extract accurate size information from elastic data using the Glauber framework.

On the other hand structure information about the wave function of the struck valence neutron can be obtained from quasi free scattering of the ^{11}Be halo nucleus. This information relies often on the validity of the single scattering approximation, or in incomplete and truncated multiple scattering expansions.

We investigate the importance of multiple scattering effects in elastic and breakup by comparing Faddeev/AGS multiple scattering calculations to n order at intermediate energies with full calculations using Padé summation.[9]

As an example we consider proton scattering from ^{11}Be at 150 MeV.

2. The Faddeev/AGS equations

The Faddeev/AGS reaction formalism is a multiple scattering framework for 3-body problems which treats all open channels in the same footing. Let us consider 3 particles (1,2,3) interacting by means of two-body potentials. We use the odd man out notation which means, for example, that the interaction between the pair (2,3) is denoted as v_1. We assume that the system is non-relativistic and write its total Hamiltonian as

$$H = H_0 + \sum_\gamma v_\gamma \,, \tag{1}$$

where H_0 is the kinetic energy operator and v_γ the interaction for the pair γ. One then defines a transition amplitude t_γ for each interacting pair according to

$$t_\gamma = v_\gamma + v_\gamma G_0 t_\gamma \tag{2}$$

with the free resolvent

$$G_0(z) = (z - H_0)^{-1}. \tag{3}$$

The three-body Faddeev/AGS[5] equations are given in terms of the operators $U^{\beta\alpha}$

$$U^{\beta\alpha} = \bar{\delta}_{\beta\alpha} G_0^{-1} + \sum_\gamma \bar{\delta}_{\beta\gamma} t_\gamma G_0 U^{\gamma\alpha} \,, \tag{4}$$

where $\bar{\delta}_{\beta\alpha} = 1 - \delta_{\beta\alpha}$; the operators $U^{\beta\alpha}$ are equivalent to the transition amplitudes on the energy shell.

The solution of the Faddeev/AGS equations can be found by iteration leading, in the elastic scattering case, to

$$U^{\alpha\alpha} = \sum_{\gamma} t_\gamma \bar{\delta}_{\gamma\alpha} + \sum_{\gamma} \bar{\delta}_{\alpha\gamma} t_\gamma \sum_{\xi} G_0 \bar{\delta}_{\gamma\xi} t_\xi \bar{\delta}_{\xi\alpha} + \sum_{\gamma} \bar{\delta}_{\alpha\gamma} t_\gamma \sum_{\xi} G_0 \bar{\delta}_{\gamma\xi} t_\xi \sum_{\eta} G_0 \bar{\delta}_{\xi\eta} t_\eta \bar{\delta}_{\eta\alpha} + \cdots .$$

For breakup one has

$$U^{0\alpha} = \sum_{\gamma} t_\gamma \bar{\delta}_{\gamma\alpha} + \sum_{\gamma} t_\gamma \sum_{\xi} G_0 \bar{\delta}_{\gamma\xi} t_\xi \bar{\delta}_{\xi\alpha} + \sum_{\gamma} t_\gamma \sum_{\xi} G_0 \bar{\delta}_{\gamma\xi} t_\xi \sum_{\eta} G_0 \bar{\delta}_{\xi\eta} t_\eta \bar{\delta}_{\eta\alpha} + \cdots .$$

The successive terms of this series can be considered up to first order (single scattering), second order (double scattering) and so on in the transition operators. One can either take into account the contribution of the terms of the series up to n order or, for the full solution, sum it using the Padé method.[9]

3. Results

We now proceed to calculate the elastic and breakup observables for proton scattering from ^{11}Be.

In order to solve the Faddeev/AGS equations we need to provide the pair interactions p-n, p-^{10}Be and n-^{10}Be. We take the realistic nucleon-nucleon CD Bonn interaction.[10] The interaction between the valence neutron and the ^{10}Be core is assumed to be L-dependent as described in Crespo et al.[7] For the potential between the proton and ^{10}Be core we take a phenomenological optical model with parameters taken from the Watson global optical potential parametrization.[7,11]

We show in Fig. 1 the elastic cross section relative to Rutherford at 150 MeV/u. As shown in the figure the single scattering approximation (thin solid line) overestimates the full calculation (thick solid line). Adding the double scattering term (dashed line) does not produce a significant contribution to the elastic observable. The third scattering term (dashed dotted) decreases the cross section and brings it closer to the full calculation. Third order scattering, is absent in the Glauber scattering framework. At this energy the convergence has been practically achieved with the 3rd order contribution. The inclusion of higher order terms represented by the dots (4th order) does not contribute significantly to the calculated elastic scattering differential cross section. Therefore at this energy the series has converged at the level of the 3rd order term.

We also investigate the convergence of the multiple scattering series for breakup scattering. We show in Fig. 2 the calculated breakup cross section $d^5\sigma/dSd\hat{k}_1 d\hat{k}_2$ versus the arclength S at 150 MeV/u. Particles 1 and 2 are detected; they are, respectively, the ^{10}Be core and the neutron. We take three quasifree configurations: (conf 1) with $\hat{k}_1 = (\theta_1, \phi_1) = (0, 0)$ and $\hat{k}_2 = (\theta_2, \phi_2) = (45, 180)$; (conf 2) with $\hat{k}_1 = (\theta_1, \phi_1) = (0, 0)$ and $\hat{k}_2 = (\theta_2, \phi_2) = (30, 180)$; (conf 3) with $\hat{k}_1 = (\theta_1, \phi_1) = (0, 0)$ and $\hat{k}_2 = (\theta_2, \phi_2) = (15, 180)$. The kinematic angles are represented in Fig. 3.

In each configuration, the solid line represents the observable obtained from the full solution of AGS equations. The single scattering approximation (SSA) represented by the dashed solid lines overestimates the full calculated breakup observable

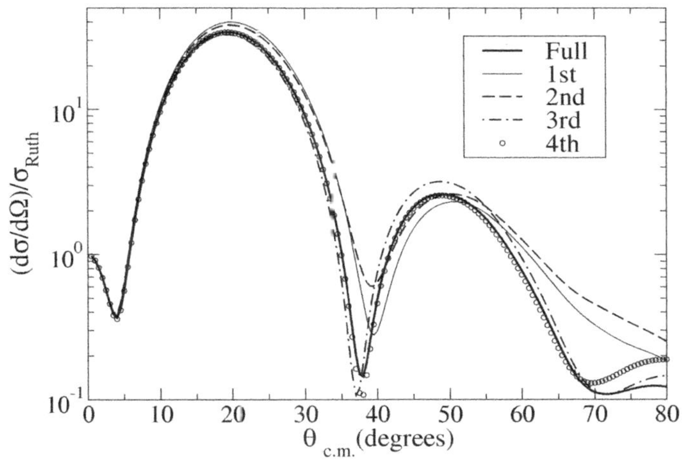

Fig. 1. Cross section for the elastic p(^{11}Be,^{11}Be)p at 150 MeV.

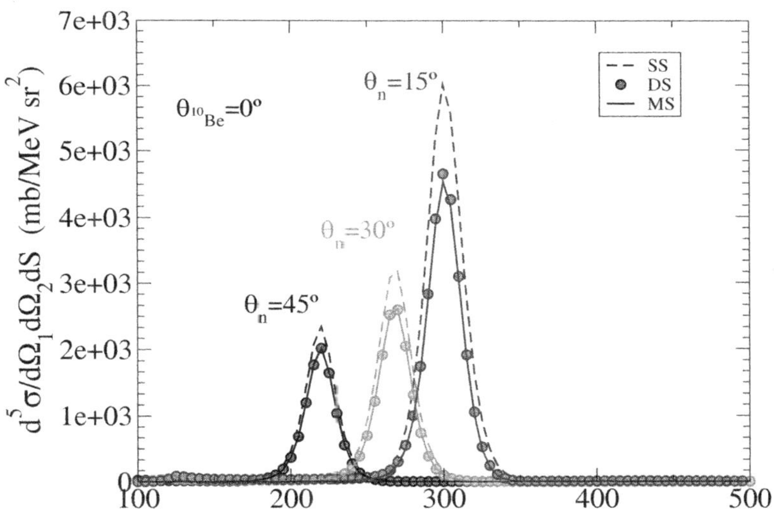

Fig. 2. Cross section for the breakup p(^{11}Be,^{10}Be n)p at 150 MeV.

in any configuration at this energy. The double scattering approximation represented by the dotted points reproduces well the full calculation and thus at this energy the series has converged to second order.

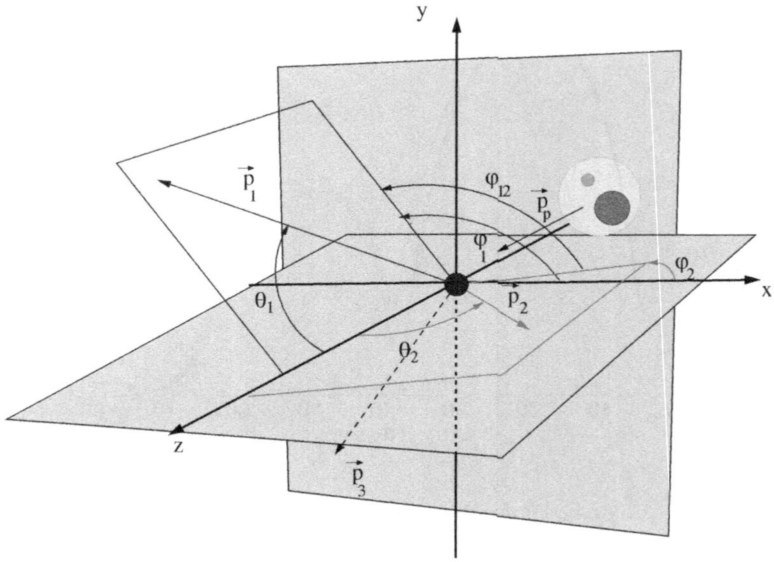

Fig. 3. Kinematic angles for breakup.

4. Conclusions

We have performed Faddeev/AGS multiple scattering calculations for elastic and breakup of ^{11}Be from protons at 150 MeV/u. We have studied the convergence of the multiple scattering series and the validity of using truncated multiple scattering frameworks.

At these incident energies convergence for the elastic scattering has been achieved using the multiple scattering expansion up to 3rd order. Therefore we need to go beyond double scattering to achieve convergence. Terms of order greater than two are naturally excluded from the Glauber multiple scattering expansion and therefore these findings provide evidence for the breakdown of this scattering formalism.

As for the breakup reaction, the series has converged to 2nd order showing that one needs to go beyond the single scattering approximation.

Acknowledgments

The work of R.C. is supported by Fundação para a Ciência e Tecnologia (FCT), through grant No. POCTI/FNU/43421/2001. A.D. is supported by the FCT grant SFRH/BPD/14801/2003 and all other authors by FCT grant POCTI/ISFL/2/275.

References

1. K.M. Watson, *Phys. Rev.* **105**, 1338 (1957).
2. R. Crespo and R.C. Johnson, *Phys. Rev. C* **60**, 034007 (1999).

3. R.J. Glauber, *Lectures in theoretical Physics* vol 1, ed W.E. Brittin (New York: Interscience, 1958).
4. J.S. Al-Khalili and J.A. Tostevin, *Phys. Rev. C* **57**, 1846 (1998).
5. E.O. Alt, P. Grassberger and W. Sandhas, *Nucl. Phys. B* **2**, 167 (1967).
6. A. Deltuva, A.C. Fonseca and P.U. Sauer, *Phys. Rev. C* **71**, 054005 (2005); *C***72**, 054004 (2005).
7. R. Crespo, E. Cravo, A. Deltuva, M. Rodríguez-Gallardo, A.C. Fonseca, *Phys. Rev. C* **76**, 014620 (2007).
8. J.S. Al-Khalili, R. Crespo, R.C. Johnson, A.M. Moro and I.J. Thompson, *Phys. Rev. C* **75**, 024608 (2007).
9. G.A.Baker and J.L. Gammel eds., *The Padé Approximation in Theoretical Physics* (Academic, New York, 1970).
10. R. Machleidt, *Phys. Rev. C* **63**, 024001 (2001).
11. B.A. Watson, P.P. Singh, and R.E Segel, *Phys. Rev.* **182**, 978 (1969).

THE DIPOLE RESPONSE OF NEUTRON HALOS AND SKINS

THOMAS AUMANN

GSI - Gesellschaft für Schwerionenforschung mbH,
D-64291 Darmstadt, Germany
E-mail: t.aumann@gsi.de

The dipole response of neutron-rich nuclei has been investigated in a series of experiments at the R3B/LAND facility at GSI utilizing the heavy-ion induced electromagnetic excitation process at high energy. Pronounced changes of the dipole response have been observed compared to stable nuclei. The present article summarizes the experimental results covering a wide mass range.

Keywords: Radioactive Beams; Dipole Response; Pygmy and Giant Resonances.

1. Introduction

The question how the excitation spectra of nuclei evolve with increasing asymmetry with respect to the neutron-to-proton ratio has been studied theoretically and experimentally in recent years by various approaches. Changes of the multipole response are expected in neutron-proton asymmetric nuclei due to the different properties of asymmetric nuclear matter and due to specific properties of these nuclei. The filling of different single-particle orbits for protons and neutrons might affect the excitation spectra as well as the resulting asymmetry in the binding energies for protons and neutrons. The differences in the neutron and proton density distributions, i.e., the development of neutron skins or halos for neutron-rich nuclei manifest itself in the collective response of the nucleus. In particular, a new dipole mode corresponding to the vibration of less bound valence neutrons forming the neutron skin against a isospin-saturated core has been predicted by various models for medium-heavy neutron-rich nuclei and experimental evidence has been accumulated meanwhile, which shall be summarized in this paper. This excitation mode is often referred to as 'soft dipole mode' or 'pygmy' resonance.

Giant-resonance data have been very valuable in the past providing constraints to the effective nucleon-nucleon interaction in the medium. The isovector properties of the forces, however, are not constrained well by the experimental data available for stable nuclei. The study of giant resonances in nuclei with asymmetric neutron-to-proton ratio may provide the missing information. The multipole response of nuclei might also be utilized to constraining fundamental properties of nuclear matter. From the measurement of isoscalar giant monopole excitations of nuclei, for instance, one can extract the compressibility of nuclear matter. Similarly, from the isoscalar dipole excitation. Recently, it has been pointed out, that also the low-lying dipole excitations observed in neutron-rich nuclei can be analyzed to constrain the isospin-asymmetric part of the equation of state of nuclear matter. It has been shown in mean-field calculations, that the dipole strength located at low excitation energy (which is related to the less bound valence neutrons responsible for the neutron skin) is directly linked to the symmetry energy (as the neutron-skin thickness is). A first attempt extracting parameters describing the density dependence of the nuclear symmetry energy has been undertaken using the GSI data for nuclei around ^{132}Sn, wich will be discussed in section 2.4.

For a detailed discussion of the theoretical approaches to study the multipole response of exotic nuclei we refer to a recent review by Paar et al.[1] A more general overview on giant resonances can be found in Ref. [2] and on experiments with high-energy neutron-rich beams in Ref. [3]

2. The dipole response of neutron-rich nuclei

2.1. *Coulomb breakup of light halo nuclei*

The large cross sections observed for the electromagnetic dissociation of halo nuclei can be explained by non-resonant transitions to the continuum due to a large overlap between the tail of the neutron wave function and continuum wave functions with large wavelength, i.e., small relative momenta q (direct-breakup model). Since the effective charge $Z_{eff} \sim A^{-\lambda}$ (with λ being the multipolarity) gets smaller for higher multipolarities, the breakup process is dominated by dipole transitions. Typel and Baur[4] estimated the $E2$ contribution for the Coulomb breakup of ^{19}C, for instance, to be more than three orders of magnitude smaller than the $E1$ cross section. Quadrupole and higher multipolarities can thus safely be neglected and the differential cross section can be factorized into the number $N_{E1}(E^*)$ of equivalent dipole photons with energy E^* associated with the rapidly varying Coulomb field of the target, and the square of the dipole matrix elements:[5]

$$\frac{d\sigma}{dE^*}(I_c^\pi) = (\frac{16\pi^3}{9\hbar c})N_{E1}(E^*)\sum_{nlj} C^2 S(I_c^\pi, n\ell j) \times \sum_m |\langle q|(Ze/A)rY_m^1|\phi_{n\ell j}(r)\rangle|^2. \quad (1)$$

The photon flux $N_{E1}(E^*)$ can be calculated in the semiclassical approximation.[6] The Coulomb breakup cross sections are calculated for individual ground-state single-particle configurations of the neutron with a relative-motion wave function $\psi_{nlj}(r)$, with r being the relative distance between the core and the valence neutron, and corresponding core state (I_c^π). The final state with relative momentum q is represented either by plane or distorted waves describing the neutron in the continuum. As one can see from the above equation, the $E1$-strength distribution is very sensitive to the spatial extension of the single-particle wave function. A pronounced halo-like tail of the wave function leads to large transition probabilities at very low relative energies. The shape of the strength distribution is thus also sensitive to

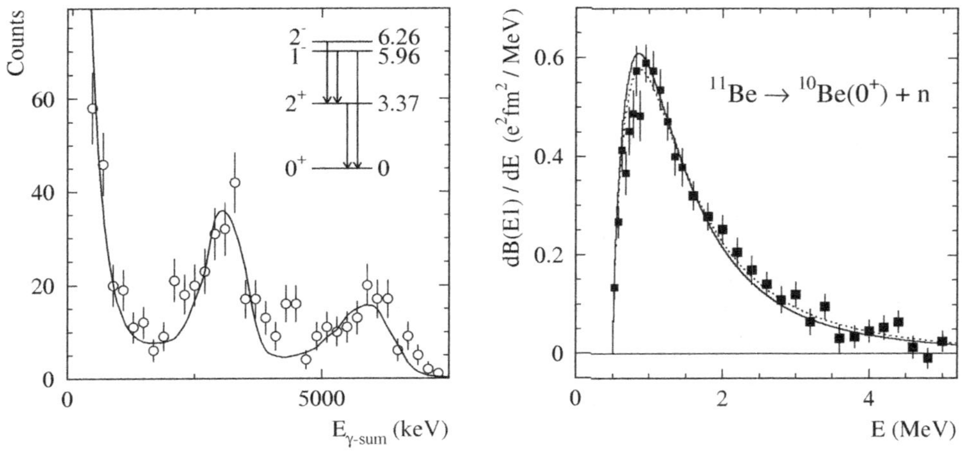

Fig. 1. Breakup of 520 MeV/nucleon ^{11}Be into ^{10}Be and one neutron measured with a lead target.[7] Left: Doppler corrected γ-sum energy spectrum measured in coincidence with a ^{10}Be fragment and a neutron. The inset shows a partial level scheme for ^{10}Be. Right: Dipole-strength distribution of ^{11}Be yielding the ^{10}Be fragment in its ground state. From Palit *et al.*[8]

the angular momentum of the valence neutron. This characteristic behavior can be utilized to probe the ground-state single-particle structure by measuring the differential electromagnetic dissociation cross section: the l value of the valence nucleon is determined by the shape of the distribution, the core state I_c^π to which the nucleon is coupled can be identified by the characteristic γ-decay (in case of excited states), while the associated spectroscopic factors $CS^2(I_c^\pi, nlj)$ can be deduced from the absolute cross sections.

As an example, the results obtained for the halo nucleus ^{11}Be[7] are shown in Figure 1. The γ-ray sum-energy spectrum as recorded in coincidence with ^{10}Be fragments and a neutron (left panel) reveals contributions to the breakup cross section involving the 2^+ first excited state at 3.37 MeV but also higher lying states of ^{10}Be at around 6 MeV excitation energy originating from the removal of more deeply bound core neutrons from the p shell populating 1^- and 2^- states. After subtracting the excited-state contributions, the differential cross sections $d\sigma/dE^*$ for ground state transitions, i.e., excitations solely related to the $2s_{1/2}\otimes^{10}\text{Be}(0^+)$ halo state, are obtained. The extracted B(E1) distribution is shown in the right frame of Figure 1 and compared to model calculations as described above. A theoretical analysis by Typel and Baur[9] results in a spectroscopic factor of 0.7 for the $2s_{1/2}\otimes^{10}\text{Be}(0^+)$ configuration. Consistent results have been obtained from a Coulomb breakup experiment at lower incident energy[10] as well as from nuclear knockout reactions.[11]

The integrated B(E1) strength is related to the spatial extension of the neutron relative wavefunction via the non-energy weighted sum rule. In case of $2n$ halo nuclei, this quantity is related to the rms distance between the center-of-mass of the two neutrons and the core, and thus contains information on the neutron-neutron spatial correlation. From the Coulomb breakup of ^{6}He, for instance, a value of $r_{\alpha-2n} = 3.36(39)$ fm, has been deduced.[12] This value is in close agreement with a recent three-body calculation.[13] Hagino and Sagawa[13] deduced an average angle between the two neutrons of 75(13) degree for ^{6}He and 65(13) degree for ^{11}Li using the data of Ref.[12] and Ref.,[14] respectively. These values suggest the presence of spatial correlations among the two halo neutrons, being more pronounced in ^{11}Li than ^{6}He.

2.2. *Low-lying dipole strength in light neutron-rich nuclei*

The evolution of the dipole-strength distribution as a function of neutron-to-proton ratio was investigated systematically for the oxygen isotope chain ranging from a neutron excess of $N - Z = 1$ up to $N - Z = 7$ at GSI by Leistenschneider *et al.*[15] ^{16}O is a strongly bound doubly magic nucleus with a neutron separation energy of $S_n = 16$ MeV, while, for the heavier isotopes, the separation energies decrease to $S_n \approx 7$ MeV for the even isotopes with $A = 18$ to 22, and $S_n \approx 4$ MeV for the odd isotopes. Thus, one may expect a decoupling of the valence neutrons from the inert ^{16}O core and the appearance of a collective soft-dipole excitation.

Fig. 2 shows the (γ, n) cross section for the unstable isotopes ^{20,22}O in comparison with that for the stable nucleus ^{16}O.[16] Evidently, the dipole response changes significantly due to the presence of the valence neutrons. Most noticeable is the sizeable dipole absorption cross section below the GDR energy region. A clear separation into two energy domains associated with GDR excitations of the core and a soft-dipole mode involving the valence neutrons is not observed. For ^{22}O, the data are compared to a large-scale shell-model calculation.[17] Qualitatively, the shell-model calculations reproduce the experimental observation of a redistribution of the $E1$ strength towards excitation energies below the GDR. For ^{20}O, cumulation of dipole strength below the neutron threshold in the range from 5 to 7 MeV was observed in a virtual-photon scattering experiment performed at MSU[19,20] at around 100 MeV/nucleon. The integrated value of about 0.1 e^2fm^2[20] for the dipole strength corresponds to much less than 1% of the TRK sum rule, which has to be compared to the low-lying strength above the threshold of about 10%, see Figure 2.

The right part of Figure 2 shows the energy-weighted experimental low-lying dipole strength S_{exp} (integrated from the particle threshold up to 15 MeV excitation energy) in units of the energy-weighted dipole sum rule $S_{\text{TRK}} \sim NZ/A$ (upper frame) and the cluster sum rule (lower frame). While no dipole strength below 15 MeV is observed for the $N = Z$ nucleus ^{16}O, about 10% of the sum-rule value is exhausted for the neutron-rich isotopes. The results for the integrated low-lying dipole strength are compared to a large-scale shell-model calculation[17] (stars) and a calculation using the quasi-particle random-phase approximation including phonon coupling (QRPA-PC).[18] Both calculations reproduce the amount of dipole strength as found experimentally at low excitation energies. Also, the experimentally ob-

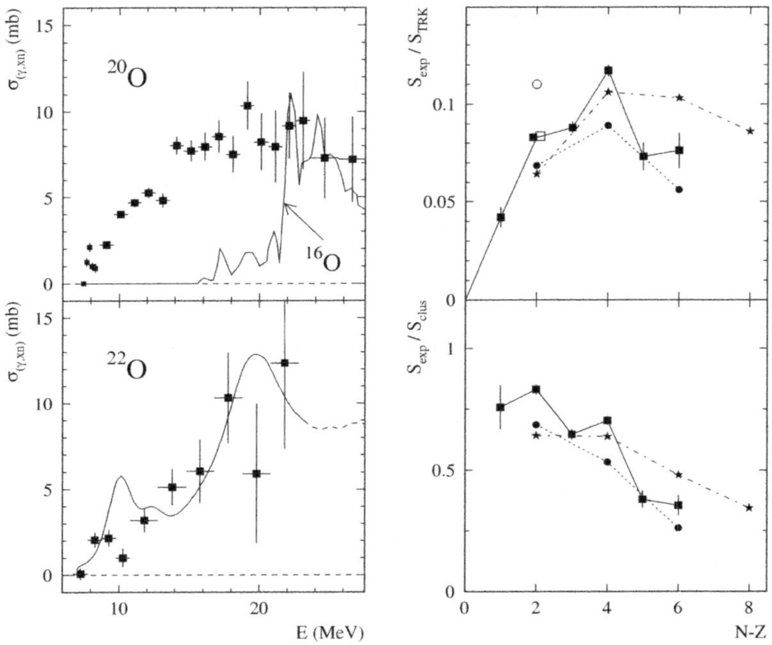

Fig. 2. Left panels: Photoneutron cross sections $\sigma_{(\gamma,xn)}$ for the unstable isotopes ^{20}O and ^{22}O (symbols)[15] and for the stable ^{16}O (solid curve in the upper panel). Right panels: Evolution of integrated (up to 15 MeV) dipole strength with neutron excess $N - Z$ in units of the TRK sum rule (upper panel) and of the cluster-sum-rule (lower panel). The data are compared to shell-model calculations by Sagawa and Suzuki[17] (stars) and the QRPA-PC calculation by Colò and Bortignon[18] (filled circles). From Palit *et al.*[8]

served trend of an increase of the strength as a function of neutron excess up to ^{20}O followed by a decrease for isotopes heavier than ^{20}O is reproduced qualitatively in both calculations. The additional strength at low excitation energy is, according to the theoretical models, not related to a collective soft dipole mode but rather to non-resonant single-particle excitations of the less bound valence neutrons, see also the review on theoretical approaches by Paar *et al.*[1]

2.3. *Collective dipole excitations of medium-mass nuclei*

A first experiment investigating the electromagnetic excitation of heavier neutron-rich nuclei was recently performed at GSI utilizing a mixed secondary beam including ^{132}Sn isotopes produced by fission of a ^{238}U beam with an energy of around 600 MeV/nucleon.[21] The differential cross sections for electromagnetic excitation of ^{130,132}Sn on a lead target are displayed in Fig. 3. The right frames display the corresponding photo-absorption (γ, n) cross sections. The spectrum is dominated by the excitation of the GDR, in contrast to the excitation of lighter nuclei discussed in section 2.2. This is due to the fact that the GDR strength in heavier nuclei is less fragmented, and concentrated at lower excitation energies, yielding larger cross sections for the electromagnetic excitation. A fit of a Lorentzian plus a gaussian parametrization for the photo-absorption cross section (solid curves) yields a position and width of the GDR comparable to those known for stable nuclei in this mass region. The GDR exhausts almost the energy-weighted dipole sum rule. It should be noted that the spectrum shown in Fig. 3 was obtained from a measurement with a rather low beam intensity of about 10 ^{132}Sn ions/s only.

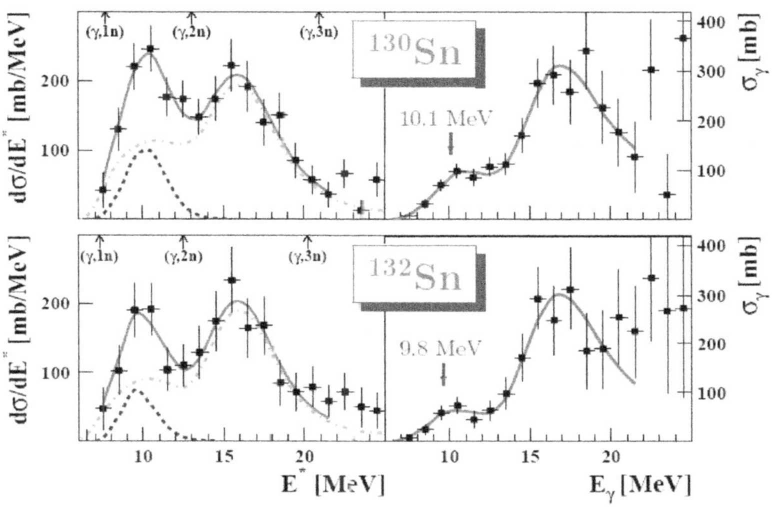

Fig. 3. Left frames: Differential cross section for the electromagnetic excitation of ^{130,132}Sn on a lead target at around 500 MeV/nucleon[21] (left). Corresponding photo-neutron cross sections are shown in the right panels. From Ref.[21]

An additional peak structure is clearly visible below the GDR energy region. The position of around 10 MeV is close to the theoretical predictions for the soft mode (Pygmy resonance). The experimentally observed strength (about 5% of the TRK sum rule) is in good agreement with the relativistic QRPA prediction by Paar *et al.*[22] as well as with the non-relativistic QRPA calculations of Sarchi *et al.*[23] The collective character of the low-lying structure is, however, still an open question: While the strength is attributed to a few single-particle excitations in the non-relativistic calculation,[23] a coherent superposition of many neutron quasi-particle configurations is found in the relativistic approach.[22] Further theoretical and experimental investigations are called for in order to shed light on the nature of the low-lying dipole strength.

Cumulated dipole strength below the neutron threshold was also observed in real-photon scattering experiments on stable nuclei at the S-DALINAC, e.g., in ^{208}Pb[25] and $N = 82$ isotopes,[26] which was discussed in the context of a Pygmy dipole resonance. However, the states observed exhaust only a rather small fraction of the energy-weighted dipole sum rule. A direct comparison with the data for

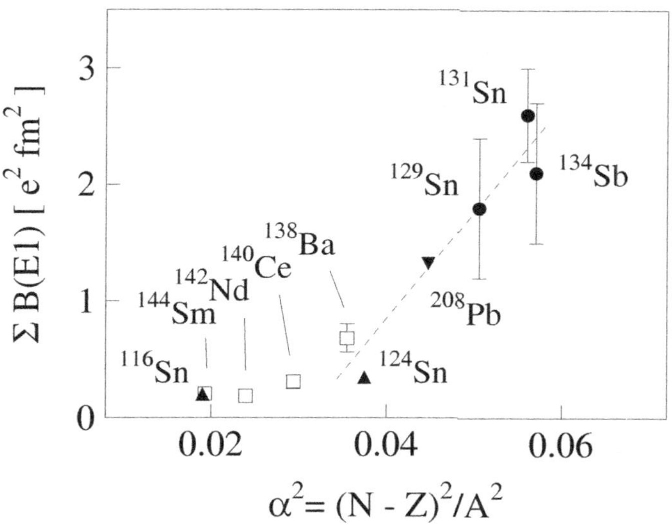

Fig. 4. Pygmy dipole strength as a function of neutron-proton asymmetry for unstable ^{129,130,131,132}Sn and ^{133,134}Sb nuclei (solid symbols)[24] in comparison with data (open symbols) available for stable nuclei.[25–28] From Ref.[24]

neutron-rich Sn isotopes is hindered by the fact that the experiments discussed above measure the strength only above the neutron threshold, which is located for the even Sn isotopes around 7 MeV. New results on low-lying dipole strength for odd-mass nuclei[24] from the same experiment[21] allow a direct comparison since the neutron threshold is much lower, at around 5 MeV, thus covering the excitation-energy region where dipole strength has been observed in stable nuclei. Figure 2.3 shows this comparison as a function of the asymmetry parameter $\alpha^2 = (N-Z)^2/A^2$. A clear tendency of increasing dipole strength is observed with increasing neutron excess.

2.4. *Low-lying dipole excitations and the equation of state of asymmetric nuclear matter*

As recently discussed in the literature, the neutron-skin thickness in nuclei is related to the isospin asymmetric part of the equation of state of nuclear matter, see, for instance, Refs.[29–31] The measurement of the neutron skin is rather difficult and so far known only for few cases with relatively poor accuracy. Since related to the neutron skin, the pygmy dipole strength might play an equivalent role in constraining the nuclear symmetry energy, as pointed out in a recent paper by Piekarewicz.[32] Relativistic mean field calculations[24] support a one-to-one correlation of low-lying $E1$ strength and the parameters a_4 and p_0 of the symmetry energy, similar as observed for the relation to the neutron-skin thickness. Klimkiewicz *et al.*[24] have analyzed the experimentally observed pygmy dipole strength in Sn isotopes (see section 2.3

and extracted, by employing the RQRPA approach, parameters for the symmetry energy ($a_4 = 32.0 \pm 1.8$ MeV and $p_o = 2.3 \pm 0.8$ MeV/fm^3) as well as neutron-skin thicknesses $R_n - R_p$ of 0.24 ± 0.04 fm for ^{132}Sn and of 0.18 ± 0.035 fm for ^{208}Pb.[24] Further theoretical and experimental investigations are called for in order to establish this method and to evaluate systematic uncertainties. This method, however, looks very promising and will certainly contribute to constraining parameters of the nuclear equation of state. In particular, the experimental difficulty to extract these quantities or the neutron-skin thickness calls for different independent approaches in order to get the systematic uncertainties inherent to the different approaches under control.

3. Conclusion

Experiments with fast fragmentation beams utilizing the heavy-ion induced electromagnetic excitation process have uncovered the dipole response of neutron-proton asymmetric nuclei. Pronounced differences compared to stable nuclei have been established: For light loosely bound neutron-rich nuclei, in particular halo nuclei, a complete decoupling of the dipole strength related to the valence neutrons compared to that related to core excitations has been observed. In general, a redistribution of $E1$ strength towards lower excitation energies below the GDR region is found for neutron-rich nuclei. For heavier nuclei around ^{132}Sn experimental evidence has been obtained a for a new collective excitation mode associated with a dipole vibration of valence neutrons forming the neutron skin versus a isospin-saturated core. The low-lying dipole strength is thereby related to the neutron-skin thickness as well as to the parameters describing the isospin-asymmetric part of the equation of state of nuclear matter. Clearly, more theoretical and experimental investigations are called for in order to establish these findings and to uncover the nature and properties of the pygmy dipole strength.

References

1. N. Paar, D. Vretenar, E. Khan, G. Colò, *Rep. Prog. Phys.* **70** (2007) 691.
2. M.N. Harakeh and A.M. Van Der Woude, *Giant Resonances: Fundamental High-Frequency Modes of Nuclear Excitations* (Oxford University Press, 2001).
3. T. Aumann, *Eur. Phys. J. A* **26** (2005) 441.
4. S. Typel and G. Baur, *Phys. Rev. C* **64** (2001) 024601.
5. G. Baur and C.A. Bertulani, *Nucl. Phys. A* **480** (1988) 615.
6. C.A. Bertulani and G. Baur, *Phys. Rep.* **163** (1988) 299.
7. R. Palit *et al.*, *Phys. Rev. C* **68** (2003) 034318.
8. R. Palit *et al.*, *Nucl. Phys. A* **731** (2004) 235.
9. S. Typel and G. Baur, *Phys. Rev. Lett.* **93** (2004) 142502.
10. N. Fukuda *et al.*, *Phys. Rev. C* **70** (2004) 054606.
11. T. Aumann *et al.*, *Phys. Rev. Lett.* **84** (2000) 35.
12. T. Aumann *et al.*, *Phys. Rev. C* **59** (1999) 1252.
13. K. Hagino, H. Sagawa, *Phys. Rev. C* (2007), to be published.
14. T. Nakamura et al, *Phys. Rev. Lett.* **96** (2006) 252502.

15. A. Leistenschneider *et al.*, *Phys. Rev. Lett.* **86** (2001) 5442.
16. E.G. Fuller, Phys. Rep. **127** (1985) 187; B.L. Berman, *At. Data Nucl. Data Tables* **15** (1975) 319.
17. H. Sagawa and T. Suzuki, *Phys. Rev. C* **59** (1999) 3116.
18. G. Colò and P.F. Bortignon, Nucl. Phys. A **696** (2001) 427.
19. E. Tryggestad *et al.*, *Phys. Rev. C* **67** (2003) 064309.
20. E. Tryggestad *et al.*, *Phys. Lett. B* **541** (2002) 52.
21. P. Adrich *et al.*, *Phys. Rev. Lett.* **95** (2005) 132501.
22. N. Paar, P. Ring, T. Nikšić, D. Vretenar, *Phys. Rev. C* **67** (2003) 034312.
23. D. Sarchi, P.F. Bortignon, G. Colò, *Phys. Lett. B* **601** (2004) 27.
24. A. Klimkiewicz *et al.*, submitted to *Phys. Rev.* (2007).
25. N. Ryezayeva *et al.*, *Phys. Rev. Lett.* **89** (2002) 272502.
26. A. Zilges *et al.*, *Phys. Lett. B* **542** (2002) 43.
27. S. Volz *et al.*, *Nucl. Phys. A* **779** (2006) 1.
28. K. Govaert *et al.*, *Phys. Rev. C* **57** (1998) 2229.
29. S. Typel and B.A. Brown, *Phys. Rev. C* **64** (2001) 027302.
30. R.J. Furnstahl, *Nucl. Phys. A* **706** (2002) 85.
31. S. Yoshida and H. Sagawa, *Phys. Rev. C* **69** (2004) 024318.
32. J. Piekarewicz, *Phys. Rev. C* **73** (2006) 044325.

GIANT AND PYGMY RESONANCES WITHIN AXIALLY-SYMMETRIC-DEFORMED QRPA WITH THE GOGNY FORCE

S. PÉRU* and H. GOUTTE

CEA/DAM Ile de France, DPTA,
B.P. 12, 91680 Bruyères-le-Châtel, France
**E-mail: sophie.peru-desenfants@cea.fr*

Fully consistent axially-symmetric-deformed HFB+QRPA calculations have been performed, where the same Gogny D1S effective force has been used both in the HFB and QRPA approaches. Results on dipole pygmy states in ^{26}Ne are shown and compared with recent experimental data. Giant dipole resonances calculated in the s-d shell nuclei $^{26-30}$Si and $^{22-28}$Mg are presented and the splitting of the resonance in the deformed nuclei is discussed.

Keywords: Pygmy resonances, Giant resonances, Nuclear structure

1. Introduction

In order to describe giant and pygmy resonances in both spherical and deformed deformed nuclei, a fully consistent Axially-Symmetric-Deformed (ASD) Quasi-Particle Random-Phase-Approximation (QRPA) approach has been developped. Such a QRPA approach enables us to describe small amplitude collective vibrations of the nuclear mean field and pairing field together with the induced ground state correlations.[1] An outline of the method is first presented, with some details concerning the projection onto good angular momentum. Then, results obtained in the neutron-rich spherical ^{26}Ne nucleus are discussed. Finally, dipole responses obtained in s-d shell nuclei,$^{22-28}$Mg and $^{26-30}$Si, are presented.

2. Formalism

In the present study, QRPA calculations are made on top of axially-symmetric-deformed Hartree-Fock-Bogolyubov (HFB) results. The only input of this microscopic approach is the Gogny D1S force,[2] which is introduced both in the HFB and QRPA descriptions. Let us note that, to maintain consistency, the two-body center of mass correction neglected in the HFB fields, is not taken into account in the QRPA equations. Calculations are performed imposing the time reversal symmetry T, the left-right symmetry, and the $T\pi_2$ symmetry, where π_2 is the reflection with respect to the (x0z) plane. As a consequence, the projection K of the angular momentum on the symmetry axis and the parity π are good quantum numbers.

In deformed nuclei, states $|JMK\rangle$ of good angular momentum J are built from the QRPA phonon states $|\theta_K\rangle$ using standard angular momentum projection techniques:

$$|JMK\rangle = \frac{\sqrt{2J+1}}{4\pi}\int d\omega \left({D^J_{MK}}^*(\omega)\, R(\omega)\, |\theta_K\rangle + (-)^{J-K}\, {D^J_{M-K}}^*(\omega)\, R(\omega)\, |\overline{\theta_K}\rangle \right), \quad (1)$$

where D^J_{MK} is a Wigner rotation matrix element, $R(\omega)$ the three dimensional rotation operator, and $|\overline{\theta_K}\rangle$ the time reversed state of $|\theta_K\rangle$.

Response functions for a given J are obtained by calculating $\langle\, 000|\hat{Q}_{\lambda\mu}|JMK\rangle$ with the multipole operators $\hat{Q}_{J\mu}{=}\sum_i r_i^J \mathcal{Y}_{J\mu}$, where $\mathcal{Y}_{J\mu}$ are the spherical harmonics, and $|000\rangle$ is the projected ground state. Using Eq.(1), one gets:

$$\langle 000|\hat{Q}_{\lambda\mu}|JMK\rangle = \sum_{\nu,|\nu|\leq J} C_{JK}\langle\tilde{0}|\hat{Q}_{J\nu}|\theta_K\rangle\delta_{K,\nu}, \quad (2)$$

where $|\tilde{0}\rangle$ is the axially-symmetric-deformed QRPA vacuum and C_{JK} are coefficients expressed by use of straightforward algebraic methods. Let us mention that in deformed nuclei many calculations corresponding to $K^\pi = 0^\pi, \pm 1^\pi, ... \pm J^\pi$ are needed to obtain the total response function.

The collectivity of the QRPA states is obtained from their contributions to the Energy-Weighted-Sum-Rule (EWSR) $M_1(Q_{\lambda\mu}) = \sum_N \mathcal{W}_N|\langle JMK|Q_{\lambda\mu}|000\rangle|^2$.

3. Pygmy dipole state in ^{26}Ne

A dipole pygmy state in the ^{26}Ne was first predicted by Cao and Ma[3] at an energy below the two-neutron threshold, and then observed in RIKEN experiments[4] around 9 MeV and exhausting 5% of the EWSR.

We have performed QRPA calculations using spherical HFB states as trial states, since a previous study[5] using the same Gogny force has shown that ^{26}Ne is spherical at the mean field level. (In Ref. 5 the ground state was found deformed in quadrupole configuration mixing calculations).

In the theoretical response of the isovector dipole mode, a pygmy state is found at 10.64 MeV with a transition probability to the ground state B(E1) = 0.1737 e^2fm^2.[6] This value is three times smaller than the experimental one, B(E1)=0.49± 0.16 e^2fm^2.[4] The calculated transition densities of neutron and of proton for this state are shown in Fig. 1. These results are in agreement with the usual picture of a neutron skin vibrating against a nuclear core: near the surface of the nucleus one observes a finite contribution of the neutrons and a negligible one from the protons. However, a large isovector response is found near the center of the nucleus. Let us note that the larger 2qp contribution in this state is identified as (2p$\frac{3}{2}$,2s$\frac{1}{2}$) which is in agreement with Ref. 3.

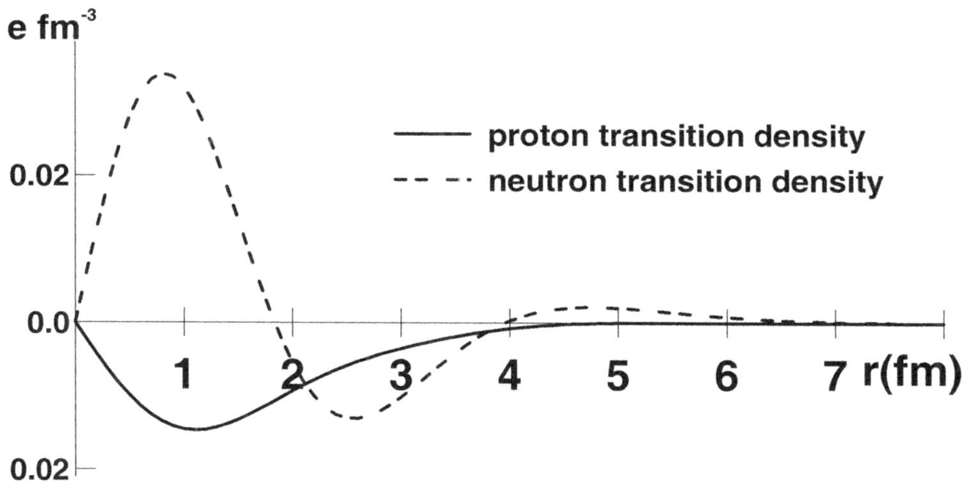

Fig. 1. Proton and neutron transition densities associated to the 10.64 MeV QRPA 1^- state in the ^{26}Ne as functions of distance.

4. Giant resonances in s-d shell nuclei

As discussed in Ref. 6, the s-d shell nuclei ^{24}Mg and ^{28}Si are good candidates to apply the ASD-QRPA approach: they are deformed and many experimental data are available. Predictions are also made for other magnesium and silicon isotopes to study the impact of intrinsic nuclear deformation on giant resonances.

Within constrained HFB calculations, the N=16 nuclei, $^{30}_{14}\mathrm{Si}_{16}$ and $^{28}_{12}\mathrm{Mg}_{16}$, are found to be spherical, the two $^{22}_{12}\mathrm{Mg}_{10}$ and $^{24}_{12}\mathrm{Mg}_{12}$ isotopes are predicted to be strongly prolate, and the $^{26}_{14}\mathrm{Si}_{12}$, $^{28}_{14}\mathrm{Si}_{14}$ and $^{26}_{12}\mathrm{Mg}_{14}$ nuclei are oblate deformed. These HFB ground states have been used as trial states in the QRPA calculations.

Theoretical isovector dipole responses are shown in Fig. 5 for $^{22-28}$Mg and $^{26-30}$Si. In spherical nuclei, responses in ^{28}Mg and ^{30}Si, are similar: a highly collective state is found at 23 MeV and two less collective fragments around 20 MeV and 25 MeV. In deformed nuclei, the strength is fragmented and is split into two components as expected.[7] The two components correspond to two different projections of the angular momentum. More precisely, in the prolate $^{22-24}$Mg isotopes, low-energy components are made of K = 0 states whereas high-energy ones contain $|K|$ = J = 1 states. The same feature is found in oblate nuclei, $^{26-28}$Si and ^{26}Mg, but the low-energy components contain $|K|$ = J = 1 states and the high-energy ones K = 0 states. The energy splitting between K = 0 and $|K|$ = 1 states appears to be related to the deformation of the ground state. Keeping in mind that for prolate shape the long axis is the symmetry axis whereas for oblate shape the symmetry axis is the shorter one, the alignment of the angular momentum along the long axis gives $|K|$ = J for prolate shape and K = 0 for oblate one. Finally, let us mention that the theoretical responses for ^{24}Mg are in good agreement with experimental

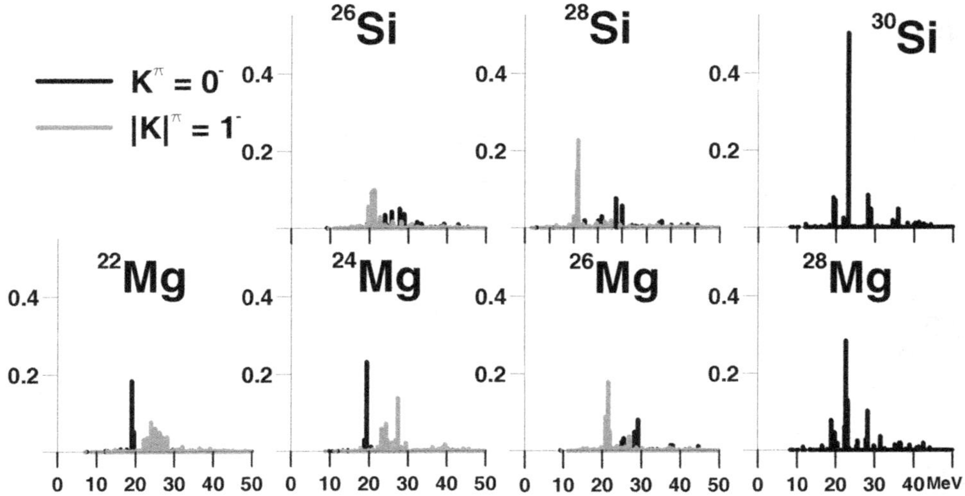

Fig. 2. Dipole response as fractions of the EWSR in $^{22-28}$Mg and $^{26-30}$Si. $K^\pi = 0^-$ and $|K|^\pi = 1^-$ components are indicated in black and grey respectively.

data.[8]

Studies of isoscalar modes, in particular monopole and quadrupole resonances, in $^{22-28}$Mg and $^{26-30}$Si nuclei are in progress.[9] As already mentioned in Ref. 6, theoretical Giant Monopole Resonances (GMRs) in ^{24}Mg and ^{28}Si are in good agreement with measurements.[10,11] By integrating the data from 9 MeV to 41 MeV the experimental mean value of the energy of the GMR is found to be 21.0 $\pm$0.6 MeV in ^{24}Mg and 21.25 $\pm$ 0.38 MeV in ^{28}Si. Our results give 20.57 MeV (^{24}Mg) and 21.15 MeV (^{28}Si).

Theoretical mean energies of the Giant Quadrupole Resonances are found to be 19.84 MeV in ^{24}Mg and 19.95 MeV in ^{28}Si. These values are larger than the experimental ones: 16.9$\pm$0.6 MeV[10] and 18.54 $\pm$ 0.25 MeV[11] in ^{24}Mg and ^{28}Si, respectively. Keeping in mind that our theoretical description does not include any adjustable parameters, the agreement we find with experimental data can be considered quite satisfactory.

5. Summary

A fully microscopic QRPA approach in axial symmetry has been applied to the exotic ^{26}Ne nucleus. The theoretical results for the dipole resonance show a pygmy state with an energy close to the experimental one. This state is explained as an oscillation of the neutron skin against a - non-inert -core.

Using the same formalism, calculations have been made in different s-d shell nuclei to analyse the main features of the giant resonance in deformed and spherical nuclei. Theoretical results on dipole resonance show a splitting of the strength in deformed nuclei in agreement with measurements. Mean values of the energy calculated in ^{24}Mg and in ^{28}Si for isoscalar monopole and quadrupole modes are in agreement with experimental data and encourage us to perform calculations of isocalar modes in other magnesium and silicon isotopes.

References

1. P. Ring and P. Schuck, *The Nuclear Many Body Problem*, (Springer-Verlag, New-York, USA, 1980).
2. J. Dechargé and D. Gogny, *Phys. Rev. C* **21**, 1568 (1980).
   J.F. Berger, M. Girod, and D. Gogny, *Comp. Phys. Comm.* **63**, 365 (1991).
3. Li-Gang Cao and Zhong-Yu Ma, *Phys. Rev. C* **71**, 034305 (2005).
4. J. Gibelin,et al, *Nucl. Phys. A* **788**, 153c (2007).
5. A. Obertelli, S. Péru et al, *Phys. Rev. C* **71**, 024304 (2005).
6. S. Péru, H. Goutte and J.-F. Berger, *Nucl. Phys. A* **788**, 44c (2007).
7. M.N. Harakeh, A. van der Woude, Giant Resonances: *Fundamental High-energy Modes of Nuclear Excitation*, (Oxford Un. Press, Oxford, 2001).
8. B.S. Ishkhanov et al,*Nucl. Phys. A* **186**, 438 (1972).
9. S. Péru and H. Goutte, *in preparation*.
10. D.H. Youngblood, Y.-W. Lui, and H.L. Clark, *Phys. Rev. C* **60**, 014304 (1999).
11. D.H. Youngblood, Y.-W. Lui, and H.L. Clark, *Phys. Rev. C* **65**, 034302 (2002).

SOFT $K^\pi = 0^+$ MODES UNIQUE TO DEFORMED NEUTRON-RICH UNSTABLE NUCLEI

K. YOSHIDA* and K. MATSUYANAGI

Department of Physics, Graduate School of Science, Kyoto University
Kyoto 606-8502, Japan
**E-mail: kyoshida@ruby.scphys.kyoto-u.ac.jp*

M. YAMAGAMI

RIKEN Nishina Center, RIKEN, Hirosawa 2-1, Wako, Saitama 351-0198, Japan

Low-frequency collective $K^\pi = 0^+$ modes in deformed neutron-rich unstable nuclei are investigated by means of the quasiparticle-random-phase approximation (QRPA) based on the coordinate space Hartree-Fock-Bogoliubov (HFB) formalism. It is found that the spatially extended structure of neutron quasiparticle wave-functions brings about a striking enhancement of the transition strengths. It is also found that the quadrupole pairing correlation plays an important role in generating coherence among two-quasiparticle excitations of neutrons.

Keywords: Neutron rich nuclei; ^{34,36,38,40}Mg, ^{62,64,66,68,70}Cr and ^{64,66,68,70,72}Fe; Collective modes; Deformed quasiparticle-random-phase approximation; Hartree-Fock-Bogoliubov method

1. Introduction

Quest for new kinds of collective motion in nuclei far from $\beta-$stability is one of the most interesting issues in nuclear structure physics and has been actively studied both experimentally and theoretically.[1] Because low-frequency collective excitations are quite sensitive to shell structure near the Fermi level, we can expect unique excitation modes to emerge associated with neutron skin and novel shell structure in unstable nuclei. In neutron rich nuclei, breaking of the spherical magic number $N = 20, 28$ and striking enhancement of $B(E2; 0_1^+ \to 2_1^+)$ in Mg isotopes towards drip line are under lively discussions in connection with onset of the quadrupole deformation and continuum coupling.[2–11] Furthermore in heavier mass region, collectivities and breaking of the $N = 40$ sub-shell have been recently discussed for neutron-rich Cr and Fe isotopes.[12–14] Although many investigations[15–28] on multipole responses of neutron-rich nuclei are available, they are largely restricted to spherical systems. Quite recently, low-frequency RPA modes in deformed neutron-rich unstable nuclei have been investigated by several groups .[29–33] These calculations, however, do not take into account the pairing correlation, or rely on the BCS pairing, which is inappropriate for describing the pairing correlation in drip line nuclei due to the unphysical nucleon gas problem.[34] In Ref. 35, we constructed a new computer code that carries out the deformed QRPA calculation based on the coordinate-space HFB formalism. In this talk, we discuss possible emergence of collective low-frequency (soft) $K^\pi = 0^+$ modes in deformed neutron-rich nuclei.

2. Formalism

We briefly summarize our approach (see Ref. 35 for details). In order to simultaneously take into account effects of nuclear deformation and pairing correlations including the continuum, we solve the HFB equation[34]

$$\begin{pmatrix} h^\tau(\boldsymbol{r}\sigma) - \lambda^\tau & \tilde{h}^\tau(\boldsymbol{r}\sigma) \\ \tilde{h}^\tau(\boldsymbol{r}\sigma) & -(h^\tau(\boldsymbol{r}\sigma) - \lambda^\tau) \end{pmatrix} \begin{pmatrix} \varphi^\tau_{1,\alpha}(\boldsymbol{r}\sigma) \\ \varphi^\tau_{2,\alpha}(\boldsymbol{r}\sigma) \end{pmatrix} = E_\alpha \begin{pmatrix} \varphi^\tau_{1,\alpha}(\boldsymbol{r}\sigma) \\ \varphi^\tau_{2,\alpha}(\boldsymbol{r}\sigma) \end{pmatrix} \tag{1}$$

directly in the cylindrical-coordinate space assuming axial and reflection symmetry, where $\boldsymbol{r} = (\rho, z, \phi)$. For the mean-field Hamiltonian h, we employ the deformed Woods-Saxon potential with the parameters used in Ref. 29, except the isovector potential strength for Mg isotopes for which a slightly smaller value, 30 MeV in stead of 33 MeV, is adopted in order to describe ^{40}Mg as a drip-line nucleus in accordance with the Skyrme-HFB[9] and Gogny-HFB calculations.[10] The pairing field is treated self-consistently by using the density-dependent contact interaction[36]

$$v_{pp}(\boldsymbol{r}, \boldsymbol{r}') = V_0 \frac{1 - P_\sigma}{2} \left[1 - \frac{\varrho^{\mathrm{IS}}(\boldsymbol{r})}{\varrho_0} \right] \delta(\boldsymbol{r} - \boldsymbol{r}'). \tag{2}$$

We use the lattice mesh size $\Delta\rho = \Delta z = 0.8$ fm and the box boundary condition at $\rho_{\max} = 10.0$ fm and $z_{\max} = 12.8$ fm. The quasiparticle energy is cut off at 50 MeV and the quasiparticle states up to $\Omega^\pi = 13/2^\pm$ for Mg isotopes, and $\Omega^\pi = 15/2^\pm$ for Cr and Fe isotopes are included.

Using the quasiparticle basis obtained by solving the HFB equation (1), we solve the QRPA equation in the matrix formulation

$$\sum_{\gamma\delta} \begin{pmatrix} A_{\alpha\beta\gamma\delta} & B_{\alpha\beta\gamma\delta} \\ B_{\alpha\beta\gamma\delta} & A_{\alpha\beta\gamma\delta} \end{pmatrix} \begin{pmatrix} f^\lambda_{\gamma\delta} \\ g^\lambda_{\gamma\delta} \end{pmatrix} = \hbar\omega_\lambda \begin{pmatrix} 1 & 0 \\ 0 & -1 \end{pmatrix} \begin{pmatrix} f^\lambda_{\alpha\beta} \\ g^\lambda_{\alpha\beta} \end{pmatrix}. \tag{3}$$

The residual interactions in the particle-particle channel appearing in the QRPA matrices A and B are the density-dependent contact interaction (2). On the other hand, for residual interactions in the particle-hole channel, we use the Skyrme-type interaction[37]

$$v_{ph}(\boldsymbol{r}, \boldsymbol{r}') = \left[t_0(1 + x_0 P_\sigma) + \frac{t_3}{6}(1 + x_3 P_\sigma)\varrho^{\mathrm{IS}}(\boldsymbol{r}) \right] \delta(\boldsymbol{r} - \boldsymbol{r}'), \tag{4}$$

with $t_0 = -1100$ MeV·fm^3, $t_3 = 16000$ MeV·fm^6, $x_0 = 0.5$, and $x_3 = 1.0$. Because the deformed Wood-Saxon potential is used for the mean field, we renormalize the residual interaction in the particle-hole channel by multiplying a factor f_{ph} to get the spurious $K^\pi = 1^+$ mode (representing the rotational mode) at zero energy

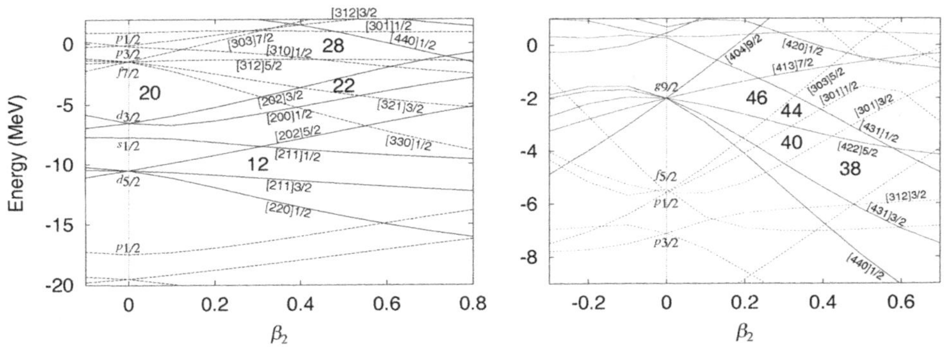

Fig. 1. Single-particle energies in the deformed WS potential for neutrons in ^{40}Mg (left) and ^{64}Cr (right), plotted as functions of the quadrupole deformation parameter β_2. Solid and dotted lines denote positive- and negative-parity levels, respectively. Single-particle levels are labelled with the asymptotic quantum numbers $[Nn_3\Lambda]\Omega$.

($v_{ph} \to f_{ph} \cdot v_{ph}$). We cut the model space by $E_\alpha + E_\beta \leq 30$ MeV due to the excessively demanding computer memory for the model space which is consistent with that adopted in the HFB calculation. Accordingly, we need another factor f_{pp} for the particle-particle channel. We determine this factor such that the spurious $K^\pi = 0^+$ mode associated with the number fluctuation appears at zero energy ($v_{pp} \to f_{pp} \cdot v_{pp}$).

3. Results and Discussion

3.1. *Mg isotopes close to the neutron drip line*

Figure 1 shows the single-particle energy diagram for the WS potential as a function of deformation parameter β_2. As β_2 increases, a level crossing between the up-sloping [303]7/2 level and the down-sloping [310]1/2 level takes place and a deformed shell gap is formed at $N = 28$ around $\beta_2 = 0.4$ (see the left panel of Fig. 1). This deformed closed shell approximately corresponds to the $(f_{7/2})^{-2}(p_{3/2})^2$ configuration in the spherical shell model representation.[8] The highest occupied level in this deformed closed shell is situated very near to the continuum threshold, so that there is no bound level above it. Neutron two-quasiparticle (2qp) excitations may take place into resonance levels like $\nu[303]7/2$, $\nu[301]1/2$ and $\nu[312]3/2$ lying just above the continuum threshold.

In Fig. 2, we show the isoscalar quadrupole transition strengths for the $K^\pi = 0^+$ excitations in Mg isotopes close to the drip line. The lowest excited state with large transition strength in ^{40}Mg is located above the neutron threshold. As discussed in Ref. 29, this $K^\pi = 0^+$ mode in ^{40}Mg is generated by coherent superposition of neutron 2qp excitations in loosely bound and resonant states. Among them, the neutron 2qp excitations, $(\nu[310]1/2)^2$, $(\nu[312]3/2)^2$ and $(\nu[303]7/2)^2$, play the major role.

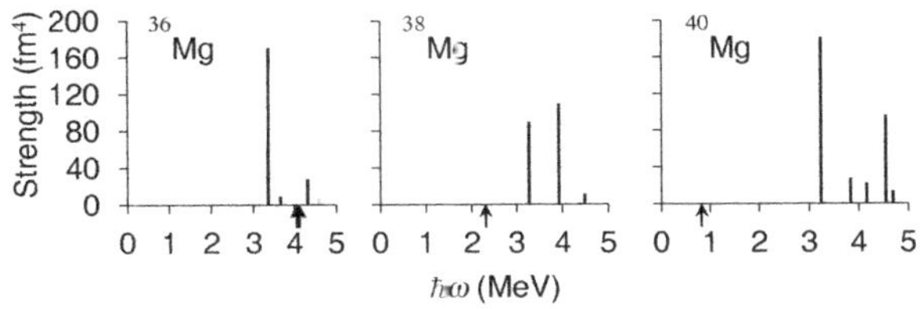

Fig. 2. Isoscalar quadrupole transition strengths $B(Q^{IS}2)$ for the $K = 0^+$ excitations built on the prolately deformed ground states of ^{36,38,40}Mg. The arrows beside the abscissa axes indicate the neutron threshold energies, $E_{th} = 4.06$ MeV (1qp continuum; $|\lambda| + \min E_\alpha$), 4.12 MeV (2qp continuum; $2|\lambda|$) for ^{36}Mg, 2.31 MeV (2qp continuum) for ^{38}Mg and 0.82 MeV (2qp continuum) for ^{40}Mg. The QRPA calculations are made by using the strength of $V_0 = -450$ MeV·fm^3 and $\beta_2 = 0.28$ for both protons and neutrons, and $f_{ph} = 0.37$, $f_{pp} = 1.54$.

3.2. ^{34}Mg

At around $\beta_2 = 0.5$ the deformed shell gap is formed at $N = 22$ due to the crossings between the up-sloping [202]3/2 and the down-sloping [330]1/2 and [321]3/2 levels (see the left panel of Fig. 1). In Fig. 3, we show the excitation spectrum for ^{34}Mg obtained by the QRPA calculation. Here excitation energies are evaluated by

$$E(I,K) = \hbar\omega_{\rm RPA} + \frac{\hbar^2}{2\mathcal{J}_{\rm TV}}(I(I+1) - K^2), \tag{5}$$

with the vibrational frequencies, $\omega_{\rm RPA}$, and the Thouless-Valatin moment of inertia, $\mathcal{J}_{\rm TV}$, calculated by the QRPA. The left panel of Fig. 4 shows the $K^\pi = 0^+$ isoscalar quadrupole transition strength. The intrinsic transition strength for the lowest $K^\pi = 0^+$ excited state is about 43 Weisskopf units (1 W.u. $\simeq 6.5$ fm^4 for ^{34}Mg). This state is created by coherent superposition of 2qp excitations among the up-sloping ν[202]3/2 and the down-sloping ν[330]1/2 and ν[321]3/2 levels, together with high-lying $2\hbar\omega$ excitations of protons.[39] In the middle and right panels of Fig. 4, we show two kinds of strength for neutron-pair transitions; one is the monopole-pair

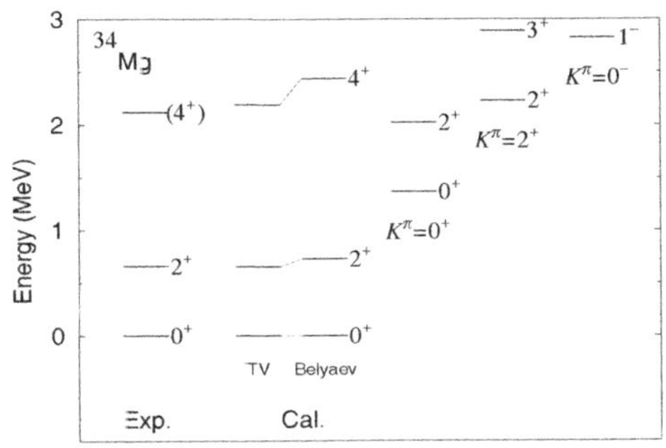

Fig. 3. Excitation spectrum for ^{34}Mg obtained by the QRPA calculation using $\beta_2 = 0.4$ and $V_0 = -400$ MeV·fm^3 and available experimental data.[4]

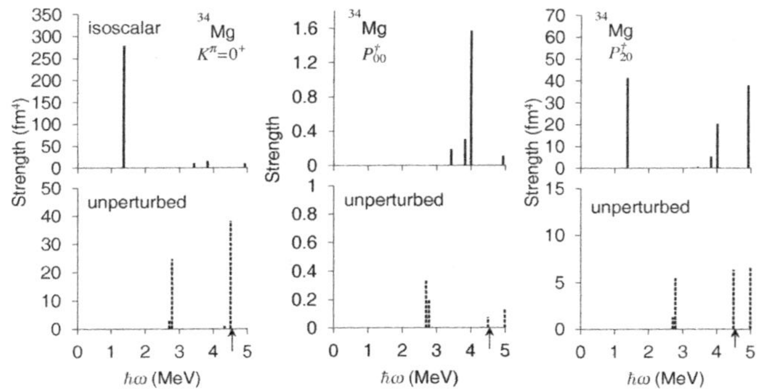

Fig. 4. QRPA strength distributions for the $K^\pi = 0^+$ isoscalar quadrupole particle-hole excitations (left) and the monopole- and quadrupole-pair excitations (middle and right) in ^{34}Mg, calculated using $f_{ph} = 0.74$ and $f_{pp} = 1.37$. For comparison, unperturbed 2qp transition strengths are shown in the lower panels. The arrow indicates the neutron threshold energy $E_{\rm th} = 4.55$ MeV for 1qp continuum.

transition and another the quadrupole-pair transition,

$$\hat{P}^\dagger_{00} = \int d\boldsymbol{r}\hat{\psi}^\dagger(\boldsymbol{r}\uparrow)\hat{\psi}^\dagger(\boldsymbol{r}\downarrow), \qquad \hat{P}^\dagger_{20} = \int d\boldsymbol{r}r^2Y_{20}(\hat{r})\hat{\psi}^\dagger(\boldsymbol{r}\uparrow)\hat{\psi}^\dagger(\boldsymbol{r}\downarrow). \tag{6}$$

For the quadrupole-pair transitions, we see a prominent peak at 1.37 MeV, whereas no peak is seen at this energy for the monopole-pair transitions. The origin of this contrasting behavior may be understood as follows: The quadrupole (monopole) pairing matrix elements for individual 2qp excitations are in opposite phase (in phase) between the down-sloping prolate-type levels and the up-sloping oblate-type levels. On the other hand, because the major component of this QRPA eigenmode is associated with pairing fluctuation, the phase of the QRPA amplitudes, $f^\lambda_{\alpha\beta}$ and $g^\lambda_{\alpha\beta}$, are opposite between these hole-like down-sloping levels and the particle-like up-sloping levels (due to the $(u^2 - v^2)$ factor; see *Appendix J* of Ref. 38). As the transition matrix elements for the QRPA eigenmodes are determined by the sum of products of the QRPA amplitudes and the individual 2qp transition matrix elements, they contributes coherently and destructively for the quadrupole and monopole pair transitions, respectively. As a consequence, the quadrupole pair transition strength to the lowest $K^\pi = 0^+$ collective excited state is greatly enhanced, whereas the monopole pair transition to this state almost vanishes. Furthermore, we have found that the coherent coupling effect between the particle-hole type quadrupole-shape fluctuations and the particle-particle (hole-hole) type pairing fluctuations play an indispensable role in generating the collectivity of the low-frequency $K^\pi = 0^+$ modes in deformed neutron-rich nuclei. Detailed discussion on microscopic mechanism generating such soft $K^\pi = 0^+$ collective modes will be given in a forthcoming publication.[39]

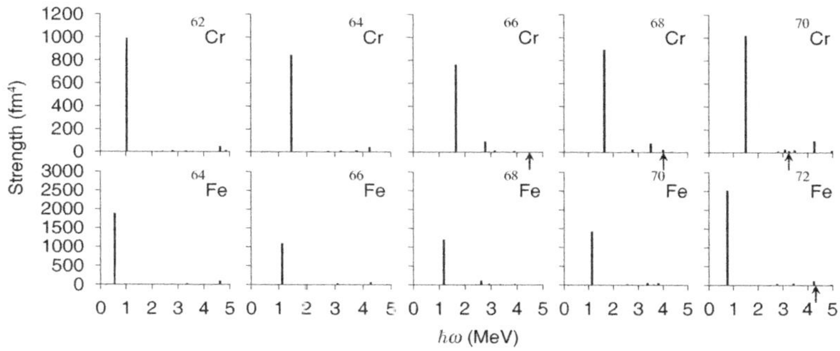

Fig. 5. Isoscalar quadrupole transition strengths for the $K^\pi = 0^+$ excitations in $^{62-70}$Cr and $^{64-72}$Fe.

3.3. *Cr and Fe isotopes with $N \approx 40$*

To show generic features of the soft $K^\pi = 0^+$ modes discussed above, we here present some results of calculation for neutron-rich Cr and Fe isotopes with $N \approx 40$. As seen in the right panel of Fig. 1, deformed shell gaps associated with different β_2 values appear at various neutron numbers due to the crossings between the down-sloping $\nu[440]1/2$, $\nu[431]1/2$ and $\nu[431]1/2$ levels and the up-sloping $\nu[301]3/2$, $\nu[301]1/2$ and $\nu[303]5/2$ levels.

Figure 5 shows the isoscalar quadrupole transition strengths for the $K^\pi = 0^+$ excitations. We can see a prominent peak at around 1 MeV in all nuclei under consideration. All low-lying states obtained here have extremely enhanced transition strengths (note that 1 W. u. is 14.6–17.1 fm^4 for $^{62-70}$Cr and 15.2–17.8 fm^4 for $^{64-72}$Fe). These excitation modes are generated by coherent superposition of neutron 2qp excitations among the up-sloping and the down-sloping levels.[39]

4. Concluding remarks

We have studied soft $K^\pi = 0^-$ modes in neutron-rich unstable nuclei by means of the QRPA simultaneously taking account of the effects of nuclear deformation, pairing correlation and continuum coupling. It is found that the spatially extended structure of neutron quasiparticle wave-functions brings about a striking enhancement of the transition strengths. It is also found that the coherent coupling effect between the quadrupole pairing fluctuation and the quadrupole shape fluctuation of neutron-skin brings forth the soft $K^\pi = 0^+$ modes in the deformed ^{34}Mg region. To demonstrate the generic feature of the soft $K^\pi = 0^+$ modes in deformed neutron-rich nuclei, we have presented several examples of the QRPA calculation also for neutron-rich Cr and Fe isotopes with $N \approx 40$. It is confirmed that the coherent coupling between the quadrupole-pairing fluctuation and the quadrupole shape fluctuation of neutron skin play the major role in generating their collectivity in the deformation region where the up-sloping and the down-sloping neutron levels cross each other.

Acknowledgments

One of the authors (K.Y) is supported by Research Fellowships of the Japan Society for the Promotion of Science for Young Scientists.

References

1. P. von Neumann-Cosel and T. Aumann (ed.), *Proceedings of the 2nd International Conference on Collective Motions in Nuclei under Extreme Conditions*, Nucl. Phys. **A788**, (2007).
2. T. Motobayashi, *et al.*, *Phys. Lett. B* **349**, 9 (1995).
3. H. Iwasaki, *et al.*, *Phys. Lett. B* **522**, 227 (2001).
4. K. Yoneda, *et al.*, *Phys. Lett. B***499**, 233 (2001).
5. J. A. Church, *et al.*, *Phys. Rev. C* **72**, 054320 (2005).
6. Z. Elekes, *et al.*, *Phys. Rev. C* **73**, 044314 (2006).
7. T. Otsuka, Y. Utsuno, M. Honma, and T. Mizusaki, *Prog. Part. Nucl. Phys.* **46**, 155 (2001).
8. E. Caurier, F. Nowacki, A. Poves, *Nucl. Phys. A* **742**, 14 (2004).
9. J. Terasaki, H. Flocard, P.-H. Heenen, and P. Bonche, *Nucl. Phys. A* **621**, 706 (1997).
10. R. Rodoríguez-Guzmń, J. L. Egido, L. M. Robledo, *Nucl. Phys. A* **709**, 201 (2002).
11. P.-G. Reinhard, *et al.*, *Phys. Rev. C* **60**, 014316 (1999) and references therein.
12. M. Hannawald *et al.*, *Phys. Rev. Lett.* **82**, 1391 (1999).
13. O. Sorlin *et al.*, *Eur. Phys. J. A* **16**, 55 (2003).
14. E. Caurier, F. Nowacki, and A. Poves, *Eur. Phys. J. A* **15**, 145 (2002).
15. I. Hamamoto, H. Sagawa and X. Z. Zhang, Phys. Rev. C **53**, 765 (1996); ibid. **55**, 2361 (1997); ibid. **56**, 3121 (1997); ibid. **57**, R1064 (1998); ibid. **64**, 024313 (2001).
16. S. Shlomo and B. Agrawal, *Nucl. Phys. A* **722**, 98c (2003).
17. M. Matsuo, *Nucl. Phys. A* **696**, 371 (2001).
18. K. Hagino, H. Sagawa, *Nucl. Phys. A* **695**, 82 (2001).
19. E. Khan, N. Sandulescu, M. Grasso, N. Van Giai, *Phys. Rev. C* **66**, 024309 (2002).
20. M. Yamagami, N. Van Giai, *Phys. Rev. C* **69**, 034301 (2004).
21. M. Matsuo, K. Mizuyama, Y. Serizawa, *Phys. Rev. C* **71**, 064326 (2005).
22. J. Terasaki, *et al.*, *Phys. Rev. C* **71**, 034310 (2005) and references therein.
23. J. Terasaki, J. Engel, *Phys. Rev. C* **74**, 044301 (2006).
24. L. G. Cao, Z. Y. Ma, *Phys. Rev. C* **71**, 034305 (2005).
25. D. Vretenar, A. V. Afanasjev, G. A. Lalazissis, P. Ring, *Phys. Rep.* **409**, 101 (2005).
26. G. Giambrone, *et al.*, *Nucl. Phys. A* **726**, 3 (2003).
27. D. Sarchi, P. F. Bortignon, G. Colò, *Phys. Lett. B* **601**, 27 (2004).
28. S. Péru, J. F. Berger, P. F. Bortignon, *Eur. Phys. J. A* **26**, 25 (2005).
29. K. Yoshida, M. Yamagami, K. Matsuyanagi, *Prog. Theor. Phys.* **113**, 1251 (2005).
30. T. Nakatsukasa and K. Yabana, *Phys. Rev. C* **71**, 024301 (2005).
31. T. Inakura, *et al.*, *Nucl. Phys. A* **768**, 61 (2006).
32. P. Urkedal, X. Z. Zhang and I. Hamamoto, *Phys. Rev. C* **64**, 054304 (2001).
33. K. Hagino, N. Van Giai and H. Sagawa, *Nucl. Phys. A* **731**, 264 (2004).
34. J. Dobaczewski, H. Flocard and J. Treiner, *Nucl. Phys. A* **422**, 103 (1984).
35. K. Yoshida, M. Yamagami, K. Matsuyanagi, *Nucl. Phys. A* **779**, 99 (2006).
36. G. F. Bertsch, H. Esbensen, *Ann. Phys.* **209**, 327 (1991).
37. S. Shlomo and G. F. Bertsch, *Nucl. Phys. A* **243**, 507 (1975).
38. D.M. Brink and R.A. Broglia, *Nuclear Superfluidity, Pairing in Finite Systems* (Cambridge University Press, 2005).
39. K. Yoshida, *et al.*, in preparation.

SYNTHESIS, DECAY PROPERTIES, AND IDENTIFICATION OF SUPERHEAVY NUCLEI PRODUCED IN ^{48}CA-INDUCED REACTIONS

YU.TS. OGANESSIAN, V.K. UTYONKOV, YU.V. LOBANOV, F.SH. ABDULLIN, A.N. POLYAKOV, R.N. SAGAIDAK, I.V. SHIROKOVSKY, YU.S. TSYGANOV, A.A. VOINOV, S. ILIEV, V.G. SUBBOTIN, A.M. SUKHOV, G.G. GULBEKIAN, S.L. BOGOMOLOV, B.N. GIKAL, A.N. MEZENTSEV, K. SUBOTIC, V.I. ZAGREBAEV, AND M.G. ITKIS

Joint Institute for Nuclear Research, 141980 Dubna, Russian Federation

K.J. MOODY, R.A. HENDERSON, J.B PATIN, D.A. SHAUGHNESSY, M.A. STOYER, N.J. STOYER, P.A. WILK, J.M. KENNEALLY, J.H. LANDRUM, J.F. WILD, AND R.W. LOUGHEED

University of California, Lawrence Livermore National Laboratory, Livermore, California 94551, USA

Thirty-four new nuclides with Z=104-116, 118 and N=161-177 have been synthesized in the complete-fusion reactions of ^{238}U, ^{237}Np, ^{242,244}Pu, ^{243}Am, ^{245,248}Cm, and ^{249}Cf targets with ^{48}Ca beams. The masses of evaporation residues were identified through measurements of the excitation functions of the xn-evaporation channels and from cross bombardments. The decay properties of the new nuclei agree with those of previously known heavy nuclei and with predictions from different theoretical models. A discussion of self-consistent interpretations of all observed decay chains originating from the parent isotopes ^{282,283}112, ^{282}113, $^{286\text{-}289}$114, ^{287,288}115, $^{290\text{-}293}$116, and ^{294}118 is presented. Decay energies and lifetimes of the neutron-rich superheavy nuclei as well as their production cross sections indicate a considerable increase in the stability of nuclei with the approach to the theoretically predicted nuclear shells with N=184 and Z=114.

1 Introduction

The existence of a region of superheavy nuclei located beyond the domain of the heaviest known nuclei has been hypothesized for about 40 years. Calculations performed with different versions of the nuclear model predict a substantial enhancement of the stability of heavy nuclei when approaching the closed spherical shells at Z=114 and N=184.

In the reactions of the doubly magic ^{48}Ca projectile with isotopes of heavy actinide elements, e.g., ^{244}Pu or ^{248}Cm, the resulting compound nuclei should have excitation energies of about 30 MeV at the Coulomb barrier. Nuclear shell effects are still expected to persist in the excited nucleus, thus increasing the survival probability of the evaporation residues (ER), as compared to "hot fusion" reactions ($E^*\approx$45-55 MeV), which were used for the synthesis of isotopes of elements Sg-Ds. Additionally, the high mass asymmetry in the entrance channel should decrease the dynamic limitations on nuclear fusion that arise in more symmetrical "cold fusion" reactions.

In spite of the advantages of ^{48}Ca-induced reactions in comparison with hot or cold complete-fusion reactions, past attempts to synthesize new elements in the reactions of

^{48}Ca projectiles with actinide targets resulted only in upper limits on their production cross sections [1]. Our present experiments are designed to attempt the production of elements 112-116 and 118 in reactions of ^{233,238}U, ^{237}Np, ^{242,244}Pu, ^{243}Am, ^{245,248}Cm, and ^{249}Cf with ^{48}Ca at the picobarn cross-section level, thus exceeding the sensitivity of the previous experiments by at least two orders of magnitude.

2 Experimental Technique

The ^{48}Ca ion beam was accelerated by the U400 cyclotron at the Flerov Laboratory of Nuclear Reactions. The typical beam intensity at the target was 1.2 pμA. The beam energy was determined with a precision of 1 MeV by a time-of-flight technique. The 32-cm^2 rotating targets consisted of the enriched (≥ 97.3%) isotopes of U to Cf deposited as oxides onto 1.5-μm Ti foils to thicknesses of about 0.34-0.40 mg cm^{-2}.

The ERs recoiling from the target were spatially separated in flight from ^{48}Ca beam ions, scattered particles and transfer-reaction products by the Dubna Gas-filled Recoil Separator. The transmission efficiency of the separator for Z=112 to 118 nuclei was estimated to be about 35-40%. Evaporation residues recoiling from the target passed through a time-of-flight system and were implanted in a 4×12-cm^2 semiconductor detector array with 12 vertical position-sensitive strips, located at the separator's focal plane. This detector was surrounded by eight 4×4-cm^2 side detectors, forming a box of detectors open from the beam side. The position-averaged detection efficiency for full-energy α particles from the decay of implanted nuclei was 87%. For detection of sequential decays of synthesized nuclides in the absence of beam-associated background, the beam was switched off automatically after a recoil was detected with an implantation energy expected for complete-fusion ERs, followed by an α-like signal with an energy expected for α decays of the parent and sometimes the daughter nuclei. Thus, the decays of the daughter nuclides were observed under very low-background conditions.

3 Experimental Results and Discussion

The most neutron-rich nuclei with the magic proton number 114 that is predicted by macroscopic-microscopic (MM) theory can be produced in the fusion reaction ^{244}Pu+^{48}Ca. The decay properties of nuclei observed in this reaction and their corresponding excitation functions [2] are shown in Figs. 1 and 2, respectively. At the three lowest ^{48}Ca energies, we synthesized an isotope with E_α=9.82 MeV and $T_{1/2}$=2.6 s that underwent two consecutive α decays followed by SF. At the three higher projectile energies, the neighboring isotope was observed, with E_α=9.95 MeV and $T_{1/2}$=0.80 s; its α decay was followed by SF of the daughter nuclide with $T_{1/2}$=97 ms. Finally, a third isotope with an even greater α-particle energy and a correspondingly lower half-life was produced at the highest bombarding energy.

As seen in Fig. 2, the measured excitation functions for the three isotopes are in agreement with empirical expectations and calculations [4] for the complete-fusion reaction ^{244}Pu+^{48}Ca followed by evaporation of 3, 4 and 5 neutrons. It is reasonable to assume that the first of the aforementioned isotopes was produced in the $[x]n$-evaporation channel leading to $^{289}114$ and the other two isotopes were produced in the $[x+1]n$ and $[x+2]n$ channels ($^{288}114$ and $^{287}114$, respectively).

Together with cross-section measurements, another approach for the mass and atomic-number identification of unknown nuclei is the method of cross bombardments, which was widely applied in previous experiments [5]. For instance, a more solid mass identification is possible if the mass number of the target nucleus is varied. Thus, if the suggested mass assignment is correct, the isotopes $^{288}114$ and $^{287}114$ could be observed in the $2n$ and $3n$ channels of the reaction $^{242}Pu+^{48}Ca$, and another even-even isotope, $^{286}114$, with a higher α-particle energy and lower lifetime could be produced via the $4n$-evaporation channel. Indeed, in the experiments aimed at the measurement of cross sections in the reaction $^{242}Pu(^{48}Ca,xn)^{290-x}114$, we observed $^{288}114$ at the lowest ^{48}Ca energy and $^{287}114$ at the three lowest energies. The α-decay of the new isotope, $^{286}114$, which undergoes SF and α decay with about equal probability (E_α=10.19 MeV, $T_{1/2}$=0.13 s), was followed by a SF nuclide with $T_{1/2}$=0.82 ms. This nuclide was produced at the two highest energies. Thus, following the previous considerations, these isotopes should be the products of the $[x-1]n$, $[x]n$ and $[x+1]n$-reaction channels.

Considering the decay properties of the observed nuclei, particularly their decay modes and the number of observed α decays before a SF is encountered, one can conclude that the first and third of the four consecutive element-114 isotopes should possess an odd number of neutrons while the second and fourth ones should be even-N

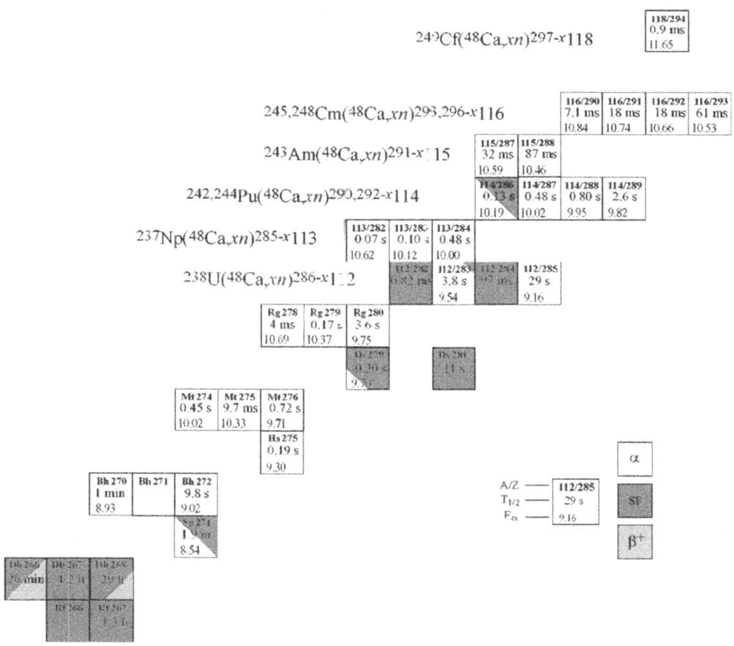

Figure 1. Half-lives, α-particle energies and decay modes of nuclei produced in corresponding reactions of ^{48}Ca ions with actinide targets [2].

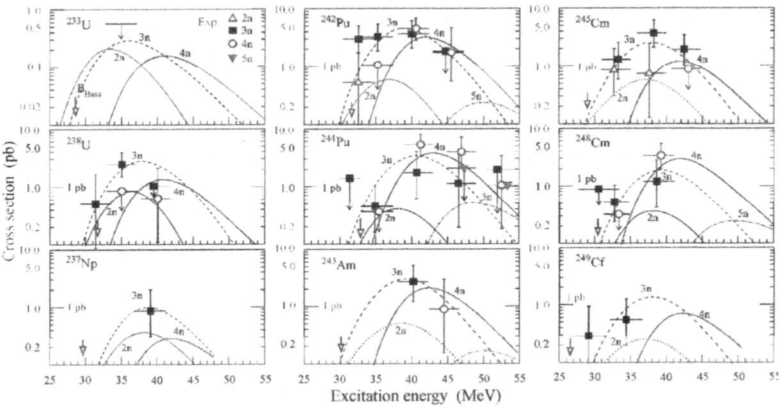

Figure 2. Excitation functions for the 2n to 5n evaporation channels from different complete-fusion reactions. The Bass barrier [3] is shown by an open arrow. Lines show the results of calculations of xn channel [4].

isotopes. The unpaired neutron in the odd-N isotopes increases the SF partial half life by 3-5 orders of magnitude; therefore, the value x can only be equal to 1, 3, 5 etc.

Another implementation of the method of cross bombardments in our case could be the production of the same isotopes of element 114 as the daughter nuclei following the α decays of heavier parent nuclei with Z=116. This method was used in our experiment aimed at the synthesis of superheavy nuclei with Z=116 in the reaction ^{248}Cm+^{48}Ca [2] in which two isotopes were observed. All the decays following the first α emission agree well with the decay chains of the [x]n and [x+1]n channels, ^{289}114 and ^{288}114, previously observed in the ^{244}Pu+^{48}Ca reaction. Thus, it was reasonable to assign the observed decays to the nuclides ^{293}116 and ^{292}116, produced via evaporation of the same number of neutrons in the reaction ^{248}Cm+^{48}Ca.

Two lighter neighboring isotopes of element 116 were produced in the reaction ^{245}Cm+^{48}Ca [2]. As in the previous case, the decay chains following the first α particle of the isotope observed at the two lowest projectile energies agree with those observed for the ^{287}114 parent nuclide from the reactions ^{244}Pu+^{48}Ca at the highest ^{48}Ca energy and ^{242}Pu+^{48}Ca at the three lowest energies. The decay properties of the descendant nuclei of the lighter isotope observed at three energies agree with those of ^{286}114 produced in the reaction ^{242}Pu+^{48}Ca at the two highest energies. Moreover, this lighter isotope was also observed in the reaction ^{249}Cf+^{48}Ca after the α decay of the parent element-118 nucleus [2]. The granddaughter nuclei from the parent nuclides of the reaction ^{245}Cm+^{48}Ca were also produced in the direct reaction ^{238}U+^{48}Ca, ^{282}112 and ^{283}112 [2]. Therefore, the isotopes observed in the reaction ^{245}Cm+^{48}Ca should be the products of the [x-1]n and [x]n channels.

The results of all of these experiments allow us to consider a possible value for x more definitely. Assuming x=1, we would conclude that the 0n channel was observed in the reaction ^{245}Cm+^{48}Ca at excitation energies E^*=33-43 MeV. However, the γ channel was

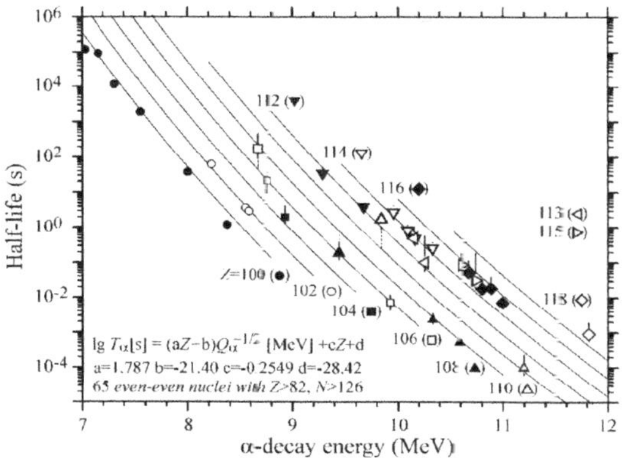

Figure 3. The dependence of T_α *vs.* Q_α for known heavy even-even isotopes with Z=100-110 and nuclei shown in Fig. 1. The lines are drawn according to the formula of Viola and Seaborg. The experimental data are taken from Refs. [2,6,8] and Refs. therein.

not observed even in cold fusion reactions at much lower excitation energies of the compound nuclei (see Ref. [6,7] and Refs. therein). The value x=5 would result in the conclusion that the parent isotopes of the reaction ^{244}Pu+^{48}Ca are the products of the 5-7n channels, which is unreasonable based both on the compound nucleus excitation energies and the competition with de-excitation by fission. Thus, we conclude that the only reasonable value for x is x=3.

Now one can discuss the possibility of other reaction channels accompanied by emission of light charged particles. In the reaction ^{243}Am+^{48}Ca, two different isotopes ^{287,288}115 were synthesized [2]. In both cases, the parent isotopes underwent five consecutive α decays followed by SF. The isotope ^{282}113 produced recently in the reaction ^{237}Np+^{48}Ca [2] demonstrates similar behavior. Comparison of the decay properties of nuclei produced in the reactions with even-Z and odd-Z target nuclei indicates that all synthesized nuclides cannot originate from the pxn channel because hindrance against SF of nuclei possessing an odd number of protons increases their SF stability by orders of magnitude. Indeed, the three isotopes of element 113, $^{282\text{-}284}$113, undergo four consecutive α decays whereas the neighboring isotopes ^{286}114, ^{282,284}112 and ^{279,281}Ds decay by SF.

Comparison of the decay properties of previously known heavy nuclides and those of the new superheavy nuclei also supports their assignment to the products of the xn-reaction channels. This can be seen in Fig. 3 where the dependences of T_α on Q_α for known even-even nuclei with Z=100-110 are shown along with the data for nuclides produced in ^{48}Ca-induced reactions. The lines are drawn in accordance with the formula by Viola and Seaborg with parameters fit to the T_α values of 65 even-even nuclei with Z>82 and N>126. The measured $T_{1/2}$ *vs.* Q_α values for all superheavy nuclei with Z=112-118, including ^{283,284}113 and ^{287,288}115, are in agreement with values expected for *allowed* α decays of isotopes of the corresponding elements. Thus, the assumption that these nuclei were produced in reactions accompanied by emission of charged particles (αxn etc.) would demand a change of assignment of all 15 $T_{1/2}$ *vs.* Q_α values to lower Z-values. The correlation between the experimental data and the empirical systematics for the heaviest nuclei with Z=112-118 indicates rather low hindrance factors, if any, for α decay. For the lighter isotopes of elements 105-113, the difference between measured and calculated T_α values results in hindrance factors of 3-10 which is consistent with values that can be extracted for the deformed nuclei located near the neutron shell N=162 (see, e.g., [6,7]).

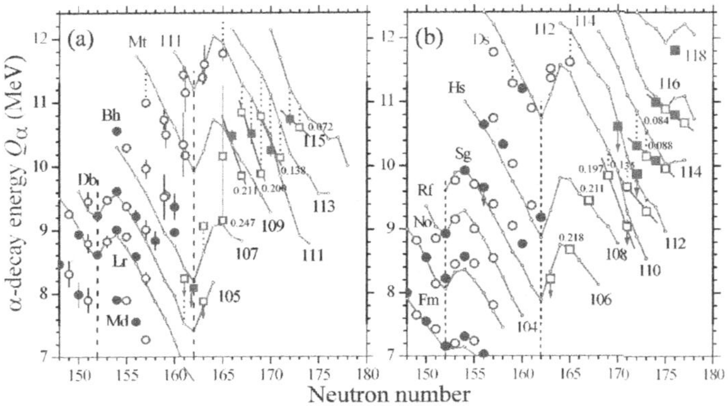

Figure 4. (a) α-decay energy *vs.* neutron number for isotopes of odd-*Z* elements (solid symbols – odd-even isotopes, open symbols – odd-odd isotopes [8], squares – data from [2] and present work). Solid lines show the theoretical Q_α values [9] for odd *Z*=103-115 elements. (b) The same for even-*Z* elements.

The experimental α-decay energies of the newly synthesized isotopes as well as previously known nuclei are plotted in Fig. 4. The α-decay energies attributed to isotopes of Mt and Bh coincide well with theoretical values [9]. The same is true for the last nuclei in the decay chain $^{275}\text{Hs} \xrightarrow{\alpha} {}^{271}\text{Sg} \xrightarrow{\alpha}$. For the isotopes $^{278\text{-}280}$Rg and $^{282\text{-}284}$113 the difference between theoretical and experimental Q_α values is 0.5-0.9 MeV. Part of this energy can be accounted for by γ-ray emission from excited levels populated during α decay. For the even-*Z* nuclei as well, the agreement between theory and experiment becomes somewhat worse as one moves from the deformed nuclei in the vicinity of neutron shells *N*=152 and *N*=162 to the more neutron-rich nuclides with *N*≥169. However, the trends of the predictions are in good agreement for the 27 nuclides with *Z*=106-118 and *N*=163-177, especially considering that the theoretical predictions of the MM model match the experimental data over a broad, previously unexplored region of nuclides. One should note that the predictions of other models for even-*Z* and odd-*Z* nuclei within the Skyrme-Hartree-Fock and the relativistic mean field theories also compare well with the experimental results (see [2] and Refs. therein). These models predict the same spherical neutron shell at *N*=184 but different proton shells, *Z*=114 (MM) and *Z*=120, 124 or 126 (SHF, RMF), yet all describe the experimental data equally well.

All of the decay chains presented in Fig. 1 end in spontaneous fission. In cases when the detector array registered both fission fragments, their sum energy E_{tot} could be used to evaluate the total kinetic energy (TKE) of the fission fragments. The value of TKE for all of the SF isotopes obtained from the ^{48}Ca-induced reactions is plotted in Fig. 5, together with the previously known data for isotopes with *Z*≥96 [10]. One can observe that, with the transition *Z*≥110, the TKE increases with increasing *Z* in agreement with the previously established dependence of TKE vs. $Z^2/A^{1/3}$ typical for the asymmetric fission of lighter nuclei. We speculate that this is due to the influence of the spherical shells at *Z*=50 and *N*=82 on the formation of the light fission fragment in the scission of superheavy nuclei. In the spontaneous fission of the lighter nuclides ^{267,268}Db, ^{271}Sg, and ^{267}Rf, the shell effect in the complementary fragment apparently results in symmetric

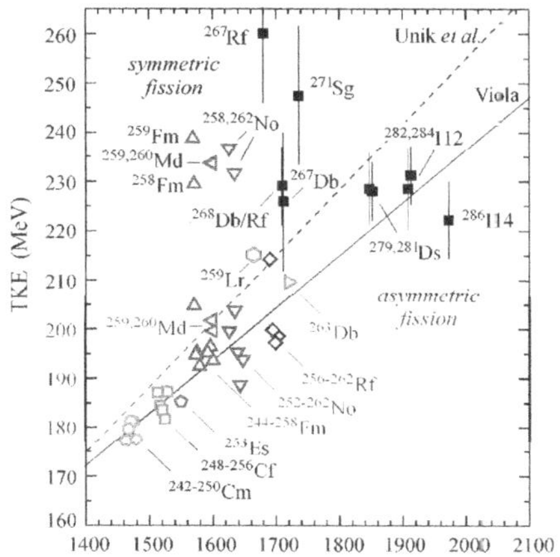

Figure 5. Experimental values of TKE *vs.* $Z^2/A^{1/3}$ (data from [10] and Refs. therein) – open symbols; data from the present work - solid squares. The lines are the linear fit [11].

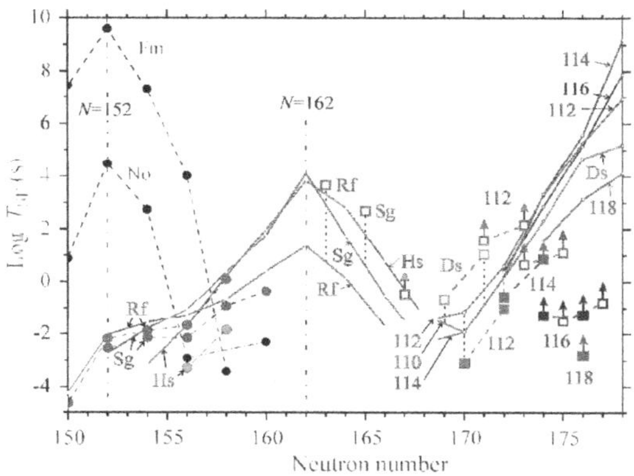

Figure 6. Common logarithm of partial spontaneous fission half-life vs. neutron number for isotopes of even-*Z* elements with $Z \geq 100$ (circles – even-even isotopes [8]). Data at $N>162$ are from [2] and the present work (open squares – odd-*N* isotopes; solid squares – even-*N* isotopes). Solid lines show the theoretical TSF values [12] for even-even Z=104-118 isotopes.

fission with high kinetic energy, characteristic of the close scission configuration that takes place in the symmetric fission of the heavy isotopes of Fm, Md, and No.

The calculations of SF half-lives are more complicated and correspondingly far less definite than those for α decay. Nevertheless, reasonable agreement between experiment and theory can be seen for terminal spontaneously fissioning isotopes in the decay chains of superheavy nuclei (see Fig. 6). The measured partial SF half-lives for three even-even isotopes $^{282,284}112$ and $^{286}114$ are lower than calculated values [12] by less than two orders of magnitude. The same can be found for the isotopes ^{267}Rf, ^{271}Sg, and ^{279,281}Ds if one assumes some hindrance factor, say 3 orders of magnitude, for SF of even-odd nuclei.

Therefore, we conclude that the superheavy nuclei produced in the ^{48}Ca-induced reactions [2] are the result of complete fusion followed by evaporation of two to five neutrons from the excited compound nuclei. Independent confirmation of these results was recently obtained in experiments aimed at the determination of the chemical properties of isotopes of ^{268}Db [13] and $^{283}112$ [14]. Therefore, confirmation of the decay properties and the determination of the chemical properties of these isotopes simultaneously signify the independent identification of all of the other superheavy nuclei observed in ^{48}Ca-induced reactions.

The nuclides with the largest neutron and proton numbers $^{293}116$ and $^{294}118$ were synthesized in the ^{48}Ca-induced reactions with the heaviest available for such experiments isotopes ^{248}Cm and ^{249}Cf and further exploration of the region of superheavy nuclei demands the use of heavier projectiles. The theoretical predictions of cross sections for production of the isotopes of element 120 substantially depend on the exact proton and neutron numbers of the double-magic superheavy nucleus, which determines the fission barriers of surrounding nuclei. For the MM barriers (Z_{shell}=114), the calculated cross section value is far below the sensitivity of reasonable experiment (σ_{4n}~0.02 pb) [4]. At the same time, according to SHF or RMF models that suggest Z=120-126 for the magic proton shell, the fission barriers of nuclei, as well as their survival probabilities, should increase as one approaches Z=120. In the experiment aimed at the synthesis of the element 120 isotopes in the ^{244}Pu(^{58}Fe,xn)$^{302-x}$120 reaction no decay chains of the fusion-evaporation reaction products were observed during an irradiation with a beam dose of 7.1×10^{18} 330-MeV ^{58}Fe projectiles. The sensitivity of the experiment corresponds to 0.4 pb for detection of one decay (1.1 pb an upper cross-section limit). The low cross section together with that for the reaction ^{249}Cf(^{48}Ca,3n)^{294}118 in comparison with corresponding values for Z=114 and 116 nuclei could indicate receding from the stability region at Z=114.

Acknowledgements

This work was performed with the support of the Russian Ministry of Atomic Energy and grants of RFBR Nos. 04-02-17186 and 07-02-00029. Much of the support for the LLNL authors was provided through the U.S. DOE under Contract No. W-7405-Eng-48. These studies were performed in the framework of the Russian Federation/U.S. Joint Coordinating Committee for Research on Fundamental Properties of Matter.

References

1. G.N. Flerov and G.M. Ter-Akopian, *Rep. Prog. Phys.* **46**, 817 (1983); P. Armbruster *et al.*, *Phys. Rev. Lett.* **54**, 406 (1985).
2. Yu.Ts. Oganessian *et al.*, *Phys. Rev.* C **69**, 021601(R) (2004); **69**, 054607 (2004); **70**, 064609 (2004); C **72**, 034611 (2005); **74**, 044602 (2006); **76**, 011601(R) (2007).
3. R. Bass, *Proceedings of the Symposium on Deep Inelastic and Fusion Reactions with Heavy Ions, West Berlin, 1979*, Lecture Notes in Physics, Vol. 117 (Springer-Verlag, Berlin, 1980), p. 281.
4. V.I. Zagrebaev, M.G. Itkis and Yu.Ts. Oganessian, *Phys. At. Nucl.* **66**, 1033 (2003); V.I. Zagrebaev, *Nucl. Phys.* **A734**, 164c (2004).
5. R.C. Barber *et al.*, *Prog. Part. Nucl. Phys.* **29**, 453 (1992).
6. S. Hofmann and G. Münzenberg, *Rev. Mod. Phys.* **72**, 733 (2000); S. Hofmann *et al.*, *Eur. Phys. J.* A **10**, 5 (2001); **14**, 147 (2002).
7. K. Morita *et al.*, *J. Phys. Soc. Jpn.* **73**, 1738 (2004); **73**, 2593 (2004); **76**, 043201 (2007); **76**, 045001 (2007).
8. Evaluated Nuclear Structure Data File (ENSDF), Experimental Unevaluated Nuclear Data List (XUNDL). http://www.nndc.bnl.gov/ensdf.
9. I. Muntian, Z. Patyk and A. Sobiczewski, *Phys. At. Nucl.* **66**, 1015 (2003); I. Muntian *et al.*, *Acta Phys. Pol.* B **34**, 2073 (2003).
10. D.C. Hoffman and M.R. Lane, *Radiochim. Acta* **70/71**, 135 (1995).

11. V.E. Viola, Jr., *Nucl. Data Tables* A **1**, 391 (1996); J.P. Unik *et al.*, *Proceedings of the Third International Atomic Energy Symposium on the Physics and Chemistry of Fission, Rochester, 1973,* (IAEA, Vienna, 1974) Vol. II, p. 19.
12. R. Smolańczuk, J. Skalski and A. Sobiczewski, *Phys. Rev.* C **52**, 1871 (1995); R. Smolańczuk, *Phys. Rev.* C **56**, 812 (1997).
13. S.N. Dmitriev *et al.*, *Mendeleev Commun.* 1 (2005); N.J. Stoyer *et al.*, *Nucl. Phys. A* **787**, 388c (2007).
14. R. Eichler *et al.*, *Nucl. Phys. A* **787**, 373c (2007).

HIGHLIGHTS OF THE BRAZILIAN RIB FACILITY AND ITS FIRST RESULTS AND HINDRANCE OF FUSION CROSS SECTION INDUCED BY ^{6}HE

P.R.S. GOMES

Instituto de Física, Universidade Federal Fluminense, Av. Litorânea s/n, Gragoatá, Niterói, R.J., 24210-340, Brazil

R. LICHTENTHÄLER, E.CREMA, L.C. CHAMON, E.A. BENJAMIM, A. LÉPINE-SZILY, V. GUIMARÃES, D.R. MENDES JUNIOR, J. ALCÂNTARA NUÑEZ, M. ASSUNÇÃO, A. BARIONI, O. CAMARGO JR., R.Z. DENKE, P.N. DE FARIA, K.C.C. PIRES, M.S.HUSSEIN

Departamento de Física Nuclear, Universidade de São Paulo, Caixa Postal 66318, 05315-970, São Paulo, S.P., Brazil

We present the main characteristics of the recently installed Brazilian radioactive beam facility (RIBRAS) and its first results on the elastic scattering and total reaction cross sections for the ^{6}He + ^{27}Al system, at near barrier energies, where the role of the nuclear break-up of the halo nucleus is studied. We also present the analysis of fusion cross sections induced by the ^{6}He halo projectile in the framework of the double-folding São Paulo potential. The results show that the fusion cross sections of ^{6}He + ^{209}Bi, ^{238}U are suppressed at energies above the barrier, when compared with coupled channel calculations that do not take into account the break-up projectile, whereas they are in agreement with the calculations at sub-barrier energies. The results for the ^{6}He + ^{209}Bi system are contradictory to the ones already reported.

1- The RIBRAS facility at São Paulo

The double superconducting solenoid system called RIBRAS (Radioactive Ion Beam Brazil) is the first facility in the South hemisphere for the production of secondary radioactive beams. It has been recently installed at the Pelletron Laboratory of the Universidade de São Paulo (USP) [1,2]. This system is similar to the TWINSOL facility at Notre Dame University [3,4]. RIBRAS consists basically of two superconducting solenoids with 6.5 T maximum central field. A 30 cm clear warm bore leads to angular acceptance within the range from 2 deg to 6 deg. This facility is installed in one of the beam lines of the 8 UD Tandem accelerator of USP, and it will be used in the near future with a LINAC post-accelerator, which will increase the present maximum beam energy by a factor three.

The first solenoid allows an in-flight selection of the reaction products emerging from the primary target in the forward direction, but the secondary beam can still be rather contaminated. So far we have been using only the first solenoid and performing elastic scattering measurements. When the two solenoids are operating, it will be possible to use differential energy loss in an energy degrader foil, located at the crossover point between the two magnets, in order to select the ion of interest by moving the non desirable ions out of its bandpass.

The production system consists of a gas cell, mounted in a ISO chamber. Its entrance window is a 2.2 μm Havar foil. The exit window is a 16 μm ^{9}Be foil, which also acts as the primary target. The gas inside the cell cools the berilium foil heated by the primary beam, and can also be used as a production target. Behind the gas cell there is a tungsten rod (Faraday cup) with a cylindrical hole, to stop the primary beam particles. The charge incident in the Faraday cup is integrated providing a measurement of the total number of particles incident in the primary target. A negative polarized collimator, placed in front of the tungsten Faraday cup, suppress the secondary electrons emitted from the cup. The secondary beam profile is measured by a X-Y Parallel Plate Avalanche Counter - PPAC - placed in the crossover point and a ΔE-E silicon telescope placed behind the PPAC. Figure 1 shows a schematic view of the RIBRAS system.

Secondary ^{8}Li beams have been produced from the neutron pickup reaction ^{9}Be (^{7}Li, ^{8}Li) ^{8}Be, with Q = +0.37 MeV, with production rate of the order of 10^{6} pps/μAe, and maximum energy of 30 MeV. Secondary ^{6}He and ^{7}Be beams have been produced from the proton stripping reaction ^{9}Be (^{7}Li, ^{6}He) ^{10}B, with Q = - 3.39 MeV, and the charge exchange reaction ^{3}He (^{7}Li, ^{7}Be) t with Q = - 0.88 MeV respectively with production rates of the order of 2.0 X10^{5} pps/μAe, and maximum energy of about 27 MeV.

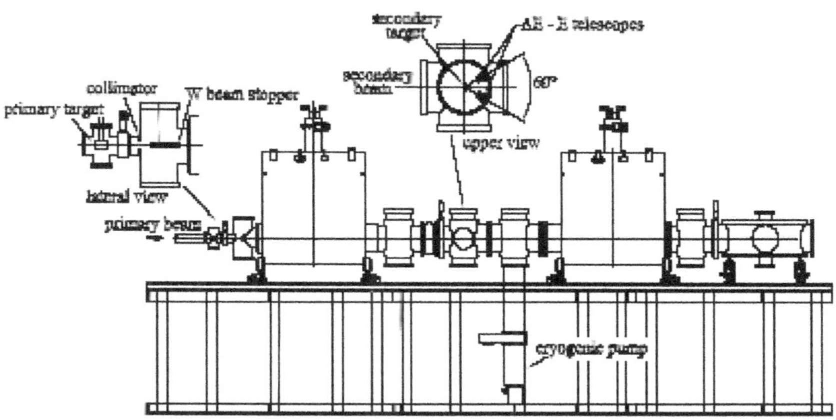

Figure 1- Schematic view of the RIBRAS double solenoid

2- What is the effect of break up on fusion and total reaction cross sections?

Nuclei that have small separation energies have a large probability of breaking-up when the colliding nuclei approach each other. Many efforts have been made, both theoretically and experimentally, on the investigation of the effect of the break-up on the fusion and total reaction cross sections [5]. So far, conflicting predictions have been made about whether the fusion of weakly bound nuclei is enhanced or hindered due to the strong coupling to the break-up channel. Studies with stable weakly bound nuclei, like ^{6}Li, ^{7}Li and ^{9}Be, are important references for the understanding of reactions induced by radioactive halo structure projectiles but important features of the reactions induced by halo radioactive beams are not present on those reactions, such as the very long range of nuclear interactions, with the possible lowering of the Coulomb barrier and consequently making easier the fusion process, and possible couplings to the soft dipole resonance. On the other hand, as the breakup populates continuum states, this process may be an irreversible way which takes flux from the fusion process and consequently suppresses the fusion cross section. The effect of the break-up on fusion may vary from sub-barrier to above-barrier energy regimes and also for light and heavy targets, since, depending on whether the break-up is dominated by the Coulomb or nuclear interactions, different effects on the fusion cross section could result. It is also very important to know, although not always taken into account, what one means by enhancement or suppression of the fusion cross section: Does one talk about complete fusion or total fusion, the last corresponding to the situation when the fusion of part of the projectile break-up fragments with the target is also considered in the fusion cross section? Does one talk about enhancement and suppression in relation to what?

There is another important open question: Is an effect of the break-up to increase the total reaction cross section, instead of affecting the fusion cross section? Therefore, it is essential to investigate the dependence of the break-up on total reaction cross sections with the break-up threshold energy of different projectiles and with the target mass.

3- The São Paulo Potential as a reliable bare potential to study reactions induced by weakly bound nuclei.

A well established fact is the strong dependence of theoretical calculations on the choice of the interacting potential to describe nuclear reactions. So, the choice of an appropriate nuclear bare interaction between two colliding nuclei, although it is a difficult task, is a crucial step in the data interpretation. As one has not direct access to the "true" bare potential, some procedures have been developed in order to constrain the choice of the bare potential to be used. Among the widely used procedures to derive interaction potentials one finds the analysis of the elastic scattering data within the context of an optical potential, where an imaginary potential is added to the real potential, in order to take into account the flux deviated from the elastic channel. The optical potential that gives the best fit to the elastic scattering data is then used to describe other reaction channels, even in coupled channel calculations including inelastic and transfer reactions in the coupling matrix. Another procedure that has been widely used to obtain the bare potential is the fit of the high-energy fusion data by a single barrier penetration model (BPM). However, both procedures suffer from several inconsistencies, as pointed out in ref [6,7].

The derivation of experimental barrier distributions probably is the best way to constrain the choice of the bare potential to be used, since their heights and shapes should be reproduced by the calculations. However, so far this can not be achieved with low-intensity radioactive beams.

In order to overcome these difficulties, we have shown recently [6] that the parameter-free São Paulo potential (SPP) [8,9] is a reliable bare interaction for studying fusion of systems involving stable weakly bound nuclei in situations in which experimental barrier distributions cannot be obtained. As it is deduced from fundamental grounds, the SPP does not contain the limitations of the other methods and does not contain any adjustable parameter. In this paper we propose to adopt the SPP as the bare potential in the study of fusion reactions induced by the radioactive weakly bound nucleus ^{6}He.

The SPP is a model for the interaction between two nuclei, based on the effects of the Pauli non-locality [8,9] and strongly supported on experimental evidences. Within this model the nuclear interaction is connected with the folding potential V_F through the expression $V_N(R) = V_F(R)\, exp(-4v^2/c^2)$, where c is the speed of light and v is the local relative velocity between the two nuclei. With the aim of providing a parameter-free description of the nuclear interaction, the SPP model includes an extensive systematic of nuclear densities. The two-parameter Fermi (2pF) distribution is assumed to be a good approximation to describe the densities. The radii of the 2pF distributions are well represented by $R_0 = 1.31\, A^{1/3} - 0.84$ fm, where A is the number of nucleons of the nucleus. The values obtained for the diffuseness of the matter distributions are similar throughout the periodic table and present small variations around the average value $a = 0.56$ fm. Within the context of this realistic systematic, the SPP does not contain any adjustable parameter. So, this model is a powerful tool to make predictions for quite different systems and energies. For bombarding energies around the Coulomb barrier, the energy-dependent term of the SPP can be ignored since the relative velocity between the interacting nuclei is close to zero in this energy regime, and the SPP becomes, basically, a double-folding potential.

When the SPP is used in the analysis of elastic scattering, an imaginary part is included. This imaginary potential has the same form as the real part and only one free parameters, its strength.

4- Near-barrier fusion of ^{6}He revised

Due to the success of the SPP as a bare potential to describe fusion of stable weakly bound systems, in the context of coupled channel calculation, we decided to analyze the available fusion data for radioactive weakly bound projectiles [7]. The present conclusions about the role of the ^{6}He on the fusion reaction mechanism on heavy targets are contradictory. Kolata et al [10] claim that there is a large sub-barrier fusion enhancement for the ^{6}He + ^{209}Bi system, whereas Raabe et al [11] say that there is no enhancement in the sub-barrier fusion of the ^{6}He + ^{238}U system. It is not very reasonable to think that these opposite conclusions are due to target structure differences.

The basis for the calculations with the SPP is a good knowledge of the matter distributions of the interacting nuclei. As the ground state density of ^{6}He has been determined from several experimental and theoretical studies [8,12-15], one can calculate the SPP for the ^{6}He+^{238}U and ^{6}He+^{209}Bi systems. For comparison, we also performed the calculation of the SPP for the ^{4}He projectile on the same targets. It is important to mention that the potential diffuseness is much larger for the ^{6}He system than for the ^{4}He, what should have a strong influence on the reaction mechanism.

The aim of the present work is not to improve the coupled channel calculations performed in refs. [10] and [11]. On contrary, in order to evidence the effect of the potential choice on the conclusions, we coupled exactly the same channels as it was done in those references: for the ^{6}He projectile, the resonance state at the continuum at $E^* = 1.8$ MeV, with $\beta_2 = 0.90$ and $r_0 = 1.06$ fm; for the ^{238}U and ^{209}Bi targets, the same inelastic excited states, with the same deformation parameters and r_0 values. We used the FRESCO code[16] in the CC calculations, with the SPP interaction supplied numerically.

The comparison of the SPP prediction and the ^{4}He + ^{238}U fusion data [17] is shown in figure 2a. As we can see in the figure, the predictions slightly underestimate the data. This can be understood by the fact that the data presented in the figure is not the fusion cross section, but rather fission cross sections, which should be slightly larger than fusion, due to presence of transfer channels.

In figure 2b we show the results of the SPP predictions for the ^{6}He + ^{238}U system. One can observe fusion cross section suppression at energies above the barrier, whereas the behavior at sub-barrier energies is predicted by the SPP. Therefore, from the present analysis one observes a suppression of the fusion cross section for the ^{6}He induced reaction of the order of 15%, similar to the values obtained for stable weakly bound nuclei. In the original work by Raabe et al [11], coupled channel calculations were performed with a bare potential calculated by a double-folding procedure (M3Y). Similar results as the ones presented here for ^{4}He and ^{6}He were obtained.

Figure 2c shows that the behavior of the ^{4}He + ^{209}Bi fusion excitation function is predicted by the SPP. In figure 2d, for the ^{6}He + ^{209}Bi system, one can observe a similar behavior as the one obtained for the ^{238}U target: fusion cross section suppression around 15% at energies above the barrier, whereas at sub-barrier energies the predictions are in agreement with the data. In the original paper by Kolata et al [10], the authors use the data above the barrier to derive the barrier height and deduce that this value was reduced by around 5 MeV, comparing with the systematic. Then they conclude that there is a large sub-barrier fusion enhancement for this system. This apparent disagreement between the sub-barrier fusion cross sections for the ^{6}He + ^{209}Bi and ^{238}U systems does not exist in our present analysis.

Thus, we conclude that, as for stable weakly bound projectiles, the hindrance of the fusion cross section induced by the halo structure ^{6}He nucleus is believed to be due to the effect of the break-up process, which produces a strong coupling between the elastic channel and the continuum states. At sub-barrier energies, there is no enhancement of the fusion cross sections, probably due to competition between break-up and other effects that may enhance fusion, as it occurs for tightly bound projectiles. Hence, we have established a very coherent and systematic behavior for the fusion induced by both, stable and radioactive weakly bound nuclei. For the ^{6}He+^{238}U system, our results are coincident with those obtained from Raabe's analysis [11], which also used a bare potential calculated by a double-folding procedure (M3Y). On the other hand, our results for the ^{6}He+^{209}Bi system are in contradiction with the results of the Kolata's analysis [10], where the bare potential was deduced from a BPM fit to the highest energies data.

Our message is that, in order to analyze fusion data for systems involving stable and radioactive weakly bound nuclei, taken around the Coulomb barrier, one needs to use a bare potential obtained from procedures based on fundamental grounds, such as double-folding methods. However, in order to do that for radioactive nuclei, it is necessary to know, at least, the ground state density of the radioactive nucleus involved, as it is nowadays available for ^{6}He.

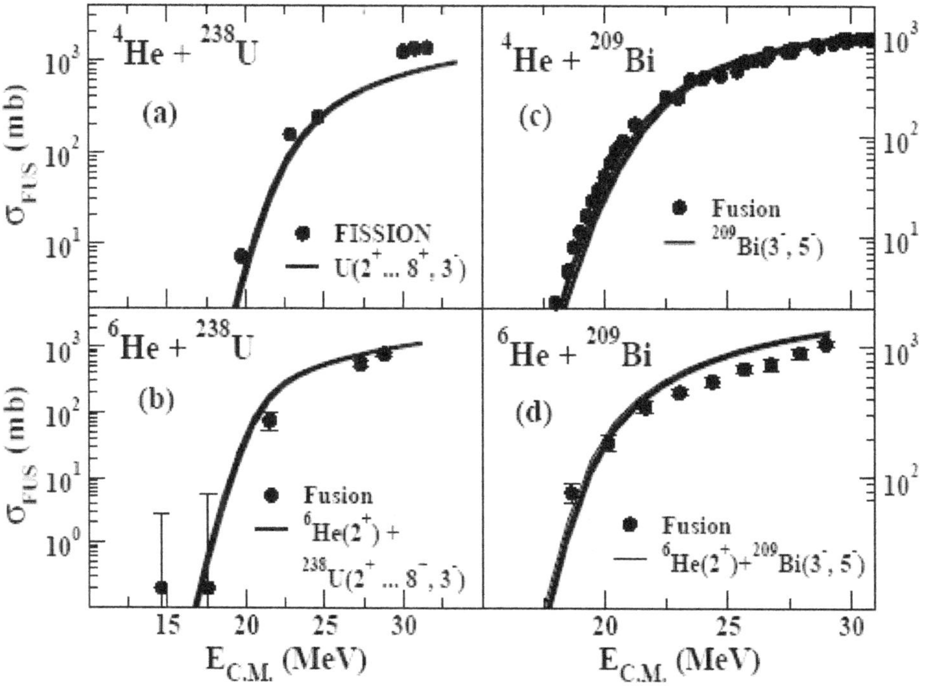

Figure. 2 - Fusion excitation functions for (a) ^{4}He + ^{238}U, (b) ^{6}He + ^{238}U, (c) ^{4}He + ^{209}Bi, (d) ^{6}He + ^{209}Bi. The bands shown in the figure represent the variation of the theoretical cross section obtained when the strength of the optical potential is varied by ±10% relative to the standard SPP in the coupled channel calculations.

5- First results with the RIBRAS facility: total reaction cross section for ^{6}He + ^{27}Al system.

Finally, we report very briefly the first results taken with the RIBRAS facility, on the elastic scattering of the ^{6}He + ^{27}Al system [18]. Our aim was to investigate the behavior of the total reaction cross section for the ^{6}He + ^{27}Al system and to compare it with the available data using the stable weakly bound ^{6,7}Li [19,20] and ^{9}Be [21] and the tightly bound ^{16}O [22] projectiles on the same target. Similar study was previously performed with the heavier ^{64}Zn target [23, 24], where the Coulomb field is stronger than for ^{27}Al, and therefore the ^{6}He break-up process for ^{27}Al should have a predominant contribution from the nuclear break-up, as compared with the heavier target. Therefore, the present study aims to contribute to the understanding of the importance of the nuclear break-up due to the long range nuclear interaction of the colliding nuclei that involves the ^{6}He halo nucleus. The study of light systems involving very weakly bound and neutron rich exotic nuclei is particularly important, since there are reactions of great astrophysical interest with these characteristics. For reactions of the ^{6}He projectile with very heavy targets, like ^{209}Bi and ^{238}U, where the Coulomb break-up predominates due to the high long range Coulomb field, very large total reaction cross sections have been derived around and below the Coulomb barrier, and the transfer plus break-up cross sections were found to correspond to most of the total reaction cross section.

Primary ^{7}Li beams were used at four energies from 19.0 to 22.0 MeV, producing ^{6}He secondary beams with energies from 9.5 to 13.4 MeV in the laboratory system, at the center of the 7.2 mg/cm^2 thick ^{27}Al secondary target, since the ^{6}He energy loss from the front to the center of the Al target was 1.6 MeV. The nominal Coulomb barrier for this system is around 8.0 MeV at the laboratory system. The energy resolution of the elastic peak was of the order of 400 keV, good enough to separate it from the inelastic excitation of the ^{27}Al first excited state, at 843 keV. The secondary beam intensities were calculated by assuming that ^{6}He scattering on a 5 mg/cm^2 thick ^{197}Au target is purely Rutherford. The production efficiency was derived from this intensity and from the integrated charge of the primary beam, which was stopped in a Faraday cup. The secondary scattered particles were detected using two ΔE-E silicon telescopes with anti-scattering collimators. The total angular uncertainty was calculated by Monte Carlo simulations as ± 3.2 degrees.

The elastic scattering angular distributions were fitted by optical model calculations, using the São Paulo Potential (SPP). Here, in addition to the usual free parameter of the normalization of the imaginary potential, another free parameter was used, the diffuseness of the nuclear density of the ^{6}He projectile, assumed to have a unique value for all energies. The reaction cross sections were calculated using the best fit optical potentials. The procedure was used in the analysis of the other angular distributions and in the determination of reaction cross sections.

In order to compare total reaction cross sections of different systems, we used the procedure suggested in ref [25], dividing the cross sections by $(A_p^{1/3} + A_T^{1/3})^2$ and the center of mass energy by $Z_p\, Z_T\, /\, (A_p^{1/3} + A_T^{1/3})$, where Z_p (Z_T) and A_p (A_T) are the charge and mass of the projectile (target), respectively. In this way, the geometrical effects are removed and the eventual strange values of the reduced radii r_0 and r_{eff}, which should be related to the physical processes to be investigated, are not washed out.
From the results of the "reduced" total reaction cross sections for the halo ^{6}He, the stable weakly bound ^{6,7}Li, ^{9}Be and the tightly bound ^{16}O projectiles on the same ^{27}Al target, one can observe that the smallest cross section is for the tightly bound projectile ^{16}O, for which the break-up process is not expected to occur. The reaction cross sections for all four weakly bound nuclei, both the stable and ^{6}He, are much larger than for the ^{16}O, but are similar between themselves, within the error bars. This result is somehow different from heavier systems involving the same projectiles and ^{64}Zn as target, since Gomes et al [24] have shown that the reaction cross section is largest for the ^{6}He [26], followed by the group of the three stable weakly bound projectiles [27-29], and finally the tightly bound ^{16}O [30] projectile produces the smallest total reaction cross section.

For the ^{64}Zn target, it has been shown [24] that the total fusion cross sections induced by weakly bound nuclei are not affected by the break-up process. For the ^{9}Be + ^{27}Al, the total fusion cross section is also not affected by the break-up, since the fusion excitation function is well described by the São Paulo Potential without any parameter fit procedure [21]. Therefore, it is reasonable to conclude that the break-up plus neutron transfer processes increase the total reaction cross section of systems involving weakly bound nuclei even when light targets are involved. This effect is larger for the halo ^{6}He nucleus than for the stable weakly bound nuclei when the target is not very light, but it is similar, within the error bars, when a target like ^{27}Al is used.

In summary, we report the first results obtained with the new radioactive beam facility RIBRAS at São Paulo, with the measurement of elastic scattering between ^{6}He and the light ^{27}Al target, at energies slightly above the Coulomb barrier. The derived total reaction cross section was compared with other systems, three stable weakly bound and one tightly bound projectiles and the same target. The larger reaction cross section for the weakly bound systems shows evidence of the importance of the nuclear break-up for these light systems.

Acknowledgements

The authors thank FAPERJ, FAPESP and CNPq for their financial support.

References

1- R. Lichtenthäler et al., *Eur. Phys. Journal A* **25**, 733 (2005).
2- A. Lepine-Szily, R. Lichtenthäler, *Nucl. Phys. A* **787,** 94c (2007).
3- J.J. Kolata et al., *Nucl. Instr. Meth. B* **40-41**, 503 (1989).
4- F.D. Becchetti et al., *Nucl. Instr. Meth. A* **505**, 377 (2003).
5- L.F. Canto, P.R.S. Gomes, R. Donangelo, M.S. Hussein, *Phys. Rep.* **424**, 1 (2006).
6- E. Crema, L.C. Chamon, P.R.S. Gomes, *Phys. Rev. C* **72**, 034610 (2005).
7- E. Crema, P.R.S. Gomes, L.C. Chamon, *Phys. Rev. C* **75**, 037601 (2007).
8- L.C. Chamon et al., *Phys. Rev. C* **66**, 014610 (2002).
9- M.A. G. Alvarez et al., *Nucl. Phys. A* **723**, 93 (2003).
10- J.J. Kolata et al., *Phys. Rev. Lett.* **81**, 4580 (1998) ; *Phys. Rev. C* **57**, R6 (1998).
11- R. Raabe et al., *Nature* **431**, 823 (2004).
12- G. D. Alkazov et al., *Phys. Rev. Lett.* **78**, 2313 (1997).
13- L. R. Gasques et al., *Phys. Rev. C* **67,** 024602 (2003).
14- M. V. Zukhov et al., *Nucl. Phys. A* **552**, 353 (1993).
15- J. S. Al-Khalili et al., *Phys. Rev. C* **54**, 1843 (1996).
16- I. Thompson, *Comput. Phys. Rep.* **7,** 167 (1988).
17- M. Trotta et al., *Phys. Rev. Lett.* **84**, 2342 (2000).
18- E. A. Benjamin et al., *Phys. Lett. B* **647**, 30 (2006).
19- J.M. Figueira et al., *Phys. Rev. C* **73**, 054603 (2006).
20- J.M. Figueira et al., *Phys. Rev. C* **75**, 017602 (2007).
21- G.V. Marti, P.R.S. Gomes et al., *Phys. Rev. C* **71**, 027602 (2005).
22- E. Crema – Master Degree Thesis, USP (1979), unpublished.
23- P. R. S. Gomes et al., *Phys. Lett. B* **601**, 20 (2004).
24- P. R. S. Gomes et al., *Phys. Rev. C* **71**, 034608 (2005).
25- P. R. S. Gomes, J. Lubian, I. Padron, R.M. Anjos, *Phys. Rev. C* **75**, 017601 (2007).
26- A. di Pietro et al., *Phys. Rev. C* **69**, 044613 (2004).; *Europhys. Lett.* **64,** 309 (2003).
27- P. R. S. Gomes et al., *Phys. Rev. C* **70**, 054605 (2004).
28- I. Padron, P. R. S. Gomes et al., *Phys. Rev. C* **66**, 044608 (2002).
29- S. B. Moraes, P. R. S. Gomes et al., *Phys. Rev. C* **61**, 064608 (2000).
30-.C. Tenreiro et al., *Phys. Rev. C* **53**, 2870 (1996).

SEARCH FOR LONG FISSION TIMES OF SUPER-HEAVY ELEMENTS WITH $Z = 114$

M. MORJEAN*, A. CHBIHI, J. D. FRANKLAND, A. MARCHIX, M. PARLOG**, C. STODEL

GANIL, DSM/CEA, IN2P3/CNRS, BP 55027,
F-14076 Caen, France
**E-mail: morjean@ganil.fr*
***On leave from National Institute for Physics and Nuclear Engineering*
RO-76900 Bucharest-Magurele, Romania

D. JACQUET, M. LAGET, L. TASSAN-GOT

IPN Orsay, BP 1
F-91406 Orsay Cedex, France

J. L. CHARVET, R. DAYRAS, A. DROUART, L. NALPAS, C. VOLANT

DSM/DAPNIA/SPhN, CEA Saclay
F-91191 Gif/Yvette Cedex, France

By contrast to composite systems with $Z = 120$ and $Z = 124$, no evidence for long lifetime components could be found for composite systems with $Z = 114$ formed in the reaction ^{208}Pb+Ge at 6.16 MeV/u.

Keywords: Super-heavy elements; Fusion; Fission time.

1. Introduction

Very high shell effects are expected to enhance the stability of super-heavy elements (SHEs) in the vicinity of closed shell structures.[1–5] However, the magic numbers are quite uncertain in this domain, especially the proton one(s): depending on the calculations considered, they can be found between $Z = 114$ and $Z = 128$. A new insight into the stability island(s) can be reached through the fission times that are expected to be highly sensitive to the fission barriers involved. The latter result essentially, for SHEs, from shell effects. At high excitation energy, the shell effects are washed out and very short fission times are thus expected. Nevertheless, competition between fission and evaporation still takes place, and neutron emission can cool down the excited nucleus, increasing thus the shell effects in the residual nucleus and therefore the fission time. The presence of a tail at very long fission times for SHE is therefore a signature of high shell effects stabilizing the nuclei after neutron emission. Such tails have indeed been measured for $Z = 120$ and 124[6–8] . In order to check if long lifetime components actually result from high fission barriers, a more neutron deficient nucleus in the same mass and charge region (having thus a lower fission barrier) has been studied in the present work.

2. Experimental results

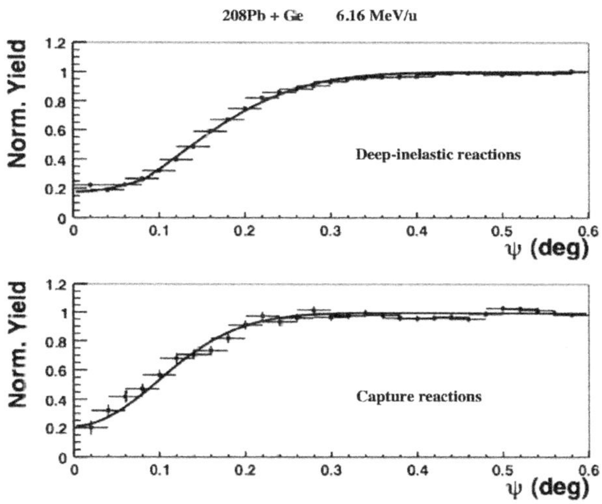

Fig. 1. Upper part: Blocking dip for projectile-like fragments. Lower part: Blocking dip for capture reactions. The full curves are just to guide the eyes.

We have studied the reaction ^{208}Pb+Ge at 6.16 MeV/u. The experimental set-up making use of the 4π charged product detection array, INDRA[9] has been previously described.[6–8] Composite systems with $Z = 114$ have been selected by retaining events for which the multiplicity of heavy fragments ($Z > 6$) is equal to 2, the total detected charge is close to 114 (the experimental overall atomic number resolution was about 2) and is shared between two fission-like fragments, with negligible light charged particle and intermediate mass fragment multiplicities. Among the selected capture reactions, reaction times longer than the experimental minimum sensitivity limit (about 10^{-18}s in the present experiment) were searched for using the blocking technique in single crystal.

Fig. 1 presents in its upper part the blocking dip measured at 20° for projectile-like fragments arising from deep-inelastic collisions. The angular distributions associated with the projectile-like fragments clearly indicate reaction times much shorter than the minimum sensitivity limit; this dip can thus be considered as a reference for reaction times shorter than 10^{-18}s. Channeling and blocking studies[10] have shown that the minimum yield measured at $\psi = 0°$, usually called χ_{min}, is increased in a sensitive way with respect to a short time reference dip (here the deep-inelastic dip) if reaction times longer than the minimum sensitivity limit (10^{-18}s) are present. Therefore, from measured different χ_{min} for deep-inelastic reactions and for capture reactions, the presence of long lifetime components could be inferred for $Z = 120$[6,7] and 124[8] . By contrast, Fig. 1 indicates quite similar χ_{min} values for fast reactions (upper part of the figure) and for capture reactions (lower part) in the case of the composite systems formed with $Z = 114$ in the present experiment.

3. Conclusion

No evidence for long lifetime components (reaction times longer than 10^{-18}s) could be found among the composite systems formed with $Z = 114$. Such a behavior can be explained either if the fusion cross-section is small as compared to the quasi-fission one or if the lifetime of the compound nuclei formed is short. Considering that long lifetimes have been actually observed for the much heavier system ^{238}U+Ge[8] for which a much weaker fusion cross-section is expected, this result rather indicates low fission barriers for the compound nucleus formed with $Z = 114$ and for its possible daughters if fast cooling by evaporation takes place. Such low fission barriers are indeed expected due to the rather high neutron deficiency of the isotopes involved with respect to the expected neutron shell closures. More data on fission times of super-heavy elements in this mass and charge region are now needed to determine the highest fission barriers, most likely associated with doubly magic spherical nuclei.

References

1. P. Møller and al., *Atomic Data and Nuclear Data Tables* **59**, 185 (1995).
2. R. Smolanczuk, *Phys. Rev. C* **56**, p. 812 (1997).
3. S. Cwiok *et al.*, *Nucl. Phys. A* **611**, p. 211 (1996).
4. M. Bender *et al.*, *Phys. Rev. C* **60**, p. 034304 (1999).
5. J. F. Berger, D. Hirata and M. Girod, *Acta Phys. Pol. B* **34**, p. 1909 (2003).
6. A. Drouart *et al.*, Exon04, in *Int. Symp. on Exotic Nuclei, PeterHof, 2004*, ed. S. World Scientific2005.
7. D. Jacquet *et al.*, Fusion06, in *Int. Conf. on Reaction Mechanisms and Nuclear Structure at the Coulomb Barrier, March 19-23 2006, Venezia, Italy*, ed. AIP2006.
8. M. Morjean *et al.*, *Eur. Phys. J. D* **45**, p. 27 (2007).
9. J. Pouthas *et al.*, *NIM A* **357**, p. 41 (1995).
10. D. S. Gemmel, *Rev. Mod. Phys.* **46**, p. 129 (1974).

MICROSCOPIC DYNAMICS OF SHAPE COEXISTENCE PHENOMENA AROUND ^{68}Se AND ^{72}Kr

NOBUO HINOHARA*, KENICHI MATSUYANAGI

Department of Physics, Graduate School of Science, Kyoto University, Kyoto 606-8502, Japan
**E-mail: hinohara@ruby.scphys.kyoto-u.ac.jp*

TAKASHI NAKATSUKASA

Theoretical Nuclear Physics Laboratory, RIKEN Nishina Center, Wako 351-0198, Japan

MASAYUKI MATSUO

Department of Physics, Faculty of Science, Niigata University, Niigata 950-2181, Japan

The adiabatic self–consistent collective coordinate (ASCC) method is applied to the pairing-plus-quadrupole (P + Q) model Hamiltonian including the quadrupole pairing, and the oblate–prolate shape coexistence phenomena in proton-rich nuclei, ^{68}Se and ^{72}Kr, are investigated. It is shown that the collective path connecting the oblate and prolate local minima runs along a triaxial valley in the (β, γ) plane. Quantum collective Hamiltonian is constructed and low-lying energy spectra and E2 transition probabilities are calculated for the first time using the ASCC method. Basic properties of the shape coexistence/mixing are well reproduced. We also clarify the effects of the time-odd pair field on the collective mass (inertial function) for the large–amplitude vibration and on the rotational moments of inertia about three principal axes.

Keywords: Nuclear structure; large–amplitude collective motion; shape coexistence.

1. Introduction

In proton rich nuclei around ^{68}Se and ^{72}Kr, oblate–prolate shape coexistence phenomena are observed.[1–4] From the mean-field point of view, the oblate and prolate shapes are mixed by the many-body tunnelling through the potential barrier between the two local minima in the potential energy landscape. To describe the shape mixing dynamics, we have to determine the collective degrees of freedom in which direction the large–amplitude shape dynamics takes place. The adiabatic self–consistent collective coordinate (ASCC) method[5] has been proposed as a microscopic theory of large–amplitude collective motion such as the shape coexistence phenomena. This theory is based on the time-dependent Hartree–Fock–Bogoliubov (TDHFB) theory and enables us to extract the collective degrees of freedom in a self–

consistent way. In the previous work,[6] we have solved the pairing-plus-quadrupole (P + Q) Hamiltonian by means of the ASCC method for ^{68}Se and ^{72}Kr nuclei, and successfully extracted the one-dimensional collective path connecting the oblate and prolate local minima. Since the two local minima are mainly connected by the triaxial degrees of freedom, we have suggested that the triaxial deformation plays an essential role in the shape coexistence dynamics of these nuclei.

In this presentation, we report results of the first calculation of low-lying energy spectra and E2 transition probabilities by means of the ASCC method. We derive the quantum collective Hamiltonian that describes the coupled collective motion of the large–amplitude vibration responsible for the oblate–prolate shape mixing and the three-dimensional rotation of the triaxial shape. The calculation yields the excited prolate rotational band as well as the oblate ground-state band. It also indicates that the shape mixing decreases as the angular momentum increases.

We also clarify the effect of the time-odd mean-field on the collective mass. This effect is ignored in the cranking mass,[7] but is included in the ASCC mass. The time-odd mean-field effect generated by the particle-hole residual interaction was investigated in a few decades ago,[8] but those generated by the pairing interaction has not been discussed so far. Quite recently we have shown, using the schematic model Hamiltonian,[9] that the time-odd component associated with the quadrupole-type pair field significantly increases the collective mass. In the present calculation, we thus include the quadrupole pairing interaction to the P + Q Hamiltonian to clarify the effect of the time-odd pair field on the collective mass and rotational moments of inertia.

2. The ASCC method

We first recapitulate the basic equations of the ASCC method. The moving-frame TDHFB state $|\phi(q,p)\rangle = e^{ip\hat{Q}(q)}|\phi(q)\rangle$ is written in terms of the collective coordinate q, and collective momentum p. The TDHFB state is expanded in terms of the collective momentum p under the adiabatic assumption. The collective Hamiltonian is expanded up to the lowest order in p as

$$\mathcal{H}(q,p,\vec{I}) = V(q) + \frac{1}{2}B(q)p^2 + \sum_{i=1}^{3}\frac{I_i^2}{2\mathcal{J}_i(q)}, \tag{1}$$

where the $V(q)$ and $B(q)$ represents the collective potential and the inverse collective mass. We add the rotational energy term with three moments of inertia $\mathcal{J}_i(q)$ to the collective Hamiltonian in order to take into account the rotational motion.

The basic equations of the ASCC method are derived from the adiabatic expansion of the time-dependent variational principle, and are summarized as

$$\delta\langle\phi(q)|\hat{H}_M(q)|\phi(q)\rangle = 0, \tag{2}$$

$$\delta\langle\phi(q)|[\hat{H}_M(q),\hat{Q}(q)] - \frac{1}{i}B(q)\hat{P}(q)|\phi(q)\rangle = 0, \tag{3}$$

$$\delta \langle\phi(q)| [\hat{H}_M(q), \frac{1}{i}\hat{P}(q)] - C(q)\hat{Q}(q) - \frac{1}{2B(q)}[[\hat{H}_M(q), \frac{dV}{dq}(q)\hat{Q}(q)], \hat{Q}(q)] - \sum_\tau \frac{\partial\lambda^{(\tau)}}{\partial q}(q)\hat{N}^{(\tau)} |\phi(q)\rangle = 0. \quad (4)$$

Equation (2) is called the moving-frame HFB equation, while Eqs. (3) and (4) are called the moving-frame QRPA equations. where $\hat{Q}(q)$ and $\hat{P}(q)$ denote the infinitesimal generators of the collective path,

$$\hat{H}_M(q) = \hat{H} - \sum_\tau \lambda^{(\tau)}(q)\hat{N}^{(\tau)} - \frac{dV}{dq}(q)\hat{Q}(q), \quad (5)$$

is the Hamiltonian in the moving-frame, and $C(q) = \frac{\partial^2 V}{\partial q^2} + \frac{1}{2B(q)}\frac{\partial B}{\partial q}\frac{\partial V}{\partial q}$ is the local stiffness parameter. The operators $\hat{H}$, $\hat{N}^{(\tau)}$ and the quantity $\lambda^{(\tau)}(q)$ represents the microscopic Hamiltonian, the particle number operators, and the chemical potential, respectively. The QRPA eigenmodes about the HFB equilibrium points, as well as static solutions of the HFB equation, are always special solutions of the ASCC equations. Therefore, the collective path can be constructed by means of the local progression procedure starting from a static HFB state. The collective path at $q+\delta q$ is calculated using the constraint $\langle\phi(q+\delta q)| \hat{Q}(q) |\phi(q+\delta q)\rangle = \delta q$ derived from the canonical variable conditions, if the collective path at q is already known. Repeating this procedure, we can find the collective path and obtain all quantities appearing in the ASCC equations and collective Hamiltonian except the rotational moments of inertia. Three rotational moments of inertia are evaluated by extending the QRPA equations for rotation at the HFB minima to the general HFB states $|\phi(q)\rangle$ on the collective path,

$$\delta \langle\phi(q)| [\hat{H}_M(q), i\hat{\Psi}_i(q)] - \mathcal{J}_i^{-1}(q)\hat{J}_i |\phi(q)\rangle = 0, \quad (6)$$

where $\hat{\Psi}_i(q)$ and $\hat{J}_i$ are the angle and angular momentum operators in the principal frame.

In the present calculation, we use the P + Q + quadrupole pairing Hamiltonian as a microscopic Hamiltonian $\hat{H}$. We adopt the same single-particle energies and the P + Q interaction strengths as in Ref. 6 and the self-consistent quadrupole-pairing strength $G_2^{\rm self}$ proposed by Sakamoto and Kishimoto.[10]

3. Collective Path

For both ^{68}Se and ^{72}Kr, we have found that the lowest HFB state possesses the oblate shape, while second lowest HFB state the prolate shape. We start from the oblate state ($q = 0$) and determine the collective path connecting the two local minima. Figure 1 shows the collective paths projected onto the (β, γ) potential energy surfaces. The path connects the two local minima via the triaxially deformed region. Figure 2 shows various quantities appearing in the collective Hamiltonian and the basic equations of the ASCC method. We define the scale of collective coordinate q

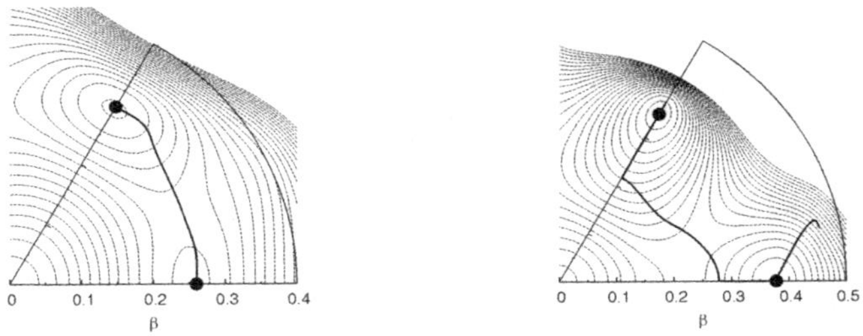

Fig. 1. Collective paths for ^{68}Se (*left*) and ^{72}Kr (*right*) calculated using the P + Q Hamiltonian including the quadrupole pairing interaction. The paths are projected onto the (β, γ) potential energy surface. The dots in the figures indicate the HB local minima. The equipotential contour lines are drawn every 100 keV.

by setting $B(q)^{-1} = 1$. The collective mass with respect to the conventional (β, γ)-coordinates is then given by $M(q) = \sqrt{(d\beta/dq)^2 + \beta^2(d\gamma/dq)^2}^{-1}$. We have found that the collective mass $M(q)$ and three rotational moments of inertia $\mathcal{J}_i(q)$ are enhanced by the time-odd pair field generated by the quadrupole pairing.

4. Energy Spectrum and Transition Probabilities

We re-quantize the collective Hamiltonian (1) and solve the collective Schrödinger equation

$$\left(-\frac{1}{2}\frac{\partial^2}{\partial q^2} + \frac{1}{2}\sum_{i=1}^{3}\mathcal{J}_i^{-1}(q)\hat{I}_i^2 + V(q)\right)\Psi_{IM,k}(q,\Omega) = E_{I,k}\Psi_{IM,k}(q,\Omega), \tag{7}$$

to obtain the excitation energies and collective wave functions. The collective wave functions $\Psi_{IM,k}(q,\Omega)$ takes the following form:

$$\Psi_{IM,k}(q,\Omega) = \sum_{K=0}^{I}\Phi_{IK,k}(q) < \Omega|IMK >, \tag{8}$$

where $\Phi_{IK,k}(q)$ and $< \Omega|IMK >$ denote the vibrational wave functions and the rotation matrices, respectively. The conventional boundary conditions and symmetry requirements for solving the Bohr collective Hamiltonian[11] are adopted.

Figures 3 display the energy spectrum and B(E2) values for ^{68}Se and ^{72}Kr. The calculation yields the excited prolate rotational band as well as the oblate ground-state band. It is seen that the inter-band E2 transitions are weaker than the intra-band E2 transitions, indicating that the oblate–prolate shape coexistence picture holds. The result of calculation also indicates that the oblate–prolate shape mixing decreases as the angular momentum increases. Thus, the basic patterns of oblate–prolate shape coexistence are successfully reproduced in the calculation. The calculation suggest the existence of excited 0^+ state in ^{68}Se, which is not yet found in experiment. By comparing two calculations, ($G_2 = 0$) and ($G_2 = G_2^{\rm self}$), we

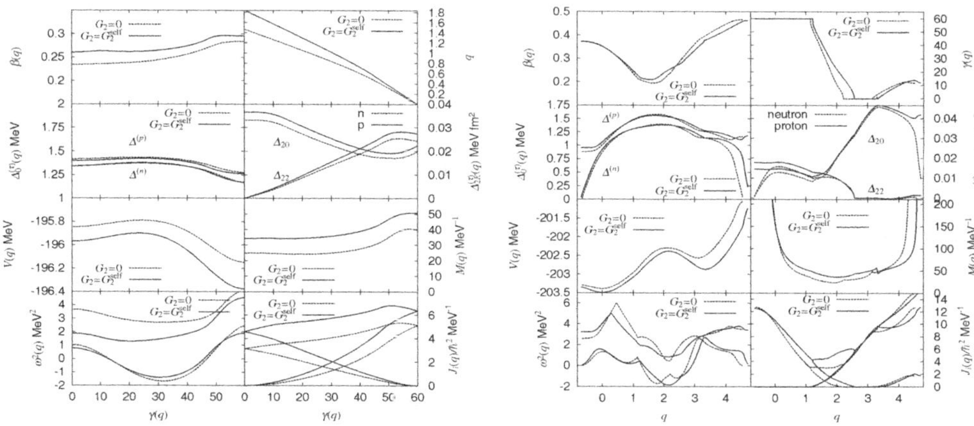

Fig. 2. The monopole pairing gaps, $\Delta_0^{(\tau)}(q)$, the quadrupole pairing gaps, $\Delta_{20}^{(\tau)}(q)$ and $\Delta_{22}^{(\tau)}(q)$, the collective potential $V(q)$, the collective mass $M(q)$, the rotational moments of inertia $\mathcal{J}_i(q)$, the lowest two moving-frame QRPA frequencies squared, $\omega^2(q) = B(q)C(q)$, and the quadrupole deformations, $\beta(q)$ and $\gamma(q)$, are plotted as functions of $\gamma(q)$. Results of two calculations using the P + Q Hamiltonian with ($G_2 = G_2^{\text{self}}$) and without ($G_2 = 0$) the quadrupole pairing interaction are compared.

found that the energy of the 0_2^- state is much lowered by including the quadrupole pairing interaction. This is due to the enhancement of the collective mass. The energies of the other members of the rotational bands are also lowered, because the quadrupole pairing also enhances the rotational moments of inertia. Experimental data for $B(\text{E2})$ are not available except $B(\text{E2}; 2_1^+ \to 0_1^+)$=1000 e^2 fm^4 in ^{72}Kr.[4] Theoretical $B(\text{E2})$ values are calculated using a polarization charge $e_{\text{pol}} = 0.881$ that reproduces this data, so that only relative magnitudes should be regarded as theoretical estimates.

In order to estimate the oblate–prolate shape mixing in quantum eigenstates, we define the oblate and prolate probabilities as follows:

$$P_{\text{ob}}(I,k) = \int_{q_{\min}}^{q_0} dq \sum_{K=0}^{I} |\Phi_{IK,k}(q)|^2, \quad P_{\text{pro}}(I,k) = \int_{q_0}^{q_{\max}} dq \sum_{K=0}^{I} |\Phi_{IK,k}(q)|^2, \quad (9)$$

where we assume that $q_{\min} \le q_{\text{ob}} < q_0 < q_{\text{pro}} \le q_{\max}$. The "boundary" between the oblate and the prolate regions is set to the top of the potential barrier or at $\gamma = 30°$. Figure 4 shows these probabilities for ^{68}Se and ^{72}Kr. It is clearly seen that the shape mixing rapidly decreases as the angular momentum increases.

5. Summary

For the first time, we have reported the result of calculation of low-lying energy spectra and E2 transition probabilities in ^{68}Se and ^{72}Kr by means of the ASCC method.

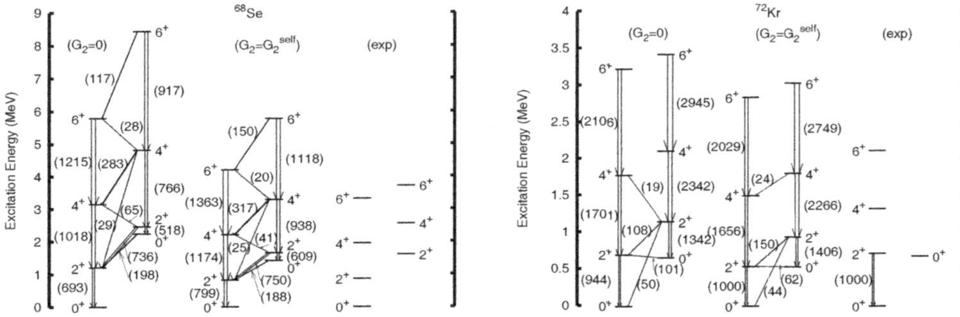

Fig. 3. Excitation energies and $B(\mathrm{E2})$ values for low-lying states of ^{68}Se and ^{72}Kr calculated by the ASCC method. In the left (middle) panel, the quadrupole pairing is ignored (included). Experimental data[1–4] are displayed in the right panel. The $B(\mathrm{E2})$ values are given in parentheses beside the arrows in unit of e^2 fm^4.

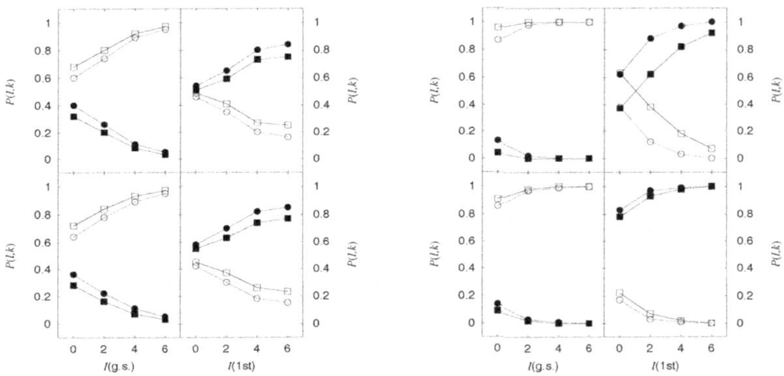

Fig. 4. The oblate and prolate probabilities evaluated for individual eigenstates in ^{68}Se (*left*) and ^{72}Kr (*right*). For each nucleus, the left (right) panel shows values for the lowest (the second lowest) state of each angular momentum. The open (closed) symbols indicate the oblate (prolate) probabilities. The probabilities defined by setting the boundary at the barrier top ($\gamma = 30°$) are shown by squares (circles) In each figure, the upper (lower) panel shows the probabilities calculated without (with) including the quadrupole pairing interaction.

We have derived the quantum collective Hamiltonian that describes the coupled collective motion of the large–amplitude vibration responsible for the oblate–prolate shape mixing and the three-dimensional rotation of the triaxial shape. The calculation yields the excited prolate rotational band as well as the oblate ground–state band. It also indicates that the oblate–prolate shape mixing decreases as the angular momentum increases.

It is surprising that basic pattern of the shape coexistence/mixing phenomena is well reproduced using the one-dimensional collective path running on the two-dimensional (β, γ) plane. Speaking more exactly, this collective path is self-consistently extracted from the huge dimensional TDHB manifold. Namely, the result of calculation indicates that the TDHB collective dynamics of the shape coexistence phenomena in these nuclei is essentially controlled by the single collective variable microscopically derived by means of the ASCC method.

We have also shown that the time-odd pair field enhances collective mass of the large-amplitude vibrational motion and the rotational moments of inertia. This finding is important in understanding the shape coexistence dynamics, because, together with the collective potential energy, these inertial functions associated with the collective kinetic energies determine the degree of localization of the collective wave function in the (β, γ) plane.

Acknowledgement

This work is supported by the Grant-in-Aid for the 21st Century COE "Center for Diversity and Universality in Physics" from the Ministry of Education, Culture, Sports, Science and Technology (MEXT) of Japan.

References

1. S. M. Fischer, *et al.*, *Phys. Rev. Lett.* **84** (2000), 4064.
2. S. M. Fischer, *et al.*, *Phys. Rev. C* **67** (2003), 064318.
3. E. Bouchez, *et al.*, *Phys. Rev. Lett.* **90** (2003), 082502.
4. A. Gade, *et al.*, *Phys. Rev. Lett.* **95** (2005), 022502; *Phys. Rev. Lett.* **96** (2006), 189901.
5. M. Matsuo, T. Nakatsukasa and K. Matsuyanagi, *Prog. Theor. Phys.* **103** (2000), 959.
6. M. Kobayasi, T. Nakatsukasa, M. Matsuo and K. Matsuyanagi, *Prog. Theor. Phys.* **113** (2005), 129.
7. P. Ring and P. Schuck, *The Nuclear Many-Body Problem* (Springer-Verlag, 1980).
8. J. Dobaczewski and J. Skalski, *Nucl. Phys. A* **369** (1981), 123.
9. N. Hinohara, T. Nakatsukasa, M. Matsuo and K. Matsuyanagi, *Prog. Theor. Phys.* **115** (2006), 567.
10. H. Sakamoto and T. Kishimoto, *Phys. Lett. B* **245** (1990), 321.
11. K. Kumar and M. Baranger, *Nucl. Phys. A* **92** (1967), 608.

α-CLUSTER STATES AND 4α-PARTICLE CONDENSATION IN ^{16}O

Y. FUNAKI

Nishina Center for Accelerator-Based Science, The Institute of Physical and Chemical Research (RIKEN), Wako 351-0098, Japan
**E-mail: funaki@riken.jp*

T. YAMADA

Laboratory of Physics, Kanto Gakuin University, Yokohama 236-8501, Japan

P. SCHUCK

Institut de Physique Nucléaire, 91406 Orsay Cedex, France

Université Paris-Sud, Orsay, F-91505, France

H. HORIUCHI and A. TOHSAKI

Research Center for Nuclear Physics (RCNP), Osaka University, Osaka 567-0047, Japan

G. RÖPKE

Institut für Physik, Universität Rostock, D-18051 Rostock, Germany

The 4α-particle condensate state in ^{16}O is discussed by using the 4α condensate model wave function (THSR wave function) and the 4α Orthogonality Condition Model (4αOCM). The experimental situation regarding the 4α condensate state is also discussed. In the OCM approach, we obtained the 0^+ state near the 4α break-up threshold with a dilute density structure, whose r.m.s. radius is about 4 fm. With the comparison of the 4α condensate state previously given by the THSR wave function, this 0^+ state is shown to have 4α condensate components.

For several years, special attention has been paid to whether or not in finite nuclei exists the Bose condensate composed of weekly interacting α particles. It is considered to be an analogue to Bose condensation for finite number of dilute bosonic atoms at very low temperature. Since the idea was first proposed several years ago by some of the present authors,[1] the possibility of the realization has been widely examined.

The recent works by the present authors have revealed that the Hoyle state, the second 0^+ state in ^{12}C has the 3α condensate character, in which 3α particles are loosely bound and condensed into an identical S-orbit.[2,3]

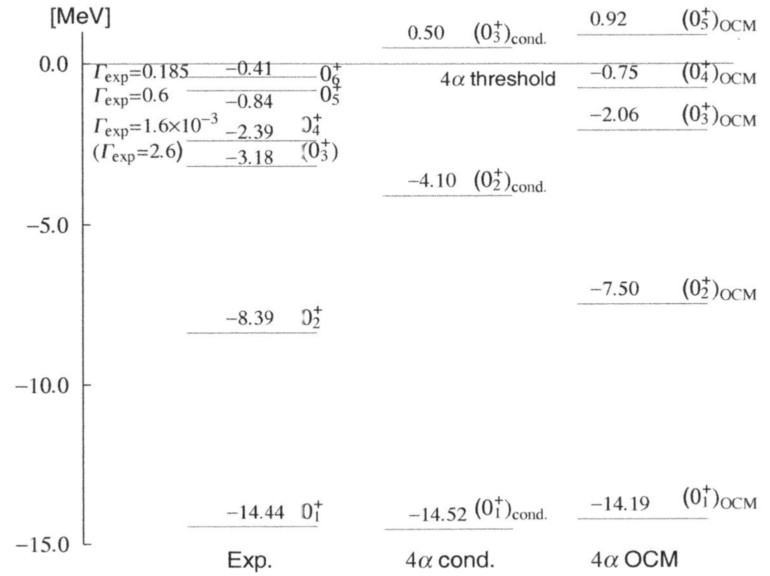

Fig. 1. Comparison of the energy spectra of the three 0^+ states given in Ref. 4 and five 0^+ states obtained via 4α OCM with the observed 0^+ levels[5] up to the 4α threshold. Schmidt-Wildermuth force is adopted as an effective nucleon-nucleon interaction, from which an α-α interaction is constructed by folding procedure, in the calculation of 4α OCM. Γ_{exp} denotes the mesured decay width. All attached numbers are in unit of MeV.

While this interpretation of the Hoyle state now seems to be confirmative, we have much less information for the 4α system. However, there is a theoretical calculation which predicts the existence of the 4α condensate state.[1] It was done by α-condensate type of cluster wave function, which hereafter we call THSR wave function and which is expressed as follows:

$$\Phi_{n\alpha} \propto \mathcal{A}\left[\prod_{i=1}^{n} \exp\Big(-2\frac{X_i^2}{B^2}\Big)\phi(\alpha_i)\right]. \tag{1}$$

Here $\boldsymbol{X}_i = \sum_j^4 \boldsymbol{r}_{j(i)}/4$, with $j(i) \equiv 4(i-1)+j$ is center-of-mass coordinates of the i-th alpha cluster. $B^2 = b^2 + 2\beta^2$, where b is the size parameter of i-th α cluster in $(0s)^4$ configuration which is common to all i's. This wave function has the characteristic feature that all center-of-mass motions of the constituent alpha clusters occupy the same spherical S-wave orbit, $\exp(-2X^2/B^2)$. It also distinguishes itself from other α-cluster models by the fact that the alpha clusters do not have any definite geometric configuration, but are allowed to move freely with small inter-alpha correlations or interactions, forming a dilute gas-like structure. The above mentioned characteristics can be put forward only as far as the variational parameter B takes a sufficiently large value so that the action of the antisymmetrizer, $\mathcal{A}$, is weak and almost negligible.

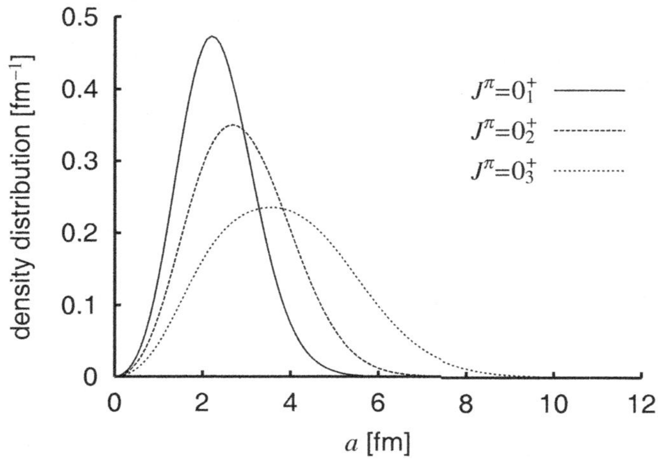

Fig. 2. Density distributions of the three 0^+ states of ^{16}O given in Ref. 4, which are the expectation values of the density operator defined as $\widehat{\rho}(a) = 1/16 \sum_{i=1}^{16} \delta(|\vec{r}_i - \vec{X}_G| - a)$.

By solving the following Hill-Wheeler equation, based on the THSR wave function Eq. (1),

$$\sum_{\beta'} \langle \Phi_{4\alpha}(\beta) \big| (H - E) \big| \Phi_{4\alpha}(\beta') \rangle f^{\lambda}_{4\alpha}(\beta') = 0, \tag{2}$$

three 0^+-states are obtained:[4] the ground state at $E_0 = -124.8$ MeV (experimental value: -127.62 MeV), a second state at excitation energy $E_{0_2^+} = 10.4$ MeV and a third one at $E_{0_3^+} = 15.0$ MeV (see Fig. 1). We show the density distribution for these three states in Fig. 2. The dilute density structure of the $(0_3^+)_{\rm cond.}$ shown in this figure indicates that this state has a quite different structure from that of the ground state with a compact density and supports the idea that this state has a gas-like structure with the 4α condensate character, as was shown in the case of the Hoyle state.[6]

Unfortunately the experimental situation in ^{16}O is by far not so complete as the one in ^{12}C. However, as one of the strong candidates of the 4α condensate state, the 0^+ state with relatively large width was recently observed at 13.6 MeV near the 4α breakup threshold energy 14.44 MeV.[7] The calculated cross section of α inelastic scattering to the $(0_3^+)_{\rm cond.}$ state is compared with the experimental data for the 13.6 MeV state in Fig. 3. Both theory and experiment are in good agreement with one another. The α-decay width for the $(0_3^+)_{\rm cond.}$ state is calculated based on the R-matrix theory. The obtained value $\Gamma = 1.5$ MeV is also consistent with the measured one, $\Gamma_{\rm exp.} = 0.6$ MeV.[8] Therefore, the newly observed state can be assigned to the theoretically predicted 4α condensate state.

However, there also exists a 0^+ state at 14.04 MeV[9] with a strong $E0$ transition probability of $M = 3.3$ fm^2, whose magnitude is close to the $E0$ value of the

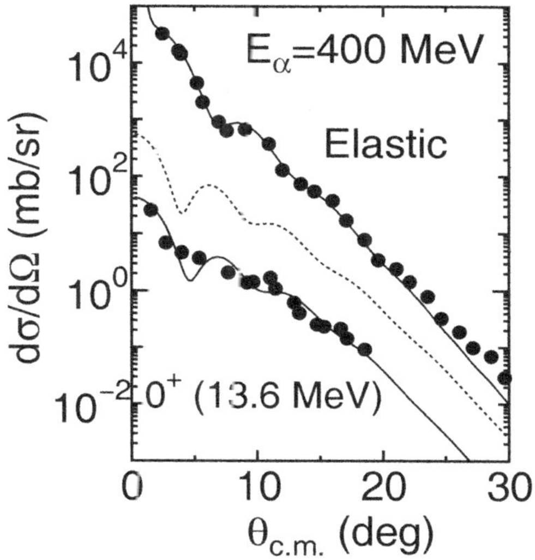

Fig. 3. Angular distribution of the elastic and $\alpha+^{16}O(0^+, E_x=13.6$ MeV) channels at the incident energy of $E_\alpha=400$ MeV. The dots are the experimental data. The solid and dotted curves are the results of microscopic coupled-channels calculation with the use of the THSR wave function, in which the ground and $(0_3^+)_{cond.}$ states are included. The dotted curve shows the result in which the $(0_2^+)_{cond.}$ is assumed to correspond to the 13.6 MeV state. This figure is taken from Ref. 10.

calculated state, that is $M = 2.5$ fm^2. We should keep in mind that the reproduction of the strong monopole transition is by no means trivial, as explained in a recent work by Yamada et al.[11] This state is therefore another candidate to be the analog to the Hoyle state in ^{16}O.

Table 1. Calculated r.m.s. radii $R_{r.m.s.}$ of the five 0^+ state obtained in the 4α OCM calculation. The monopole transition matrix elements $M(E0)$ are also shown, together with the observed values.

	$R_{r.m.s.}$ (fm)	$M(E0)$ (fm^2)	$M(E0)_{exp.}$ (fm^2)
$(0_1^+)_{OCM}$	2.7		
$(0_2^+)_{OCM}$	3.1	4.2	0_2^+: 3.55
$(0_3^+)_{OCM}$	2.9	4.1	0_4^+: 4.03
$(0_4^+)_{OCM}$	3.9	2.4	0_5^+: no data
$(0_5^+)_{OCM}$	3.1	2.0	0_6^+: 3.3

In order to clear up entirely the situation in ^{16}O, we next solve the four-body equation of motion based on the 4α Orthogonality Condition Model (4α OCM) in a large model space which is spanned by Gaussian basis functions. In this approach, α-

α relative motions are solved without any model assumption, while only the essential part of Pauli principle on constituent nucleons is taken into account. This approach will allow us to simultaneously discuss the $\alpha+^{12}$C and 4α cluster structures, as well as the shell-model-like structure of the ground state. We should note that the 4α condensate model wave function cannot reproduce all of the observed excited spectra, due to its smaller functional space with respect to α-α relative motions than that by the present OCM approach.

The calculated energy spectra are shown in Fig. 1. We obtained the five 0^+ states up to and just above the 4α break-up threshold, unlike the result using the 4α THSR wave function. The calculated r.m.s. radii $R_{\text{r.m.s.}}$ and monopole transition matrix elements $M(E0)$ for these states are shown in Table 1. The calculated $M(E0)$ values are all large and in good agreement with those of the corresponding observed 0^+ levels, respectively. We should keep in mind that these strong monopole transition strengths, being comparable with the single particle strength, indicate that the states have α cluster structures.[11,12]

The state we should focus on is the $(0_4^+)_{\text{OCM}}$ state, which is obtained just below the 4α threshold and whose r.m.s. radius 3.9 fm is larger than any other state. The large r.m.s. radius is necessary for a state to have non-negligible α condensate components since α condensate state should have a loosely bound α-particles structure like a gas. The wave function of this state is further found to be similar to the 4α THSR wave function obtained in Ref. 1, as a result of the RWA (Reduced Width Amplitude) analysis, the details of which will be given in another paper. These results mean that this state is a strong candidate of the 4α condensate state with dilute density structure. Also from the RWA analysis, we can conclude that the $(0_5^+)_{\text{OCM}}$ can be described by the mixture of components of $\alpha+^{12}\text{C}(1^-)$ and $\alpha+^{12}\text{C}(3^-)$ cluster structures and smaller component of $\alpha+^{12}\text{C}(0_2^+)$ structure. The estimated value of α-decay width of the $(0_5^+)_{\text{OCM}}$ state is four times larger than that of the $(0_6^+)_{\text{OCM}}$ state. This was calculated based on the R-matrix theory. Therefore, one suggestion we can make for the experimental situation is to assign the observed 0_4^+ and 0_5^+ states to $(0_5^+)_{\text{OCM}}$ and $(0_6^+)_{\text{OCM}}$ states, respectively. We can also explain that the $(0_2^+)_{\text{OCM}}$ and $(0_3^+)_{\text{OCM}}$ states have $\alpha+^{12}\text{C}(0_1^+)$ and $\alpha+^{12}\text{C}(2_1^+)$ structures, respectively, as is completely consistent with the other OCM calculations, where the $\alpha+^{12}$C structure in the excited spectra of ^{16}O was discussed.[13,14] More details will be given in future publication.

One of the authors (Y. F.) is grateful for financial assistance from the Special Postdoctoral Researchers Program of RIKEN. Numerical computation in this work was carried out at the RCNP Computer Facility.

References

1. A. Tohsaki *et al.*, *Phys. Rev. Let.* **87**, 192501 (2001).
2. Y. Funaki *et al.*, *Phys. Rev. C* **67**, 051306 (2003).
3. T. Yamada and P. Schuck, *Euro. Phys. J. A* 26, 185 (2005).
4. A. Tohsaki, H. Horiuchi, P. Schuck and G. Röpke, *Nucl. Phys. A***738**, 259 (2004).

5. F. Ajzenberg-Selove, *Nucl. Phys.* *A***460**, 1 (1986).
6. Y. Funaki, A. Tohsaki, H. Horiuchi, P. Schuck and G. Röpke, *Eur. Phys. J. A*, **24** (2005), 321.
7. T. Wakasa *et al.*, *Phys. Lett. B* **653**, 173 (2007).
8. Y. Funaki, T. Yamada, A. Tohsaki, H. Horiuchi, P. Schuck and G. Röpke, *Mod. Phys. Lett. A* **21** (31-33), 2331 (2006).
9. M. Stroetzel and A. Goldmann, *Phys. Lett. B* **29** , 306 (1969).
10. M. Takashina, private communications.
11. T. Yamada, H. Horiuchi, K. Ikeda, Y. Funaki and A. Tohsaki, arXive: nucl-th/0703045.
12. T. Kawabata *et al.*,*Phys. Lett. B* **646**, 6 (2007).
13. Y. Suzuki, *Prog. Theor. Phys.* **55**, 1751 (1976); **56**, 111 (1976).
14. K. Fukatsu and K. Katō, *Prog. Theor. Phys.* **87**, 151 (1992).

EVOLUTION OF THE $N = 28$ SHELL CLOSURE FAR FROM STABILITY

O. SORLIN

Grand Accélérateur National d'Ions Lourds (GANIL), CEA/DSM - CNRS/IN2P3
Bvd Henri Becquerel, BP 55027, F-14076 Caen Cedex 5, France
E-mail: sorlin@ganil.fr

The present contribution focuses on three recent experimental achievements obtained at the GANIL facility on the study of the $N = 28$ shell closure south to the doubly magic nucleus $^{48}_{20}$Ca. First, the single particle energies of the neutron fp states in ^{47}Ar have been determined by transfer (d,p) reaction using a radioactive beam of $^{46}_{18}$Ar$_{28}$. A reduction of the $N = 28$ gap and of the SO splittings of the f and p states is found. Second, an $E0$ isomer has been discovered in $^{44}_{16}$S$_{28}$ which is suggestive of a shape mixing. Third, the energy of the first 2^+ state of $^{42}_{14}$Si$_{28}$ has been obtained through in-beam γ-ray emission induced by a 2 protons knock-out reaction. Its low value, 770(19) keV, points to a large deformation there. The gradual change from spherical to deformation in the $N = 28$ isotones from $Z = 20$ to $Z = 14$ is ascribed to specific proton-neutron interactions and to enhanced quadrupole correlations. A global description of all the physics involved at $N = 28$ is derived along with experimental results.

Keywords: Shell closure, transfer (d,p), spectroscopic factor, single particle energy, isomer E0, in-beam spectroscopy, tensor force, spin-orbit interaction

1. Introduction

The magic number 28, revealed for instance in the doubly magic nucleus $^{48}_{20}$Ca, is the first originating from the spin-orbit (SO) coupling in atomic nuclei. The SO interaction lowers the $f_{7/2}$ neutron orbit just into the middle of the gap between the sd and fp oscillator shells, resulting in a neutron magic number at $N = 28$ between the $f_{7/2}$ and $p_{3/2}$ orbits. The sequence of the neutron orbits around $N = 28$ exhibits 2 pairs of spin-orbit partners $f_{7/2}$, $p_{3/2}$, $p_{1/2}$ and $f_{5/2}$, ranked as a function of decreasing binding energies. The evolution of the $N = 28$ shell gap is therefore potentially linked to changes in the SO splitting. Variation of these neutron orbits south to ^{48}Ca as protons are progressively removed from the quasi-degenerate $s_{1/2}$ and $d_{3/2}$ orbits can be traced back to the nuclear forces V^{pn} involving protons and neutrons in the sd and fp shells, respectively.

In case of a reduction of the $N = 28$ shell gap quadrupole collectivity (E2) is likely to be favored across it, i.e. between the occupied ($f_{7/2}$; $\ell = 3$) and valence states ($p_{1/2}$, $p_{3/2}$; $\ell = 1$) which differ by two units of angular momentum. As this peculiar quadrupole symmetry exists as well for protons in the sd shells, the rigidity of the ${}^{44}_{16}S_{28}$ and ${}^{42}_{14}Si_{28}$ isotones against quadrupole deformation depends on the size of the spherical proton ($Z = 16, 14$) and neutron $N = 28$ shell gaps.

The present review focuses on recent experimental achievements obtained at the GANIL facility the last few years on the evolution of the $N = 28$ shell closure. Different methods have been used, such as (i) the neutron stripping reaction ${}^{46}_{18}$Ar(d,p) to determine the neutron single particle energies of the f and p states in ${}^{47}_{18}$Ar,[1] (ii) the delayed electron spectroscopy to find a shape isomer in ${}^{44}_{16}$S,[2] (iii) the in-beam γ-ray spectroscopy to determine the energy of the first 2^+ state in ${}^{42}_{14}$Si.[3] The salient features from these experiments are described in the following paragraphs.

2. Study of the neutron single particle states in ^{47}Ar

A pure ^{46}Ar radioactive beam of 10.2(1) A·MeV was delivered by the SPIRAL facility[4] at a mean intensity of $2{\cdot}10^4$ pps. It was guided with a parallel optics onto a 0.38(6) mg/cm^2 thick CD_2 target in which the ^{46}Ar(d,p)^{47}Ar transfer reaction took place. The position of the beam on the target was obtained event by event with a resolution (FWHM) of 1 mm using a position-sensitive Multi Wire Proportional Chamber (MWPC)[5] placed 11 cm downstream from the target. The energy and angle of the reaction protons were measured with the MUST detector array[6] comprising eight highly segmented double-sided Si detectors covering polar angles ranging from 110 to 170° with respect to the beam direction. The transfer-like products ^{47}Ar(^{46}Ar) were selected and identified by the SPEG[7] spectrometer in the case of a neutron pick-up to bound(unbound) states in ^{47}Ar.

The excitation energy spectrum of Fig. 1(a) was obtained using the proper kinematics relations. It has been fitted using a sum of 9 gaussian curves of width σ=175 keV. The Q value of the transfer reaction to the ground state (GS) of ^{47}Ar is Q=1.327(80) MeV, leading to a mass excess Δm(^{47}Ar)=-25.20(9) MeV. The present mass excess value gives rise to an N=28 gap of 4.47(9) MeV in ^{46}Ar, which is 330(90) keV smaller than in ^{48}Ca. Experimental proton angular distributions are shown in Fig. 1 for the ground state (top) and for the second excited state (bottom). From a comparison to Distorted Wave Born Approximation (DWBA) calculations using the DWUCK4[8] code and the optical potentials of Ref.,[9,10] the transferred angular momentum ℓ and the vacancy $(2J+1)C^2S$, [C^2S being the Spectroscopic Factor (SF), J the total spin value] of individual orbitals in ^{47}Ar have been determined. Similar procedure has been applied to all other peaks,[1] which are not shown here. The first two peaks were attributed to $\ell = 1$ transfer to p states, and the third one to an $\ell = 3$ transfer to the $f_{7/2}$ state. The angular distribution of the three peaks around 3.4 MeV (Fig. 1) was reproduced with a combination of ℓ=3 and ℓ=4 components. The ℓ=3 part is most likely contained in the first two peaks as the ℓ=4 transfer to the $g_{9/2}$ state is located at $\simeq$4 MeV in other N=29 isotones. An ℓ=4 value was unambiguously attributed to the 5500(85) keV state.

For a closed-shell nucleus, the total vacancy is expected to be equal to $2J + 1$ for the valence orbitals (such as the GS), and 0 for occupied ones (such as the $f_{7/2}$

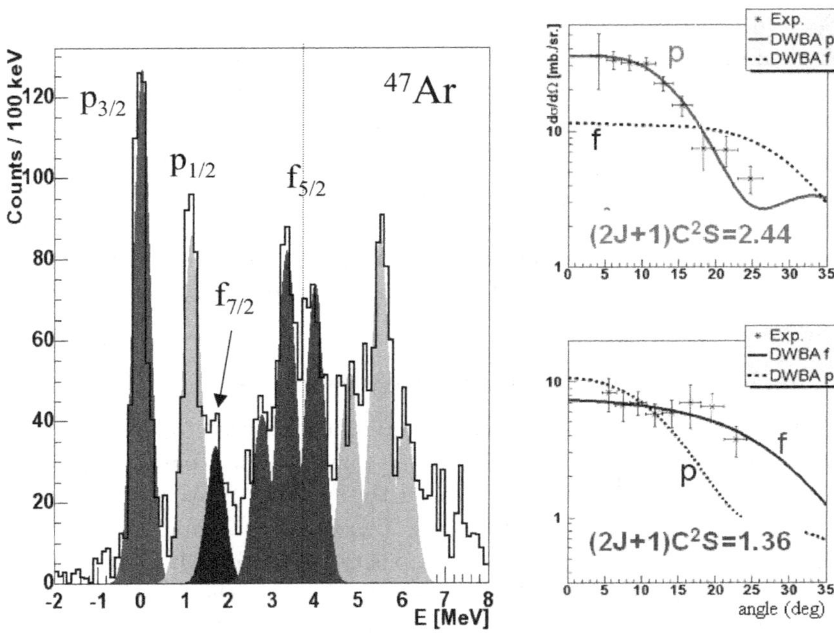

Fig. 1. Left: Background subtracted inclusive spectrum fitted by means of 9 gaussian, Right : Experimental proton angular distributions of the ground state (top) and of the second excited state (bottom) in ^{47}Ar. The curves correspond to DWBA calculations assuming transfer to p and f states are shown for comparison.

orbit). The GS vacancy value of ^{47}Ar, (2.44(20)), is significantly smaller than its maximum value of 4. As a matter of comparison, that of ^{49}Ca is 3.36(14) using similar global optical potentials. This means that a fraction of neutrons is already occupying the valence $p_{3/2}$ orbit in the ^{46}Ar nucleus core. It turns out that the missing vacancy value (1.56) of the $p_{3/2}$ GS counterbalances the excess (1.36 instead of 0) of the 1740 keV state, meaning that some of the $f_{7/2}$ neutrons have moved to the $p_{3/2}$ orbital in ^{46}Ar. This confirms that some amount of neutron particle-hole excitation is already occurring across $N = 28$ in ^{46}Ar. This process is likely due to the reduction of the $N = 28$ shell gap and to the increase of correlations. It will be reinforced if further reduction of the $N = 28$ shell gap occurs for lower Z values.

The structure of ^{47}Ar was calculated in the Shell Model (SM) framework using the ANTOINE code[11] and the *sdpf* interaction[12] to which the relevant neutron-proton monopole interactions V^{pn} were modified[13] to reproduce the experimental binding energies and spectroscopic factors of the known p states in ^{45,47}Ar[13–15] and ^{47,49}Ca.[16] By applying this constraint, the full particle strength of the $\nu f_{7/2}$, $\nu p_{3/2}$ and $\nu p_{1/2}$ and $\nu f_{5/2}$ orbits has been determined in ^{47}Ar. The resulting Single Particle Energies (SPE) are compared to those of the $^{49}_{20}$Ca isotone[17,18] in Fig. 2. Between the two isotones 2 protons have been removed, leading to a 'natural' decrease of the binding energies of all fp states due to 'missing' proton-neutron interactions.

Bearing in mind that the $d_{3/2}$ and $s_{1/2}$ orbitals are quasi-degenerate in ^{48}Ca,[19] the removal of 2 protons is equiprobable from these orbits. If the nuclear forces between protons and neutrons involved for each orbits were similar in strength this decrease would have been the same for all the f and p states. Rather their binding energy vary differently. The orbits in which the angular momentum is aligned ($\ell_{\uparrow}$) with the intrinsic spin, such as $f_{7/2}$ and $p_{3/2}$, become relatively much less bound than the $f_{5/2}$ and $p_{1/2}$ orbits where the angular momentum and intrinsic spin are anti-aligned ($\ell_{\downarrow}$). From these experimental trends, it is found that the proton-neutron monopole matrix elements involving the $d_{3/2}$ and $s_{1/2}$ protons with the neutron fp orbits should contain a spin-dependent term $\tilde{V}$ in addition to the central part of the interaction. Quantitatively, values of $\tilde{V}^{pn}_{d_{3/2}f_{\downarrow,\uparrow}}$ have been determined to be +280 keV (repulsive) and -210 keV (attractive) for the $f_{5/2}$ and $f_{7/2}$ orbitals, respectively. In this case, the change of the f SO splitting was tentatively attributed to the action of tensor forces between the $d_{3/2}$ protons and the f neutrons.[1,20] The change of the p SO splitting was ascribed to the removal of a certain fraction of $s_{1/2}$ protons, and to a much weaker extent to the removal of $d_{3/2}$ protons. This latter constraint is provided by the fact that the p SO splitting remains constant ($\simeq$1.7 MeV) between $^{41}_{20}$Ca$_{21}$[18] and $^{37}_{16}$S$_{21}$[21] after the removal of four protons from the $d_{3/2}$ orbit. As the protons $s_{1/2}$ occupy the center of the nucleus, the change of the neutron p SO is due to depletion of central density of the nucleus. This points to the central density dependence of the SO interaction which was usually thought to be a surface term exclusively. Details on this interaction can be found in Refs.[1,22] For the $p_{1/2}$ and $p_{3/2}$ orbitals, we have extracted $\tilde{V}^{pn}_{s_{1/2}p_{\downarrow,\uparrow}}$ of +170 keV and -85 keV, respectively. With these effects, the spin-orbit splittings for the f and p states have been reduced by about 10%[1,13] between ^{49}Ca and ^{47}Ar, by the removal of only two protons. The variation of the single particle energies in term of monopole interactions can be pursued towards the $^{43}_{14}$Si nucleus in which, as compared to $^{49}_{20}$Ca, 4 protons have been removed from the $d_{3/2}$ and 2 from the $s_{1/2}$ orbits. Using the monopole matrix elements derived in the interaction, a global shrink between the fp orbits is foreseen, such as reductions of the (i) $N = 28$ gap by $330 \times 3 \simeq 1000$ keV, (ii) f SO splitting by $4(0.28 + 0.21) \simeq 2000$ keV and (iii) p SO splitting by $2(0.17 + 0.085) \simeq 500$ keV. This global shrink of SPE is expected to reinforce particle-hole excitations across $N = 28$, which are of quadrupole nature. Consequently a progressive increase of collectivity, or onset of quadrupole deformation would develop below $Z = 18$. A search for a shape coexistence or mixing between a spherical and deformed state has been searched for in $^{44}_{16}$S$_{28}$ to ascertain the hypothesis discussed above.

3. Electron spectroscopy of ^{44}S

In $^{44}_{16}$S$_{28}$ the 2^+_1 energy and the $B(E2)$ value were found to be intermediate between a deformed and a spherical nucleus.[23] This situation is suggestive of a possible shape mixing between two configurations, an assumption which can be confirmed by the search for a low lying 0^+_2 state. A secondary beam of ^{44}S was produced by the

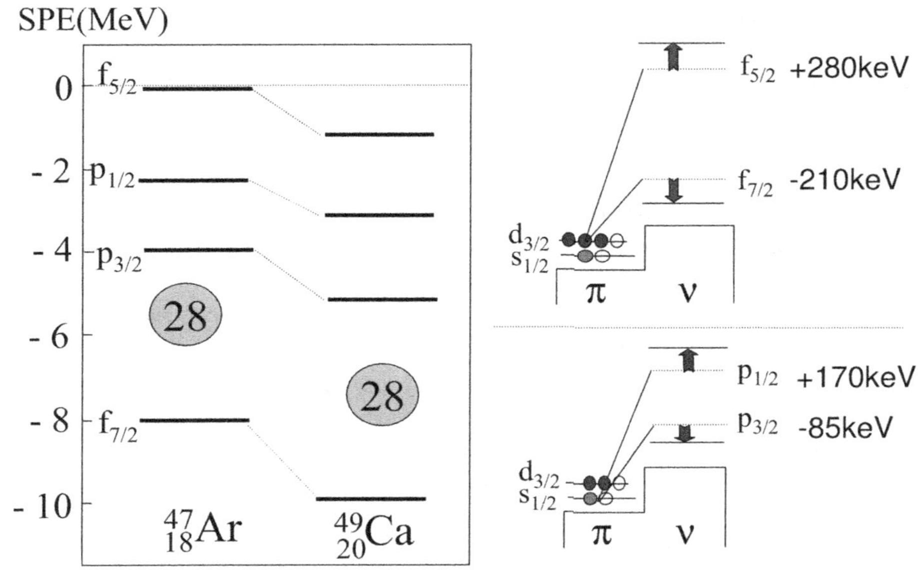

Fig. 2. Left : Neutron single particle energies (SPE) of the fp orbitals for the ${}^{47}_{18}Ar_{29}$ and ${}^{49}_{20}Ca_{29}$ nuclei. Right : Schematic view of the proton-neutron interactions invloved to change the f (top) and p (bottom) SO splittings.

fragmentation of a ^{48}Ca beam at 60A·MeV into a Be target, and selected by the LISE achromatic spectrometer. During this reaction mechanism part of the nuclei can be produced in a 0_2^+ state, which can be isomer or not. To search for an isomer transition, these nuclei were implanted in a thin mylar foil located at the center of an array of four Si-Li and two Ge crystals detectors which aimed at detecting conversion electrons and γ-rays, respectively. Such a 0_2^+ state was discovered through the detection of a delayed electron transition at an energy of 1362.5(1)keV[2] with a lifetime of 2.3 $\pm$ 0.5 μs. Combining the conversion-electron and γ-ray spectroscopy it was found that the 0_2^+ state decays both directly to the ground state via internal conversion, and to the 2_1^+ state at 1329 keV. This experiment has been repeated few months ago to obtain more statistics and to be able to determine the ratio between the two reduced transition probabilities, B(E2;$0_1^+ \rightarrow 2_1^+$) [obtained in Ref.[23]] and B(E2;$0_2^+ \rightarrow 2_1^+$) [determined in the present experiment]. The analysis is in progress. Similar and large B(E2) values would reveal that the 0_1^+ and 0_2^+ levels have mixed wave functions. A large reduced strength $\rho(E0)$ would in turn point to the existence of a spherical and strongly deformed configurations in ^{44}S which co-exist before mixing.

4. Determination of the 2^+ energy in ^{42}Si

Excited states in the neutron-rich $_{14}$Si isotopes have been recently discovered at the NSCL [24] and GANIL laboratories.[3] These two experiments used nucleon removal reactions from secondary beams centered around ^{42}P and ^{44}S at intermediate energy to produce the ^{40}Si and ^{42}Si nuclei and study the γ-rays from their de-excitation in flight. The detection of the γ-rays was achieved by arrays of detectors which surrounded the production target in which the reaction occurred. In the case of Ref.[24] a segmented array of Ge detectors was used, leading to a photo-peak γ-ray efficiency of about 3% at 1 MeV. An array composed of 70 BaF_2 detectors was used in Ref.[3] in order to obtain a higher photo-peak efficiency of about 30% at 1 MeV and 20% at 2 MeV. Campbell et al.[24] have determined the energy of the 2^+ state in $^{40}_{14}Si_{26}$ at 986(5) keV. While the 2^+ energies increase in Ca isotopes from $N = 24$ to reach a maximum around 4 MeV at $N = 28$, the 2^+ energies in the Si isotopes start to deviate from those of the Ca isotopes at $N = 26$ (see Fig. 3) which hints for a reduced N=28 shell gap.[24] As 2^+ neutron excitations occurring before the complete filling of the $\nu f_{7/2}$ shell can be generated inside the shell, they are not that sensitive to the size of the $N = 28$ gap. Conversely, in $^{42}_{14}Si_{28}$ the neutron $f_{7/2}$ shell is *a priori* filled, and the 2^+ state comes mainly from particle-hole excitations across the $N = 28$ gap. It is therefore extremely sensitive to any reduction of the $N = 28$ shell gap. Bastin et al.[3] have established a 2^+ at 770(19) keV for the $N = 28$ nucleus ^{42}Si.

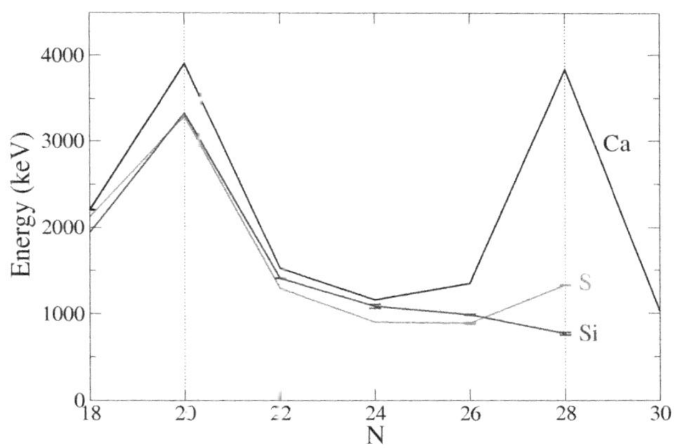

Fig. 3. Experimental $E(2_1^+)$ values in the $_{14}$Si, $_{16}$S and $_{20}$Ca isotopic chains as a function of the neutron number N. Data are taken from the compilation of Ref.,[25] except for ^{38}Si,[26] ^{40}Si,[24] ^{42}Si,[3] and ^{40}S.[27]

As shown in Fig. 3 this dramatic decrease of 2^+ energy ^{42}Si is a proof of the disappearance of the spherical $N = 28$ shell closure at $Z = 14$. This extremely low energy of 770 keV -actually one of the smallest among nuclei having a similar atomic mass - cannot be obtained solely from neutron excitations. Proton core excitations should play an important role, which could in principle be evidenced by measuring the B(E2) values in the Si isotopic chain while reaching $N = 28$.

The behaviors of the $N = 20$ to $N = 28$ isotones south of Ca isotopes are significantly different and instructive. The $N = 20$ shell gap persists with up to the removal of 4 and 6 protons to reach the quasi doubly magic $^{36}_{16}S_{20}$ and $^{34}_{14}Si_{20}$ nuclei, respectively. For these nuclei the existence of significant proton sub-shell closures at $Z = 16$ and $Z = 14$ and the persistence of the $N = 20$ gap are warrants of their rigidity against quadrupole excitations. Conversely, at $N = 28$ the studies of the proton[19] and neutron single particle energies [1,17,18] have revealed a much different situation. The combined compression of the energy of the proton and neutron orbits, plus the favored quadrupole excitations across $N = 28$, produce a rich variety of behaviors and shapes in the even $N = 28$ isotones; spherical ^{48}Ca; oblate non-collective ^{46}Ar; coexistence in ^{44}S, and rotor oblate ^{42}Si .[3,28] This variety of shapes is also supported by mean field calculations, relativistic[31] or non-relativistic.[29,30,32,33] In these models subtle changes in the potential energy surfaces arise from the choice of the effective interactions, in particular for the ^{44}S nucleus. However the global trend to deformation is found as the quadrupole symmetry between the occupied and valence protons and neutrons orbits dominates over effects of the residual proton-neutron interaction.

5. Summary

The spectroscopy of the neutron-rich $N = 28$ isotones has been investigated at GANIL by various complementary methods, such as the neutron stripping (d,p), electron spectroscopy, and in-beam γ-ray spectroscopy. By gathering all pieces of information, a coherent description of the evolution of the $N = 28$ shell gap with the physics processes involved has been proposed.

The $N = 28$ shell gap, as well as the neutron f and p SO splittings are significantly reduced. Reasons for this were attributed to a subtle combination of tensor, central and SO forces. This global shrink of SPE favors the development of quadrupole deformation south to ^{48}Ca. These reductions of SO splittings arise from new effects that have been put forward these last few years both from theoretical and experimental point of views. In intensity these reductions of SO splittings are much larger than the ones expected by an increased diffuseness of the nuclear matter far from the valley of stability. They are opening up a new vista on the various components giving rise to changes of the SO splittings in hitherto unknown parts of the chart of nuclides. Noteworthy is the fact that similar physics processes (tensor force and $E2$ symmetry) are expected to be at play in other doubly magic nuclei formed by the SO interaction, such as $^{78}_{50}Ni_{28}$ and $^{100}_{50}Sn_{50}$.

Acknowledgements

I wish to thank all my colleagues (physicists and technicians) who have participated in the mounting, analysis and theoretical interpretation of the experiments presented in this review. Special thanks are addressed to the students, post-docs and researchers, L. Gaudefroy, S. Grévy, B. Bastin, D. Sohler and I. Stefan who have been taking a decisive part in particular for the data analysis. The MUST collaboration have played a key role for the success of the (d,p) experiment, thanks to the diverse and complementary competences from the IPN Orsay, SPhN Saclay and GANIL laboratories. I have appreciated and learnt very much from efficient collaborations with F. Nowacki, A. Poves, T. Otsuka and J. Pieckarewicz. I also thank few other collaborators who keep on bringing efforts and ideas on the subject, F. Azaiez, Z. Dombràdi, D. Guillemaud-Mueller, K. L. Kratz, M. G. Saint-Laurent and M. Stanoiu.

References

1. L. Gaudefroy *et al.*, *Phys. Rev. Lett.* **97**, 092501 (2006)
2. S. Grévy *et al.*, *Eur. Phys. J. A* **25**, s01-111 (2005)
3. B. Bastin *et al.*, *Phys. Rev. Lett.* **99**, 022503 (2007)
4. A.C.C. Villari *et al.*, *Nucl. Instr. Meth. B* **204**, 173 (2003)
5. S. Ottini-Hustache *et al.*, *Nucl. Instr. Meth. A* **431**, 476 (1999)
6. Y. Blumenfeld *et al.*, *Nucl. Instr. Meth. A* **421**, 471 (1999)
7. L. Bianchi *et al.*, *Nucl. Instr. Meth. A* **276**, 509 (1989)
8. P. D. Kunz, University of Colorado, (unpublished)
9. G. L. Wales and R. C. Johnson, *Nucl. Phys. A* **274**, 168 (1976)
10. R. L. Varner *et al.*, *Phys. Rep.* **201**, 57 (1991)
11. E. Caurier, ANTOINE code, IReS, Strasbourg 1989-2002; E. Caurier and F. Nowacki, *Act. Phys. Pol. B* **30**, 705 (1999)
12. S. Nummela *et al.*, *Phys. Rev. C* **63**, 044316 (2001)
13. L. Gaudefroy *et al.*, *Phys. Rev. Lett.* **99**, 099202 (2007)
14. L. Gaudefroy *et al.*, *Eur. Phys. J. A* **27** s01, 309 (2006)
15. A. Gade *et al.*, *Phys. Rev. C* **71**, 051301(R) (2005)
16. ENSDF database, http://www.nndc.bnl.gov/ensdf/
17. R. Abegg *et al.*, *Nucl. Phys. A* **303**, 121 (1978)
18. Y. Uozumi *et al. Nucl. Phys A* **576**, 123 (1994)
19. P. D. Cottle *et al*, *Phys. Rev. C* **58**, 3761 (1998)
20. T. Otsuka *et al.*, *Phys. Rev. Lett.* **95**, 232502 (2005)
21. C. E. Thorn *et al.*, *Phys. Rev. C* **30**, 1442 (1984)
22. B. G. Todd-Rutel *et al.*, *Phys. Rev. C* **69**, 021301(R) (2004)
23. T. Glasmacher *et al*, *Phys. Lett. B* **395**, 163 (1997)
24. C.M. Campbell *et al.*, *Phys. Rev. Lett.* **97**, 112501 (2006)
25. S. Raman, C.W. Nestor, and P. Tikkanen, *At. Data Nucl. Data Tables* **78**, 1 (2001)
26. X. Liang *et al.*, *Phys. Rev. C* **67**, 024302 (2003)
27. J.A. Winger *et al.*, *Phys. Rev. C* **64**, 064318 (2001)
28. E. Caurier F. Nowacki and A. Poves, *Nucl. Phys. A* **742**, 14 (2004)
29. T.R. Werner *et al.*, *Nucl. Phys. A* **597**, 327 (1996)
30. P.-G. Reinhardt *et al.*, *Phys. Rev. C* **60**, 014316(1999)
31. G.A. Lalazissis *et al.*, *Phys. Rev. C* **60**, 014310 (1999)
32. S. Péru *et al.*, *Eur. Phys. J. A* **49**, 35 (2000)
33. R. Rodriguez-Guzman *et al.*, *Phys. Rev. C* **65**, 024304 (2002)

CONTINUUM QRPA APPROACH AND THE SURFACE DI-NEUTRON MODES IN NUCLEI NEAR THE NEUTRON DRIP-LINE

MASAYUKI MATSUO

Department of Physics, Faculty of Science, Niigata University, Niigata, 950-2181, Japan
**E-mail: matsuo@nt.sc.niigata-u.ac.jp*

YASUYOSHI SERIZAWA, KAZUHITO MIZUYAMA

Graduate School of Science and Technology, Niigata University, Niigata, 950-2181, Japan

By means of the continuum QRPA calculations applied to excitation modes in medium and heavy mass neutron-rich nuclei we demonstrate that a surface di-neutron mode emerges as a fundamental mode of excitation in superfluid neutron-rich nuclei near the drip-line. We report also a recent development of the continuum QRPA theory in which the Skyrme energy functional is utilized up to the velocity dependent terms and the energy weighted sum-rule is satisfied.

Keywords: Continuum QRPA; Skyrme-HFB theory, Di-neutron correlation

1. Introduction

Collective excitation modes in neutron-rich nuclei have been attracting many attentions in the last two decades. It is expected that the presence of weakly bound neutrons, the neutron skin and the neutron halo in neutron-rich nuclei may bring about exotic modes of excitation. It is therefore one of the most important goals of nuclear structure theories to describe/predict the excitation modes under these new circumstances. If one wants to describe n-rich nuclei in the medium and heavy mass regions, the quasiparticle random phase approximation (QRPA) is one of the most promising method, and indeed new schemes of the QRPA have been developed recently.[1–5] Among those the continuum QRPA theory plays a special role to describe nuclei near the neutron drip-line as it enables us to treat precisely neutrons in the continuum orbits as well as in the weakly bound orbits.[1,2] In this presentation we shall report recent two achievements of the continuum QRPA. Namely we shall first discuss that the surface di-neutron mode, which is predicted by the continuum QRPA, is regarded as a fundamental mode of excitation in n-rich nuclei near the neutron drip-line. We also discuss a recent theoretical development which enables us to describe multipole responses embedded in the continuum spectrum of unstable nuclei on the basis of the Skyrme energy functional while preserving the energy weighted sum rule.

2. Surface di-neutron mode in the Skyrme-HFB plus continuum QRPA approach

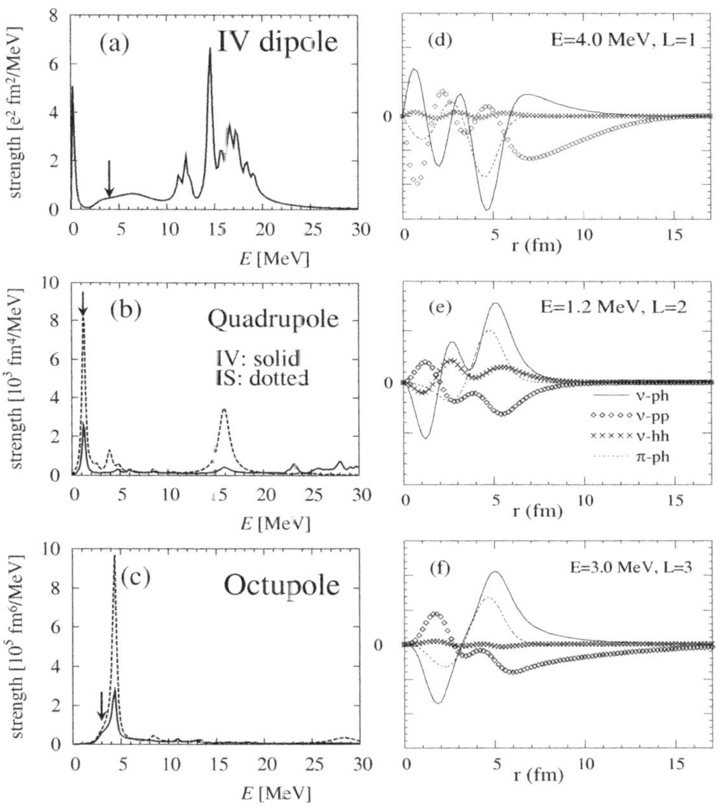

Fig. 1. The dipole, quadrupole and octupole responses in ^{84}Ni, calculated using the SLy4 parameter set for the Skyrme functional and the mixed-type DDDI.

We employ the Skyrme-Hartree-Fock-Bogoliubov theory to describe the ground state. Within this framework the Hartree-Fock potential and the pair potential relevant for the pair correlation are derived simultaneously from the Skyrme energy functional and the pair energy functional. For the latter we use the density dependent delta interaction (DDDI),[6,7] whose strength we parametrize as $V_0[\rho] = v_0 \left(1 - \eta \left(\frac{\rho_n(\boldsymbol{r})}{\rho_c}\right)^{\alpha}\right)$. Solving the Bogoliubov equation to obtain the quasiparticle states, one is able to construct the density and the pair density in a self-consistent way.

We would like to emphasize here that this approach is quite versatile in describing various situations of the pair correlation. Besides the weak coupling pairing (the conventional pairing in stable nuclei), the method is applicable also to the strong coupling pairing where, due to strong ^1S pair interaction between the like-Fermions, the Bose-Einstein condensation of bound Fermion pairs (bosons) may be formed.[8] Because of this feature the method is able to describe the spatial di-neutron correlation which arises from the strong coupling features of the pairing. Indeed it is found recently that the Cooper pairs (the correlated pairs) described within the HFB framework exhibit a significant spatial di-neutron correlation in the ground states of medium and heavy mass neutron-rich nuclei especially in the surface region.[9,10] The physics behind this is that the strength of the nn interaction in the ^1S channel increases with decreasing momentum as it is characterized by the large scattering length $a = -18$ fm at the lowest momentum. The DDDI incorporates this feature in terms of a rather large density dependence (e.g.,$\eta = 0.7 - 0.8$[11]).

We can describe also di-neutron correlations in the excitation modes by means of the QRPA based on the HFB mean-field methods. The QRPA using the Skyrme functional and the DDDI enables us to evaluate directly three kinds of the transition densities $\rho_i^{ph}(\boldsymbol{r}) = \langle\Phi_i|\sum_\sigma \psi^\dagger(\boldsymbol{r}\sigma)\psi(\boldsymbol{r}\sigma)|\Phi_0\rangle$, $P_i^{pp}(\boldsymbol{r}) = \langle\Phi_i|\psi^\dagger(\boldsymbol{r}\uparrow)\psi^\dagger(\boldsymbol{r}\downarrow)|\Phi_0\rangle$, and $P_i^{hh}(\boldsymbol{r}) = \langle\Phi_i|\psi(\boldsymbol{r}\downarrow)\psi(\boldsymbol{r}\uparrow)|\Phi_0\rangle$. In addition to the standard particle-hole transition density $\rho_i^{ph}(\boldsymbol{r})$, $P_i^{pp}(\boldsymbol{r})$ represents the amplitude for motion of singlet pair of nucleons and we call it particle-pair amplitude. $P_i^{hh}(\boldsymbol{r})$ represents also another amplitude for nucleon pairs but for populating the excited state by pair removal process from the $A+2$ ground state.

An example of the continuum QRPA calculation is shown in Fig.1, where we plot the calculated responses of ^{84}Ni for multipole excitations caused by the electric dipole (isovector dipole), the isoscalar/isovector quadrupole and octupole fields. The transition densities evaluated at excitation energy marked with arrows are shown also in the right column. Looking at the transition densities of the soft dipole excitation (Fig.1(d)), we interpret this excitation as motion of surface neutrons which accompanies the strong di-neutron correlation since the significant neutron-pair amplitudes (ν-pp) dominates in the exterior region ($r > 6$ fm). The characteristics of di-neutron motion is not found in the case of the quadrupole response, neither in the low-lying 2^+ collective state (the sharp peak at $E_x = 1.2$ MeV, Fig.1(e)). The octupole case is slightly different. In the continuum bump just above the threshold energy($E_x \sim 3$ MeV, Fig.1(f)), we found the di-neutron correlation in the transition densities. It suggests that the surface di-neutron mode is not limited just to the dipole case, but it may be present also in other multipoles.

3. Surface di-neutron mode as a fundamental mode of excitation[12]

Now we would like to demonstrate that the surface di-neutron mode is a fundamental mode of excitation which originates from an interplay between the strong

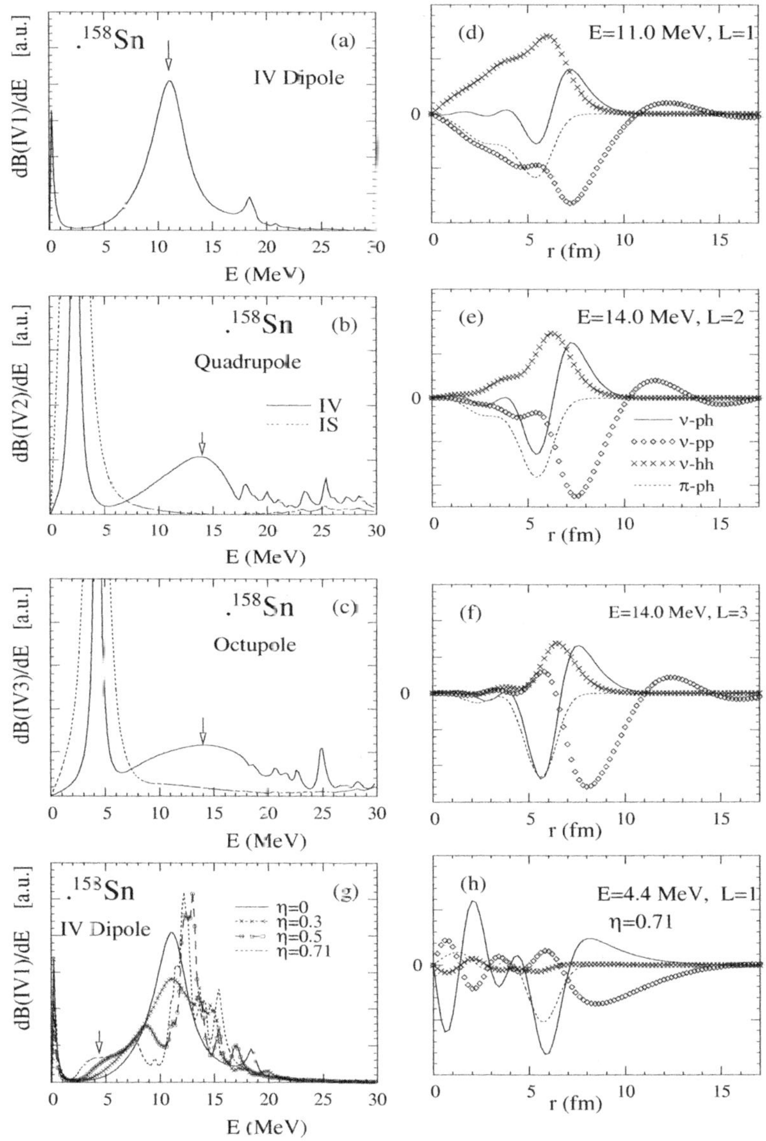

Fig. 2. (a-c): The dipole, quadrupole and octupole strength functions in ^{158}Sn under the strong pairing. (d-f): The transition densities of the surface di-neutron modes at excitation energies marked with arrows. (g): The isovector dipole strength functions for various values of the density-dependence factor η. (h): The transition densities of the soft dipole excitation (at the arrowed energy) in the realistic paring case $\eta = 0.71$. The SLy4 parameter set is used.

surface pairing and the weak binding of neutrons. For this purpose we consider here an extreme case where the pairing interaction is governed only by the large scattering length $a = -18$ fm without the density dependent part of the DDDI. Namely we put $\eta = 0$ while the overall constant v_0 is chosen to reproduce the scattering length.[6,7,11] In this case the pairing correlation becomes strong not only in the surface but also inside the nucleus where the density is close to the saturation. The pairing potential $\Delta(\boldsymbol{r})$ is evaluated as large as $\Delta \sim 15$ MeV. This is of course unrealistic, but it can be used as a magnifying glass to visualize effects of the strong pairing and the di-neutron correlation. We have calculated ^{120}Sn and ^{158}Sn (in the latter case the neutron separation energy is order of 1 MeV).

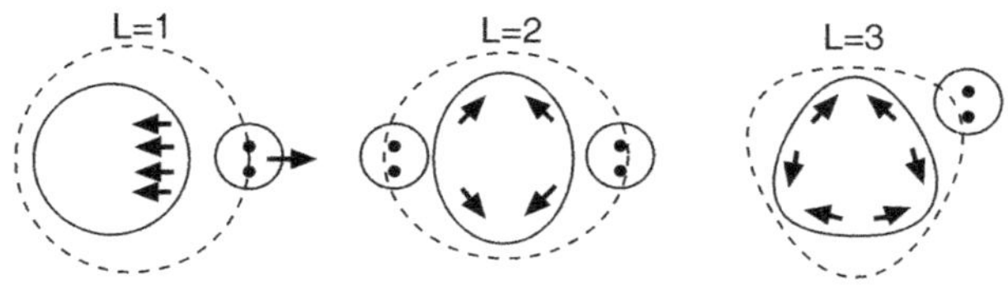

Fig. 3. Schematic picture of the representative surface di-neutron modes.

Figure 2 shows the multipole responses of "^{158}Sn" under the condition of the strong pairing. It is seen that the profiles of the strength function appears very different from those in the realistic pairing case. (We do not show the results for the realistic pairing, but it is quite similar to the case of ^{84}Ni shown in Fig.1). Interestingly very simple behaviors are seen in the strength functions. It appears that there are only two classes of states, one at the low excitation energy $E = 2-5$ MeV which carries large isoscalar strength, and a very broad distribution of isovector strength around $E = 10-15$ MeV, which are present in all the multipoles. Fig.2(d-f) shows the transition densities of the broad isovector modes, and we find a common behavior in the surface and exterior regions: the neutron-pair transition density (ν-pp) is the largest there while at radius $r \sim 4-7$ fm just around the surface the particle-hole transition densities of neutrons and protons (ν-ph and π-ph) exhibit the character of an isoscalar surface vibration. This is readily interpreted as motion of the di-neutrons coupled to the out-of-phase surface vibration (Fig.3). We regard this as representative form of the surface di-neutron mode. The transition densities of the isoscalar low-lying modes, on the other hand, display the profile known for the Bohr-Tassie surface vibration mode in which the nucleus behaves as an incompressible and irrotational one-kind fluid. It is also noted that the transition densities of the broad mode in the isovector dipole response exhibit, in the interior part of the nucleus, the typical behavior of the Steinwedel-Jensen hydrodynamic mode of the isovector density oscillation. In the case of ^{120}Sn under the strong pairing condition we see only two hydrodynamic modes of excitation (the Bohr-Tassie and the Steinwedel-Jensen modes), but in the case of ^{158}Sn the surface di-neutron mode emerges as the third fundamental mode of excitation.

We have checked also how the representative surface di-neutron mode is related to the excitation modes predicted in the realistic pairing case. By changing gradually the density-dependence factor η from 0 to the realistic value 0.71, we can trace how the surface di-neutron mode varies. Fig.2(g) shows the variation of the isovector dipole strength distribution. It is seen that the single broad peak (the surface di-neutron mode) in the strong pairing case is gradually split to two modes: the soft dipole excitation around $E_x = 2 - 6$ MeV and the high-lying resonance around $E_x = 12 - 18$ MeV (the latter is the giant dipole resonance). Looking into the transition densities of the soft dipole excitation (Fig.2(h)), it is confirmed that the soft dipole excitation inherits the character of the surface di-neutron mode. The character of the surface di-neutron mode is seen also in the octupole response, but not in the quadrupole responses as we discussed in connection with Fig.1. This can be traced also by following the change caused by the variation of η.

4. Skyrme continuum QRPA preserving the sum rule[13]

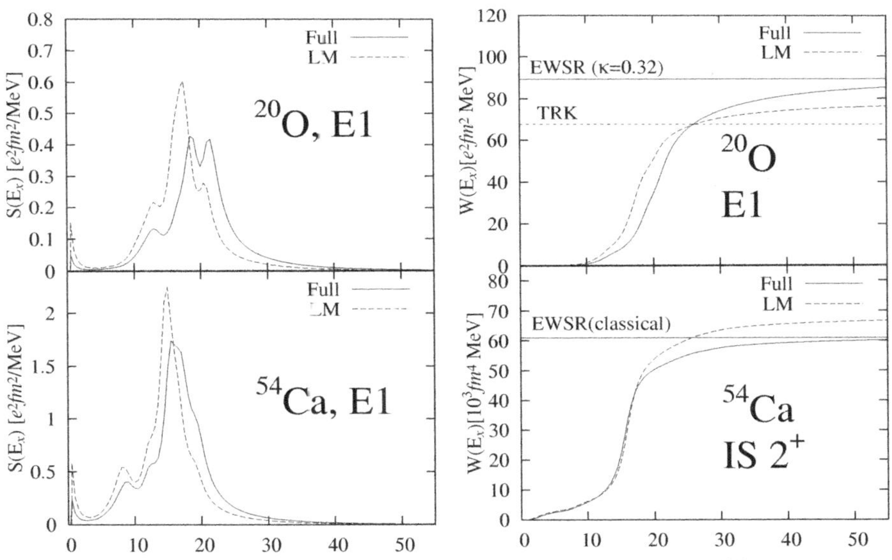

Fig. 4. (Left) The B(E1) isovector dipole strength function for ^{20}O and ^{54}Ca, calculated by means of the Skyrme continuum QRPA. The SkM* is used for the Skyrme functional while the DDDI of the mixed type is adopted for the pairing interaction. The dashed line is the results in the Landau-Migdal approximation. (Right) The running sum of the energy weighted strength for the isovector dipole in ^{20}O and for the isoscalar quadrupole in ^{54}Ca.

The Skyrme energy functional is expressed in terms of several kinds of densities such as the nucleon density $\rho(\boldsymbol{r})$, their derivatives $\Delta\rho(\boldsymbol{r})$ and $\nabla\rho(\boldsymbol{r})$, the kinetic energy density $\tau(\boldsymbol{r})$, the current $\boldsymbol{j}(\boldsymbol{r})$, the spin density, and the spin-orbit tensor.

The self-consistent formulation of the Skyrme QRPA is achieved only if the linear responses in all the densities are taken into account. The continuum QRPA requires, however, a significant amount of computation time in summing up the continuum single-particle states. Thus in the previous continuum QRPA calculations,[1,2] only the local density fluctuation $\delta\rho(\boldsymbol{r},\omega)$ has been treated on top of the pair density fluctuations $\delta\tilde{\rho}(\boldsymbol{r},\omega)$ and $\delta\tilde{\rho}^*(\boldsymbol{r},\omega)$. This, however, brings about violation of the self-consistency and the sum rule, and it causes an ambiguity in establishing a precise correspondence between the experimental data and the Skyrme functional. To improve this situation, we have recently formulated a new version of the continuum QRPA so that the theory should preserve the energy weighted sum rule for multipole density responses (i.e. not involving the spin degrees of freedom).[13] This is achieved by taking into account all the kinds of the densities in the linear response equation of the continuum QRPA except the spin-dependent densities. Numerical results are shown in Fig.4, where the calculation have been performed for the dipole and quadrupole responses in ^{20}O and ^{54}Ca.

We show the calculated isovector dipole strengths in the left column of Fig.4. For the sake of comparison, we have also evaluated the strength function obtained by using the Landau-Migdal (LM) approximation to the residual interaction.[2] (The calculations in the previous sections also use the LM approximation.) Here the LM approximation means that among the densities relevant to the Skyrme functional we consider only the local density $\delta\rho(\boldsymbol{r},\omega)$ for the linear responses while the residual interaction for $\delta\rho(\boldsymbol{r},\omega)$ is evaluated in terms of the Landau-Migdal parameter F_0 and G_0. A part of the velocity dependent terms of the Skyrme interaction is taken into account through F_0 and G_0, but the linear responses in the other kinds of densities of the Skyrme functional (e.g. the current $\delta\boldsymbol{j}(\boldsymbol{r},\omega)$) are neglected. The difference between the LM approximation and the full calculation is not negligible although the basic features of the excitation modes are not affected. The influence is smaller in the case of SLy4 parameter than in the case using the SkM* parameters.

The new formulation of the Skyrme continuum QRPA satisfies the energy weighted sum rule. To show this, we plot the running sum $W(E_x) = \int_0^{E_x} dEEdB(IS(IV)L)/dE$ of the energy weighted strength in the right column of Fig.4. This quantity should converge to the energy weighted sum rule (EWSR) value at sufficiently large E_x. Here the EWSR value is equivalent to the classical sum rule for the isoscalar strength while for the isovector strength there exists the enhancement factor ($\kappa = 0.32$ for the dipole in ^{20}O) which arises from the velocity dependent terms of the Skyrme effective interaction. In the ^{20}O dipole case the running sum evaluated at the highest calculated energy $E_x = 55$ MeV reaches 94% of the EWSR. This suggests that the EWSR is satisfied in the present calculation. In the case of the Landau-Migdal approximation, however, the sum is only 85 % of the EWSR. The same is true for the isoscalar quadrupole response, where the sum approaches to the 99% of the EWSR. This satisfactory result indicates the power of this Skyrme continuum QRPA. We note however that this is a very time-consuming calculation. To perform calculations within a reasonable time, we here adopted a rather large smoothing parameter $\epsilon = 1$ MeV. Because of this limit we here demonstrate only the basic aspects of the method.

5. Conclusions

We have studied excitation modes of neutron-rich nuclei in the medium and heavy mass regions by means of the continuum QRPA method formulated on the basis of the Skyrme-Hartree-Fock-Bogoliubov theory. Employing the density-dependent delta interaction which gives rise to stronger attraction in the surface and exterior regions, we predict that the soft dipole excitation accompanies the character of surface di-neutron motion. To clarify the nature of this mode, we considered a hypothetical situation where the pairing interaction is described only by the large scattering length of the nn interaction. In such a strong pairing case and in neutron-rich nuclei close to the drip-line the representative surface di-neutron modes emerge in all the multipoles, and hence this mode can be regarded as a fundamental mode of excitation. The character of the representative surface di-neutron mode is inherited to the soft dipole excitation in the realistic pairing case, and also to the soft octupole excitation. We have demonstrated also a recent development of the continuum QRPA in which the energy weighted sum rule is satisfied up to the enhancement factor. This is achieved by including the velocity dependent terms of the Skyrme interaction within the continuum QRPA.

Acknowledgments

The work is supported by the Grant-in-Aid for Scientific Research (No. 17540244) from the Japan Society for the Promotion of Science, and also by the JSPS Core-to-Core Program, International Research Network for Exotic Femto Systems (EFES).

References

1. M. Matsuo, *Nucl. Phys. A* **696**, 371 (2001).
2. E. Khan, N. Sandulescu, M. Grasso, and Nguyen Van Giai, *Phys. Rev. C* **66**, 024309 (2002).
3. N. Paar, P. Ring, T. Nikšić, and D. Vretenar, *Phys. Rev. C* **67**, 034312 (2003).
4. J. Terasaki, J. Engel, M. Bender, J. Dobaczewski, W. Nazarewicz, and M. Stoitsov, *Phys. Rev. C* **71**, 034310 (2005).
5. N. Paar, D. Vretenar, E. Khan, G. Colo *Rep. Prog. Phys.* **70**, 691 (2007).
6. G. F. Bertsch and H. Esbensen, *Ann. Phys.* **209**, 327 (1991).
7. E. Garrido, P. Sarriguren, E. Moya de Guerra, and P. Schuck, *Phys. Rev. C* **60**, 064312 (1999).
8. A. J. Leggett, in *Modern Trends in the Theory of Condensed Matter*, Lecture Note in Physics 115, (Springer-Verlag, Berlin, 1980); *J. de Phys.* **41**, C7-19 (1980).
9. M. Matsuo, K. Mizuyama, and Y. Serizawa, *Phys. Rev. C* **71**, 064326 (2005).
10. N. Pillet, N. Sandulescu, and P. Schuck, *Phys. Rev. C* **76**, 024310 (2007).
11. M. Matsuo, *Phys. Rev. C* **73**, 044309 (2006).
12. M. Matsuo, Y. Serizawa, K. Mizuyama, *Nucl. Phys. A* **788**, 307c (2007).
13. K. Mizuyama, M. Matsuo, Y. Serizawa, preprint arXiv:0706.1115 (nucl-th).

DEFORMED RELATIVISTIC HARTREE-BOGOLIUBOV MODEL FOR EXOTIC NUCLEI

S.G. ZHOU*

Institute of Theoretical Physics, Chinese Academy of Sciences, Beijing 100080, China
Center of Theoretical Nuclear Physics, National Laboratory of Heavy Ion Accelerator, Lanzhou 730000, China
**E-mail: sgzhou@itp.ac.cn*

J. MENG

School of Physics, Peking University, Beijing 100871, China
Institute of Theoretical Physics, Chinese Academy of Sciences, Beijing 100080, China
Center of Theoretical Nuclear Physics, National Laboratory of Heavy Ion Accelerator, Lanzhou 730000, China

P. RING

Department of Physics, Technische Universität München, 85748 Garching, Germany

A deformed relativistic Hartree-Bogoliubov (DRHB) model is developed aiming at a proper description of exotic nuclei, particularly deformed ones with large spatial extension. In order to give an adequate description of both the contribution of the continuum and the large spatial distribution in exotic nuclei, the DRHB equations are solved in a Woods-Saxon basis in which the radial wave functions have proper asymptotic behaviors at large distance from the nuclear center which is crucial for the formation of halo. The formalism and the numerical procedure of the DRHB model in a Woods-Saxon basis are briefly presented.

Keywords: Relativistic mean field model; Bogoliubov transformation; Deformed halo; Woods-Saxon basis.

1. Introduction

One of the exotic phenomena observed in nuclei close to drip lines is the halo in which the extremely weakly binding property leads to many new features, e.g., the coupling between the bound states and the continuum due to pairing correlations and the large spacial density distribution. In order to give an adequate theoretical description of the halo phenomenon, the asymptotic behavior of nuclear densities at large r must be considered properly and the discrete bound states, the continuum and the coupling between them must be treated self consistently. This could be achieved by solving the non-relativistic Hartree-Fock-Bogoliubov (HFB)[1–3] or relativistic Hartree-Bogoliubov (RHB)[4–6] equations in coordinate space which can fully take into account the mean-field effects of the coupling to the continuum.

In Refs.,[2–6] the spherical symmetry is assumed. Since most of the known nuclei are deformed, whether or not there exist deformed halos and what new features are expected in deformed exotic nuclei are very interesting questions[7–12] which could be answered by the deformed counterparts of the HFB or RHB models. Nevertheless for deformed nuclei, to solve the HFB or RHB equations in coordinate space becomes much more sophisticated and numerically very time consuming. Although many efforts have been made to develop non-relativistic HFB models either in (discretized) coordinate space or in a basis with improved asymptotic behavior,[13–20] the deformed relativistic Hartree-Bogoliubov model has been developed only in conventional harmonic oscillator basis.[21–24]

The deformed relativistic Hartree equations have been solved in a basis in which the basis wave functions are calculated from solving the Dirac equation with spherical Hartree potentials.[25,26] In Ref.,[27] the Woods-Saxon basis was proposed as a reconciler between the harmonic oscillator basis and the coordinate space. The Woods-Saxon wave functions have more realistic asymptotic behavior at large r. One can also use a box boundary condition to discretize the continuum. It has been shown that to solve the spherical relativistic Hartree equations in a Woods-Saxon basis is almost equivalent to to do it in coordinate space.[27] The Woods-Saxon basis may be used in more complicated situations, e.g., for the description of exotic nuclei where both deformation and pairing have to be taken into account.[28] In this proceeding paper we will present briefly the formalism, numerical procedure and some results of the deformed relativistic Hartree Bogoliubov model in a Woods-Saxon basis. Throughout the paper, this model will be labelled as the DRHBWS model.

The paper is organized as follows. In Sec. 2, we give the formalism of the DRHBWS model. The numerical procedure and some preliminary results are presented in Sec. 3. A summary is given in Sec. 4.

2. Formalism

The starting point of the relativistic Hartree theory is a Lagrangian density where nucleons are described as Dirac spinors which interact via the exchanges of several mesons (σ, ω, and ρ) and the photon,[29–33]

$$\begin{aligned}\mathcal{L} = {} & \bar{\psi}_i \left(i\partial\!\!\!/ - M\right) \psi_i + \frac{1}{2}\partial_\mu\sigma\partial^\mu\sigma - U(\sigma) - g_\sigma\bar{\psi}_i\sigma\psi_i \\ & - \frac{1}{4}\Omega_{\mu\nu}\Omega^{\mu\nu} + \frac{1}{2}m_\omega^2\omega_\mu\omega^\mu - g_\omega\bar{\psi}_i\omega\!\!\!/\psi_i \\ & - \frac{1}{4}\vec{R}_{\mu\nu}\vec{R}^{\mu\nu} + \frac{1}{2}m_\rho^2\vec{\rho}_\mu\vec{\rho}^\mu - g_\rho\bar{\psi}_i\vec{\rho\!\!\!/}\vec{\tau}\psi_i \\ & - \frac{1}{4}F_{\mu\nu}F^{\mu\nu} - e\bar{\psi}_i\frac{1-\tau_3}{2}A\!\!\!/\psi_i, \end{aligned} \tag{1}$$

with the summation convention used, $x\!\!\!/ \equiv \gamma^\mu x_\mu = \gamma_\mu x^\mu$, M the nucleon mass, and m_σ, g_σ, m_ω, g_ω, m_ρ, g_ρ masses and coupling constants of the respective mesons. The nonlinear self-coupling for the scalar mesons is crucial for a satisfactory description of the surface properties,[34]

$$U(\sigma) = \frac{1}{2}m_\sigma^2\sigma^2 + \frac{g_2}{3}\sigma^3 + \frac{g_3}{4}\sigma^4, \tag{2}$$

and field tensors for the vector mesons and the photon fields are defined as

$$\begin{cases} \Omega_{\mu\nu} = \partial_\mu\omega_\nu - \partial_\nu\omega_\mu, \\ \vec{R}_{\mu\nu} = \partial_\mu\vec{\rho}_\nu - \partial_\nu\vec{\rho}_\mu - g_\rho(\vec{\rho}_\mu \times \vec{\rho}_\nu), \\ F_{\mu\nu} = \partial_\mu A_\nu - \partial_\nu A_\mu. \end{cases} \tag{3}$$

For the ground state of nuclei with time reversal symmetry, the nucleon spinors are the eigenvectors of the stationary Dirac equation,

$$[\boldsymbol{\alpha}\cdot\boldsymbol{p} + V(\boldsymbol{r}) + \beta(M + S(\boldsymbol{r}))]\,\psi_i(\boldsymbol{r}) = \epsilon_i\psi_i(\boldsymbol{r}), \tag{4}$$

and equations of motion for the mesons and the photon are

$$\begin{cases} (-\Delta + \partial_\sigma U(\sigma))\,\sigma(\boldsymbol{r}) = -g_\sigma\rho_s(\boldsymbol{r}), \\ (-\Delta + m_\omega^2)\,\omega^0(\boldsymbol{r}) = g_\omega\rho_v(\boldsymbol{r}), \\ (-\Delta + m_\rho^2)\,\rho^0(\boldsymbol{r}) = g_\rho\rho_3(\boldsymbol{r}), \\ -\Delta A^0(\boldsymbol{r}) = e\rho_p(\boldsymbol{r}), \end{cases} \tag{5}$$

where ω^0 and A^0 are time-like components of the vector ω and the photon fields and ρ^0 the 3-component of the time-like component of the iso-vector vector ρ meson. Equations (4) and (5) are coupled by the vector and scalar potentials,

$$\begin{cases} V(\boldsymbol{r}) = g_\omega\omega^0(\boldsymbol{r}) + g_\rho\tau_3\rho^0(\boldsymbol{r}) + e\frac{1-\tau_3}{2}A^0(\boldsymbol{r}), \\ S(\boldsymbol{r}) = g_\sigma\sigma(\boldsymbol{r}), \end{cases} \tag{6}$$

and various densities

$$\begin{cases} \rho_s(\boldsymbol{r}) = \sum_{i=1}^A \bar{\psi}_i(\boldsymbol{r})\psi_i(\boldsymbol{r}), \\ \rho_v(\boldsymbol{r}) = \sum_{i=1}^A \psi_i^\dagger(\boldsymbol{r})\psi_i(\boldsymbol{r}), \\ \rho_3(\boldsymbol{r}) = \sum_{i=1}^A \psi_i^\dagger(\boldsymbol{r})\tau_3\psi_i(\boldsymbol{r}), \\ \rho_c(\boldsymbol{r}) = \sum_{i=1}^A \psi_i^\dagger(\boldsymbol{r})\frac{1-\tau_3}{2}\psi_i(\boldsymbol{r}). \end{cases} \tag{7}$$

The pairing correlation is included via the Bogoliubov transformation and the relativistic Hartree-Bogoliubov equation reads,[35]

$$\sum_{\sigma'p'}\int d^3\boldsymbol{r}' \begin{pmatrix} h_D(\boldsymbol{r}\sigma p, \boldsymbol{r}\sigma' p') - \lambda & \Delta(\boldsymbol{r}\sigma p, \boldsymbol{r}'\sigma' p') \\ -\Delta^*(\boldsymbol{r}\sigma p, \boldsymbol{r}'\sigma' p') & -h_D(\boldsymbol{r}\sigma p, \boldsymbol{r}\sigma' p') + \lambda \end{pmatrix} \begin{pmatrix} U_k(\boldsymbol{r}'\sigma' p') \\ V_k(\boldsymbol{r}'\sigma' p') \end{pmatrix} = E_k \begin{pmatrix} U_k(\boldsymbol{r}\sigma p) \\ V_k(\boldsymbol{r}\sigma p) \end{pmatrix}, \tag{8}$$

where E_k is the quasiparticle energy, h_D the Dirac Hamiltonian in (4), and λ the Fermi energy. Here $p = 1, 2$ (or $\pm$) is used to represent the particle-antiparticle degree of freedom.

If in the pp channel, we use a zero range density dependent force,

$$V_{p_1p_2p_1'p_2'}(\boldsymbol{r}_1, \boldsymbol{r}_2; \sigma_1\sigma_2\sigma_1'\sigma_2') = \frac{1}{4}V_0\delta(\mathbf{r}_1 - \mathbf{r}_2)\left[1 - 4\vec{\sigma}_{11'}\cdot\vec{\sigma}_{22'}\right]\left[\mathbf{I}_{11'}^p\cdot\mathbf{I}_{22'}^p\right], \tag{9}$$

for axially deformed nuclei with spacial reflection symmetry, we may expand the potentials, $S(\boldsymbol{r})$, $V(\boldsymbol{r})$, and $\Delta(\boldsymbol{r})$. and various densities in terms of the Legendre polynomials,

$$f(\boldsymbol{r}) = \sum_\lambda f_\lambda(r) P_\lambda(\cos\theta), \ f_\lambda(r) = \frac{2\lambda+1}{2}\int d\cos\theta f(\boldsymbol{r}) P_\lambda(\cos\theta), \ \lambda = 0, 2, 4, \cdots . \tag{10}$$

The quasi particle wave function is expanded in the Woods-Saxon basis $\{\epsilon_{i\kappa m}, \varphi_{n\kappa m}(\boldsymbol{r}\sigma p)\}$ as,

$$U_k(\boldsymbol{r}\sigma p) = \sum_{i\kappa} \begin{pmatrix} u^{(m)}_{k,(i\kappa)} \varphi_{i\kappa m}(\boldsymbol{r}\sigma p) \\ u^{(\bar{m})}_{k,(i\tilde{\kappa})} \tilde{\varphi}_{i\kappa m}(\boldsymbol{r}\sigma p) \end{pmatrix},$$

$$V_k(\boldsymbol{r}\sigma p) = \sum_{i\kappa} \begin{pmatrix} v^{(m)}_{k,(i\kappa)} \varphi_{i\kappa m}(\boldsymbol{r}\sigma p) \\ v^{(\bar{m})}_{k,(i\tilde{\kappa})} \tilde{\varphi}_{i\kappa m}(\boldsymbol{r}\sigma p) \end{pmatrix}, \tag{11}$$

where the single particle energy $\epsilon_{i\kappa m}$ and wave function $\varphi_{i\kappa m}(\boldsymbol{r}\sigma p)$ are obtained from solving the spherical Dirac equation with Woods-Saxon-like potentials,[36]

$$\varphi_{i\kappa m}(\boldsymbol{r}\sigma) = \frac{1}{r}\begin{pmatrix} iG_{i\kappa}(r) Y^l_{jm}(\theta,\phi,\sigma) \\ -F_{i\kappa}(r) Y^{\tilde{l}}_{jm}(\theta,\phi,\sigma) \end{pmatrix}, \quad j = l \pm \frac{1}{2}, \tag{12}$$

with $G^\kappa_\alpha(r)/r$ and $F^\kappa_\alpha(r)/r$ the radial wave functions for the upper and lower components and $Y^l_{jm}(\theta,\phi,\sigma)$ the spin spherical harmonics where $\kappa = (-1)^{j+l+1/2}(j+1/2)$ and $\tilde{l} = l + (-1)^{j+l-1/2}$. The states both in the Fermi sea and in the Dirac sea should be included in the basis for the completeness. In the Woods-Saxon basis, for each m-block the RHB equation (8) turns out to be,

$$\begin{pmatrix} \mathcal{A} & \mathcal{B} \\ \mathcal{C} & \mathcal{D} \end{pmatrix} \begin{pmatrix} \mathcal{U} \\ \mathcal{V} \end{pmatrix} = E \begin{pmatrix} \mathcal{U} \\ \mathcal{V} \end{pmatrix}, \tag{13}$$

where

$$\mathcal{U} = \left(u^{(m)}_{k,(i\kappa)} \right), \ \mathcal{V} = \left(v^{(m)}_{k,(i\tilde{\kappa})} \right), \tag{14}$$

$$\mathcal{A} = \left(A^{(m)}_{(i\kappa)(i'\kappa')} \right), \ \mathcal{D} = \left(-A^{(m)}_{(i\tilde{\kappa})(\widetilde{i'\kappa'})} \right), \tag{15}$$

$$\mathcal{B} = \left(\Delta^{(m)}_{(i\kappa)(\widetilde{i'\kappa'})} \right), \ \mathcal{C} = \left(-\Delta^{(m)}_{(i\tilde{\kappa})(i'\kappa')} = \Delta^{(m)}_{(i'\kappa')(i\tilde{\kappa})} \right). \tag{16}$$

The derivation of the matrix elements will be given in a detailed paper.[37]

3. Numerical procedure and results

There are several parameters which have to be introduced for numerical reasons, e.g., the mesh size Δr, the box size $R_{\max}$, and the cut off parameter on λ in the expansion (10), $\lambda_{\max}$. Instead of a cut off on the radial quantum number n in the

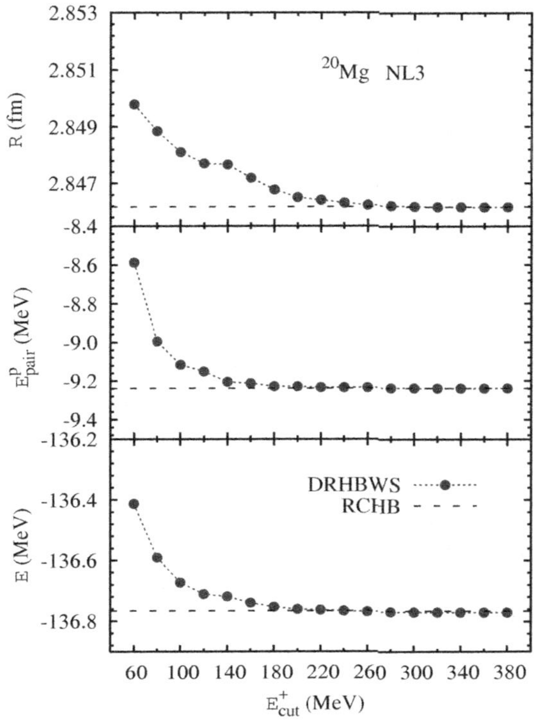

Fig. 1. The total binding energy E (the lower panel), the proton pairing energy $E^{\mathrm{p}}_{\mathrm{pair}}$ (the middle panel), and the rms radius R (the upper panel) of ^{20}Mg versus the cutoff energy E^{+}_{cut} in the Woods-Saxon basis for the DRHBWS model (solid circles). The spherical RCHB results (dashed lines) are also included for comparison.

expansion (11), an energy cutoff E^{+}_{cut} is introduced for positive energy states in the Woods-Saxon basis and in each κ-block, the number of negative energy states in the Dirac sea is the same as that of positive energy states in the Fermi sea. We have investigated the dependence of our results on some of these parameters in spherical and deformed relativistic Hartree models.[27,28] It's found that a box of the size $R_{\mathrm{max}} = 4r_0A^{1/3}$ with $r_0 = 1.2$ fm, a step size $\Delta r = 0.1$ fm, $\lambda_{\mathrm{max}} = 4$, and $E^{+}_{\mathrm{cut}} = 100$ MeV give relative deviations of the binding energy, the rms radius, and the quadrupole moment from the standard ones smaller than 0.1 % for light nuclei. In order to reduce the computational time, a small cutoff $\lambda_{\mathrm{max}} = 3$ is used which would not introduce sizable errors. The parameter set NL3[38] is used for the Lagrangian density.

There are two parameters (ρ_0 and V_0) in the phenomenological pairing force (9). Since we are using a zero range force, a cutoff $E^{\mathrm{q.p.}}_{\mathrm{cut}}$ must be introduced to define the pairing window. We take the empirical value 0.152 fm^{-3} for the saturation density ρ_0. The pairing strength $V_0 = 374$ MeV fm^3 and the cutoff $E^{\mathrm{q.p.}}_{\mathrm{cut}} = 60$ MeV reproduce the proton pairing energy in the spherical nucleus ^{20}Mg from the spherical relativistic Hartree Bogoliubov theory in a harmonic oscillator basis in which the Gogny force is used in the pp channel.

As the first application of the DRHBWS method, we study a spherical nucleus ^{20}Mg so that a comparison can be made between the DRHBWS and the spherical RCHB[4,6] results. The ground state properties of ^{20}Mg are calculated with the DRHBWS and the RCHB codes. The total binding energy E, the proton pairing energy $E^{\rm p}_{\rm pair}$, and the rms radius R of ^{20}Mg are plotted versus $E^{+}_{\rm cut}$ in the lowest, middle and top panels of Fig. 1, respectively. The spherical RCHB results, regarded as exact ones, are shown as horizontal dashed lines. When the basis size increases, E, $E^{\rm p}_{\rm pair}$, and R all converge to the corresponding exact values. In practical calculations, one can choose $E^{+}_{\rm cut}$ according to the balance between the desired accuracy and the computational cost. For light nuclei, one can safely use $E^{+}_{\rm cut} = 100$ MeV which results in accuracies in the total binding energy and the proton pairing energy of about a hundred keV and in the rms radius of around 0.002 fm.

We are investigating some light deformed nuclei close to the neutron drip line by using the DRHBWS model. The results will be presented elsewhere.[37]

4. Summary

In order to give a proper description of exotic nuclei, particularly deformed ones with large spatial extension, the deformed relativistic Hartree-Bogoliubov (DRHB) equations are solved in a Woods-Saxon basis in which the radial wave functions have proper asymptotic behaviors at large distance from the nuclear center which is crucial for the formation of halo.

In this contribution, the formalism and the numerical procedure of the DRHB model in a Woods-Saxon basis are briefly presented. Some preliminary results, namely, the results for a spherical nucleus ^{20}Mg is also given and compared with the spherical relativistic Hartree-Bogoliubov model.

Acknowledgments

This work was supported by the National Natural Science Foundation of China under Grant Nos. 10435010, 10475003 and 10575036, the Major State Basic Research Development Program of China under contract No. 2007CB815000, the Knowledge Innovation Project of Chinese Academy of Sciences under contract Nos. KJCX3-SYW-N02 and KJCX2-SW-N17, and Asia Link Programme CN/Asia-Link 008 (94 791). The computation of this work was performed on the HP-SC45 Sigma-X parallel computer of ITP and ICTS and supported by Supercomputing Center, CNIC, Chinese Academy of Sciences.

References

1. A. Bulgac, *Preprint FT-194-1980*, Central Institute of Physics, Bucharest, 1980 [arXiv: nucl-th/9907088].

2. J. Dobaczewski, H. Flocard, and J. Treiner, *Nucl. Phys. A* **422**, 103 (1984).
3. J. Dobaczewski, W. Nazarewicz, T. R. Werner, J.-F. Berger, C. R. Chinn, and J. Dechargé, *Phys. Rev. C* **53**, 2809 (1996).
4. J. Meng and P. Ring, *Phys. Rev. Lett.* **77**, 3963 (1996).
5. W. Pöschl, D. Vretenar, G. A. Lalazissis, and P. Ring, *Phys. Rev. Lett.* **79**, 3841 (1997).
6. J. Meng, *Nucl. Phys. A* **635**, 3 (1998).
7. X. Li and P.-H. Heenen, *Phys. Rev. C* **54**, 1617 (1996).
8. T. Misu, W. Nazarewicz, S. Åberg, *Nucl. Phys. A* **614**, 44 (1997).
9. L. Guo, E. G. Zhao, and F. Sakata *Commun. Theor. Phys.* **40**, 573 (2003).
10. J. Meng, H.-F. Lü, S. Q. Zhang and S. G. Zhou, *Nucl. Phys. A* **722c**, 366 (2003).
11. F. M. Nunes, *Nucl. Phys. A***757**, 349 (2005).
12. J. C. Pei, F. R. Xu, and P. D. Stevenson, *Nucl. Phys. A* **765**, 29 (2006).
13. J. Terasaki, P.-H. Heenen, H. Flocard, and P. Bonche, *Nucl. Phys. A* **600**, 371 (1996).
14. J. Terasaki, H. Flocard, P.-H. Heenen, and P. Bonche, *Nucl. Phys. A* **621**, 706 (1997).
15. M. Yamagami, K. Matsuyanagi, and M. Matsuo, *Nucl. Phys. A* **693**, 579 (2001).
16. N. Tajima, *Phys. Rev. C* **69**, 034305 (2004).
17. M. V. Stoitsov, J. Dobaczewski, P. Ring, and S. Pittel, *Phys. Rev. C* **61**, 034311 (2000).
18. M.V. Stoitsov, J. Dobaczewski, W. Nazarewicz, S. Pittel, D.J. Dean, *Phys. Rev. C* **68**, 054312 (2003).
19. E. Terán, V. E. Oberacker, and A. S. Umar, *Phys. Rev. C* **67**, 064314 (2003).
20. V. E. Oberacker, A. S. Umar, E. Terán, and A. Blazkiewicz, *Phys. Rev. C* **68**, 064302 (2003).
21. D. Vretenar, G. A. Lalazissis, and P. Ring, *Phys. Rev. Lett.* **82**, 4595-4598 (1999).
22. G. A. Lalazissis, D. Vretenar, and P. Ring, *Nucl. Phys. A* **650**, 133-156 (1999).
23. T. Niksic, D. Vretenar, P. Ring, and G. A. Lalazissis, *Phys. Rev. C* **65**, 054320 (2002).
24. T. Niksic, D. Vretenar, G. A. Lalazissis, and P. Ring, *Phys. Rev. C* **69**, 047301 (2004).
25. J.-K. Zhang and D. S. Onley, *Phys. Lett. B* **209**, 145 (1988).
26. J.-K. Zhang and D. S. Onley, *Nucl. Phys. A* **526**, 245 (1991).
27. S. G. Zhou, J. Meng, and P. Ring, *Phys. Rev. C* **68**, 034323 (2003).
28. S. G. Zhou, J. Meng, and P. Ring, *AIP Conf. Proc.* **865**, 90 (2006).
29. B. D. Serot and J. D. Walecka, *Adv. Nucl. Phys.* **16**, 1 (1986).
30. P.-G. Reinhard, *Rep. Prog. Phys.* **52**, 439 (1989).
31. P. Ring, *Prog. Part. Nucl. Phys.* **37** (1996) 193; *ibid*, **46**, 165 (2001).
32. D. Vretenar, A. V. Afanasjev, G. A. Lalazissis, and P. Ring, *Phys. Rep.* **409**, 101 (2005).
33. J. Meng, H. Toki, S. G. Zhou, S. Q. Zhang, W. H. Long, and L. S. Geng, *Prog. Part. Nucl. Phys.* **57**, 470 (2006).
34. J. Boguta and A. R. Bodmer, *Nucl. Phys. A* **292**, 413 (1977).
35. H. Kucharek and P. Ring, *Z. Phys. A* **339** (1991) 23.
36. W. Koepf and P. Ring, *Z. Phys. A* **339**, 81 (1991).
37. S. G. Zhou, J. Meng, and P. Ring, *in preparation.*
38. G. A. Lalazissis, J. König, and P. Ring, *Phys. Rev. C* **55** (1997) 540.

TWO- AND THREE-BODY CORRELATIONS IN THREE-BODY RESONANCES AND CONTINUUM STATES

K. KATŌ

Department of Physics, Faculty of Science, Hokkaido University, Sapporo 060-0810, Japan.
E-mail: kato@nucl.sci.hokudai.ac.jp

K. IKEDA

The Nishina accellerator center, The Institute of Physical and Chemical Research (RIKEN), Wako, Saitama 351-0198, Japan.
E-mail. k-ikeda@postman.riken.go.jp

Two- and three-body correlations in weakly bound states and continuum states of the Borromean systems such as ^{6}He and ^{11}Li are discussed. In these systems, a two-neutron decay channel comes down to the lowest threshold, and the pairing correlation between valence neutrons is very important together with couplings with continuum states. In addition to those correlations, the tensor correlation is investigated for the halo structure and its excited states. Concerning the problem how to observe the correlation in the halo state experimentally, the final interaction between ^{4}He and n is shown to dominate the Coulomb-breakup cross section of ^{6}He.

Keywords: Borromean system; Pairing correlation; Tensor correlation; Coulomb breakup.

1. Introduction

In drip-line nuclei, much interest has been concentrated on the exotic properties of the halo structure. Two-neutron halo nuclei have been successfully studied as the Borromean system with the core+n+n three-body models.[1] The nuclei ^{6}He and ^{11}Li are typical examples of the Borromean systems, which have no bound states of any sub-system in three-body core+n+n systems. Two-neutron halo structures of ^{6}He and ^{11}Li are investigated in great detail.[1] However, we have not a full understanding of the neutron-neutron and neutron-core correlations in the halo state and its excited states observed in the Coulomb-breakup reaction.

The halo neutrons are weakly bound to the core nucleus, and largely include continuum components. Recently, the Gamow shell model calculations[2–4] have succeeded in description of the single-particle continuum components. Here, we employ the cluster orbital shell model (COSM)[5–7] to describe single-particle resonant states and continua. The neutron-neutron correlation such as the di-neutron clustering, however, cannot appropriately be described by the Gamow shell model and COSM. We discuss that the di-neutron correlation can be treated with the extended cluster model (ECM).[8,9] To describe both continuum and di-neutron correlations, we carried out the hybrid-TV model[10,11] for $n+n+^4$He (^{6}He) and $n+n+^9$Li (^{11}Li).

We discuss the mechanism to control the two-neutron halo structure in ^{6}He and ^{11}Li. The di-neutron correlation is induced by the neutron-neutron pairing interaction. From the point of view of the pairing correlation in ^{11}Li, the p-shell neutrons in the valence part of the ^{9}Li core should be treated on the same footing as the halo neutrons. We here carry out an couple-channel $n+n+^9$Li model calculation including the pairing correlation in ^{9}Li.[12] Furthermore, we discuss that the tensor correlation in the ^{9}Li core plays an important role in a appearance of the di-neutron correlation together with the Pauli-blocking effects.[13]

To observe the correlation in the halo state, the Coulomb breakup reactions are investigated by solving the three-body continuum states. The Borromean system has three-body continuum states starting from the lowest threshold energy and two-body resonant states in its sub-systems. We solve the three-body continuum states and the resonant states within the complex scaling method (CSM).[14] Applying the CSM to the Lippmann-Schwinger solutions, we show that the direct breakup term can be separated from the total strength. From the result of the small direct contribution to the cross section, we can understand that the final state interaction dominates the total strength of the ^{6}He Coulomb breakup.

2. Di-neutron correlation in ^{6}He

The nuclei ^{6}He and ^{11}Li have the two-neutron hallo structure in which the two-neutron correlation plays an important role. The motion of two neutrons above the *core* (^{4}He and ^{11}Li) is expressed by the cluster orbital shell model (COSM) basis functions.[5–7] The center-of-mass motion is completely removed and single-particle continuum states are described by linear combinations of different-size Gaussian basis functions. By taking into account the continuum coupling, the two-neutron correlation has been investigated as an important ingredient of mechanism of the hallo structure.[11]

The ^{6}He nucleus dominantly has a $(p_{3/2})^2$ two-neutron configuration. The $S=0$ components in this simple $(p_3/2)^2$ configuration is 66.6 % and gives two kinds of structure in the two-neutron density distribution; the so-called cigar-type and di-neutron type. The di-neutron distribution is represented by the peak at a small angle θ between n-^{4}He coordinates as shown in Fig. 1. This angle corresponds to the peak of $|P_\ell(\theta)|^2$, and is approximated as $\theta \sim 1/\ell$. Therefore, larger ℓ configurations are necessary in order to describe the density distribution including the di-neutron correlation at the small angle.[9] By our calculations, in the case of ^{6}He, angular momentum configurations of $\ell > 15$ are needed. The recent Gamow shell model calculations,[2] although single-particle continuum components are well described, include only the single configuration of $\ell = 1$, where the di-neutron correlation cannot be described appropriately.

To describe the di-neutron correlation, the so-called T-type basis functions in

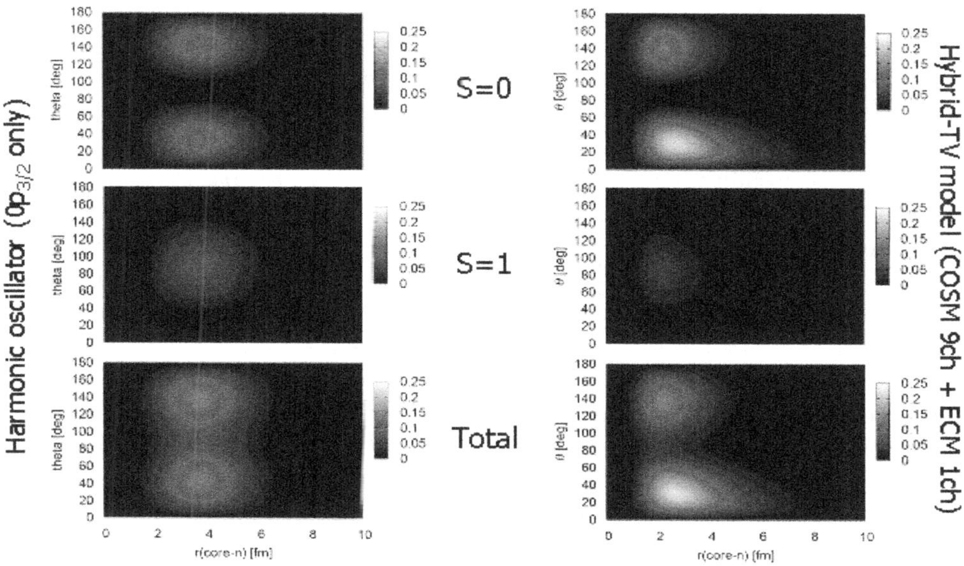

Fig. 1. Density distributions of two neutrons in the $n + n + {}^4$He model for ^{6}He. Right and left hand panels present density distributions of the harmonic oscillator $(0p_{3/2})^2$ configuration and the hybrid-TV model wave functions with 9 channels for COSM plus 1 channel for ECM, respectively. The coordinates (θ, r) are the angle between two n-^{4}He relative vectors and the distance between n and ^{4}He.

the extended cluster model (ECM) are needed.[8,9] When we use the hybrid-TV (COSM+ECM) basis functions to describe two-neutron motion above the core, it is not necessary to employ large angular momentum configurations in the COSM basis.[11] In Fig. 1, we show the two-neutron density distribution in the hybrid-TV calculations. The di-neutron correlation increases clearly by employing the T-type basis states with the single channel of $[0 \otimes 0]_0$ $S = 0$ components, and we can see in the right of Fig. 1 that the di-neutron cluster configuration is enhanced. The T-type basis functions efficiently describe higher angular momentum configurations of COSM. In the case of ^{6}He, its mixing is about 10 %, while the dominant configuration is $(p_{3/2})^2$ of 85 %.

3. Di-neutron correlation in ^{11}Li

In the case of ^{11}Li, the formation mechanism of the di-neutron cluster is very different from the case of ^{6}He. The ^{11}Li nucleus is considered to be described approximately by a $n + n + core(^9\text{Li})$ model, where the ^{9}Li core dominantly has a $(p_{3/2})^4$ neutron closed configuration.[8–10] Therefore, the two neutrons above the ^{9}Li *core* are expected to occupy the $p_{1/2}$ orbit and their pairing correlation has to be treated together with the p-shell neutrons in the ^{9}Li core. From this point of view, the di-neutron correlation in ^{11}Li must be reproduced by taking account of the pairing correlation in ^{9}Li, which is expressed by $(p_{3/2})^4 + (p_{3/2})^2(p_{1/2})^2$ approximately. We performed a coupled three-body $[n + n + ^9\text{Li}((p_{3/2})^4)] + [n + n + ^9\text{Li}((p_{3/2})^2(p_{1/2})^2)]$ calculation, and obtained a large $(s_{1/2})^2$ mixing for weakly-coupled halo two neutrons due to the Pauli-blocking effect.[12] Recently, it has been shown that the analogous Pauli-blocking effects are richly produced by the tensor correlation in the ^{9}Li core.[13] In Fig. 2, we illustrate the Pauli-blocking effects with the pairing and tensor correlations schematically.

The neutron density distribution of ^{11}Li, calculated by solving the coupled-channel $n + n + ^9$Li model taking the pairing and tensor correlations, is displayed in Fig. 3. The $(p_{1/2})^2$ amplitude of two halo neutrons is dominant, approximately 50 %, but the $(s_{1/2})^2$ amplitude holds a large component of $\sim 40\%$. The two-peak structure, which is seen in the ^{6}He case, of the two-neutron density distribution disappears in ^{11}Li, and the di-neutron configuration dominates. This results indicates an importance of s-wave components differently from the ^{6}He case. The s-wave components expands the neutron spatial distribution, and then the pairing interaction between neutrons increases.

Figure 3 also presents the calculation results[13] of the Coulomb break-up cross section of ^{11}Li comparing with the experimental data by Nakamura et al.[15] In this calculation, we used the ^{11}Li wave function of the coupled-channel $n + n + ^9$Li model taking account of the pairing and tensor correlations and a density correction of an exponential behavior in calculations of the neutron-^{9}Li interaction. The experimental data are well explained by this calculation, but the pairing interaction alone cannot reproduce the low-energy peak.

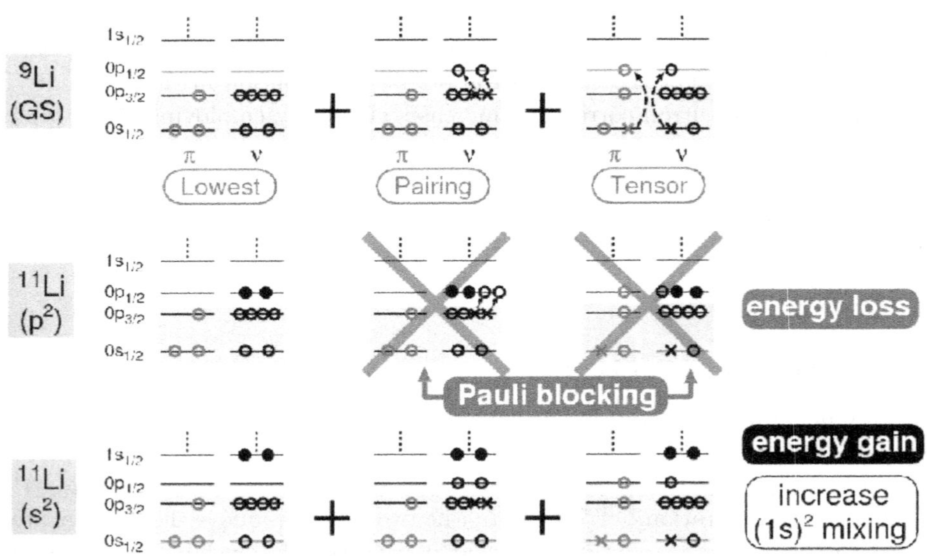

Fig. 2. Expected effects of pairing and tensor correlations in ^{11}Li.

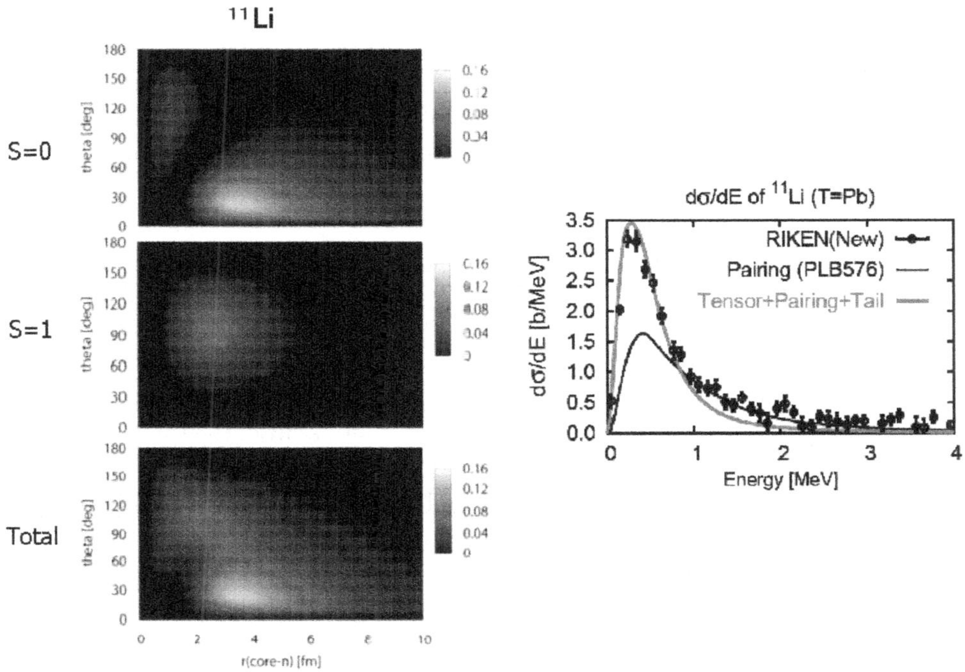

Fig. 3. Left panels: density distributions of two neutrons in the coupled-channel $n+n+{}^9$Li model for ^{11}Li. The coordinates (θ, r) are the angle between two n-^{9}Li relative vectors and the distance between n and ^{9}Li. Right panels: the Coulomb breakup cross sections of ^{11}Li. The experimental data were taken from Ref.15.

4. Observation of correlations in Coulomb breakup reactions

Weakly bound states such as halo states easily breakup by action of an external force. Therefore, it is expected that we can investigate the properties of the weakly bound state from observation of the response to the external force. Since the Coulomb interaction is simple as the external force, the Coulomb breakup reaction is considered as a good probe for such purpose. However, when a final state interaction works strongly in the breaking process, it may be difficult to observe the di-neutron correlation in the initial state directly.

The final state, in the case of the present Borromean system, is a three-body continuum state including the case of a resonant state in any two-body subsystem. Such a case must be treated by the Faddeev method usually. However, without help of the Faddeev method, we here try to apply the complex scaling method[14] and to describe the continuum states within an L^2 basis functions. The complex scaled three-body continuum state is expressed as

$$|\Psi^\theta(\vec{k}, \vec{K})\rangle = |\vec{k}^\theta, \vec{K}^\theta\rangle + \frac{1}{E - H^\theta} V^\theta |\vec{k}^\theta, \vec{K}^\theta\rangle, \tag{1}$$

where the first and second terms in right hand are the plane-wave with the relative momenta ($\vec{k}$, $\vec{K}$) and the interacting solution, respectively. The complex scaled Green function of the second term automatically satisfies out-going boundary conditions for all decaying channels. This Green function is approximately described by using a finite number of eigen-solutions $\{E_i, \phi_i^\theta\}$ of the complex scaling Hamiltonian H^θ. Thus, we have

$$|\Psi^\theta(\vec{k}, \vec{K})\rangle \approx |\vec{k}^\theta, \vec{K}^\theta\rangle + \sum_i |\phi_i^\theta\rangle \frac{1}{E - E_i} \langle \phi_i^\theta | V^\theta | \vec{k}^\theta, \vec{K}^\theta \rangle . \tag{2}$$

We apply this expression (Eq. (2)) to calculations of the final states in the Coulomb breakup process, and then we can extract the direct breakup contribution of $< \vec{k}^\theta, \vec{K}^\theta | \hat{E}1^\theta | \Psi_{gr}^\theta >$, which includes no final state interaction, from the total strength $< \Psi^\theta(\vec{k}, \vec{K}) | \hat{E}1^\theta | \Psi_{gr}^\theta >$.

In Fig. 4, we show the results of the Coulomb breakup strength for ^{6}He. To see the effects from di-neutron and n-^{4}He correlations, we display the breakup strength in T-type and Y-type coordinate representations, respectively. From this result, we can see a very small strength and no prominent structure in the direct breakup. The total strengths in Y- and T-type coordinates have large peaks corresponding to the final state interactions of the virtual n-n resonance and the n-^{4}He ($3/2^-$) resonance, respectively. This result indicates that it is rather difficult to observe di-neutron correlations in the ground state of ^{6}He through the Coulomb breakup.

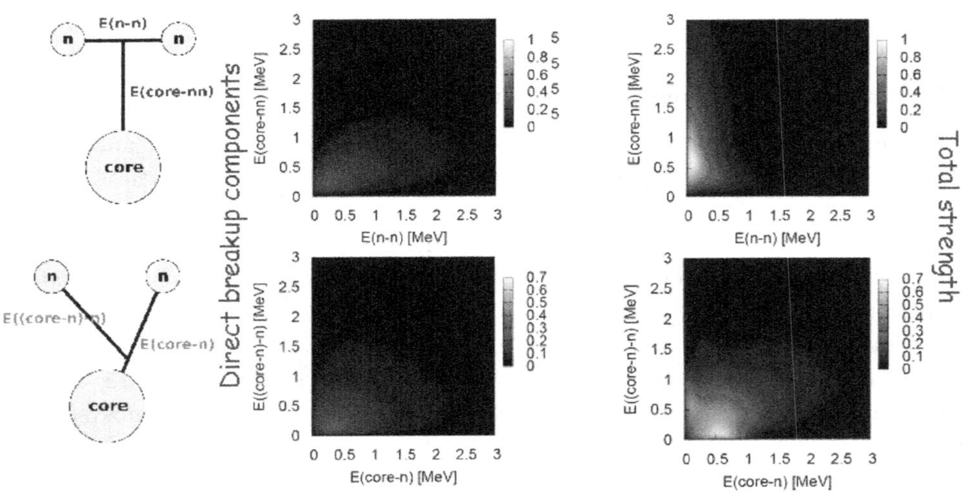

Fig. 4. Direct components and the total strengths in Y- and T-type coordinates of the ^{6}He Coulomb breakup.

5. Summary

It was shown that higher partial waves between n and *core* are needed to reproduce the di-neutron correlation in ^{6}He and ^{11}Li. This higher partial wave components in the $n + n + core$ model are appropriately reproduced by the ECM basis functions in the T-type coordinate system. The distinct difference between ^{6}He and ^{11}Li is the mixing of s-waves: its amplitude is close to 50 % in ^{11}Li, but the $(p_{3/2})^2$ configuration is dominant ($\approx$ 85 %) in ^{6}He. The large mixing of the s-waves in ^{11}Li is explained by the pairing and tensor correlations in ^{9}Li and the Pauli-blocking effects dominantly for the $p_{1/2}$ neutrons. We proposed a new approach of the complex-scaled Lippmann-Schwinger equation for three-body continuum states, and applied this method to calculations of the Coulomb breakup strength. The interesting result indicates that the final state interaction dominates and the di-neutron correlation in the ground state cannot be distinguished in the breakup cross section.

Acknowledgments

We would like to thank Dr. T. Myo, Dr. C. Kurokawa, Dr. R. Suzuki, Mr. Y. Kikuchi, and Prof. H. Masui for their collaboration. This work is supported by the JSPS Core-to-Core Program, International Research Network for Exotic Femto Systems (EFES).

References

1. S. Aoyama, T. Myo, K. Katō and K. Ikeda, *Prog. Theor. Phys.* **116**, 1 (2006).
2. G. Hagen, M. Hjorth-Jensen and J. S. Vaagen, *Phys. Rev. C* **71**, 044314 (2005).
3. J. Rotureau, N. Michel, W. Nazarewicz, M. Ploszajczak and J. Dukelsky *Phys. Rev. Lett.*.**97**, 110603 (2006).
4. N. Michel, W. Nazarewicz and M. Ploszajczak *Phys. Rev. C* **75**, 031301 (2007).
5. Y. Suzuki and K. Ikeda, *Phys. Rev. C***38**, 410 (1988).
6. H. Masui, K. Katō, K. Ikeda, *Phys. Rev. C* **73**, 034318 (2006).
7. H. Masui, K. Katō, K. Ikeda, *Phys. Rev. C* **75**, 034316 (2007).
8. Y. Tosaka, Y. Suzuki and K.Ikeda, *Prog. Theor. Phys.* **83**, 1140 (1990); Y. Tosaka and Y. Suzuki, *Nucl. Phys. A* **512**, 46 (1990).
9. K. Ikeda, *Nucl. Phys. A* **538** (1992), 355c.
10. S. Mukai, S. Aoyama, K. Katō and K. Ikeda, *Prog. Theor. Phys.* **99**, 381 (1998).
11. S. Aoyama, S. Mukai, K. Katō and K. Ikeda, *Prog. Theor. Phys.* **93**, 99 (1995); *ibid.* **94**, 343 (1995).
12. T. Myo, S. Aoyama, K. Katō and K. Ikeda, *Prog. Theor. Phys.* **108**, 133 (2002).
13. T. Myo, K. Katō, H. Toki and K. Ikeda, *Phys. Rev. C* **76**, 024305 (2007).
14. J. Aguilar and J.M.Combes, *Commun. Math. Phys.* **22**, 269 (1971); E. Balslev and J.M. Combes, Commun. Math. Phys. **22**, 280 (1971).
15. T. Nakamura *et al.*, *Phys. Rev. Lett.* **96**, 252502 (2006).

PION- AND RHO-MESON EFFECTS IN RELATIVISTIC HARTREE-FOCK AND RPA

NGUYEN VAN GIAI*

Institut de Physique Nucléaire, IN2P3 and Université Paris-Sud, 91405 Orsay, France
** E-mail: nguyen@ipno.in2p3.fr*

W.H. LONG, H.Z. LIANG, J. MENG

School of Physics, Peking University, Beijing 100871, China

H. SAGAWA

Center for Mathematical Sciences, University of Aizu, Aizu-Wakamatsu, 965-8580 Fukushima, Japan

This talk summarizes the present status of the relativistic Hartree-Fock (RHF) approach based on density-dependent effective Lagrangians. This approach keeps all the success of the widely used relativistic Hartree models, but it can also treat the missing effects due to the neglect of the exchange (Fock) terms. These effects are recovered by explicitly including the coupling of nucleons to effective pions and to the tensor component of rho mesons. On some selected examples it will be shown that these effects can be important for nuclear properties such as single-particle spectra in isotopic and isotonic chains, or spin-isospin excitation modes.

Keywords: relativistic Hartree-Fock; relativistic RPA; effective Lagrangian; charge-exchange excitations; spin-isospin modes

1. Introduction

Within the relativistic scheme, mean field theory has achieved great success in the description of finite nuclei and nuclear matter during the past years. One of the most outstanding models is the relativistic Hartree approach with the no-sea approximation, namely the relativistic mean field (RMF) theory.[1,2]

In the framework of the RMF approach, however, there exist two serious defects. One is the missing of one-pion exchange process. Namely, with the relativistic Hartree approach, the one-pion exchange has zero contribution in mediating nuclear interactions. It is expected that the one-pion exchange may have minor effects in the spin-saturated system, whereas in the spin-unsaturated systems it plays an essential role in determining the isospin dependence of the shell evolutions. Another problem is related to the tensor correlations coming from the ρ-tensor couplings. In the relativistic Hartree approximation, the contributions from the tensor couplings are practically negligible. In recent non-relativistic[3–5] and relativistic studies,[6] it is shown that the tensor forces have distinct effects on the shell evolution of nuclei.[7] In fact these two defects are mainly due to the absence of Fock terms which are dropped in RMF. We will discuss here some physical properties specific to the relativistic Hartree-Fock (RHF) approach.

2. The model

2.1. *Effective Lagrangian*

The starting point is an effective Lagrangian $\mathcal{L}$ for a system of nucleons interacting via meson exchanges. The nucleon fields are denoted by ψ whereas the fields of the 4 types of mesons are $\sigma, \omega, \vec{\pi}, \vec{\rho}$ (the arrows of $\vec{\pi}, \vec{\rho}$ indicate that these fields are vectors in isospin space). The photon field A_μ induces the Coulomb interaction between protons. The general form of $\mathcal{L}$ is:

$$
\begin{aligned}
\mathcal{L} = \bar{\psi}\Big[& i\gamma^\mu\partial_\mu - M - g_\sigma\sigma - g_\omega\gamma^\mu\omega_\mu - g_\rho\gamma^\mu\vec{\tau}\cdot\vec{\rho}_\mu + \frac{f_\rho}{2M}\sigma_{\mu\nu}\partial^\nu\vec{\rho}^\mu\cdot\vec{\tau} \\
& - \frac{f_\pi}{m_\pi}\gamma_5\gamma^\mu\partial_\mu\vec{\pi}\cdot\vec{\tau} - e\gamma^\mu\frac{1-\tau_3}{2}A_\mu\Big]\psi + \frac{1}{2}\partial^\mu\sigma\partial_\mu\sigma - \frac{1}{2}m_\sigma^2\sigma^2 - \frac{1}{4}\Omega^{\mu\nu}\Omega_{\mu\nu} \qquad (1) \\
& + \frac{1}{2}m_\omega^2\omega_\mu\omega^\mu - \frac{1}{4}\vec{R}_{\mu\nu}\cdot\vec{R}^{\mu\nu} + \frac{1}{2}m_\rho^2\vec{\rho}^\mu\cdot\vec{\rho}_\mu + \frac{1}{2}\partial_\mu\vec{\pi}\cdot\partial^\mu\vec{\pi} - \frac{1}{2}m_\pi^2\vec{\pi}\cdot\vec{\pi} - \frac{1}{4}F^{\mu\nu}F_{\mu\nu} \,.
\end{aligned}
$$

where $\Omega^{\mu\nu} \equiv \partial^\mu\omega^\nu - \partial^\nu\omega^\mu$, $\vec{R}^{\mu\nu} \equiv \partial^\mu\vec{\rho}^\nu - \partial^\nu\vec{\rho}^\mu$, $F^{\mu\nu} \equiv \partial^\mu A^\nu - \partial^\nu A^\mu$. In Eq.(1) $\vec{\tau}$ stands for the Pauli matrices representing the isospin, and the γ^μ are the Dirac matrices. The term $\bar{\psi}[\ \ ]\psi$ describes the nucleon fields interacting through meson and photon exchanges whereas the rest represents a system of free mesons and photons.

The Hamilton operator $\mathcal{H}$ is the $(0,0)$-component of the energy-momentum tensor $T^{\mu\nu} \equiv \frac{\partial\mathcal{L}}{\partial(\partial_\mu\phi_i)}\partial^\nu\phi_i - g^{\mu\nu}\mathcal{L}$

$$
\mathcal{H} \equiv T^{00} = \frac{\partial\mathcal{L}}{\partial\dot{\phi}_i}\dot{\phi}_i - \mathcal{L} \,. \qquad (2)
$$

It is convenient to work with a Hamiltonian acting only in the subspace of nucleons. To this end, one uses the Euler-Lagrange equations to obtain the Klein-Gordon and Proca equations which are obeyed by the meson and photon fields. Thus, it is possible to formally express the latter fields in terms of the nucleon fields. One can also make the simplifying assumption of neglecting retardation effects because the nuclear excitation energies are small compared to the meson masses. Then, one can obtain the nucleonic Hamiltonian in the following form[8] :

$$
\begin{aligned}
H = & \int d^3x[\bar{\psi}(-i\vec{\gamma}\cdot\vec{\nabla} + M)\psi] \\
& + \frac{1}{2}\int d^3x_1 d^4x_2 \sum_i \bar{\psi}(x_1)\bar{\psi}(x_2)\Gamma_i(1,2)D_i(x_1,x_2)\psi(x_2)\psi(x_1) \,. \qquad (3)
\end{aligned}
$$

In this work, we introduce a density-dependence into the effective Lagrangian by allowing the meson-nucleon couplings to depend on the baryonic density at each point of space, $g_i = g_i(\rho(\mathbf{r}))$, and similarly for f_π and f_ρ. Then, the interaction vertices of the Hamiltonian (3) are generalizations of those of Ref.8:

$$\begin{aligned}
\Gamma_\sigma(1,2) &\equiv -g_\sigma(1)g_\sigma(2),\\
\Gamma_\omega(1,2) &\equiv +g_\omega(1)\gamma_\mu(1)g_\omega(2)\gamma^\mu(2),\\
\Gamma_\rho^V(1,2) &\equiv +g_\rho(1)\gamma_\mu(1)\vec{\tau}(1)\cdot g_\rho(2)\gamma^\mu(2)\vec{\tau}(2),\\
\Gamma_\rho^T(1,2) &= \frac{1}{4M^2} f_\rho(1)\sigma_{\nu k}(1) f_\rho(2)\sigma^{\nu l}(2)\partial^k(1)\partial_l(2)\vec{\tau}(1)\cdot\vec{\tau}(2),\\
\Gamma_\rho^{VT}(1,2) &= \frac{1}{2M}\Big[f_\rho(1)\sigma^{k\nu}(1)g_\rho(2)\gamma_\nu(2)\partial_k(1) + g_\rho(1)\gamma_\nu f_\rho(2)\sigma^{k\nu}(2)\partial_k(2)\Big]\vec{\tau}(1)\cdot\vec{\tau}(2),\\
\Gamma_\pi(1,2) &\equiv -[\frac{f_\pi}{m_\pi}\vec{\tau}\gamma_5\gamma_\mu\partial^\mu]_1\cdot[\frac{f_\pi}{m_\pi}\vec{\tau}\gamma_5\gamma_\nu\partial^\nu]_2,\\
\Gamma_A(1,2) &\equiv +\frac{e^2}{4}[\gamma_\mu(1-\tau_3)]_1[\gamma^\mu(1-\tau_3)]_2.
\end{aligned} \tag{4}$$

The meson propagators have the usual Yukawa form,

$$D_i(\mathbf{x}_1,\mathbf{x}_2) = \frac{1}{4\pi}\frac{e^{-m_i|\mathbf{x}_1-\mathbf{x}_2|}}{|\mathbf{x}_1-\mathbf{x}_2|}\ . \tag{5}$$

2.2. *Density-dependent couplings*

The determination of the parameters of the model is explained in Refs. 9,10 and 6. They are obtained by a least square fit of bulk properties of nuclear matter plus the binding energies of a set of selected nuclei. Several combinations of the coupling strengths (g_σ and g_ω for the two isoscalar mesons, f_π for the pseudo-vector pion, g_ρ and f_ρ for the vector and tensor couplings of the ρ meson) are thus obtained. In Fig.1 we show the density dependence of the coupling strengths for two of our RHF models: the model PKO1 contains the scalar σ-, vector ω- and ρ-, pseudo-vector π-couplings, but as a simplification the tensor ρ-coupling has been dropped. Recently, a new version called PKA1 which contains the additional tensor ρ-coupling f_ρ has been constructed.[6] For PKA1, two more nuclei (^{140}Ce and ^{146}Gd) have been added to the list of selected nuclei used for determining PKO1, and new constraints on single-particle energies around the Z=58 and 64 subshells are imposed. This is because the results of PKO1, although quite satisfactory in general still suffer of an artificial shell closure at Z=58 common to most relativistic models and the new model PKA1 aims at solving this problem. In Fig.1 we show the density dependence of the couplings of the two models, PKO1 and PKA1. For comparison, the coupling strengths of the density-dependent RMF model DD-ME2 of Ref. 11 are also shown.

3. Single-particle spectra

In this section we briefly discuss single-particle spectra in ^{132}Sn and ^{208}Pb calculated with the two RHF models PKO1 and PKA1, in comparison with results obtained

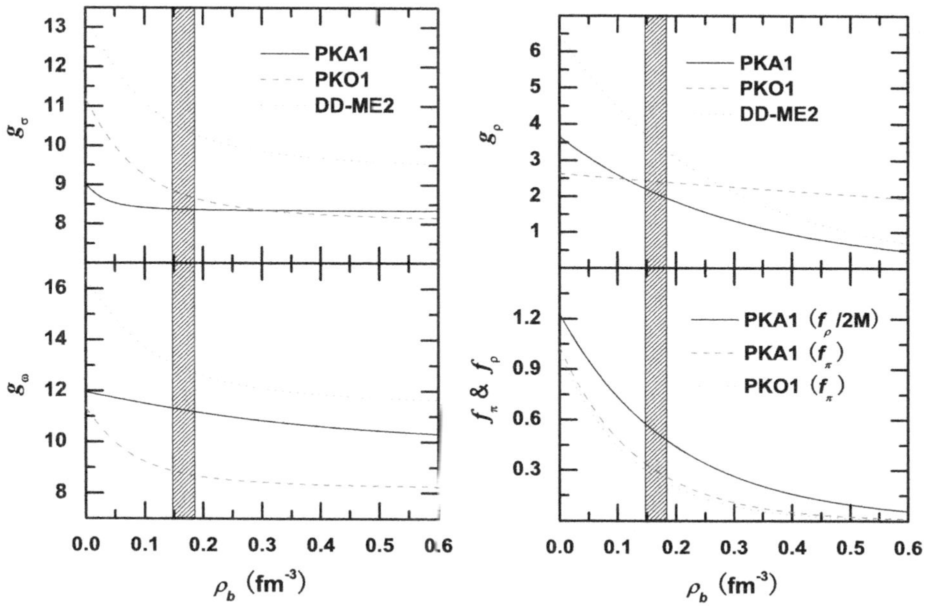

Fig. 1. (color online) The density-dependent meson-nucleon couplings in the isoscalar (left panel: g_σ and g_ω) and isovector (right panel: g_ρ, f_ρ and f_π) channels as functions of density for the DDRHF effective interactions PKA1 and PKO1, and the RMF model DD-ME2. The shadowed area denotes the empirical saturation density region.

using two representative RMF models, the PK1 parametrization[12] with non-linear self-couplings of mesons and the density-dependent DD-ME2 model already mentioned.

The proton numbers of the two N=82 nuclei ^{140}Ce and ^{146}Gd correspond to the artificial shell closure Z=58 of RMF models (and also of PKO1) and to the experimental sub-shell closure Z=64. respectively. Since the model with rho tensor-coupling has more freedom it is possible to impose in the parameter search the condition that the shell gaps at N=58 and Z=58 be small in ^{140}Ce and ^{146}Gd. Then, the sub-shell closures at Z=64 and N=64 are well reproduced by PKA1.[6] This improvement in the single-particle spectra can be seen in another N=82 nucleus, namely ^{132}Sn whose single-particle spectrum is not part of the parameter search procedure. In Fig. 2 are shown the neutron (left panel) and proton (right panel) single-particle levels calculated with PKA1, PKO1, PK1, and DD-ME2. The experimental data from Ref.[13] are also shown for comparison. In this figure, one can also find the spurious shell structures Z or $N = 58$ appearing in the results of PKO1, PK1, and DD-ME2. Compared to these three results, PKA1 shows distinct improvement on this problem. Namely, the spurious shell gaps do not appear any more and fairly good agreement with the data is obtained with PKA1.

In addition to N (or Z) $= 58$, another spurious shell structure at the nucleon number 92 appears in RMF calculations.[14] In Fig. 3, it is clearly shown in both

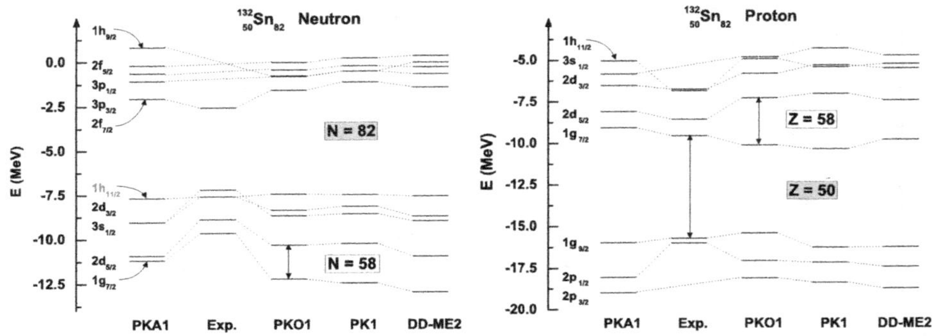

Fig. 2. Single-particle energies of ^{132}Sn. The results are calculated by DDRHF with PKA1 and PKO1, and RMF with PK1 and DD-ME2. Experimental data are taken from Ref. 13.

neutron and proton single-particle spectra that N (or Z) $= 92$ becomes a spurious shell structure in the results of PKO1, PK1, and DD-ME2. With the inclusion of the ρ-tensor correlations, the artificial shell closures at $N = 92$ and $Z = 92$ disappear in both neutron and proton spectra in the results of PKA1. PKA1 also shows another improvement in the order of single-particle levels, e.g., for the neutron states $2g_{9/2}$ and $1i_{11/2}$ in ^{208}Pb. Thus, the RHF with PKA1 provides the same ordering as the data while the other three cases fail to predict the correct ordering.

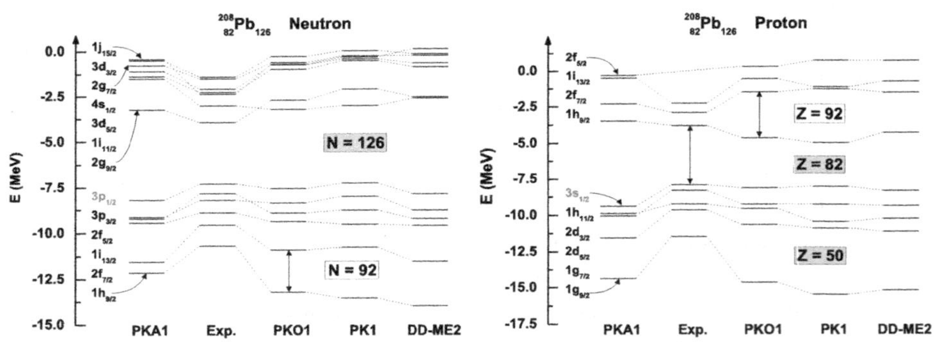

Fig. 3. Single-particle energies of ^{208}Pb. The results are calculated by DDRHF with PKA1 and PKO1, and RMF with PK1 and DD-ME2. The experimental data are taken from Ref. 13.

4. Spin-isospin excitations

A natural extension of the RHF model is the relativistic random phase approximation (RRPA) which enables one to describe excited nuclear states on the same footing as the ground states, and we shall now briefly describe the first results obtained in this new approach.[15]

The RRPA based on the RMF is well known (see, e.g., Ref. 16) and the most recent developments of the RMF-RRPA are presented in P. Ring's contribution to these proceedings.[17] The RMF-RRPA can describe very well nuclear excitations in the non-spin flip channels, but the case of spin flip excitations poses a problem because the mesons entering the model (scalar σ, vector ω, isovector-vector ρ) cannot contribute to the particle-hole (p-h) residual interaction in the spin flip channel by their direct contribution only. They would contribute to the p-h interaction by their exchange terms but these terms are excluded in RMF. The only way out is to add to the p-h interaction the direct contribution of a pion-exchange term. The π-nucleon coupling strength is fixed at its empirical value, $f_\pi^2 = 0.08$, and a zero-range counter term of the form

$$V_{\pi\delta}(1,2) = g'[\frac{f_\pi}{m_\pi}\vec{\tau}\gamma_0\gamma_5\boldsymbol{\gamma}]_1 \cdot [\frac{f_\pi}{m_\pi}\vec{\tau}\gamma_0\gamma_5\boldsymbol{\gamma}]_2\delta(\boldsymbol{r}_1 - \boldsymbol{r}_2). \tag{6}$$

must be added in order to cancel the contact interaction due to the pion pseudovector coupling. For this cancellation to be realized the value of g' must be exactly equal to 1/3.[8] However, in RMF-RRPA it is not possible to obtain satisfactory results with this theoretical value of g' and one has to treat it as an adjustable new parameter of the model. To obtain an agreement with the empirical excitation energy of the Gamow-Teller (G-T) giant resonance in ^{208}Bi the value $g' \simeq 0.55$ is adopted[18] when calculating spin flip modes with the RMF parameter set DD-ME1.

Here, we present the first results of a fully self-consistent RHF-RRPA calculation of the G-T resonance (GTR) and the spin-dipole resonance (SDR) excited from the parent nucleus ^{208}Pb in (p,n) and (n,p) reactions. The parameter set PKO1 has been used together with $g' = 1/3$. In the left panel of Fig.4 are shown the G-T strength distributions calculated with different approximations: a) RHF without residual interaction, b) RRPA with only σ and ω in the residual interaction, c) RRPA with σ, ω and ρ in the residual interaction, d) full RRPA with $g' = 1/3$. These results show that the isoscalar mesons σ and ω have an important effect on the G-T mode through their exchange contributions. This is in contrast to the RMF-RRPA where the isoscalar mesons play no role in the charge-exchange channels. The full calculation predicts the G-T resonance at 18.3 MeV in ^{208}Bi, to be compared with the measured value of 19.2 MeV. It remains to see in the future what would be the contribution to the G-T resonance of a tensor-ρ component in the model. In the right panel of Fig.4 we show the calculated SDR strength distributions in the nuclei ^{90}Nb and ^{90}Y, in comparison with the ^{90}Zr(p,n) and ^{90}Zr(n,p) data of Yako et al.[19] It can be seen that the strength distributions in both isospin channels are correctly described by the self-consistent RHF-RRPA approach. It is expected that the RRPA strength distributions at energies above the main peaks must underestimate the empirical distributions because spreading effects caused by the coupling to configurations more complex than 1 particle-1 hole configurations are still missing in the model. This feature can be seen in Fig.4 .

It is particularly important to have a model which can describe simultaneously the ground state properties (binding energies, radii) and the SDR, in view of the issue of neutron density distributions in nuclei. Indeed, a precise measurement of the SDR sum rule $S_- - S_+$ can give a direct information on the ground state quantity $N\langle r^2\rangle_n - Z\langle r^2\rangle_p$. In the present calculations, we find $S_- - S_+$=150.2 fm^2 and $r_n - r_p$=0.092 fm in ^{90}Zr to be compared to the experimental values of Ref. 19, 148±12 fm^2 and 0.07±0.04 fm .

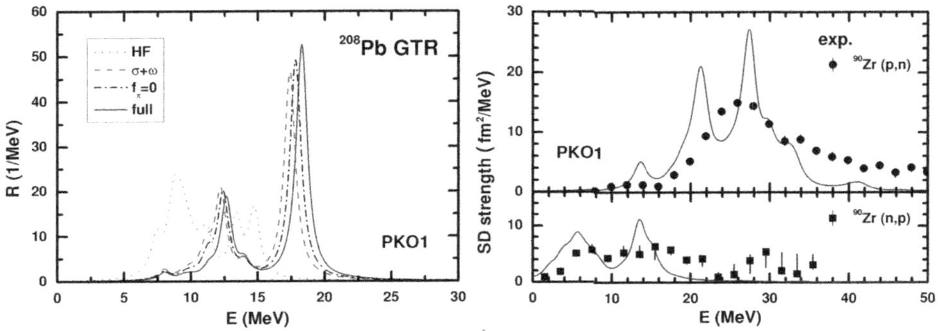

Fig. 4. Left panel: Strength distribution for GTR of ^{208}Pb calculated by RHF+RPA with PKO1. The dotted line shows the unperturbed (HF) excitation, the dashed line shows the calculation with only σ and ω p-h residual interaction, the dash-dotted line shows the calculation setting $f_\pi = 0$ in the p-h residual interaction, and the solid line shows the fully self-consistent calculation. Right panel: Charge exchange SD strength for τ_- channel (upper part) and τ_+ channel (lower part) calculated by RHF+RPA with PKO1. The experimental data are taken from Ref. 19 .

5. Conclusion

In this paper we have briefly presented a density-dependent effective Lagrangian suitable for a relativistic description of nuclear ground states and excited states. The specificity of this model is that it does not neglect the exchange contributions (the Fock terms in the mean field, the exchange terms in the particle-hole interaction) of the meson-induced N-N interactions. These exchange contributions are generally dropped on the belief that they can be compensated by adopting renormalized values of the coupling strengths in the direct terms. However, it can be seen that this is not always possible. For instance, the absence of pion contributions to the mean field due to the Hartree approximation cannot be compensated by readjusting other coupling strengths, as one can see by examining the evolution of single-particle spectra calculated in isotopic or isotonic chains. Furthermore, we have seen in the example of charge-exchange excitations that, isoscalar mesons like σ and ω mesons can give important exchange contributions to the particle-hole residual interaction while their direct contributions are zero. Thus, the mechanism leading to the description of charge-exchange excitations is different depending on whether exchange contributions are kept or not.

Of course, including the exchange contributions in a relativistic approach like RHF is a major complication as compared to the RMF and RMF-RRPA cases. More work is needed to improve the present state: at the mean field level, pairing correlations should be better described in a RHF-Bogoliubov framework, and deformed cases have to be treated; at the RPA level, quasiparticle RRPA studies must be undertaken.

Acknowledgments

This work is partly supported by the National Natural Science Foundation of China under Grant No. 10435010, and 10221003, and the Japanese Ministry of Education, Culture, Sports, Science and Technology by Grant-in-Aid for Scientific Research under the program number (C(2)) 16540259, and the European Community project Asia-Europe Link in Nuclear Physics and Astrophysics CN/Asia-Link 008(94791).

References

1. J.D. Walecka, *Ann. Phys. (N.Y.)* **83**, 491 (1974).
2. B.D. Serot and J.D. Walecka. *Adv. Nucl. Phys.* **16**, 1 (1986).
3. T. Otsuka, T. Suzuki, R. Fujimoto, H. Grawe and Y. Akaishi, *Phys. Rev. Lett.* **95**, 232502 (2005).
4. G. Colò, H. Sagawa, S. Fracasso and P.F. Bortignon, *Phys. Lett. B* **646**, 227 (2007).
5. D.M. Brink and Fl. Stancu, *Phys. Rev. C* **75**, 064311 (2007).
6. W.H. Long, H. Sagawa, N. Van Giai and J. Meng, *Phys. Rev. C* **76**, 034314 (2007).
7. J.P. Schiffer *et al.*, *Phys. Rev. Lett.* **92**, 162501 (2004).
8. A. Bouyssy, J.-F. Mathiot, N. Van Giai, and S. Marcos *Phys. Rev. C* **36**, 380 (1987).
9. W.H. Long, N. Van Giai and J. Meng, *Phys. Lett. B* **640**, 150 (2006).
10. W.H. Long, N. Van Giai and J. Meng, arXiv:nucl-th/0608009; W.H. Long, Ph.D thesis, Université Paris-Sud (2005), unpublished.
11. T. Nikšić, D. Vretenar, P. Finelli and P. Ring, *Phys. Rev. C* **66**, 024306 (2002).
12. W.H. Long, J. Meng, N. Van Giai, and S.G. Zhou, *Phys. Rev. C* **69**, 034319 (2004).
13. A.-M. Oros, Ph. D. thesis, University of Köln, unpublished (1996).
14. L.S. Geng, J. Meng, H. Toki, W.H. Long, and G. Shen, *Chin. Phys. Lett.* **23**,1139 (2006).
15. H.Z. Liang, N. Van Giai, and J. Meng, *in preparation.*
16. M.L'Huillier and N. Van Giai, *Phys.Rev. C* **39**, 2022 (1989).
17. P. Ring and D. Pena, these Proceedings.
18. N. Paar, T. Nikšić, D. Vretenar, and P. Ring, *Phys. Rev. C* **69**, 054303 (2004).
19. K. Yako, H. Sagawa and H. Sakai, *Phys. Rev. C* **74**, 051303 (R) (2006).

STUDY OF THE STRUCTURE OF NEUTRON RICH NUCLEI BY USING β-DELAYED NEUTRON AND GAMMA EMISSION METHOD

YANLIN YE*, ZHIHUAN LI, JIANLING LOU , ZAN KONG, HUI HUA, TAO ZHENG, YUCHENG GE, DONGXING JIANG

School of Physics and State Key Laboratory of Nuclear Physics and Technology, Peking University, Beijing 100871 China
**E-mail: yeyl@pku.edu.cn*

β-delayed neutron emission is often the dominant decay mode for neutron-rich nuclei near the drip-line. The large decay energy (Q value) allows to populate the highly excited states of the daughter nucleus, which are of special importance to study the nuclear structure of the unstable nuclei. So far many experiments of this kind have been carried out in several laboratories.

We present here briefly the experimental method and basic detection setups for the experiment of this kind. The major detector systems in the world are outlined and their performances are compared to each other. The latest experimental results are presented, including the study of the β-delayed neutron and gamma emission of the isotopes $^{18,21}N$ and the excited states of the daughter nuclei $^{18,21}O$, respectively.

Keywords: β decay; neutron and γ emission; highly excited states.

1. Introduction

β-delayed neutron emission is often the dominant decay mode for light neutron-rich nuclei. The large decay energy (Q value) allows to populate the highly excited states of the daughter nucleus, which are of special importance to study the nuclear structure of the unstable nuclei. So far many experiments of this kind have been carried out for nucleus such as $^{49,51,52,53}K$, $^{34,35}Al$, ^{32}Mg used TONNERRE (neutron ball) at GANIL [1–4], for nucleus ^{23}O, $^{17,18,19,20}N$, $^{17,18}C$, ^{15}B used neutron ball at MSU,[5–9] for nucleus ^{14}Be, ^{17}B, ^{19}C used neutron wall at RIKEN,[10–12] and for nucleus $^{18,21}N$ used a combined setup composed of neutron walls and neutron ball by Peking University (PU) group.[13–15]

The experimental method, the major detection setups and their performance will be described in section 2. As an example, the β-delayed neutron and γ emission for the nucleus ^{18}N, the the excited states of the daughter nuclei ^{18}O , and the comparison between the OXBASH calculation and all the existed data for ^{18}N decay will be presented in section 3. The latest preliminary results for ^{21}N will also be shown in section 3. The perspectives of β-delayed neutron and γ emission experiments will be given in the last section.

2. Experimental Facilities

The main characteristics of experimental facilities used in the β-delayed neutron emission experiments are summarized in Table 1. The intrinsic efficiency is related to the thickness of the scintillator used and the signal threshold of each detector. Compared to other arrays listed in Table 1, the PU neutron wall has a moderate solid angle coverage but the lowest neutron detection threshold, namely only 100 keV, which is important to the measurement of the emitted neutrons near the threshold.

Table 1. Comparison of the main experimental facilities in the world

Array	Scintillator	Size[a] cm^3	Pieces	R[b] cm	$\Delta\Omega$[c] %	Threshold keV(n)	ξ_n[d] %	Ref
TONNERRE	BC400	160*20*4	32	120	45	200	44	[1-4]
MSU	BC412	157*7.6*2.54	16	100	15	800	20	[5-9]
RIKEN	BC408	100*6*6	15	125	11	200	55	[10-12]
PU Ball	BC408	157*40(20)*2.5	8	100	30	300	25	new
PU Walls	BC408	40*4.5*2.5	20	62	7.6	100	35	new

Note: [a] *size*: length*width*thickness, for PU ball, the width is 40 cm at the middle of each detector and reduced to 20 cm at both ends.

Note: [b] *R*: the flight path (in the case of the RIKEN wall and PU wall, the flight path is measured with respect to the center of the detector).

Note: [c] $\Delta\Omega$: solid angle as a fraction of 4π

Note: [d] ξ_n: the intrinsic detection efficiency for 2 MeV neutrons.

3. Study of $^{18,21}N$

3.1. *the beta decay of* ^{18}N

For ^{18}N, the β-decay was initially studied by Chase et al. in 1964.[16] Since the β-decay energy of ^{18}N is 13.899 MeV, which lie above the α threshold at S_α=6.227 MeV and the neutron threshold at S_n=8.044 MeV, delayed emissions of neutrons or alpha particles can occur.

The attempt to measure the energy spectrum of the β-delayed neutrons was first made by Scheller *et al*,[7] and nine neutron peaks were observed at energies from 0.99 to 3.26 MeV with a total branching ratio of 2.2 $\pm$ 0.4%. However this only accounts for a small portion of the total neutron-emission branching ratio (Pn) of 14.3 $\pm$ 2.0% determined by Reeder *et al.*[17] The missing probability is attributed to the relatively high neutron detection threshold and the poor statistics in that experiment. We have recently reported some new results for the beta-delayed neutron emission of ^{18}N.[13] In the experiment special effort was made to reduce the energy threshold for neutron detection. In addition to the previously used neutron sphere,[15] which has an advantage of covering a large solid angle but a disadvantage of having a high neutron energy threshold, a neutron wall composed of short scintillation bars was used for detecting low energies neutrons. Two low-energy neutron groups at 0.58 ± 0.02 MeV and 0.79 ± 0.04 MeV and with branching ratios of $5.14 \pm 1.12\%$ and $0.28 \pm 0.06\%$, respectively, were observed by the neutron wall.[13] Another new transition is also measured by neutron ball in the same experiment.[14] The observed neutron energy is $3.78 \pm\ 0.05$ MeV with a branching ratio of $0.05 \pm 0.03\%$, which corresponds to an excitation energy of 12.05 ± 0.05 MeV in the daughter nucleus ^{18}O. The logft value of this transition is 5.24 ± 0.3, which, together with the previously reported electron scattering data[18], determines the J^{π} value of this level to be 1^-. The large solid angle (30% of 4π sr of the whole neutron sphere) and the relatively low background counting are the main reasons for which we were able to observe the high energy peak at 3.78 MeV by the neutron sphere but not by the neutron wall. Therefore, for neutron detectors, it is also important to have a high signal to background ratio.

The β-delayed α-spectrum of ^{18}N was firstly studied by Z.Zhao et al and the absolute beta decay branching ration of ^{18}N to the I_3^- and I_4^- states in ^{18}O had been given in Ref.[19] The β-delayed α-spectrum of ^{18}N had also been measured by L. Buchmann *et al.*[20] resolving the structure of a broad α-particle distribution at α-energies of 1.5-5.5 MeV into several states that show a distinct interference pattern. R-matrix fits assuming a solely $J^{\pi} = 1^-$ partial wave confirm this finding.

The β-delayed γ-rays of ^{18}N had also been studied in PU's experiment.[13] Many peaks were found and identified to be the known transitions in the daughter nucleus ^{18}O. From the work of Olness *et al.* ,[21] the 1652 keV transition in ^{18}O has a relative γ branching ratio of $60.48 \pm 1.80\%$. In the PU's work, the 1652 keV transition in ^{18}O was found to have an absolute branching ratio of $42.5 \pm 3.9\%$ which results in absolute branching ratio for all γ-ray emission of $70.3 \pm 6.8\%$. This value is consistent with the measurement of $76.7 \pm 7.2(\text{stat}) \pm 5.8(\text{normd})\%$ measured by France III *et al.* [22] It is also in good agreement with the result of $70.9 \pm 2.1\%$ which was deduced from the total branching ratio for non-γ-ray decay.

The shell-model calculation was carried out by Zhang Yu-mei *et al.*[23] using the code OXBASH in the *psd* and *spsdpf* model spaces. The WBT interaction was used, and only the allowed β decay of ^{18}N was considered. For the first time, Fig. 1 shows experimental $B(\text{GT})$ values of all the allowed β-decay from ^{18}N to the neutron unbound states ,[13] α unbound states [19] and γ bound states [22] of ^{18}O together with the theoretical calculations in *psd* and *spsdpf* model spaces, respectively. It can be seen that the $B(\text{GT})$ distribution obtained from the present experiment and previous experiment is qualitatively in agreement with that calculated in psd model space. For large $B(\text{GT})$ the calculated energies seem systematically exceed those of the measured ones by a few hundred keV. The lack of quantitative agreement between the theoretical predictions and the measurements suggests further development of the appropriate structure model.

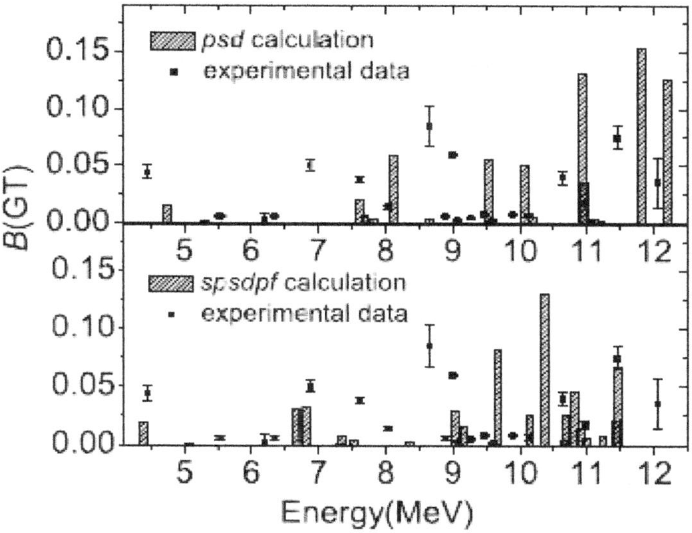

Fig. 1. Comparison of shell model calculation of $B(GT)$ values for all the allowed beta-decay of ^{18}N to neutron unbound states, α unbound states, and γ bound states of ^{18}O.

3.2. *The beta decay of ^{21}N*

For neutron-rich nucleus ^{21}N, the β-decay energy is 17.189 MeV, which lies above the single neutron separation energy at S_n=3.806 MeV and the two neutron separation energy at S_{2n}=11.41 MeV, and delayed emissions of single neutron or two neutrons can occur. There exists little experimental information about the beta decay of ^{21}N, excepting the half-time with larger error and the total β-delayed neutron emission branching ratio,[17] and the level schemes lower than the single neutron separation energy obtained by nuclear reaction method.[24]

We have upgraded our neutron detection facilities in order to reduce the neutron detection threshold further and improve the detection efficiency after the experiment for ^{18}N. The beta decay of ^{21}N using the upgraded neutron facilities was carried-out in Lanzhou in January of 2007. Both ^{16}C beam and ^{17}N beams were used to calibrate neutron detection efficiency in this experiment. The experimental setup was shown in Fig.2, in which Fig.2(a) gives the whole detector used in this experiment and Fig.2(b) presents its inside part. Some differences can be seen from this setup in comparison to that for the previous experiment for ^{18}N .[13] Four clover γ detectors, are used in this experiment. The distance between the center of each clover detector to the that of the beta detector is about 12 cm. Radioactive sources such as ^{60}Co,^{207}Bi,^{152}Eu as well as the radioactive beams of ^{17}N and ^{19}O were used to calibrate the efficiency of the γ detectors. The total efficiency is about 1% for each clover detector at 1.33 MeV.

The primary beam composed of ^{26}Mg ions at 68.8 MeV/nucleon was provided by the HIRFL in Lanzhou and impinged on a ^{9}Be primary target. The produced fragments were separated, purified, and collected using Radioactive Ion Beam Line in Lanzhou (RIBLL).[25] First, the kind of ions contained in secondary beam was identified by a dE detector together with the time-of-flight (tof) from terminal T1 to T2 at RIBLL. Once a ^{21}N was identified, the beam was stopped by a chopper for 300 ms, and the β-delayed neutron and γ rays resulted from ^{21}N decay were detected within this 300 ms. The beam intensity of ^{21}N is only about 0.5 pps. Fig.3(a) and (b) present the primary result of neutron time-of-flight spectrum obtained by neutron wall and neutron ball, respectively. Neutron ball, with large solid angle coverage, provides better statistics than neutron walls, but have higher neutron detection threshold, so the neutron peak at about 0.44 MeV can not be clearly seen by the neutron ball, but by the neutron wall. It seems that a group of neutrons at about

(a) (b)

Fig. 2. the experimental setup in the experiment with ^{21}N, (a) the whole detector system; (b) the central implantation detector and the surrounded clover detectors

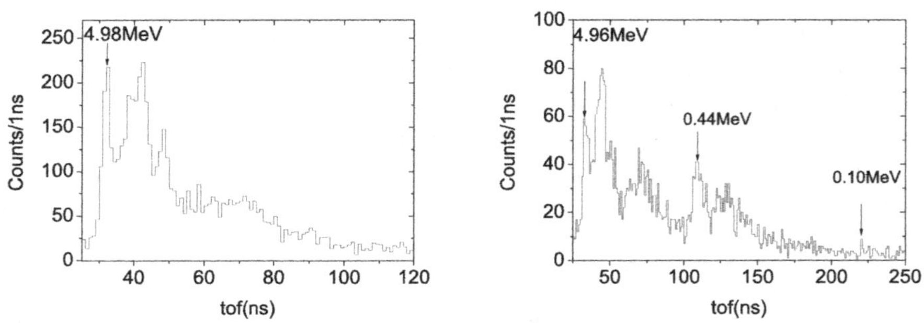

(a) (b)

Fig. 3. Neutron Time-of-flight spectra resulted from decay of ^{21}N measured by the neutron wall (a) and neutron ball (b).

0.10MeV could be identified by the neutron wall. Neutron ball has a better energy resolution for neutrons with energies higher than 2 MeV (tof less than 50ns) than that for neutron wall, due to the shorter flight distance (only about 62 cm) for the latter. At least four groups of neutrons at energies higher than 2 MeV can be identified from the tof spectrum of the neutron ball. Fig.4 presents the preliminary result of β-delayed γ spectrum obtained by all four clovers, with (a) showing the γ rays with lower energies, whereas (b)the high energy part. The 1218 keV transition belongs to the β-delayed γ emission of ^{21}N, and 351 keV and 280 keV transitions are related to the β decay of ^{21}O and ^{21}F, respectively, as identified by the gating analysis. The γ-rays at energies of 1674 keV, 2397 keV, 2780 keV, 3175 keV and 4072 keV are emitted from ^{20}O, the granddaughter nucleus of the ^{21}N. The γ transitions at energies of 1057 keV and 1634 keV are from the β decay of ^{20}O and its daughter nucleus ^{20}F, respectively. More analysis are carrying on and more accurate results will be reported later on.

4. Summary and Perspectives

β-delayed particle emission has large probability when approaching the drip-line, and can provide rich spectroscopic information for the excited states. The particle decay measurement, together with the γ spectroscopy, provides a promising tool to study these highly excited states, and is applicable to very low beam intensity, such as ^{21}N. Systematic measurements are needed and theoretical description should be improved.

Nearly all the light neutron-rich Li, Be, B, C, N, O isotopes have been largely studied using β-delayed neutron and γ emission method, but only very few work done for Na, Mg, Al, K and heavier isotopes. Therefore, this method can still be used for large number of nuclei closed to the drip line and provide us with rich spectroscopic information.

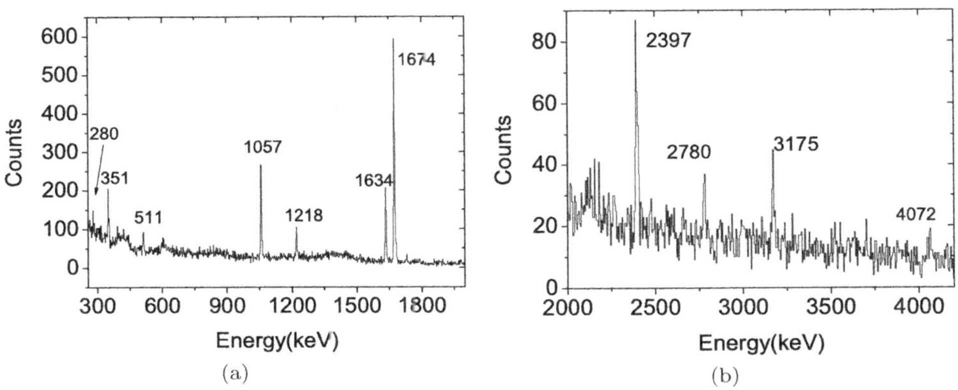

Fig. 4. β-delayed γ spectrum from ^{21}N: (a) low energy part and (b) high energy part.

Acknowledgements

We would like to acknowledge the staffs of HIRFL (Heavy Ion Research Facility in Lanzhou) for providing the ^{22}Ne and ^{26}Mg primary beam and the staffs of RIBLL for all kinds of assistance. This work is supported by the National Basic Research Program (973 program) of China (2007CB815000,2005CB724800), the National Natural Science Foundation of China (10775003,10475004, 10405001, 10221003).

References

1. A. Buta, T. Martin, C. Timis et al, *Nucl. Instrum. Methods A* **455**, 412 (2000).
2. F. Perrot, F. Marechal, C. Jollet et al, *Phys. Rev. C* **74**, 014313 (2006).
3. S. Nummela, P. Baumann, E. Caurier et al, *Phys. Rev. C* **63**, 044316 (2001).
4. S. Grevy, S. Pietri, L. Achouri, J.C.Ange lique et al, *Nucl.Phys.A* **734**, 369 (2004).
5. C. S. Sumithrarachchi, D. J. Morrissey, B. A. Brown et al, *Phys. Rev. C* **75**, 024305 (2007).
6. C. S. Sumithrarachchi, D. W. Anthony, P. A. Lofy et al, *Phys. Rev. C* **74**, 024322 (2006).
7. K.W.Scheller, J.Gorres, J.G.Ross et al, *Phys. Rev. C* **49**, 1 (1994).
8. K.W.Scheller, J.Gorres, S.Vouzoukas et al, *Nucl.Phys. A* **582**, 109 (1995).
9. R.Harkewicz, D.J.Morrissey et al, *Phys. Rev. C* **44**, 6 (1991).
10. A.Ozawa, G.Raimann, R.N.Boyd et al, *Nucl.Phys. A* **592**, 244 (1995).
11. G. Raimann, A. Ozawa, R. N. Boyd et al, *Phys. Rev. C* **53**, 1 (1996).
12. N. Aoi, K. Yoneda et al, *Phys. Rev. C* **66**, 014301 (2002).
13. Z.H. Li, Y.L. Ye, H. Hua, D.X. Jiang et al, *Phys. Rev. C* **72**, 064327 (2005).
14. J.L. Lou, Z.H. Li, Y.L. Ye et al, *Phys. Rev. C* **75**, 057302 (2007).
15. Q. Y. Hu, Y. L. Ye, Z. H. Li et al, *IEEE on Nucl Scie* **52**, 1 (2005).
16. L. F. Chase et al,*Phys. Rev. Lett.* **13**, 665 (1964).
17. P. L. Reeder, R.A.Warnar, W.K.Hensley, *Phys. Rev. C* **44**, 1435(1991).
18. R. M. Sellers et al, *Phys. Rev. C* **51**, 1926 (1995).
19. Z. Zhao et al, *Phys. Rev. C* **39**, 1985 (1989).
20. L. Buchmann,1,* J. DAuria,2 M. Dombsky et al, *Phys. Rev. C* **75**, 012804(R) (2007).
21. J. W. Olness, E. K. Warburton, D. E. Alburger et al, *Nucl. Phys. A* **373**, 13 (1982).
22. R. H. France III, Z. Zhao, and M. Gai, *Phys. Rev. C* **68**, 057302 (2003).
23. Zhang Yu-Mei et al, *Chin. Phys. Lett.* **23**, 2046 (2006).
24. M. Stanoiu, F. Azaiez, Zs. Dombrdi et al, *Phys. Rev. C* **69**, 034312 (2004).
25. Z. Sun et al, *Nucl. Instrum. Methods A* **503**, 496 (2003).

PRODUCTION OF SECONDARY RADIOACTIVE ^{21}Na BEAM FOR THE STUDY OF ^{21}Na(α,p)^{24}Mg STELLAR REACTION*

DAM NGUYEN BINH, LE HONG KHIEM

Institute of Physics and Electronics, Vietnam Academy of Science and Technology
Hoang Quoc Viet Rd. Nghia Do, Hanoi, Vietnam
dnbinh@grad.iop.vast.ac.vn

S. KUBONO, H. YAMAGUCHI, Y. WAKABAYASHI, S. HAYAKAWA

Center for Nuclear Study, University of Tokyo,
2-1 Hirosawa, Wako, Saitama, 351-0198, Japan

A. KIM

Ewha Womans University,
11-1 Daehyun-Dong, Seodamun-gu, Seoul,120-750,Korea

The availability of radioactive beams has produced great opportunities for advances in our understanding of the nucleosynthesis occurring in stellar explosions such as novae, X-ray burst and supernovae. By using an in-flight low-energy radioisotope beam separator (CRIB) at Center for Nuclear Study (CNS), University of Tokyo, we have successfully produced the ^{21}Na proton-rich beam for the study of ^{21}Na(α,p)^{24}Mg reaction which is related to the astrophysically important production of ^{22}Na in the stellar explosive environments. Since it is the first time when this reaction is studied experimentally, we have performed a test experiment to produce the ^{21}Na beam and to estimate the feasibility of the experimental study of ^{21}Na(α,p)^{24}Mg reaction.

1. Introduction

The synthesis of ^{22}Na in the explosive hydrogen burning has been extensively investigated in the past two decades. Proton-capture reactions on the "seed" nuclei ^{20}Ne are considered to be responsible for the synthesis of significant amounts of the unstable nucleus ^{21}Na.[1] This is followed either by its β^+ decay into ^{21}Ne ("cold" NeNa cycle), which leads to ^{22}Na by means of ^{21}Ne$(p,\gamma)^{22}$Na, or, at high enough temperatures, by additional proton captures on ^{21}Na ("hot" NeNa cycle) leading to ^{22}Mg that subsequently decays into ^{22}Na. ^{22}Na could be observable by the characteristic 1.275 MeV gamma ray with the half-life of 2.6 yr. Considerable efforts were undertaken in the search for a galactic 1.275 MEV γ line by satellite in the INTEGRAL Project, but found no flux of 1.275 MeV gamma rays above the background

*This research has been partially supported by CNS, University of Tokyo.

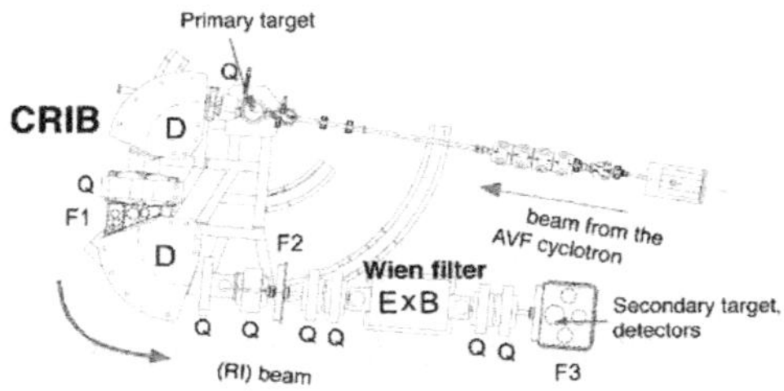

Fig. 1. The schematic plane view of CRIB.

from a survey of several ONe type novae.[2] People believe that the unknown nuclear reaction rates involved in the synthesis of this radioisotope, such as ^{21}Na(p,γ) and ^{22}Mg(p,γ), are the main source of uncertainty associated with calculating the amount of ^{22}Na in the explosive stellar events, although the reaction of ^{22}Na(p,γ) has been studied precisely at TRIUMF. The ^{22}Na abundance might be reduce by the ^{21}Na(α,p)^{24}Mg reaction. Therefore, the purpose of our study is to perform a direct measurement of the rates of ^{21}Na(α,p)^{24}Mg reaction. There is no experiment that measured the reaction ^{21}Na(α,p)^{24}Mg directly at the temperatures of interest. This is the breakout reaction from the NeNa cycle to the next cycle. Since this is the first time this reaction is studied experimentally, we need to perform a test experiment before performing the final one. The main purpose of the test experiment is to produce a high-quality and high-intensity ^{21}Na beam using the CRIB facility of CNS, the University of Tokyo. In this report, the production of ^{21}Na beam by a transfer reaction ^{20}Ne(d,p)^{21}Na in inverse kinematics and the feasibility for studying the ^{21}Na(α, p)^{24}Mg are presented. These tests are very important to decide the experimental plan for the final experiment that will be performed in near future.

2. Experiment and results

The ^{21}Na beam was produced by ^{20}Ne(d,p)^{21}Na transfer reaction with the primary beam of accelerated ^{20}Ne^{6+} from the AVF cyclotron of the Institute of Physics and Chemicals Research (RIKEN). Figure 1 shows a schematic plane view of CRIB and a connection with the RIKEN AVF cyclotron. The primary beam ^{20}Ne^{6+} is accelerated by the AVF cyclotron and delivered to a production gas target at F0 chamber of CRIB. The deuterium gas target at pressure of 1 atm and was cooled down to 77 K was confined in a cylinder chamber of 80 mm length with an entrance and exit windows made by 2.2 μm thick Havar foils and installed at F0 chamber.

The proton-rich beam ^{21}Na is produced by (d,p) transfer reaction in inverse kinematics. The composition of beam obtained after the production target includes some kind of ion as shown in a particle identification figure at F2, measured by SSD (Figure 2). Due to the difference of the product ions after the F0 target, the ^{21}Na beam was separated by a slit at the momentum dispersive focal plane F1.

We have made the optimization for ^{21}Na^{10+} and ^{21}Na^{11+}. We found that the

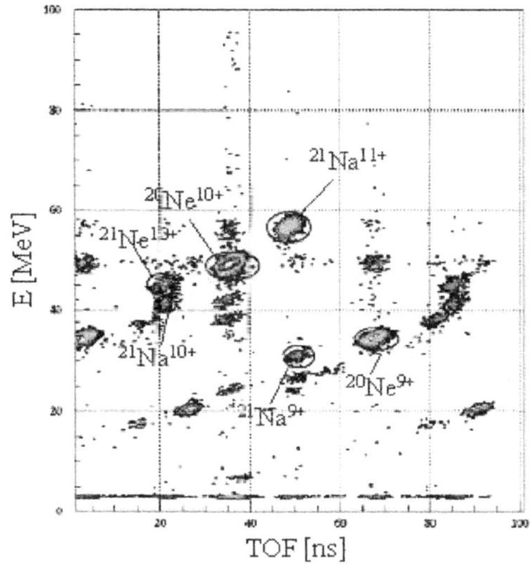

Fig. 2. Particle identification at F2 before Wien filter.

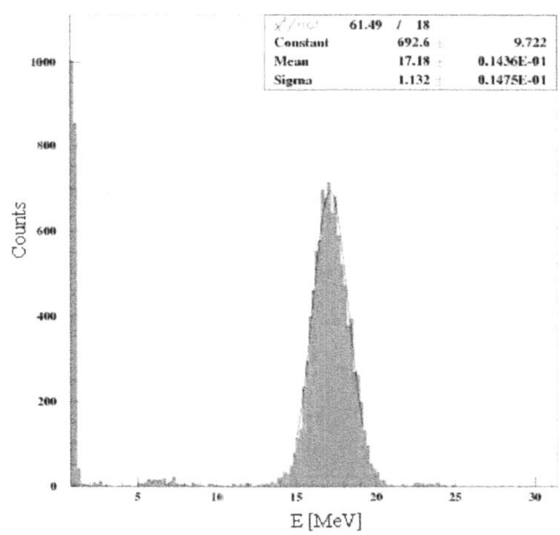

Fig. 3. Beam energy spectrum obtained at F3 chamber.

secondary $^{21}Na^{11+}$ beam is better than the other. Thus, we decided to use this beam for the stellar $^{21}Na(\alpha,p)^{24}Mg$ reaction. After selecting at momentum focal plane F1, the beam was still contaminated by several several nuclear species. The beam can be made with high purity by the Wien filter that is installed in front of F3 chamber where the reaction will be studied. In fact, the purity of ^{21}Na beam obtained at F3 was 98.6%, and the energy was 17.18 MeV with a standard deviation of 1.13 MeV, as shown in figure 3.

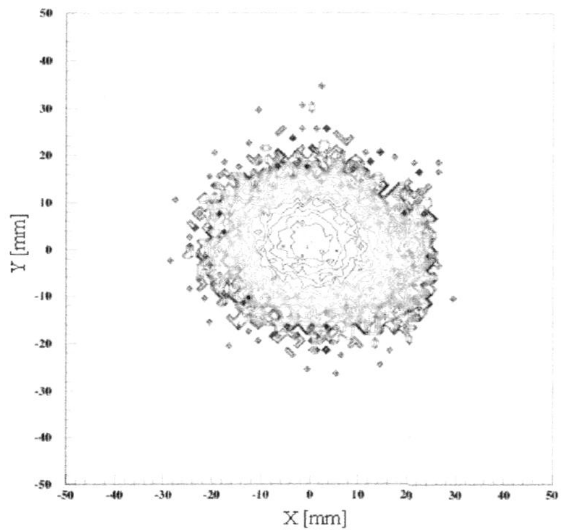

Fig. 4. Profile of the ^{21}Na beam at F3.

Figure 4 shows the profile of the ^{21}Na beam obtained at the Helium gas target position at F3. The intensity of the $^{21}Na^{11+}$ beam that was obtained was 9.5×10^3 pps for the primary beam current of 30 enA. By extrapolation the beam intensity to the beam current, we could get the beam intensity of 10^6 pps at primary beam current of 0.52 pμA.

The test experiment of the ^{21}Na(α,p)^{24}Mg reaction was performed with the ^{21}Na beam produced here, and the proton spectrum from the ^{21}Na(α,p) reaction has been taken. The data analysis is still in progress.

3. Summary

We have succeeded to produce the low-energy beam of ^{21}Na to be used for the study of ^{21}Na(α,p)^{24}Mg stellar reaction which is important for production of ^{22}Na at high temperature condition like in X-ray burst. This reaction is a breakout process from NeNa cycle to MgAl cycle in the nucleosynthesis occurring in the explosive environments, which bypasses the waiting point at ^{22}Mg. The final experiment of ^{21}Na(α,p) will be performed in near future.

References

1. C.J. Lister et al., *Phys. Rev. Lett.* **92**, 3 (2004).
2. A.F. Iyudin et al., *Astron. Astrophys.* **300**, 422.
3. J.M. D'Auria et al., *Phys. Rev. C* **69**, 065803 (2004).
4. C. Ruiz et al., *Phys. Rev. C* **71**, 025802 (2005).
5. M. Wiescher and J. Gorres et al., *Nucl. Phys. A* **484**, 90 (1988).
6. F. Stegmuller et al., *Nucl. Phys. A* **601**, 168 (1996).

ASYMMETRIC NUCLEAR MATTER PROPERTIES WITHIN THE BRUECKNER THEORY

W. ZUO*

Institute of Modern Physics, Chinese Academy of Sciences, Lanzhou 730000, China

U. LOMBARDO and H.-J. SCHULZE

Department of Physics and Astrophysics, Catania University, Via Santa Sofia 64, I-95123, Italy

We investigate the equation of state (EOS) and single particle properties of isospin asymmetric nuclear matter by using the Brueckner theory. We have concentrated on the effect of microscopic three-body forces and studied especially the isospin dependence of the EOS and the properties of asymmetric nuclear matter.

Keywords: Properties of asymmetric nuclear matter; Symmetry energy; Brueckner theory, Three-body force.

1. Introduction

The equation of state (EOS) and single particle (s.p.) properties of isospin asymmetric nuclear matter (ANM), especially their isovector components, play an important role in nuclear physics, heavy ion physics and astrophysics. In heavy ion collisions (HIC) induced by radioactive beams, a transient state of dense and highly isospin-asymmetric nuclear matter can be formed and the isospin dynamics and observables have been shown to be quite sensitive to the density dependence of symmetry energy.[1] In the study of nuclear structure, the properties of neutron-rich nuclei far from the nuclear stable line, such as the radius, the neutron-skin thickness and the density distribution turn out to depend sensitively on the isospin behavior of the nuclear EOS at densities around and below the normal nuclear matter density.[2] Besides its general interest in nuclear physics and heavy ion physics, the properties of ANM at high densities is expected to be extremely significant for understanding many phenomena in astrophysics. In fact, the EOS of ANM is a fundamental input of the stellar structure model (TOV equation) which determines the properties, such as the maximum mass and the density profile, of neutron stars.

Microscopically the properties of ANM have been extensively investigated by adopting various many-body approaches. In the present paper, we shall report our research results on the properties of ANM within the Brueckner-Hartree-Fock

*E-mail: zuowei@impcas.ac.cn

(BHF) framework.

2. Theoretical approaches

Our investigation is based on the Brueckner-Bethe-Goldstone (BBG) theory for ANM.[3,4] The extension of the BBG scheme to include three-body forces can be find in Ref.[5,6] The starting point of the BBG scheme is the Brueckner reaction G matrix, which satisfies the Bethe-Goldstone (BG) equation. In solving the BG equation for the G-matrix, we adopt the continuous choice for the s.p. potential $U^{\rm bhf}(k)$ since it provides a much faster convergency of the hole-line expansion than the gap choice.[7] Under the continuous choice, the s.p. potential describes physically at the BHF level the nuclear mean field felt by a nucleon in nuclear medium.

The realistic nucleon-nucleon (NN) interaction v_{NN} in the present calculation contains two parts, i.e., the Argonne V_{18} (AV_{18}) two-body interaction[8] plus the contribution of the microscopic TBF based on the meson-exchange current approach.[5] We have shown in Ref.[9] that the microscopic TBF provides a repulsive contribution to the EOS of nuclear matter and improves remarkably the predicted saturation properties. We find that the main relativistic correction to the nuclear EOS in the DBHF approach can be reproduced quantitatively by the $2\sigma - N\bar{N}$ component (i.e., the Z-diagram contribution) of the microscopic TBF. However, the TBF components from the other elementary processes becomes important at high densities and can not be completely neglected even at the saturation density.

In our BHF calculation, the TBF contribution has been included by reducing the TBF to an equivalently effective two-body interaction via a suitable average with respect to the third-nucleon degrees of freedom according to the standard scheme as described in Ref.[5] In r-space, the equivalent two-body force $V_3^{\rm eff}$ reads:

$$\begin{aligned}\langle \vec{r}_1{}'\vec{r}_2{}'|V_3^{\rm eff}|\vec{r}_1\vec{r}_2\rangle = \frac{1}{4}{\rm Tr}\sum_n \int {\rm d}\vec{r}_3{\rm d}\vec{r}_3{}'\phi_n^*(\vec{r}_3{}')(1-\eta(r_{13}'))(1-\eta(r_{23}')) \qquad (1)\\ \times\, W_3(\vec{r}_1{}'\vec{r}_2{}'\vec{r}_3{}'|\vec{r}_1\vec{r}_2\vec{r}_3)\phi_n(\vec{r}_3)(1-\eta(r_{13}))(1-\eta(r_{23})).\end{aligned}$$

A detailed description and justification of the above scheme can be found in Ref.[5] It is worth stressing that the effective force $V_3^{\rm eff}$ depends strongly on density. It is the density dependence of the $V_3^{\rm eff}$ that induces the TBF rearrangement contribution to the s.p. potential in nuclear medium within the BHF framework. The TBF rearrangement contribution to the s.p. potential can be calculated as follows:[10]

$$U^{\rm tbf}(k) \approx \frac{1}{2}\sum_{k_1k_2} n_{k_1}n_{k_2}\left\langle k_1k_2\left|\frac{\delta V_3^{\rm eff}}{\delta n_k}\right|k_1k_2\right\rangle_A \qquad (2)$$

3. Results and Discussions

3.1. *Isospin Dependence of the EOS of ANM & Symmetry Energy*

In order to see clearly the isospin dependence of the EOS of ANM, we report in Fig.1 the isovector part of the EOS as a function of β^2, where β is the isospin asymmetry of ANM. In the figure, the left and middle panels show the results for cold nuclear matter at several typical values of density in the two cases of including and not including the TBF contribution, respectively. In the right panel is given the results for ANM at finite-temperature ($T = 20$MeV). We find that the EOS of ANM fulfills satisfactorily a quadratic dependence on asymmetry parameter β in the whole asymmetry range $0 \leq \beta \leq 1$ in both cases with and without including the TBF contribution. Accordingly, the EOS of ANM may be expressed as: $E_A(\rho, T, \beta) = E_A(\rho, T, 0) + E_{sym}(\rho, T)\beta^2$. From the right panel we can see that the linear dependence of the EOS of ANM on β^2 can be extended to the finite-temperature case. The above result provides an microscopic support for the empirical β^2-law extracted from the nuclear mass table and extended its validity up to the highest isospin asymmetry. The β^2-law of the EOS of ANM leads to two important consequences. First, it indicates that the EOS of ANM at any isospin asymmetry is determined completely by the EOS of symmetric nuclear matter (SNM) and the symmetry energy E_{sym}. Second, the above β^2-law implies that the difference of the neutron and proton chemical potentials in β-stable neutron star matter is determined by the symmetry energy in an explicit way: $\mu_n - \mu_p = 4\beta E_{sym}$ and thus the symmetry energy plays a crucial role in predicting the composition of neutron stars.

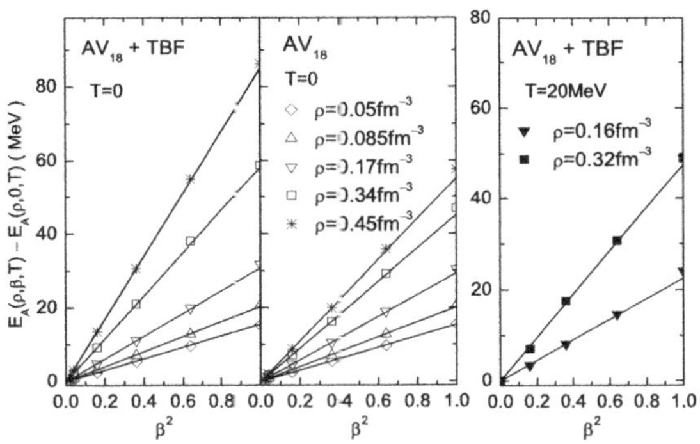

Fig. 1. Difference between the energy per nucleon of ANM and that of symmetric nuclear matter.

In Fig.2 is reported the predicted symmetry energy as a function of density. In the left panel is shown the TBF effect on symmetry energy. And the right panel gives the thermal effect on symmetry energy. It is seen that the TBF effect is reasonably small at low densities around and below the saturation density. The TBF provides a repulsive contribution to symmetry energy at high densities. The TBF repulsion increases rapidly as a function of density. As an consequence, it makes the density dependence of symmetry energy at low densities quite different from that at high

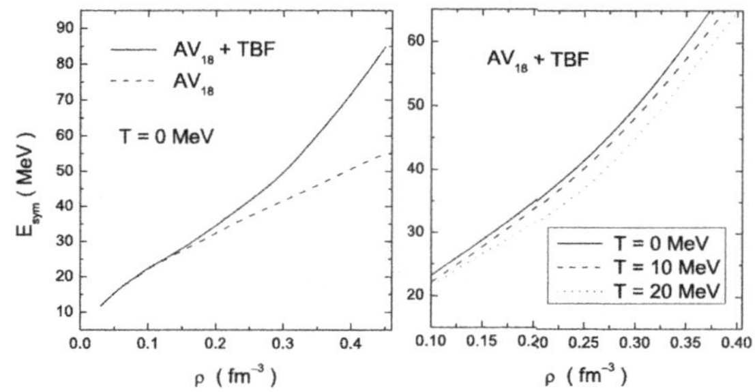

Fig. 2. TBF effect (left panel) and thermal effect (right panel) on symmetry energy.

densities and changes the stiffness of the density dependence of symmetry energy from asy-soft in low density region to asy-stiff in high density region. From the right panel we see that the symmetry energy decreases as the temperature increases.

3.2. *Isospin Dependence of Single Particle Potentials*

The isovector part of s.p. potential in neutron-rich matter is described by the isospin asymmetry potential defined as: $U_{sym} = (U_n - U_p)/2\beta$, which turns out to be crucial for predicting the isospin observables in HIC induced by radioactive beams. In the

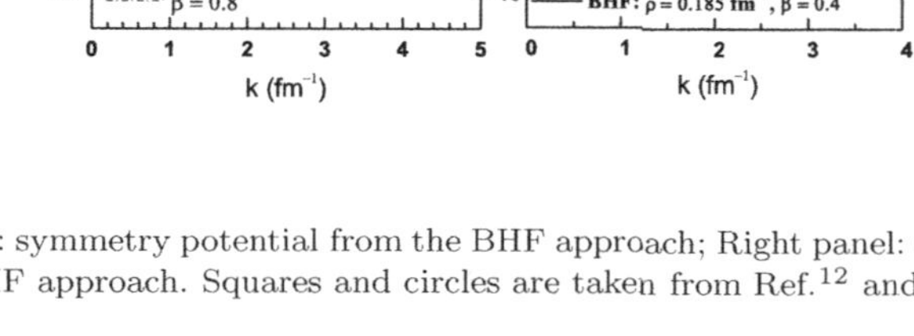

Fig. 3. Left panel: symmetry potential from the BHF approach; Right panel: comparison with the results of the DBHF approach. Squares and circles are taken from Ref.[12] and Ref.[13] respectively.

left panel of Fig.3, we report our calculated symmetry potential vs. momentum for three typical densities and several isospin asymmetries. It is seen that the symmetry potential depends strongly on both density and momentum. Above the Fermi surface, $U_{\rm sym}$ decreases rapidly as a function of momentum and saturates at high enough momenta. In the momentum region relevant to the intermediate HIC up to a beam energy about 300 MeV/A, U_{sym} is positive, implying that its effect is repulsive on neutrons and attractive on protons. We also notice that the dependence of symmetry potential on β is rather weak, which indicates a linear dependence of the neutron and proton s.p. potentials on β and provides an microscopic support for the Lane potential.[11] In the right panel, we compare our symmetry potential with the predictions from the DBHF approach.[12,13] The symmetry potential obtained by the BHF approach display an overall agreement with those by the DBHF approach. We have also compared in Ref.[14] our microscopic symmetry potential with the phenomenological parametrizations of momentum-dependent symmetry potential[15] adopted in the dynamical simulations of HIC and found our microscopic U_{sym} shows a remarkably different behavior as a function of density and momentum from the phenomenological ones, which suggests the that it is necessary to apply the microscopic symmetry potential in the calculations of HIC.

3.3. *Neutron-proton effective mass splitting in neutron-rich matter*

The neutron-proton (n-p) effective mass splitting measures the difference between the isospin-behaviors of the neutron and proton effective masses in neutron-rich matter. In the microscopic Brueckner framework, the predicted neutron effective mass is larger than the proton one in neutron-rich matter,[10,14] i.e., $m_n^* \geq m_p^*$. In the relativistic DBHF approach, it has been shown recently in Refs.[16] that the splitting of the nonrelativistic-type effective mass is consistent with the nonrelativistic BHF predictions. In order to get insight into the microscopic mechanism of the isospin

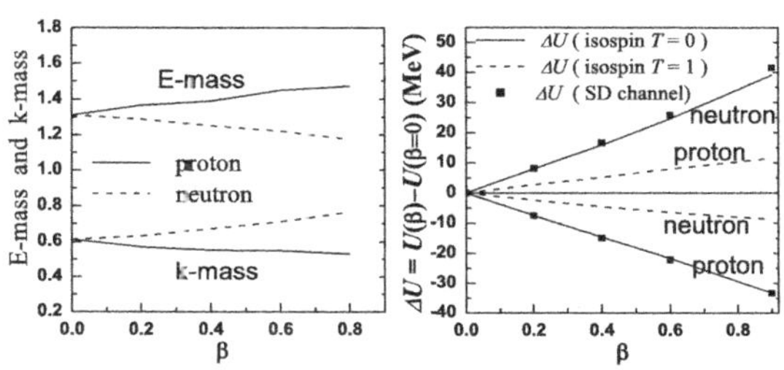

Fig. 4. Left: proton and neutron k-masses and E-masses; Right: isospin $T = 0$, $T = 1$ and SD channel contributions to the proton and neutron s.p. potentials. The density is $\rho = 0.17$ fm^{-3}.

splitting, we show in the left panel of Fig.4 the proton and neutron k-masses m_k and E-masses m_e at Fermi surfaces vs. β. It is seen that the isospin splitting is opposite for m_k and m_e. In neutron-rich matter the neutron k-mass becomes larger than the proton one, while the neutron-proton E-mass splitting is reversed. Although the absolute strengths of the splitting for the two masses are about the same, the relative splitting of m_k is much more pronounced than that of m_e due to the smaller value of m_k. As a consequence, the k-mass predominates the isospin splitting of the total effective masses m^* and leads to the result of $m_n^* \geq m_p^*$ in neutron-rich matter. In the right panel are displayed the contributions separately from the isospin $T = 0$ and $T = 1$ channels to the proton and neutron s.p. potentials at $k = 0$, as a function of β with respect to their values in SNM. It is seen that the variation vs. β of the $T = 0$ component is much larger than that of the $T = 1$ component, implying that the isospin dependence of the s.p. potentials is determined to a large extent by the $T = 0$ component. It is also seen that the net contribution of the $T = 0$ channels stems almost completely from the SD tensor channel (squares). The above result implies that the isospin behavior of the momentum dependence of the s.p. potentials which determines the n-p effective mass splitting, is essentially controlled by the nature of the NN interaction as discussed in Ref.[14]

3.4. *TBF rearrangement effect on Single Particle Properties*

It has been pointed out in Ref.[17] that the BHF s.p. potential is too attractive and its momentum dependence is too weak at large densities and high momenta for describing the high-energy experimental elliptic flow data. In order to discuss the TBF-induced rearrangement effect, we compare in Fig. 5 the s.p. potentials predicted in three different cases. It is seen that the s.p. potential without the

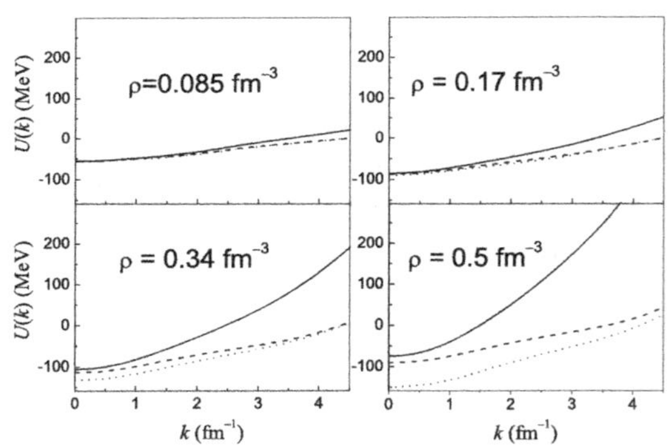

Fig. 5. The predicted s.p. potentials for SNM at several densities in three cases: without the TBF contribution (dotted curves); including the TBF effect only via the G-matrix (dashed curves); including the TBF effect via both the G-matrix and the rearrangement contribution (solid curves).

TBF is most attractive. At low densities, the TBF effects are reasonably small. The TBF effects become stronger rapidly as the density increases. The TBF effect via the G-matrix alone provides an moderate repulsion at high densities which is more pronounced at lower momenta and weakens the momentum dependence of the s.p. potential. The TBF-induced rearrangement effect in nuclear medium provides an additional repulsive and strongly momentum-dependent contribution at high densities and high momenta. The repulsion due to the TBF rearrangement is shown to be much stronger at high momenta as compared to that via the G-matrix. The rearrangement repulsion induced by the TBF reduces largely the attraction and enhances strongly the momentum-dependence of the s.p. potential at high densities and high momenta. Therefore, it improves substantially the agreement between our microscopic s.p. potential and the parametrized one of Ref.[17]

4. Summary

In summary, we report our recent investigation on the properties of ANM within the Brueckner theory. We have shown that the empirical parabolic law for the EOS of ANM can be extended to the highest asymmetry and to the finite-temperature case in both cases of including and not including the TBF. The TBF leads to a strong enhancement of symmetry energy at high densities and makes the density dependence of symmetry energy at high densities much stiffer. The symmetry energy is found to decrease as the temperature increases. Our calculated symmetry potential turns out to be in an overall agreement with those obtained by the DBHF approach. The predicted n-p effective mass splitting is $m_n^* \geq m_p^*$ in neutron-rich matter. In the BHF framework, the n-p effective mass splitting is shown to be controlled essentially by the nature of the NN interactions. At high densities and momenta, the TBF induces a strongly repulsive and momentum-dependent rearrangement modification of s.p. potential, which improves substantially the agreement of our microscopic s.p. potential with the parametrized potential for describing the elliptic flow data.[17]

Acknowledgments

The work is supported in part by the National Natural Science Foundation of China (10575119), the Knowledge Innovation Project(KJCX3-SYW-N2) of Chinese Academy of Sciences, the Major State Basic Research Developing Program of China under No. 2007CB815004, CAS/SAFEA International Partnership Program for Creative Research Teams (CXTD-J2005-1), and the Asia-Link project (CN/ASIA-LINK/008(94791)) of the European Commission.

References

1. B.A. Li, *Phys. Rev. Lett.* **88**, 192702 (2002); L.W. Chen, C.M. Ko and B.A. Li, *Phys. Rev. Lett.* **94**, 032701 (2005).
2. K. Oyamatsu, I. Tanihata, Y. Sugahara *et al.*, *Nucl. Phys. A* **634**, 3 (1998).
3. I. Bombaci and U. lombardo, *Phys. Rev. C* **44**, 1892 (1991).
4. W. Zuo, I. Bombaci, and U. Lombardo, *Phys. Rev. C* **60**, 024605 (1999).
5. P. Grangé, A. Lejeune, M. Martzoff, and J.-F.Mathiot, *Phys. Rev. C* **40**, 1040 (1989).
6. W. Zuo, A. Lejeune, U. Lombardo, and J.-F. Mathiot, *Eur. Phys. J. A* **14**, 469 (2002).

7. H.Q. Song, M. Baldo, G. Giansiracusa *et al.*, *Phys. Rev. Lett.* **81**, 1584 (1998).
8. R.B. Wiringa, V.G.J. Stoks, and R. Schiavilla, *Phys. Rev. C* **51**, 38 (1995).
9. W. Zuo, A. Lejeune, U. Lombardo, and J.-F. Mathiot, *Nucl. Phys. A* **706**, 418 (2002).
10. W. Zuo, U. Lombardo, H.-J. Schulze *et al.*, *Phys. Rev. C* **74**, 014317 (2006).
11. A. M. Lane, *Nucl. Phys.* **35**, 676 (1962).
12. E.N.E. van Dalen, C. Fuchs and A. Faessler, *Nucl. Phys. A* **744**, 227 (2004).
13. F. Sammarruca, W. Barredo and P. Krastev, *Phys. Rev. C* **71**, 064306 (2005).
14. W. Zuo, L.G. Cao, B.A. Li *et al.*, *Phys. Rev. C* **72**, 014005 (2005).
15. C.B. Das, S. Das Gupta, C. Gale *et al.*, *Phys. Rev. C* **67**, 034611 (2003).
16. E.N.E. van Dalen, C. Fuchs and A. Faessler, *Phys. Rev. Lett.* **95**, 022302 (2005); Z.Y. Ma, J. Rong, B.Q. Chen *et al.*, *Phys. Lett. B* **604**, 170 (2004).
17. P. Danielewicz, *Nucl. Phys. A* **673**, 375 (2000).

PART B

Poster Presentations

STUDY OF GIANT DIPOLE RESONANCE IN CONTINUUM RELATIVISTIC RANDOM PHASE APPROXIMATION

DING YANG*, ZHONG-YU MA

China Institute of Atom Energy
Beijing 102413, P.R. of China
**E-mail: yd9955@sina.com*

LIGANG CAO

Institute of Modern Physics, Chinese Academy of Science
Lanzhou 730000, P.R. of China

The isovector giant dipole resonance(IVGDR) of finite nuclei is studied in the framework of the continuum relativistic random phase approximation(CRRPA). In this method the contribution of the continuum spectrum to nuclear excitations is treated exactly by the single particle Green's function technique. We study the strength distribution and non energy and energy-weighted moments m_0 and m_1 of IVGDR in stable nucleus ^{120}Sn . The results are compared to those obtained by commonly used relativistic random phase approximation with a discretized spectrum in the continuum.

Keywords: Relativistic Random Phase Approximation(RRPA); Continuum RRPA; Single particle Green's function.

The relativistic mean field (RMF) method with non-linear isocalar meson self-interactions has been very successful in describing the ground state properties not only spherical, deformed nuclei but also nuclei far from the β-stability line.[1,2] The experimental mass of all measured nuclei can well be reproduced with an overall error less than 0.5 percentage. Furthermore the self-consistent relativistic random phase approximation (RRPA) has been built on the RMF ground state to describe various collective excitations.[3–6] In those methods mentioned above the nucleon wave functions in the continuum are discretized by a basis expansion method or box approximation. Therefore the escaping width of the collective states was missing, especially the continuum contribution near the threshold might not be treated properly. Due to the closeness of Fermi surface to the particle continuum in exotic nuclei the coupling between bound states and the continuum becomes important. The excitation near particle threshold is more important in describing the Pygmy Dipole Resonance(PDR) in exotic nuclei, so it is required to consider the contribution of continuum rigorously when one works on the properties of nuclei far from the β-stable line.

In this work we extend our RRPA calculation to treat the continuum exactly by employing Green's function technique.[7] The single particle Green's function is calculated numerically in the coordinate space. The RRPA calculation is fully consistent in the sense that the nucleon wave functions and the RRPA particle-hole residual interaction are calculated on the same footing. We shall apply the continuum RRPA method to study the properties of collective excitations in finite nuclei. The response functions of IVGDR in stable nucleus ^{120}Sn as well as the non energy weighted and energy weighted moments m_0, m_{-1} are investigated.

We first briefly present the theoretical framework of the continuum RRPA(CRRPA) in terms of the Green's function method. The response function of a quantum system to an external field is given by the imaginary part of the polarization operator,

$$R(P,P;\mathbf{k},\mathbf{k}';E) = \frac{1}{\pi} Im\Pi^R(P,P;\mathbf{k},\mathbf{k}';E) \ , \tag{1}$$

where P is an external field operator. The RRPA polarization operator is obtained by solving the Bethe-Salpeter equation,

$$\Pi(P,P;\mathbf{k},\mathbf{k}',E) = \Pi_0(P,P;\mathbf{k},\mathbf{k}',E) - \sum_i g_i^2 \int d^3k_1 d^3k_2 \Pi_0(P,\Gamma^i;\mathbf{k},\mathbf{k}_1,E) \\ D_i(\mathbf{k}_1,\mathbf{k}_2,E)\Pi(\Gamma_i,P;\mathbf{k}_2,\mathbf{k}',E) \ , \tag{2}$$

To be consistent, the residual particle-hole interactions in RRPA calculations must be obtained from the same Lagrangian as in the description of the nuclear ground state. They are generated by exchanging various mesons: the isoscalar scalar meson σ, the isoscalar vector meson ω and the isovector vector meson ρ. Therefore, in Eq.(2) the sum i runs over σ, ω and ρ mesons, and g_i and D_i are the corresponding coupling constants and meson propagators. The vertex functions are $\Gamma^i = 1$ for σ meson and $\Gamma^i = \gamma^\mu, \gamma^\mu\tau_3$ for ω and ρ mesons, respectively. The meson propagators in the non-linear model are non-local in the momentum space, and therefore have to be calculated numerically.[8] The detailed expressions of non-linear meson propagators $D_i(\mathbf{k}_1,\mathbf{k}_2,E)$ can be found in Ref.[8] Π_0 is the unperturbed polarization operator, which has the following retarded form,

$$\Pi_0^{RMF}(P,P;r,r';E) \\ = \sum_{a,b=A,\bar{\alpha}} \left[\frac{\langle \bar{f}_a(r)\,|P|\,h_b(r)\rangle \langle \bar{h}_b(r')\,|P|\,f_a(r')\rangle}{E_b - (E_a + E) - i\eta} \right. \\ \left. + \frac{\langle \bar{f}_a(r')\,|P|\,h_b(r')\rangle \langle \bar{h}_b(r)\,|P|\,f_a(r)\rangle}{E_b - (E_a - E) + i\eta} \right] , \tag{3}$$

where the indices a, A, $\bar{\alpha}$ represent the hole, particle and Dirac sea states, respectively. The $(f_a(r),\ E_a)$ and $(h_b(r),\ E_b)$ are the corresponding wave functions and energies, respectively.

Defining the single particle Green's function,

$$G(r,r';E) = \sum_{b=c,A,\bar{\alpha}} \frac{h_b(r)h_b^+(r')}{E_b - E \mp i\eta} , \tag{4}$$

and rewriting Eq.(3) in term of the single particle Green's function, we can get

$$\begin{aligned}
&\Pi_0^{RMF}(P,P;r,r';E) \\
&= \sum_a \left[\bar{f}_a(r)P \sum_{b=a,A,\bar{\alpha}} \frac{h_b(r)\bar{h}_b(r')}{E_b-(E_a+E)-i\eta} P f_a(r') \right. \\
&\quad \left. + \bar{f}_a(r')P \sum_{b=a,A,\bar{\alpha}} \frac{h_b(r')\bar{h}_b(r)}{E_b-(E_a-E)+i\eta} P f_a(r) \right] \\
&= \sum_a \left[\langle \bar{f}_a(r)PG(r,r';\omega^+)\gamma^0 P f_a(r')\rangle \right. \\
&\quad \left. + \langle \bar{f}_a(r')PG(r',r;\omega^-)\gamma^0 P f_a(r)\rangle \right] ,
\end{aligned} \tag{5}$$

where $\omega^+ = E_a + E + i\eta$ and $\omega^- = E_a - E - i\eta$, E_a is the eigenvalue of hole states, E is the excitation energy. We shall emphasize that the internal sum over b in Eq.(3) extends over only unoccupied particle and Dirac sea states, while the sum over b in the single particle Green's function runs over all states. One could add and subtract a sum over all occupied states in the internal sum over b in the first and second terms of Eq.(3). It is interesting to find that those two subtracted terms cancel each other exactly.

To simplify the problem, the Bethe-Salpeter function is usually solved in momentum space, the Fourier transformation of the above equation is

$$\Pi_0^{RMF}(P,P;k,k';E) = \int e^{-ik\cdot r}\Pi_0^{RMF}(P,P;r,r';E)e^{ik'\cdot r'}drdr' . \tag{6}$$

After multipole expansions the final expression of the L component of the unperturbed polarization operator in momentum space is

$$\begin{aligned}
&\Pi_0^L(P,P;k,k';E) \\
&= (4\pi)^2 \sum_a \left[\langle \bar{f}_a(r)Pj_L(kr)Y_{LM}(\hat{r})G(r,r';\omega^+)\gamma^0 Pj_L(k'r')Y_{LM}^*(\hat{r}')f_a(r')\rangle \right. \\
&\quad \left. + \langle \bar{f}_a(r')Pj_L(k'r')Y_{LM}^*(\hat{r}')G(r',r;\omega^-)\gamma^0 Pj_L(kr)Y_{LM}(\hat{r})f_a(r)\rangle \right] .
\end{aligned} \tag{7}$$

The single particle Green's function represents as a solution to the system of equations,

$$[\alpha\cdot\mathbf{p} + V(r) + \beta M^* - E]\, G(\mathbf{r},\mathbf{r}';E) = \delta(\mathbf{r}-\mathbf{r}') , \tag{8}$$

where $M^* = M + S(r)$ is the nucleon effective mass, $S(r) = -g_\sigma\sigma(r)$ is the scalar potential. $V(r) = g_\omega\omega^0(r) + g_\rho\rho_3^0(r)\tau_3 + \frac{1}{2}(1+\tau_3)A^0(r)$ is the vector potential.

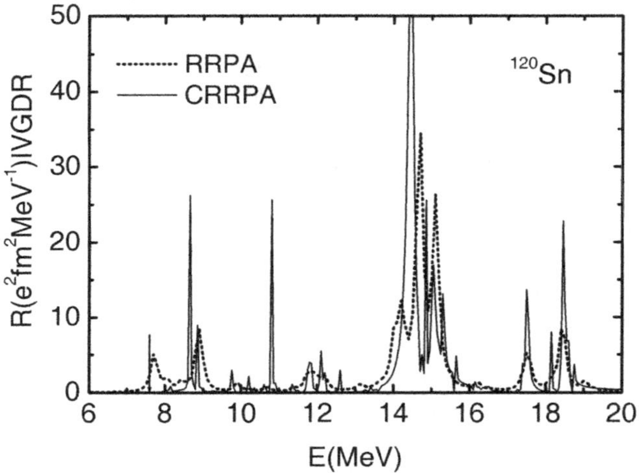

Fig. 1. The IVGDR response functions in ^{120}Sn

The single particle Green's function can be decomposed into radial function $g_{ij}(r,r')$ and spin-spherical harmonics,

$$G(\mathbf{r},\mathbf{r}';E)=\sum_{\kappa m}\frac{1}{rr'}\begin{pmatrix} g_{11}^{\kappa}\mathscr{Y}_{\kappa m}(\Omega_r)\mathscr{Y}^*_{\kappa m}(\Omega_{r'}) & -ig_{12}^{\kappa}\mathscr{Y}_{\kappa m}(\Omega_r)\mathscr{Y}^*_{-\kappa m}(\Omega_{r'}) \\ ig_{21}^{\kappa}\mathscr{Y}_{-\kappa m}(\Omega_r)\mathscr{Y}^*_{\kappa m}(\Omega_{r'}) & g_{22}^{\kappa}\mathscr{Y}_{-\kappa m}(\Omega_r)\mathscr{Y}^*_{-\kappa m}(\Omega_{r'}) \end{pmatrix}, \quad (9)$$

where the spin-spherical harmonics stand for

$$\mathscr{Y}_{\kappa m}(\Omega)=\sum_{m'}Y_{lm'}(\Omega)\chi_{1/2\ m-m'}\langle l,m',1/2,m-m'\,|j,m\rangle\ , \quad (10)$$

where $j=|\kappa|-1/2$ and $l=j\pm1/2$ for $\kappa>0$ and $\kappa<0$, respectively. $Y_{lm'}(\Omega)$ is the spherical harmonic function and $\chi_{1/2\ m-m'}$ two-component Pauli spinor . Inserting Eq.(9) into Eq.(8) yields the radial equation,

$$\begin{pmatrix} -M^*-V+E & d/dr-\kappa/r \\ d/dr+\kappa/r & -M^*+V-E \end{pmatrix}\begin{pmatrix} g_{11}^{\kappa}\ g_{12}^{\kappa} \\ g_{21}^{\kappa}\ g_{22}^{\kappa} \end{pmatrix}=\delta(r-r')\begin{pmatrix} -1\ 0 \\ 0\ \ 1 \end{pmatrix}. \quad (11)$$

The radial Green's matrix is calculated numerically as proper combinations of the regular and irregular solutions of the homogeneous part of Eq.(11). The regular solutions are

$$u^{\kappa}=\begin{pmatrix} u_1^{\kappa} \\ u_2^{\kappa} \end{pmatrix}, \quad (12)$$

which are normalizable at the origin, and irregular solutions are

$$v^{\kappa}=\begin{pmatrix} v_1^{\kappa} \\ v_2^{\kappa} \end{pmatrix}, \quad (13)$$

which are normalizable at infinity, exist for any value of the energy E. The irregular solutions are determined by the boundary condition at infinity.

The radial Green's functions are given in terms of u and v by

$$g_{ij}^{\kappa}(r,r') = \frac{1}{\Delta^{\kappa}} \begin{cases} u_i^{\kappa}(r)v_j^{\kappa}(r'), & r \le r' \\ v_i^{\kappa}(r)u_j^{\kappa}(r'), & r > r' \end{cases} \quad i,j = 1,2 \ . \tag{14}$$

The normalization of the Green's function is determined by the Wronskian determinant $\Delta^{\kappa} = v_1^{\kappa}u_2^{\kappa} - u_1^{\kappa}v_2^{\kappa}$ of the regular and irregular solutions.

In this work, we study the IVGDR in the stable nucleus ^{120}Sn in terms of CRRPA built on the RMF ground state, where NL3 parameter set[9] is adopted. The response strengths of IVGDR in ^{120}Sn is plotted in Fig. 1. The results are compared to those

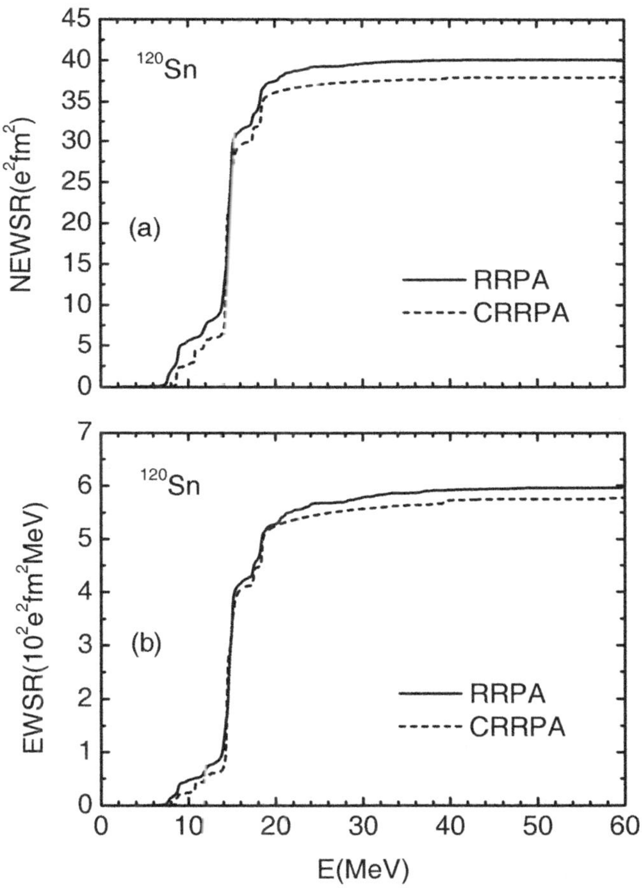

Fig. 2. The non energy-weighted moment m_0 (a) and energy-weighted moment m_{-1} (b) of IVGDR response function in ^{120}Sn.

calculated in the RRPA approach with a discretized continuum spectrum, where a small averaging parameter $\Delta = 0.2$MeV is introduced by hand. It is shown in Fig.1 that the strength distributions of the response function given by two approaches (CRRPA and RRPA) are very similar. The results can be also validated by the non-energy weighted and energy weighted sum rules. We calculate various moments of the response function as a function of the interval energy,

$$m_k(E) = \int_0^E R^L(E^{'})E^{'k}dE^{'}. \tag{15}$$

The non-energy-weighted moment m_0 and energy weighted moment m_1 as functions of the energy are shown in Fig. 2(a) and Fig. 2(b), where the solid and dashed curves represent the results of RRPA and CRRPA, respectively. It is observed that both non-energy weighted and energy weighted sum rules in CRRPA are slightly smaller than those obtained in RRPA. This is due to the fact that in our CRRPA calculations the discrete excitations from bound states to bound states are missing, which have to be treated specially.

In summary, we have extended the discretized relativistic random phase approximation to take account the contribution of the continuum in a proper way. The method is based on the Green's function technique. In this method, the summation of the continuum is exactly taken account by a single particle Green's function. The negative states in the Dirac sea are also included in the single particle Green's function in the no sea approximation. Single particle Green's function is calculated numerically by the product of the regular and irregular solutions of the Dirac equation.

The CRRPA is used to study the properties of isovector giant dipole resonance in stable nucleus ^{120}Sn. The calculation is fully consistent: the effective nuclear interaction NL3 is used both to determine the bound states and the Green's function in the full energy space, and the meson propagator which presents the particle-hole residual interaction in nuclear collective excitation. Preliminary results on IVGDR in stable nucleus ^{120}Sn are obtained.

References

1. G. A. Lalazissis, S. Raman and P. Ring, *At. Data Nucl. Data Tables* **71**, 1 (1999).
2. L. S. Geng, H. Toki, and J. Meng, *Prog. Theor.Phys.* **113**, 785 (2005).
3. Zhong-yu Ma, N. V. Giai, A. Wandelt, D. Vretenar and P. Ring, *Nucl. Phys. A* **686**, 173 (2001); Zhong-yu Ma, *Commun. Theo. Phys.* **32**, 493 (1999).
4. P. Ring, Zhong-yu Ma, N. V. Giai, A. Wandelt, D. Vretenar, and Li-gang Cao, *Nucl. Phys. A* **694**, 249 (2001).
5. Zhong-yu Ma, A. Wandelt, N. V. Giai, D. Vretenar, P. Ring, and Li-gang Cao,*Nucl. Phys. A* **703**, 222 (2002).
6. J. Piekarewicz, *Phys. Rev. C* **64**, 024307 (2001).
7. Li-Gang Cao, Ding Yang and Zhong-yu Ma, submitted to *Phys. Lett. B* .
8. Zhong-yu Ma, H. Toki, N. V. Giai, *Phys. Rev. C* **55**, 2385 (1997).
9. G. A. Lalazissis, J.König and P. Ring, *Phys. Rev. C* **55**, 540 (1997).

CHIRAL BANDS FOR QUASI-PROTON AND QUASI-NEUTRON COUPLING WITH A TRIAXIAL ROTOR

B. QI AND S.Q. ZHANG*

School of Physics, Peking University,
Beijing 100871, China
**E-mail: sqzhang@pku.edu.cn*

S.Y. WANG

Department of Space Science and Applied Physics, Shandong University at Weihai,
Weihai 264209, China

J. MENG

School of Physics, Peking University,
Beijing 100871, China
Institute of Theoretical Physics, Chinese Academy of Sciences,
Beijing 100080, China
Center of Theoretical Nuclear Physics, National Laboratory of Heavy Ion Accelerator,
Lanzhou 730000, China
E-mail: mengj@pku.edu.cn

A particle rotor model (PRM) with a quasi-proton and a quasi-neutron coupled with a triaxial rotor is developed and applied to study chiral doublet bands with configurations of an $h_{11/2}$ proton and an $h_{11/2}$ quasi-neutron. With pairing treated by the BCS approximation, the present quasi-particle PRM is aimed at simulating one proton and many neutron holes coupled with a triaxial rotor. It is found that aplanar rotation or equivalently chiral geometry exists beyond the usual one proton and one neutron hole coupled with a triaxial rotor. After including the pairing correlation, the model describes the candidate chiral bands in ^{126}Cs successfully, which supports the interpretation of chirality geometry.

Keywords: chirality; particle rotor model; quasiparticle; aplanar rotation.

1. Introduction

Chirality is a common property in chemistry, biology and particle physics, etc. In nuclear physics, Frauendorf and Meng pointed out the rotation of triaxial nuclei may attain the chiral character and give rise to chiral doublet bands,[1] i.e., a pair of separated $\Delta I = 1$ bands (normally regarded as nearly degenerate) with the same parity.

So far, although quite a number of candidate chiral bands have been observed in odd-odd and odd-A nuclei in A ~100, 130 and 190 mass region, their identification is still an open question.[2–6] The essential starting point for understanding their properties is based on one particle and one hole coupled with a rigid triaxial rotor. Based on this scenario, a set of observable signatures have been suggested as fingerprints of chiral bands.[1,3,6,7] However, in a realistic nucleus, it is more natural that there will be more than one nucleon in a high j-shell, e.g., the candidate chiral doublet bands reported for $N = 75$ isotones with $Z = 55$ (^{130}Cs), 57 (^{132}La), 59 (^{134}Pr), 61 (^{136}Pm) and 63 (^{138}Eu), and for $Z = 55$ (Cs) isotopes with $N = 69$, 71, 73, 75, and 77. The Fermi energy of a proton (neutron) will undoubtedly change with Z (N) in these isotones (isotopes). Therefore it is interesting and necessary to investigate the doublet bands with valence nucleons sitting in the middle of a high j-shell, or alternatively multi-particles sitting in a high j-shell.

To address these issues, in Ref. 8, we developed a particle rotor model with a quasi-proton and a quasi-neutron coupled with a triaxial rotor to study chiral doublet bands. With the pairing correlations taken into account by the BCS approximation, the configuration of multi-particles sitting in a high j-shell can be simulated by adjusting the Fermi energy.

2. Model

The particle rotor model (PRM) has been extensively used in the investigation of the chiral rotation due to the advantages of the good angular momentum and simple picture. In Refs.,[1,7,10] the PRM with one particle and one hole coupled with a triaxial rotor has been developed and used in the analysis of chiral bands. The PRM Hamiltonian is written as

$$\hat{H} = \hat{H}_{\text{core}} + \hat{H}_{\text{p}} + \hat{H}_{\text{n}}. \tag{1}$$

$\hat{H}_{\text{core}}$ represents the Hamiltonian of the rotor,

$$\hat{H}_{\text{core}} = \sum_{\nu=1}^{3} \frac{(\hat{I}_\nu - \hat{j}_{\text{p}\nu} - \hat{j}_{\text{n}\nu})^2}{2\mathcal{J}_\nu}, \tag{2}$$

where $\hat{I}_\nu, \hat{j}_{p\nu}, \hat{j}_{n\nu}$ respectively denote the angular momentum operators for the nucleus, the valence proton and neutron. The moments of inertia for irrotational flow are adopted, i.e., $\mathcal{J}_\nu = \mathcal{J}\sin^2(\gamma - \frac{2\pi}{3}\nu)$. H_{p} and H_{n} describe the Hamiltonians of the single proton and neutron outside the rotor which for a single-j model can be given as,

$$H_{\text{p(n)}} = \pm\frac{1}{2}C\left\{\cos\gamma\left(j_3^2 - \frac{j(j+1)}{3}\right) + \frac{\sin\gamma}{2\sqrt{3}}\left(j_+^2 + j_-^2\right)\right\} \tag{3}$$

where the plus sign refers to a particle and the minus to a hole and the coupling parameter C is proportional to the quadrupole deformation β.

Including the pairing by the standard BCS quasiparticle approximation, the PRM can be generalized to the two quasiparticles coupled with a triaxial rotor cases. The single-particle energies ϵ_ν obtained by diagonalizing the single proton (neutron)

Hamiltonian in PRM[1,10] are replaced with the corresponding quasiparticle energies E_ν,

$$E_{\nu i} = \sqrt{(\epsilon_{\nu i} - \lambda_i)^2 + \Delta_i^2}, \qquad i = \mathrm{n, p} \tag{4}$$

where $\varepsilon_{\nu i}$ is the single particle energies, Δ_i the pairing gaps and λ_i the Fermi energies. With inclusion of pairing, the present model goes beyond the one particle one hole coupled with the triaxial rotor model, and the configuration of multi-particles sitting in a high j-shell can be simulated by adjusting the Fermi level. The detailed formalism and numerical calculation can be seen in Ref. 8.

3. Results

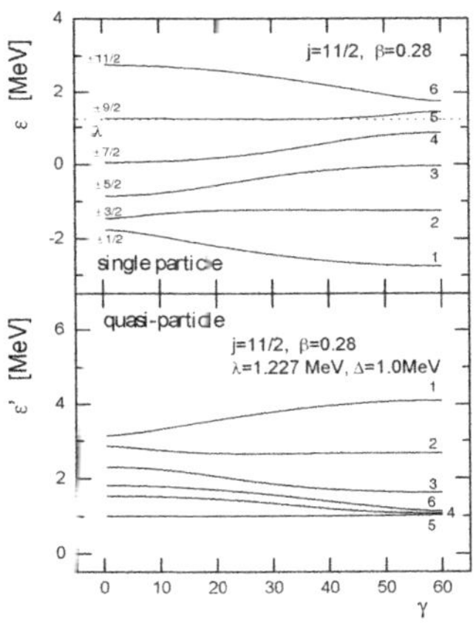

Fig. 1. Upper panel: The single particle energy ε with single-j Hamiltonian ($j = 11/2, \beta = 0.28$) as a function of γ deformation. The dashed line indicates the Fermi energy λ, which is used to obtain the quasiparticle energy ε' in the lower panel. Lower panel: Quasiparticle energy ε' for the same parameters as a function of γ deformation. The pairing parameters are $\lambda = 1.227$ MeV, $\Delta = 1$ MeV. Each level $(1, 2, \cdots, 6)$ corresponds to that with the same number in the upper panel.

The single particle energy ε with $1h_{11/2}$ j-shell and $\beta = 0.28$ is plotted in the upper panel of Fig. 1 as a function of the γ deformation of the deformed well. The six degenerate levels are respectively indicated by $1, 2, \cdots, 6$, as well as the corresponding third angular momentum components $\pm 1/2, \pm 3/2, \cdots, \pm 11/2$ at $\gamma = 0°$ (which is good quantum number only for $\gamma = 0°$). With pairing taken into account by the BCS calculation, the quasiparticle energy ε' with $\lambda = 1.227$ MeV and $\Delta = 1$ MeV is given in the lower panel of Fig. 1. Since the Fermi energy is $\lambda \simeq \varepsilon_5$, the state ε_5 is now the lowest quasiparticle state located at ~ 1 MeV due to the pairing gap Δ. With the pairing correlations taken into account by the BCS approximation, the configuration of multi-particles sitting in a high j-shell can be simulated by adjusting the Fermi energy.

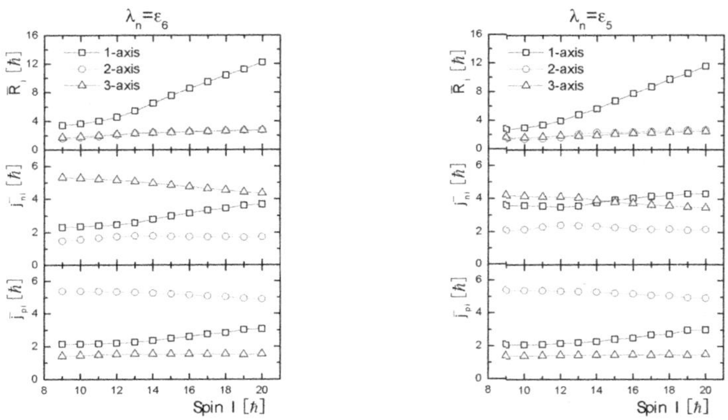

Fig. 2. The expectation values in body-fixed reference frame for the three components of the collective, odd-neutron, and odd proton angular momenta — defined by $\bar{R}_i = \sqrt{\langle \hat{R}_i^2 \rangle}$, $\bar{j_{pi}} = \sqrt{\langle \hat{j}_{pi}^2 \rangle}$, and $\bar{j_{ni}} = \sqrt{\langle \hat{j}_{ni}^2 \rangle}$, (i=1, 2, 3) — are plotted as functions of spin I. In the calculations, $\beta = 0.28$, $\gamma = 30°$, $\mathcal{J} = 30$ MeV^{-1} and the odd proton is fixed to be a pure $h_{11/2}$ particle, while the odd neutron is treated as a BCS quasiparticle with $\Delta = 1$ MeV, and $\lambda_n = \varepsilon_6$ (left panel), ε_5 (right panel). Open squares, open circles and open triangles correspond to the 1-axis, 2-axis and 3-axis, respectively.

The key to the formation of chiral bands in triaxial nuclei is the existence of aplanar rotation. Using wave functions obtained from the particle-rotor model, one can calculate the expectation values of angular momenta. The expectation values of the total angular momentum $\vec{I}$, collective angular momentum $\vec{R}$, the valence-proton and -neutron angular momentum $\vec{j}_p$, $\vec{j}_n$ have been investigated for $\lambda_n = \varepsilon_6, \cdots, \varepsilon_1$ as functions of spin I. In Ref. 8, these values of doublet bands have been given. When chiral geometry is realized, the observed states of chiral doublet bands may be written as[7] $|I+\rangle = \frac{1}{\sqrt{2}}(|IL\rangle + |IR\rangle)$, $|I-\rangle = \frac{i}{\sqrt{2}}(|IL\rangle - |IR\rangle)$, where $|IL\rangle$, $|IR\rangle$ are left- and right-handed geometry states in body-fixed reference frame. The expectation values of the angular momentum for $|IL\rangle$, $|IR\rangle$ states are the same, and can be obtained from observed states. For simplicity, the cases for $\lambda_n = \varepsilon_6, \varepsilon_5$ are shown in Figs. 2. For $\lambda_n = \varepsilon_6$, the collective angular momentum, and valence-proton and -neutron angular momentum align along the intermediate axis (1-), the short axis (2-) and the long axis (3-) respectively, i.e., a chiral picture results. For $\lambda_n = \varepsilon_5$, the configuration is similar to one proton plus three neutron holes in a single $h_{11/2}$ shell. In this case, the orientations of $\vec{R}$ and $\vec{j}_p$ are similar to the left panel of Fig. 2, while the third component of the angular momentum $\vec{j}_n$ is reduced. The total angular momentum $\vec{I}$ is still aplanar, but its inclination angle to the 1-2 plane becomes smaller.

More results and detail analysis of the angular momentum orientations, energy separation between the partner bands, and behavior of electromagnetic transitions can be found in Ref. 8. It shows: (i) The chiral geometry holds even for the valence nucleons deviating from a pure particle-hole configuration; (ii) The near constant energy separation ($\sim$ 200 keV) between the partner bands, which has been observed in many candidate chiral bands experimentally, can be obtained for $\lambda_n = \varepsilon_6$ and $\lambda_n = \varepsilon_5$ for certain spin and deformation γ intervals; (iii) Either a deviation of the core shape from $\gamma = 30°$ or a deviation of the Fermi energy surface from a particle-hole configuration will hinder the level degeneracy and prefer a near constant energy difference; (iv) The odd-even staggering of $B(M1)/B(E2)$ values is strongly influenced by the deformation γ as well as the Fermi surface λ.

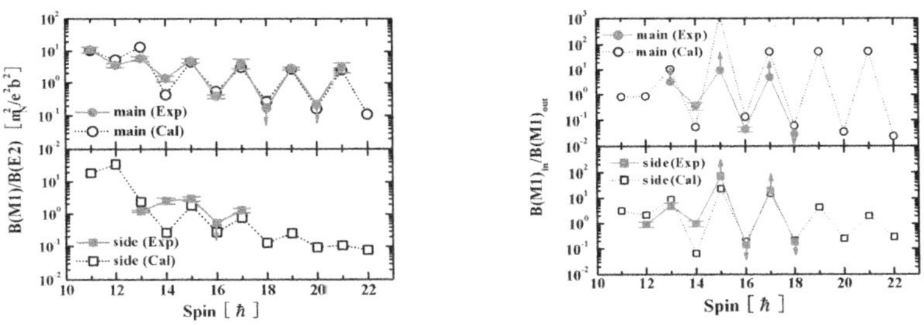

Fig. 3. Comparisons between the calculated $B(M1)/B(E2)$, $B(M1)_{in}/B(M1)_{out}$ and experimental data available for the main and side band in ^{126}Cs. The data are taken from Ref. 11.

The observation of the chirality in ^{126}Cs, ^{128}Cs, ^{130}Cs, ^{132}Cs provides a good opportunity to investigate the dependence of its occurrence on the number of the valence neutron in high-j shell. In particular, for Cs isotopes, the proton Fermi levels are supposed to lie in the lower $\pi h_{11/2}$ subshell, and compared with ^{130}Cs and ^{132}Cs where the neutron Fermi levels are supposed to lie in the upper $\nu h_{11/2}$ subshell, the neutron Fermi level in ^{126}Cs is likely close to be in the middle of the $\nu h_{11/2}$ subshell. Using the above two quasiparticles coupled with a triaxial rotor model, the doublet bands in ^{126}Cs has been investigated. The detailed parameters can be seen in Ref. 9. The calculated $B(M1)/B(E2)$ and $B(M1)_{\rm in}/B(M1)_{\rm out}$ values together with their corresponding experimental results for the doublet bands in ^{126}Cs are presented in the Fig. 3. It can be seen that the agreement for the $B(M1)/B(E2)$ ratios at the whole spin region is excellent. The magnitude, staggering and the decreasing trend of the ratios with spin are reproduced quite well. Furthermore the experimental staggering phase is exactly reproduced in the calculation, i.e., the value at odd spin is larger than the one at even spin. Similarly, the calculated $B(M1)_{\rm in}/B(M1)_{\rm out}$ ratios also reproduce the experimental magnitude, staggering and the trend pattern quite well. To further confirm the picture of the chirality in ^{126}Cs, the orientation of the angular momentum for the rotor as well as the valence proton and neutron are investigated, which indicates a remarkable aplanar rotation in this nucleus. This provides additional support for the existence of chiral bands in ^{126}Cs.

4. Summary

With pairing treated by the BCS approximation, the present quasi-particles PRM is aimed at simulating one proton and many neutron holes coupled with a triaxial rotor. After a detailed analysis of the angular momentum orientations, energy separation between the partner bands, and behavior of electromagnetic transitions, it is demonstrated that aplanar rotation or equivalently chiral geometry, does exist beyond the simple one proton and one neutron hole coupled with a triaxial rotor. For the candidate chiral doublet bands in ^{126}Cs, not only the energy spectra but also the electromagnetic transition ratios can be well reproduced by the above model, which supports the existence of chiral bands in this nucleus. The PRM with multi-proton particles (holes) and multi-neutron holes (particles) coupled with a triaxial rotor is in progress.

Acknowledgments

This work is partly supported by the National Natural Science Foundation of China under Grant No. 10605001, 10435010, 10221003, and 10505002.

References

1. S. Frauendorf and J. Meng, *Nucl. Phys. A* **617**, 131 (1997).
2. D. Tonev *et al.*, *Phys. Rev. Lett.* **96**, 052501 (2006).
3. C. M. Petrache, G. B. Hagemann, I. Hamamoto and K. Starosta, *Phys. Rev. Lett.* **96**, 112502 (2006).
4. E. Grodner *et al.*, *Phys. Rev. Lett.* **97**, 172501 (2006).
5. P. Joshi, M. P. Carpenter, D. B. Fossan, T. Koike, E. S. Paul, G. Rainovski, K. Starosta, C. Vaman, and R. Wadsworth, *Phys. Rev. Lett.* **98**,102501 (2007).
6. S. Y. Wang, S. Q. Zhang, B. Qi and J. Meng, *Chin. Phys. Lett.* **24**, 664 (2007).
7. T. Koike, K. Starosta and I. Hamamoto, *Phys. Rev. Lett.* **93**, 172502 (2004).
8. S. Q. Zhang, B. Qi, S. Y. Wang, and J. Meng, *Phys. Rev. C* **75**, 044307 (2007).
9. S. Y. Wang, S. Q. Zhang, B. Qi, and J. Meng, *Phys. Rev. C* **75**, 024309 (2007).
10. J. Peng, J. Meng, and S. Q. Zhang, *Phys. Rev. C* **68**, 044324 (2003).
11. S. Y. Wang, Y. Z. Liu, T. Komatsubara, Y. J. Ma, and Y. H. Zhang, *Phys. Rev. C* **74**, 017302 (2006).

CONTINUUM PROPERTIES OF THE HARTREE-FOCK MEAN FIELD WITH FINITE-RANGE INTERACTIONS

HOANG SY THAN* and DAO TIEN KHOA

Institute for Nuclear Science & Technique, VAEC
P.O. Box 5T-160, Nghia Do, Hanoi, Vietnam
**E-mail: hsthan@vaec.gov.vn*

NGUYEN VAN GIAI and ELIAS KHAN

Institut de Physique Nucléaire, IN2P3-CNRS/ Université Paris-Sud
91406 Orsay, France.

A method of solving in coordinate representation the Hartree-Fock (HF) equations with finite-range interactions is presented. The calculations are performed with the effective Gogny interaction D1S. Results for the continuum properties of the single-particle spectra in Oxygen isotopes are discussed. The present method can be extended to treat the HF-BCS problem in neutron-rich nuclei with inclusion of the resonant continuum.

Keywords: Hartree-Fock equations; Finite-range interactions; Resonant continuum.

1. Introduction

Mean field theories have successfully been applied to nuclear structure problems, in particular for stable nuclei. They are also useful to investigate basic characters of the effective interactions. The Skyrme interaction[1] has been popular in the Hartree-Fock (HF) calculations, since the zero-range form is easy to handle and the self-consistent equations can be conveniently solved in coordinate space. For the finite range interaction, the Gogny force[2] is widely applied to the mean field calculations and the M3Y interaction (Michigan three-range Yukawa) has also been applied for HF studies,[3] but the calculations are usually done by adopting the harmonic-oscillator basis-set. One notes that the M3Y interaction derives from the bare NN interaction through the G-matrix and its parametrization by Yukawa functions.[4] With a certain number of modifications, M3Y-type interactions have successfully been applied to nuclear reactions.[5,6]

Physical single-particle resonances appear naturally as the outgoing-wave solutions of the Schrödinger equation corresponding to a finite depth potential. Unlike the case of the familiar Skyrme-HF potentials which are essentially local (the non-locality being simply described by a position-dependent effective mass[7]), the HF potential generated by a finite-range interaction is fully non-local. Thus, the aim of this paper is to show how the coordinate space HF equations for a non-local self-consistent potential can be actually solved with correct boundary conditions. We restrict our study to spherical symmetry.

We study here the exact continuum of the neutron spectra in neutron-rich systems for which one expects that the continuum plays an important role close to the drip lines. The calculations are done for the Oxygen isotopes with the Gogny D1S[8] interaction.

2. Hartree-Fock equations in coordinate representation

We now take as an example a Gogny force which contains a sum of two gaussians, a zero-range density-dependent part and a zero-range spin-orbit part. For such an interaction, the HF potential consists of two local terms: $U^H(\mathbf{r}_1)$ containing the direct (Hartree) contributions of the finite-range components and the contributions (direct + exchange) of the spin-orbit force, $U^{t_0}(\mathbf{r}_1)$ corresponding to the direct plus exchange contributions of the density-dependent component, and a non-local term $U_i^F(\mathbf{r}_1, \mathbf{r}_2)$ due to the exchange (Fock) contributions of the finite-range components. In spherical symmetry the single-particle wave functions can be separated as

$$\varphi_i(\mathbf{r}, \sigma, \tau) = \frac{u_i(r)}{r} \sum_{m_l m_s} \left\langle l m_l \frac{1}{2} m_s | j m \right\rangle Y_{l m_l}(\hat{r}) \chi_{m_s}(\sigma) \chi_q(\tau) , \qquad (1)$$

where the index i stands for the following set of quantum numbers: the charge q, the principal quantum number n, the orbital angular momentum l, the total angular momentum j, and the magnetic quantum number m.

In the context of HF equations with finite-range interactions, Vautherin and Vénéroni[9] have introduced a method called "*trivially equivalent local potential*" where one can transform the HF integro-differential equation into a purely differential equation. In fact, a drawback of the method is that this equivalent local potential can have poles at the nodes of the wave functions $u_i(r)$[9,10] . In Ref.[9] an iterative method was introduced to overcome this problem, based on the linearization of the local equivalent potential around the poles. However, in this work we would like to use another method for solving the HF equations, the so-called BGW method.[11] Three functions $D_i(r)$, $G_i(r)$ and $H_i(r)$ are introduced as

$$\begin{aligned} D_i(r_1) &= u_i^2(r_1) + a^2[u_i'(r_1)]^2 \\ G_i(r_1) &= \frac{1}{D_i(r_1)} \int U_i^F(r_1, r_2)[u_i(r_1)u_i(r_2) + a^2 u_i'(r_1) u_i'(r_2)]dr_2 \\ H_i(r_1) &= \frac{a^2}{D_i(r_1)} \int U_i^F(r_1, r_2)[u_i(r_2)u_i'(r_1) - u_i(r_1)u_i'(r_2)]dr_2 , \end{aligned} \qquad (2)$$

where a can take an arbitrary value. The case $a = 0$ corresponds to the trivially equivalent local potential. With the help of the definitions (2) it is straightforward to transform the integro-differential HF equations for the radial wave functions $u_i(r)$

into

$$-u_i''(r_1) + \frac{l_i(l_i+1)}{r_1^2}u_i(r_1) + \frac{2m}{\hbar^2}[U^H(r_1) + U^{to}(r_1) + G_i(r_1) - \epsilon_i]u_i(r_1) + \frac{2m}{\hbar^2}H_i(r_1)u_i'(r_1) = 0 \,. \quad (3)$$

Note that the potentials $U^H(r_1)$ and U^{to} do not exhibit any state dependence. One can see that Eq. (3) becomes formally a differential equation where the potentials depend on the solutions and therefore they must be solved iteratively. At each iteration of the HF scheme we evaluate the potentials of all states by using the wave functions of the previous iteration and we repeat this procedure until convergence. An additional difficulty comes from the fact that the first derivative of the wave function $u_i(r_1)$ appears in Eq. (3). This is not a major problem since the wave function $u_i(r)$ and its derivative can be determined together by a Numerov method.

3. Resonant continuum of the Hartree-Fock field

In general, the HF equations are usually solved by imposing to the radial wave functions the condition that they vanish beyond a given distance R (box radius). Thus, the continuous spectrum is replaced by a set of discrete energies, whose density depends on the box radius. In this section we discuss how the HF can be solved by keeping the correct asymptotic conditions for the bound and unbound neutron wave functions.

The asymptotic behaviour of the HF wave function is determined by the physical condition that, at large distances the nuclear mean field vanishes. The condition requires an effective interaction of finite range and finite-range non-locality and we can readily find the asymptotic behaviour of the physical solutions at infinity. Thus, for a negative chemical potential λ, i.e., for a bound system, there are two well separated regions in the single-particle spectrum. At negative energies the spectrum is discrete and the radial HF wave functions decay exponentially at infinity. For $E > 0$ the spectrum is continuous and the solutions[12] are

$$u_{l_i}(k_i, r)|_{r\to\infty} \longrightarrow Ak_i r[cos(\delta_{l_i})j_{l_i}(k_i, r) + sin(\delta_{l_i})n_{l_i}(k_i, r)] \,, \quad (4)$$

where j_{l_i} and n_{l_i} are the spherical Bessel and Neumann functions, respectively, δ_{l_i} is the phase shift corresponding to the angular momentum (l, j) and $k = (\frac{2mE}{\hbar^2})^{1/2}$. One can see that the HF wave functions have the standard form of a scattering state. In Eq. (4) A is an unknown coefficient which is determined by a normalization condition. The phase shift δ_{l_i} can be found by integrating Eq.(3) out to large distances and using the relation

$$tg(\delta_{l_i}) = \frac{r_1 n_{l_i}(k_i, r_1)u_{l_i}(k_i, r_2) - r_2 n_{l_i}(k_i, r_2)u_{l_i}(k_i, r_1)}{r_1 j_{l_i}(k_i, r_1)u_{l_i}(k_i, r_2) - r_2 j_{l_i}(k_i, r_2)u_{l_i}(k_i, r_1)} \,. \quad (5)$$

The problem now is how to choose the values of radius r_1 and r_2 in Eq. (5). Normally, the selected values of r_1 and r_2 are of the order of a few times the nuclear radius.

Table 1. Results of HF calculations for Oxygen isotopes: neutron single-particle energies ϵ, resonance energies E_R and widths Γ in the d3/2 and f7/2 channels.

	1d3/2			1f7/2		
	ϵ (MeV)	E_R (MeV)	Γ (keV)	ϵ (MeV)	E_R (MeV)	Γ (keV)
^{16}O	0.896	0.841	110	2.900	6.010	1663
^{18}O	0.746	0.685	69	2.916	5.770	1559
^{20}O	0.472	0.408	21	2.925	5.440	1398
^{22}O				2.928	5.060	1205
^{24}O				2.919	4.630	1039
^{26}O				2.881	4.000	769

The radial wave functions are calculated by integrating the HF equations outwards starting from the initial conditions at $r = 0$, and inwards starting from (4) if $E > 0$, and from an exponentially decreasing function if $E < 0$. The solutions are propagated by a modified Numerov method towards the matching point, where the continuity conditions for the wave functions and their derivatives are imposed. In the case of a bound state the continuity conditions and the normalization condition determine the coefficients A and the energy E. The difficulty of an exact continuum calculation, with asymptotic solutions given by Eq. (4), is to identify the energy regions where the localization of the wave function changes quickly with the single-particle energy. These are the regions of single-particle resonant states. The positions and the widths of these HF resonances are related to the total phase shift, calculated from the matching conditions, as

$$\delta(E) \simeq arctg\frac{\Gamma}{2(E_R - E)} \tag{6}$$

where E_R and Γ are the energy and the width of the resonant state. In our calculations we study for each (l, j) channel the behaviour of the phase shift and we estimate the energy (width) of the resonant state from the energies where the derivative of the phase shift is maximum (half of its maximum). Then we choose for exploring the energy region of a resonant state an energy grid with a small step.

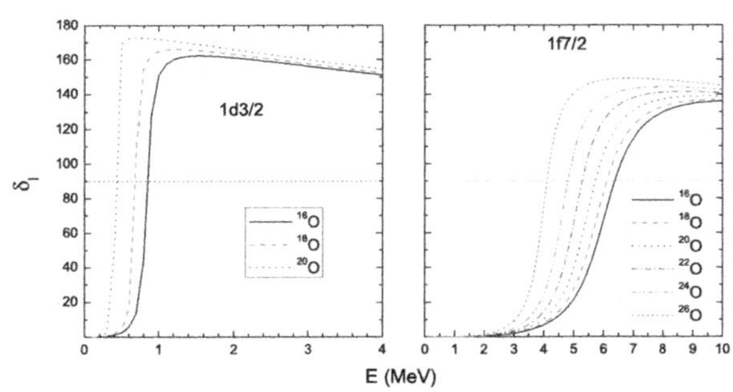

Fig. 1. Phase shifts in the $d3/2$ and $f7/2$ neutron channels for Oxygen isotopes.

We now apply the above resonant HF approach to the case of Oxygen isotopes using the finite range Gogny interaction D1S.[8] All calculations are carried out up to a distance of 20 fm, but the numerical results discussed in the following do not depend sensitively on the precise choice of the values r_1 and r_2 (see Eq. (5)) in the range (3-4)R, where R is the nuclear radius. In Table 1 we show the resonance energies and the widths calculated from the derivatives of the phase shift in the $d3/2$ and $f7/2$ neutron channels for $^{16-26}$O. In both channels they are related to unbound states with energies smaller than 4 MeV and they are single-particle resonances.

In Fig. 2 we show the phase shifts in the $d3/2$ and $f7/2$ channels for Oxygen isotopes in the energy region of the resonant states. For the $d3/2$ channel we only show the phase shift for $^{18-20}$O, because the phase shift δ_{l_i} changes so quickly around the value of $\pi/2$ that the shape of its derivative is too thin at this position. The results for the $f7/2$ channel are illustrated in the right panel and we can see that the phase shifts are not symmetric around the resonance energy in some cases. Thus, the position of some resonances may appear for values of the total phase shift which are quite far from $\pi/2$. For more details, we have plotted the phase shifts (full line) and their derivatives (dashed line) in the $f7/2$ channel for ^{16}O and ^{26}O in Fig. 3. In the case of ^{16}O, the resonance energy and the width estimated from the derivative of the phase shift are $E = 6.010$ MeV and $\Gamma = 1663$ keV while the energy corresponding to this phase shift when $\delta_{l_i} \simeq \pi/2$ is $E = 6.371$ MeV. For the ^{26}O case, the results of calculations are $E = 4$ MeV and $\Gamma = 769$ keV. In this case the total phase shift cross 90^0 at the resonance energy.

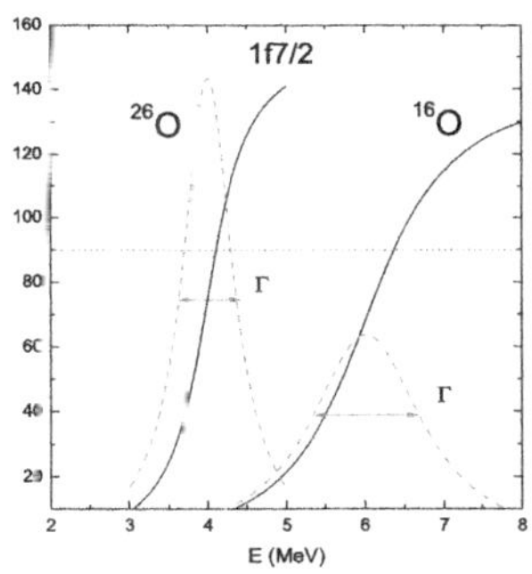

Fig. 2. Phase shifts (full line) and their derivatives (dashed line) in the $f7/2$ neutron channel for ^{16}O and ^{26}O.

4. Summary and Perspectives

In this work we have presented a new method for solving HF equations in coordinate space with finite range interactions. An iterative scheme is used for solving the integro-differential HF equations via this method. This may be useful because we can avoid the problem of the equivalent local potential of the Vautherin's method.[9] The first application is to predict the single-particle resonances in Oxygen isotopes calculated with the finite-range Gogny interaction D1S.[8] The results of the phase shifts and widths of the resonant states in the $d3/2$ and $f7/2$ neutron channels for $^{16-26}$O are presented.

From this paper, one can see that our method can be easily used with finite range interactions and we plan to apply the present technique to the density-dependent M3Y[5] interaction. This would allow one to study both nuclear structure and reaction problems[6] self-consistently with the same starting effective interaction which is itself based on a G-matrix approach. On the other hand, the physics of exotic nuclei close to the drip lines has triggered a new interest for the study of paring correlations in finite systems. Thus, a further step toward full self-consistency with the same finite-range force for calculating the mean field and the pairing field can be made, using the same technique. We should note that the effect of the resonant continuum upon pairing correlations has been studied in the framework of the BCS approximation[13] with zero-range forces, and we plan to use the density-dependent M3Y force to perform this type of study. Work in this direction is in progress.

Acknowledgements

The authors wish to thank N. Sandulescu for fruitful discussions. This research project has been supported, in part, by the Natural Science Council of Vietnam, EU Asia-Link Program CN/Asia-Link/008 (94791) and Vietnam Atomic Energy Commission (VAEC).

References

1. D.G. Vautherin, D.M. Brink, *Phys. Rev. C* **5**, 626 (1972).
2. J. Dechargé and D. Gogny, *Phys. Rev. C* **21**, 1568 (1980).
3. H. Nakada, *Phys. Rev. C* **68**, 014316 (2003).
4. N. Anantaraman, H. Toki and G.F. Bertsch, *Nucl. Phys. A* **398**, 269 (1983).
5. Dao T. Khoa, W. von Oertzen and A. A. Ogloblin, *Nucl. Phys. A* **602**, 98 (1996).
6. Dao T. Khoa, Hoang Sy Than, Do Cong Cuong, *Phys. Rev. C* **76**, 014603 (2007).
7. C.B. Dover and N. Van Giai, *Nucl. Phys. A* **190**, 373 (1972).
8. L.M. Robledo, J.L. Egido, J.F. Berger, M. Girod, *Phys. Lett. B* **187**, 223 (1987).
9. D.G. Vautherin, M. Vénéroni, *Phys. Lett. B* **25**, 175 (1967).
10. M. Grasso, N. Van Giai, N. Sandulescu, *Phys. Lett. B* **535**, 103 (2002).
11. K. A. Brueckner, J. L. Gammel, H. Weitzner, *Phys. Rev.* **110**, 431 (1958).
12. M. Grasso, N. Sandulescu, N. Van Giai, R.J. Liotta, *Phys. Rev. C* **64**, 064321 (2001).
13. N. Sandulescu, N. Van Giai, R.J. Liotta, *Phys. Rev. C* **61**, 061301(R) (2000).

A STUDY OF PAIRING INTERACTION IN A SEPARABLE FORM

YUAN TIAN[1] and ZHONGYU MA[2]
China Institute of Atom Energy
Beijing 102413, China
E-mail: [1]tiany@ciae.ac.cn
[2]mazy12@ciae.ac.cn

P. RING
Physik department, Technische Universität München,
D-85747 Garching, Germany
E-mail: ring@ph.tum.de

The pairing force of Gogny in the $^1S_0$ channel has been applied with great success in many applications in finite nuclei. It is of finite range and therefore rather complicated. Based on this force, we propose a new rather simple pairing interaction, which has basically the same properties as the Gogny force. It is separable in momentum space, translational invariant and, as the Gogny force of finite range. Using techniques of Talmi and Moshinsky, we show, that it can be represented as a sum of separable terms in oscillator space, which converges quickly. Because of its separability it is very easy to handle. It is fully determined by the Gogny force. It is used for the calculation of pairing properties of finite nuclei in the Relativistic Hartree Bogoliubov (RHB) approach. It is found that this new force can well depict the pairing properties on almost the same footing as the original pairing interaction not only in nuclear matter, but also in finite nuclei. This simple force can be easily applied in the calculation of deformed nuclei, and the investigation of pairing properties in nuclei close to the drip lines as well as for further investigations going beyond mean field.

1. Introduction

The structure of the nucleus and the properties of exotic nuclear systems strongly depend on their possible superfluid nature. In finite nuclei, pairing correlation constitutes an important part of the residual interaction and it has a strong influence on most of the low-energy properties of the system. Despite its major role, the present knowledge of the pairing force and the nature of pairing correlations in nuclei are quite poor. Properties such as the range of the effective pairing interaction, its link to the bare force, its possible surface character in finite nuclei and its density dependence still have to be clarified.

Recently Duguet proposed a microscopic effective interaction to treat pairing correlations in the $^1S_0$ channel.[1] He starts in nuclear matter and introduces a separable pairing force of a product of two Gaussians adjusting the parameters to the gap at the Fermi surface $\Delta_F(\rho)$ as a function of the density, calculated with the bare nucleon-nucleon interaction. This effective interaction is tailored in such a way that reproduces the pairing properties of the bare force. Starting from this simple ansatz and summing up *pp*- and *hh*-ladders he derives after using several approximations an effective pairing force in the nuclear medium for Hartree+BCS calculations. This effective force is non-local and of finite range. It depends on the momentum and on the density, but it contains no additional phenomenological parameters. In a further approximation he derives from this effective force a density dependent zero range force with a smooth cut-off parameter proportional to the BCS occupation numbers v_k^2. This is a very ambitious concept because it not only starts with the nucleon-nucleon interaction, which is well known, but it takes into account in addition correlations within the medium. On the other side it contains a number of approximations and assumptions which may be valid in nuclear matter, but which are hard to control in finite nuclei. In addition this force is tailored for Hartree+BCS calculations and it is not clear to what extend it can be used in case where one needs Hartree-Bogoliubov theory for the description of pairing.

Our new concept is much more modest. We start from the fact, that the phenomenological Gogny force[2] provides an excellent description of pairing properties in finite nuclei. This has been shown over the years in numerous applications of this force by the group of Gogny but also by the Madrid group.[3] In addition the pairing part of this force has been used with great success in the framework Relativistic Hartree-Bogoliubov (RHB) theory (for a review see Ref.[4]) On the other side this force is of finite range and not separable. Therefore the numerical effort is considerable and this is certainly the reason why this pairing force is not applied more frequently by other groups.

We start with the same separable ansatz in nuclear matter as Duguet. However, we do not consider this as a bare force from one has to derive an effective pairing interaction by summing up *pp*- and *hh*-ladders. On the contrary we determine this simple separable force from the pairing properties of the Gogny force in nuclear matter and we use it for extensive calculations in finite nuclei. First we transform it to r-space and to oscillator space. We show that is has a relatively simple form in both cases too. In particular we find that the matrix elements of this force can be represented in oscillator space by a sum of separable terms, which converges very quickly. Therefore this force can be without any difficulties in full Hartree-Bogoliubov calculations.

The purpose of this paper is to apply this force for Relativistic Hartree-Bogoliubov (RHB) calculations in finite nuclei, where it has been shown that the pairing channel can be well described by a non-relativistic force[5] , in order to investigate its validity and its equivalence with the full Gogny force.

2. Separable Pairing Interaction in Nuclear Matter

We start from the ansatz of Duguet for the pairing force in the $^1S_0$ channel of nuclear matter in momentum space: In general the antisymmetric matrix element of the pairing interaction $V^{^1S_0}$ is expressed in the plane-wave basis as,

$$\langle \mathbf{k}_1 s_1 \mathbf{k}_2 s_2 | \frac{1-P_\sigma}{2} V^{^1S_0} | \mathbf{k}_3 s_3 \mathbf{k}_4 s_4 \rangle_a = \frac{1}{2} \langle \mathbf{k}_1 \mathbf{k}_2 | V^{^1S_0} | \mathbf{k}_3 \mathbf{k}_4 \rangle_s (\delta_{s_1 s_3} \delta_{s_2 s_4} - \delta_{s_1 s_4} \delta_{s_2 s_3}), \tag{1}$$

where $\mathbf{K} = \mathbf{k}_1 + \mathbf{k}_2$, $\mathbf{K}' = \mathbf{k}_3 + \mathbf{k}_4$ are the total momenta and $\mathbf{k} = \frac{1}{2}(\mathbf{k}_1 - \mathbf{k}_2)$, $\mathbf{k}' = \frac{1}{2}(\mathbf{k}_3 - \mathbf{k}_4)$ are the relative momenta of the particle pairs respectively.

$$\langle \mathbf{k}_1 \mathbf{k}_2 | V^{^1S_0} | \mathbf{k}_3 \mathbf{k}_4 \rangle = \delta(\mathbf{K} - \mathbf{K}') \langle \mathbf{k} | V^{^1S_0} | \mathbf{k}' \rangle. \tag{2}$$

It has been pointed that the bare interaction in the $^1S_0$-channel is to a good approximation separable and nonlocal at low energies.[6] In the center of mass the matrix element is approximated by a separable form,

$$\langle \mathbf{k} | V_{sep}^{^1S_0} | \mathbf{k}' \rangle = \lambda v(k) v(k'). \tag{3}$$

The isospin quantum number has not been specified since we consider only the $T = 1$ channel of the pairing force. The $T = 0$ channel plays only a role for proton-neutron pairing in the vicinity of the $N = Z$ line. λ is a strength parameter. The shape of the function $v(k)$ is adjusted to the gap parameter at the Fermi surface $\Delta_F(k_F)$ as a function of the density expressed by k_F. In order to do this we have to solve the gap equation for the separable force.

Inserting the separable interaction $\lambda v(k) v(k')$ into Gap Equation, the solution of the gap equation takes the form $\Delta(k) = \Delta_0 v(k)$, where Δ_0 is the gap at zero momentum and satisfies the non-linear equation

$$1 = -\int_0^\infty \frac{k^2 dk}{4\pi^2} \frac{\lambda v^2(k)}{\sqrt{(\epsilon(k) - \mu)^2 + \Delta_0^2 v^2(k)}} \tag{4}$$

The gap at the Fermi surface is then obtained as $\Delta_F(k_F) = \Delta_0 v(k_F)$. Of course, the solution depends on the parameter λ and on the function $v(k)$. This function is determined by a numerical mapping of the curve $\Delta_F(k_F)$ to the function $v(k)$, in such a way that the resulting function $\Delta_F(k_F)$ represents exactly the curve derived from the solution of the gap equation for the Gogny force.[7]

For simplicity we first use a Gaussian shape for $v(k)$

$$v(k) = \exp(-\alpha^2 k^2) \tag{5}$$

When comparing with various Gogny force, we obtain sets of parameters $\lambda = -738$ MeV fm^3 and $\alpha^2 = 0.405$ fm^2 for the Gogny force D1 and $\lambda = -728$ MeV fm^3 and $\alpha^2 = 0.415$ fm^2 for the Gogny force D1S. In the left panel of Fig. 1. we show the gaps in the $^1S_0$ channel obtained from Gogny forces and from the corresponding separable forces. The separable forces reproduce the gap curves almost perfectly, especially in the case of Gogny force D1S. We also adjusted the bare potential Bonn A and we find the parameters: $\lambda = -877$ MeV fm^3 and $\alpha^2 = 0.53$ fm for the Bonn A potential. The corresponding curves are shown in right panel of Fig. 1. we find that the pairing gap can be well reproduced at low densities and we observe slight deviations at densities with $k_F > 1.0 fm^{-1}$. That means, the short range part of the

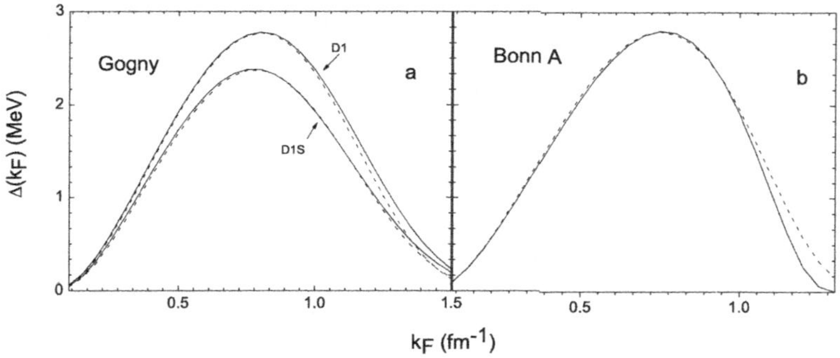

Fig. 1. Left panel: Comparison of $^1S_0$ pairing gaps in nuclear matter for the Gogny forces D1 and D1S and corresponding separable forces with λ=-738 MeV fm^3 α^2=0.405 fm^2 for D1 and λ=-728 MeV fm^3 α^2=0.415 fm^2 for D1S. Solid and dashed curves are for Gogny forces and corresponding separable forces, respectively. Right panel: Comparison of $^1S_0$ pairing gaps obtained from the potential Bonn A and the separable force with λ=-877MeV fm^3, α^2 = 0.53fm. The solid and dashed curves refer to Bonn A and corresponding separable force.

bare force, which is strongly repulsive cannot be reproduced so well by the Gaussian ansatz. To improve the agreement one probably needs a more general ansatz for the separable force going beyond the Gaussian form.

3. Separable Pairing Interaction in Finite Nuclei

Now we have to investigate how these separable forces describe the pairing properties in finite nuclei. For this purpose we investigate the properties of pairing correlations in the framework of RHB theory in some spherical isotope chains, as for instance. the Sn- and Pb-isotopes.

In coordinate space, pairing interaction 2 can be written as:

$$V(\mathbf{r}_1, \mathbf{r}_2, \mathbf{r}_3, \mathbf{r}_4) = \frac{\lambda}{(4\pi\alpha^2)^3}\delta(\mathbf{R} - \mathbf{R}')e^{-\frac{r^2}{4\alpha^2}}e^{-\frac{r'^2}{4\alpha^2}} \tag{6}$$

and the pairing matrix element in a spherical oscillator basis with oscillator parameter b is obtained as:

$$V_{1234} = C\sum_N V_{12}^N \cdot V_{34}^N \tag{7}$$

with $C = (2\pi)^{-3}\pi\lambda b^{-3}\sqrt{2}$ and the single particle matrix element

$$V_{12}^N = \sqrt{\frac{2j_1+1}{2(2l_1+1)}}M_{n_1l_1n_2l_2}^{N0n0}\int R_{n0}(\frac{r}{b})(\frac{1}{\sqrt{2\alpha^2}})^3 e^{-\frac{r^2}{4\alpha^2}}r^2dr \tag{8}$$

In Fig. 2 we show the pairing energy $E_{pair} = \frac{1}{2}\mathrm{Tr}(\kappa V\kappa)$ as a function of the mass number for the isotope chains ^{100}Sn$\sim^{160}$Sn and ^{164}Pb$\sim^{264}$Pb.

As we see from the upper part of Fig.2 good agreement of the pairing energy calculated with Gogny pairing force and its separable form is observed, where the

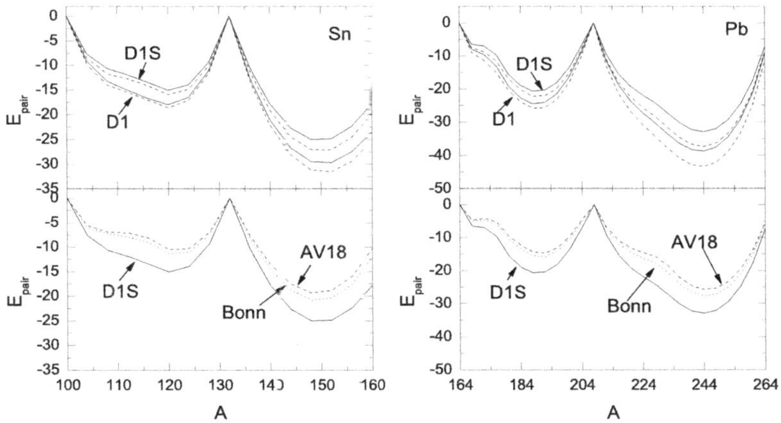

Fig. 2. Pairing energies in Sn-isotopes $^{100}Sn\sim^{160}Sn$ and Pb-isotopes $^{164}Pb\sim^{264}Pb$. Upper figure shows the result of RHB calculation with Gogny(D1 and D1S) pairing interactions(solid curves) and their corresponding separable forms pairing interaction(dotted curves). The lower figure shows the result of RHB calculation with the Gogny D1S pairing interaction (solid curves), and the separable pairing interactions corresponding to the potentials Bonn A (dotted curves) and AV18 (dashed curves).

largest discrepancy is less than 10%. Having in mind that the pairing energy depends in these nuclei extremely sensitively on the strength parameter of the pairing force, we can conclude that the separable pairing interaction can depict the paring properties in finite nuclei on almost same footing as its corresponding pairing interaction. Therefore, we can use the separable pairing interaction as a replacement for of its complicated counterpart of Gogny. We further calculate the pairing energies in those two isotope chains with the separable form of the pairing interaction obtained from the bare potential Bonn A and the potential AV18 potential. It is seen in the lower part of Fig.2, that the pairing energy obtained with the bare NN interaction, such as Bonn A and AV18 potential is considerably weaker than those of Gogny force. From this fact we can conclude, that there might be an additional influence of an induced force resulting from the exchange of phonons corresponding to low-lying surface vibrations.[8,9]

4. Summary

We present a simple separable form of the pairing interaction in the $^1S_0$ channel. Its shape is, without and additional phenomenological parameters, adjusted to reproduces either the pairing part of the Gogny force in the a bare or effective NN interaction by reproducing the gap in nuclear matter at the Fermi surface as a function of the density. It is found that a Gaussian ansatz works rather well for the Gogny force and at not too high densities also for the bare potential Bonn A. By comparing the matrix element of the separable form and its corresponding pairing interactions, the physical content of the separable force is clarified as an effective interaction at low-energy. Then we used this separable forces for RHB calculations to determine the pairing energies for chains of Sn- and Pb-isotopes and compare the results with those of full Gogny calculations. . Good agreement is observed, the largest discrepancy is less than 10%. The therefore can conclude that it is a good approximation to use the separable form instead a complicated full Gogny force in the pairing channel. Finally we calculate pairing energies for the same chains with the separable form deduced from the bare potential Bonn A. In comparison with the pairing energies obtained with Gogny pairing force, it is found that the bare NN interactions, such as Bonn interactions produce weaker pairing energies than those of Gogny force. This simple separable force can be easily applied in the calculation of deformed nuclei, and the investigation of pairing properties in nuclei close to the drip lines as well as for further extensions in beyond the mean field approximation.

Acknowledgements

This work has been supported by the National Natural Science Foundation of China under Grant Nos 10235030, 10475116 and 10535010, the European Community project Asia-Europe Link in Nuclear Physics and Astrophysics, CN/ASIA-LINK/008(94791)and by the Bundesministerium für Bildung und Forschung, Germany, under the project MT 246.

References

1. T. Duguet, *Phys. Rev. C* **69**, 054317 (2004).
2. J. Dechargé and D. Gogny, *Phys. Rev. C* **21**, 1568 (1980).
3. J. L. Egido and L. M. Robledo, in *Lecture Notes in Physics*, edited by G. Lalazissis, P. Ring, and D. Vretenar (Springer-Verlag, Heidelberg, 2004), Vol. 641, p. 269.
4. D. Vretenar, A. V. Afanasjev, G. A. Lalazissis, and P. Ring, *Phys. Rep.* **409**, 101 (2005).
5. M. Serra and P. Ring, *Phys. Rev. C* **65**, 064324 (2002).
6. G. E. Brown and A. D. Jackson, *The Nucleon-Nucleon Interaction* (North Holland, Amsterdam, 1976).
7. H. Kucharek, P. Ring, P. Schuck, R. Bengtsson, and M. Girod, *Phys. Lett. B* **216**, 249 (1989).
8. F. Barranco, R. A. Broglia, G. Gori, E. Vigezzi, P. F. Bortignon, and J. Terasaki, *Phys. Rev. Lett.* **83**, 2147 (1999).
9. J. Terasaki, F. Barranco, R. A. Broglia, E. Vigezzi, and P. F. Bortignon, *Nucl. Phys. A* **697**, 127 (2002).

MICROSCOPIC STUDY OF THE INELASTIC α+^{12}C SCATTERING

DO CONG CUONG and DAO TIEN KHOA

Institute for Nuclear Science and Technique, VAEC, P.O. Box 5T-160, Nghia Do, Hanoi, Vietnam.

Microscopic folding model analysis of the inelastic α+^{12}C scattering to 0_2^+ and 2_1^+ states in ^{12}C at E_{lab} = 104 to 240 MeV has been done using the 3-α resonating group wave functions and a complex density-dependent M3Y interaction deduced from the Brueckner Hartree Fock results for nuclear matter. In difference from 2_1^+ excitation at 4.44 MeV, the DWBA or CC analyses show that the strength of the 0_2^+ state at 7.65 MeV (the Hoyle state) exhausts only 7% of the monopole energy weighted sum rule (EWSR), compared to about 15% of the EWSR extracted from inelastic electron scattering data. The full $E0$ transition strength predicted by microscopic α-cluster models of the Hoyle state can be shown to exhaust up to 22% of the EWSR. Our present study has found that the missing $E0$ strength of the Hoyle state is likely due to the short lifetime and weakly bound structure of 0_2^+ state which significantly enhance absorption in the exit α+^{12}C*(0_2^+) channel.

Keywords: Inelastic α-scattering, 0_2^+ and 2_1^+ states in ^{12}C, folding model analysis

1. Introduction

Given a vital role in the stellar synthesis of Carbon, the isoscalar 0_2^+ state at 7.65 MeV in ^{12}C (known as the Hoyle state) has been studied over the years in numerous experiments. Although this state was clearly identified long ago in the inelastic α+^{12}C scattering at medium energies[1–4] and inelastic electron scattering[5] as an isoscalar $E0$ excitation, our knowledge about its unique structure is still far from complete.[6] Since the Hoyle state lies slightly above the α-decay threshold of 7.27 MeV, its wave function should have a dominant α cluster component. In fact, the Hoyle state has been well described in terms of three α clusters by the Resonating Group Method (RGM) some thirty years ago.[7,8] Very interesting is the condensate scenario[9,10] where three α clusters were shown to condense into the lowest (s-state) of their potential, and thus, forming a Bose-Einstein condensate (BEC). A somewhat more complicated structure of the Hoyle state was shown by the Fermionic Molecular Dynamics (FMD) calculation.[11] To validate conclusion made in the structure calculation, the wave functions must be carefully probed in the nuclear reactions study. In this aspect, the model wave functions of the Hoyle state[8,11,12] have been shown to give a reasonable description of the inelastic (e, e') data. The electric monopole transition moment

$$M(E0, 0_1^+ \to 0_2^+) = e\sqrt{4\pi} \int_0^\infty \rho^{\rm proton}_{0_1^+ \to 0_2^+}(r) r^4 dr \tag{1}$$

predicted by the RGM,[8] BEC[10] and FDM[11] calculations is around 6.62, 6.45 and 6.53 e fm^2, respectively, which are about 20% larger[a] than the experimental moment $M(E0) \approx 5.37 \pm 0.22$ e fm^2 deduced from the (e, e') data.[5] The (e, e') scattering probes, however, only the charge transition density and it is necessary to study also hadron scattering which probes the nuclear density. Such experiments have been done, e.g., for inelastic ^{3,4}He,^{6}Li+^{12}C scattering[13–15] and the data analyses, either in the distorted wave Born approximation (DWBA) or coupled-channel (CC) formalism, were done using the collective model form factor (FF) for the Hoyle state. These analyses indicated that the observed $E0$ strength of the Hoyle state exhausts about 7 to 9% of the isoscalar $E0$ energy weighted sum rule (EWSR). In particular, the DWBA analysis of the inelastic α+^{12}C data, measured recently at $E_{\rm lab}$ = 240 MeV with high precision,[15] has found that the Hoyle state exhausts 7.6 ± 0.9% of the EWSR. We recall that the $E0$ sum rule fraction S_0 of a monopole excitation is determined[16] as

$$S_0 = E_{\rm x} |M(IS0, 0_1^+ \to 0_2^+)|^2 / \left(\frac{2\hbar^2 A < r^2 >}{m}\right), \tag{2}$$

where $E_{\rm x}$, m and A are the excitation energy, nucleon mass and target mass number, respectively, $< r^2 >$ is the mean square radius and $M(IS0, 0_1^+ \to 0_2^+)$ is the isoscalar monopole transition moment which is determined by the same Eq. (1) but using the nuclear transition density instead of the charge transition density. If we assume $M(IS0, 0_1^+ \to 0_2^+) \approx 2M(E0, 0_1^+ \to 0_2^+)/e$ and take $< r^2 > \approx 5.78$ fm^2 (estimated using the ground state density of ^{12}C given by the RGM[8]), then we obtain $S_0 \approx$ 22.8, 21.7 and 22.2% of the EWSR from the RGM,[8] BEC[10] and FDM[11] results, respectively, and $S_0 \approx 15.0 \pm 1.3$% of the EWSR from the experimental $E0$ moment, in a perfect agreement with experimental value.[5] Given a good description of the (e, e') data by the cluster models[8–11] we conclude that the $E0$ strength of the Hoyle state should exhaust 15 to 20% of the EWSR and a puzzle remains why at least half of the monopole strength is missing in the inelastic α+^{12}C scattering.

2. Results of the DWBA and CC analyses

To investigate the missing monopole strength of the Hoyle state, we have performed a consistent folding model analysis of the inelastic α+^{12}C scattering at $E_{\rm lab}$ = 104 to 240 MeV using microscopic nuclear densities obtained from the 3-α RGM wave functions by Kamimura,[8] and a new complex density-dependent M3Y interaction. Before discussing our results, we recall that the same RGM nuclear densities[8] were

[a]a direct comparison of (1) with the observed $M(E0)$ can be made if the small difference between the charge- and proton transition densities is neglected.

used in a number of folding model studies (see, e.g., Refs.[17–21]) of the inelastic α+^{12}C scattering. Excepting the early work by Bauhoff[17] where a density independent Gaussian has been used as αN interaction in the single-folding calculation, other studies mentioned here have used the well-known DDM3Y[22] interaction in the double-folding calculation of the α+^{12}C potentials. Since the DDM3Y interaction is real, the imaginary parts of both the optical potential (OP) and inelastic FF were chosen phenomenologically in these studies. For example, in the CC analysis of the α+^{12}C scattering by Ohkubo and Hirabayashi[18,19] the imaginary inelastic FF was neglected and parameters of the Woods-Saxon imaginary OP were adjusted separately for each exit channel to obtain a good CC fit to the measured cross sections. Although one could achieve a reasonable description of the inelastic α+^{12}C scattering data in such a CC analysis, an arbitrary choice of the imaginary potentials makes it difficult to estimate accurately the absolute $E0$ transition strength. Up to now, the $E0$ strength of the Hoyle state has been deduced from the DWBA or CC analyses of the inelastic ^{3,4}He.^{6}Li+^{12}C scattering[13–15] using the breathing mode (BM) model[16] for the nuclear transition density

$$\rho_{0_1^+\to 0_2^+}(r) = -\beta\ [3\rho_0(r) + r\frac{d\rho_0(r)}{dr}], \tag{3}$$

where $\rho_0(r)$ is the ground-state (g.s.) density of ^{12}C. In this case, $S_0 = [\beta/\beta_{\rm max}]^2$, where $\beta_{\rm max}$ is the deformation parameter required for a monopole excitation to exhaust 100% of the EWSR.[23] The inelastic scattering FF is then obtained by the folding calculation using (3) and some effective nucleon-nucleon (NN) or α-nucleon (αN) interactions.[24–26] In general, the (complex) strength of the αN or NN interaction is first adjusted to the best optical model or CC description of elastic scattering and then is used without any further renormalization to calculate the inelastic FF. As a result, the only parameter in the inelastic channel is the deformation parameter β which must be determined from the best DWBA or CC fit to the inelastic scattering data. In the present work, we have constructed a new *complex*, density dependent M3Y-Paris interaction for the folding model study of inelastic α+^{12}C scattering. The parameters of the complex density dependence were carefully calibrated against the Brueckner Hartree-Fock results[27] for the nucleon OP in nuclear matter at each considered energy. The new density dependent interaction, dubbed hereafter as CDJLM interaction, was used to calculate the complex OP and inelastic FF for the DWBA and CC analyses of the elastic and inelastic α+^{12}C scattering data at $E_{\rm lab}$ = 104,[1,2] 139,[3] 172.5,[4] and 240 MeV.[15] The generalized double-folding method[25] was used to calculate the complex α-nucleus potential as the HF-type matrix element of the CDJLM interaction

$$U_{A\to A^*} = \sum_{i\in\alpha;j\in A,j'\in A^*} [< ij'|u_{\rm D}|ij > + < ij'|u_{\rm EX}|ji >], \tag{4}$$

where A and A^* are states of the target in the entrance- and exit channel of the α-nucleus scattering, respectively. Thus, Eq. (4) gives the elastic (diagonal) OP if A^* =

A and inelastic (transition) FF if otherwise. All the DWBA and CC calculations have been performed using the code ECIS97 written by Raynal.[31]

To fine tune the strength of complex CDJLM interaction, the optical model (OM) analysis of the measured elastic $\alpha+^{12}$C scattering data was done at each energy using the following complex OP

$$U_0(R) = N_{\rm R}{\rm Re}[U_{0_1^+\to 0_1^+}(R)] + iN_{\rm I}{\rm Im}[U_{0_1^+\to 0_1^+}(R)] + V_{\rm C}(R), \tag{5}$$

where $V_{\rm C}(R)$ is the elastic Coulomb potential between a point charge and a uniform charge distribution of radius $R_{\rm C} = 1.3 \times (4^{1/3} + 12^{1/3})$ fm. The renormalization coefficients $N_{\rm R}$ and $N_{\rm I}$ were adjusted first to the best OM fit of the elastic scattering data and then used further to scale the real and imaginary inelastic folded FF for the DWBA calculation.[15,17,25,26] Since $N_{\rm R}$ and $N_{\rm I}$ are an approximate way to take into account the higher-order (dynamic polarization) contributions to the microscopic OP,[25,33] they need to be readjusted again (to fit the elastic data) in the CC calculation, to account for the remaining nonelastic channels which were not included into the CC scheme. For consistency, these new $N_{\rm R}$ and $N_{\rm I}$ factors are also used to scale the complex inelastic folded FF for the CC calculation. We found that the $N_{\rm R}$ coefficient given by the best OM fit of the elastic scattering data is slightly above unity which indicates that the CDJLM interaction is quite a realistic choice for the double-folding calculation of the α-nucleus potential. The best fit $N_{\rm I}$ was found around 1.3 to 1.5.

The DWBA results obtained with the complex folded OP and inelastic folded FF are compared with the measured cross section for inelastic $\alpha+^{12}$C scattering to the Hoyle state in Figs. 1 and 2. We found that the calculated DWBA cross sections systematically overestimate data at all energies if the inelastic folded FF is obtained with the (*unrenormalized*) RGM transition density.[8] Given the accurate choice of the density dependent NN interaction and nuclear densities, this discrepancy is clearly not associated with usual uncertainties of the folding model analysis of α-nucleus scattering.[26] Since the RGM transition density has been proven to reproduce nicely the (e,e') data,[8,12] the folding + DWBA results shown in Figs. 1 and 2 might indicate a "suppression" of the monopole $E0$ strength occurred in the inelastic $\alpha+^{12}$C scattering. To further probe this effect, we have used a realistic Fermi distribution[30] for the g.s. density of ^{12}C to generate the BM transition density (3) for the DWBA analysis of inelastic scattering to the Hoyle state. We have fixed the monopole deformation parameter β so that (3) gives exactly the same $E0$ transition strength as that given by the RGM transition density and, hence, exhausts 22.8% of the isoscalar monopole EWSR as discussed above. As can be seen from Figs. 1 and 2, the DWBA cross sections given by the inelastic FF folded with the BM transition density for the Hoyle state are very close to those given by the RGM transition density.

To match the calculated DWBA cross section to the data points, we need to scale both transition densities by a factor of 0.55 which reduces the sum rule strength of the Hoyle state to $S_0 \approx 6.9\%$ of the EWSR. Such a small sum rule strength is

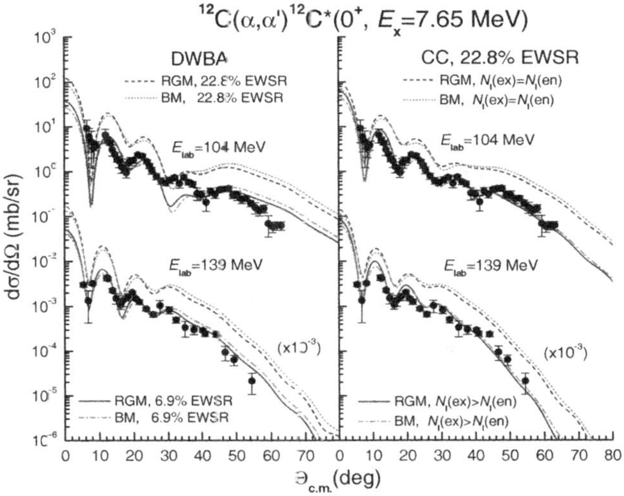

Fig. 1. Inelastic $\alpha+^{12}$C scattering data at $E_{lab} = 104^{1,2}$ and 139^3 MeV for the 0_2^+ state of ^{12}C in comparison with the DWBA and CC results given by the complex folded OP and inelastic FF obtained with two choices of the transition density for the 0_2^+ state. A good DWBA description of the data is reached only if the the transition densities are scaled by a factor of 0.55 which reduces $E0$ sum rule strength to $S_0 \approx 6.9\%$ of the monopole EWSR. If S_0 is kept at 22.8% of the EWSR, as given by the RGM result,[8] the data can be reproduced in the CC calculation only by using a more absorbing OP for the $\alpha+^{12}C^*(0_2^+)$ channel. See more details in text.

significantly below the empirical range of 15 - 20% of the EWSR discussed above. The sum rule strength of the Hoyle state found in our DWBA analysis is comparable to that ($7.6 \pm 0.9\%$) found in the folding + DWBA analysis of the inelastic $\alpha+^{12}$C scattering at 240 MeV,[15] where a Gaussian has been used for the αN interaction in the folding calculation of the OP and inelastic FF.

A widely used prescription for the DWBA calculation of inelastic hadron scattering is to assume the same complex OP for the entrance and exit channels[16] and the only difference in the elastic waves is a slight shift in kinematics caused by a lower energy in the exit channel. For such a subtle nuclear state like the Hoyle state which has unusually dilute and extended structure,[6] the diagonal potential $U_{0_2^+\to 0_2^+}$ should be quite different from $U_{0_1^+\to 0_1^+}$ which was used as the OP in both the entrance and exit channels of the DWBA calculation presented above. Actually, the importance of using explicitly different OP for different exit channels in the CC description of $\alpha+^{12}$C scattering at 104 and 139 MeV was pointed out earlier by Bauhoff.[17] Thus, one must go beyond DWBA and do the CC analysis of inelastic $\alpha+^{12}$C scattering to the Hoyle state before making any conclusion on the missing monopole strength. Given the importance of $0_2^+ \to 2_1^+$ transition in the stellar Carbon production, we have considered following coupling scheme in our CC calculation

$$0_1^+ \leftrightarrow 2_1^+ \leftrightarrow 0_2^+ \leftrightarrow 0_1^+. \qquad (6)$$

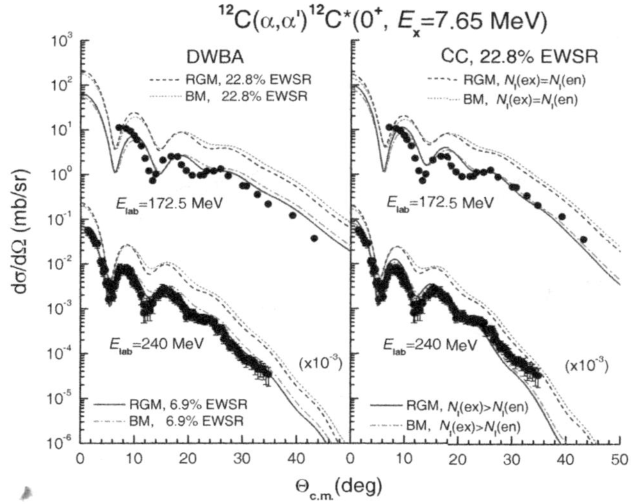

Fig. 2. The same as Fig. 1 but for the data measured at $E_{\rm lab} = 172.5$[4] and 240[15] MeV.

In the CC scheme (6), the two-step excitation ($0_1^+ \to 2_1^+ \to 0_2^+$) of the Hoyle state is treated in equal footing with the direct excitation ($0_1^+ \to 0_2^+$). Thus, the CC description of the inelastic $\alpha+{}^{12}$C scattering to 0_2^+ state should be more realistic compared to the DWBA calculation which takes into account the direct excitation only.[34] For the inputs of the CC calculation, in addition to three inelastic FF's for transitions between 0_1^+, 2_1^+ and 0_2^+ states, three complex OP's have been calculated using the diagonal nuclear densities of these states given by the RGM calculation.[8] For consistency, the $N_{\rm R(I)}$ factors have been readjusted again to obtain a good CC description of the elastic scattering data. While the best-CC-fit $N_{\rm R}$ factor remains more or less the same, $N_{\rm I}$ factor becomes significantly smaller and closer to unity, especially at the two low energies. This result shows explicitly how the inelastic channels in the CC scheme (6) contribute to the enhanced $N_{\rm I}$ factor found in the OM analysis of elastic scattering. All the CC results are shown in the right panels of Figs. 1 and 2, and it can be seen that the CC cross sections strongly overestimate the inelastic data for 0_2^+ state, in the same way as found in the DWBA calculation, when the inelastic FF is obtained with the original RGM transition density.[8] The two-step excitation ($0_1^+ \to 2_1^+ \to 0_2^+$) was found to be negligible at all considered energies and the CC cross sections shown in Figs. 1 and 2 are practically given by the direct (one-step) excitation of 0_2^+ state only. We note that the original RGM density for the $2_1^+ \to 0_2^+$ transition could not reproduce the observed $B(E2)$ transition probability and we have scaled this transition density, as recommended in Ref.,[8] by a factor of 1.53 to reproduce the experimental transition rate $B(E2, 2_1^+ \to 0_2^+) \approx 2.6 \pm 0.8\ e^2{\rm fm}^4$.[35] Nonetheless, the folded $2_1^+ \to 0_2^+$ transition FF gives still a very weak contribution to the CC cross section for the 0_2^+ excitation. An agreement with data can be reached only if the $E0$ transition density is reduced by about half as found in the DWBA analysis. It is clear that the missing monopole strength in the inelastic $\alpha+^{12}$C scattering to the Hoyle state is *not* associated with uncertainties of the DWBA or CC analyses, and there should be some physics effect that damps the $E0$ transition strength. We further mention a recent folding + CC calculation of inelastic $\alpha+^{12}$C scattering at $E_{\rm lab} = 172.5, 240$ and 386 MeV by Takashina,[36] where coupling to all the low-lying excited states of ^{12}C as well as 3-α breakup channel was taken into account, but the CC cross section for the 0_2^+ excitation remains lying significantly higher than the data points, in about the same manner as found in our CC analysis.

Such a "damping" of the transition strength in inelastic nucleus-nucleus scattering has also been observed in our recent folding model studies of inelastic ^{16}O+^{16}O scattering to the 2_1^+ and 3_1^- states of ^{16}O[37] and one-neutron transfer ^{16}O(^{16}O,^{17}O)^{15}O* reaction to the $3/2^-$ excited state of ^{15}O.[38] The DWBA or CC description of high-precision inelastic scattering or one-neutron transfer data was possible only if the absorption in the exit channel is significantly increased.[37] The enhanced absorption in the exit channel is a direct consequence of the suppression of nuclear refraction in nonelastic channels.[29] Namely, a short-lived (excited) and loosely bound cluster like ^{16}O$^*_{2^+}$ or ^{15}O$^*_{3/2^-}$ has a shorter mean free path in the nuclear medium which implies a weaker refraction or stronger absorption[39,40] in the exit channel compared to the entrance channel where both nuclei are in their (stable) ground states.[29,37,38] In a similar scenario, the 0_2^+ state of ^{12}C is very weakly bound and particle unstable (with a mean lifetime $\tau \approx 10^{-16}$ s) and it is natural to expect that the OP in the $\alpha+^{12}$C$^*(0_2^+)$ exit channel is more absorbing than the OP in the entrance $\alpha+^{12}$C channel. Since the elastic α scattering on the Hoyle state cannot be measured, we have adjusted $N_{\rm I}$ of the imaginary OP of the $\alpha+^{12}$C$^*(0_2^+)$ channel to achieve a reasonable CC description of the inelastic data for 0_2^+ state, while keeping $N_{\rm R}$ factor and complex strength of the transition FF unchanged. From Figs. 1 and 2 one can see that a very good CC description of the inelastic 0_2^+ data can be reached by using $N_{\rm I} \approx 2.5 - 3.4$ for the imaginary OP in the $\alpha+^{12}$C$^*(0_2^+)$ channel.

To illustrate whether the enhanced absorption is also required for some other excited states of ^{12}C or is it a unique effect associated with fragile structure of the Hoyle state, we have done the same analysis for inelastic $\alpha+^{12}$C scattering to the 2_1^+ state at 4.44 MeV and the results are plotted in Fig. 3. Like the Hoyle state, the RGM transition density of the 2_1^+ state has also been used to successfully reproduce the (e,e') data.[8] The electric transition strength predicted by the RGM transition density is $B(E2, 0_1^+ \to 2_1^+) = 46.5\ e^2$fm^4 which agrees fairly with the measured $B(E2)$ transition rate of $40 \pm 4\ e^2$fm^4.[35,41] We note that 2_1^+ state is a very strong excitation and the coupling strength by the 2_1^+ excitation is strongest in our CC scheme (6). From Fig. 3 one can see that a consistent folding model description of

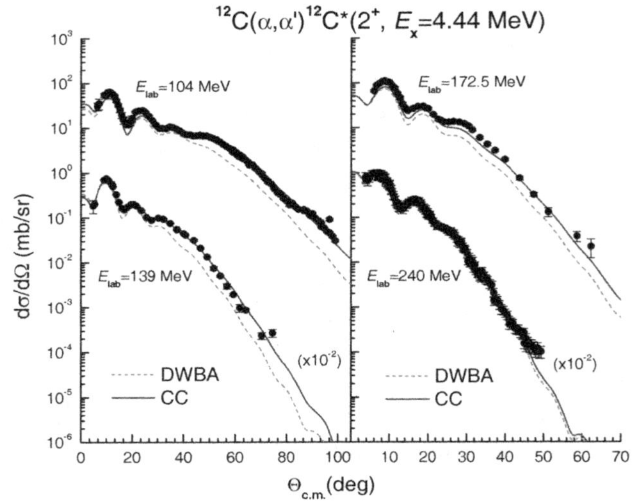

Fig. 3. Inelastic $\alpha+^{12}$C scattering data measured at $E_{\rm lab} = 104,$[1,2] 139,[3] 172.5[4] and 240[15] MeV for the 2_1^+ excitation at 4.44 MeV in ^{12}C in comparison with the DWBA and CC results obtained with the complex folded OP and inelastic folded FF. See more details in text.

the inelastic $\alpha+^{12}$C scattering to the 2_1^+ state at α-energy up to 172.5 MeV can be reached only in the CC calculation. The measured inelastic $\alpha+^{12}$C scattering cross section for the 2_1^+ state is nicely reproduced by the CC calculation at all energies without the need to enhance the imaginary OP in the $\alpha+^{12}\mathrm{C}^*(2_1^+)$ channel, i.e., the same $N_{\rm R(I)}$ factors were used for the entrance and exit $\alpha+^{12}\mathrm{C}^*(2_1^+)$ channels. This indicates that the enhanced absorption found above for the $\alpha+^{12}\mathrm{C}^*(0_2^+)$ channel seems to be associated particularly with the fragile structure and short lifetime of the Hoyle state. We recall that the 2_1^+ state of ^{12}C is more robust and particle stable, and it decays via γ-emission only. With a mean lifetime $\tau \approx 6 \times 10^{-14}$ s[41] the 2_1^+ state lives about 600 times longer than the Hoyle state and 10 times longer than the 2_1^+ state of ^{16}O for which the absorption enhancement has been found in the folding model study of inelastic ^{16}O+^{16}O scattering.[37] We note that such an absorption enhancement has also been found in the semi-microscopic CC analysis of the same $\alpha+^{12}$C data by Ohkubo and Hirabayashi.[19] These authors have used phenomenological Woods-Saxon (WS) potentials of the same geometry for the imaginary OP in all exit channels of the inelastic $\alpha+^{12}$C scattering and adjusted the WS depth each case to reproduce the inelastic data. From Table I of Ref.[19] one can find that the WS imaginary OP for the $\alpha+^{12}\mathrm{C}^*(0_2^+)$ channel is nearly 4 times deeper than that for the entrance $\alpha+^{12}\mathrm{C}_{\rm g.s.}$ channel, while the best-fit WS depth found for the $\alpha+^{12}\mathrm{C}^*(2_1^+)$ channel is only slightly larger than that of the $\alpha+^{12}\mathrm{C}_{\rm g.s.}$ channel. For the 3_1^- state of ^{12}C the CC calculation of Ref.[19] has shown also a sizable enhancement of absorption, in an agreement with the results of our

CC analysis using the CDJLM interaction which will be presented elsewhere. We note further that small fraction ($S_0 \approx 9\%$) of the monopole EWSR found in the DWBA analyses of inelastic ^{3}He,^{6}Li+^{12}C scattering to the Hoyle state[13,14] could well be associated with the same absorption effect as that found in the present work.

Our results of the CC description of inelastic $\alpha+^{12}$C scattering as well as those of inelastic scattering and one-neutron transfer reaction measured with the ^{16}O+^{16}O system[37,38] show clearly that there is a correlation between the weak binding and/or short lifetime of the *excited* target-like cluster and the *absorption* in the exit channel of a quasi-elastic scattering reaction. Although it is still difficult to establish a systematics for this "absorption" phenomenon, the results obtained sofar indicate the need to have a realistic choice for the OP not only in the entrance but also in the exit channel of the inelastic nucleus-nucleus scattering. The standard method of choosing the same complex OP in both the entrance and exit channels of nucleus-nucleus scattering might lead to a large uncertainty in the deduced transition strength if one simply scales the inelastic FF to match the DWBA or CC results to the measured angular distributions. In particular, this effect must be accurately taken into account in the study of the nucleus-nucleus scattering measured with unstable beams where nuclei in the entrance and exit channels are very differently bound.

In conclusion, a realistic (complex) density dependence was introduced into the M3Y-Paris interaction, based on the Brueckner Hartree Fock calculation of nuclear matter, for the folding model study of the $\alpha+^{12}$C scattering at $E_{\rm lab}$ = 104 to 240 MeV. Given an accurate estimation of the bare $\alpha+^{12}$C optical potential, our folding model analysis has shown consistently that there should be an enhancement of absorption in the exit $\alpha+^{12}$C$^*(0_2^+)$ channel due to the short lifetime and weakly bound structure of the Hoyle state, which accounts for the "missing" monopole strength of the Hoyle state observed earlier in the DWBA analysis of inelastic $\alpha+^{12}$C scattering.

Acknowledgments

The authors are grateful to Prof. W. von Oertzen, Prof. S. Ohkubo, Dr. M. Takashina and Dr. X. Chen for their helpful communications. We also thank Prof. M. Kamimura for providing us with the revised parametrization of the RGM densities from Ref.[8] The present research has been supported, in part, by Natural Science Council of Vietnam, EU Asia-Link Program CN/ASIA-LINK/008 (94791) and Vietnam Atomic Energy Commission (VAEC).

References

1. G. Hauser, R. Löhken, H. Rebel, G. Schatz, G.W. Schweimer, and J. Specht, *Nucl. Phys. A* **128**, 81 (1969).
2. J. Specht, G.W. Schweimer, H. Rebel, G. Schatz, R. Löhken, G. Hauser, *Nucl. Phys. A* **171**, 65 (1971).

3. S.M. Smith, G. Tibell, A.A. Cowley, D.A. Goldberg, H.G. Pugh, W. Reichart, N.S. Wall, *Nucl. Phys. A* **207**, 273 (1973).
4. S. Wiktor, C. Mayer-Böricke, A. Kiss, M. Rogge, P. Turek, and H. Dabrowski, *Acta Phys. Pol. B* **12**, 491 (1981); A. Kiss, C. Mayer-Böricke, M. Rogge, P. Turek, and S. Wiktor, *J. Phys. G* **13**, 1067 (1987).
5. P. Strehl, *Z. Phys.* **234**, 416 (1970).
6. M. Freer, *J. Phys. G* **34**, 789 (2007); M. Freer, contribution to this volume.
7. E. Uegaki, S. Okabe, Y. Abe, H. Tanaka, *Prog. Theor. Phys.* **57**, 1262 (1977).
8. M. Kamimura, *Nucl. Phys. A* **351**, 456 (1981); M. Kamimura, private communication.
9. A. Tohsaki, H. Horiuchi, P. Schuck, G. Röpke, *Phys. Rev. Lett.* **87**, 192501 (2001).
10. Y. Funaki, A. Tohsaki, H. Horiuchi, P. Schuck, G. Röpke, *Phys. Rev. C* **67**, 051306(R) (2003).
11. M. Chernykh, H. Feldmeier, T. Neff, P. von Neumann-Cosel, A. Richter, *Phys. Rev. Lett.* **98**, 032501 (2007).
12. Y. Funaki, A. Tohsaki, H. Horiuchi, P. Schuck, G. Röpke, *Eur. Phys. J. A* **28**, 259 (2006).
13. D. Lebrun, M. Buenerd, P. Martin, P. de Saintignon, G. Perrin, *Phys. Lett. B* **97**, 358 (1980).
14. W. Eyrich, A. Hofmann, A. Lehmann, B. Mühldorfer, H. Schlösser, H. Wirth, H. J. Gils, H. Rebel, S. Zagromski, *Phys. Rev. C* **36**, 416 (1987).
15. B. John, Y. Tokimoto, Y.W. Lui, H.L. Clark, X. Chen, D.H. Youngblood, *Phys. Rev. C* **68**, 014305 (2003).
16. G.R. Satchler, *Direct Nuclear Reactions* (Clarendon Press: Oxford, 1983).
17. W. Bauhoff, *Phys. Lett. B* **139**, 223 (1984).
18. Y. Hirabayashi, S. Ohkubo, *Phys. At. Nuclei* **65**, 683 (2002).
19. S. Ohkubo, Y. Hirabayashi, *Phys. Rev. C* **70**, 041602(R) (2004).
20. M. Takashina, Y. Sakuragi, *Phys. Rev. C* **74**, 054606 (2006).
21. S. Ohkubo, Y. Hirabayashi, *Phys. Rev. C* **75**, 044609 (2007).
22. A.M. Kobos, B.A. Brown, P.E. Hodgson, G.R. Satchler, A. Budzanowski, *Nucl. Phys. A* **384**, 65 (1982).
23. D.J. Horen, J.R. Beene, G.R. Satchler, *Phys. Rev. C* **52**, 1554 (1995).
24. D.T. Khoa, G.R. Satchler, W. von Oertzen, *Phys. Rev. C* **56**, 954 (1997).
25. D.T. Khoa, G.R. Satchler, *Nucl. Phys. A* **668**, 3 (2000).
26. G.R. Satchler, D.T. Khoa, *Phys. Rev. C* **55**, 285 (1997).
27. J.P. Jeukenne, A. Lejeune, C. Mahaux, *Phys. Rev. C* **16**, 80 (1977).
28. N. Anantaraman, H. Toki, and G.F. Bertsch, *Nucl. Phys. A* **398**, 269 (1983).
29. D.T. Khoa, W. von Oertzen, H.G. Bohlen, S. Ohkubo, *J. Phys. G* **34**, R111 (2007).
30. D.T. Khoa, *Phys. Rev. C* **63**, 034007 (2001).
31. J. Raynal, Computing as a Language of Physics (IAEA, Vienna, 1972) p.75; J. Raynal, coupled-channel code ECIS97 (unpublished).
32. C.J. Batty, E. Friedman, H.J. Gils, H. Rebel, Adv. Nucl. Phys. 19 (1989) 1.
33. H. Feshbach, *Theoretical Nuclear Physics* Volume II (Wiley-Interscience: New York, 1992).
34. W. von Oertzen, private communication.
35. P.M. Endt, At. *Data and Nucl. Data Tables* **23**, 3 (1979); F. Ajenberg-Selove, *Nucl. Phys. A* **506**, 1 (1990).
36. M. Takashina, talk presented at DREB07 Workshop, June 2007.
37. D.T. Khoa, H.G. Bohlen, W. von Oertzen, G. Bartnitzky, A. Blazevic, F. Nuoffer, B. Gebauer, W. Mittig, P. Roussel-Chomaz, *Nucl. Phys. A* **759**, 3 (2005).
38. H.G. Bohlen, D.T. Khoa, W. von Oertzen, B. Gebauer, F. Nuoffer, G. Bartnitzky, A.

Blazevic, W. Mittig, P. Roussel-Chomaz, *Nucl. Phys. A* **703**, 573 (2002).
39. J.P. Jeukenne, A. Lejeune, C. Mahaux, *Phys. Rep.* **25**, 83 (1976); *Phys. Rev. C* **16**, 80 (1977).
40. J.W. Negele, K. Yazaki, *Phys. Rev. Lett.* **47**, 71 (1981).
41. S. Raman, C.W. Nestor Jr., P. Tikkanen, At. *Data and Nucl. Data Tables* **78**, 1 (2001).

PROBING THE HIGH DENSITY BEHAVIOR OF THE SYMMETRY ENERGY

F. ZHANG*

Institute of theoretical Department , Lanzhou University,
Lanzhou, 730000, China
Institute of Modern Physics, Chinese Academy of Science, Lanzhou 730000, China

**E-mail: zhangfang@impcas.ac.cn*
www.impcas.ac.cn

W.ZUO and GAO-CHAN YONG

Institute of Modern Physics, Chinese Academy of Science, Lanzhou 730000, China
Graduate School, Chinese Academy of Science, Beijing 100039, P.R. China

E-mail: zuowei@impcas.ac.cn

Using the isospin- and momentum-dependent hadronic transport model IBUU04. Effects of the symmetry energy on π^-/π^+ ratio and neutron to proton differential flow were studied. It was found that these two observables are sensitive to the symmetry energy, especially at incident energy about per nuclei 400 MeV. We also using the neutron to proton ratio to probe the symmetry energy, and found that the effects of the symmetry energy is more notable in central neutron-rich collisions.

Keywords: Transport model, π^-/π^+ ratio, neutron to proton differential flow, neutron to proton ratio.

1. Introduction

The density dependence of nuclear symmetry energy $Esym(\rho)$ is very important for understanding many interesting questions in astrophysics,[1,2] the novel structures of radioactive nuclei[3] and many observables of heavy-ion reactions.[4,5] However, it is still very poorly known both theoretically and experimentally, especially at supra-normal densities. Fortunately, nuclear reactions induced by radioactive beams provide a unique opportunity to pin down the symmetry energy in a broad density range. In present work, based on isospin-dependent transport model calculations, we studied the effects of symmetry energy on π^-/π^+ ratio and neutron to proton differential flow, then we using neutron to proton ratio as a probe to study the density dependence of the symmetry energy for nuclear reactions induced by neutron-rich nuclei.

2. Brief introduction of the transport model IBUU04

Our study is based on the isospin- and momentum-dependent hadronic transport model IBUU04. Guided by a Hartree-Fock calculation using the Gogny effective interaction, the single nucleon potential was parameterized as[6]

$$\begin{aligned} U(\rho,\delta,\vec{p},\tau) &= A_u \frac{\rho_{\tau'}}{\rho_0} + A_l \frac{\rho_\tau}{\rho_0} \\ &+ B(\frac{\rho}{\rho_0})^\sigma (1 - x\delta^2) - 8\tau x \frac{B}{\sigma+1} \frac{\rho^{\sigma-1}}{\rho_0^\sigma} \delta\rho_{\tau'} \\ &+ \frac{2C_{\tau,\tau}}{\rho_0} \int d^3p' \frac{f_\tau(\vec{r},\vec{p'})}{1+(\vec{p}-\vec{p'})^2/\Lambda^2} \\ &+ \frac{2C_{\tau,\tau'}}{\rho_0} \int d^3p' \frac{f_{\tau'}(\vec{r},\vec{p'})}{1+(\vec{p}-\vec{p'})^2/\Lambda^2} \end{aligned} \tag{1}$$

In the above $\tau = 1/2$ $(-1/2)$ for neutrons (protons) and $\tau \neq \tau'$; $\sigma = 4/3$; $f_\tau(\vec{r},\vec{p})$ is the phase space distribution function at coordinate $\vec{r}$ and momentum $\vec{p}$. The parameters $A_u, A_l, B, C_{\tau,\tau}, C_{\tau,\tau'}$ and Λ were obtained by fitting the momentum-dependence of the $U(\rho,\delta,\vec{p},\tau)$ predicted by the Gogny Hartree-Fock and/or the Brueckner-Hartree-Fock calculations, saturation properties of symmetric nuclear matter and the symmetry energy of 30 MeV at normal nuclear matter density $\rho_0 = 0.17/fm^3$. The compressibility of symmetric nuclear matter K_0 is set to be 211 MeV. The momentum-dependence of the symmetry potential steams from the different interaction strength parameters $C_{\tau,\tau'}$ and $C_{\tau,\tau}$ for a nucleon of isospin τ interacting, respectively, with unlike and like nucleons in the background fields,and more $C_{unlike} = -103.4$ MeV while $C_{like} = -11.7$ MeV.

The parameters $A_u(x)$and $A_l(x)$ depend on the x parameter according to $Au(x) = -95.98 - x\frac{2B}{\sigma+1}$ and $A_l(x) = -120.57 + x\frac{2B}{\sigma+1}$.The parameter x was introduced to mimic predictions on the density dependence of the symmetry energy $Esym(\rho)$ by microscopic and/or phenomenological many-body theories. The symmetry energy as a function of density was shown in Fig. 1.

3. Results and discussions

3.1. *Beam energy dependence of the π^-/π^+ ratio*

We know most of the pion mesons come from delta resonances in intermediate energy heavy-ion collisions, the contribution of N^* is negligible,the direct pion production is also neglected in the present calculations due to its small cross sections and large theoretical uncertainty. Fig. 2 shows the transverse momentum distribution of the π^-/π^+ ratio at different beam energies in the reaction of $^{132}Sn +^{124} Sn$ with the parameter $x = 1$ and $x = -1$ in semi-central collisions. It was seen the π^-/π^+ ratio is rather sensitive to the symmetry energy. The magnitude of π^-/π^+ ratio obtained by using the soft energy ($x = 1$) than that by using the stiff energy ($x = -1$). It's interesting to notice that the plot show a prominent feature, i.e. the ratios show a peak at about 150 MeV/c. These peaks are so-called Coulomb peak. One can

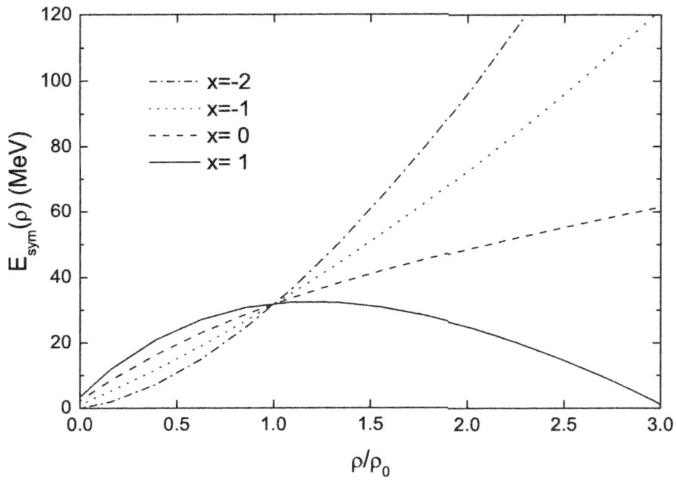

Fig. 1. Density dependence of nuclear symmetry energy with two parameter x settings in Eq.1

seen that the positions of the Coulomb peaks remains nearly unchanged with the increasing of the beam energies. From Fig. 2 we also can find for the two beam energies E_{baem}/A = 400MeV and 600 MeV, the π^-/π^+ ratio are quite sensitive to the symmetry energy, especially around Coulomb peaks. However, as the beam energy increases from E_{beam}/A =600 MeV to 800 MeV, the π^-/π^+ ratio become less sensitive to the symmetry energy. This is because the coulomb potential among protons may leads to the matter in reaction region more neutron-rich. Whereas the effects of the symmetry potential, which is repulsive to neutrons and attractive to protons, is to make the reaction regions less neutron-rich. The competition between the above-mentioned two effects results in lowering of Coulomb peaks and the reduction of the difference between the π^-/π^+ ratios by using the two different symmetry energies as the beam energy goes from 600/A MeV to 800/A MeV.

3.2. *Beam energy dependence of the neutron to proton differential flow*

It is well known that the collective flow is also sensitive to the symmetry energy. In present work we use the neutron to proton differential flow to probe the symmetry energy, which has a virtue that minimize influences of the isoscalar potential but maximize the effects of the symmetry energy potential, this type of nucleon flow combines constructively contributions of the symmetry potential to the collective flow of both neutrons and protons.

Fig. 3 presents the reduced rapidity distribution of neutron to proton differential flow for the reactions of neutron-rich $^{132}Sn+^{124}Sn$ system with three different beam energies in semi-central collisions. Because of the symmetry energy is repulsive to neutrons while respect to protons. The stiff symmetry energy ($x = -1$)

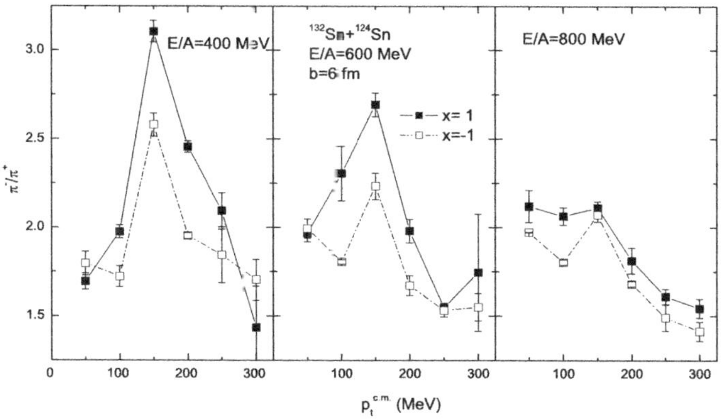

Fig. 2. Kinetic energy distribution of the charged pion mesons with the two symmetry energies in the reactions of $^{132}Sn + ^{124}Sn$ and $^{112}Sn + ^{112}Sn$ at $E_{beam}/\mathrm{A} = 400$ MeV and an impact parameter of 6 fm.

gives a higher magnitude of the neutron to proton differential flow as one expected. At higher beam energies, the magnitude of the neutron to proton differential flow increases because of the dominating role of the collisions and their fast growth with beam energy.[7] From Fig. 3, it's also found the effect of the symmetry energy decreased with the beam energy. This is because the role of the nucleon-nucleon collisions is important while the effects of the symmetry potential reduced at high incident energies. Thus the difference between the neutron to proton differential flow by adopting two symmetry energy decreased at high beam energy.

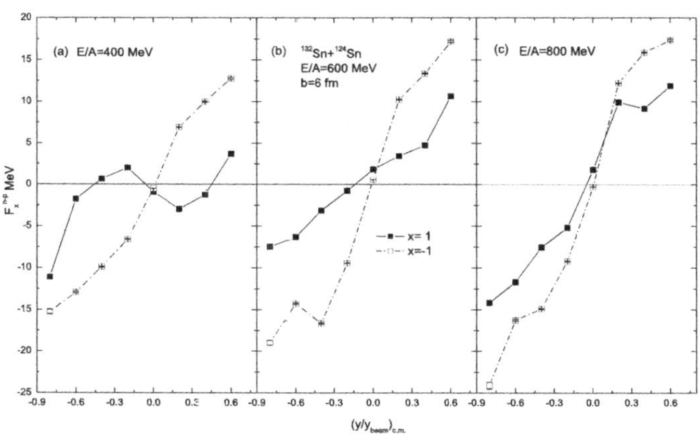

Fig. 3. Neutron to proton differential flow as a function of reduced rapidity by adopting two symmetry energies in the same reactions as in Fig. 2.

3.3. *Impact parameter and isospin dependence of the neutron to proton ratio*

We studied the reaction of $^{132}S_n +^{124}S_n$ system with various impact parameters at beam energy per nucleon 400 MeV. Shown in Fig. 4 is the neutron to proton ratio as a function of kinetic energy by adopting two symmetry energy $x = 0$ and $x = -1$ in central and semi-peripheral collisions.

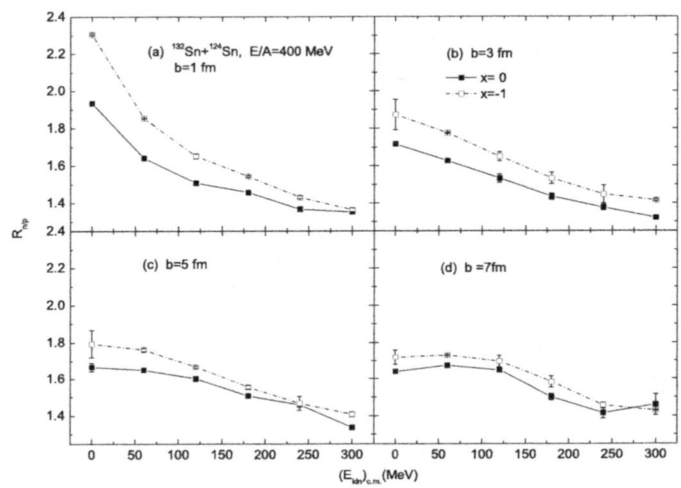

Fig. 4. Neutron to proton ratio as a function of kinetic energy with different impact parameter at incident energy $E/A = 400$ MeV by adopting two symmetry energy

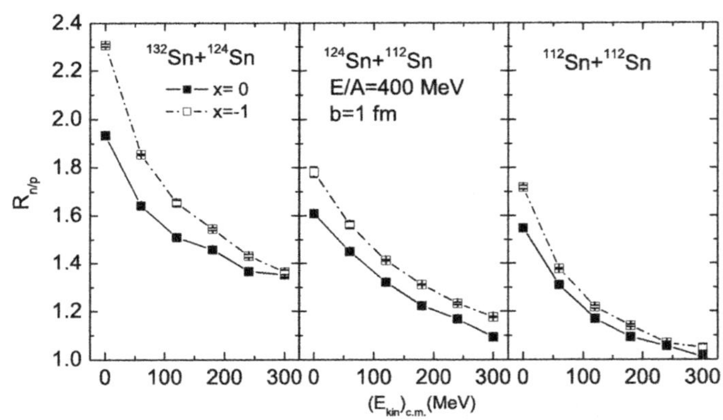

Fig. 5. Kinetic energy distributions of neutron to proton ratio in different reactions stages in the reaction at $E_{beam}/\text{A} = 400$ MeV and an impact parameter of 1 fm.

From Fig. 4 it's seen that a higher neutron/proton ratio of free nucleons with stiff symmetry energy of $x = -1$ due to stronger repulsive(attractive) symmetry potential for neutrons(protons). The rise of the neutron/proton ratio at low kinetic energies due to the Coulomb force which pushes protons away from low energy to high energy. We also can find the sensitivity of this ratio to the symmetry energy is more evidence in central heavy ion collisions. With the increasing of the impact parameter, the sensitivity of the neutron to proton ratio to the symmetry energy reduced. This because a very useful concept in heavy ion physics is that of spectators and participants. In the reaction region the participants are more in central heavy ion collisions than in semi-peripheral collisions. Therefor, the maximum density the system can be reached is more high in central collisions. Thus the effects of the density dependent symmetry energy is more notable in central collision.

We examine in Fig. 5 the kinetic energy distribution of the neutron/proton ratio dN_n/dN_p in the two symmetry energies case of in three reaction systems. It is interesting to notice that the sensitivity to of neutron the proton ratio to the symmetry energy increased with the isospin asymmetry in the system. This is because the symmetry potential is isospin asymmetry dependent, It implies the larger isospin asymmetry is, the stronger symmetry effect is. For the more asymmetric system^{132}Sn+^{124}Sn, effects of symmetry energy are more prominent than the less asymmetry system ^{124}Sn+^{112}Sn due to larger isospin asymmetry in the system, for the more symmetric system effects of the symmetry energy are almost can negligible because of small isospin asymmetry in the system. Therefor, the measurement of the ratio $dN_n/dN_p(E_{kin})$ in neutron-rich, central heavy-ion collisions for nucleons is suitable for probing the high density behavior of symmetry energy.

4. Summary

In summary, using IBUU04 transport model, effects of beam energy on symmetry energy was studied by using π^-/π^+ ratio and neutron to proton differential flow. We find that sensitivity of these to observables to the symmetry energy reduced as the increasing of the beam energy. We also studied the effects of impact parameter and isospin on symmetry energy by using neutron to proton ratio. It was showed the maximum symmetry energy effects in central neutron-rich collisions.

References

1. H.A. Bethe, *Rev. Mod. Phys.* **62**, 801 (1990).
2. J.M. Lattimer *et al.*, *Astr. Phys. Jour.* **550**, 426 (2001).
3. R.J. Furnstahl, *Nucl. Phys. A* **706**, 85 (2002).
4. P.Danielewicz, R. Lacey and W.G. Lynch, *Science* **298**, 1592 (2002).
5. V.Baran, M.Colonna, M.Di.Toro *et al.*, *Phys. Rev. C* **72**, 064620 (2005).
6. C. B. Das *et al.*, *Phys. Rev. C* **67**, 034611 (2003).
7. B. A. Li , *Nucl. Phys. A* **708**, 365-390 (2002).

MICROSCOPIC CALCULATIONS BASED ON A SKYRME FUNCTIONAL PLUS THE PAIRING CONTRIBUTION

JUN LI[1,2*] GIANLUCA COLÒ[2†] and JIE MENG[1‡]

1 School of Physics, Peking University, Beijing, 100871, China
2 Dipartimento di Fisica, Università degli Studi and INFN Sez. di Milano, 20133, Milano, Italy
*E-mail: *jun.li@mi.infn.it, †gianluca.colo@mi.infn.it, ‡mengj@pku.edu.cn*

A self-consistent Quasiparticle Random-Phase-Approximation (QRPA) model which employs the canonical Skyrme-Hartree-Fock-Bogoliubov (HFB) basis, is used to study the monopole collective excitations of tin isotopes. The influence of the spurious state on the strength function of the isoscalar monopole excitations is clearly assessed. We compare the effect of different kinds of pairing forces (volume pairing, surface pairing and mixed pairing) on the monopole excitation strength function. The energy of the Isoscalar Giant Monopole Resonance (ISGMR) is calculated for tin isotopes with the Skyrme force SLy5 and the volume pairing force, and compared with the results of Random-Phase-Approximation (RPA).

Keywords: Hartree-Fock-Bogoliubov; Quasiparticle Random-Phase-Approximation; Isoscalar Giant Monopole Resonance.

1. Introduction

The nuclear structure community have made great achievements in understanding the structure of the ground state and of the excited states of stable atomic nuclei. Moreover, these has been a recent blooming both on the experimental and theoretical side of the studies concerning the exotic properties of the nuclei far from the stability valley. A better knowledge of the Hamiltonian or density functional is also relevant for exotic systems as well as to solve open problems for ordinary (not exotic) nuclear matter. One of the open problems concerns the nuclear incompressibility which is related to the monopole excitation energy.

Among the various approaches for calculating the monopole excitation energy are: 1) Standard RPA method based on stable Hartree-Fock (HF) mean field solutions.[1,2] 2) QRPA method based on the HF mean field and Bardeen-Cooper-Schriffer (BCS)[3] pairing field solutions.[1,2] 3) QRPA method on top of the HFB basis, especially on top of canonical Skyrme-HFB basis.[4] Recently, many works appeared in which the HF-RPA,[5,6] the HF-BCS-QRPA,[7–9] and the HFB-QRPA,[4] have been used to study the nuclear collective excitations. Moreover, the relativistic mean field (RMF)-RPA[10,11] or RMF-BCS-QRPA methods[12] also have been applied to the analysis of the monopole excitation. If the density functionals are characterized by a nuclear incompressibility K_∞ around 230~240 MeV (or 250~270 MeV in the case of RMF), they give the right ISGMR centroid energies compared with the experimental data in ^{208}Pb.[9,10] However, most of them also overestimate the centroid energies in the tin isotopes.[10,11] The questions arise about why is tin soft? And whether there is need of more detailed considerations on the nuclear pairing in tin? To understand the second question, in the present work we study the ground state properties of the tin isotopes using the HFB theory with a Skyrme mean field and different kinds of pairing force, and we analyze the problems of the ISGMR by means of a self-consistent QRPA calculation.

The paper is organized as follows. Section II discusses the HFB-QRPA employed in this study. Since the QRPA equation is solved on the canonical HFB basis, in this section we also present the canonical HFB theory in brief. In section III, we give the numerical details and the results. In this section, we also discuss the method for projecting out the spurious state. In some detail, we study the energy of ISGMR in tin isotopes. The last section contains a short summary.

2. QRPA with canonical HFB basis

The standard approach for describing collective excitations in open-shell nuclei is the so-called QRPA on top of stable solutions, i.e., on top of HFB solutions. Therefore, the first step in solving the self-consistent QRPA is to obtain the ground state properties by solving the HFB equations. Since the HFB theory is discussed in detail in many textbooks and articles,[1,13,14] here we just present briefly the HFB theory and the features of the canonical basis.

Starting from the effective Hamiltonian of the nuclear system, and using the variational principle, we can obtain the HFB equations in the coordinate representation (Eqs. (4.6) of Ref.[14]). Solving the HFB equations in coordinate representation, we obtain the quasiparticle wave functions and quasiparticle energies. According to the Eqs. (4.7a) of Ref.,[14] we can also obtain the density matrices. Diagonalizing the density matrices, we get the canonical basis. Since the canonical basis wave functions are consequently localized, it is instructive to consider the HFB equations in this particular basis. They read

$$(h-\lambda)_{\mu\nu}(u_\mu v_\nu + u_\nu v_\mu) + \tilde{h}_{\mu\nu}(u_\mu u_\nu - v_\nu v_\mu) = 0, \tag{1}$$
$$(h-\lambda)_{\mu\nu}(u_\mu u_\nu - v_\nu v_\mu) - \tilde{h}_{\mu\nu}(u_\mu v_\nu + u_\nu v_\mu) = E_{\mu\nu}, \tag{2}$$

where h and $\tilde{h}$ are the particle-hole and particle-particle Hamiltonian in the canonical basis, respectively. λ is the Fermi energy, and $E_{\mu\nu}$ represents the HFB Hamiltonian in the canonical basis. The square of v_μ and u_μ represent the probability that a certain state μ is occupied or empty and are solely determined by the diagonal matrix elements of particle-hole and particle-particle Hamiltonians, $v_\mu = -\mathrm{sgn}(\tilde{h}_{\mu\mu})\sqrt{\frac{1}{2} - \frac{h_{\mu\mu}-\lambda}{2E_{\mu\mu}}}, \quad u_\mu = \sqrt{\frac{1}{2} + \frac{h_{\mu\mu}-\lambda}{2E_{\mu\mu}}}$.

The QRPA is aimed at the description of small amplitude, collective excitations. Defining the phonon excitation operators in the angular momentum coupled representation,

$$Q_\nu^\dagger(JM) = \sum_{\alpha\geq\beta} \frac{1}{\sqrt{1+\delta_{\alpha\beta}}} \sum_{m_\alpha m_\beta} \langle j_\alpha m_\alpha j_\beta m_\beta | JM\rangle (X_{\alpha\beta}^\nu \alpha_\alpha^\dagger \alpha_\beta^\dagger - Y_{\alpha\beta}^\nu \alpha_{\bar\beta}\alpha_{\bar\alpha}), \quad (3)$$

where $\alpha_\alpha^\dagger$ and α_α are the quasiparticle creation and annihilation operators in canonical basis, respectively. The QRPA equation can be easily obtained from the linearization of Schrödinger equation,

$$\sum_{\alpha<\beta} \begin{pmatrix} A_{\alpha\beta,\gamma\delta} & B_{\alpha\beta,\gamma\delta} \\ -B_{\alpha\beta,\gamma\delta}^* & -A_{\alpha\beta,\gamma\delta}^* \end{pmatrix} \begin{pmatrix} X_{\gamma\delta}^\nu \\ Y_{\gamma\delta}^\nu \end{pmatrix} = E_\nu \begin{pmatrix} X_{\alpha\beta}^\nu \\ Y_{\alpha\beta}^\nu \end{pmatrix}. \quad (4)$$

Actually, in the canonical basis, the the matrix elements $A_{\alpha\beta,\gamma\delta}$ and $B_{\alpha\beta,\gamma\delta}$ in the equations of HFB-QRPA are almost same as those of BCS-QRPA, except for the presence of off-diagonal terms of the unperturbed HFB Hamiltonian in the former case.[15]

3. Results and discussions

In this work, in the particle-hole channel, we use a Skyrme force, i.e., SLy5,[16] and in the particle-particle channel, we use different kind of zero-range pairing forces (volume pairing, surface pairing and mixed pairing). The zero-range density-dependent pairing force

$$v(\mathbf{r}_1, \mathbf{r}_2) = v_0 \left[1 - \eta \left(\frac{\rho(\frac{\mathbf{r}_1+\mathbf{r}_2}{2})}{\rho_0}\right)^\alpha\right] \delta(\mathbf{r}_1 - \mathbf{r}_2), \quad (5)$$

where $\alpha = 1$, $\rho_0 = 0.16$ fm^{-3}; the values of η are 0, 1, and 0.5 for volume, surface and mixed pairing forces, respectively; the value of v_0 is fixed by fitting the experimental neutron gap of ^{120}Sn (Δ_n = 1.32 MeV). In the case of the Skyrme force SLy5, v_0= -170.92, -537.05 and -272.99 MeV fm^3 for volume, surface and mixed pairing forces, respectively. There are some numerical parameters in the actual calculations: (i) We use a high-quasiparticle-energy cutoff (200 MeV) and a maximum angular momentum $j_{max} = 15/2$. (ii) The radius of box is fixed at 20 fm with a small mesh (0.05 fm).

The self-consistency of the HFB+QRPA calculations requires the use in QRPA of a residual force derived from the HFB fields. At the same time for a real self-consistent calculation, all the quasiparticle states produced by the HFB calculation must be used to build the matrices A and B in the QRPA equation. In actual calculations, there are about 500-600 quasiparticle states in the HFB calculation up to 200 MeV which is the HFB cutoff. Some states with very small values of occupation probability in canonical basis or some two quasiparticle excitations with very high energy give little contribution to the QRPA spectrum, so we cut off the canonical-basis wave functions by excluding very small values of occupation probability and very high values of single particle energy, and check the convergence of the QRPA solution. Fig. 1 shows the isoscalar 0^+ strength function (averaged

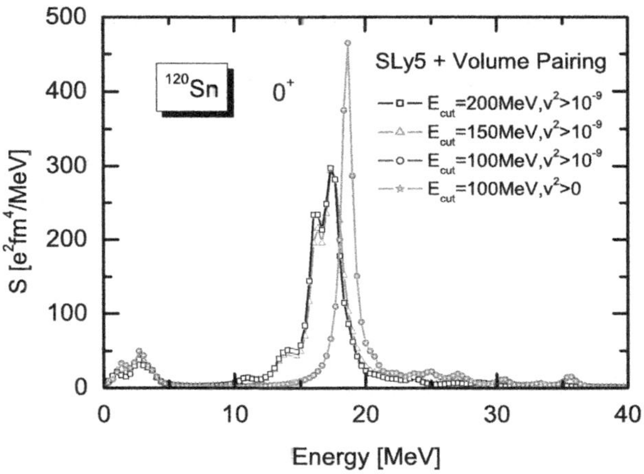

Fig. 1. (Color online)The isoscalar 0^+ strength function (averaged with Lorentzians having 1 MeV width) in ^{120}Sn with Skyrme force SLy5 and the volume pairing force for different values of the single particle energy cutoff and the occupation probabilities.

with Lorentzians having 1 MeV width) in ^{120}Sn with the Skyrme force SLy5 and the volume pairing force defined above, for different values of cutoff in single particle energy and occupation probability. Comparing the curve with circles with the curve with stars in Fig. 1, we find that the levels with occupation probabilities smaller than 10^{-9} have little influence on the isoscalar 0^+ strength function. Similarly, from the curves with squares, up-triangle, and circles in Fig. 1, we find $E_{cut} = 150 \sim 200$ MeV is a suitable single particle energy cutoff in calculations to make our results stable within 300 KeV of the centroid energy. Therefore, in the calculations which are shown below, the value of the cutoff on occupation probability is fixed at 10^{-9}, and the single particle energy cutoff is selected as 200 MeV.

In the spherical QRPA solution, there exists a problem of the spurious state in the monopole channel due to the symmetry broken by the HFB solution.[1] The spurious state should be orthogonal to all other physical states, and should appear at zero energy. However, in actual calculations the spurious state is at low but not zero energy for numerical inaccuracies, and contributes to the total strength. Therefore, the spurious state must be projected out from the real physical states. In this work, we use the same method as in Ref.[17] to project out the spurious state.

In the case of the isoscalar monopole excitation, we deal with the operator which does not carry any angular momentum and spin, $F^{IS}_{monopole} = \sum_{i=1}^{A} r_i^2$. The macroscopic picture for a state excited by this operator is so-called "breathing mode". The strength function related to this operator is defined as $S(E) = \sum_n |\langle n|F^{IS}_{monopole}|0\rangle|^2 \delta(E - E_n)$. The moments of the strength function, or sum rules, are $m_k = \int E^k S(E) dE$, and the centroid energy can be defined as the ratio between the energy-weighted sum rule m_1 and the non-energy weighted sum rule m_0: $E_0 = m_1/m_0$.

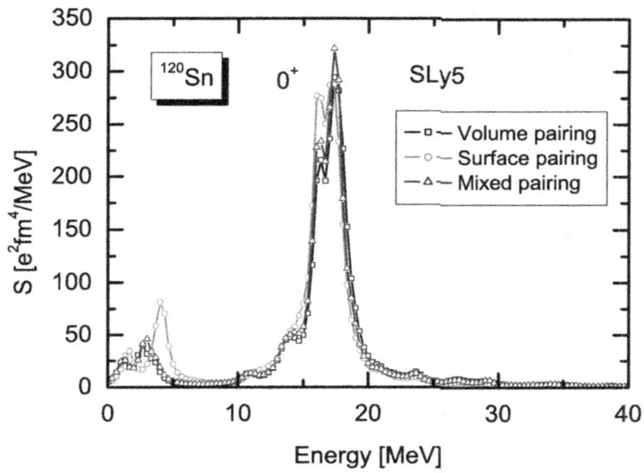

Fig. 2. (Color online)The isoscalar 0^+ strength function (averaged with Lorentzians having 1 MeV width) in ^{120}Sn with the Skyrme force SLy5 for different kinds of pairing force.

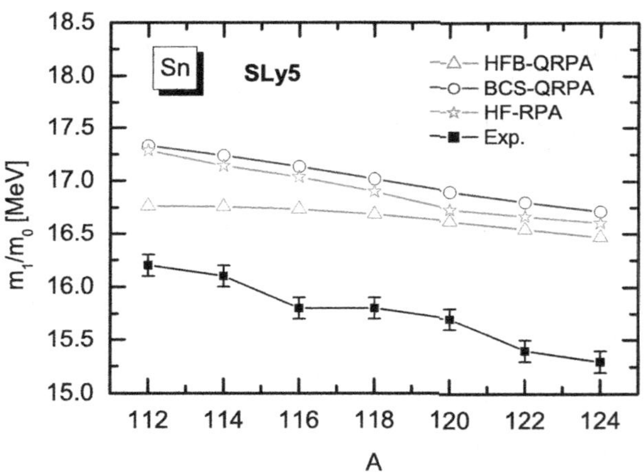

Fig. 3. (Color online)Systematics of the moment ratios m_1/m_0 for the ISGMR strength distributions in the tin isotopes as a function of mass number. They are evaluated in the energy interval between 10.5 and 20.5 MeV. The experimental data are extracted from the Ref.[18]

Fig. 2 shows the isoscalar 0^+ strength function (averaged with Lorentzians having 1 MeV width) in ^{120}Sn with Skyrme force SLy5 for different kinds of pairing force. The results with volume and mixed pairing force are almost the same, while the results with surface pairing force is different from the others, especially in the low energy region.

Fig. 3 shows the centroid energy in the tin isotopes with the Skyrme force SLy5 and the volume pairing force. For the sake of companion, we also give the results of HF-RPA and HF-BCS-QRPA. The moment ratios are evaluated in the energy interval between 10.5 and 20.5 MeV, and the experimental data come from Ref.[18] HF-RPA and HF-BCS-QRPA overestimate the centroid energy about 1 MeV in tin isotopes. One possible guess that theoretical models overestimate the centroid energies in tin isotopes is that tin is more sensitive to pairing correlation and it is need of a more appropriate method to deal with the pairing correlation than HF-BCS plus QRPA, i.e., of a HFB-QRPA. As shown in Fig. 3, HFB-QRPA makes the theoretical results significantly close to experiment, especially in ^{112,114,116}Sn. However, the trend of the energies is not as in experiment.

4. Conclusions

In summary, we have developed a fully self-consistent QRPA on top of HFB. The model uses canonical basis and it has been applied to the isoscalar monopole excitation. Since the particle number symmetry is broken by the HFB approximation, and this results in the problem of the spurious state, we project out the spurious state from real physical states completely. We check the influence of pairing correlations on the isoscalar 0^+ strength function and the centroid energy in tin isotopes. Pairing improves the absolute values of energy compared to experiment in isotopes the ^{112,114,116}Sn. However, the theoretical results are still overestimated with respect to experiment. Possible reasons are presently under study.

References

1. P. Ring and P. Schuck, *The Nuclear Many-Body Problem* (Springer-Verlag, Berlin, 1980).
2. D. J. Rowe, Nuclear collective motion (Methuen, London, 1970).
3. L. Cooper, J. Bardeen, and J. Schrieffer, *Phys. Rev.* **108**, 1175 (1957).
4. J. Terasaki *et al.*, *Phys. Rev. C* **71**, 034310 (2005); J. Terasaki and J. Engel, *Phys. Rev. C* **74**, 044301 (2006).
5. B. K. Agrawal, S. Shlomo, and A. I. Sanzhur, *Phys. Rev. C* **67**, 034314 (2003); B. K. Agrawal and S. Shlomo, *Phys. Rev. C* **70**, 014308 (2004).
6. T. Sil *et al.*, *Phys. Repv. C* **73**, 034316 (2006).
7. G. Colò *et al.*, *Phys. Rev. C* **70**, 024307 (2004); G. Colò *et al.*, *Nucl. Phys. A* **722**, 111c-116c (2006).
8. S. Fracasso and G. Colò, *Phys. Rev. C* **72**, 064310 (2005).
9. N. Paar *et al.*, *Rep. Prog. Phys.* **70** 691 (2007), and references therein.
10. J. Piekarewicz, *Phys. Rev. C* **76**, 031301(R) (2007).
11. H. Sagawa *et. al.*, nucl-th/07060966 (2007).
12. Cao Li-Gang and Ma Zhong-Yu, *Chin. Phys. Lett.* **21**, 810 (2004).
13. W. Nazarewicz, T. R. Werner, and J. Dobaczewski, *Phys. Rev. C* **50**, 2860 (1994).
14. J. Dobaczewski *et al.*, *Phys. Rev. C* **53**, 2809 (1996).
15. Jun Li, Gianluca Colò, and Jie Meng, To be published.
16. E. Chabanat *et al.*, *Nucl. Phys. A* **643**, 441 (1998).
17. G. Colò *et al.*, *Phys. Lett. B.* **485**, 362 (2000)
18. T. Li, U. Garg *et al.*, To be published.

IN-MEDIUM CROSS SECTIONS IN DIRAC-BRUECKNER-HARTREE-FOCK APPROACH*

LUO PEIYAN and ZUO WEI

Institute of Modern Physics, Chinese Academy of Sciences, Lanzhou, 730000, P.R.C.

U.LOMBARDO

Dipartimento di Fisica e Astronomia, viale Andrea Doria6,95125 Catania,Italy

We calculate the in-medium nucleon-nucleon scattering cross sections from the G-matrix in the framework of the Dirac-Brueckner-Hartree-Fock (DBHF) approach. With the two representations of the G-matrix, the differential and total cross sections are different, which are mainly from the different nucleon effective masses.

1. Introduction

The transport models, which are employed to describe the central reactions in heavy ion collisions(HIC), are powerful approaches to reproduce the collective observable in HIC and probe the equation of state(EOS) of nuclear matter.[1-3] Those models, such as the quantum molecular dynamics (QMD) and Boltzmann-Uehling-Uhlenbeck (BUU) equation as well as their relativistic extensions (RBUU and RQMD), treat the nucleon mean fields and nucleon-nucleon scattering cross sections as the basic input ingredients. Up to now, most caculations of particle producing in nucleus-nucleus reaction are using the free cross sections or the approximations $\sigma^*_{NN} = 0.8\sigma^{free}_{NN}$, $\sigma^*_{NN}/\sigma^{free}_{NN} = (m^*/m)^2$ instead of the in-medium cross sections.[4-6]

Many recent papers have devoted to the in-medium N-N scattering problem within different methods, such as the collectivistic Brueckner approach,[7-9] and Dirac-Brueckner-Hartree-Fock(DBHF) approach.[10,11] Here we apply the DBHF approach for its successfully describing the mechanism of the EOS, which meets the empirical region of saturation.[12] However, because of the uncertainty in determining the nucleon self-energy in DBHF approach, Horowitz and Serot[13] have developed a projection technique in which projects the G-matrix onto five covariant amplitudes. This set of five covariant amplitudes is not unique, corresponding to different representations, and we adopt two different representations (pseudo-scalar(ps) and complete pseudo-vector(pv) representations).

*This work is partially supported by Asia-Europe Link Program [CN/ASIA-LINK/008 (094-791)].

2. Relativistic Brueckner approach

In the relativistic Brueckner approach, the essential point is using Dirac equation to describe the single-particle motion in the nuclear matter.

$$[\gamma_\mu \widetilde{k}^{*\mu} - \widetilde{M}^*]u_\lambda(k) = 0 \tag{1}$$

by introducing the reduced kinetic momentum $\widetilde{k}^{*\mu} = \frac{k^\mu + Re\Sigma^\mu(k)}{1+\Sigma_v(k)}$ and the reduced effective mass $\widetilde{M}^* = \frac{M+Re\Sigma_s(k)}{1+\Sigma_v(k)}$, the scalar Σ_s and vector $\Sigma^\mu = (\Sigma_0, k\Sigma_v)$ are the three components of the self-energy. The solution of Dirac equation is

$$u_\lambda(k) = \sqrt{\frac{\widetilde{E}^*(\vec{k}) + \widetilde{M}^*}{2\widetilde{M}^*}} \begin{pmatrix} 1 \\ \frac{2\lambda|\vec{k}|}{\widetilde{E}^*(\vec{k})+\widetilde{M}^*} \end{pmatrix} \chi_\lambda \tag{2}$$

where $\widetilde{E}(\vec{k}) = \sqrt{\vec{k}^2 + \widetilde{M}^{*2}}$, χ_λ denotes a two-component Pauli spinor with $\chi = \pm\frac{1}{2}$. The normalization of Dirac spinor is $\widehat{u}_\lambda(k, k_F)u_\lambda(k, k_F) = 1$. The basic quantity in DBHF approach is the G-matrix, satisfying the Thompson equation which is most easily solved in the two-nucleon c.m. frame,[10,14]

$$G(\vec{p}, \vec{q}, \vec{P})|_{c.m.} = V(\vec{p}, \vec{q}) + \int \frac{d^3\vec{k}}{(2\pi)^3} \frac{M^{*2}}{E^{*2}(\vec{k})} \frac{Q(\vec{k}, \vec{P})}{2E^*(\vec{q}) - 2E^*(\vec{k}) + i\epsilon} G(\vec{k}, \vec{q}, \vec{P}) \tag{3}$$

where $\vec{q}$, $\vec{k}$, $\vec{p}$ are the relative momenta of the initial, intermediate and final states, respectively, $\vec{P}$ is the total center of mass(c.m.) momentum. The starting energy in Eq.(3)is fixed to $\sqrt{s^*} = 2E^*(\vec{q})$. The Pauli operator Q, preventing the nucleons scattering onto occupied intermediate states, usually is replaced by an angle-averaged Pauli operator $\overline{Q}$ for simplicity. In order to calculate the components of the self-energy in rest frame from the G-matrix in c.m. frame, we need to construct the G-matrix covariantly using the ps and pv representations. Thus starting from reasonable initial values for every part of self-energy, one obtains the G-matrix by solving the in-medium Thompson equation in momentum space, then leading to a new set of values for self-energy to be used in the next iteration; this procedure is continued until convergence is achieved.

3. On-shell in-medium nucleon-nucleon cross section

For on-shell nucleon nucleon interaction, there are only five independent G-matrix elements, one can directly calculate the cross sections from those elements[15,16] ($|\vec{p}| = |\vec{q}|$). The differential cross section is given as,[11]

$$d\sigma = \frac{(M^*)^4}{s^* 4\pi^2} |\widehat{G}(q, q, \theta)|^2 d\Omega \tag{4}$$

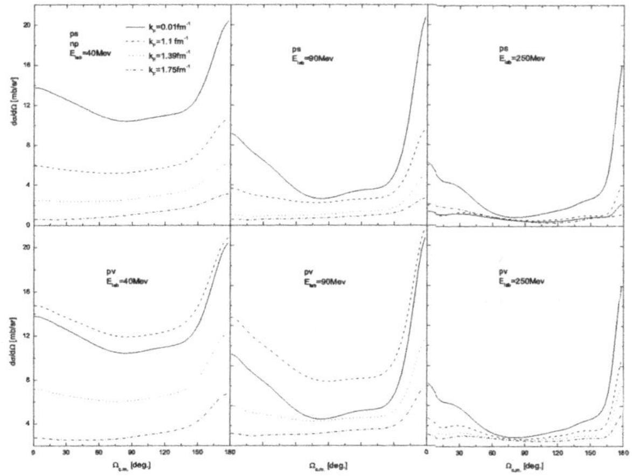

Fig. 1. In-medium np differential cross sections obtained at various laboratory energies and densities are functions of angles. The upper three with ps representation and the lower three with pv representation.

and

$$|\widehat{G}(q,q,\theta)|^2=\sum_{i=1}^{6}\beta_i[(\sum_J\frac{2J+1}{4\pi}d^J_{\lambda_i\lambda_i'}(\theta)ReG_i^J(q,q))^2+(\sum_J\frac{2J+1}{4\pi}d^J_{\lambda_i\lambda_i'}(\theta)ImG_i^J(q,q))^2] \tag{5}$$

The weighting factors β_i =2, i=1, ..., 4, and $\beta_5=\beta_6=4$ arise from the sum over all helicity states. Using the orthogonality relation for the rotation matrices,

$$\int d\cos(\theta)d^J_{\lambda_i\lambda_i'}(\theta)d^{J'}_{\lambda_i\lambda_i'}(\theta)=\frac{2}{2J+1}\delta_{JJ'} \tag{6}$$

We can get the total cross section

$$\sigma_{tot}=\int d\Omega\frac{(M^*)^4}{s^*4\pi^2}|\widehat{G}(q,q,\theta)|^2=\frac{(M^*)^4}{s^*4\pi^2}\sum_{i=1}^{6}\beta_i\sum_J\frac{2J+1}{4\pi}([ReG_i^J(q,q)]^2+[ImG_i^j(q,q)]^2]) \tag{7}$$

where $G_i^J=0.5(G_i^{J,I=0}+G_i^{J,I=1})$ and $G_i^J=G_i^{J,I=1}$ are the average neutron-proton and proton-proton on-shell G-matrix elements respectively.

4. Numerical results and conclusions

In this section we present the numerical results to investigate the influence of these two representations of the G-matrix to the cross sections. For the bare nucleon-nucleon potential we adopt Bonn A. First we calculate the G-matrix in the c.m. frame of the two interaction nucleons, i.e., we use Eq.(3) with $\overrightarrow{P}=0$. The relative momentum q is related to the kinetic energy of the incident nucleon in the "laboratory system" $E_{lab}=2q^2/m$, in which the other nucleon is at rest. In the following we consider the in-medium cross sections at three different Fermi momenta k_F =1.1,

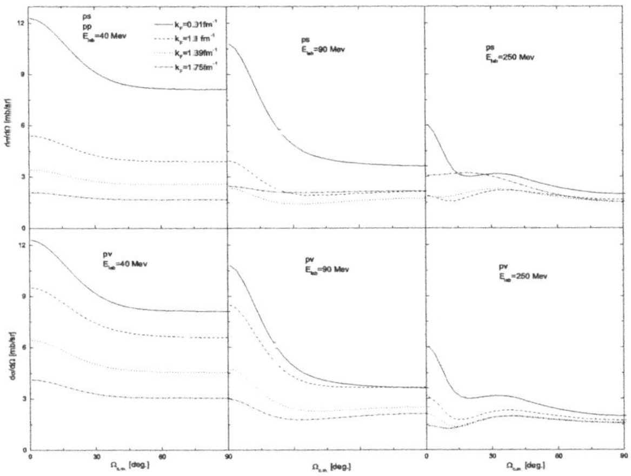

Fig. 2. In-medium pp differential cross sections, same as Fig.1.

1.39 and 1.75 fm^{-1}, corresponding to densities ρ =0.09, 0.18 and 0.36 fm^{-3} with the nuclear saturation density ρ_0 =0.18 fm^{-3}.

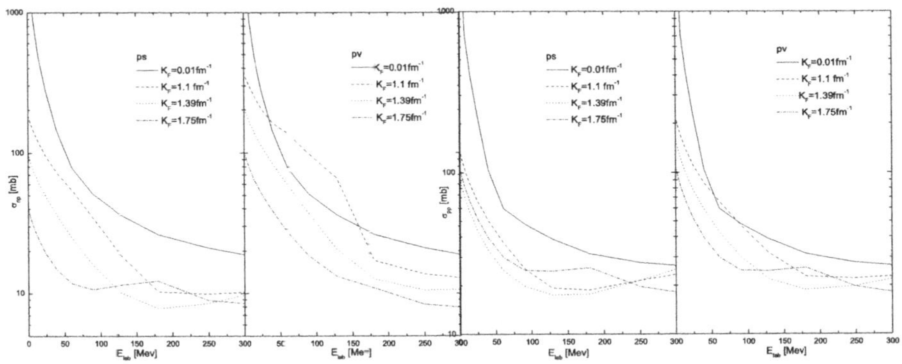

(a) (b)

Fig. 3. (a)In-medium np total cross sections are functions of laboratory energies, left one with ps representation, right one with pv representation. (b) In-medium pp total cross sections with two representations, same as (a).

In Fig.1, the differential np cross sections at different densities are shown at E_{lab} =40, 90 and 250 Mev with the two representations of the G-matrix for comparison. The vacuum results are also presented. With the ps presentation, the np differential cross section tends to decrease with the increasing of density at fixed laboratory energy and also decrease with the increasing of laboratory energy at fixed density. With the pv presentation, the results are similar except that at the density $\frac{1}{2}\rho_0$ the lines are greatly enhanced at low energies. We also present the pp differential cross section in Fig.2, the results of which are much more isotropic compared with Fig.1 with two representations. At last, we give the results of the np and pp total cross sections as functions of laboratory energies in Fig.3. For np channel, the total cross section with ps representation decreases with the increasing of density, while with pv representation there is a bomb at moderate energy and density $\frac{1}{2}\rho_0$; for pp channel, the bomb is not evident. For every figure, the result with pv representation is less changed with density or laboratory energy than that with ps representation, which is mainly because of the difference of effective masses with these two representations, at different Fermi momenta $k_F = 1.1, 1.39$and $1.75\ fm^{-1}$, with ps representation M^* =592.4, 479.2, 339.0 Mev, while with pv representation M^* =766.6, 660.2, 527.6 Mev.

References

1. M. Demoulins, D. L'Hôte, J. P. Alard *et al.*, *Phys. Lett. B* **241**, 476 (1990).
2. B. A. Li and A. Sustich, *Phys. Rev. Lett.* **82**, 5004 (1999).
3. C. Hartnack, H. Oeschler and J.Aichelin, *Phys. Rev. Lett.* **96**, 012302 (2006).
4. George F. Bertsch, Gerald E. Brown, Volker Koch and Bao-An Li, *Nucl. Phys. A* **490**, 745 (1988)
5. Declan Persram and Charles Gale, *Phys. Rev. C* **65**, 064611 (2002)
6. Bao-An Li, Lie-Wen Chen, *Phys. Rev. C* **72**, 064611 (2005)
7. G. Giansiracusa, U. Lombardo and N.Sandulescu, **Phys. Rev. C 53**, R1478 (1996).
8. J. Jaenicke, J. Aichelin, N. Ohtsuka, R. Linden and A. Faessler, *Nucl. Phys. A* **536**, 201 (1992).
9. T. Alm, G. Röpke and M. Schmidt, *Phys. Rev. C* **50**, 31 (1994)
10. G. Q. Li and R. Machleidt, *Phys. Rev. C* **48**, 1702 (1993)
11. C. Fuchs, Amand Faessler and M. El-Shabshiry, *Phys. Rev. C* **64**, 024003 (2001)
12. T. Gross-Boelting, C. Fuchs and Amand Faessler, *Nucl. Phys. A* **648**, 105 (1999)
13. C. J. Horowitz, B. D. Serot, *Nucl. Phys. A* **464**, 613 (1987)
14. C. Fuchs, arXiv: nucl-th/ 0309003
15. A. Bohnet, N. Ohtsuka, J. Aichelim, R. Linden, and A. Faessler, *Nucl. Phys. A* **494**, 349 (1989).
16. B. ter Haar and R. Malfliet, *Phys. Rev. C* **36**, 1611 (1987)

THE EFFECT OF THE TENSOR FORCE ON SINGLE-PARTICLE STATES AND ON THE ISOTOPE SHIFT

WEI ZOU, GIANLUCA COLÒ and PIER FRANCESCO BORTIGNON

Dipartimento di Fisica, Università degli Studi and INFN, Sezione di Milano, 20133 Milano, Italy
**E-mail: weizou2048@gmail.com, gianluca.colo@mi.infn.it, pierfrancesco.bortignon@mi.infn.it*

ZHONGYU MA

China Institute of Atomic Energy (CIAE), Beijing 102413, China
E-mail: mazy12@ciae.ac.cn

HIROYUKI SAGAWA

Center for Mathematical Sciences, University of Aizu, Aizu-Wakamatsu, Fukushima 965-8560, Japan
E-mail: sagawa@u-aizu.ac.jp

In the attempt to reach a global view of the effects on single-particle states caused by the tensor interaction, we analyze the evolution of the spin-orbit splittings in the Ca isotopes and in the N=28 isotones. We also study the isotope shift in the charge radii of the Pb isotopes. Our framework is Hartree-Fock (HF) plus Bardeen-Cooper-Schrieffer (BCS) theory, and we employ a Skyrme interaction complemented by an effective zero-range tensor force, plus a density-dependent interaction in the pairing channel.

Keywords: Tensor force; Spin-orbit splitting; Isotope shift; Hartree-Fock theory

1. Introduction

The Skyrme forces are widely used in nuclear structure calculations, but most of the parameter sets do not include a tensor component. In the study of Fl. Stancu, D.M. Brink and H. Flocard,[1] who added this component to the SIII parameterization, it was pointed out that some spin-orbit splittings in magic nuclei can be improved with the inclusion of a tensor force, but the improvements are considered "minor" by the authors themselves. Until very recently, not so much effort has been put on the study of the effects of the tensor force on the evolution of the shell structure of atomic nuclei. In the framework of mean-field calculations the issue has been addressed in Ref.[2,3] using the Gogny interaction, and in Ref.[4] using the Skyrme interaction. While in the former work it has been pointed out for the first time that the tensor force is responsible for a specific attraction (repulsion) between protons and neutrons lying in $j_>$ and $j_<$ ($j_>$ and $j_>$, or $j_<$ and $j_<$) orbitals, in the latter it has been shown that the effects of the tensor force can be understood in a very transparent way also by inspecting the Skyrme Hartree-Fock (SHF), or SHF plus Baardeen-Cooper-Schrieffer (SHF plus BCS), formulas. Moreover, in Ref.,[4] the detailed study of the Sn isotopes and N=82 isotones has performed by taking into account the effects of the triplet-even and triplet-odd tensor contributions separately. We should point out that the effect of the tensor force on the evolution of single-particle states has been examined by other groups as well.[5,6]

In the present paper, which is a follow-up of Ref.,[4] we extend our study of the tensor correlations to lighter systems. The goal is to see whether the same conclusions reached before, can be confirmed by looking at empirical data in another mass region. Moreover, we analyze another observable which should be sensitive to the tensor correlation, namely the charge radii of the Pb isotopic chain.

The zero-range triplet-even (T) and triplet-odd (U) tensor terms are

$$\begin{aligned} v_T = \frac{T}{2}\{[(\sigma_1 \cdot k')(\sigma_2 \cdot k') - \frac{1}{3}(\sigma_1 \cdot \sigma_2)k'^2]\delta(r_1 - r_2) \\ + [(\sigma_1 \cdot k)(\sigma_2 \cdot k) - \frac{1}{3}(\sigma_1 \cdot \sigma_2)k^2]\delta(r_1 - r_2)\} \\ + U\{(\sigma_1 \cdot k')\delta(r_1 - r_2)(\sigma_2 \cdot k) - \frac{1}{3}(\sigma_1 \cdot \sigma_2)[k' \cdot \delta(r_1 - r_2)k]\} \end{aligned} \tag{1}$$

In this expression, the operator $k = (\nabla_1 - \nabla_2)/2i$ acts on the right and $k' = -(\nabla_1 - \nabla_2)/2i$ acts on the left. The coupling constants T and U denote the strength of the triplet-even and triplet-odd tensor interactions, respectively; we treat these coupling constants as free parameters in the following study.

The zero-range tensor force contributes to the spin-orbit potential, which is

$$U_{s.o.}^{(q)} = \frac{1}{2r}(2W_0\frac{d\rho_q}{dr} + W_0'\frac{d\rho_{q'}}{dr}) + (\alpha\frac{J_q}{r} + \beta\frac{J_{q'}}{r}), \tag{2}$$

where q=0 (1) labels neutrons (protons), ρ is the usual density and J the so-called spin-orbit density. W_0 and W_0' are parameters of the spin-orbit part of the usual Skyrme forces. On the other hand, α and β get contributions both from the exchange part of the Skyrme force and from the tensor force: $\alpha = \alpha_C + \alpha_T$ and $\beta = \beta_C + \beta_T$. The central exchange contributions read

$$\begin{aligned} \alpha_C &= \frac{1}{8}(t_1 - t_2) - \frac{1}{8}(t_1x_1 + t_2x_2), \\ \beta_C &= -\frac{1}{8}(t_1x_1 + t_2x_2), \end{aligned} \tag{3}$$

while the tensor contributions can be expressed as

$$\begin{aligned} \alpha_T &= \frac{5}{12}U, \\ \beta_T &= \frac{5}{24}(T + U). \end{aligned} \tag{4}$$

In the spin-saturated systems, the quantity J is negligibly small, while it gets positive (negative) contribution when the $j_>$ ($j_<$) orbitals are filled by particles.

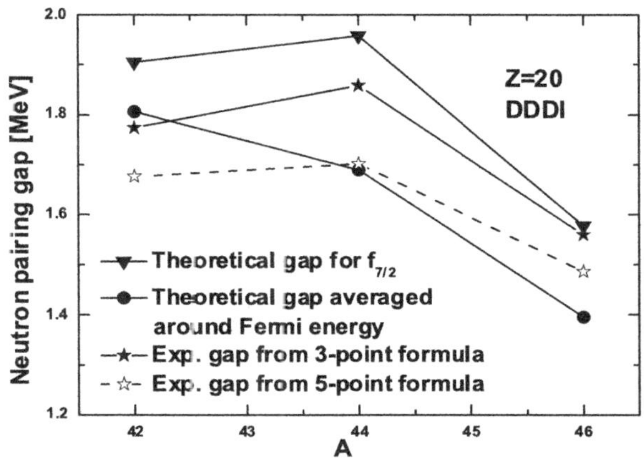

Fig. 1. Theoretical neutron pairing gaps for the Ca isotopes obtained by using HF-BCS, with SLy5 plus the density-dependent δ interaction (DDDI). The comparison with the empirical values allows fitting the parameter $V_0 = 740$ MeV·fm^3.

All the above formulas are valid within the SHF framework. In the present work, except for the double-magic systems, we perform HF-BCS in order to take into account the pairing correlations. In the pairing channel, we employ a zero-range, density-dependent delta interaction (DDDI) whose strength is fitted to reproduce the empirical pairing gaps. The results which are shown below correspond to the choice of a surface pairing force, or DDDI, as in Ref.[4] We have checked that the results are the same if we choose a volume-type pairing, e.g., a simple delta-force $V_0 \cdot \delta(\vec{r}_1 - \vec{r}_1)$.

2. Results for the single-particle states

We have performed a study of the valence nucleon states in the Ca isotopes using the Skyrme interaction SLy5[8]and two different pairing interactions. It is shown in Fig. 1 that, by comparing the pairing gap near the Fermi surface (i.e., $\Delta(f_{7/2})$) and the average pairing gap with the result of the three-point and five-point formulas respectively, we can fit a suitable parameter $V_0 = 740$ MeV·fm^3 in the case of the DDDI. If we wish to use the simpler DF, we must use $V_0 = 305$ MeV·fm^3.

The single-proton states in the Ca isotopes are displayed in Fig. 2. It can be immediately noticed that the original SLy5 interaction, without tensor terms, fails to reproduce the trend of the experimental data. On the other hand, a substantial improvement is induced by the the inclusion of the tensor interaction with $(\alpha_T, \beta_T) = (-170, 100)$ MeV·fm^5. In particular, the experimental results in the magic isotopes ^{40}Ca and ^{48}Ca are very well reproduced. As a matter of fact, these results can be qualitatively understood in terms of simple arguments, by looking at Eq. (2). The present nuclei are spin-saturated as far as the protons are concerned, so that the proton spin density J_p is negligibly small. Consequently, the term in α

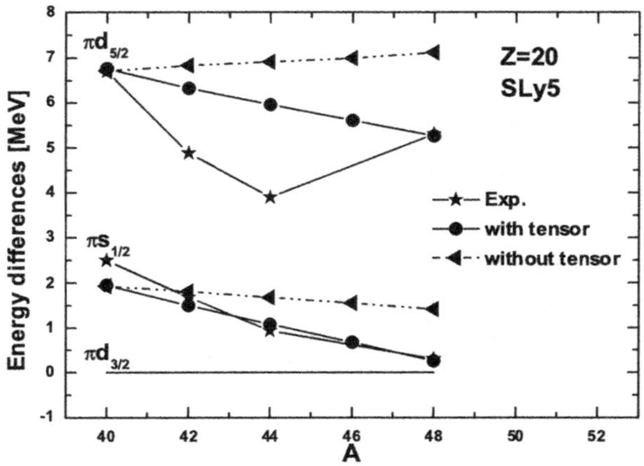

Fig. 2. Energy differences between the $1d_{3/2}$ single-proton state and, respectively, the $2s_{1/2}$ and $1d_{5/2}$ states, plotted along the Ca isotopes. The calculations are performed within HF-BCS using SLy5 and the DDDI for pairing, without and with the tensor term. The experimental data are taken form Ref.[7]

does not give any contribution to the spin-orbit potential and only the term in β can be responsible for the isospin dependence of the spin-orbit splitting. In going from $A = 42$ to $A = 48$, the $f_{7/2}$ neutron orbit is gradually filled and J_n is increased. Then, the positive value of α_T shrinks in absolute value the spin-orbit potential and decreases the spin-orbit splitting. This explains why in the figure the energies of the $1d_{5/2}$ and $2s_{1/2}$ decrease with respect to the $1d_{5/2}$ energy, when the tensor is included. Without the tensor term, the trend of the spin-orbit splitting is the opposite as in the experiment. We can conclude that it is quite satisfactory that the tensor interaction, with the same parameters already employed for the Sn isotopes and the N=82 isotones in Ref.,[4] lead to a better description of the states in the Ca isotopes as compared to the original Skyrme force SLy5.

3. The isotope shift in the charge radii of Pb isotopes

As we discussed above, the tensor terms have sizable effect on the spin-orbit splitting, and affect the position of the energy orbits. Since this has obvious and direct relationship with the charge radii, we expect that the tensor terms can modify them. In particular, in the case of the Pb isotopes, the so-called isotope shifts defined as

$$\delta r^2 = [\langle r^2 \rangle - \langle r^2 \rangle (^{208}Pb)] \qquad (5)$$

are experimentally well-known. In the past, usual Skyrme parameter sets have been shown to be unable to explain the experimental trend of the isotope shifts. This trend can be reproduced by the parameter sets introduced in Ref.,[13] in which the ratio W_0'/W_0 (cf. Eq. (2) is different from one. Relativistic mean field models also perform well as far as the reproduction of the isotope shifts is concerned.

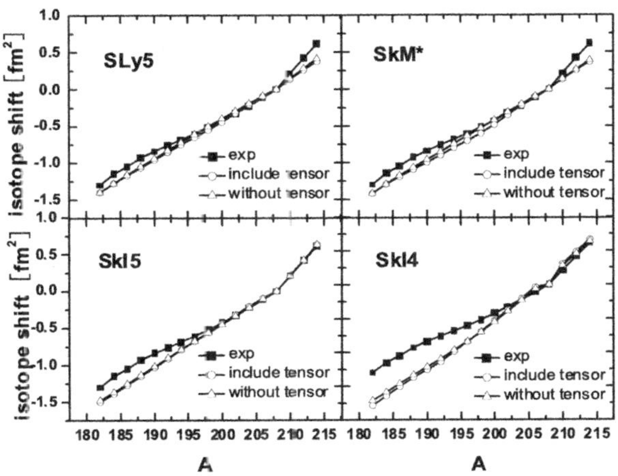

Fig. 3. Isotope shifts in the Pb isotopes. The experimental data are compared with results obtained by using the parameter sets SLy5, SkM*, SkI5, SkI4, with and without the tensor terms. The experimental data are taken from Refs.[9–12]

In this work, we have checked the effect of the inclusion of tensor terms on top of existing Skyrme forces, on the isotope shifts in the case of Pb. Fig. 3 present our results. In all these calculations, we take the pairing correlations into account by means of a simple constant gap approximation, Δ_n being given by $11.2/\sqrt{A}$ MeV. At shell closure, that is, at N=126, the data show a clear kink. This change of slope is not reproduced by the parameter sets SkM* and SLy5, but is reproduced if one employs the forces SkI4 and SkI5 introduced in Ref.[13] This pattern is not markedly changed if we include the tensor force. If we wished to reproduce the experimental kink by changing the tensor force parameters, we would need values of α_T and β_T which are quite different from those used in our previous discussion. Moreover, we should point out that data for the very neutron-rich (and neutron-deficient) isotopes are not easy to reproduce. Whether this can be improved by major changes in the Skyrme functionals, or by the inclusions of correlations beyond mean field (that is, particle-vibration coupling), should be the one of the goals of future works.

Acknowledgments

We acknowledge the financial support from the the Asia-Link project [CN/ASIA-LINK/008(94791)] of the European Commission.

References

1. F. Stancu, D.M. Brink and H. Flocard, *Phys. Lett. B* **68**, 108 (1977).
2. T. Otsuka, T. Suzuki, R. Fujimoto, H. Grawe and Y. Akaishi, *Phys. Rev. Lett.* **95**, 232502 (2005).
3. T. Otsuka, T. Matsuo and D. Abe, *Phys. Rev. Lett.***97**, 162501 (2006).
4. G. Colò, H. Sagawa, S. Fracasso, P.F. Bortignon, *Phys. Lett. B* **646**, 227 (2007).

5. D.M. Brink and F. Stancu,*Phys. Rev. C* **75**, 064311 (2007).
6. J. Dobaczewski, in Proceedings of the Third ANL/MSU/JINA/INT RIA Workshop, edited by T. Duguet, H. Esbensen, K. M. Nollett, and C. D. Roberts (World Scientific, Singapore, 2006).
7. P. Doll, G.J. Wagner, K.T. Knopfle, *Nucl. Phys. A* **263**, 270 (1976).
8. E. Chabanat, P. Bonche, P. Haensel, J. Meyer, R. Schaeffer, *Nucl. Phys. A* **635**, 231 (1998).
9. S.B.Dutta *et al.*,*Z. Phys. A* **341**, 39 (1991).
10. M.Anselment *et al.*,*Nucl. Phys. A* **451**, 471 (1986).
11. U.Dinger *et al.*, *Z. Phys. A* **328**, 253 (1987).
12. H.De Witte *et al.*, *Z. Phys. A* **328**, 253 (1987).
13. P.-G. Reinhard, H. Flocard, *Nucl. Phys. A* **584**, 467 (1995).

^{18}Ne EXCITED STATES TWO-PROTON DECAY

M. DE NAPOLI[*,1)], E. RAPISARDA[1)], G. RACITI[1)], G. CARDELLA[1)], F. AMORINI[1)], F. GIACOPPO[1)] and C. SFIENTI[2)]

[1)] *Dipartimento di Fisica, Universitá degli studi di Catania and LNS-INFN, 64, Via S. Sofia, I-95123 Catania, Italy*

[2)] *Gesellschaft für Schwerionenforchung, D-64291 Darmstadt, Germany*

[*] *E-mail: denapoli@lns.infn.it*

Two-proton radioactivity studies have been performed on excited states of ^{18}Ne produced by ^{20}Ne fragmentation at the FRS of the Laboratori Nazionali del Sud and excited via Coulomb excitation on a ^{209}Pb target. The ^{18}Ne levels decay has been studied by complete kinematical reconstruction. In spite of the low statistic, the energy and angular correlations of the emitted proton pairs indicate the presence of ^{2}He emission toghether with the *democratic* decay.

Keywords: Reaction induced by unstable nuclei, Coulomb excitation, Decay by proton emission, Properties of nuclei-nuclear energy levels, Charged-particle spectroscopy.

1. Introduction

The simultaneous emission of two protons was proposed more than 40 years ago as a possible decay mode for even-Z-nuclei beyond or close the proton drip line.[1] This decay mode is, in general, in competition with the sequential 2p decay, where the two protons are emitted sequentially via an intermediate state. Moreover, the two-proton decay may proceed via an uncorrelated (*democratic*) emission of the two protons, or a correlated (*diproton*) emission, where a ^{2}He cluster is emitted.
In many experiments, such as the breakup of ^{6}Be[2] and ^{12}O[3,4] or the β-delayed 2p emissions,[5–7] the results were consistent with a sequential decay description. Recently, the first observation of two-proton decay came from ^{45}Fe at ground states[8,9] and from a metastable state in ^{94}Ag.[10] Simultaneous emission of two protons has also been observed from the decay of the 6.15 MeV, $I^{\pi}=1^{-}$ level in ^{18}Ne, produced in the fusion reaction ^{17}F+p.[11] However, the energy and angular distribution data do not distinguish between *diproton* and *democratic* emission. Recently, in order to discriminate between these two decay mechanisms, an experiment was performed by using a ^{18}Ne beam at about 30 AMeV, produced with the FRIBs facility[12,13] at the Laboratori Nazionali del Sud (LNS). The ^{18}Ne levels were populated by Coulomb excitation on a ^{209}Pb target and their decay has been studied by complete kinematical reconstruction.

2. Experimental Setup

The secondary beam has been produced by the fragmentation of a primary stable ^{20}Ne beam at 45 AMeV, delivered by the Superconducting Cyclotron (SC), on a ^{9}Be, 500 μm thick, production target. The secondary ions have been separated in-flight in the fragment separator of the LNS [12,13] operated at a Bρ setting optimized for the ^{18}Ne. A primary current of 300 enA produced a total RIBs rate of 10^5 ions/sec at the exit of the fragment separator. The secondary beam has been transported with a 60% of transmission up to the TRASMA scattering chamber.[12,13] The ^{18}Ne rate was about 9% of the total RIBs mixture rate, i.e. about 5.5×10^3 pps.
The measurement of energy, charge and mass, via ΔE-ToF technique, and position of each ion of the transported RIBs was derived, event-by-event, before the interaction with the secondary ^{209}Pb target. The time-of-flight was measured with respect to the radio frequency (RF) signal provided by the SC and the energy loss was measured by a 16×16 X-Y Si-Strip detector 5×5 cm^2 of area and 300μm thick[12,13] (Fig. 1). The ^{18}Ne incident energy on the ^{209}Pb target was in the range of 31 ÷ 35 AMeV as evaluated from the measured energy loss in the Si-Strip.

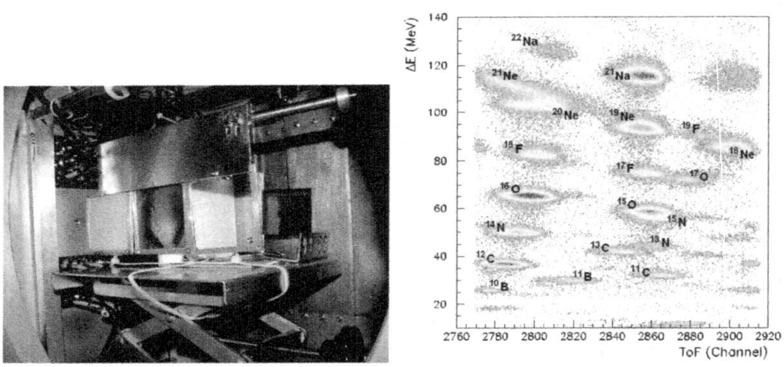

Fig. 1. Left: the Si-Strip *tagging* detector in front of the target holder at the entrance of the TRASMA scattering chamber. Right: energy-loss in the Si-Strip *tagging* detector versus ToF.

The detection system consisted on two Si-CsI hodoscopes with different granularity (Fig. 2):

- 81 two-fold telescopes: 300 μm Silicon detectors 1×1 cm^2 of active area followed by a 1×1 cm^2 and 10 cm long CsI(Tl);
- 89 three-fold telescopes 50 μm + 300 μm Silicon detectors both having 3×3 cm^2 surface followed by a 6 cm long CsI(Tl) of the same surface.

Due to its high granularity and low threshold, the array is a suitable detector for momentum and angular correlation measurements. The two-folds telescopes cover the angular range $\theta_{lab} = \pm5^\circ$ in the step of $\pm0.6^\circ$ while the three-folds telescopes cover the angular range between $\pm5^\circ$ and $\pm21.5^\circ$ in step of $\pm1.5^\circ$.

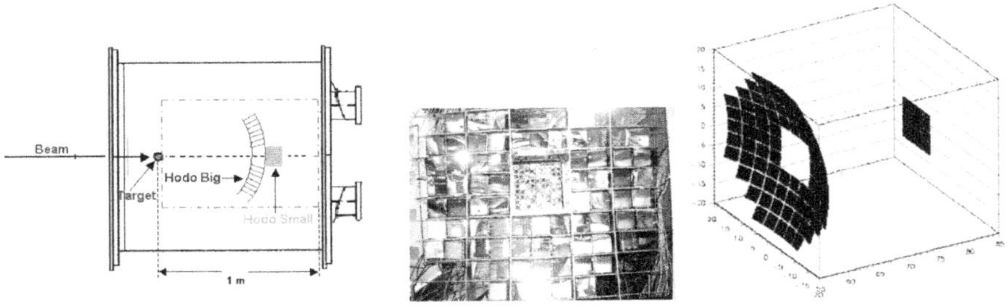

Fig. 2. Left: schematic view of the experimental setup. Middle: both hodoscopes inside the TRASMA scattering chamber. Right: 3D reconstruction of the hodoscopes geometry: the covered solid angle is 0,34 str and the geometrical efficiancy is 72%

3. Kinematic reconstruction procedure

The projectiles have been selected by gating on the ΔE-ToF plot provided by the *tagging* detector and then a complete reconstruction of the decay kinematic in the center-of-mass system (CM) was performed. The X-Y coordinates of the interaction point on the target as well as the energies and angles of all outgoing decay particles were measured event-by-event. From the velocities and angles of the decay products measured in the hodoscopes, the CM velocity, i.e. the velocity of the decaying nucleus, was determined. The CM energy, including q-value and energy loss in the target, must be almost equal to the projectile energy, reconstructed from the ΔE measured in the Si-Strip. With an iterative procedure it was possible to identify the target slice where the energy conservation is satisfied, i.e. where the reaction takes place.[14] Therefore it was possible to reconstruct, event-by-event, the invariant mass in the CM system and thus the excitation energy spectrum. Background produced by reactions on the Si-Strip *tagging* detector has been measured with the empty frame runs and subtracted from the data.
The whole procedure was checked for the well-known ^{16}O stable nucleus. Figure 3 (left panel) shows the excitation energy spectrum as determined from the fully measured ^{12}C+α events produced by the ^{16}O. The well-known 1^- and 2^+ excited levels of ^{16}O[15,16] are recognized despite the poor energy resolution. The experimental energy resolution is about 500 keV, mainly dominated by the error in the determination of the depth of the interaction point in the thick ^{209}Pb target.[14] The spectrum presents only the 1^- and 2^+ levels as expected since only these levels can be populated via Coulomb excitation.

In Ref.[17] the Coulomb excitation cross section of the 13.05 MeV state in ^{16}O projectiles incident on ^{209}Pb targets was calculated as a function of the incident energy. For an incident energy of about 30 AMeV the calculated Coulomb excitation cross section is about 0.9 mb. In order to measure this value, the peak corresponding to the 13.05 MeV level in the experimental spectra (Fig. 3, left panel) was integrated and the extracted value was corrected for the contribution of the other neighboring levels and for the hodoscopes efficiency. Having measured the number of incoming

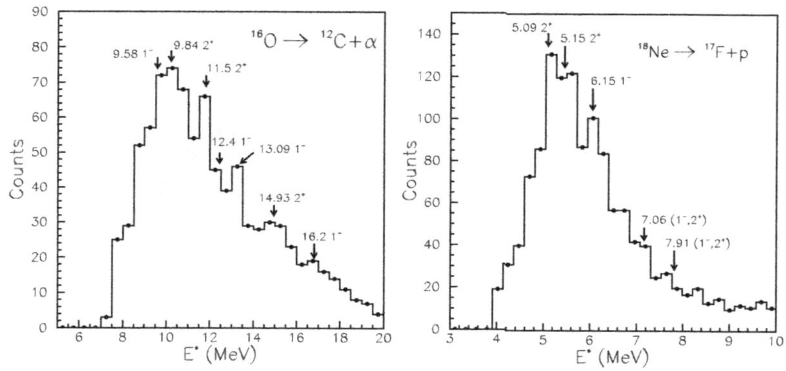

Fig. 3. Left: Experimental excitation energy spectrum of ^{16}O from the ^{12}C+α events (left panel) and of ^{18}Ne from the ^{17}F+p events (right panel). The indicated levels are listed in Ref.[16]

^{16}O projectiles by the Si-Strip and considering the branching ratio of this level in ^{12}C+α to be about 20%,[16] an experimental cross section of 0.87±0.22 mb has been estimated, which is quite near to the calculated value.[17]
The excitation energy spectrum of the ^{17}F+p events produced by selecting ^{18}Ne projectile is shown in the same figure (right panel). The peaks corresponding to the decay of the 5.09 MeV(2^+), 5.15 MeV(2^+) and 6.15 MeV(1^-) excited states[16] are recognized although partially superimposed.

4. ^{18}Ne two-proton decay

The same procedure was then applied to the fully measured ^{16}O+2p events among the ones produced by the selected ^{18}Ne secondary beam. The excitation energy spectrum is shown in Fig. 4 (top-left panel). The presence of the 6.15 MeV peak confirms the evidence for simultaneous two-proton decay of this ^{18}Ne state reported in Ref.[11] In fact, for this level, the sequential 2p decay is energetically forbidden (Fig. 4, top-right panel). Whether the presence of two protons in the final state of the the known 7.06 MeV(1^-,2^+), 7.91 MeV(1^-,2^+) and 8.5 MeV (1^-,2^+) levels indicates a two-proton decay mechanism, deserve further investigations since, for these levels, the sequential 2p decay channel is available. Moreover, the Coulomb excitation allowed to populate unknown high-lying states around 10.7 MeV, 11.2 MeV, 12.5 MeV, 13.7 MeV and 14.5 MeV, possible future candidates for two-proton radioactivity. Indeed, recently, in the study of the decay of excited states of ^{17}Ne, performed at GANIL, the observed data seem to indicate that one or several high-lying states could decay by a correlated 2p emission, despite the fact that the sequential branch is widely open.[18]

In order to understand whether the two-proton decay of the 6.15 MeV level is *diproton* or *democratic*, the relative energy and angular correlation of the two emitted protons have been studied. Figure 4 (bottom panels) shows the relative energy and the relative CM angle of the two protons for events selected inside the excitation

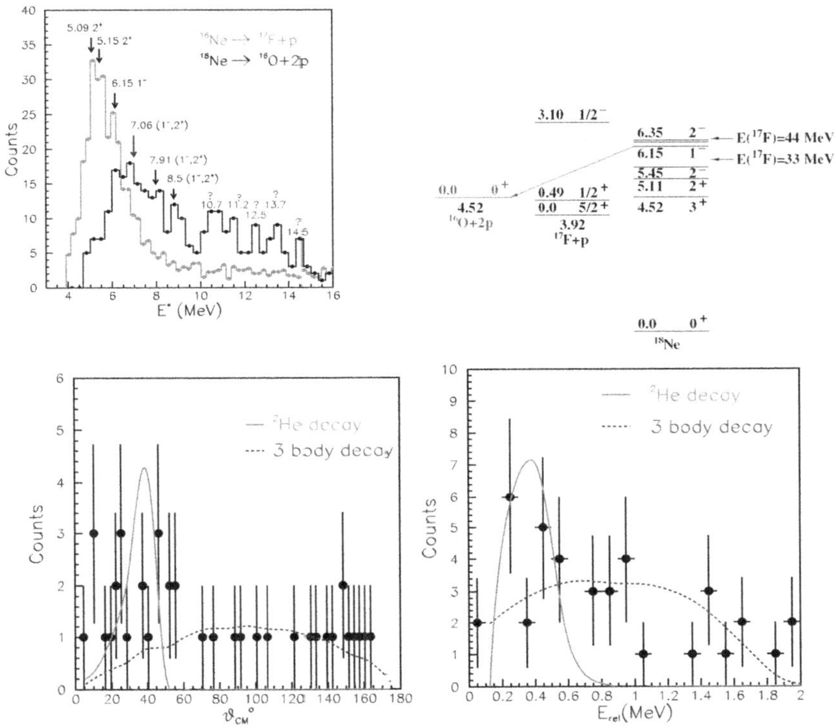

Fig. 4. Top-Left: experimental excitation energy spectrum of ^{18}Ne obtained from ^{16}O+2p events compared to that of the ^{17}F+p channel (scaled). The indicated levels are listed in Ref.[16] Top-Right: decay scheme of ^{18}Ne from Ref.[11] Bottom: relative energy (left) and CM relative angle (right) of the two protons emitted from the 6.15 MeV level, compared to Monte Carlo simulations.

energy range $5.9 \leq E^* \leq 6.5$ MeV. Data are compared to Monte Carlo simulations, filtered for the geometry and detector constrains, assuming ^2He emission (solid line) and three-body decay neglecting final state interaction (dashed line). As shown by the simulations, the two possible mechanisms for simultaneous two-proton emission lead to different energy and angular correlation between the two protons, provided the correlations are studied over a large enough angular range.[11]

Despite the poor statistic, the relative energy and angle indicate the presence of events where the two protons are emitted both uncorrelated and in the form of a ^2He. To extract the branching ratios the simulation curves, normalized to the data, have been integrated: the ^2He emission is about 40% of the total two-proton decay whereas the three-body emission is about 60%.

5. Conclusions

Recently, an experiment to search for and study two-protons emission from excited states of ^{18}Ne was performed at the Laboratori Nazionali del Sud. The beam was produced by ^{20}Ne fragmentation on a ^{9}Be target at 45 AMeV at the LNS-FRS.[12,13] The ^{18}Ne levels were populated by Coulomb excitation on a ^{209}Pb target. Several levels were identified in the excitation energy spectrum built by kinematical reconstruction from the ^{17}F+p and ^{16}O+2p fully measured decay events. The presence of the 6.16 MeV peak in the ^{16}O+2p energy spectrum confirms the observed two-protons decay of such level.[11] In particular, the relative energy and angle of the two protons, analysed in the excitation energy range 5.9≤(E*)≤6.5 MeV, indicate the presence of both *diproton* (40%) and *democratic* (60%) decay mechanisms. However, for a precise measure of the branching ratio more statistic is needed. Finally, the Coulomb excitation allowed to populate high-lying known and unknown states in ^{18}Ne. Whether the presence of two protons decay of such states is the result of a simultaneous two-protons emission is still under investigation.

References

1. V.I. Goldansky, *Nucl. Phys.* **19**, 482 (1960).
2. D.F. Geesaman *et al.*, *Phys. Rev. C* **15**, 1835 (1977).
3. R.A. Kryger *et al.*, *Phys. Rev. Lett.* **74**, 860 (1995).
4. A. Azhari *et al.*, *Phys. Rev. C* **58**, 2568 (1998).
5. M.D. Cable *et al.*, *Phys. Rev. Lett.* **50**, 404 (1983).
6. C. Detraz *et al.*,*Nucl. Phys. A* **514**, 529 (1990).
7. I. Mukha *et al.*, *Nucl. Phys. A* **630**, 394c (1999).
8. J. Giovinazzo *et al*, *Phys. Rev. Lett.* **89**, 102501 (2002).
9. M. Pfützner *et al.*, *Eur. Phys. Jour.* **14**, 279 (2002).
10. M. I. Mukha *et al.*, *Nature* **439**, 298 (2006).
11. J. Gomez del Campo *et al*, *Phys. Rev. Lett.* **86**, 43 (2001).
12. G. Raciti *et al.*, *Proceeding of the 10th International Conference on Nuclear Reaction Mechanisms*, Varenna, **11** (2003).
13. E. Rapisarda *et al.*, *EPJ A in press.*
14. G. Cardella *et al.*, *Proceeding of the International Conference on Proron Emitting Nuclei, Lisbon* (2007), in press.
15. F. Ajzenberg, *Rev. of Mod. Phys.* **27**, 77 (1955).
16. *www.nndc.bnl.gov*
17. C.A. Bertulani *et al.*, *Phys. Lett. B* **650**, 233 (2007).
18. B. Blank, *Proceeding of the International Workshop XXXI on Gross Properties of Nuclei and Nuclear Excitations*, Hirschegg, Austria (2003)

THE ISOMERIC RATIO AND ANGULAR MOMENTUM OF FRAGMENT ^{135}Xe IN PHOTOFISSION OF HEAVY NUCLEI

TRAN DUC THIEP*[5], IU. GANGRSKI[1], V.I. JEMENHIK[2], G.V. MINSHINSKI[1], IU.E. PENHIONZKEVITCH[1], N.N. KOLESNHIKOV[3], V.G. LUCASHIK[4], TRUONG THI AN[5], NGUYEN THE VINH[5], PHAN VIET CUONG[5], NGUYEN TUAN KHAI[5]

[1]*Joint Institute for Nuclear Research, Dubna, Russia.*
[2]*Institute for Nuclear Research, Kiev, Ucraina*
[3]*Moscow State University M.B. Lomonosov, Russia*
[4]*Institute for Cosmic Research, Moscow, Russia*
[5]*Institute of Physics and Electronics, VAST, Hanoi, Vietnam*
**E-mail: tdthiep@iop.vast.ac.vn*

In this work we present results of our study of isomeric ratio using the method of gaseous flow and the angular momentum of fragment ^{135}Xe produced in photofission of heavy elements ^{232}Th, ^{237}Np, ^{238}U, ^{243}Am and ^{248}Cm induced by 25 MeV bremsstrahlung produced at Microtron MT-25 of the Flerov Laboratory of Nuclear Reaction, JINR, Dubna, Russia. On the basis of the statistical method proposed by Huizenga and Vandenbossch the root mean square angular momentum of the fragment corresponding to the measured isomeric ratio was deduced. The results are discussed and compared with those by other authors.

Keywords: Photofission, fragment, isomeric ratio, angular momentum, bremsstrahlung, gaseous flow, Microtron.

1. Introduction

The study on the isomeric ratio and the angular momentum of the fission fragment provides information on the scission configuration as well as mechanism of energy dissipation, leading to better understanding of the dynamics of the fission process.[1,2] On the other hand, the angular momentum is closely connected to the ratio of the probabilities of population of the isomeric and ground states of the fragment, so called isomeric ratio. The study of this ratio in nuclei formed in states with different angular momenta presents one of directions in studying nuclear structure. This allows to obtain important information about the dependence of level density on angular momentum and the probabilities of radiation transitions between the level. Simultaneously the isomeric ratio in fission fragment allows to discuss about characteristics of the fission process, in part the role of different kinds of collective movements, affecting on increasing the angular momentum of fragments at the scission moment of fissioning nuclei.[3,4] Up to now most of investigations were concentrated on the study of angular momentum of fission fragments in low energy fission of actinides with neutrons, but the data for photofission are very few in the literature. The study led to very interesting effects such as the effects of nuclear structure and scission-point deformation on the fragment angular momentum and the correlation of fragment angular momentum (J_{rms}) with average neutron number and elemental yield.[5]

In this work we present our results on study of the isomeric ratio and angular momentum of fragment ^{135}Xe formed in photofission of heavy elements ^{232}Th, 237Np, ^{238}U, ^{243}Am, and ^{248}Cm induced by 25 MeV bremsstrahlung (i.e. in the dipole resonance region).

2. Experimental

Fission fragment ^{135}Xe is belonging to fission chain with mass number A=135. The scheme of formation ^{135}I and ^{135m,g}Xe is shown in Fig. 1 and the decay characteristics of ^{135m}Xe and ^{135g}Xe are taken from[6,7] and presented in Table 1.

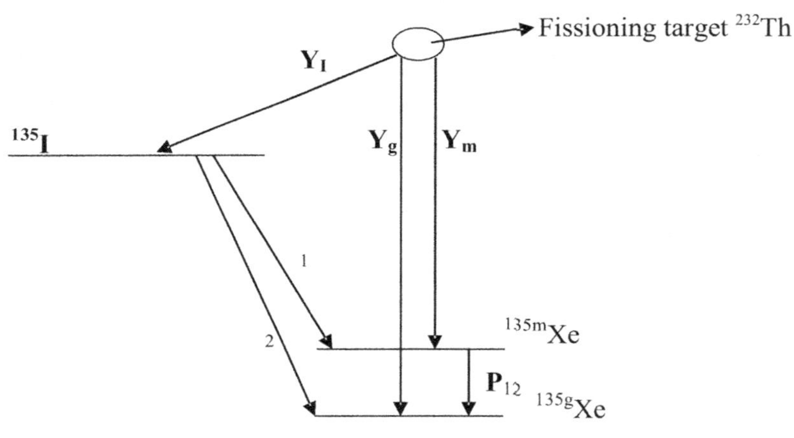

Fig. 1. Scheme of formation of ^{135}I and ^{135m,g}Xe.

As it is seen from Fig. 1 ^{135}Xe is in both isomeric and ground states and formed directly from the photofission and from decay of ^{135}I during the reaction process with independent yields Ym and Yg. When the reaction process finished ^{135}Xe is formed only from the decay of ^{135}I.

Usually the photopeaks characterizing the isomeric and ground states of a fission fragment are observed on high background of other fragments and it makes the determination of the isomeric ratio less accurate or impossible. For effective decrease of the mentioned background different techniques were applied, for example catcher foil technique,[8] fission fragment transportation[9] and chemical isotope

Table 1. The decay characteristics of ^{135m}Xe and ^{135}Xe

Xe^{135g}(T_{12} = 9.104 hours)			Xe^{135m}(T_{12} = 15.65 minutes)		
Spin [h]	Energy [keV]	Intensity [%]	Spin [h]	Energy [keV]	Intensity [%]
$\frac{3}{2}^+$	158.2	0.2300	$\frac{11}{2}^-$	526	81.20
	358.3	0.2000			
	407.9	0.3200			
	608.1	2.5800			
	249.7	90.600			

separations.[10,11] In this work as ^{135}Xe is a gaseous state fragment, for determination of its isomeric ratio we used the inert gaseous flow method described in[12] for transporting this fragment from the reaction camera to the fragment collectors. The experimental set-up is shown in Fig. 2. The inert gaseous flow used was He of 2atm pressure. Under this pressure the inert gaseous flow will enter the reaction camera and carry all gaseous fragments (Xe and Kr) through the filter to the collectors. All other fission fragments produced are in solid state will be catched on the wall of the reaction camera or remain in the filter.

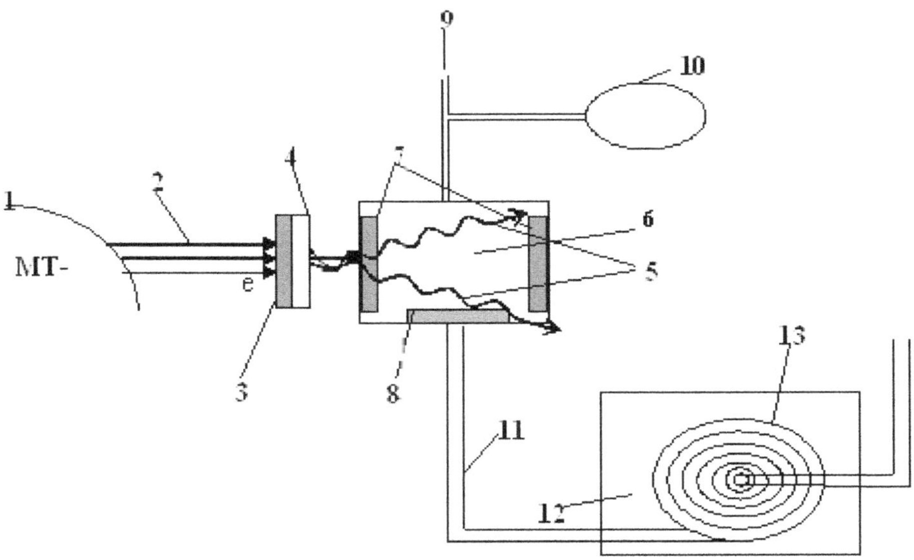

Fig. 2. The experimental setup.

Electron accelerator MT-25; 2. Accelerated electron beam; 3. W-convertor of electron to photon; 4. Al-absorber of low energy electron; 5. Bremsstrahlung photon flux; 6. Reaction camera; 7.Fissioning target; 8. Fragment filter; 9. Inert gaseous flow; 10. Manometer; 11. Capillary; 12. Cryostat; 13. Fragment collector.

In the experiment, two collectors were used . The first one works when the photofission process is happening, it means that this collector collects all gaseous fragments ^{135}Xe from ^{134}I and directly from ^{237}Np target. After the photofission stopped the second collectors begins to work and it collects gaseous fragment $^{135}_{54}Xe$ only from $^{134}_{53}I$.

Target preparation and irradiation.

The targets were prepared by thin ^{237}Np, ^{238}U, ^{243}Am, and ^{248}Cm layers of 0.1 $\div$ 0.3mg.cm^{-2}thickness on two aluminium foils of 0.1mm thickness (for ^{232}Th the target was prepared by thorium foils of 50 μm thickness) and placed at two sides of reaction camera. They were irradiated by 25 MeV bremsstrahlung produced by Microtron MT-25 of the Flerov Laboratory of Nuclear Reaction, JINR, Dubna. The description of this accelerator and its characteristics are presented in.[13] The average electron beam was 14-15μA. The time of fragment collecting (i.e the time of irradiation) for the first collector and the time of fragment collecting for the second collector were 45 min. The time between two collections was about 10 min.

Measurement of fragment gamma spectra:

The gamma spectra of the first and the second collectors were measured with HPGe – semiconductor detectors, produced by CANBERRA company. The detectors has volumes of 80 and 150 cm^3 with 2.2 and 1.8 keV energy resolution at 1332 keV gamma ray of ^{60}Co respectively

Determination of isomeric ratio:

The isomeric ratio was determined from different combinations of gamma spectra of the first and the second collectors for different times of cooling and measurement by procedure shown in.[12] In determination of the isomeric ratio of fission fragment ^{135}Xe only two gamma ray of 249.7 and 526 keV were used due to their high intensities and it makes the results more accurate.

3. Results and discussions

It is known that the spectrum of gamma radiation from fission fragment is very complicated and consists of hundreds peaks, in part photopeak of different isotopes cover each other, and simultaneously the dead time and the Compton background are very high. This leads to less accurate results even for pure photopeaks of interest. In this work by using the method of inert gas flow all solid fragments are catched in the reaction camera, except gaseous fragments Xe and Kr, therefore the gamma spectra is much simpler because the background from other fragments is eliminated. Fig. 3 shows part of the spectrum photofission of ^{237}Np induced by 25 MeV bremsstrahlung, measured for the second collector. Here the photopeaks characterizing the isomeric and ground states of ^{135}Xe are seen very clear. Other photopeaks appeared in the spectrum are from different isotopes of Xe and Kr. This spectrum is very simple in comparison with total spectrum of photofission and makes the data processing easier, with a higher accuracy.

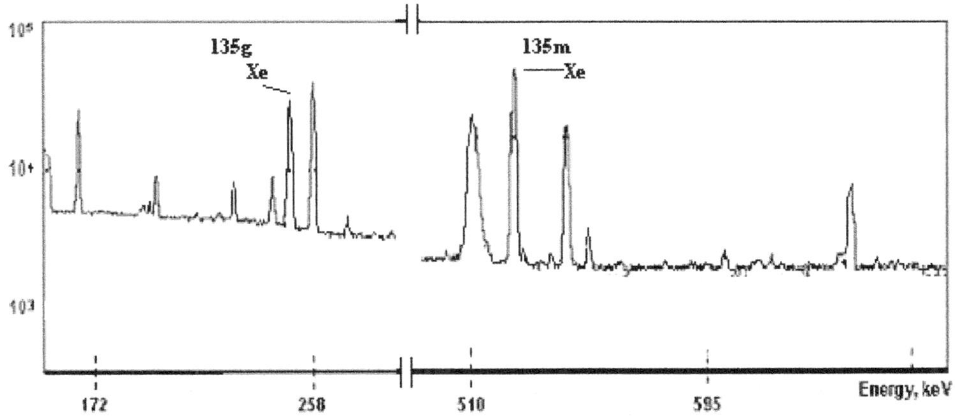

Fig. 3. Spectrum for the case of photofission of Np^{237} at 25 MeV bremsstrahlung

The choice of ^{135}Xe was due to the condition that this nucleus can be formed not only directly from fission but also in one-step reactions such as (n, γ) of (γ, n). The mechanism of these reactions are well-known and the angular momentum of absorbed γ – quantum and the energy of projectile neutron are fixed. Therefore in the case of these reactions it is possible to perform the detail calculation of change on the angular momentum of the nucleus at population of the isomeric and ground states and then from the isomeric ratio are obtained the parameters of level density, which can be used for calculation of the isomeric ratio in photofission fragment ^{135}Xe.

The results are shown in Table 2. For ^{232}Th in error limit our value coincides with that shown in[14] and for ^{238}U and ^{237}Np the obtained results are higher than the known ones in.[15] The angular momentum were calculated by the statistical method proposed by Huizenga-Vandenbosh[16] and are shown in Fig. 3. The most important parameters in the statistical model are parameters of level density a and spin cut-off σ. In order to obtain these parameters was performed the procedure of calculation of the isomeric ratios for ^{134}Xe (n, γ) and ^{136}Xe(γ, n) reaction. The isomeric ratios in these reactions are presented in[17] (IR= 0.013 for ^{134}Xe(n, $\gamma)^{135m,g}$Xe and IR= 0.110 for ^{136}Xe(γ, n)^{135m,g}Xe. The obtained parameters a and σ are in agreement between theoretical and experimental isomeric ratio values were used for calculation of the angular momentum J of fragment ^{135}Xe. In order to determine the average excitation energy of the fragment and the number of neutrons, evaporated from it, the similar procedure shown in[18] was used.

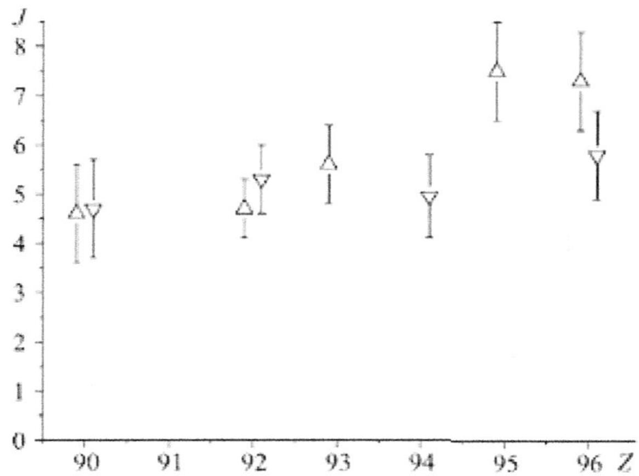

Fig. 4. The dependence of angular momentum J of photofission fragment ^{135}Xe on the atomic number of fissioning nuclide. ∇ - fission induced by neutron. Δ – photofission.

One can see in Table 2 and Fig. 3 that the angular momenta are high (5-7 h) and in the same order of that in fission by thermal and fast neutron.[5,19] A growth of the average angular momenta with the increase of atomic number Z of fissioning nuclei was observed. This fact is connected with high excitation energy of the fragment and, therefore, with big number of evaporation neutrons which lead to decreasing fluctuation of the angular momenta as well as are connected to processes accompanying the fission leading to filling the spins of primary fragments.

Table 2. Yield, average neutron numbers $\bar{\nu}$, isomeric ratio IR and average angular momentum of the fragment ^{135}Xe, $\bar{J}$

Parameter	$^{232}Th(\gamma, f)$	$^{238}U(\gamma, f)$	$^{237}Np(\gamma, f)$	$^{243}Am(\gamma, f)$	$^{248}Cm(\gamma, f)$
$Y_i(^{135}Xe)$, $1/f$	0.0012(8)	0.0033(8)	0.023(3)	0.015(2)	0.024(5)
$Y_c(^{135}I)$, $1/f$	0.040(5)	0.047(6)	0.030(4)	0.040(5)	0.028(5)
$\bar{\nu}$	1.0(3)	1.3(3)	1.5(4)	1.8(4)	2.5(5)
IR	0.75(10)	0.76(12)	1.41(18)	2.73(1.0)	2.68(35)
$\bar{J}, \hbar$	4.6(1.0)	4.9(6)	5.6(8)	7.5(1.0)	7.3(1.0)

It is also possible the influence on the isomeric ratio of the final spin values of odd nuclei ^{237}Np and ^{243}Am. More detail information about these effects can be obtained from future experiments to study the dependence of the isomeric ratios on mass number of fragments, their kinetic energy and angular momentum of fissioning nuclei.

Acknowledgement

In conclusion the author would like to express their sincere thanks to Iu. T. Oganessian, M.G. Itkis and S.N. Dmitriev for their kind interest in this work, J. Kliman and A. Daniel for useful discussions, G.B. Buclanov for the preparation of the targets and A.G. Belov for the operation of the irradiation facility.

The supports from RFFR (Grant N^o: 04-02-16955) and the Vietnam National Program for Fundamental Research in Natural Science are gratefully acknowledged.

The investigation has been performed at the Flerov Laboratory of Nuclear Reaction, JINR, Dubna.

References

1. H.O. Denschlag *et. al*, Proceeding of the International Symposium on the Physics and Chemistry of Fission, Julich 1979, Inter. Atomic Energy Agency, Vienna, 1980, V.II, p.153.
2. J.P. Bocquet *et. al*, Proceeding of the International Symposium on the Phys. and Chem. of Fission, Julich 1979, IAEA, Vienna, 1980 V.II, p.159.
3. Nix J.R., Swiatecki W.J. *Nuclear Physics* , V.71, p.1 (1965).
4. Rasmusse J.O., Nrenberg W., Mang H.J. *Nucl. Phys. A* **136**, 465 (1969).
5. H. Naik, *Phys. Rev. C* **71**. 014304 (2005).
6. C. M. Lederer, V. S. Shirley, Table of Isotope, 7th Edition, John Wiley, 1978.
7. J. Blachot and C. Fiche At Data and Nuc. Data Tab. **20**, N3, 242 (1977).
8. H. Thierens *et. al*, *Nucl. Instrum. Methods* **134**, 299 (1976).
9. Yu.P. Gangrski *et. al*, Preprint of the Joint Institute for Nuclear Research, Dubna, Russia, P13, 66 (2001).
10. H. Thierens *et. al*, *Phys. Rev. C* **25**, 1546 (1982).
11. A.P. Tonchev *et. al*, *J. Radioanal. Nucl. Chem., Letters* 155(5), 299 (1991).
12. Tran Duc Thiep *et. al*, *J. of Communication in Physics*, **16**, 177 (2006).
13. A.G.Belov *et. al*, Print of JINR, D15-93-80, Dubna, Russia (1993).
14. O.A.Bezshiuko *et. al*, Coll. of Sci pub. of the Kiev Inst. for Nucl. Res., N2(8), 46 (2002).
15. O.A. Bezshiyko *et. al*, Bull. of Uni. of Kiev, Ser. Phys. &Math. 2 (2004).
16. R.Vandenbossch and J.Huizenga, *Phys. Rev,* **120**, 1305 (1960) .
17. A.G. Belov *et. al*, *Russian J. of Nucl. Phys.,* **64**, 1987 (2001).
18. D.De Frenne *et. al*, *Phys. Rev. C*, 1356 (1982).
19. H.Naik *et. al*, *Z. Phys. A,* **331**, 335 (1988).

SEARCH FOR CORRELATED TWO-NUCLEON SYSTEMS IN ^{6}Li AND ^{6}He NUCLEI VIA ONE-NUCLEON EXCHANGE REACTION

N.T. KHAI[1,2], T. SUDA[1], A. YOSHIDA[1], T. OHNISHI[1], H. TAKEDA[1], T. SUZUKI[3], T. SUGAWARA[3], S. NAKAJIMA[3], K. OGATA[3], K. KOBAYASHI[3], Y. SUZUKI[3], L.H. KHIEM[2,6], P.Q. HUNG[4] and A. OZAWA[5]

[1] *RIKEN (Institute of Chemical and Physical Research), 2-1 Hirosawa, Wako, Saitama 351-0198, Japan*

[2] *Institute of Physics and Electronics, NCST of Vietnam, P.O. Box 429, Boho, Hanoi 10000, Vietnam*

[3] *Saitama University, Japan*

[4] *Hanoi University of Science, 334 Nguyen Trai, Thanh Xuan, Hanoi, Vietnam*

[5] *Tshukuba University, Ibaraki 305-8571, Japan*

[6] *Center for Nuclear Study, University of Tokyo, RIKEN campus, 2-1 Hirosawa, Wako, Saitama 351-0198, Japan*

The one-nucleon exchange (ONE) reactions of p(^{6}He, ^{4}He+n)d and p(^{6}Li, ^{4}He+p)d were studied experimentally with the secondary beams of ^{6}He at energy of 70.5 AMeV and ^{6}Li at 74.4 AMeV at the RIKEN Accelerator Research Facility (RARF). The experiment is to consider nucleon-nucleon correlation in the di-nucleon configuration of the ^{6}He and ^{6}Li nuclei. In addition, another reaction with same products as the exchange reaction of interest was also observed for the ^{6}He + p and ^{6}Li + p systems.

Keywords: One-nucleon exchange reaction, di-neutron correlation, solid-H2 target, backward scattering.

1. Introduction

The ^{6}Li and ^{6}He nuclei are the typical six-nucleon systems consisting of α core and two loosely bound nucleons. ^{6}He is the lightest halo nucleus showing Borromean binding mechanism, which suggests a strong di-neutron correlation for the formation of the two neutron halo (2n-halo) in this nucleus.[1–3] ^{6}Li is a stable nucleus whose cluster properties have been well investigated through the (e, e'd) experiment, in which the neutron-proton pair in the nucleus is found to have a similar wave function as that of a free deuteron.[4] For the ^{6}He nucleus, correlation between two neutrons is still an open problem. Recently, the experimental studies on 1n- and 2n-transfer reactions in the ^{6}He + p system, and 2n exchange in the ^{6}He + ^{4}He reaction have revealed the "di-neutron" and "cigar-like" configurations for the ground state structure.[2,3] These predictions await further experimental verification. The (d, p) elastic-scattering experiment showed an increase in cross section at very backward-scattering angles in the center of mass (CM) system.[5] This characteristics is mainly attributed to the nucleon exchange mechanism. This suggests that if a similar reaction can be observed for the 2n system in ^{6}He, the measurement will then provide a direct verification of the existence of the correlated two-neutron structure in this nucleus, because the exchange mechanism requires a rather high relative momentum between neutrons in the 2n system, typically 150 MeV/c.

2. Experimental setup and data analysis

We report the experimental search for the 2n correlation in the ^{6}He nucleus via the one-nucleon exchange (ONE) reaction p(^{6}He, ^{4}He+n)d in inverse kinematics, using an incident ^{6}He beam at an energy of 70 AMeV delivered by the RIKEN Accelerator Research Facility (RARF) and a newly developed solid-H2 target (SHT).[6] For reference, a similar reaction was also measured for the ^{6}Li + p system. The reaction products, which are an α particle, a slow deuteron and a high-energy neutron for the ^{6}He + p system (and high-energy proton for the ^{6}Li + p system) have been measured in coincidence. The detector systems were designed to determine the kinematical characteristics of the particles at backward-scattering angles in the CM system.

The kinetic energy and scattering angle of the deuteron were measured by a two-strip-array system consisting of front and rear arrays placed at 350 mm from the target. Each array consists of 12 semi-circular strips divided into 6 pairs. The strip is 10 mm thick and 14.5 mm wide. The radii of the innermost and outermost strips are 30 and 205.5 mm, respectively. In order to use in high-intensity beam, the strips are made from plastic scintillator. In such a design the front array is used to detect the backward-scattering deuterons produced in an angular range from 5° to 30° in the laboratory system with a resolution of $\pm 2^o$, whereas the rear one plays a role as veto for high-energy particles which are able to be produced in interaction of the beam with either the target or support materials. In this experiment the source of high-energy particles mainly comes from fragmentation of the beam particles.

In addition, a He-gas bag is fittingly arranged between the target and the detector in order to reduce the energy loss and multiple scattering of the slow deuterons.

For the strip-detector system the particle identification was done by analysis: (1) Time of flight (TOF) - Energy loss (ΔE_F) correlation for the front strips and (2) energy loss (ΔE_R) - (ΔE_F) in the front and rear ones. Fig. 1 shows the particle identification given by TOF - (ΔE_F) correlation, where the positions of low-energy particles stopped in the front strips and high-energy particles passing through or stopped in the rear ones could be detected.

On the other side of reaction plane, a thick plastic-wall detector of 600x400 mm^2 consisting of front and rear walls of same thickness of 100 mm was located at a distance of 700 mm from the target for measurement of the high-energy neutron (proton). The system covers the same range of angles as the deuteron detector. In addition, a thin plastic layer, called charge veto, was placed in front of the detector to identify charged particles

Fig. 2 shows the particle identification given by the front wall using the energy

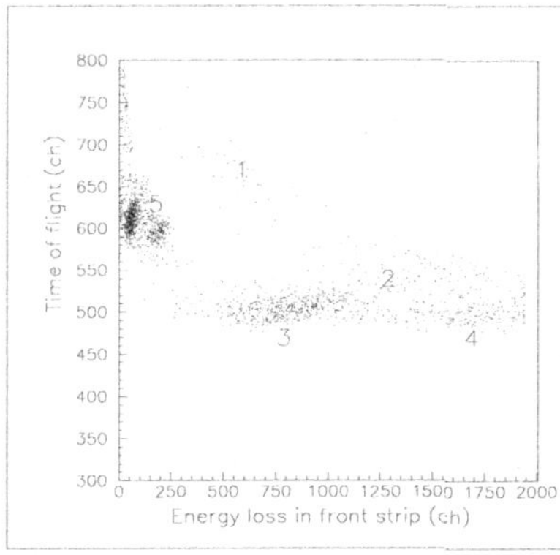

Fig. 1. Particle identification given by strips: 1-Low-energy particles stopped in the front strips; 2-The light particles (p,d,t) stopped in the rear strips; 3,4-The high-energy particles (p,d,t,^{3,4}He) passing through both strips; 5-Cross-talk effects between strips

loss - time-of-flight correlation, where locations of the particles are well separated.

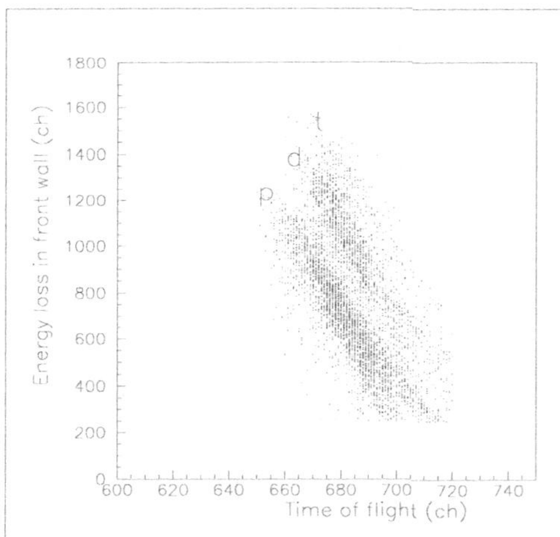

Fig. 2. Particle identification given by front wall

Located at 1000 mm downstream the target, in the beam direction, is a detector system for α-particle measurement. It consists of three plastic layers of 140x140 mm^2 with the corresponding thicknesses of 20, 40 and 20 mm in order to cover an angular range of $\pm$ 4°. A multi-wire proportional chamber (MWPC) with 140 vertical and 140 horizontal wires of 1 mm spacing is placed in front of the system for position measurement. The identification for α particle was performed using the energy-loss measurements in first two layers. In this system the third layer plays a role as veto for high-energy particles. Fig. 3 shows the particle identification using the energy-loss correlations in the layers, where the positions of the particles are clearly visible. However, in order to select the α particles produced by the ONE reaction we need to do a coincidence analysis with the particle identifications given by the strips and wall detector.

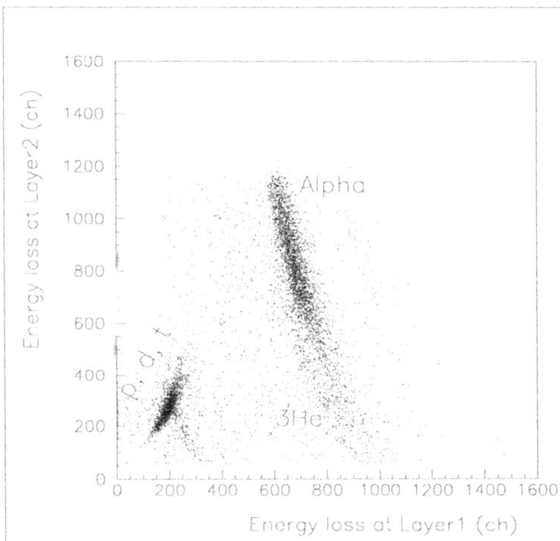

Fig. 3. α-particle identification using energy-loss correlation in first two layers

The above-mentioned results showed a good possibility of particle identification from the structure of the detector systems used in the experiment. This allows us to make the strictly coincident analysis to select the products of the studied reaction, i.e., low-energy deuteron, α particle and high-energy proton for ^{6}Li + p system, or neutron for ^{6}He + p system. In the data analysis an especially important interference was identified. That is 1n-transfer reaction which is a two-step process going through unbound nuclei ^{5}Li decaying into α particle and proton for ^{6}Li + p system: $^6\text{Li} + \text{p} \rightarrow \text{d} + {}^5\text{Li} \rightarrow \text{d} + \alpha + \text{p}$

and ^{5}He decaying into α particle and neutron for ^{6}He + p system: $^6\text{He} + \text{p} \rightarrow \text{d} + {}^5\text{He} \rightarrow \text{d} + \alpha + \text{n}$ The final products of this process are identical with those produced by the one-step reaction. Hence, kinematically, it should be assumed that the events obtained from the coincident analysis can be originated from the ONE reaction as well as from the 1n-transfer one. The kinematical calculations showed that under the present geometry of the experiment the energies of deuterons produced by the ONE and 1n-transfer reactions are almost similar. The key used to identify two these reactions are based on energy difference of α particles and protons for ^{6}Li + p system, and of α particles and neutrons for ^{6}He + p system.

Fig. 4 shows a correlation between α-particle and proton energies obtained experimentally. We did a simulation to determine the kinematical characteristics of α

particle and proton produced in ^{6}Li + p interaction. The simulation is based on two possible configurations of the ^{6}Li nucleus: α-np configuration which is assumed to be dominant in producing the ONE reaction whereas 1n-transfer reaction is mainly from n-^{5}Li configuration. The simulation allowed us to determine the kinematical location and some possible overlapping between two these reaction channels. This is very useful for evaluation of the event separation from the correlation obtained experimentally.

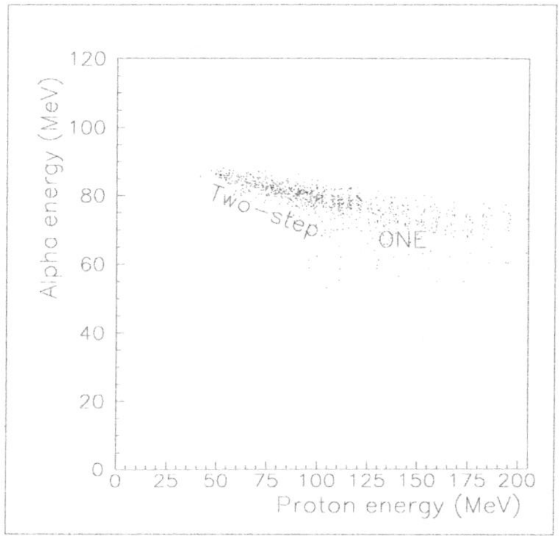

Fig. 4. Correlation between α particle and proton energies

Fig. 5 shows the momentum and angular distributions of α particles produced by the ONE reaction for ^{6}Li + p system, where the angular distribution is obtained from the position measurement given by MWPC:

i. For the momentum distribution fitting with Gaussian function gives a width $\sigma \simeq 40$ MeV/c.

ii. The maximum value in the angular distribution is about 1.5^o.

These results are well consistent with the simulation, in which the nucleon exchange mechanism is assumed in inverse kinematics collision between the two-nucleon system and proton. α particle is then predicted to conserve the momentum vector. Consequently, its angular distribution should be consistent with the momentum determination in α - di-nucleon configuration.

Related to the structure of ^{6}He the study on di-neutron elastic transfer in the ^{4}He(^{6}He, ^{6}He)^{4}He reaction has revealed a small contribution (up to 15 %) of the P-wave part corresponding to the n-^{5}He configuration, so called cigar-like, in the ^{6}He wave function.[7] However, shell-model calculations of relevant spectroscopic amplitudes (SA) predicted a large probability of the n-^{5}He configuration.[8,9] In addition the experiment on the three-body breakup of ^{6}He at energy 240 AMeV showed the larger cross section of single-neutron knockout (σ_{-1n} = 127 ± 14 mb for C tar-

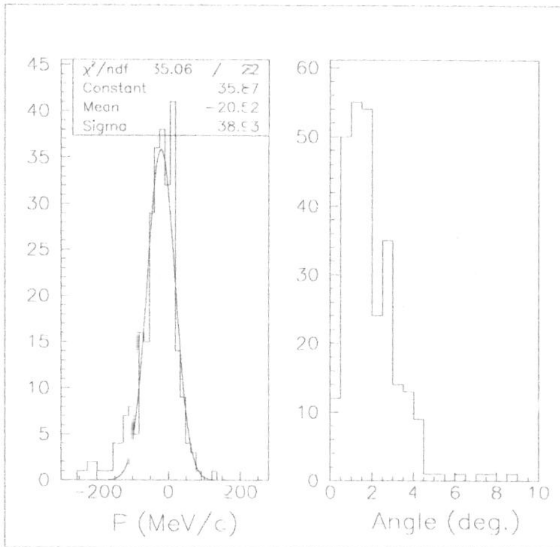

Fig. 5. Momentum and angular distributions of α-particles

get and 320 ± 90 mb for Pb target) compared with two-neutron knockout (σ_{-2n} = 33 ± 23 mb for C target and 180 ± 100 mb for Pb target) [10]. This experiment concluded about a large extent of the ^{5}He ground-state resonance following single-neutron knockout.[10] These researches can suggest some relation between the predicted configurations and mechanism of the reactions with participation of the ^{6}He nucleus.

3. Conclusion

The nucleon exchange reactions in inverse kinematics for ^{6}Li + p and ^{6}He + p systems were measured for the first time at the energy of about 70 AMeV at the backward-scattering angles in the CM system. Kinematically, a consistence in correlation between relative momentum of two-nucleon system and scattering angle of α particle was obtained. The component of the wave function corresponding to the two-neutron structure in ^{6}He is similar to that of p-n pair in ^{6}Li. This suggests that the two neutrons in ^{6}He is spatially close each other as p-n pair in ^{6}Li. We have observed the events probably attributed to the other reaction mechanism in both the ^{6}He + p and ^{6}Li + p systems. Kinematically, they are consistent with one-neutron transfer process although the beam energy is quite high. At present, theoretical considerations are being carried out to reach the conclusions from the physics viewpoint.

Acknowledgement

The authors gratefully acknowledge the support of the cyclotron staff for a smooth accelerator operation. One of authors (N.T. Khai) is grateful for the APA program between RIKEN and Hanoi University of Science.

References

1. T. Kobayashi *et al.*, *Phys. Rev. Lett.* **60**, 2599 (1988).
2. G. M. Ter-Akopian *et al.*, *Phys. Lett. B* **426**, 251 (1998).
3. Yu Oganessian, V.T. Zagrebaev, and J. Vaagen,*Phys. Rev. Lett.* **82**, 4996 (1999).
4. R. Ent *et al.*, *Nucl. Phys. A* **578**, 93 (1994).
5. K. Sekiguchi *et al.*, *Phys. Rev. C,* **65**, 034003 (2002).
6. S. Ishimoto *et al.*, *Nucl. Instr. and Meth. A* **480**, 304 (2002).
7. Dao T. Khoa, W. von Oertzen, *Phys. Lett. B* **595**, 193 (2004).
8. R. Wolski *et al.*, *Phys. Lett. B* **467**, 8 (1999).
9. Yu. F. Smirnov, Yu. M. Tchuvilsky, *Phys. Rev. C* **15**, 84 (1977).
10. T. Aumann *et al.*, *Phys. Rev. C* **59**, 1252 (1999).

PART C

Summary of ISPUN07 and List of Participants

SUMMARY TALK OF ISPUN07

N. ALAMANOS*
DSM/DAPNIA/SPhN, CEA-Saclay,
Gif-sur-Yvette, F-91191, France
**E-mail: nalamanos@cea.fr*

It is always extremely difficult to give the summary talk of a conference where the number and quality of the presentations is as great as it has been at ISPUN07. We have had sixty six talks which renders the job of the person giving the summary talk rather difficult. In order to carry out this task I have decided to group the subjects into three categories:
i) Construction of new facilities ii) Experiments iii) Theory.
Naturally, the choice of subjects within each category reflects my personal tastes and corresponds without doubt to an unfair selection. I would nevertheless like to excuse myself on the grounds of the large number of subjects covered during this conference which consequently imposes a rather drastic selection.

1. Construction of new facilities

We have had the pleasure – and it has indeed been a pleasure – to hear a number of presentations of new facilities which have either commenced operation this year or will do so in the very near future. The new isotope ^{125}Pd has been produced for the first time at the RIKEN Big-Rips facility (T. Motobayashi), whereas the commissioning of SHARAQ should take place next year (S. Kubono). SHARAQ opens the possibility of studying spin-isospin excitations in exotic nuclei. Magnex at Catania in Italy has commenced operation and should very soon form part of the "standard" experimental installations of the accelerator at Catania (F. Cappuzzello). Another important project, presented by P. Van Duppen, is the upgrade to ISOLDE, HIE-ISOLDE. Piet placed the accent not only on the production of radioactive ion beams but also on the purification of these beams, an essential requirement for the success of experiments in this area and often neglected in the design phase of new radioactive beam facilities. The first feasibility study for EXL with prototype detectors in the actual experimental storage ring was presented by N. Kalantar-Nayestanaki. The SPIRAL2 project, currently under construction in France, was presented in detail by S. Gales. Finally, P. Gomes presented the Sao-Paulo facility, the only radioactive ion beam facility in the southern hemisphere.

It is with great pleasure that I have dwelt on the blossoming of all these new projects, as I am convinced that these new facilities will contribute to renewing the excitement concerning research in nuclear physics. They will contribute to the continued teaching of nuclear physics in universities and the selection and training of high quality students and post-doctoral researchers. They will also contribute to both the authorities and the public viewing nuclear physics as a leading domain of science. Furthermore, they will allow a body of well-educated scientists with opinions in the domain of nuclear energy to be constituted, who will be able to participate in the debates, foreseen in the years to come, concerning the world energy problem.

2. Experiments

2.1. *Sub-barrier fusion*

The possibility of an increase in the fusion cross section induced by weakly-bound nuclei at incident energies below the Coulomb barrier has been the object of several studies in recent years. We have had several talks on this subject by D.J. Hinde, P.R.S. Gomes and V. Scuderi. The picture which seems to emerge from these studies is that:

In all systems studied, the fusion cross section induced by weakly-bound projectiles is hindered at above barrier-energies. No sub-barrier fusion enhancement was observed.

I would like to take this opportunity to point out the excellent study by N. Keeley et al., [*Prog. Part. Nucl. Phys.* **59**, 579 (2007)] where the most recent results on sub-Coulomb fusion are discussed in a very compact fashion.

2.2. *Low and high energy direct reactions*

A new phenomenon in the study of the structure of exotic nuclei is the observation of a decoupling between the excitation of protons and neutrons ($M_n \gg M_p$). This behaviour has been clearly demonstrated for ^{16}C and ^{14}Be. According to T. Motobayashi this indicates that ^{16}C should have an egg-like structure. Similar behaviour has been observed in ^{14}Be by T. Nakamura.
Among other results we cite:

A bound $4n$ was not observed; ^{7}H seems to exist as a resonance close to the $t + 4n$ threshold, and some strong evidence for parity inversion $(2s_{1/2} - 1p_{1/2})$ in ^{9}He was reported by D. Beaumel. A detailed discussion of the structure of ^{8}He was given by S. Gales and myself and new experimental data for proton inelastic scattering on ^{7}Be, ^{10}Be, ^{11}Be, ^{12}Be, ^{14}Be and ^{8}B, aiming at obtaining their matter density distributions, were presented by P. Egelhof. Obviously it will be interesting to dispose of matter density distributions for halo and non-halo nuclei. Among the nuclei studied ^{11}Be, ^{14}Be and ^{8}B are candidates for a halo structure.

In his talk P. Van Duppen presented new results for the Coulomb excitation of Cu and Zn isotopes, using for these studies the first post-accelerated, laser ionized isomeric beam from REX-ISOLDE. I would also like to dwell on the results of a novel experimental technique presented by S. Sakaguchi concerning polarisation measurements of exotic nuclei. Although this technique currently suffers from poor statistics it is important to continue the development of such alternative approaches providing complementary information on the structure of exotic nuclei. The structure of light neutron-rich nuclei, for instance ^{12}Be, has also been studied by alpha induced reactions. The results presented by S. Shimoura prefigure a part of the experimental programme of the SHARAQ device.

The dependence of the nucleon removal strengths at the Fermi surface as a function of the asymmetry of the two Fermi surfaces was presented by J.A. Tostevin. From this presentation we can conclude that $1N$ removal analyses indicate a systematic dependence of the suppression of shell model single-particle strength on the asymmetry of the two Fermi surfaces. All (sd-shell, test) data sets are consistent with shell model spectroscopy and a suppression [0.50(5)] of $2N$ shell model strength. The cases studied so far all have similar Fermi surface asymmetries so any $2N$ asymmetry dependence is not yet evident. These results were obtained at fragmentation beam energies with the appropriate techniques. Obviously it will be important to complement these studies with low energy dedicated transfer reaction measurement.

The previous remark is a natural link with the talk of O. Sorlin who discussed the first proof of weakening of the $N = 28$ gap between ^{48}Ca and ^{46}Ar, derived from experimentally-constrained monopole variations. A shrinking of the single particle energies gradually occurs when $N \gg Z$ due to two-body $p - n$ interactions, central, tensor and spin orbit which favour particle-hole excitations and E2 collectivity. It will be important to extend these results with appropriate new experiments. Finally, it was demonstrated in this talk that ^{26}Ne is indeed a deformed nucleus.

I do not wish to leave this domain without saying a few words on the results of M. Morjean and V.K. Utyonkov. M. Morjean reported on new evidence for survival with large probabilities of highly excited nuclei with $Z = 120$ and $Z = 124$ at times longer than 10-18*sec*. V. K. Utyonkov made a nice review of the super heavy elements research programme followed by the Dubna group. The ensemble of the analysis of the Dubna results allows one to plot the fission barrier height as a function of the neutron number for nuclei with atomic numbers between $Z = 102$ and $Z = 116$. The question which one can ask is the following. What is the link between the two experimental approaches? Is it possible to combine the life-times observed by M. Morjean and the fission barriers observed by the latter experiments?

2.3. *Pygmy resonances*

The search for and study of pygmy resonances has recently become an industry. Personally, I was first confronted with pigmy resonances when analysing twenty years ago ^{208}Pb heavy ion inelastic scattering. Indeed ^{208}Pb presents a well marked pigmy resonance clearly observed in photo absorption measurements, which manifests its presence with a clear bump in the excitation spectrum of ^{208}Pb. I remember that at the time this bump was an inconvenient complication in the analysis.

In studies of exotic nuclei the observation of these modes may shed some light on the symmetry energy and eventually on the neutron proton skin thickness, since the energy weighted sum rule exhausted by these modes is related to this quantity. Some nice work concerning the excitation and decay characteristics (γ,γ), (γ,n) of these modes was presented by T. Aumann and G. Benzoni.

3. Theory

3.1. *Direct reactions*

Several subjects were also discussed on the theory side during this conference, beginning with a new multi-channel T-matrix theory for describing low energy capture reactions by K. Amos. With this approach it is possible to calculate several observables without adjustable parameters, for example the excitation function for the $\alpha(^3\text{He},^3\text{He})\alpha$ reaction. The agreement between the calculation and the experimental results seems to be excellent something to keep an eye on. We were also informed by Z.Y. Ma of new attempts to obtain a complex microscopic potential to describe the interaction between heavy ions. It should be recalled that although we currently dispose of several theoretical approaches for calculating the complex proton-nucleus potential, this is not the case for the interaction between heavy ions. The description of the interactions induced by exotic nuclei is even more difficult than for stable nuclei. Beyond the interaction potential, which is still not well evaluated theoretically, in the case of weakly-bound nuclei couplings to the continuum must also be included. Indeed, it was noted that a description of the scattering of weakly-bound and unbound states requires calculations including couplings to the continuum. The accuracy of these calculations was investigated by A. Vitturi, who indicated that coupled channels descriptions based on discretisation of the continuum can be accurate when a proper choice is made of the number of discrete states, of the energy mesh and of the energy cut-off.

3.2. *Shell model and cluster physics*

I believe that the cluster model wave function is part of a large basis shell model wave function. For this reason I think there is a strong link between the shell model and cluster physics. From the recent shell model calculations by T. Otsuka it becomes obvious that the tensor force of the nucleon-nucleon interaction is important for predicting the properties of nuclei far from stability. I don't want to insist on a recent debate, largely discussed in O. Sorlin's talk, concerning the microscopic structure of ^{42}Si, doubly-magic or strongly oblate. I think that the interested reader can find the answer for himself. Now, concerning clustering in nuclei, I would like to remind you of the nice experimental review by M. Freer and the calculations presented by Y. Kanada-En'yo, M. Kimura and Y. Funaki. Y. Kanada-En'yo's calculations for the ^{8}He ground state favour a $(p_{3/2})^4 + (p_{3/2})^2(p_{1/2})^2$ structure in agreement with the experimental results presented.

3.3. *The QRPA*

It is beyond the scope of this summary talk to enter into details concerning the status of the different QRPA calculations presented at this conference. In his calculations G. L. Colò includes two-body Coulomb and two-body spin-orbit interactions which are essential ingredients for successfully describing the properties of nuclei. I was also pleased to see that in his relativistic QRPA calculations P. Ring, although he reproduces a huge amount of the structure properties of nuclei, also reproduces the E1 low energy bump in Pb which as I said previously was a nightmare for me when analysing inelastic scattering experiments on this nucleus.

T. Nakatsukasa has applied his linear response with time-dependent Skyrme density functional type calculations in studying the structure of ^{26}Ne and in particular the low-lying peak at $E_x = 8$ MeV. It is not obvious, at least to me, that the predicted properties of this peak are sensitive to the microscopic interaction used for performing the calculations. M. Matsuo and S.G. Zhou have made a tremendous effort to include the continuum in their calculations whereas S. Peru has presented truly self-consistent deformed QRPA calculations, which as far as I know are a real achievement in this domain.

I am coming to the end of my summary talk, the aim of which was to recall some of the physics results presented at this conference. The conference has been very lively and led to many exchanges among the participants. We have seen the emergence of new results of very high quality on the structure of exotic nuclei. The study of exotic nuclei also obliges us from time to time to revisit results obtained in the past with stable nuclei and to observe that we lack data on certain properties of these nuclei. Indeed, I note that to investigate pairing in finite nuclei P. F. Bortignon has asked for spectroscopic factors in Ca, Ni, Sm, etc. and B. Tsang has undertaken to satisfy this request.

Concerning theory, we have seen that low-Energy nuclear theory is currently going through an unprecedented revival. I gave some hints concerning microscopic calculations. Furthermore, A. Schwenk has convinced us that on a more fundamental level we are in the process of obtaining the first explicit link between Quantum Chromo Dynamics and two/three-nucleon interactions.

4. End

The conference was a success due to the quality of the organization and I would like to take this opportunity to thank the organizing committee and especially Dao Tien Khoa and Nguyen Van Giai. It was decided that the next ISPUN conference will be organised in 2010. We are all looking forward to it!

LIST OF PARTICIPANTS

Nicolas Alamanos
CEA-Saclay, DAPNIA-SPhN
91191 Gif-sur-Yvette cedex
France
alamanos@cea.fr

Ken Amos
School of Physics
University of Melbourne
Victoria 3010, Australia
amos@physics.unimelb.edu.au

Thomas Aumann
GSI
Planckstrasse 1. D-64291, Darmstadt
Germany
t.aumann@gsi.de

Didier Beaumel
Institut de Physique Nucleaire
Université Paris-Sud 11
91406 Orsay Cedex, France
beaumel@ipno.in2p3.fr

Giovanna Benzoni
University of Milan,
Dipartimento di Fisica, Via Celoria 16,
I-20133 Milano, Italy
Giovanna.Benzoni@mi.infn.it

Hans Gerhard Bohlen
Hahn-Meitner-Institut,
Glienicker Strasse 100,
14109 BERLIN, Germany
bohlen@hmi.de

P. F. Bortignon
University of Milan,
Dipartimento di Fisica, Via Celoria 16,
I-20133 Milano, Italy
pierfrancesco.bortignon@mi.infn.it

Francesco Cappuzzello
University di Catania and INFN-LNS
Via S.Sofia 62, I 95125, Catania
Italy
cappuzzello@lns.infn.it

Richard Francis Casten
Yale University
Physics Dept., WNSL, P. O. Box 208124,
New Haven, CT 06520-8124 USA
rick@riviera.physics.yale.edu

Gianluca Colò
Dipartimento di Fisica
Università degli Studi and INFN,
20133 Milano, Italy
gianluca.colo@mi.infn.it

Raquel Crespo
Dept. de Física, Inst. Superior Tecnico
Taguspark, 2780-990, Porto Salvo,
Oeiras, Portugal
raquel.crespo@tagus.ist.utl.pt

Angelo Cunsolo
Universit di Catania and INFN-LNS
Via S.Sofia 62, I 95125, Catania
Italy
cunsolo@lns.infn.it

Dam Nguyen Binh
Institute of Physics and Electronics
10 DaoTan, CongVi, BaDinh, HaNoi
Vietnam
dnbinh@grad.iop.vast.ac.vn

Dao Tien Khoa
Institute for Nuclear Science & Tech.
P.O. Box 5T-160 Hoang Quoc Viet
Nghia Do, Hanoi, Vietnam
khoa@vaec.gov.vn

Pierre Darriulat
Institute for Nuclear Science & Tech.
P.O. Box 5T-160 Hoang Quoc Viet
Nghia Do, Hanoi, Vietnam
darriulat@vaec.gov.vn

Marzio De Napoli
Dipartimento di Fisica,
Universitá di Catania & LNS-INFN
64, Via S. Sofia, I-95123, Catania, Italy
denapoli@lns.infn.it

Massimo Di Toro
Laboratori Nazionali del Sud
INFN, Via S.Sofia, 62
I-95123 Catania, Italy
ditoro@lns.infn.it

Do Cong Cuong
Institute for Nuclear Science & Tech.
P.O. Box 5T-160 Hoang Quoc Viet
Nghia Do, Hanoi, Vietnam
dccuong@vaec.gov.vn

Piet Van Duppen
Inst. voor Kern- en Stralingsfysica
University of Leuven (K.U.Leuven)
B-3001 Leuven Belgium
piet.vanduppen@fys.kuleuven.be

Peter Egelhof
GSI
Planckstrasse 1. D-64291, Darmstadt
Germany
p.egelhof@gsi.de

Martin Freer
School of Physics and Astronomy
University of Birmingham
Birmingham B15, 2TT, UK
M.Freer@bham.ac.uk

Yasuro Funaki
RIKEN
2-1, Hirosawa, Wako, Saitama 351-0198
Japan
funaki@riken.jp

Sydney Galès
GANIL
BP 5027, 14076 CAEN cedex 5
France
gales@ganil.fr

Paulo Roberto Silveira Gomes
Inst. de Fisica, UFF, Av. Litoranea
Campus da Praia Vermelha, Niteroi,
R.J.,24210-340, Brazil
paulogom@if.uff.br

Magdalena Jolanta Gorska
GSI Darmstadt
Planckstrasse 1. D-64291, Darmstadt
Germany
m.gorska@gsi.de

Fairouz Hammache
Institut de Physique Nucleaire
Université Paris-Sud 11
91406 Orsay Cedex, France
hammache@ipno.in2p3.fr

David John Hinde
Department of Nuclear Physics,
Australian National University,
Canberra, ACT 0200, Australia
david.hinde@anu.edu.au

Nobuo Hinohara
Department of Physics
Kyoto University, Kyoto 606-8502
Japan
hinohara@ruby.scphys.kyoto-u.ac.jp

Hoang Hoa Mai
Institute for Nuclear Science & Tech.
P.O. Box 5T-160 Hoang Quoc Viet
Nghia Do, Hanoi, Viet Nam
hhmai@vaec.gov.vn

Hoang Sy Than
Institute for Nuclear Science & Tech.
P.O. Box 5T-160 Hoang Quoc Viet
Nghia Do, Hanoi, Viet Nam
hsthan@vaec.gov.vn

Aksel S. Jensen
Department of physics and Astronomy
University of Aarhus,
DK-8000 Aarhus C, Denmark
asj@phys.au.dk

Nasser Kalantar-Nayestanaki
KVI, University of Groningen
Zernikelaan 25, 9747 AA Groningen
The Netherlands
nasser@kvi.nl

Yoshiko Kanada-En'yo
Yukawa Institute for Theoretical Physics
Kyoto University, 606-8502 Kyoto
Japan
yenyo@yukawa.kyoto-u.ac.jp

Kiyoshi Kato
Dept. of Physics, Faculty of Science
Hokkaido University,
Sapporo 060-0810, Japan
kato@nucl.sci.hokudai.ac.jp

Masaaki Kimura
University of Tsukuba
Tsukuba 305-8571
Japan
masaaki@nucl.ph.tsukuba.ac.jp

Shigeru Kubono
CNS, University of Tokyo
Hirosawa 2-1, Wako
Saitama, 351-0198 Japan
kubono@cns.s.u-tokyo.ac.jp

Le Hong Khiem
Institute of Physics and Electronics
10 DaoTan, CongVi, BaDinh, HaNoi
Vietnam
lhkhiem@iop.vast.ac.vn

Haozhao Liang
School of Physics
Peking University, Beijing 100871
China
hzliang@pku.edu.cn

Umberto Lombardo
Dipartimento di Fisica e Astronomia
viale A.Doria 6 I 95125 Catania
Italy
Lombardo@lns.infn.it

Peiyan Luo
Institute of Modern Physics
509 Nanchang Road., Lanzhou 730000
China

Zhong-yu Ma
China Institute of Atomic Energy
P.O.Box 275/18, Beijing 102413
China
Mazy12@ciae.ac.cn

Masayuki Matsuo
Department of Physics,
Niigata University
Niigata 950-2181, Japan
matsuo@nt.sc.niigata-u.ac.jp

Maurice Morjean
GANIL
BP 55027, F-14076 Caen Cedex 5
France
morjean@ganil.fr

Tohru Motobayashi
RIKEN Nishina Center
2-1 Hirosawa, Wako,
Saitama 351-0198, Japan
motobaya@riken.jp

Takashi Nakamura
Tokyo Institute of Technology
Dept. of Physics, 2-12-1 O-Okayama
Meguro, Tokyo 152-8551, Japan
nakamura@ap.titech.ac.jp

Takashi Nakatsukasa
University of Tsukuba
Institute of Physics, Tsukuba 305-8571
Japan
nakatsukasa@nucl.ph.tsukuba.ac.jp

Ngo Van Luyen
University of Natural Sciences
Dept. of Physics, 334 Nguyen Trai
Thanh Xuan, Hanoi, Vietnam

Nguyen Dang Chien
Dalat University
01 Phu Dong Thien Vuong Street, Dalat
Vietnam
ndchien@grad.iop.vast.ac.vn

Nguyen Dinh Dang
INST & RIKEN
Hirosawa 2-1, Wako,
Saitama 351-0198, Japan
dang@postman.riken.go.jp

Nguyen Van Giai
Institut de Physique Nucleaire
Université Paris-Sud 11
91406 Orsay Cedex, France
nguyen@ipno.in2p3.fr

Nguyen Tuan Khai
Institute of Physics and Electronics
10 DaoTan, CongVi, BaDinh, HaNoi
Vietnam
lhkhiem@iop.vast.ac.vn

Nguyen Duy Ly
Dalat University
01 Phu Dong Thien Vuong Street, Dalat
Vietnam
ngduyly@yahoo.com

Nguyen Tran Tho
University of Natural Sciences
227 Nguyen Van Cu Street,
Ho Chi Minh City, Vietnam
nguyentrantho2005@yahoo.co.uk

Shigeo Ohkubo
Kochi Women's University
5-15 Eikokuji-cho, Kochi 780-8515
Japan
shigeo@cc.kochi-wu.ac.jp

Takaharu Otsuka
University of Tokyo
Hongo, Bunkyo-ku, Tokyo, 113-0033
Japan
otsuka@phys.s.u-tokyo.ac.jp

Sophie Péru
CEA/DAM Ile de France, DPTA,
B.P. 12, 91680 Bruyères-le-Châtel,
France
sophie.peru-desenfants@cea.fr

Pham Ngoc Diep
Institute for Nuclear Science & Tech.
P.O. Box 5T-160 Hoang Quoc Viet
Nghia Do, Hanoi, Vietnam
dieppn@vaec.gov.vn

Pham Ngoc Dong
Institute for Nuclear Science & Tech.
P.O. Box 5T-160 Hoang Quoc Viet
Nghia Do, Hanoi, Vietnam
dongpn@vaec.gov.vn

Zsolt Podolyák
University of Surrey
Dept. of Physics, Guildford GU2 7XH
United Kingdom
Z.Podolyak@surrey.ac.uk

Bin Qi
School of Physics
Peking University, Beijing 100871
China
bqi@jcnp.pku.edu.cn

Peter Ring
Physics Department
Technical University Munich
Garching, D85748, Germany
ring@ph.tum.de

Zhongzhou Ren
Department of Physics
Nanjing University, Hankou Road,
Nanjing 210093, China
zren@nju.edu.cn

Hiroyuki Sagawa
Center for Mathematical Sciences
Univ. of Aizu, Aizu-Wakamatsu
Fukushima 965-8580, Japan
sagawa@u-aizu.ac.jp

Satoshi Sakaguchi
CNS, University of Tokyo
Hirosawa 2-1, Wako
Saitama, 351-0198 Japan
satoshi@cns.s.u-tokyo.ac.jp

Hideyuki Sakai
Department of Physics
University of Tokyo, Hongo 7-3-1,
Bunkyo, Tokyo 113-0033, Japan
sakai@phys.s.u-tokyo.ac.jp

Achim Schwenk
TRIUMF
4004 Wesbrook Mall, Vancouver,
BC, V6T 2A3, Canada
schwenk@triumf.ca

Valentina Scuderi
Laboratori Nazionali del Sud
INFN, Via S.Sofia, 62
I-95123 Catania, Italy
scuderiv@lns.infn.it

Susumu Shimoura
CNS, University of Tokyo,
2-1 Hirosawa, Wako
Saitama 351-0198, Japan
shimoura@cns.s.u-tokyo.ac.jp

Olivier Sorlin
GANIL
BP 5027 14076 CAEN cedex 5
France
sorlin@ganil.fr

Toshio Suzuki
Department of Physics
Nihon University, Setagaya-ku,
Tokyo 156-8550, Japan
suzuki@phys.chs.nihon-u.ac.jp

Yuan Tian
China Institute of Atomic Energy
P.O. Box 275/18, Beijing 102413
China
tiany@ciae.ac.cn

Jeff A. Tostevin
Department of Physics
University of Surrey, Guildford
Surrey GU2 7XH, UK
j.tostevin@surrey.ac.uk

M. Betty Tsang
Michigan State University
Cyclotron Laboratory East Lansing
MI 48824, USA
tsang@nscl.msu.edu

Vladimir K. Utyonkov
Joint Inst. for Nuclear Research
141980, Dubna, Moscow region
Russia
utyonkov@sungns.jinr.ru

Andrea Vitturi
Dipartimento di Fisica
Via Marzolo, 8 I-35131 PADOVA
Italy
vitturi@pd.infn.it

Vo Thi Ngoc Tho
University of Natural Sciences
227 Nguyen Van Cu Street,
Ho Chi Minh City, Vietnam
ngoctho_vo8@yahoo.com

Vuong Kim Au
Cyclotron Institute,
Texas A&M University, College Station,
Texas 77840, USA
auvuong@comp.tamu.edu

Qi Wang
Institute of Modern Physics
509 Nanchang Road., Lanzhou 730000
China
wangqi@impcas.ac.cn

Chang Xu
Department of Physics
Nanjing University, Hankou Road,
Nanjing 210093, China
changxu99@gmail.com

Ding Yang
China Institute of Atomic Energy
P.O.Box 275/18, Beijing 102413
China
yd9955@sina.com

Yanlin Ye
School of Physics
Peking University, Beijing 100871
China
yeyl@pku.edu.cn

Kenichi Yoshida
Department of Physics
Kyoto University, Kyoto 606-8502
Japan
kyoshida@ruby.scphys.kyoto-u.ac.jp

Fang Zhang
Institute of Modern Physics,
Chinese Academy of Science
Lanzhou 730000, China
zhangfang@impcas.ac.cn

Ying Zhang
School of Physics
Peking University, Beijing 100871
China
yzhang@jcnp.pku.edu.cn

Shan-Gui Zhou
Institute of Theoretical Physics
Chinese Academy of Sciences
Beijing 100080, China
sgzhou@itp.ac.cn

Wei Zuo
Institute of Modern Physics
509 Nanchang Road, Lanzhou 730000
China
zuowei@impcas.ac.cn

Made in the USA
Monee, IL
17 September 2026

63471866R00306